Collected Papers of S.R.S. Varadhan

Collected Papers of S.R.S. Varadhan

Volume I: Limit Theorems, Review Articles

Volume II: PDE, SDE, Diffusions, Random Media

Volume III: Large Deviations

Volume IV: Particle Systems and Their Large Deviations

S.R.S. Varadhan

Collected Papers III

Large Deviations

Editors
Rajendra Bhatia
Abhay Bhatt
K. R. Parthasarathy

HINDUSTAN BOOK AGENCY
Springer

Author:
Professor S.R.S. Varadhan
New York University
New York, USA

Volume Editors:
Professor Rajendra Bhatia
Indian Statistical Institute
New Delhi, India

Professor Abhay Bhatt
Indian Statistical Institute
New Delhi, India

Professor K. R. Parthasarathy
Indian Statistical Institute
New Delhi, India

A co-publication with the Hindustan Book Agency, New Delhi, licensed for sale in all countries outside of India. Sold and distributed within India by the Hindustan Book Agency, P 19 Green Park Extn., New Delhi 110 016, India

HBA ISBN 978-93-80250-39-7 (4 volume set)

ISBN 978-3-642-33231-9 (4 volume set)
ISBN Volume III 978-3-642-33546-4
Springer Heidelberg Dordrecht London New York

Library of Congress Control Number: 2012947217

Math. Subj. Class (2010): 60-06, 60F99, 60J60, 60G44, 60F10, 60K35

Cover design: WMXDesign GmbH, Heidelberg, Germany

Printed on acid-free paper

Springer is part of Springer Science+Business Media (www.springer.com)

Printed and bound by Gopsons Papers Ltd., Noida

Contents

Volume 1

Limit Theorems

Review Articles

Volume 2

PDE, SDE, Diffusions

Random Media

Volume 3

Large Deviations

Volume 4

Particle Systems and their Large Deviations

Large Deviations

COMMUNICATIONS ON PURE AND APPLIED MATHEMATICS, VOL. XIX, 261–286 (1966)

Asymptotic Probabilities and Differential Equations*

S. R. S. VARADHAN
Indian Statistical Institute and Courant Institute of Mathematical Sciences

1. Recently Donsker [1] considered the behavior of the solution u_ε for the initial value problem

$$u_t = uu_x + \frac{\varepsilon^2}{2} u_{xx} + p(x, t) , \tag{1.1}$$

$u(0, x) = u(x)$; as $\varepsilon \to 0$. It was shown that, under mild conditions, $\lim_{\varepsilon \to 0} u_\varepsilon = u$ exists and that u is the natural solution for (1.1) with $\varepsilon = 0$. The method used was to transform (1.1) by a transformation of Hopf [3] to a linear equation of the form

$$V_t = \frac{\varepsilon^2}{2} V_{xx} + \frac{P(x, t)}{\varepsilon^2} V . \tag{1.2}$$

Since (1.2) can be solved as a function space integral, the problem reduces to the analysis of the asymptotic behavior of some function space integrals. This was carried out by Schilder [8] and the development resembles the "saddle point" method of one dimensional integrals.

One can consider variations of (1.1) of the form

$$u_t = [f(u)]_x + p(t, x) , \tag{1.3}$$

where $f(u)$ is a convex function of u. The question that can be asked is whether one can somehow alter (1.3) slightly so that the resulting equation can be made linear, by a simple transformation, in such a way that it can be solved easily and naturally. It turns out as will be shown here that it is possible to do this in such a way that the solution to (1.3) is approximated by the solution to the altered equations. The approach is similar to that of Donsker and Schilder. But one encounters more general families of measures on function space whose asymptotic behavior is to be investigated.

---

* This paper represents results obtained at the Courant Institute of Mathematical Sciences, New York University, under the sponsorship of the Sloan Foundation Grant for Probability and Statistics. Reproduction in whole or in part is permitted for any purpose of the United States Government.

A similar approach in some special cases was carried out by Lax [5], [6]. Sections 2 and 3 deal with the asymptotic properties of measures and integrals based on these measures in function space. Sections 4 and 5 give the application to the problem described earlier.

2. Let $f(\theta)$ be a convex function of θ. We shall suppose that $|f(\theta)| < \infty$ for all θ. We further assume that $f(\theta)$ is not a linear function of θ and that $f(0) = 0$. Define the conjugate convex function $h(y)$ by the relation

$$h(y) = \max_{\theta} [\theta y - f(y)] ; \tag{2.1}$$

$h(y)$ has the following properties:

(i) $h(y)$ is a convex function of y,

(ii) $h(y) \geqq 0$,

(iii) $h(y) < \infty$ for y in some nondegenerate interval,

(iv) $f(\theta) = \max_{y} [\theta y - h(y)]$,

(v) $\lim_{|y| \to \infty} \frac{|h(y)|}{|y|} = +\infty$,

(vi) if f^+ and f^- are the left and right derivatives of $f(\theta)$ at $\theta = 0$, then $h(y)$ is nonincreasing for $y \leqq f^-$, equals zero on $f^- \leqq y \leqq f^+$ and is nondecreasing for $y \geqq f^+$.

Let $\{a_n\}$ be any sequence of non-negative numbers chosen and fixed in advance such that

$$\lim_{n \to \infty} \frac{a_n}{n} = \infty .$$

Define a distribution F_n by the formula

$$dF_n = c_n \exp\left\{-\frac{a_n}{n} h(y)\right\} dy , \qquad -\infty < y < \infty ; \tag{2.2}$$

c_n is chosen such that

$$c_n^{-1} = \int_{-\infty}^{\infty} \exp\left\{-\frac{a_n}{n} h(y)\right\} dy .$$

Lemma 2.1. *c_n is well defined and* $\lim_{n \to \infty} \frac{n}{a_n} \log c_n = 0$.

Proof: Using properties (ii)–(vi) listed above this can be derived by standard estimation.

Lemma 2.2.

$$\lim_{n \to \infty} \frac{n}{a_n} \log\left[\int c_n \exp\left\{\frac{a_n}{n} [\theta y - h(y)]\right\} dy\right] = f(\theta) .$$

Proof: This can be proved by familiar saddle point considerations.

For each n, $\{X_{n,j}\}$ is a sequence of independent and identically distributed random variables all having the common distribution F_n. For each n, the partial sums $S_{n,k}$ are defined as

$$S_{n,0} = 0,$$

$$S_{n,k} = \sum_{j=1}^{k} X_{n,j}.$$

Let $[0, T]$ be some finite time interval. Ω is the space of functions on $[0, T]$ which have left and right limits at each point, vanish at 0 and are right continuous at every point $t \in [0, T]$. Ω is endowed with the topology of uniform convergence on the interval $[0, T]$. For each n a measure P_n is constructed on Ω as follows. Let us define a stochastic process $z_n(t)$ by

$$z_n(t) = \frac{1}{n} S_{n,[nt]}, \qquad 0 \leqq t \leqq T, \tag{2.3}$$

$[nt]$ denoting the integral part of nt. P_n is the measure on Ω corresponding to $z_n(t)$.

Our interest is in the asymptotic behavior of P_n. We shall show that in some suitable sense

$$dP_n \simeq \exp\left\{-a_n \int_0^T h(z'(t))\, dt\right\} \prod dz(t).$$

The precise formulation of the above statement is as follows. Let $\Omega_0 \subseteq \Omega$ be the set of those functions z in Ω, which are absolutely continuous with respect to t and for which

$$\int_0^T h(z'(t))\, dt < \infty.$$

Define a function $I(z)$ for $z \in \Omega$ by

$$I(z) = \begin{cases} \int_0^T h(z'(t))\, dt & \text{for} \quad z \in \Omega_0, \\ \infty & \text{for} \quad z \in \Omega - \Omega_0, \end{cases}$$

with the uniform topology on Ω; $I(z)$ is lower semicontinuous. We shall prove in this section the following theorem.

THEOREM 2.1. *If C is a closed set in Ω and G is an open set in Ω, then*

$$\limsup_{n\to\infty} \frac{1}{a_n} \log P_n(C) \leqq -\inf_{z\in C} I(z),$$

$$\liminf_{n\to\infty} \frac{1}{a_n} \log P_n(G) \geqq -\inf_{z\in G} I(z).$$

We shall prove the theorem by establishing several lemmas.

LEMMA 2.3. *Let C be a subset of Ω of the form*

$$C = [z\colon z(t_2) - z(t_1) \geqq x], \qquad t_2 > t_1, \delta = t_2 - t_1 .$$

Then

$$\limsup_{n\to\infty} \frac{1}{a_n} \log P_n(C) \leqq -\inf_{x\geqq y} \delta h\left(\frac{y}{\delta}\right).$$

Proof:

$$z_n(t_2) - z_n(t_1) = \xi_n = \frac{S_n[nt_2] - S_n[nt_1]}{n} .$$

If $\lambda_n = [nt_2] - [nt_1]$, ξ_n has a distribution which is identical with that of $(S_n, \lambda_n)/n$. Hence

$$Ee^{\theta\xi_n} = \left[c_n \int \exp\left\{\frac{\theta y}{n} - \frac{a_n}{n} h(y)\right\}\right]^{\lambda_n},$$

$$P_n(C) = P\{\xi_n \geqq x\} \leqq e^{-\theta x} Ee^{\theta\xi_n} \qquad \text{for} \qquad \theta \geqq 0$$

$$= e^{-\theta x}\left[c_n \int \exp\left\{\frac{\theta y}{n} - \frac{a_n}{n} h(y)\right\} dy\right]^{\lambda_n} .$$

Replacing θ by $a_n\theta$, we have for every $\theta > 0$

$$P_n(C) \leqq e^{-a_n\theta x}\left[c_n \int \exp\left\{\frac{a_n}{n}[\theta y - h(y)]\right\} dy\right]^{\lambda_n},$$

or setting

$$\phi_n(\theta) = \frac{n}{a_n} \log\left[\int c_n \exp\left\{\frac{a_n}{n}[\theta y - h(y)]\right\} dy\right],$$

$$\log P_n(C) \leqq a_n\theta x + \frac{\lambda_n}{a} \cdot \frac{a_n}{n} \phi_n(\theta) ,$$

(2.4)

$$\frac{1}{a_n} \log P_n(C) \leqq \frac{\lambda_n}{n} \phi_n(\theta) - \theta x .$$

Using Lemma 2.2, we see that, since $\lambda_n/n \to \delta$ as $n \to \infty$,

$$\limsup_{n\to\infty} \frac{1}{a_n} \log P_n(C) \leqq \delta f(\theta) - \theta x . \tag{2.5}$$

Since (2.5) is true for all $\theta \geqq 0$,

$$\limsup_{n\to\infty} \frac{1}{a_n} \log P_n(C) \leqq \inf_{\theta\geqq 0} [\delta f(\theta) - \theta x] = -\delta \sup_{\theta\geqq 0}\left[\theta \frac{x}{\delta} - f(\theta)\right]. \tag{2.6}$$

CASE i: $x > \delta f^+(0)$.

If $\theta < 0$,

$$f(\theta) \geqq \theta f^- \geqq \theta f^+ \geqq \frac{\theta x}{\delta},$$

or

$$\frac{\theta x}{\delta} - f(\theta) \leqq 0 .$$

Hence

$$\sup_{\theta \geqq 0} \left[\theta \frac{x}{\delta} - f(\theta)\right] = \sup_{\theta} \left[\theta \frac{x}{\theta} - f(\theta)\right] = h\left(\frac{x}{\delta}\right).$$

Therefore (2.6) reduces to

$$\limsup_{n \to \infty} \frac{1}{a_n} \log P_n(C) \leqq -\delta h\left(\frac{x}{\delta}\right).$$

Since $h(y)$ is nondecreasing for $y \geqq f^+$,

$$\inf_{y \geqq x} h\left(\frac{y}{\delta}\right) = h\left(\frac{x}{\delta}\right)$$

if $x > \delta f^+$, and the lemma is proved.

CASE ii: $x \leqq \delta f^+(0)$.

Since $h(y) = 0$ for $f^- \leqq y \leqq f^+$,

$$\inf_{y \geqq x} h\left(\frac{y}{\delta}\right) = 0$$

if $a \leqq \delta f^+$, and the lemma is valid trivially.

LEMMA 2.4. *Let C be a closed set of the form*

$$C = [z(t_2) - z(t_1) \leqq x] , \qquad t_2 > t_1 , \delta = t_2 - t_1 .$$

Then

$$\limsup_{n \to \infty} \frac{1}{a_n} \log P_n(C) \leqq -\inf_{y \leqq x} \delta h\left(\frac{y}{\delta}\right).$$

The proof is similar to that of Lemma 2.3. We now prove a lemma to be used later.

LEMMA 2.5. *Let $\varepsilon > 0$ be given. Then there exists a $\delta_0 > 0$ such that for all t_1, t_2, with $0 \leqq t_2 - t_1 \leqq \delta_0$,*

$$P_n[z : | z(t_2) - z(t_1)| \geqq \varepsilon] \leqq \tfrac{1}{2}$$

for all $n \geqq n_0$.

Proof: From equation (2.4), for $\theta \geqq 0$,

$$P_n[z\colon z(t_2) - z(t_1) \geqq \varepsilon] \leqq \exp\left\{-a_n\theta\varepsilon + \lambda_n \frac{a_n}{n}\,\theta_n(\theta)\right\}. \tag{2.7}$$

Take $\theta = 1$. Since $\theta_n(1) \to f(1)$, $\theta_n(1) \leqq M$ for all n, $\lambda_n \leqq n\delta_0 + 1$. Using these in (2.7), we obtain

$$P_n[z\colon z(t_2) - z(t_1) \geqq \varepsilon] \leqq \exp\left\{-\varepsilon a_n + a_n\left(\delta_0 + \frac{1}{n}\right)M\right\}.$$

If n is such that $1/n \leqq \delta_0$, we have

$$P_n[z\colon z(t_2) - z(t_1) \geqq \varepsilon] \leqq \exp\{-\varepsilon a_n + 2M\delta_0 a_n\}.$$

For any $\varepsilon > 0$, choosing δ_0 small enough and n large enough, we can achieve that

$$P_n[z\colon z(t_2) - z(t_1) > \varepsilon] \leqq \tfrac{1}{4}.$$

Similarly we can make

$$P_n[z\colon z(t_2) - z(t_1) \leqq -\varepsilon] \leqq \tfrac{1}{4}.$$

Combining the two inequalities we see that the lemma is true.

LEMMA 2.6. *Let C be a closed set of the form*

$$C = [z\colon x_1 \leqq z(t_2) - z(t_1) \leqq x_2], \quad t_2 > t_1,\, t_2 - t_1 = \delta.$$

Then

$$\limsup_{n\to\infty} \frac{1}{a_n} \log P_n(C) \leqq -\inf_{x_1 \leqq y \leqq x_2} \delta h\left(\frac{y}{\delta}\right).$$

Proof: CASE i: $x_1 \geqq \delta f^+$.

Since

$$P_n(C) \leqq P_n[z\colon z(t_2) - z(t_1) \geqq x_1],$$

$$\begin{aligned}\limsup_{n\to\infty} \frac{1}{a_n} \log P_n(C) &\leqq \limsup_{n\to\infty} \frac{1}{a_n} \log P_n[z\colon z(t_2) - z(t_1) \geqq x_1] \\ &\leqq -\inf_{y \geqq x_1} \delta h\left(\frac{y}{\delta}\right) \\ &= -\inf_{x_2 \geqq y \geqq x_1} \delta h\left(\frac{y}{\delta}\right).\end{aligned}$$

We use Lemma 2.3 and the fact that $h(y)$ is nondecreasing for $y \geqq f^+$.

CASE ii: $x_2 \leqq \delta f^-$.

This is treated similar to case i using Lemma 2.4.

CASE iii: $[\delta f^-, \delta f^+]$ and $[x_1, x_2]$ have a point in common.

In this case if y is a common point, $y/\delta \in [f^-, f^+]$. Consequently

$$\inf_{x_1 \leqq y \leqq x_2} \delta h\left(\frac{y}{\delta}\right) = 0,$$

and the lemma is trivially true.

We now introduce some notation. Let k points $0 < t_1 < t_2 < \cdots < t_k \leqq T$ be chosen. Let Π denote the map that sends $z \in \Omega$ to the point $\{z(t_1), z(t_2) - z(t_1), \cdots, z(t_k) - z(t_{k-1})\}$ of R_k. We denote by y a point $(y_1, \cdots, y_k)$ of R_k and by $\psi(y)$ the function

$$\psi(y) = \sum_{j=1}^{k} (t_j - t_{j-1}) h\left(\frac{y_j}{t_j - t_{j-1}}\right).$$

For any set $A \subseteq R_k$, $\Pi^{-1} A$ is the subset of Ω defined by

$$\Pi^{-1} A = [z: z(t_1), z(t_2) - z(t_1), \cdots, z(t_k) - z(t_{k-1}) \in A].$$

LEMMA 2.7. *Let A be a closed subset of R_k. Then*

$$\limsup_{n\to\infty} \frac{1}{a_n} \log P_n(\Pi^{-1} A) \leqq -\inf_{y \in A} \psi(y).$$

Proof: Let first A be a rectangle of the form

$$\begin{aligned} A &= [a_1 \leqq y_1 \leqq b_1 \leqq \cdots \leqq a_k \leqq y \leqq b_k] \\ &= A_1 \cap A_2 \cap \cdots \cap A_k. \end{aligned}$$

Then

$$\begin{aligned} \limsup_{n\to\infty} \frac{1}{a_n} \log P_n\left(\Pi^{-1} \bigcap_{j=1}^{k} A_j\right) &= \limsup_{n\to\infty} \frac{1}{a_n} \log P_n\left(\bigcap_{j=1}^{k} \delta\, \Pi^{-1} A_j\right) \\ &= \limsup_{n\to\infty} \frac{1}{a_n} \sum \log P_n(\Pi^{-1} A_j) \\ &\leqq \sum \limsup_{n\to\infty} \frac{1}{a_n} \log P_n(\Pi^{-1} A_j) \\ &\leqq -\sum \inf_{a_j \leqq y \leqq b_j} \delta_j h\left(\frac{y_j}{\delta_j}\right) \\ &= -\inf_{y \in A} \psi(y). \end{aligned}$$

Next let A be a closed bounded set. Let A^ε be the sphere of radius ε around A. We can find closed rectangles $A_1, \cdots, A_N$ such that

$$A \subseteq A_1 \cup \cdots \cup A_N \subseteq A^\varepsilon.$$

Then

$$\begin{aligned}\limsup_{n\to\infty}\frac{1}{a_n}\log P_n(\Pi^{-1}A) &\leqq \limsup_{n\to\infty}\frac{1}{a_n}\log P_n(\cup\,\Pi^{-1}A_j)\\ &\leqq \limsup_{n\to\infty}\frac{1}{a_n}\log \sum P_n(\Pi^{-1}A_j)\\ &\leqq \limsup_{n\to\infty}\frac{1}{a_n}\log\,[N\max P_n(\Pi^{-1}A_j)]\\ &= \limsup_{n\to\infty}\frac{1}{a_n}\max\log\,[P_n(\Pi^{-1}A_j)] \qquad (2.8)\\ &= \max\limsup_{n\to\infty}\frac{1}{a_n}\log\,[P_n(\Pi^{-1}A_j)]\\ &\leqq \max\,[-\inf_{y\in A_j}\psi(y)]\\ &= -\inf_{y\in A_j}\psi(y)\\ &\leqq -\inf_{y\in A^\varepsilon}\psi(y)\,.\end{aligned}$$

Since $\psi(y)$ is lower semicontinuous and since A is a closed bounded set,

$$\liminf_{\varepsilon\downarrow 0\;\; y\in A^\varepsilon}\psi(y)=\inf_{y\in A}\psi(y)\,.$$

Since ε is arbitrary in (2.8) we have, letting $\varepsilon\downarrow 0$,

$$\limsup_{n\to\infty}\frac{1}{a_n}\log P_n(\Pi^{-1}A)\leqq -\inf_{y\in A}\psi(y)\,.$$

Now let A be an arbitrary closed set. Define

$$\begin{aligned} E_N &= \{y\colon |y_j|\leqq N \text{ for all } j\}\,,\\ F_N &= \{y\colon |y_j|\geqq N \text{ for some } j\}\,,\\ A_N &= A\cap E_N\,,\\ F^{(1)}_{N,j} &= \{y\colon y_j\geqq N\}\,,\\ F^{(2)}_{N,j} &= \{y\colon y_j\leqq -N\}\,,\\ \delta_j &= t_j-t_{j-1}\,,\qquad t_0=0\,,\end{aligned}$$

$$\limsup_{n\to\infty}\frac{1}{a_n}\log P_n(\Pi^{-1}F^{(1)}_{n,j})\leqq -\inf_{y\geqq N}\delta_j h\left(\frac{y}{\delta_j}\right),$$

$$\limsup_{n\to\infty}\frac{1}{a_n}\log P_n(\Pi^{-1}F^{(2)}_{n,j})\leqq -\inf_{y\leqq -N}\delta_j h\left(\frac{y}{\delta_j}\right).$$

Since $F_N = \bigcup_j F_{n,j}^{(1)} \bigcup_j F_{n,j}^{(2)}$, we can get as before

$$\limsup_{n\to\infty} \frac{1}{a_n} \log P_n[\Pi^{-1} F_N] \leqq -\inf_j \inf_{|y|\geqq N} \delta_j \delta h\left(\frac{y}{\delta_j}\right) = -\rho_N . \tag{2.9}$$

Since $h(y) \to \infty$ as $|y| \to \infty$, we have

$$\lim_{N\to\infty} \rho_N = +\infty . \tag{2.10}$$

Now if A is an arbitrary closed set in R_k, it follows that for any N

$$\begin{aligned} \limsup_{n\to\infty} \frac{1}{a_n} \log P_n[\Pi^{-1} A] &\leqq \limsup_{n\to\infty} \frac{1}{a_n} \log P_N[\Pi^{-1} A_N \cup F_N] \\ &\leqq \max\left[-\inf_{y\in AN} \psi(y), -\rho_N\right] \\ &\leqq \max\left[-\inf_{y\in A} \psi(y), -\rho_N\right] . \end{aligned} \tag{2.11}$$

Since $\rho_N \to \infty$ as $N \to \infty$, we have, by letting $N \to \infty$ in (2.11),

$$\limsup_{n\to\infty} \frac{1}{a_n} \log P_n[\Pi^{-1} A] \leqq -\inf_{y\in A} \psi(y)$$

which proves the lemma.

Let τ be the map which sends a point $z \in \Omega$ to a point $[z(t_1), \cdots, z(t_k)]$ in R_k. Let $\Phi(y)$ be the function

$$\Phi(y) = \sum (t_j - t_{j-1}) h\left(\frac{y_j - y_{j-1}}{t_j - t_{j-1}}\right) ,$$

$y_0 = t_0 = 0$. Then the following lemma holds.

LEMMA 2.8. *For any closed set A in R_k*

$$\limsup_{n\to\infty} \frac{1}{a_n} \log P_n[\tau^{-1} A] \leqq \inf_{y\in A} \Phi(y) .$$

Proof: Let f be the map of R_k onto R_k defined by

$$f(y_1, \cdots, y_k) = y_1, y_1 + y_2, \cdots, y_1 + y_2 \cdots + y_k .$$

Then

$$\tau = f\pi \qquad \text{and} \qquad \tau^{-1} A = \pi^{-1} f^{-1} A .$$

Moreover, f is continuous and $f^{-1}A$ is closed if A is closed. Hence

$$\begin{aligned}\limsup_{n\to\infty}\frac{1}{a_n}\log P_n[\tau^{-1}A] &= \limsup_{n\to\infty}\frac{1}{a_n}\log P_n[\pi^{-1}f^{-1}A]\\ &\leqq -\inf_{y\in f^{-1}A}\psi(y)\\ &= -\inf_{y\in A}\psi(f^{-1}y)\\ &= -\inf_{y\in A}\Phi(y)\,.\end{aligned}$$

We now prove an elementary lemma about sums of independent random variables. Let $x_1, x_2, \cdots$ be independent random variables and $S_j = x_1 + \cdots + x_j$.

LEMMA 2.9. *Let $P\{|S_n - S_j| \geqq T\} \leqq \frac{1}{2}$ for $j = 1, 2, \cdots, n$. Then*

$$P\{\sup_{1\leqq j\leqq n}|S_j| \geqq 2T\} \leqq 2P\{|S_n| \geqq T\}\,.$$

Proof: Define

$$E = \{\sup_{1\leqq j\leqq n}|S_j| \geqq 2T\}\,,$$

$$E_j = \{\sup_{1\leqq i\leqq j-1}|S_i| < 2T, |S_j| \geqq 2T\}\,.$$

Then E is the disjoint union of E_j, and

$$\begin{aligned}P\{E, |S_n| \leqq T\} &= \sum P\{E_j, |S_n| \leqq T\}\\ &\leqq \sum P\{E_j, |S_n - S_j| \geqq T\}\\ &= \sum P(E_j)P\{|S_n - S_j| \geqq T\} \qquad (2.12)\\ &\leqq \tfrac{1}{2}\sum P(E_j)\\ &= \tfrac{1}{2}P(E)\,.\end{aligned}$$

On the other hand,

$$P\{E, |S_n| > T\} \leqq P\{|S_n| \geqq T\}\,. \qquad (2.13)$$

Adding (2.12) and (2.13), we have

$$P(E) \leqq \tfrac{1}{2}P(E) + P\{|S_n| \geqq T\}\,,$$

or

$$P(E) = P\{\sup_{1\leqq j\leqq n}|S_j| \geqq 2T\} \leqq 2P\{|S_n| \geqq T\}\,.$$

We are now ready to estimate $P_n(C)$ for an arbitrary closed set C in Ω. For $z \in \Omega$ let us define a map θ_n of $\Omega \to \Omega$ as follows.

The interval $[0, T]$ is divided into N equal parts; the points of division are $T/N, 2T/N, \cdots, dT/N, \cdots, NT/N$. I_j is the interval $(j-1)T/N \leqq t \leqq jT/N$. For z in Ω, $\theta_N z$ is that element ω of Ω which equals z at the points of division and is linear on each I_j. $\|z\|$ denotes the maximum on $[0, T]$ of the function $|z(t)|$.

LEMMA 2.10. *For any* $\varepsilon > 0$,

$$\lim_{N\to\infty} \limsup_{n\to\infty} \frac{1}{a_n} \log P_n[z: \|\theta_N z - z\| \geqq \varepsilon] = -\infty .$$

Proof:

(2.14)
$$\|\theta_N z - z\| \leqq \max_j \sup_{t\in I_j} |z(t_j) - z(t)| ,$$

where t_j is the left end point of I_j. Let, in view of Lemma 2.5, N be chosen such that

$$\sup_{t\in I_j} P\{|z(t_{j+1}) - z(t)| \geqq \tfrac{1}{4}\varepsilon\} \leqq \tfrac{1}{2} .$$

Then using Lemma 2.9,

$$P_n\{z: \sup_{t\in I_j} |z(t_j) - z(t)| \geqq \tfrac{1}{2}\varepsilon\} \leqq 2P_n\{z: |z(t_{j+1}) - z(t_j)| \geqq \tfrac{1}{4}\varepsilon\} .$$

Using (2.14) for N large enough,

$$\begin{aligned}
&\limsup_{n\to\infty} \frac{1}{a_n} \log P_n[z: \|\theta_N z - z\| \geqq \varepsilon] \\
&\leqq \limsup_{n\to\infty} \frac{1}{a_n} \log P_n\left[z: \max_j \sup_{t\in I_j} |z(t_j) - z(t)| \geqq \tfrac{1}{2}\varepsilon\right] \\
&\leqq \limsup_{n\to\infty} \frac{1}{a_n} \log \left[\sum_j 2P_n\{z: |z(t_{j+1}) - z(t_j)| \geqq \tfrac{1}{4}\varepsilon\}\right] \\
&= \limsup_{n\to\infty} \frac{1}{a_n} \log \left[\max_j P_n[z: |z(t_{j+1}) - z(t_j)| \geqq \tfrac{1}{4}\varepsilon\right] \\
&\leqq -\inf_{|y|\geqq\frac{1}{4}\varepsilon} \frac{T}{N} h\left(\frac{\varepsilon N}{4T}\right) .
\end{aligned}$$

Property (v) of $h(y)$ now ensures that

$$\lim_{N\to\infty} \limsup_{n\to\infty} \frac{1}{a_n} \log P_n[z: \|\theta_N z - z\| \geqq \varepsilon] = -\infty .$$

LEMMA 2.11. *Let C be any closed subset of Ω. Let C^ε be the sphere around C of radius ε. Then*

$$\limsup_{n\to\infty} \frac{1}{a_n} \log P_n(C) \leqq -\lim_{\varepsilon\downarrow 0} \inf_{z\in C^\varepsilon} I(z) .$$

Proof:

(2.15)
$$P_n(C) \leqq P_n[z: \|\theta_N z - z\| \geqq \varepsilon] + P_n[z: \theta_N z \in C^\varepsilon] .$$

We denote by L_N the range of θ_N and by τ_N the map of $\Omega \to R_N$ sending ω

to the vector $\omega(jT/N), j = 1, 2, \cdots, N$; τ_N maps L_N onto R_N in a bicontinuous manner. Further $\tau_N\theta_N = \tau_N$. Hence

$$\begin{aligned} P_n[z\colon \theta_N z \in C^\varepsilon] &\leqq P_n[z\colon \theta_N z \in \overline{C^\varepsilon}] \\ &= P_n[z\colon \theta_N z \in \overline{C^\varepsilon} \cap L_N] \\ &= P_n[z\colon \tau_N\theta_N z \in \tau_N(\overline{C^\varepsilon} \cap L_N)] \\ &= P_n[z\colon \tau_N z \in \tau_N(\overline{C^\varepsilon} \cap L_N)] , \end{aligned}$$

and therefore,

$$\begin{aligned} \limsup_{n\to\infty} \frac{1}{a_n} &\log P_n[z\colon \theta_N z \in C^\varepsilon] \\ &\leqq \limsup_{n\to\infty} \frac{1}{a_n} \log P_n[\tau_N^{-1}\tau_N(\overline{C^\varepsilon} \cap L_N)] \\ &\leqq -\inf_{y\in\tau_N(\overline{C^\varepsilon}\cap L_N)} \Phi(y) \\ &= -\inf_{z\in\overline{C^\varepsilon}\cap L_N} \Phi(\tau_N z) . \end{aligned}$$

But for z in L_N, $\Phi(\tau_N z) = I(z)$. Hence

$$\begin{aligned} \limsup_{n\to\infty} \frac{1}{a_n} \log P_n[z\colon \theta_N z \in C^\varepsilon] &\leqq -\inf_{z\in C^\varepsilon\cap L_N} I(z) \\ &\leqq -\inf_{z\in C^\varepsilon} I(z) . \end{aligned}$$

In view of Lemma 2.10 and (2.15),

$$\limsup_{n\to\infty} \frac{1}{a_n} \log P_n(C) \leqq -\inf_{z\in\overline{C^\varepsilon}} I(z) .$$

Since this is true for every $\varepsilon > 0$ and since $\overline{C^\varepsilon} \subseteq C^{2\varepsilon}$,

$$\limsup_{n\to\infty} \frac{1}{a_n} \log P_n(C) \leqq -\liminf_{\varepsilon\downarrow 0}\, \inf_{z\in C^\varepsilon} I(z) .$$

LEMMA 2.12. *The set $[z\colon I(z) \leqq M]$ is a compact subset of continuous functions in Ω.*

Proof: Since $I(z) < \infty$ implies that z is continuous, it suffices to show that the set in question is an equicontinuous family of functions. Let us denote the set by A. For $z \in A$ write

$$z_1(t) = \int_0^t \alpha(z'(s))\, ds ,$$

$$z_2(t) = \int_0^t \beta(z'(s))\, ds ,$$

where

$$\alpha(x) = \begin{cases} 0 & \text{if} \quad |x| > \lambda, \\ x & \text{if} \quad |x| > \lambda. \end{cases}$$

Clearly $\alpha + \beta \equiv x$ and $z_1 + z_2 \equiv z$. As z varies over A the first derivative of z_1 is bounded by λ and hence is equicontinuous. We shall complete the proof by showing that , by choosing λ large, $\|z_2\|$ can be made uniformly small over A.

Using property (v) of $h(y)$, let λ be chosen such that

$$\frac{h(y)}{|y|} \geqq \frac{1}{\varepsilon} \quad \text{for} \quad |y| \geqq \lambda .$$

Then

$$\begin{aligned} \|z_2\| &\leqq \int_0^T |\beta(z'(s))|\, ds \\ &\leqq \varepsilon \int_0^T h(z'(s))\, ds \\ &\leqq \varepsilon M . \end{aligned}$$

$I(z)$ is lower semicontinuous; therefore if the $[z\colon I(z) \leqq M]$ are compact sets,

$$\liminf_{\varepsilon \downarrow 0} \inf_{z \in C^\varepsilon} I(z) = \inf_{z \in C} I(z) \tag{2.16}$$

for any closed C. Lemmas 2.11 and 2.12 complete the proof of one part of Theorem 2.1.

Lemma 2.13. *Let G be the subset of Ω defined by*

$$G = [z\colon |z(t) - a| < \varepsilon] ,$$

where ε is some positive number. Then

$$\liminf_{n \to \infty} \frac{1}{a_n} \log P_n(G) \geqq -h\left(\frac{a}{t}\right) .$$

Proof: Let $\lambda_n = [nt]$. Then

$$\begin{aligned} P_n[G] &= P\left[\left|\frac{x_{n1} + \cdots + x_{n\lambda_n}}{n} - a\right| < \varepsilon\right] \\ &\geqq P\left[\left|x_{nj} - \frac{an}{\lambda_n}\right| < \varepsilon \text{ for all } j\right] \\ &\geqq P\left[\left|x_{nj} - \frac{a}{t}\right| < \tfrac{1}{2}\varepsilon \text{ for all } j\right] \text{ if } n \text{ is large enough} \\ &= \left\{c_n \int_{|y - a/t| < \frac{1}{2}\varepsilon} \exp\left[-\frac{a_n}{n} n(y)\right] dy\right\}^{\lambda_n} . \end{aligned}$$

The function $h(y)$ is continuous except for jumps to $+\infty$; therefore, if $h(a/t) < \infty$, $h(y)$ is continuous at that point either from the left or from the right so that, for small enough Δ, there is a one-sided interval around a/t, where $h(y) \geqq h(a/t) - \Delta$. Hence

$$P_n[G] \geqq c_n^{\lambda_n}\delta^{\lambda_n} \exp\left\{-\frac{a_n}{n}\left[h\left(\frac{a}{t}\right) - \Delta\right]\lambda_n\right\}.$$

Since

$$\lim_{n\to\infty} \frac{\lambda_n}{n} = t\,,$$

$$\lim_{n\to\infty} a\frac{n}{a_n} \log c_n = 0\,,$$

$$\liminf_{n\to\infty} \frac{1}{a_n} \log P_n[G] \geqq -t\left[h\left(\frac{a}{t}\right) - \Delta\right]. \tag{2.17}$$

Now letting $\Delta \to 0$, we have

$$\liminf_{n\to\infty} \frac{1}{a_n} \log P_n[G] \geqq -th\left(\frac{a}{t}\right).$$

The following lemmas are derived easily from Lemma 2.13 in a manner which is similar to that used in the derivation of Lemmas 2.7 and 2.8.

LEMMA 2.14. *Let A be a sphere around a point y in R, τ the map in Lemma 2.8, Φ the function defined there. Let $G = \tau^{-1}A$, then*

$$\liminf_{n\to\infty} \frac{1}{a_n} \log P_n[G] \geqq -\Phi(y)\,.$$

LEMMA 2.15. *Let G be the sphere around a point f in Ω of radius ε. Then*

$$\liminf_{n\to\infty} \frac{1}{a_n} \log P_n[G] \geqq -I(f)\,.$$

Proof: If f is not continuous, $I(f) = +\infty$ and there is nothing to prove. Let f be continuous, then

$$\lim_{N\to\infty} \|\theta_N f - f\| = 0\,.$$

Let N be chosen large enough so that

$$\|\theta_N f - f\| < \tfrac{1}{3}\varepsilon\,.$$

Then

$$\begin{aligned} P_n[G] &= P_n[z: \|z - f\| < \varepsilon] \\ &\geqq P_n[z: \|\theta_N z - \theta_N f\| < \tfrac{1}{3}\varepsilon, \|\theta_N z - z\| < \tfrac{1}{3}\varepsilon] \qquad (2.18) \\ &\geqq P_n[z: \|\theta_N z - \theta_N f\| < \tfrac{1}{3}\varepsilon] - P_n[z: \|\theta_N z - z\| \geqq \tfrac{1}{3}\varepsilon]\,, \end{aligned}$$

and hence

$$\begin{aligned}\liminf_{n\to\infty}\frac{1}{a_n}\log P_n[z: \|\theta_N z-\theta_N f\|<\tfrac{1}{3}\varepsilon] &= \liminf_{n\to\infty}\frac{1}{a_n}\log P_n[z: \|\tau_N z-\tau_N f\|<\tfrac{1}{3}\varepsilon]\\ &= \liminf_{n\to\infty}\frac{1}{a_n}\log P_n[\tau_N^{-1}(y: \|y-\tau_N f\|<\tfrac{1}{3}\varepsilon]\\ &\geqq -\Phi(\tau_N f)\\ &= -I(\theta_N f)\\ &\geqq -I(f)\,.\end{aligned}$$

On the other hand, Lemma 2.10 implies that if N is large enough, the second term of (2.18) with a negative sign is negligible compared to the first term. Therefore,

$$\liminf_{n\to\infty}\frac{1}{a_n}\log P_n[G]\geqq -I(f)\,.$$

Now let G be any open set. If $f\in G$, there exists a sphere around f contained in G. If S is the sphere, then

$$\begin{aligned}\liminf_{n\to\infty}\frac{1}{a_n}\log P_n[G] &\geqq \liminf_{n\to\infty}\frac{1}{a_n}\log P_n[S]\\ &\geqq -I(f)\,;\end{aligned}$$

since this is true for every f in G,

$$\liminf_{n\to\infty}\frac{1}{a_n}\log P_n[G]\geqq -\inf_{z\in G} I(z)\,.$$

This completes the proof of Theorem 2.1.

COROLLARY 2.1. *If $\bar{A}$, A^0 are the closure and interior of a set A in Ω and if*

$$\inf_{z\in A^0} I(z)=\inf_{z\in\bar{A}} I(z)=m\,,$$

then

$$\lim_{n\to\infty}\{P_n[A]\}^{1/a_n}=e^{-m}\,.$$

3. Let Ω be a regular topological space, and P_n a sequence of probability measures on Ω. The σ-field of subsets of Ω on which P_n is defined is assumed to include all open and closed subsets of Ω. Let a_n be a sequence of non-negative numbers such that

$$a_n \to \infty,$$

(3.1)
$$\limsup_{n\to\infty} \frac{1}{a_n} \log P_n(C) \leqq -\inf_{z\in C} I(z),$$

$$\liminf_{n\to\infty} \frac{1}{a_n} \log P_n(G) \geqq -\inf_{z\in G} I(z),$$

where G is open in Ω and C is closed in Ω. $I(z)$ is assumed to have the following properties:

(i) $0 \leqq I(z) \leqq \infty$,

(ii) $I(z)$ is lower semicontinuous everywhere in Ω,

(iii) $[z: I(z) \leqq M]$ is a compact subset of Ω for every finite M.

Let $F_n(z)$ be a suitable function on Ω. By $J_n(A)$ we denote the integral

(3.2)
$$J_n(A) = \int_A \exp\{a_n F_n(z)\}\, dP_n,$$

$$J_n = J_n(\Omega) = \int_\Omega \exp\{a_n F_n(z)\}\, dP_n.$$

We shall investigate the asymptotic behavior of J_n. We make the following assumptions on the sequence $F_n(z)$.

(i) Given $\varepsilon > 0$, there exists, for any z_0 such that $I(z_0) < \infty$, a neighborhood N_{z_0} of z_0 in Ω such that, for all sufficiently large n and z in N_{z_0},

$$F_n(z) \leqq F(z_0) + \varepsilon.$$

(ii) $F_n(z) \leqq L$ for all n and z in Ω.

THEOREM 3.1. *Under conditions* (i) *and* (ii) *on* $F_n(z)$,

$$\limsup_{n\to\infty} \frac{1}{a_n} \log J_n \leqq \sup_{z\in\Omega} [F(z) - I(z)].$$

Proof: Let

$$\Lambda = \sup_{z\in\Omega} [F(z) - I(z)].$$

Let $f \in \Omega$ be an arbitrary point in Ω such that $I(f) < \infty$. Then for any $\varepsilon > 0$ there exists (from property (ii) of $I(z)$ and property (i) of $F_n(z)$) a neighborhood N_f of f such that

$$F_n(z) \leqq F(f) + \varepsilon \qquad \text{for} \qquad z \in N_f,$$

$$I(z) \geqq I(f)) - \varepsilon \qquad \text{for} \qquad z \in N_f.$$

Let G_f be another neighborhood of f such that

$$\bar{G}_f = C_f \subset N_f ,$$

then

$$\begin{aligned} J_n(G_f) &\leqq \exp \{\sup_{z\in G_f} a_n F_n(z)\} P_n(G_f) \\ &\leqq \exp \{a_n \sup_{z\in N_f} a_n F_n(z)\} P_n(C_f) \\ &\leqq \exp \{a_n[F(f) + \varepsilon]\} P_n(C_f) . \end{aligned}$$

Therefore,

$$\begin{aligned} \limsup_{n\to\infty} \frac{1}{a_n} \log J_n(G_f) &\leqq F(f) + \varepsilon - \inf_{z\in C_f} I(z) \\ &\leqq F(f) - I(f) + 2\varepsilon \qquad (3.3) \\ &\leqq \Lambda + 2\varepsilon . \end{aligned}$$

Let K be the set $\{z\colon I(z) \leqq M\}$. For a given $\varepsilon > 0$, and any $f \in K$, there exists a neighborhood G_f of f such that

$$\limsup_{n\to\infty} \frac{1}{a_n} \log J_n(G_f) \leqq \Lambda + 2\varepsilon .$$

Since by property (iii) of $I(z)$, K is compact in Ω, there exists a finite union of $\{G_f\}, f \in K$, which covers K. Let us denote by G this finite union. Then

$$\begin{aligned} \limsup_{n\to\infty} \frac{1}{a_n} \log J_n(G) &\leqq \max_{f=f_1,\cdots,f_N} \limsup_{n\to\infty} \frac{1}{a_n} \log J_n(G_f) \\ &\leqq \Lambda + 2\varepsilon . \end{aligned}$$

Let $C = \Omega - G$. Then

$$J_n(C) \leqq \exp \{a_n L\} P_n(C) .$$

$$\limsup_{n\to\infty} \frac{1}{a_n} \log J_n(C) \leqq L - \inf_{z\in C} I(z) \leqq L - M .$$

If we choose M large enough so that $L - M \leqq \Lambda$,

$$\limsup_{n\to\infty} \frac{1}{a_n} \log J_n \leqq \max [\Lambda + 2\varepsilon, L - M] = \Lambda + 2\varepsilon .$$

Since $\varepsilon > 0$ is arbitrary,

$$\limsup_{n\to\infty} \frac{1}{a_n} \log J_n \leqq \Lambda .$$

We now replace the assumption that $\{F_n(z)\}$ is uniformly bounded from above by the less restrictive condition

$$\lim_{L\to\infty} \limsup_{n\to\infty} \frac{1}{a_n} \log \left[\int_{F_n(z)\geqq L} \exp \{a_n F_n(z)\}\, dP_n \right] = -\infty . \qquad (3.4)$$

THEOREM 3.2. *If $\{F_n(z)\}$ satisfies condition* (i) *and* (3.4), *then*

$$\limsup_{n\to\infty} \frac{1}{a_n} \log J_n \leqq \sup_z [F(z) - I(z)] .$$

Proof: Let us denote by $\bar{F}_n(z)$ the function defined to be

$$\bar{F}_n(z) = \min [F_n(z), L] .$$

Then $\bar{F}_n(z)$ satisfies the conditions of Theorem 3.1 and, hence, if

$$\bar{J}_n = \int_\Omega \exp \{a_n \bar{F}_n(z)\} dP_n ,$$

$$\limsup_{n\to\infty} \frac{1}{a_n} \log \bar{J}_n \leqq \sup [\bar{F}(z) - I(z)] \leqq \sup [F(z) - I(z)] .$$

Condition (3.4) insures that for large L

$$\limsup_{n\to\infty} \frac{1}{a_n} \log J_n = \limsup_{n\to\infty} \frac{1}{a_n} \log \bar{J}_n .$$

This proves the theorem.

Remark. Condition (i) is equivalent to the following:
Whenever $z_n \to z$ and $I(z) < \infty$,

$$\limsup_{n\to\infty} F_n(z_n) \leqq F(z) .$$

This is always true if $F_n = F$ for all n, and F is upper semicontinuous.

Let $f \in \Omega$ be such that for any $\varepsilon > 0$ there exists a neighborhood G_f of f in Ω such that, for n suitably large and z in G_f,

$$F_n(z) \geqq F(f) - \varepsilon . \tag{3.5}$$

Then

$$J_n \geqq J_n(G_f) \geqq \exp \{a_n[F(f) - \varepsilon]\} P_n(G_f) ,$$

$$\liminf_{n\to\infty} \frac{1}{a_n} \log J_n \geqq F(f) - \varepsilon - \inf_{z \in G_f} I(z) \geqq F(f) - I(f) - \varepsilon .$$

Since $\varepsilon > 0$ is arbitrary,

$$\liminf_{n\to\infty} \frac{1}{a_n} \log J_n \geqq F(f) - I(f) . \tag{3.6}$$

If Ω_0 is defined as the set of points f in Ω for which, given any $\varepsilon > 0$, there exists a neighborhood G_f such that (3.5) holds for suitably large n, we have from (3.6) the following theorem.

THEOREM 3.3.

$$\liminf_{n\to\infty} \frac{1}{a_n} \log J_n \geqq \sup_{z\in\Omega_0} [F(z) - I(z)] .$$

We can now summarize Theorems 3.2 and 3.3 in the following manner. Let $\{F_n(z)\}$ be a sequence of functions satisfying condition (3.4). Assume further:

(i) For every z in Ω such that $I(z) < \infty$ and $z_n \to z$,

$$\limsup_{n\to\infty} F_n(z_n) \leqq F(z) .$$

(ii) For every z in Ω_0 and $z_n \to z$,

$$\liminf_{n\to\infty} F_n(z_n) \geqq F(z) .$$

(iii) $\Lambda = \sup_{z\in\Omega} [F(z) - I(z)] = \sup_{z\in\Omega_0} [F(z) - I(z)]$.

Then the following theorem holds.

THEOREM 3.4.

$$\lim_{n\to\infty} \frac{1}{a_n} \log J_n = \Lambda .$$

Let A be any closed subset of Ω. Let $\{F_n(z)\}$ satisfy the conditions of Theorem 3.2. Define $\bar{F}_n(z) = F_n(z)$ for $z \in A$, $\bar{F}_n(z) = -\infty$ for $z \in \Omega - A$, $\bar{F}(z) = F(z)$ for $z \in A$ and $\bar{F}(z) = -\infty$ for $z \in \Omega - A$. Then $\{\bar{F}_n(z)\}$ satisfies the conditions and one obtains

THEOREM 3.5.

$$\limsup_{n\to\infty} \frac{1}{a_n} \log J_n(A) \leqq \sup_{z\in A} [F(z) - I(z)] .$$

Let us now suppose that $\{F_n(z)\}$ is a sequence of functions satisfying the condition (3.4) and such that:

(i) Whenever $I(z) < \infty$ and $z_n \to z$,

$$\limsup_{n\to\infty} F_n(z_n) \leqq F(z) .$$

(ii) The supremum $\Lambda = \sup_{z\in\Omega} [F(z) - I(z)]$ is attained at a point f in Ω.

(iii) $\sup_{\|z-f\|\geqq\varepsilon} [F(z) - I(z)] < \Lambda$ for every $\varepsilon > 0$.

(iv) $\liminf_{n\to\infty} F_n(z_n) \geqq F(f)$ for every sequence $z_n \to f$.

We define a measure Q_n on Ω by the formula

$$Q_n(A) = \frac{1}{J_n} J_n(A) \qquad \text{for every} \qquad A . \tag{3.7}$$

THEOREM 3.6. *If $g(z)$ is a bounded function on Ω which is continuous at $z = f$, then*

$$\lim_{n\to\infty} \int g(z)\, dQ_n = g(f) .$$

Proof: Since $g(z)$ is continuous at $z = f$, it suffices to show that

$$\lim_{n\to\infty} Q_n(A) = 0 ,$$

for every set A of the form $\|z - f\| \geqq \varepsilon$. Since A is a closed set,

$$\limsup_{n\to\infty} \frac{1}{a_n} \log J_n(A) < \Lambda$$

and

$$\liminf_{n\to\infty} \frac{1}{a_n} \log J_n = \Lambda .$$

This proves the theorem.

4. We now go back to the case considered in Section 2. Let P_n be the measures defined there. Let $\{F_n(z)\}$ be any sequence of functions such that

$$F_n(z) \leqq A + B\, \|z\|$$

for some constants A and B, independent of n. Then we can show that condition (3.4) is satisfied.

LEMMA 4.1. *If $f_n(z) \leqq A + B\, \|z\|$ for all n and z,*

$$\lim_{L\to\infty} \limsup_{n\to\infty} \frac{1}{a_n} \log\left[\int_{F_n(z)\geqq L} \exp\{a_n F_n(z)\}\, dP_n\right] = -\infty .$$

Proof:

$$\begin{aligned}\int_{F_n(z)\geqq L} \exp\{a_n F_n(z)\}\, dP_n &\leqq \int_{F_n(z)\geqq L} \exp\{a_n[A + B\, \|z\|]\}\, dP_n \\ &\leqq \int_{\|z\|\geqq (L-A)/B} \exp\{a_n[A + B\, \|z\|]\}\, dP_n .\end{aligned}$$

It remains to prove that

$$\lim_{L\to\infty} \limsup_{n\to\infty} \frac{1}{a_n} \log\left|\int_{\|z\|\geqq L} \exp\{a_n B\, \|z\|\}\, dP_n\right| = -\infty .$$

For this purpose we need to estimate $P_n[\|z\| \geqq L]$. It is easy to see that, for L suitably large, the conditions of Lemma 2.9 are satisfied and, hence,

$$P_n[\|z\| \geqq L] \leqq 2P_n[|z(T)| \geqq \tfrac{1}{2}L] .$$

Therefore it suffices to prove that, for every B,

$$\lim_{L\to\infty} \limsup_{n\to\infty} \frac{1}{a_n} \log \left[\int_{|z(T)| \geqq L} \exp \{a_n B\, |z(T)|\}\, dP_n \right] = -\infty . \tag{4.1}$$

From equation (2.4) we have

$$P_n[z(T) \geqq M] \leqq \exp \{-a_n \max_{\theta \geqq 0} [\theta M - TQ_n(\theta)]\} \leqq \exp \{-a_n[\theta_0 M - K]\} ,$$

where K is a constant depending on θ_0. Since there exists a similar estimate for $P_n[z(T) \leqq -M]$, we can ensure that for a given value of θ_0 there exists a constant K such that

$$P_n[|z(T)| \geqq M] \leqq K \exp \{-a_n \theta_0\} .$$

By choosing θ_0 bigger than B, one can establish (4.1).

Let $P(\tau, x)$ be a continuous function of x for $0 \leqq t \leqq T$ and $-\infty < x < \infty$. Let $U(x)$ be a continuous function of x in $-\infty < x < \infty$. Moreover, let

(i) $U(x) \leqq A + B\, |x|$ for all x ,

(ii) $P(\tau, x) \leqq A + B\, |x|$ for all τ and x .

Consider the functions $F_n(z)$ on Ω defined by

$$F_n(z) = \int_0^{t_n} P(\tau, z(t_n - \tau))\, d\tau + U(z(t_n)) ,$$

and let

$$F(z) = \int_0^{t} P(\tau, z(t - \tau))\, d\tau + U(z(t)) .$$

Since z is continuous in $0 \leqq t \leqq T$, it is not difficult to see that, if $z_n \to z$ and $t_n \to t$ as $n \to \infty$,

$$\lim_{n\to\infty} F_n(z_n) = F(z)$$

and

$$F_n(z) \leqq A_0 + B_0\, \|z\| ,$$

for all z with $I(z) < \infty$. We therefore have

THEOREM 4.1.

$$\lim_{n\to\infty} \frac{1}{a_n} \log \left[\int_\Omega \exp \{a_n F_n(z)\}\, dP_n \right] = \sup_{z \in \Omega} [F(z) - I(z)] ,$$

where

$$I(z) = \int_0^T h(z'(\tau))\, d\tau .$$

COROLLARY 4.1. *Since $F(z)$ depends only on the path up to time t, we can write*

$$\sup_{z\in\Omega}[F(z) - I(z)] = \sup_{z\in\Omega}\left[F(z) - \int_0^t h(z'(\tau))\,d\tau\right].$$

LEMMA 4.2. *Let $F(z) \leqq A + B\,\|z\|$ and let $F(z)$ be upper semicontinuous. Then for any closed set C the supremum*

$$\Lambda = \sup_{z\in C}[F(z) - I(z)]$$

is attained.

Proof: Since $\Lambda > -\infty$ we can assume, without loss of generality, that $z_n \in C$ is such that

$$F(z_n) - I(z_n) \to \Lambda\,.$$

If we show that z_n is compact, then, in view of the proper semicontinuity assumptions, the supremum will be attained at any limit point (which are necessarily within the closed set C). Since

$$F(z_n) - I(z_n) \to \Lambda > -\infty\,,$$

there are two possibilities. Either $I(z_n) \to +\infty$, or there exists a subsequence with $I(z_n)$ bounded, which would mean that there exists at least one limit point. So let us suppose that $I(z_n) \to +\infty$. Let $L > B$ and let M be chosen such that if $|y| \geqq M$, $|n(y)| \geqq L\,|y|$,

$$\begin{aligned} I(z_n) &= \int_0^T h(z_n'(\tau))\,d\tau \\ &\geqq L\int_{[t:|z_n'(t)|\geqq M]} |z_n'(\tau)|\,d\tau \\ &\geqq L\int_0^T |z_n'(\tau)|\,d\tau - TML, \\ &\geqq L\,\|z_n\| - TML\,, \end{aligned}$$

then

$$F(z_n) - I(z_n) \leqq (B - L)\,\|z_n\| + TML + A\,.$$

Since $F(z_n) - I(z_n) \to \Lambda > -\infty$, and $L > B$, we must have $\|z_n\|$ bounded or $F(z_n)$ bounded from above. This in turn would imply that $I(z_n)$ is bounded and hence the lemma.

Let $\{F_n(z)\}$, $F(z)$ be as described in Theorem 4.1. Moreover, let the supremum

$$\Lambda = \sup_{z\in\Omega} F(z) - \int_0^t h(z'(\tau))\,d\tau$$

be attained at a unique function f in the space of functions on the interval $[0, t]$. Let $g(z)$ be a function of z for z in Ω depending only on $z(\tau)$ for $0 \leqq \tau \leqq t$. Let g be continuous at $z = f$. Then

THEOREM 4.2.

$$\lim_{n\to\infty} \frac{\int g(z) \exp\{a_n F_n(z)\}\, dP_n}{\int \exp\{a_n F_n(z)\}\, dP_n} = g(f)\,.$$

Proof: The proof is not quite the same as that of Theorem 3.6, because the supremum over Ω may not be unique. However the conditions ensure that any function $\bar{f}$ in Ω where the sup is attained must be an extension of f defined on $0 \leqq \tau \leqq t$. The function $g(z)$ depends only on $z(\tau)$ for $0 \leqq \tau \leqq t$. Hence taking C_ε to be the closed set

$$C_\varepsilon = [z\colon \sup_{0\leqq\tau\leqq t} |z(\tau) - f(\tau)| \geqq \varepsilon]\,,$$

it suffices to show that for every $\varepsilon > 0$

$$\lim_{n\to\infty} \frac{J_n[C_\varepsilon]}{J_n} = 0\,.$$

But

$$\limsup_{n\to\infty} \frac{1}{a_n} \log J_n[C_\varepsilon] \leqq \sup_{z\in C_\varepsilon} [F(z) - I(z)] < \Lambda\,.$$

From now on the theorem is proved in the same way as Theorem 3.6.

5. Let us start with the functions (continuous in x)

$$U(x) \leqq A + B\,|x|\,,$$
$$P(t, x) \leqq A + B\,|x|\,,$$

and consider the integral

(5.1) $$V_n(t, x) = \int_\Omega H_n(t, z, x)\, dP_n\,,$$

where

$$H_n(t, z, x) = \exp\left\{a_n\left[\int_0^t P(t-\tau, z(\tau) + x)\, d\tau + U(z(t) + x)\right]\right\}.$$

It is easy to check, by taking suitable conditional expectations in (5.1), that

(5.2) $$V_n\left(\frac{j+1}{n}, x\right) = \exp\left\{a_n \int_{j/n}^{(j+1)/n} P(\tau, x)\, d\tau\right\} \int V_n\left(\frac{j}{n}, x + \frac{y}{n}\right) dF_n(y)\,,$$
$$V_n(0, x) = \exp\{a_n U(x)\}$$

Here P_n and F_n are as in Section 2. If we take log on both sides of (5.2) and put

$$U_n(t, x) = \frac{1}{a_n} \log V_n(t, x)\,,$$

then

$$U_n\left(\frac{j+1}{n}, x\right) - U_n\left(\frac{j}{n}, x\right) = \int_{j/n}^{(j+1)/n} P(\tau, x)\, d\tau$$

$$+ \frac{1}{a_n} \log \int \exp\left\{a_n \left[U_n\left(\frac{j}{n}, x + \frac{U}{n}\right) - U_n\left(\frac{j}{n}, x\right)\right]\right\} dF_n(y) .$$

Multiplying by n and taking the limit as $n \to \infty$,

$$U_t(t, x) = P(t, x) + f(U_x(t, x)) .$$

Hence, formally denoting $u = U_x$,

$$u_t = p(t, x) + [f(u)]_x .$$

We shall first examine the behavior of $U_n(t, x)$. Since the recurrence relation (5.2) yields the value of $U_n(t, x)$ for t of the form j/n, we can choose j_n such that $j_n/n \to t$.

Theorem 5.1.

$$\lim_{n\to\infty} U_n(t_n, x) = U(t, x)$$

exists, whenever $t_n \to t$ as $n \to \infty$, and the limit is given by

$$U(t, x) = \sup_{z\in\Omega} \left[\int_0^t P(t - \tau, z(\tau) + x)\, d\tau + U(z(t) + x) - \int_0^t h(z'(t))\, d\tau\right].$$

Proof:

$$V_n(t_n, x) = \int_\Omega \exp\{a_n F_n(z)\}\, dP_n ,$$

where

$$F_n(z) = \int_0^{t_n} P(t_n - \tau, z(\tau) + x)\, d\tau + U(z(t_n) + x)$$

$$= \int_0^{t_n} P(\tau, z(t_n - \tau) + x)\, d\tau + U(z(t_n) + x) .$$

Since Theorem 4.1 is applicable now (the translation by x does not affect the validity) the theorem is proved.

Theorem 5.2. *Let*

$$U(x) = \int_0^x u(y)\, dy ,$$

$$P(t, x) = \int_0^x p(t, y)\, dy ,$$

$$u_n(t, x) = \frac{d}{dx} U_n(t, x) ,$$

where u and p are continuous functions. Then, if for a given t and x the supremum in Theorem 5.1 is attained at a unique function $\omega(\tau)$, then

$$\lim_{n\to\infty} u_n(t_n, x) = \int_0^t p(t-\tau, \omega(\tau) + x)\, d\tau + u(\omega(t) + x).$$

Proof: This follows immediately from the definition of $u_n(t, x)$ and Theorem 4.2.

We can summarize the results as follows. Suppose we are interested in the equation

$$u_t = [f(u)]_x + p(t, x) \tag{5.3}$$

with initial condition $u_0(x) = u(x)$. We form the functions $U(x)$, $P(t, x)$ which are indefinite integrals of $u(x)$ and $p(t, x)$, respectively. We then construct V_n successively by the linear model (5.2). We define

$$U_n = \frac{1}{a_n} \log V_n .$$

Then under suitable conditions, $U_n \to U$, where U is given in Theorem 5.1. U_x is known to solve the equation (5.3) in the weak sense, [4], [7].

Under additional conditions we even have the convergence of dU_n/dx for specific values of t and x.

An important special case is when $f(\theta)$ is of the form

$$f(\theta) = c\left[\log \int_{-\infty}^{\infty} e^{\theta y}\, dF(y)\right].$$

In this case F_n can be taken to be equal to F for all n and a_n can be chosen to be n.

The estimates of Section 2 can be obtained in a similar way starting instead from the known results for large deviations of arithmetic means. This includes the case studied by Lax [6]. In that case F is a discrete measure on a discrete set of equally spaced points.

If $f(\theta)$ has the further special form

$$f(\theta) = c\left[\log \int_{-\infty}^{\infty} e^{\theta y}\, dF(y)\right],$$

where F is an infinitely divisible distribution, the discrete manner in which P_n is constructed could be replaced by a continuous process and instead of (5.2) we would obtain

$$\frac{dV_n}{dt} = a_n V_n P + L_n[V_n],$$

where L_n is the generator of a suitable process with independent increments. The case where $f = \frac{1}{2}\theta^2$ leads through the Gaussian distribution to Brownian motion and the second derivative for L_n.

For the detailed exposition of processes with independent increments and connections with semigroups and their generators see Hille and Phillips [2].

Bibliography

[1] Donsker, M. D., *Function space integrals and non-linear differential equations*, to appear.
[2] Hille, E., and Phillips, R. S., *Functional Analysis and Semigroups*, Rev. Ed., Amer. Math. Soc., Providence, R. I., 1957.
[3] Hopf, E., *On the partial differential equation* $u_t + uu_x = \mu u_{xx}$, Comm. Pure Appl. Math., Vol. 3, 1950, pp. 201–230.
[4] Conway, E., and Hopf, E., *Hamilton's theory and the generalized solutions of the Hamilton Jacobi equation*, J. Math. Mech., Vol. 13, 1964, pp. 939–986.
[5] Lax, P. D., *Hyperbolic systems of conservation laws, II*, Comm. Pure Appl. Math., Vol. 10, 1957, pp. 537–566.
[6] Lax, P. D., *On difference schemes for solving initial value problems for conservation laws*, Proc. Rome Symposium, Provisional International Computation Centre, Libreria Eredi Virgilio Veschi, Roma, 1958, pp. 69–78.
[7] Oleinik, O. A., *On discontinuous solutions of non-linear differential equations*, Uspekhi Mat. Nauk, Vol. 12, 1957, pp. 3–73. (In Russian.)
[8] Schilder, M., *On a Laplace asymptotic formula for Wiener integrals*, to appear in the Trans. Amer. Math. Soc.

Received November, 1965.

COMMUNICATIONS ON PURE AND APPLIED MATHEMATICS, VOL. XX, 431–455 (1967)

On the Behavior of the Fundamental Solution of the Heat Equation with Variable Coefficients*

S. R. S. VARADHAN

1. Introduction

The solution $q(t, x, y)$ of the equation

$$\frac{\partial q}{\partial t} = \frac{1}{2} \sum_{i=1}^{k} \frac{\partial^2 q}{\partial x_i^2} \tag{1.1}$$

with the boundary condition $q(t, x, y) \to \delta_x(y)$ as $t \to 0$ can of course be written explicitly as

$$q(t, x, y) = (2\pi t)^{-k/2} \exp\left\{-\frac{1}{2t} \|x - y\|^2\right\}, \tag{1.2}$$

where $\|x - y\|$ denotes the Euclidean distance. Looking at (1.2) one sees immediately that

$$\lim_{t \to 0} [-2t \log q(t, x, y)] = \|x - y\|^2 . \tag{1.3}$$

We shall consider the analogue of (1.1). Let $p(t, x, y)$ be the solution of the equation

$$\frac{\partial p}{\partial t} = \frac{1}{2} \sum_{i,j=1}^{k} a_{ij}(x) \frac{\partial^2 p}{\partial x_i \, \partial x_j} \tag{1.4}$$

with the boundary condition $p(t, x, y) \to \delta_x(y)$ as $t \to 0$. We prove a formula similar to (1.3):

$$\lim_{t \to 0} [-2t \log p(t, x, y)] = d^2(x, y) , \tag{1.5}$$

where $d(x, y)$ is the distance induced by a Riemannian metric derived from the coefficients $\{a_{ij}(x)\}$.

* This paper represents results obtained at the Courant Institute of Mathematical Sciences, New York University, under the sponsorship of the Office of Naval Research, Contract Nonr-285(46), and a Sloan Foundation Grant for Probability and Statistics. Reproduction in whole or in part is permitted for any purpose of the United States Government.

If G is a region with a boundary B, we consider the solution $\phi(x, \lambda)$ of the equation

$$\frac{1}{2}\sum a_{ij}(x)\frac{\partial^2\phi}{\partial x_i\,\partial x_j} = \lambda\phi \qquad \text{for} \qquad x \in G$$

with the boundary value $\phi = 1$ on B. We prove that

$$\lim_{\lambda\to\infty}\,[-(2\lambda)^{-1/2}\log\phi(x, \lambda)] = d(x, B)\,, \tag{1.6}$$

where x is any point of G and $d(x, B)$ is the shortest distance to the boundary B from x.

The behavior (1.5) of $p(t, x, y)$ is derived from the behavior (1.6) of $\phi(x, \lambda)$ which is proved first. Although these results are closely related to certain properties of Markov processes, the probabilistic connections will be explored separately, [3].

Section 2 covers the preliminaries. The assumptions and the main theorems are stated there. The later sections cover the actual proof.

2. Preliminaries

L stands for the following differential operator acting on smooth functions on R_k:

$$Lf = \frac{1}{2}\sum a_{ij}(x)\frac{\partial^2 f}{\partial x_i\,\partial x_j}\,. \tag{2.1}$$

The matrix $\{a(x)\}$ of coefficients is of course symmetric and positive definite and the following assumptions are made concerning $a(x)$:

(i) Uniform Hölder condition, i.e., for all i and j we have

$$|a_{ij}(x) - a_{ij}(y)| \leqq M\,\|x - y\|^{\alpha} \text{ for all } x, y, \qquad \alpha > 0.$$

(2.2) (ii) Uniform ellipticity condition, i.e., there exist constants α and A such that for any vector $\xi_1, \cdots, \xi_k$, and x

$$A\sum \xi_i^2 \geqq \sum a_{ij}(x)\xi_i\xi_j \geqq \alpha\sum \xi_i^2\,.$$

We then consider the equation

$$\frac{\partial p}{\partial t} = Lp \tag{2.3}$$

with the boundary condition $p(t, x, y) \to \delta_x(y)$ as $t \to 0$.

The following theorem concerning the existence of a solution $p(t, x, y)$ for (2.3) can be found in [1], [2].

THEOREM 2.1. *Under the hypothesis* (2.2) *on the coefficients there exists a solution* $p(t, x, y)$ *for equation* (2.3) *and it has the following properties:*

(i) $p(t, x, y) \geqq 0$,

(ii) $\int p(t, x, y)p(s, y, z)\, dy = p(t + s, x, z)$,

(iii) $p(t, x, y)$ *is continuous in* t, x, *and* y,

(iv) *there are constants* M *and* α *such that*

$$p(t, x, y) \leqq Mt^{-k/2} \exp\left\{-\frac{\alpha}{2t}\|x - y\|^2\right\},$$

(v) *there are positive constants* α_1, α_2, M_1, M_2, λ *such that*

$$p(t, x, y) \geqq M_1 t^{-k/2} \exp\left\{-\frac{\alpha_1}{2t}\|x - y\|^2\right\}$$

$$-M_2 t^{-k/2+\lambda} \exp\left\{-\frac{\alpha_2}{2t}\|x - y\|^2\right\}.$$

Further $p(t, x, y)$ *is unique.*

We now introduce the metric d. Let $p(\tau)$, $0 \leqq \tau \leqq 1$, be a smooth path in R_k. Then the length of such a path is defined as

$$l(p) = \int_0^1 [\dot{p}(\tau)a^{-1}(p(\tau))\dot{p}(\tau)]^{1/2}\, d\tau,$$

where $\dot{p}(\tau)$ stands for $dp(\tau)/d\tau$, $a^{-1}(p(\tau))$ for the matrix inverse to $a(p(\tau))$ and $(\theta a \theta)$ for the quadratic form $\Sigma a_{ij}\theta_i\theta_j$; $l(p)$ is the natural length in a metric defined locally as

$$ds^2 = \sum a^{ij}(x)\, dx_i\, dx_j\,.$$

The operator L modified by a first order term is Laplacian in this metric. The global distance $d(x, y)$ induced by this metric is defined as

(2.4) $$d(x, y) = \inf_{\substack{p:\ p(0)=x \\ p(1)=y}} l(p)\,.$$

We can now state the first main theorem concerning the behavior of $p(t, x, y)$ introduced in Theorem 2.1 for small t.

THEOREM 2.2.

$$\lim_{t\to 0} [-2t \log p(t, x, y)] = d^2(x, y)$$

uniformly over x *and* y *such that* $d(x, y)$ *is bounded.*

We consider an open set G in R_k. Let B be the boundary of G. It is assumed that every point b on B is a limit point of the exterior of G.

Let $\phi(x, \lambda)$ be the solution of the equation

$$L\phi = \lambda\phi \quad \text{for} \quad x \in G,$$
$$\phi = 1 \quad \text{for} \quad b \in B. \tag{2.5}$$

The solution exists for $\lambda \geqq 0$, if the boundary value is properly interpreted. The next theorem is concerned with the behavior of the solution $\phi(x, \lambda)$ for large λ.

THEOREM 2.3. *With the assumptions stated above,*

$$\lim_{\lambda\to\infty}\left[-\frac{1}{\sqrt{2\lambda}}\log\phi(x,\lambda)\right] = d(x, B)$$

uniformly over compact subsets of $G \cup B$.

We state some notation and a few elementary consequences concerning the distance $d(x,y)$. The distance $[(x-y)\Lambda(x-y)]^{1/2}$ is denoted by $\|x-y\|_\Lambda$, where Λ is a positive definite matrix. If Λ is chosen as $a^{-1}(z)$, then $\|x-y\|$ is denoted by $\|x-y\|_z$; $\|x-y\|$ is of course the Euclidean distance. It is clear from the assumptions that d, $\| \quad \|$, $\| \quad \|_z$ are all equivalent in the sense that the ratio of any two is bounded by a universal constant. We shall denote by $\Delta(x,y)$ any one of these metrics. If a and b are positive definite matrices, we shall mean by $a \geqq b$ that $a - b$ is positive semidefinite. The following is a consequence of the Hölder condition in (2.2).

LEMMA 2.4. *There exists a function $\delta(\varepsilon) \downarrow 0$ as $\varepsilon \downarrow 0$ such that if $\Delta(x,y) \leqq \varepsilon$, then*

$$[1-\delta(\varepsilon)]a(x) \leqq a(y) \leqq [1+\delta(\varepsilon)]a(x),$$
$$[1-\delta(\varepsilon)]a^{-1}(x) \leqq a^{-1}(y) \leqq [1+\delta(\varepsilon)]a^{-1}(x).$$

From Lemma 2.4 and the definition of $d(x,y)$ we deduce

LEMMA 2.5. *There exists a function $\delta(\varepsilon) \downarrow 0$ as $\varepsilon \downarrow 0$ such that if $\Delta(x,y) \leqq \varepsilon$ and $\Delta(x, z) \leqq \varepsilon$, then*

$$[1-\delta(\varepsilon)]\,\|x-y\|_z \leqq d(x,y) \leqq [1+\delta(\varepsilon)]\,\|x-y\|_z\,.$$

3. Elliptic Equation

In this section we prove Theorem 2.3. The method consists of comparing our solution with the solution of an equation having constant coefficients in a small domain and then piecing together the estimate with the help of the maximum principle.

Let y be an arbitrary point in R_k. Let the set S_ε and its boundary B_ε be defined as follows:

$$S_\varepsilon = [x: \|x - y\|_y < \varepsilon] ,$$

$$B_\varepsilon = [x: \|x - y\|_y = \varepsilon] .$$

We denote by L_y the operator

$$L_y f = \tfrac{1}{2} \sum a_{ij}(y) \frac{\partial^2 f}{\partial x_i \, \partial x_j} .$$

It is an operator with constant coefficients. Let $\psi_\varepsilon(x, \lambda)$ be the solution to the equation

$$(3.1) \qquad \begin{aligned} L_y \psi_\varepsilon &= \lambda \psi_\varepsilon \qquad \text{for} \qquad x \in S_\varepsilon , \\ \psi_\varepsilon &= 1 \qquad\quad \text{for} \qquad x \in B_\varepsilon . \end{aligned}$$

THEOREM 3.1. *For every $\lambda > 0$, $\psi(x, \lambda)$ is explicitly given by the formula below for $x \in S_\varepsilon$:*

$$\psi_\varepsilon(x, \lambda) = \frac{\cosh(\sqrt{2\lambda}\, r)}{\cosh(\sqrt{2\lambda}\, \varepsilon)} \qquad \textit{for} \qquad k = 1 ,$$

$$\psi_\varepsilon(x, \lambda) = \frac{\int_0^\pi \cosh(\sqrt{2\lambda}\, r \cos\theta) \sin^{k-2}\theta \, d\theta}{\int_0^\pi \cosh(\sqrt{2\lambda}\, \varepsilon \cos\theta) \sin^{k-2}\theta \, d\theta} \qquad \textit{for} \qquad k \geqq 2 ,$$

where $r = \|x - y\|_y$. For all k, $\psi_\varepsilon(x, \lambda)$ is a convex function of x.

Proof: The proof is elementary and consists of direct verification.

THEOREM 3.2. *For every $\rho > 0$ there exists a constant $M_\rho < \infty$ depending only on the dimension k such that, for all y, ε and λ,*

$$\psi_\varepsilon(y, \lambda) \leqq M_\rho \exp\{-\varepsilon(1 - \rho)\sqrt{2\lambda}\} .$$

Proof: When $x = y$ or $r = 0$ the numerator in Theorem 3.1 reduces to a dimension constant. As for the denominator, it can be estimated from below by limiting the range of integration to the region $0 \leqq \theta \leqq \cos^{-1}(1 - \rho)$.

We now consider the sets

$$T_\varepsilon = [x: d(x, y) < \varepsilon] ,$$

$$D_\varepsilon = [x: d(x, y) = \varepsilon] .$$

Then in view of Lemma 2.5, there exists a function $\delta(\varepsilon) \downarrow 0$ as $\varepsilon \downarrow 0$ such that the set

$$S = [x: \|x - y\|_y < \varepsilon(1 - \delta(\varepsilon))]$$

is contained in T_ε. Let B be the boundary of S:

$$B = [x: \|x - y\|_y = \varepsilon(1 - \delta(\varepsilon))] .$$

Obviously B is contained in $T_\varepsilon \cup D_\varepsilon$.

Let $f_\varepsilon(x, \lambda)$, $g_\varepsilon(x, \lambda)$, and $h_\varepsilon(x, \lambda)$ denote the solutions of the following equations:

$$\begin{aligned} Lf &= \lambda f && \text{for} \quad x \in T_\varepsilon , \\ f &= 1 && \text{for} \quad x \in D_\varepsilon , \end{aligned} \tag{3.2}$$

$$\begin{aligned} Lg &= \lambda g && \text{for} \quad x \in S , \\ g &= 1 && \text{for} \quad x \in B , \end{aligned} \tag{3.3}$$

$$\begin{aligned} L_y h &= \lambda h && \text{for} \quad x \in S , \\ h &= 1 && \text{for} \quad x \in B . \end{aligned} \tag{3.4}$$

Lemma 3.3.

$$f(y, \lambda) \leqq g(y, \lambda) .$$

Proof: From equation (3.2) it is obvious that $f \leqq 1$ in $T_\varepsilon \cup D_\varepsilon$. Since $B \subset T_\varepsilon \cup D_\varepsilon$, it follows that $f \leqq 1$ on B. But f and g satisfy the same equation in S and on the boundary B of S, $g = 1$ and $f \leqq 1$. Hence $f \leqq g$ everywhere in S. Therefore in particular, $f(y, \lambda) \leqq g(y, \lambda)$.

Lemma 3.4. *There exists a function $\delta(\varepsilon) \downarrow 0$ as $\varepsilon \downarrow 0$ such that*

$$g(y, \lambda) \leqq h(y, \lambda(1 - \delta(\varepsilon))) .$$

Proof: By Theorem 3.1, h is convex and by Lemma 2.4

$$a(x) \leqq \frac{1}{1 - \delta(\varepsilon)} a(y) .$$

Therefore,

$$\begin{aligned} \tfrac{1}{2} \sum a_{ij}(x) \frac{\partial^2 h}{\partial x_i \, \partial x_j} &\leqq \frac{1}{2(1 - \delta(\varepsilon))} \sum a_{ij}(y) \frac{\partial^2 h}{\partial x_i \, \partial x_j} \\ &= \frac{1}{(1 - \delta(\varepsilon))} L_y h \\ &= \frac{\lambda}{(1 - \delta(\varepsilon))} h , \end{aligned}$$

or

$$Lh(x, \lambda(1 - \delta(\varepsilon))) \leqq \lambda h \qquad \text{in} \qquad S\,.$$

But

$$Lg(x, \lambda) = \lambda g \qquad \text{in} \qquad S\,,$$

and h and g have the same boundary value on B. Hence the solution $g(x, \lambda)$ is smaller than $h(x, \lambda(1 - \delta(\varepsilon)))$ in S. In particular,

$$g(y, \lambda) \leqq h(y, \lambda(1 - \delta(\varepsilon)))\,.$$

THEOREM 3.5. *The solution $f_\varepsilon(x, \lambda)$ of equation (3.2) has the following property. There exist a function $\delta(\varepsilon) \downarrow 0$ as $\varepsilon \downarrow 0$ and a constant $M_\rho < \infty$ for every $\rho > 0$ such that, for all ε, y and λ,*

$$f_\varepsilon(y, \lambda) \leqq M_\rho \exp\{-(1 - \delta(\varepsilon))(1 - \rho)\varepsilon\sqrt{2\lambda}\}\,.$$

The proof is immediate from Theorem 3.2 and Lemmas 3.3 and 3.4.

Let G be any arbitrary region with a boundary B. Let $\phi(x, \lambda)$ be the solution to the equation

$$\text{(3.5)} \qquad \begin{aligned} L\phi &= \lambda\phi \qquad \text{for} \qquad x \in G\,, \\ \phi &= 1 \qquad \text{for} \qquad x \in B\,. \end{aligned}$$

If G is unbounded, we determine ϕ by the additional restriction that $0 \leqq \phi \leqq 1$. If G is bounded, it is automatically true that $0 \leqq \phi \leqq 1$. Let

$$\text{(3.6)} \qquad d(x, B) = \inf_{b \in B} d(x, b)\,.$$

We then have the following estimate concerning $\phi(x, \lambda)$.

THEOREM 3.6. *For every $\rho > 0$, there exists a function $C_\rho(x)$ such that*

$$\phi(x, \lambda) \leqq C_\rho(x) \exp\{-(1 - \rho)d(x, B)\sqrt{2\lambda}\}\,.$$

$C_\rho(x)$ depends only on $d(x, B)$ and is uniformly bounded whenever x varies in such a way that $d(x, B)$ remains bounded.

Proof: Let $x \in G$ be arbitrary. Then the set

$$S_R = [y\colon d(x, y) < d(x, B)]$$

is obviously contained in G and its boundary is

$$B_R = [y\colon d(x, y) = d(x, B)]\,.$$

Hence if $\psi(x, \lambda)$ is the solution of

$$(3.7) \qquad \begin{aligned} L\psi &= \lambda\psi \qquad \text{for} \qquad y \in S_R\,, \\ \psi &= 1 \qquad \text{for} \qquad y \in B_R\,, \end{aligned}$$

then

$$\phi(x, \lambda) \leqq \psi(x, \lambda)\,.$$

To estimate $\psi(x, \lambda)$ we choose ε small so that $n\varepsilon = d(x, B)$ and define the sets $S_1, S_2, \cdots S_n$ as follows:

$$S_j = [y: d(x,y) < j\varepsilon]\,,$$

$$B_j = [y: d(x,y) = j\varepsilon]\,.$$

Let ψ_j be the solution to the equation

$$(3.8) \qquad \begin{aligned} L\psi_j &= \lambda\psi_j \quad \text{for } y \text{ in } S_j\,, \\ \psi_j &= 1 \quad \text{for } y \text{ on } B_j\,. \end{aligned}$$

We estimate ψ_{j+1} on B_j. It is obvious that if $b_j \in B_j$, the set $S = [y: d(b_j, y) < \varepsilon]$ is contained in S_{j+1} and hence

$$\psi_{j+1}(b_j, \lambda) \leqq f_\varepsilon(b_j, \lambda)\,,$$

where f_ε refers to the function defined in equation (3.2). Therefore,

$$(3.9) \qquad \psi_{j+1}(b_j, \lambda) \leqq M_\rho \exp\{-[1 - \delta(\varepsilon)](1 - \rho)\sqrt{2\lambda}\,\varepsilon\}\,.$$

Now we can use the inequality

$$\phi(b_j, \lambda) \leqq \psi_{j+1}(b_j, \lambda) \sup_{b_{j+1} \in B_{j+1}} \phi(b_{j+1}, \lambda)$$

which with (3.9) leads to

$$(3.10) \qquad \phi(x, \lambda) \leqq [M_\rho]^n \exp\{-[1 - \delta(\varepsilon)](1 - \rho)n\varepsilon\sqrt{2\lambda}\}\,.$$

Since $\delta(\varepsilon) \downarrow 0$ as $\varepsilon \downarrow 0$ and since $n\varepsilon = d(x, B)$, this leads immediately to the theorem.

For the rest of the section we shall be concerned with obtaining a lower bound.

Let y be an arbitrary point in R_k, let S be the set

$$S = [x: \|x - y\|_y < \varepsilon]$$

and B its boundary

$$B = [x: \|x - y\|_y = \varepsilon]\,.$$

Γ is that part of the boundary B defined by

$$\Gamma = [x: \|x - y\|_y = \varepsilon, \|x - z\|_y < \varepsilon\sqrt{2\eta}],$$

where z is a point of B and η is a positive number. So Γ is a small portion of the boundary around z.

Let $\phi(x, \lambda)$ be the solution to the equation

$$(3.11) \qquad \begin{aligned} L\phi &= \lambda\phi \quad \text{for } x \text{ in } S, \\ \phi &= 1 \quad \text{for } x \text{ on } \Gamma, \\ \phi &= 0 \quad \text{for } x \text{ on } B - \Gamma. \end{aligned}$$

We have the following theorem giving us a lower bound on $\phi(y, \lambda)$.

THEOREM 3.7. *There exists a function $\delta(\varepsilon) \downarrow 0$ as $\varepsilon \downarrow 0$ and a constant M_ρ for every $\rho > 0$ such that for all y and λ*

$$\phi(y, \lambda) \geqq \exp\{-\varepsilon(1 + \delta(\varepsilon))\sqrt{2\lambda}\}[1 - M_\rho \exp\{-\varepsilon(\eta - \rho - \delta(\varepsilon))\sqrt{2\lambda}\}].$$

Proof: Let θ be a vector in R_k. For any z in R_k, θz is the scalar product $\Sigma\, \theta_i z_i$. For x in S we have

$$\begin{aligned} Le^{\theta(x-y)} &= \tfrac{1}{2}[\theta a(x)\theta]e^{\theta(x-y)} \\ &\geqq \tfrac{1}{2}(1 - \delta(\varepsilon))(\theta\Lambda\theta)e^{\theta(x-y)}, \end{aligned}$$

where $\Lambda = a(y)$. Putting $\lambda = \frac{1}{2}(1 - \delta(\varepsilon))[\theta\Lambda\theta]$, we have

$$Le^{\theta(x-y)} \geqq \lambda e^{\theta(x-y)} \quad \text{for } x \text{ in } S.$$

Hence if $g(x, \lambda)$ is the solution to

$$(3.12) \qquad \begin{aligned} Lg &= \lambda g \quad \text{in } S, \\ g &= e^{\theta(x-y)} \quad \text{for } x \text{ on } B, \end{aligned}$$

then

$$(3.13) \qquad e^{\theta(x-y)} \leqq g(x, \lambda).$$

Let $\psi(x, \lambda)$ be the solution to

$$(3.14) \qquad \begin{aligned} L\psi &= \lambda\psi \quad \text{for } x \text{ in } S, \\ \psi &= 1 \quad \text{for } x \text{ on } B. \end{aligned}$$

Then from Theorem 3.2, there is, for every $\rho > 0$, a constant M_ρ such that

$$(3.15) \qquad \psi(y, \lambda) \leqq M_\rho \exp\{-\varepsilon(1 - \rho)\sqrt{2\lambda}\}.$$

Let us now take the vector $\theta = k\Lambda^{-1}(z - y)$, where the constant k is to be chosen later, such that

$$\lambda = \tfrac{1}{2}(1 - \delta(\varepsilon))[\theta\Lambda\theta]\,. \tag{3.16}$$

The following estimates are easily obtained:

$$\sup_{x \in B} \theta(x - y) = k\varepsilon^2\,,$$

$$\sup_{x \in B - \Gamma} \theta(x - y) = k\varepsilon^2(1 - \eta)\,.$$

Hence on B,

$$e^{\theta(x-y)} \leqq e^{k\varepsilon^2(1-\eta)+k\varepsilon^2}\chi_\Gamma(x)\,, \tag{3.17}$$

where $\chi_\Gamma(x)$ is the indicator function of Γ. Let us recall that $\phi(x, \lambda)$ satisfies the equation

$$\begin{aligned} L\phi &= \lambda\phi \quad \text{for } x \text{ in } S\,,\\ \phi &= \chi_\Gamma(x) \quad \text{for } x \text{ on } B\,, \end{aligned} \tag{3.18}$$

and compare the equations (3.12), (3.14) and (3.18). The functions g, ψ, and ϕ satisfy the same equation in S but with different boundary values $e^{\theta(x-y)}$, 1, and $\chi_\Gamma(x)$, respectively, on B. Since (3.17) is valid on the boundary, the following inequality is valid throughout S:

$$g(x, \lambda) \leqq e^{k\varepsilon^2(1-\eta)}\psi(x, \lambda) + e^{k\varepsilon^2}\phi(x, \lambda)\,. \tag{3.19}$$

Setting $x = y$ in (3.19) and using (3.13), we obtain

$$e^{k\varepsilon^2(1-\eta)}\psi(y, \lambda) + e^{k\varepsilon^2}\phi(y, \lambda) \geqq 1\,.$$

Using now the estimate (3.15) for $\psi(y, \lambda)$, we get

$$\phi(y, \lambda) \geqq e^{-k\varepsilon^2} - e^{-k\varepsilon^2\eta}M_\rho e^{-\varepsilon(1-\rho)\sqrt{2\lambda}}\,. \tag{3.20}$$

Now (3.16) leads to the value $\varepsilon k = [1 + \delta(\varepsilon)]\sqrt{2\lambda}$. Substitution of this value of k in (3.20), a little simplification and replacement of $\delta(\varepsilon)$ by a larger function which again tends to zero as $\varepsilon \to 0$ yields

$$\phi(y, \lambda) \geqq e^{-\varepsilon(1+\delta(\varepsilon))\sqrt{2\lambda}}(1 - M_\rho e^{-\varepsilon(\eta-\rho-\delta(\varepsilon))\sqrt{2\lambda}})\,.$$

This completes the proof.

Let y and z be two arbitrary points in R_k with $d(x,y) = \varepsilon$. Let S be the set

$$S = [x\colon d(x, z) \leqq 2\varepsilon\sqrt{\eta}]\,.$$

We consider the solution $\phi(x, \lambda)$ of the equation

$$(3.21)\qquad \begin{aligned} L\phi &= \lambda\phi \quad \text{for } x \text{ in the exterior of } R_k - S\,, \\ \phi &= 1 \quad \text{for } x \text{ on the boundary } B \text{ of } S\,. \end{aligned}$$

We further assume that η is small enough so that y is in the exterior of S. Then we have

THEOREM 3.8. *There exists a function $\delta(\varepsilon) \downarrow 0$ as $\varepsilon \downarrow 0$, and a constant M_ρ for every $\rho > 0$ such that, for all y and z,*

$$\phi(y, \lambda) \geqq \exp\{-\varepsilon(1 + \delta(\varepsilon))\sqrt{2\lambda}\}[1 - M_\rho \exp\{-(\eta - \rho - \delta(\varepsilon))\sqrt{2\lambda}\}]\,.$$

Proof: Let $\varepsilon' = \|y - z\|_y$. Choose η' such that the set

$$E = [x: \|x - z\|_y \leqq \varepsilon'\sqrt{2\eta'}]$$

is contained in S. Let F denote the boundary

$$F = [x: \|x - z\|_y = \varepsilon'\sqrt{2\eta'}]$$

of S. Let $\psi(x, \lambda)$ be the solution to the equation

$$(3.22)\qquad \begin{aligned} L\psi &= \lambda\psi \quad \text{for } x \text{ in the exterior of } E\,, \\ \psi &= 1 \quad \text{for } x \text{ on } F\,. \end{aligned}$$

Let S' be the set

$$S' = [x: \|x - y\|_y < \varepsilon']$$

with boundary

$$B' = [x: \|x - y\|_y = \varepsilon']$$

and Γ' the set $\Gamma' = B' \cap E$. Let $g(x, \lambda)$ be the solution of

$$(3.23)\qquad \begin{aligned} Lg &= \lambda g \quad \text{for } x \text{ in } S'\,, \\ g &= \chi_{\Gamma'}(x) \quad \text{for } x \text{ on } B'\,. \end{aligned}$$

Since $E \subset S$ and ψ is the solution to (3.22) in the exterior of E, ψ is defined on the boundary B of S and $0 \leqq \psi \leqq 1$ on B. But ϕ is the solution of (3.21) in the exterior of S and therefore for x in the exterior of S

$$(3.24)\qquad \phi(x, \lambda) \geqq \psi(x, \lambda)\,.$$

Let A be the set $S' \cap$ (exterior of E). Its boundary is given by $[B' \cap$ (exterior of E)$] \cup (F \cap S')$. Then g and ψ are both solutions of the same equation $Lf = \lambda f$ in A. As for their boundary values,

$$\begin{aligned} &\text{on } B' \cap (\text{exterior of } E), \quad g = 0\,, \quad \psi \geqq 0\,, \\ &\text{on } F \cap S', \quad g \leqq 1\,, \quad \psi = 1\,. \end{aligned}$$

Hence throughout A, $\psi \geqq g$. Since y is in A, we have with (3.24)

$$\phi(y, \lambda) \geqq g(y, \lambda) .$$

Using Theorem 3.7 to estimate $g(y, \lambda)$, we have

$$g(y, \lambda) \geqq \exp\{-\varepsilon'(1 + \delta(\varepsilon'))\sqrt{2\lambda}\}[1 - M_\rho \exp\{-(\eta' - \rho - \delta(\varepsilon'))\sqrt{2\lambda}\}] .$$

It is easy to estimate using Lemma 2.5:

$$\varepsilon' \leqq \varepsilon(1 + \delta(\varepsilon))$$

and

$$\eta' \geqq 2\eta\left(\frac{1 - \delta(\varepsilon)}{1 + \delta(\varepsilon)}\right)^2 .$$

Therefore for ε small enough, $\eta' \geqq \eta$ and this proves the theorem.

Let y, z be two points with $d(y, z) = d$. Let n be an integer such that $n\varepsilon = d$. Let G be the sphere around z of radius $4\varepsilon\sqrt{\eta}$:

$$G = [x: d(x, z) \leqq 4\varepsilon\sqrt{\eta}] .$$

Let $\phi(x, \lambda)$ be the solution of the equation

$$\text{(3.25)} \qquad \begin{aligned} L\phi &= \lambda\phi \quad \text{for } x \text{ in the exterior of } G , \\ \phi &= 1 \quad \text{on the boundary of } G . \end{aligned}$$

THEOREM 3.9. *For the solution ϕ of equation (3.25) there exist a function $\delta(\varepsilon) \downarrow 0$ as $\varepsilon \downarrow 0$ and a constant M_ρ for every $\rho > 0$ such that*

$$\phi(y, \lambda) \geqq \exp\{-d(1 + 4\sqrt{\eta})(1 + \delta(\varepsilon))\sqrt{2\lambda}\} \times [1 - M_\rho \exp\{-\varepsilon(1 - 4\sqrt{\eta})(\eta - \rho - \delta(\varepsilon))\sqrt{2\lambda}\}]^n .$$

Proof: Let the points $y = y_0, y_1, y_2, \cdots, y_n = z$ be chosen such that $d(y_i, y_{i+1}) = \varepsilon$; $n\varepsilon$ being equal to d. Let $\varepsilon > 0$. Construct the sets C_i successively as follows:

$$C_1 = \overline{S(y_1, 2\varepsilon\sqrt{\eta})} ,$$

$$C_2 = \overline{\bigcup_{z_1 \in C_1} S(y_2, 2\sqrt{\eta}\, d(z_1, y_2))} ,$$

$$C_{j+1} = \overline{\bigcup_{z_j \in C_j} S(y_{j+1}, 2\sqrt{\eta}\, d(z_j, y_{j+1}))} ,$$

and let

$$\Delta_j = \sup_{z_j \in C_j} d(z_j, y_j),$$

$$\rho_{j-1} = \sup_{z_{j-1} \in C_{j-1}} d(z_{j-1}, y_j),$$

$$\mu_{j-1} = \inf_{z_{j-1} \in C_{j-1}} d(z_{j-1}, y_j).$$

Then

$$\rho_{j-1} \leqq \varepsilon + \Delta_{j-1},$$

$$\mu_{j-1} \geqq \varepsilon - \Delta_{j-1},$$

$$\Delta_j \leqq 2\sqrt{\eta}\, \rho_{j-1},$$

$$\rho_0 = \varepsilon, \qquad \Delta_0 = 0.$$

It follows from these equations that

$$\begin{aligned}\Delta_j &\leqq \varepsilon[2\sqrt{\eta} + (2\sqrt{\eta})^2 + \cdots (2\sqrt{\eta})^j] \\ &\leqq \varepsilon 2\sqrt{\eta}\,\frac{1}{1 - 2\sqrt{\eta}} \\ &\leqq 4\varepsilon\sqrt{\eta} \qquad \text{for} \qquad 2\sqrt{\eta} \leqq \tfrac{1}{2}, \qquad\qquad j = 1, 2, \cdots n.\end{aligned}$$

If ε_j is a typical distance from C_j to y_{j+1}, then

$$\varepsilon - \Delta_j \leqq \varepsilon_j \leqq \varepsilon + \Delta_j,$$

so that

$$\varepsilon(1 - 4\sqrt{\eta}) \leqq \varepsilon_j \leqq \varepsilon(1 + 4\sqrt{\eta}).$$

Now we can apply Theorem 3.8 and obtain a uniform estimate for x in C_{n-1}. Repeated application of Theorem 3.8 leads to

$$\begin{aligned}\phi(y, \lambda) \geqq{} & \exp\{-d(1 + 4\sqrt{\eta})(1 + \delta(\varepsilon))\sqrt{2\lambda}\} \\ & \times [1 - M_\rho \exp\{-\varepsilon(1 - 4\sqrt{\eta})(\eta - \rho - \delta(\varepsilon))\sqrt{2\lambda}\}]^n.\end{aligned}$$

THEOREM 3.10. *Let z be a point in R_k, S the sphere $S = [x\colon d(x, z) < \delta]$. Let $\phi(x, \lambda)$ be the solution to the equation $L\phi = \lambda\phi$ in the exterior of S with $\phi = 1$ on the boundary of S. Then*

$$\liminf_{\lambda \to \infty} \frac{1}{\sqrt{2\lambda}} \log \phi(x, \lambda) \geqq -d(x, z)$$

uniformly over all x, z and δ, provided δ remains bounded away from zero and $d(x, z)$ remains bounded.

The proof follows at once from Theorem 3.9. One chooses η first and then ρ and ε suitably; n remains bounded for any $\varepsilon > 0$ if d is bounded.

Let G be a region with exterior E. B is the boundary of G. It is assumed that B is also the boundary of E.

LEMMA 3.11. *Let $x \in G$ and let $d = d(x, B)$. Then for any $\varepsilon > 0$ there exists a $\delta > 0$ and a point $z \in E$ such that*

(i) $d(x, z) \leqq d + \varepsilon$,

(ii) $[y: d(y, z) < \delta] \subset E$.

Moreover δ can be chosen uniformly when x varies over a bounded subset of G.

Proof: Let $x \in G \cup B$. Then there exists a y on B such that $d(x, B) = d(x, y)$. The sphere around y of radius $\frac{1}{2}\varepsilon$ has a non-empty intersection with E and there exists a sphere around some points z of some radius δ completely contained inside the intersection. Obviously $d(x, z) \leqq d(x, y) + \frac{1}{2}\varepsilon$. To show uniformity we point out that if x' is in $G \cup B$ with $d(x, x') < \frac{1}{2}\varepsilon$, then

$$\begin{aligned} d' = d(x', B) &\geqq d(x, B) - d(x, x') \\ &\geqq d - \tfrac{1}{2}\varepsilon \\ &\geqq d(x, z) - \tfrac{1}{2}\varepsilon - \tfrac{1}{2}\varepsilon \\ &= d(x, z) - \varepsilon, \end{aligned}$$

or

$$d(x, z) \leqq d' + \varepsilon .$$

In other words, the sphere of radius δ around z serves not only for x but for a neighborhood around x as well. A routine compactness argument using finite subcovers proves the lemma for compact subsets of $G \cup B$ or for bounded subsets of G.

Now we can prove the main theorem:

THEOREM 3.12. *Let G be any region with boundary B. B is assumed to be the boundary of the exterior E of G as well. If $\phi(x, \lambda)$ is the solution of the equation*

$$L\phi = \lambda\phi \quad \text{for} \quad x \quad \text{in} \quad G,$$

$$\phi = 1 \quad \text{on} \quad B,$$

then

$$\lim_{\lambda \to \infty} \left[-\frac{1}{\sqrt{2\lambda}} \log \phi(x, \lambda) \right] = d(x, B)$$

uniformly for x varying over bounded subsets of G.

The proof follows from Theorems 3.6, 3.10, and Lemma 3.11.

4. The Fundamental Solution

Let $F(t)$ be a nondecreasing function of t for $t \geqq 0$ with $0 \leqq F(t) \leqq 1$ and $F(0) = 0$. Let $\phi(\lambda)$ be the Laplace transform

$$\phi(\lambda) = \int_0^\infty e^{-\lambda\tau}\, dF(\tau) . \tag{4.1}$$

We assume that the $\{\phi_\alpha(\lambda)\}$ are Laplace transforms of $\{F_\alpha(t)\}$ and that

$$\lim_{\lambda\to\infty} \left[-\frac{1}{\sqrt{2\lambda}} \log \phi_\alpha(\lambda) \right] = d_\alpha , \tag{4.2}$$

where the d_α are such that $0 \leqq d_\alpha \leqq D < \infty$. We want to conclude that if the limit in (4.2) is uniform with respect to α, then

$$\lim_{t\to 0} \left[-2t \log F_\alpha(t) \right] = d_\alpha^2 \tag{4.3}$$

uniformly in α.

Initially we assume that (4.2) holds uniformly and that in addition $0 < d \leqq d_\alpha \leqq D < \infty$. The following lemmas are conclusions regarding $F_\alpha(t)$.

Lemma 4.1. *For every $\varepsilon > 0$ there exists a constant $A_\varepsilon < \infty$ such that, for all $t \geqq 0$,*

$$F_\alpha(t) \leqq A_\varepsilon \exp\left\{ -\frac{(1-\varepsilon)}{2t} d_\alpha^2 \right\}.$$

Proof: Since $0 < d \leqq d_\alpha \leqq D < \infty$ we can conclude, from the fact that (4.2) holds uniformly, that for every $\varepsilon > 0$ there exists a constant A_ε such that for all $\lambda \geqq 0$

$$\phi_\alpha(\lambda) \leqq A_\varepsilon \exp\{-\sqrt{1-\varepsilon}\,\sqrt{2\lambda}\, d\alpha\} . \tag{4.4}$$

It follows from (4.1) that

$$\phi_\alpha(\lambda) \geqq e^{-\lambda t} F_\alpha(t) ,$$

or

$$F_\alpha(t) \leqq e^{\lambda t} \phi_\alpha(\lambda) . \tag{4.5}$$

Choosing $\lambda = (1-\varepsilon) d_\alpha^2/2t$ and combining (4.5) and (4.4) we get

$$F_\alpha(t) \leqq A_\varepsilon \exp\left\{ -\frac{(1-\varepsilon)}{2t} d_\alpha^2 \right\} .$$

COROLLARY 4.2.

$$\limsup_{t\to 0} 2t \log F_\alpha(t) \leqq -d_\alpha^2$$

uniformly in α.

LEMMA 4.3. *For every* $k > 1$,

$$\int_{kt}^{\infty} \exp\left\{-\frac{d_\alpha^2\tau}{2t^2}\right\} F_\alpha(\tau)\, d\tau \leqq \frac{2A_\varepsilon t^2}{\varepsilon d^2} \exp\left\{-\frac{(1-\varepsilon)d_\alpha^2}{2t} - \frac{\rho_k}{t}\right\},$$

where $\rho_k = \frac{1}{2}d^2(k + 1/k - 2)$.

Proof: Using Lemma 4.1, we have

$$\int_{kt}^{\infty} \exp\left\{-\frac{d_\alpha^2\tau}{2t}\right\} F_\alpha(\tau)\, d\tau \leqq A_\varepsilon \int_{kt}^{\infty} \exp\left\{-\frac{d_\alpha^2\tau}{2t^2}\right\} \exp\left\{-\frac{(1-\varepsilon)d_\alpha^2}{2t}\right\} d\tau$$

$$= A_\varepsilon t \int_{k}^{\infty} \exp\left\{-\frac{1}{2t} d_\alpha^2\left[u + \frac{(1-\varepsilon)}{u}\right]\right\} du$$

$$= A_\varepsilon t \int_{k}^{\infty} \exp\left\{-\frac{\varepsilon d_\alpha^2}{2t}\right\} \exp\left\{-\frac{(1-\varepsilon)d_\alpha^2}{2t}\left[u + \frac{1}{u}\right]\right\} du$$

$$\leqq \frac{2A_\varepsilon t^2}{\varepsilon d^2} \exp\left\{-\frac{(1-\varepsilon)d_\alpha^2}{t} - \frac{\rho_k}{t}\right\}.$$

LEMMA 4.4. *For every* $k' < 1$,

$$\int_{0}^{k't} \exp\left\{-\frac{d_\alpha^2\tau}{2t^2}\right\} F_\alpha(\tau)\, d\tau \leqq \frac{2A_\varepsilon t^2}{\varepsilon d^2} \exp\left\{-\frac{(1-\varepsilon)d_\alpha^2}{t} - \frac{\rho_{k'}}{t}\right\},$$

where $\rho_{k'} = \frac{1}{2}d^2[k' + 1/k' - 2]$.

The proof is similar to that of Lemma 4.3.

LEMMA 4.5.

$$\lim_{t\to 0}\left[-2t \log F_\alpha(t)\right] = d_\alpha^2$$

uniformly in α.

Proof: Because of (4.2), there exists, for every $\varepsilon > 0$, a constant B_ε such that for all $\lambda \geqq 0$

$$\phi_\alpha(\lambda) \geqq B_\varepsilon \exp\{-(1+\varepsilon)\sqrt{2\lambda}\, d_\alpha\}. \tag{4.6}$$

However,

$$(4.7)\qquad \begin{aligned}\phi_\alpha(\lambda) &= \lambda \int_0^\infty e^{-\lambda\tau} F_\alpha(\tau)\, d\tau \\ &= \lambda\left[\int_0^{k't} + \int_{k't}^{kt} + \int_{kt}^\infty\right] e^{-\lambda\tau} F_\alpha(\tau)\, d\tau .\end{aligned}$$

Therefore, substituting $\lambda = d_\alpha^2/2t^2$ and combining (4.6) and (4.7) with Lemmas 4.3 and 4.4, we have

$$(4.8)\qquad \int_{k't}^{kt} \exp\left\{-\frac{d_\alpha^2\tau}{2t^2}\right\} F_\alpha(\tau)\, d\tau \geqq C_\varepsilon t^2 \exp\left\{\frac{(1+\varepsilon)\, d_\alpha^2}{t}\right\}\left(1 - k_\varepsilon \exp\left\{\frac{2\varepsilon D^2 - \rho}{t}\right\}\right),$$

where $\rho = \min(\rho_k, \rho_{k'})$.

For a given k and k' with $k' < 1 < k$ one can find ε small enough so that $2\varepsilon D^2 - \rho < 0$. Therefore, we conclude from (4.8) that, for every k, k' with $k' < 1 < k$,

$$(4.9)\qquad \liminf_{t\to 0} t \log\left[\int_{k't}^{kt} \exp\left\{-\frac{d_\alpha^2}{2t^2}\right\} F_\alpha(\tau)\, d\tau\right] \geqq -d_\alpha^2$$

uniformly in α.

On the other hand,

$$\int_{k't}^{kt} \exp\left\{-\frac{d_\alpha^2\tau}{2t^2}\right\} F_\alpha(\tau)\, d\tau \leqq F_\alpha(kt) \exp\left\{-\frac{k'd_\alpha^2}{2t}\right\}$$

which with (4.9) yields

$$\liminf_{t\to 0} t \log F_\alpha(kt) \geqq -d_\alpha^2 + \tfrac{1}{2}k'd_\alpha^2 ,$$

or

$$\liminf_{t\to 0} 2t \log F_\alpha(t) \geqq -2kd_\alpha^2 + kk'd_\alpha^2$$

uniformly in α. Since k and k' can be chosen close to 1, we have

$$\liminf_{t\to 0} 2t \log F_\alpha(t) \geqq -d_\alpha^2$$

uniformly in α. This is one part of the lemma and the other part is Corollary 4.2.

Now we drop the assumption that $d > 0$. We assume only that $0 \leqq d_\alpha \leqq D$.

Lemma 4.6. *Lemma 4.5 still holds if we assume only that* $0 \leqq d_\alpha \leqq D$.

Proof: It is of course enough to show that if, for $\lambda \geqq \lambda_0$,

$$-\varepsilon \leqq -\frac{1}{\sqrt{2\lambda}} \log \phi_\alpha(\lambda) \leqq \varepsilon,$$

then there exists a t_0 and a function $\eta(\varepsilon) \to 0$ as $\varepsilon \to 0$ such that

$$-\eta \leqq -2t \log F_\alpha(t) \leqq \eta .$$

Since $0 \leqq \phi_\alpha(\lambda) \leqq 1$ and $0 \leqq F_\alpha(t) \leqq 1$, one side is trivial. If

$$\phi_\alpha(\lambda) \geqq e^{-\varepsilon\sqrt{2\lambda}} \qquad \text{for} \qquad \lambda \geqq \lambda_0,$$

we shall show that for ε small enough there exist η and t_0 such that

$$F_\alpha(t) \geqq e^{-\eta/t} \qquad \text{for} \qquad t \leqq t_0 .$$

To prove this we notice that

$$\phi_\alpha(\lambda) \leqq F_\alpha(t) + e^{-\lambda t},$$

or

$$F_\alpha(t) \geqq \phi_\alpha(\lambda) - e^{-\lambda t}$$

$$\geqq e^{-\varepsilon\sqrt{2\lambda}} - e^{-\lambda t} \qquad \text{for} \qquad \lambda \geqq \lambda_0 .$$

Choose $\lambda = 2/t^2$. If $t \leqq t_0 = \sqrt{2/\lambda_0}$ and $\varepsilon < \frac{1}{2}$, then

$$f_\alpha(t) \geqq e^{-2\varepsilon/t}(1 - e^{-1/t})$$

which concludes the proof.

We have, therefore, proved the following theorem:

Theorem 4.7. *Let* $\phi_\alpha(\lambda) = \int_0^\infty e^{-\lambda t}\, dF_\alpha(\tau)$. *If*

$$\lim_{\lambda \to \infty} \left[-\frac{1}{\sqrt{2\lambda}} \log \phi_\alpha(\lambda) \right] = d_\alpha$$

uniformly in α *and if* $0 \leqq d_\alpha \leqq D < \infty$, *then*

$$\lim_{t \to 0} \left[-2t \log F_\alpha(t) \right] = d_\alpha^2$$

uniformly in α.

Let $G \subset R_k$ be a region with a boundary B. It is assumed that B is the boundary of the exterior of G as well. Let S be the cylinder in $R_k \times (0, \infty)$ with base G.

That is,

$$S = G \times (0, \infty) .$$

The boundary of S is $\{B \times (0, \infty)\} \cup \{G \times \{0\}\}$. We consider the solution $u(x, t)$ to the equation

$$\begin{aligned} u_t - Lu &= 0 \quad \text{for} \quad (x, t) \in G \times (0, \infty) , \\ u(x, 0) &= 0 \quad \text{for} \quad x \in G, \\ u(x, t) &= 1 \quad \text{for} \quad x \in B, \quad t > 0 . \end{aligned} \tag{4.10}$$

The solution to the equation exists and has the properties:

$u(x, t)$ is nondecreasing in t for $t \geqq 0$,

$0 \leqq u(x, t) \leqq 1$.

THEOREM 4.8.

$$\limsup_{t \to 0} 2t \log u(x, t) \leqq -d^2(x, B)$$

uniformly over all regions G and points x in G so long as d(x, B) remains bounded.

Proof: If $\phi(x, \lambda) = \int_0^\infty e^{-\lambda\tau} u(x, d\tau)$, then $\phi(x, \lambda)$ satisfies the equation

$$\begin{aligned} L\phi &= \lambda\phi \quad \text{for} \quad x \in G , \\ \phi &= 1 \quad \text{for} \quad x \in B . \end{aligned} \tag{4.11}$$

From Theorem 3.6, there exists for every $\rho > 0$, a $C_\rho < \infty$ such that

$$\phi(x, \lambda) \leqq C_\rho \exp \{-(1 - \rho)\sqrt{2\lambda}\, d(x, B)\}$$

so long as $d(x, B)$ remains bounded. From equation (4.5), we have

$$u(x, t) \leqq e^{\lambda t}\phi(x, \lambda)$$

for every $\lambda \geqq 0$. Choosing $\lambda = (1 - \rho)^2 d^2(x, B)/2t^2$,

$$u(x, t) \leqq C_\rho \exp \left\{\frac{(1 - \rho)^2 d^2(x, B)}{2t}\right\}$$

which proves the theorem.

THEOREM 4.9. *The fundamental solution $p(t, x, y)$ of Theorem 2.1 has the property*

$$\limsup_{t \to 0} 2t \log p(t, x, y) \leqq -d^2(x, y)$$

uniformally over all x, y such that d(x, y) is bounded.

Proof: Let $d(x,y) = d$. Define, for any $\varepsilon > 0$,

$$G = [z\colon d(x, z) < d - \varepsilon]\,,$$

$$B = [z\colon d(x, z) = d - \varepsilon]\,.$$

Then $p(t, x, y)$ satisfies the equation

$$\text{(4.12)}\qquad \begin{aligned} p_t - Lp &= 0 &&\text{for}\quad (x, t)\ \text{in}\ G \times (0, \infty)\,,\\ p(0, x, y) &= 0 &&\text{for}\quad x \in G\,,\\ p(t, x, y) &= p(t, b, y) &&\text{for}\quad b \in B\,. \end{aligned}$$

If $u(x, t)$ is the solution to

$$\text{(4.13)}\qquad \begin{aligned} u_t - Lu &= 0 &&\text{for}\quad (x, t)\ \text{in}\ G \times (0, \infty)\,,\\ u &= 0 &&\text{for}\quad x \in G\ \text{and}\ t = 0\,,\\ u &= 1 &&\text{for}\quad x \in B\ \text{and}\ t > 0\,, \end{aligned}$$

then clearly

$$\text{(4.14)}\qquad p(t, x, y) \leqq \Big[\sup_{\substack{0 \leqq \tau < \infty \\ b \in B}} p(\tau, b, y)\Big]\, u(x, t)\,.$$

Since $d(b, y) \geqq \varepsilon$, we have from property (iv) of Theorem 2.1

$$\text{(4.15)}\qquad \sup_{0 < \tau < \infty}\ \sup_{b \in B} p(\tau, b, y) \leqq M_\varepsilon < \infty\,.$$

Hence, combining (4.14) and (4.15) with Theorem 4.8, we obtain

$$\lim_{t \to 0}\ 2t \log p(t, x, y) \leqq -(d - \varepsilon)^2\,.$$

Since ε is arbitrary, this proves the theorem. However, apparently the uniformity holds only in regions of the form $0 < \varepsilon \leqq d(x, y) \leqq D < \infty$. But the direct estimate in Theorem 2.1,

$$p(t, x, y) \leqq Ct^{-k/2} \exp\left\{\frac{-\alpha\, \|x - y\|^2}{2t}\right\},$$

establishes the uniformity over all x, y such that $d(x, y)$ remains bounded.

We now proceed to obtain a lower bound on $p(t, x, y)$ for small t.

THEOREM 4.10. *There exists a function $\psi(\varepsilon)$ such that $\psi(\varepsilon) \to 0$ as $\varepsilon \to 0$ and*

$$\liminf_{t \to 0}\ \inf_{x, y : d(x, y) \leqq \varepsilon}\ 2t \log p(t, x, y) \geqq -\psi(\varepsilon)\,.$$

Proof: Let $s > 0$. Define G_s and its boundary B_s by

$$G_s = [z : d(y, z) > s]\,,$$
$$B_s = [z : d(y, z) = s]\,.$$

Let $u_s(x, t)$ be the solution to the equation

$$\begin{aligned} u_t - Lu &= 0 \quad \text{in} \quad G_s \times (0, \infty)\,, \\ u &= 1 \quad \text{on} \quad B_s \times (0, \infty)\,, \\ u &= 0 \quad \text{on} \quad G_s \times \{0\}\,. \end{aligned} \tag{4.16}$$

Let $\phi_s(x, \lambda) = \int_0^\infty e^{\lambda\tau} u_s(x, d\tau)$. Then $\phi_s(x, \lambda)$ satisfies the equation

$$\begin{aligned} L\phi &= \lambda\phi \quad \text{in} \quad G_s\,, \\ \phi &= 1 \quad \text{on} \quad B_s\,. \end{aligned} \tag{4.17}$$

From Theorem 3.9,

$$\begin{aligned} \phi_s(x, \lambda) &\geqq \exp\{-d(1 + 4\sqrt{\eta})(1 + \delta(\varepsilon))\sqrt{2\lambda}\} \\ &\quad \times [1 - M_\rho \exp\{-\varepsilon(1 - 4\sqrt{\eta})(\eta - \rho - \delta(\varepsilon))\sqrt{2\lambda}\}]^n\,, \end{aligned} \tag{4.18}$$

where $d = d(x, y)$, ε, η, ρ are arbitrary, $n\varepsilon = d$, with the restriction that $s \geqq 4\varepsilon\sqrt{\eta}$. Since

$$u_s(x, t) \geqq \phi_s(x, \lambda) - e^{-\lambda t}$$

for every λ, let us choose $\lambda = 8/t^2$, $\varepsilon = s$ and η small enough. We further put $s = \sqrt{t}$. Then using (4.18) we get

$$u_{\sqrt{t}}(x, t) \geqq \exp\{-\frac{4d}{t}(1 + \theta)\}(1 - M_\rho e^{-C/\sqrt{t}})^{d/\sqrt{t}} - e^{-8/t}, \tag{4.19}$$

where θ is a small number which can be chosen arbitrarily small. This function is defined for x such that $d(x, y) \geqq s = \sqrt{t}$. It is clear from (4.19) that if $\varepsilon \leqq 1$

$$\liminf_{t \to 0} \inf_{\substack{x, y : d(x, y) \geqq \sqrt{t} \\ d(x, y) \leqq \varepsilon}} t \log u_{\sqrt{t}}(x, t) \geqq -4\varepsilon\,. \tag{4.20}$$

We now consider the solution $u_s(x, t_0, t)$ to the equation

$$\begin{aligned} u_t - Lu &= 0 \quad \text{in} \quad G_s \times (0, \infty)\,, \\ u &= 0 \quad \text{on} \quad B_s \times (0, t_0)\,, \\ u &= 1 \quad \text{on} \quad B_s \times (t_0, \infty)\,, \\ u &= 0 \quad \text{on} \quad G_s \times \{0\}\,. \end{aligned} \tag{4.21}$$

It is clear that

$$u_s(x, t_0, t) = 0 \qquad \text{for} \qquad t \leqq t_0,$$

$$u_s(x, t_0, t) = u_s(x, t - t_0) \qquad \text{for} \qquad t > t_0.$$

We are interested in $u_{\sqrt{t}}(x, \frac{1}{2}t, t) = u_{\sqrt{t}}(x, \frac{1}{2}t)$. From (4.20) we deduce

$$\liminf_{t \to 0} \inf_{\substack{(x,y):d(x,y) \geqq \sqrt{t} \\ d(x,y) \leqq \varepsilon}} t \log u_{\sqrt{t}}(x, \tfrac{1}{2}t, t) \geqq -8\varepsilon. \tag{4.22}$$

Going back to $p(t, x, y)$ we see that it satisfies an equation similar to (4.21) in $G_s \times (0, \infty)$ with different boundary values. However, since the solution $u_s(x, t_0, t)$ depends only on the boundary values on $B_s \times (0, \infty)$ for $0 < \tau < t$ we conclude by direct comparison that

$$p(t, x, y) \geqq u_s(x, t_0, t) \inf_{\substack{b \in B_s \\ t_0 \leqq \tau \leqq t}} p(\tau, b, y).$$

Since t_0 and t are arbitrary, we choose $t_0 = \frac{1}{2}t$ and $s = \sqrt{t}$. Therefore,

$$p(t, x, y) \geqq u_{\sqrt{t}}(x, \tfrac{1}{2}t, t) \inf_{\substack{b \in B_{\sqrt{t}} \\ \frac{1}{2}t \leqq \tau \leqq t}} p(\tau, b, y). \tag{4.23}$$

For t sufficiently small we have from property (v) of Theorem 2.1

$$\inf_{\substack{b \in B_{\sqrt{t}} \\ \frac{1}{2}t \leqq \tau \leqq t}} p(\tau, b, y) \geqq \gamma > 0.$$

This with (4.22) and (4.23) yields

$$\liminf_{t \to 0} \inf_{\substack{x,y:d(x,y) \leqq \varepsilon \\ d(x,y) \geqq \sqrt{t}}} 2t \log p(t, x, y) \geqq -8\varepsilon.$$

On the other hand, if $d(x, y) \leqq \sqrt{t}$, again from property (v) of Theorem 2.1,

$$\liminf_{t \to 0} \inf_{x,y:d(x,y) \leqq \sqrt{t}} 2t \log p(t, x, y) \geqq 0.$$

This completes the proof.

THEOREM 4.11. *Let $\varepsilon > 0$ and let G_y be the set $[z: d(y, z) < \varepsilon]$. If $k(x, t)$ is defined as*

$$k(x, t, y) = \int_{G_y} p(t, x, z)\, dz,$$

then

$$\liminf_{t \to 0} 2t \log k(x, t, y) \geqq -d^2(x, y)$$

uniformly over all x, y such that $d(x, y)$ is bounded.

Proof: Let

$$H_y = [z: d(y, z) > \tfrac{1}{2}\varepsilon] ,$$
$$B_y = [z: d(y, z) = \tfrac{1}{2}\varepsilon] .$$

From property (iv) of Theorem 2.1 it follows that

$$k(x, t, y) \to 1 \quad \text{as} \quad t \to 0 \tag{4.24}$$

uniformly for x on B_y.

Assume that $u(x, t)$ solves the equation

$$\begin{aligned} u_t - Lu &= 0 \quad \text{in} \quad H_y \times (0, \infty) , \\ u &= 1 \quad \text{on} \quad B_y \times (0, \infty) , \\ u &= 0 \quad \text{on} \quad H_y \times \{0\} . \end{aligned} \tag{4.25}$$

Then the Laplace transform $\phi(x, \lambda)$ solves the equation

$$L\phi = \lambda\phi \quad \text{for } x \text{ in } H_y ,$$
$$\phi = 1 \quad \text{on} \quad B_y .$$

From Theorem 3.12 it follows that

$$\lim_{t\to\infty} \left[-\frac{1}{\sqrt{2\lambda}} \log \phi(x, \lambda) \right] = d(x, B_y) .$$

Moreover, from Theorems 3.6 and 3.10 it is clear that the above limit is uniform whenever $d(x,y)$ remains bounded. Of course $\phi(x, \lambda)$ is defined only when $d(x,y) \geqq \frac{1}{2}\varepsilon$. Now from Theorem 4.7 we obtain

$$\lim_{t\to\infty} 2t \log u(x, t) \geqq -d^2(x, B_y) \geqq -d^2(x,y) \tag{4.26}$$

uniformly whenever $d(x,y)$ remains bounded. Standard comparison of $k(x, t, y)$ which satisfies an equation similar to (4.25) with $u(x, t)$ shows that

$$k(x, t, y) \geqq u(x, t) \inf_{\substack{0\leqq\tau\leqq t \\ b\in B_y}} k(b, \tau, y) . \tag{4.27}$$

Inequalities (4.26) and (4.27) complete the proof insofar as the region $d(x,y) \geqq \frac{1}{2}\varepsilon$ is concerned. However, property (iv) of Theorem 2.1 shows that

$$k(x, t, y) \to 1 \quad \text{as} \quad t \to 0$$

uniformly over the set $d(x,y) \leqq \frac{1}{2}\varepsilon$.

THEOREM 4.12.

$$\liminf_{t\to 0} 2t \log p(t, x, y) \geqq -d^2(x, y)$$

uniformly over all x, y such that $d(x, y)$ is bounded.

Proof: If G_y is the set $[z: d(y, z) \leqq \varepsilon]$, then

$$p(t, x, y) = \int p([1-\delta]t, x, z)p(\delta t, z, y)\, dz$$

$$\geqq \int_{G_y} p([1-\delta]t, x, z)p(\delta t, z, y)\, dz$$

$$\geqq k(x, [1-\delta]t, y) \inf_{z\in G_y} p(\delta t, z, y)\,.$$

Therefore, for every $\varepsilon > 0$ and $\delta > 0$, applying Theorems 4.10 and 4.11 we get

$$\liminf_{t\to 0} 2t \log p(t, x, y) \geqq -\left[\frac{d^2(x, y)}{1-\delta} + \frac{\psi(\varepsilon)}{\delta}\right].$$

Since δ is arbitrary, let us choose

$$\delta = \frac{[\psi(\varepsilon)]^{1/2}}{d(x, y) + [\psi(\varepsilon)]^{1/2}}\,.$$

Then

$$\liminf_{t\to 0} 2t \log p(t, x, y) \geqq -[d(x, y) + [\psi(\varepsilon)]^{1/2}]\,.$$

Since ε is arbitrary and $\psi(\varepsilon) \to 0$ as $\varepsilon \to 0$,

$$\liminf_{t\to 0} 2t \log p(t, x, y) \geqq -d^2(x, y)\,.$$

Since $d(x, y)$ is bounded, the only possible trouble regarding uniformity can occur when $\delta \to 1$ or $d(x, y)$ is very small. But then Theorem 4.10 can be used directly to show that it is uniform even around $d(x, y) = 0$.

We have finally proved

THEOREM 4.13.

$$\lim_{t\to 0} \left[-2t \log p(t, x, y)\right] = d^2(x, y)$$

uniformly in x, y such that $d(x, y)$ is bounded.

Remarks. The assumption in Theorem 3.12 that B is the common boundary of G and of its exterior can be weakened further. It suffices to assume that, for a dense set $B_0 \subset B$, every sufficiently small sphere S around $b_0 \in B_0$ is disconnected by B.

We can consider equation (2.3) in a domain $G \subset R_k$ with the extra condition that the solution is zero on the boundary. Then there is a formula similar to Theorem 2.2, where $d(x, y)$ is replaced by $d_G(x, y)$ which is the length of the shortest path from x to y not leaving G.

These results are best proved using probabilistic methods and are therefore left to [3].

Acknowledgement. The methods used by the author in an earlier version of this paper were heavily probabilistic. The author expresses his thanks to Prof. L. Nirenberg whose suggestions were useful in replacing the probabilistic methods by analytical ones.

Bibliography

[1] Friedman, A., *Partial Differential Equations of Parabolic Type*, Prentice-Hall, Englewood Cliffs, N.J., 1964.

[2] Miranda, C., *Equazioni alle Derivate Parziali di Tipo Ellittico*, Springer, Berlin, 1955.

[3] Varadhan, S. R. S., *Diffusion processes in a small time interval*, to appear in Comm. Pure. Appl. Math., Vol. 20, 1967.

Received August, 1966.

COMMUNICATIONS ON PURE AND APPLIED MATHEMATICS, VOL. XX, 659–685 (1967)

Diffusion Processes in a Small Time Interval*

S. R. S. VARADHAN

1. Introduction

Let $X(t)$ be a diffusion process in k dimensions with diffusion coefficients $\{a_{ij}(x)\}$. The transition probability densities of the homogeneous Markov process $X(t)$ are given by the fundamental solution $p(t, x, y)$ of the equation

$$\frac{\partial p}{\partial t} = \tfrac{1}{2} \sum a_{ij}(x) \frac{\partial^2 p}{\partial x_i \, \partial x_j} . \tag{1.1}$$

Our interest is in the behavior of the process $X(t)$ for small t. For that purpose we introduce a small parameter $\varepsilon > 0$ and define a process $X_\varepsilon(t) = X(\varepsilon t)$. We consider in detail the process $X_\varepsilon(t)$ in a fixed time interval $[0, T]$. This of course corresponds to studying the behavior, as $\varepsilon \to 0$, of the diffusion processes corresponding to the equations

$$\frac{\partial p}{\partial t} = \tfrac{1}{2}\varepsilon \sum a_{ij}(x) \frac{\partial^2 p}{\partial x_i \, \partial x_j} .$$

A similar analysis for Brownian motion was carried out by M. Schilder [4] and such an analysis has been used by Donsker in [2]. The same approach in connection with processes with independent increments was considered by the author, [6].

Section 2 outlines the contents of the paper and describes some preliminary material. The later sections contain the actual proofs.

2. Preliminaries and Summary

We denote by L_ε the operator

$$L_\varepsilon f = \tfrac{1}{2}\varepsilon \sum a_{ij}(x) \frac{\partial^2 f}{\partial x_i \, \partial x_j} \tag{2.1}$$

acting on smooth functions on R_k. The following assumptions are made regarding the coefficients $\{a_{ij}(x)\}$:

(A) $\{a_{ij}(x)\}$ is a symmetric positive definite matrix for each x.

---

* This paper represents results obtained under a Sloan Foundation grant for Probability and Statistics. Reproduction in whole or in part is permitted for any purpose of the United States Government.

(B) The coefficients satisfy a uniform Hölder condition:

$$|a_{ij}(x) - a_{ij}(y)| \leqq M\,|x - y|^{\alpha} .$$

(C) The matrices $\{a_{ij}(x)\}$ satisfy a uniform ellipticity condition:

$$\alpha_1 \sum \xi_j^2 \leqq \sum a_{ij}(x)\xi_i\xi_j \leqq \alpha_2 \sum \xi_j^2$$

for some $0 < \alpha_1 \leqq \alpha_2 < \infty$.

Under these assumptions there exists a fundamental solution $p_\varepsilon(t, x, y)$ to the equation

$$\frac{\partial p}{\partial t} = \tfrac{1}{2}\varepsilon \sum a_{ij}(x) \frac{\partial^2 p}{\partial x_i\, \partial x_j} . \tag{2.2}$$

Moreover, if we denote L_ε for $\varepsilon = 1$ by L, and the corresponding $p_\varepsilon(t, x, y)$ by $p(t, x, y)$, it is obvious that

$$p_\varepsilon(t, x, y) = p(\varepsilon t, x, y) . \tag{2.3}$$

Let us choose and fix a $T < \infty$. We denote by Ω the space of all continuous functions $\omega(t)$ on $[0, T]$ with values in R_k. The topology in Ω is that of uniform convergence. The measurable sets are the Borel sets. We denote by Ω_x all functions $\omega \in \Omega$ with $\omega(0) = x$, where x is a point in R_k. For any set $A \subset \Omega$, A_x will be the intersection of A and Ω_x, $\|\omega\|$ will denote the sup norm.

For every $\varepsilon > 0$, $p_\varepsilon(t, x, y)$ satisfies the Chapman-Kolmogorov equations (semigroup property)

$$p_\varepsilon(t + s, x, y) = \int p_\varepsilon(t, x, z) p_\varepsilon(s, z, y)\, dz . \tag{2.4}$$

We can therefore use these functions as probability densities and construct a Markov process. For each $\varepsilon > 0$ they lead to a measure P_x^ε on Ω corresponding to the process starting at time $t = 0$ at the point x. In other words,

$$P_x^\varepsilon[\Omega_x] = 1 . \tag{2.5}$$

If $A \subset \Omega$ is a cylinder set of the form

$$A = \{\omega : \omega(t_1), \cdots \omega(t_n) \in B\} ,$$

where $0 < t_1 < t_2 \cdots < t_n \leqq T$ and B is a measurable subset of the n-fold product of R_k, then $P_x^\varepsilon[A]$ is defined as

$$P_x^\varepsilon[A] = \int_B p_\varepsilon(t_1, x, y_1) p_\varepsilon(t_2 - t_1, y_1, y_2) p_\varepsilon(t_n - t_{n-1}, y_{n-1}, y_n)\, dy_1 \cdots dy_n . \tag{2.6}$$

The relation (2.6) completely determines P_x^ε on Ω. The family $\{P_x^\varepsilon\}$ is referred to as the Markov process corresponding to the operator L_ε. For more details and for further properties of $\{P_x^\varepsilon\}$, such as the strong Markov property, the reader is referred to Dynkin [3].

We are interested in the behavior of P_x^ε as $\varepsilon \to 0$. For this purpose we introduce a functional $I(\omega)$ on Ω. Let us denote by $a^{-1}(x)$ the matrix inverse of the coefficient matrix $\{a_{ij}(x)\}$ at the point x. If θ is a vector, then $\theta a^{-1}(x)\theta$ will denote the quadratic form using the matrix $a^{-1}(x)$. For any $\omega \in \Omega$ which is differentiable, $\dot{\omega}(\tau)$ will be the vector $d\omega(\tau)/d\tau$. We now define $I(\omega)$ as follows:

If ω is absolutely continuous with a square integrable derivative on $[0, T]$, then

$$I(\omega) = \tfrac{1}{2}\int_0^T [\dot{\omega}(\tau)a^{-1}(\omega(\tau))\dot{\omega}(\tau)]\, d\tau\,, \tag{2.7a}$$

otherwise,

$$I(\omega) = +\infty\,. \tag{2.7b}$$

We have the following theorem concerning $I(\omega)$.

THEOREM 2.1. *The function $I(\omega)$ is lower semicontinuous on all of Ω. Moreover, for any finite constant N, the set $\{\omega : I(\omega) \leqq N\}$ is an equicontinuous family of continuous functions on $[0, T]$.*

The proof is straightforward and is omitted.

We next introduce the global Riemannian distance $d(x, y)$ on R_k. For smooth functions $\omega(\tau)$ we first define the length

$$l(\omega) = \int_0^T \sqrt{\dot{\omega}(\tau)a^{-1}(\omega(\tau))\dot{\omega}(\tau)}\, d\tau\,, \tag{2.8}$$

and then we define

$$d(x, y) = \inf_{\substack{\omega:\ \omega(0)=x \\ \omega(T)=y}} l(\omega)\,.$$

It is clear that d is independent of T and is the natural geodesic distance; $l(\omega)$ is of course the geodesic length.

There is a relation between $I(\omega)$ and $l(\omega)$. The geodesic length $l(\omega)$ is independent of the parametrization of the curve, whereas $I(\omega)$ depends on the parametrization. For, a class of ω all representing the same curve $I(\omega)$ is minimum when the curve is parametrized so as to cover equal lengths in equal time. This corresponds to the fact that for the minimizing parametrization the integrand in (2.8) is a constant. We therefore obtain

LEMMA 2.2. *For* $0 \leqq \alpha < \beta \leqq T$,

$$\inf_{\substack{\omega:\,\omega(\alpha)=x \\ \omega(\beta)=y}} I(\omega) = \frac{d^2(x, y)}{2(\beta - \alpha)} .$$

An obvious extension is

LEMMA 2.3. *For* $0 \leqq t_1 < t_2 < t_3 \cdots < t_n \leqq T$,

$$\inf_{\substack{\omega:\,\omega(t_j)=x_j \\ j=1,2,\cdots n}} I(\omega) = \frac{1}{2} \sum_{j=1}^{n-1} \frac{d^2(x_{j+1}, x_j)}{t_{j+1} - t_j} .$$

We are now ready to state the main theorem of Section 3.

THEOREM 2.4. *Let* $C \subset \Omega$ *be closed and* $G \subset \Omega$ *be open. Then*

$$\limsup_{\substack{\varepsilon \to 0 \\ y \to x}} \varepsilon \log P_y^\varepsilon[C] \leqq -\inf_{\omega \in C_x} I(\omega) ,$$

$$\liminf_{\substack{\varepsilon \to 0 \\ y \to x}} \varepsilon \log P_y^\varepsilon[G] \geqq -\inf_{\omega \in G_x} I(\omega) .$$

In proving Theorem 2.4 the following theorem proved in [7] plays an important role.

THEOREM 2.5.

$$\lim_{t \to 0} - 2t \log p(t, x, y) = d^2(x, y)$$

uniformly over all x, y *such that* $d(x, y)$ *is bounded.*

The rest of this paper, namely Sections 4, 5, 6 are applications of Theorem 2.4. In Section 4 the following problem is considered: Let $D \subset R_k$ be an open subset with a boundary B. Let us consider the solution $p_D(t, x, y)$ of equation (1.1) but with the boundary condition

$$p_D(t, x, y) \to 0 \qquad \text{as} \qquad x \to B .$$

In other words, the process is killed on reaching the boundary B of D. We want to consider the behavior of

$$\lim_{t \to 0} \frac{p_D(t, x, y)}{p(t, x, y)} \tag{2.9}$$

for points x, y in D. We introduce the distance

$$d_D(x,y) = \inf_{\substack{\omega(0)=x \\ \omega:\omega(T)=y \\ \omega(\tau)\in D \text{ for } 0\leq\tau\leq T}} l(\omega) .$$

For suitably smooth regions D it is shown that

$$\lim_{t\to 0} - 2t \log p_D(t, x, y) = d_D^2(x, y) .$$

Moreover, the ratio (2.9) behaves as $t \to 0$ as follows: If the geodesics from x to y do not leave D, the ratio goes to 1. Otherwise it goes to 0. This is a rough form of the result. The precise version will be found in Section 4.[1] Section 5 is concerned with some applications to certain differential equations. Section 6 deals with the case when the operator L has an additional first order term. Essentially the same results can be obtained in this case. In fact, they are deduced from the present case.

3. Behavior of $\{P_x^\varepsilon\}$

Let $0 = t_0 < t_1 < t_2 \cdots < t_n = T$ be some time points. We shall denote by π this partition of the interval $[O, T]$. T_π will denote the map from Ω to the $(n + 1)$-fold product of R_k defined by

$$T_\pi\omega = \{\omega(t_0), \omega(t_1), \cdots, \omega(t_n)\} . \tag{3.1}$$

For any set A in the range of T_π the inverse image $T_\pi^{-1}A$ is defined in the usual manner. We then have from (2.6)

$$P_y^\varepsilon[T_\pi^{-1}A] = \int_{A_y} \prod_{j=1}^{n} p_\varepsilon(t_j - t_{j-1}, y_{j-1}, y_j)\, dy_j , \tag{3.2}$$

where A_y is the section of A with $y_0 = y$.

LEMMA 3.1. *Let $C \subset \Omega$ be a closed set of the form $C = T_\pi^{-1}A$, where A is closed. Then*

$$\limsup_{\substack{\varepsilon\to 0 \\ y\to x}} \varepsilon \log P_y^\varepsilon[C] \leqq -\inf_{\omega\in C_x} I(\omega) .$$

Proof: First let us suppose that A is closed and bounded. Then

$$P_y^\varepsilon[C] = \int_{A_y} \prod_{j=1}^{n} p_\varepsilon(t_j - t_{j-1}, y_{j-1}, y_j)\, dy_j . \tag{3.3}$$

[1] A more detailed investigation of this behavior for the case of Brownian motion was carried out by Ciesielski [1].

From Theorem 2.5 and Lemma 2.3 we see that

$$\limsup_{\substack{y\to x\\ \varepsilon\to 0}} \varepsilon \log P_y^\varepsilon[C] \leqq \limsup_{y\to x}\left[-\inf_{y_1,y_n\in A_y}\frac{1}{2}\sum\frac{d^2(y_{j-1},y_j)}{t_j-t_{j-1}}\right]$$

$$\leqq -\inf_{y_1,y_n\in A_x}\frac{1}{2}\sum\frac{d^2(y_{j-1},y_j)}{t_j-t_{j-1}}$$

$$= -\inf_{\omega\in C_x} I(\omega)\ .$$

The first step follows from straight-forward estimation of the integral (3.3) using the uniformity assured by Theorem 2.5. Since A is closed and bounded, points chosen in A_y have limit points in A_x as y approaches x. This justifies the second step. The last step is the identification furnished by Lemma 2.3.

If A is unbounded, we choose a Λ large enough so that

(3.4) $$\Lambda > \inf_{\omega\in C_x} I(\omega)\ .$$

There is a standard estimate (Theorem 0.5, page 229, Volume II of [3]) of the form

(3.5) $$p_\varepsilon(t,x,y) \leqq \frac{M}{(t\varepsilon)^{k/2}}\exp\left\{-\frac{\alpha}{2t\varepsilon}|x-y|^2\right\}.$$

We now define

$$S_\rho = \{|y_0| \leqq \rho, \cdots, |y_n| \leqq \rho\}$$

and S_ρ^* as the complement of S_ρ. From (3.5) we see that we can choose ρ large enough so that for all y in a compact set K

(3.6) $$P_y^\varepsilon[T_\pi^{-1}S_\rho^*] \leqq Me^{-\Lambda/\varepsilon}\ .$$

Let us set

$$C_\rho = T_\pi^{-1}[A\cap S_\rho]\ ,$$

$$C_\rho^* = T_\pi^{-1}[S_\rho^*]\ .$$

Since $A \cap S_\rho$ is bounded and closed,

$$\limsup_{\substack{\varepsilon\to 0\\ y\to x}} \varepsilon\log P_y^\varepsilon[C_\rho] \leqq -\inf_{\substack{\omega\in C_\rho\\ \omega\in C_x}} I(\omega)$$

$$\leqq -\inf_{\omega\in C_x} I(\omega)\ .$$

From (3.6) we have

$$\limsup_{\substack{\varepsilon\to 0\\ y\to x}} \varepsilon \log P_y^\varepsilon[C_\rho^*] \leqq -\Lambda .$$

Since $P_y^\varepsilon[C] \leqq P_y^\varepsilon[C_\rho] + P_y^\varepsilon[C_\rho^*]$,

$$\limsup_{\substack{\varepsilon\to 0\\ y\to x}} \varepsilon \log P_y^\varepsilon[C] \leqq -\inf_{\omega\in C_x} I(\omega) .$$

We now choose for any integer n equally spaced time points $0 = t_0 < t_1 < \cdots t_n = T$. The partition has a length $h_n = T/n$. We denote by ω_π the trajectory obtained from ω by taking the points $\{\omega(t_j)\}$ and joining the successive ones by geodesics with natural parametrization. If there are several geodesics between two such points it is immaterial which one is chosen. In this way we obtain for each ω a ω_π which is piecewise geodesic and

$$\omega(t_j) = \omega_\pi(t_j) , \qquad j = 0, 1, 2, \cdots, n .$$

LEMMA 3.2. *For every* $\delta > 0$,

$$\limsup_{n\to\infty} \limsup_{\varepsilon\to 0} \sup_x \varepsilon \log P_x^\varepsilon[\|\omega - \omega_\pi\| \geqq \delta] = -\infty .$$

Proof: In [7] it was proved (Theorem 3.5) that if $\phi(\lambda, z)$ is the solution to the equation

$$L\phi = \lambda\phi \quad \text{for} \quad z \in U ,$$

$$\phi = 1 \quad \text{for} \quad z \in \delta U ,$$

where U is the set

$$U = \{z : d(x, z) < \eta\} ,$$

with a boundary δU, then for every $\mu > 0$ there is a constant M_μ such that, for all $\eta \leqq \eta_0$ and for all $x \in R_k$,

$$\phi(\lambda, x) \leqq M_\mu \exp\{-\sqrt{2\lambda}\,\eta(1 - \mu)\} . \tag{3.7}$$

Noting that if τ is the first exit time for the process starting at x for the set U, we then have

$$\phi(\lambda, x) = E_x[e^{-\lambda\tau}] .$$

We conclude that the expression

$$\phi(\varepsilon, \lambda, x) = \int e^{-\lambda\tau}\, dP_x^\varepsilon \tag{3.8}$$

satisfies the equation

$$\varepsilon L\phi = \lambda\phi \quad \text{in} \quad U,$$
$$\phi = 1 \quad \text{on} \quad \delta U.$$

Consequently, from (3.7), we have

$$\phi(\varepsilon, \lambda, x) \leqq M_\mu \exp\left\{-\sqrt{\frac{2\lambda}{\varepsilon}}\,\eta(1-\mu)\right\}.$$

Choosing $\mu = 1 - 1/\sqrt{2}$, we obtain

$$\phi(\varepsilon, \lambda, x) \leqq M \exp\left\{-\sqrt{\frac{\lambda}{\varepsilon}}\,\eta\right\}.$$

Using (3.8), a Tchebechev type of estimate yields

$$P_x^\varepsilon[\tau \leqq h] \leqq M \exp\left\{-\frac{\eta^2}{4\varepsilon h}\right\}.$$

We notice that

$$P_x^\varepsilon[\tau \leqq h] = P_x^\varepsilon\left[\sup_{0\leqq t\leqq h} d(\omega(t), \omega(0)) \geqq \eta\right]$$

and we therefore obtain, for all x and $\eta \leqq \eta_0$,

$$P_x^\varepsilon\left[\sup_{0\leqq t\leqq h} d(\omega(t), \omega(0)) \geqq \eta\right] \leqq M \exp\left\{-\frac{\eta^2}{4\varepsilon h}\right\}. \tag{3.9}$$

We want to estimate

$$P_x^\varepsilon\left[\sup_{0\leqq t\leqq T} d(\omega(t), \omega_\pi(t)) \geqq 2\eta\right].$$

Since $\omega(t_j) = \omega_\pi(t_j)$ for $j = 0, 1, \cdots, n$, and ω_π is piecewise geodesic for t in (t_j, t_{j+1}),

$$\begin{aligned} d(\omega_\pi(t), \omega_\pi(t_j)) &\leqq d(\omega_\pi(t_{j+1}), \omega_\pi(t_j)) = d(\omega(t_{j+1}), \omega(t_j)) \\ &\leqq \sup_{t_j\leqq t\leqq t_{j+1}} d(\omega(t_j), \omega(t)). \end{aligned}$$

We now have

$$\begin{aligned} \sup_{0\leqq t\leqq T} d(\omega(t), \omega_\pi(t)) &= \sup_{0\leqq j\leqq n-1}\ \sup_{t_j\leqq t\leqq t_{j+1}} d(\omega(t), \omega_\pi(t)) \\ &\leqq \sup_{0\leqq j\leqq n-1}\ \sup_{t_j\leqq t\leqq t_{j+1}} [d(\omega(t), \omega(t_j)) + d(\omega_\pi(t), \omega_\pi(t_j))] \\ &\leqq 2 \sup_{0\leqq j\leqq n-1}\ \sup_{t_j\leqq t\leqq t_{j+1}} d(\omega(t), \omega(t_j)). \end{aligned} \tag{3.10}$$

From the Markov property and the inequality (3.10), we obtain

$$P_x^\varepsilon\left[\sup_{0\le t\le T} d(\omega(t), \omega_\pi(t)) \ge 2\eta\right] \le P_x^\varepsilon\left[\sup_{0\le j\le n-1}\ \sup_{t_j\le t\le t_{j+1}} d(\omega(t), \omega(t_j)) \ge \eta\right]$$

$$\le \sum_{j=0}^{n-1} P_x^\varepsilon\left[\sup_{t_j\le t\le t_{j+1}} d(\omega(t), \omega(t_j)) \ge \eta\right]$$

$$\le n \sup_x P_x^\varepsilon\left[\sup_{0\le t\le h} d(\omega(t), \omega(0)) \ge \eta\right]$$

$$\le nM \exp\left\{-\frac{\eta^2}{4\varepsilon h}\right\}.$$

Since $|x - y| \le \beta\, d(x,y)$ for some β, we have for $\delta \le \delta_0$

$$P_x^\varepsilon[\|\omega - \omega_\pi\| \ge \delta] \le nM \exp\left\{-\frac{\delta^2}{16\beta^2\varepsilon h}\right\}.$$

Consequently,

$$\limsup_{n\to\infty}\ \limsup_{\varepsilon\to 0}\ \sup_x\ \varepsilon \log P_x^\varepsilon[\|\omega - \omega_\pi\| \ge \delta] = -\infty .$$

THEOREM 3.3. *For any closed set $C \subset \Omega$,*

$$\limsup_{\substack{\varepsilon\to 0\\ y\to x}} \varepsilon \log P_y^\varepsilon[C] \le -\inf_{\omega\in C_x} I(\omega) .$$

Proof: Let us define for $\delta > 0$ the sets C_δ and C_δ^δ as

$$C_\delta = \{\omega : \omega \in C\ |\omega(0) - x| < \delta\} ,$$

$$C_\delta^\delta = \{z : \|z - \omega\| < \delta \text{ for some } \omega \in C_\delta\} .$$

We introduce the function

$$I^\delta(\omega) = \inf_{z:\|z-\omega\|<\delta} I(z) .$$

Then $I^\delta(\omega)$ is finite everywhere. Moreover if

$$T_\delta = \inf_{\omega\in C_\delta^\delta} I(\omega) ,$$

then $\omega \in C_\delta \Rightarrow I^\delta(\omega) \ge T_\delta$. Since $y \to x$ we can assume $|y - x| < \delta$. We would therefore obtain

$$P_y^\varepsilon[C] = P_y^\varepsilon[C_\delta] \le P_y^\varepsilon[I^\delta(\omega) \ge T_\delta] .$$

Further,

$$P_y^\varepsilon[I^\delta(\omega) \geqq T_\delta] \leqq P_y^\varepsilon[\|\omega - \omega_\pi\| \geqq \delta] + P_y^\varepsilon[I(\omega_\pi) \geqq T_\delta] \tag{3.11}$$

and

$$\lim_{n\to\infty} \limsup_{\varepsilon\to 0} \sup_x \varepsilon \log P_y^\varepsilon[\|\omega - \omega_\pi\| \geqq \delta] = -\infty . \tag{3.12}$$

The set $[I(\omega_\pi) \geqq T_\delta]$ is equal to

$$\left\{\omega : \frac{1}{2}\sum_{j=0}^{n-1} \frac{d^2(\omega(t_j), \omega(t_{j+1}))}{t_{j+1} - t_j} \geqq T_\delta\right\} .$$

Hence for every n,

$$\limsup_{\substack{\varepsilon\to 0\\ y\to x}} \varepsilon \log P_y^\varepsilon[I(\omega_\pi) \geqq T_\delta] \leqq -T_\delta . \tag{3.13}$$

It follows from (3.11), (3.12) and (3.13) that

$$\limsup_{\substack{\varepsilon\to 0\\ y\to x}} \varepsilon \log P_y^\varepsilon[C] \leqq -T_\delta .$$

This is true for every $\delta > 0$ and $\lim_{\delta\to 0} T_\delta = T$, where

$$T = \inf_{\omega \in C_x} I(\omega) .$$

The proof of the theorem is now complete.

LEMMA 3.4. *Let $f \in \Omega$ and $f(0) = x$. Let V be the set $V = \{\omega : \|\omega - f\| < \delta\}$, where $\delta > 0$. Then*

$$\liminf_{\substack{\varepsilon\to 0\\ y\to x}} \varepsilon \log P_y^\varepsilon[U] \geqq -I(f) .$$

Proof: We can assume without loss of generality that δ is small and that $I(f) = a < \infty$. Let V_π denote the set

$$V_\pi = \{\omega : |\omega(t_j) - f(t_j)| \leqq \tfrac{1}{2}\delta \text{ for } j = 0, 1, 2, \cdots, n\} ,$$

then

$$P_y^\varepsilon[V_\pi] = \int_{A_\pi} \prod p_\varepsilon(t_j - t_{j-1}, y_{j-1}, y_j)\, dy_j ,$$

where A_π is the set

$$A_\pi = \{y : |y_j - f(t_j)| \leqq \tfrac{1}{2}\delta \text{ for } j = 0, 1, \cdots, n\} .$$

Using Theorem 2.3, one can obtain

$$\liminf_{\substack{\varepsilon\to 0\\ y\to x}} \varepsilon \log P_y^\varepsilon[V_\pi] \geqq -I(f_\pi) \geqq -I(f) .$$

It is therefore sufficient to prove that

$$\lim_{n\to\infty} \limsup_{\substack{\varepsilon\to 0\\ y\to x}} \varepsilon \log P_y^\varepsilon[U_\pi - U] = -\infty . \tag{3.14}$$

We see that $\omega \in V_\pi - V \Rightarrow |\omega(t_j) - f(t_j)| \leqq \frac{1}{2}\delta$ for all j but $\|\omega(t) - f(t)\| \geqq \delta$. This implies that, for some interval $t_j \leqq t \leqq t_{j+1}$,

$$|\omega(t_j) - f(t_j)| < \tfrac{1}{2}\delta , \qquad |\omega(t_{j+1}) - f(t_{j+1})| < \tfrac{1}{2}\delta ,$$

but $|\omega(\tau) - f(\tau)| \geqq \delta$ for some τ in $t_j \leqq \tau \leqq t_{j+1}$. Further, we can choose n large enough so that, for the resulting π, $|f(\tau) - f(t_j)| < \frac{1}{4}\delta$ for all τ in $t_j \leqq \tau \leqq t_{j+1}$. We conclude from all this that, if $\omega \in V_\pi - V$, then $|\omega(\tau) - \omega(t_j)| > \frac{1}{4}\delta$ for τ in (t_j , t_{j+1}) for some j. Hence $d(\omega(\tau), \omega(t_j)) \geqq \beta\frac{1}{4}\delta$ for some $\beta > 0$. This in turn implies that $I(\omega) \geqq \beta^2\delta^2/32h$. Moreover, $V_\pi - V$ is closed and therefore, by Theorem 3.3,

$$\lim_{n\to\infty} \limsup_{\substack{\varepsilon\to 0\\ y\to x}} \varepsilon \log P_y^\varepsilon[V_\pi - V] = -\infty .$$

This establishes (3.14) and hence the lemma is proved.

THEOREM 3.5. *Let $G \subset \Omega$ be open. Then*

$$\liminf_{\substack{\varepsilon\to 0\\ y\to x}} \varepsilon \log P_y^\varepsilon[G] \geqq -\inf_{\omega\in G_x} I(\omega) .$$

Proof: Since $G \subset \Omega$ is open for $f \in \Omega$, there is a sphere V around f contained in Ω. By Lemma 3.4,

$$\liminf_{\substack{\varepsilon\to 0\\ y\to x}} \varepsilon \log P_y^\varepsilon[G] \geqq -I(f) .$$

This is true for all $f \in \Omega$ with $f(0) = x$. Hence the theorem is true. Theorems 3.3 and 3.5 contain Theorem 2.4.

We conclude this section with a corollary to Theorem 3.5.

COROLLARY 3.6. *Let $A \subset \Omega$ be measurable, A° and $\bar{A}$ the interior and closure of A. Assume that for $x \in K$ (K compact)*

$$\phi(x) = \inf_{\omega\in\bar{A}\cap\Omega_x} I(\omega) = \inf_{\omega\in A^\circ\cap\Omega_x} I(\omega) .$$

Then $\phi(x)$ is continuous on K and

$$\lim_{\varepsilon \to 0} \varepsilon \log P_x^\varepsilon[A] = -\phi(x)$$

uniformly over K.

4. Fundamental Solution in a Subdomain

$D \subset R_k$ is an open set and we kill the process at the boundary B of D. We consider the fundamental solution $q_\varepsilon(t, x, y)$ of the equation

$$(4.1) \qquad \frac{\partial q}{\partial t} = \tfrac{1}{2}\varepsilon \sum a_{ij}(x) \frac{\partial^2 q}{\partial x_i \, \partial x_j} \quad \text{for} \quad x \in D ,$$
$$q_\varepsilon(t, x, y) \to 0 \quad \text{as} \quad x \to B .$$

It is clear that

$$q_\varepsilon(t, x, y) \leqq p_\varepsilon(t, x, y) .$$

Moreover, if we define the set function

$$(4.2) \qquad q_\varepsilon(t, x, A) = \int_A q_\varepsilon(t, x, y)\, dy$$

for sets $A \subset D$, then the connection between (4.2) and the measures P_x^ε is

$$(4.3) \qquad q_\varepsilon(t, x, A) = P_x^\varepsilon\{\omega : \omega(t) \in A, \omega(\tau) \in D \text{ for } 0 \leqq \tau \leqq t\} .$$

Let us introduce two subsets of Ω,

$$G = \{\omega : \omega(\tau) \in D \text{ for } 0 \leqq \tau \leqq T\} ,$$

and

$$\bar{G} = \text{closure of } G .$$

Then $\bar{G} \subset \{\omega : \omega(\tau) \in D \cup B \text{ for } 0 \leqq \tau \leqq T\}$. In some cases the inclusion is proper and that is relevant. We introduce the distances

$$\underline{d}_D(x, y) = \inf_{\substack{\omega(0)=x \\ \omega(T)=y \\ \omega \in \bar{G}}} l(\omega) ,$$

and

$$\bar{d}_D(x, y) = \inf_{\substack{\omega(0)=x \\ \omega(T)=y \\ \omega \in G}} l(\omega) .$$

The set D is called "smooth" if $\underline{d}_D(x, y) = \bar{d}_D(x, y)$. This means that if a trajectory of length l can be approximated by trajectories in D, then it can be also approximated by trajectories within D whose lengths approach l. In such a case we shall denote by $d_D(x, y)$ the common value of $\underline{d}_D$ and $\bar{d}_D$. We can check that the following relations hold:

$$\inf_{\substack{\omega \in G \\ \omega(0)=x \\ \omega(T)=y}} I(\omega) = \frac{1}{2T} \bar{d}_D^2(x, y) , \tag{4.4}$$

$$\inf_{\substack{\omega \in \bar{G} \\ \omega(0)=x \\ \omega(T)=y}} I(\omega) = \frac{1}{2T} \underline{d}_D^2(x, y) .$$

LEMMA 4.1. *For $A \subset D$ and closed.*

$$\limsup_{\substack{\varepsilon \to 0 \\ y \to x}} \varepsilon \log q_\varepsilon(T, y, A) \leqq -\inf_{z \in A} \frac{\underline{d}_D^2(x, z)}{2T} .$$

Proof: Let us take $C \subset \Omega$ as

$$C = \{\omega : \omega(T) \in A, \ \omega \in \bar{G}\} .$$

Then from relation (4.3), we have

$$\limsup_{\substack{\varepsilon \to 0 \\ y \to x}} \varepsilon \log q_\varepsilon(T, y, A) \leqq \limsup_{\substack{\varepsilon \to 0 \\ y \to x}} \varepsilon \log P_y^\varepsilon[C]$$

$$\leqq -\inf_{\omega \in C_x} I(\omega) = -\frac{1}{2T} \inf_{z \in A} \underline{d}_D^2(x, z) .$$

Lemma 4.1 is only a weak version of what we want to prove. To study the density $q_\varepsilon(t, x, y)$ we use a slightly different approach.

LEMMA 4.2. *Let x, y be points in D. Then*

$$\limsup_{\substack{\varepsilon \to 0 \\ x' \to x \\ y' \to y}} \varepsilon \log q_\varepsilon(T, y, y') \leqq -\frac{1}{2T} \underline{d}_D^2(x, y) .$$

Proof: Let us take a neighborhood S_δ around y of radius δ and a neighborhood $S_{2\delta}$ around S_δ of radius δ. Let τ be the first time the set $S_{2\delta}$ is reached starting from x'. Then for $y' \in S_\delta$,

$$q_\varepsilon(T, x', y') = \int_{\substack{\tau < T \\ \omega \in G}} q_\varepsilon(t - \tau, \omega(\tau), y') P_{x'}^\varepsilon(d\omega) . \tag{4.5}$$

Since $\omega(\tau)$ is in the boundary of $S_{2\delta}$ and y' is in S_δ, we have $|\omega(\tau) - y'| \geqq \delta$. Moreover, $q_\varepsilon(t, x, y) \leqq p_\varepsilon(t, x, y)$ and, if x and y are kept away, the functions are bounded. Hence for all $\omega(\tau)$ and y',

$$q_\varepsilon(T - \tau, \omega(\tau), y') \leqq M_\delta < \infty .$$

Continuing with (4.5) this yields

$$q_\varepsilon(T, x', y') \leqq M_\delta P^\varepsilon_{x'}[\omega : \omega \in G, \tau \leqq T] .$$

Therefore,

$$\limsup_{\substack{\varepsilon \to 0 \\ x' \to x \\ y' \to y}} \varepsilon \log q_\varepsilon(T, x', y') \leqq -\frac{1}{2T} \inf_{z \in S_{2\delta}} d^2_D(x, z) .$$

Since this is true for every $\delta > 0$, we prove the lemma by letting $\delta \to 0$.

LEMMA 4.3. *For $A \subset D$ and open,*

$$\liminf_{\substack{\varepsilon \to 0 \\ y \to x}} \varepsilon \log q_\varepsilon(T, y, A) \geqq -\inf_{z \in A} \frac{d^2_D(x, z)}{2T} .$$

Proof: We have

$$q_\varepsilon(T, y, A) = P^\varepsilon_y[\omega : \omega \in G, \omega(T) \in A] .$$

The set on the right is open. A direct application of Theorem 3.5 yields

$$\liminf_{\substack{\varepsilon \to 0 \\ y \to x}} \varepsilon \log q_\varepsilon(T, y, A) \geqq -\inf_{\substack{\omega \in G_x \\ \omega(T) \in A}} I(\omega)$$

$$= -\frac{1}{2T} \inf_{z \in A} d^2_D(x, z) .$$

LEMMA 4.4. *Let $K \subset D$ be compact. Then for $\Delta > 0$,*

$$\liminf_{\substack{\varepsilon \to 0 \\ x, y \in K \\ |x - y| \to 0}} \varepsilon \log q_\varepsilon(\Delta, x, y) \geqq 0 .$$

Proof: We use the formula

$$q_\varepsilon(\Delta, x, y) = p_\varepsilon(\Delta, x, y) - \int_{\tau \leqq \Delta} p_\varepsilon(\Delta - \tau, \omega(\tau), y) P^\varepsilon_x(d\omega) ,$$

where τ is the first exit time from D. Since $y \in K \subset D$ and K is compact, $|y - \omega(\tau)| \geqq \eta > 0$ for all $y \in K$ and $\omega(\tau) \in B$. Therefore,

$$q_\varepsilon(\Delta, x, y) \geqq p_\varepsilon(\Delta, x, y) - \sup_{\substack{b \in K \\ 0 \leqq u \leqq \Delta \\ y \in B}} p_\varepsilon(u, b, y) \tag{4.6}$$

$$\geqq p_\varepsilon(\Delta, x, y) - Me^{-c\eta^2/\varepsilon} .$$

Since by Theorem 2.5

$$\liminf_{\substack{\varepsilon \to 0 \\ |x-y| \to 0}} \varepsilon \log p_\varepsilon(\Delta, x, y) \geqq 0 ,$$

it follows from (4.6) that

$$\liminf_{\substack{\varepsilon \to 0 \\ x,y \in K \\ |x-y| \to 0}} \varepsilon \log q_\varepsilon(\Delta, x, y) \geqq 0 .$$

Lemma 4.5. *Let x, y be points in D. Then*

$$\liminf_{\substack{\varepsilon \to 0 \\ x' \to x \\ y' \to y}} \varepsilon \log q_\varepsilon(T, x', y') \geqq -\frac{1}{2T} d_D^2(x, y) .$$

Proof: Since $q_\varepsilon(t, x, y)$ satisfies the semigroup property,

$$q_\varepsilon(T, x', y') = \int_D q_\varepsilon(T - \Delta, x', z) q_\varepsilon(\Delta, z, y')\, dz$$

$$\geqq \int_{z \in S} q_\varepsilon(T - \Delta, x', z) q_\varepsilon(\Delta, z, y')\, dz$$

$$\geqq q_\varepsilon(T - \Delta, x', S) \inf_{z \in S} q_\varepsilon(\Delta, z, y') .$$

We take S to be a small sphere around y of radius δ. Then,

$$\liminf_{\substack{\varepsilon \to 0 \\ x' \to x \\ y' \to y}} \varepsilon \log q_\varepsilon(T:x', y') \geqq \liminf_{\substack{\varepsilon \to 0 \\ x' \to x}} \varepsilon \log q_\varepsilon(T - \Delta, x', S)$$

$$+ \liminf_{\substack{\varepsilon \to 0 \\ y' \to y}} \inf_{z \in S} \varepsilon \log q_\varepsilon(\Delta, z, y') .$$

Combining Lemmas 4.3 and 4.4, we have if we let $\delta \to 0$

$$\liminf_{\substack{\varepsilon \to 0 \\ x' \to x \\ y' \to y}} \varepsilon \log q_\varepsilon(T, x', y') \geqq -\frac{1}{2(T - \Delta)} d_D^2(x, y) .$$

Since $\Delta > 0$ is arbitrary, we complete the proof of Lemma 4.5 by letting $\Delta \to 0$.

Lemmas 4.2 and 4.5 yield

THEOREM 4.6. *Let D be smooth. Then*

$$\lim_{\varepsilon \to 0} \varepsilon \log q_\varepsilon(T, x, y) = -\frac{1}{2T} d_D^2(x, y)$$

uniformly for $x, y \in K$, *where* $K \subset D$ *is any compact set.*

COROLLARY 4.7. *For D smooth,*

$$\lim_{t \to 0} t \log q(t, x, y) = -\tfrac{1}{2} d_D^2(x, y) .$$

Let us set

$$R_\varepsilon(t, x, y) = p_\varepsilon(t, x, y) - q_\varepsilon(t, x, y) , \tag{4.7}$$

$$d_D^*(x, y) = \inf_{\substack{\omega(0)=x \\ \omega(T)=y \\ \omega \notin G}} l(\omega) . \tag{4.8}$$

Then

$$\inf_{\substack{\omega(0)=x \\ \omega(T)=y \\ \omega \notin G}} I(\omega) = \frac{1}{2T} d_D^{*2}(x, y) .$$

Moreover, it is easily verified that

$$d_D^*(x, y) = \inf_{z \in B} [d(x, z) + d(z, y)] .$$

LEMMA 4.8.

$$\limsup_{\substack{\varepsilon \to 0 \\ x' \to x \\ y' \to y}} \varepsilon \log R_\varepsilon(T, x', y') \leqq -\frac{1}{2T} d_D^{*2}(x, y) .$$

Proof: This is proved in a manner similar to Lemma 4.2. In order to get to y' from x' by a trajectory not in G, we have to reach the boundary first and then come to a sphere around y. From this point on the proof is exactly like that of Lemma 4.2. Since τ is defined differently, we end up with

$$R_\varepsilon(T, x', y') = \int_{\tau \leqq T} p_\varepsilon(t - \tau, \omega(\tau), y') P_{x'}^\varepsilon(d\omega) ,$$

where τ is the first time the trajectory enters a set $S_{2\delta}$ around y after reaching the boundary B of D. Hence, by suitably letting $\delta \to 0$, we have

$$\limsup_{\substack{\varepsilon\to 0\\ x'\to x\\ y'\to y}} \varepsilon \log R_\varepsilon(T, x', y') \leqq -\frac{1}{2T} d_D^*(x, y) .$$

For any region D with boundary B, Lemmas 4.8, 4.2 and Theorem 2.5 imply

$$\lim_{t\to 0} 2t \log p(t, x, y) = -d^2(x, y) ,$$

$$\limsup_{t\to 0} 2t \log q(t, x, y) \leqq d_D^2(x, y) , \tag{4.9}$$

$$\limsup_{t\to 0} 2t \log R(t, x, y) \leqq -d_D^{*2}(x, y) .$$

An immediate consequence of (4.9) is

THEOREM 4.9. *Let x, y in D be such that all the geodesics from x, y lie completely within D. Then the same is true for all a, b near x, y, respectively, and*

$$\lim_{t\to 0} \frac{q(t, a, b)}{p(t, a, b)} = 1$$

uniformly for a, b near x, y. On the other hand, let x, y in D be such that no geodesic from x to y can be approximated uniformly by paths completely in D. Then the same is true for all a, b near x, y, respectively, and

$$\lim_{t\to 0} \frac{q(t, a, b)}{p(t, a, b)} = 0$$

uniformly for a, b near x, y.

5. Some Associated Differential Equations

Let $D \subset R_k$ be a domain with boundary B. We assume throughout this section that the set of points $b \in B$, with the property that every sufficiently small neighborhood N_b around b is disconnected by B, is dense in B.

Let $\phi(x, \lambda)$ be the solution of the equation

$$\tfrac{1}{2} \sum a_{ij}(x) \frac{\partial^2 \phi}{\partial x_i\, \partial x_j} = \lambda \phi \quad \text{for} \quad x \in D ,$$
$$\phi = 1 \quad \text{on} \quad B . \tag{5.1}$$

We then have

THEOREM 5.1.

$$\lim_{\lambda\to\infty} -\frac{1}{\sqrt{2\lambda}} \log \phi(x, \lambda) = d(x, B)$$

uniformly in x over compact subsets of $D \cup B$.

Proof: Let τ be the exit time from the set D. Then the solution $\phi(x, \lambda)$ of equation (5.1) has a representation

$$\phi(x, \lambda) = \int e^{-\lambda\tau} P_x(d\omega)$$

$$= \int e^{-\lambda\tau} F(x, dt) ,$$

where

$$F(x, t) = P_x[\tau \leqq t] .$$

In order to estimate $\phi(x, \lambda)$ for large λ we need estimate $F(x, t)$ for small t:

$$F(x, \varepsilon) = P_x[\tau \leqq \varepsilon] = P_x^\varepsilon[\tau \leqq 1] .$$

Let C be the set

$$C = \{\omega : \omega(s) \in B \text{ for some } s, \quad 0 \leqq s \leqq 1\}.$$

Then C is closed and $C = \{\omega : \tau \leqq 1\}$. Therefore,

$$\limsup_{\substack{\varepsilon\to 0\\ y\to x}} \varepsilon \log F(y, \varepsilon) = \limsup_{\substack{\varepsilon\to 0\\ y\to x}} \varepsilon \log P_y^\varepsilon[C] \tag{5.2}$$

$$\leqq -\inf_{C_x} I(\omega) = -\tfrac{1}{2} d^2(x, B) .$$

Let G be defined as the set of trajectories ω for which there are two points $0 \leqq u_1 < u_2 \leqq 1$ in the time interval $0 \leqq s \leqq 1$ such that, in the interval $u_1 \leqq \tau \leqq u_2$, $\omega(\tau)$ lies completely in one of the disconnected neighborhoods N_b and $\omega(u_1)$ and $\omega(u_2)$ belong to two different components of the disconnected set N_b. It is clear that G is open and $G \subset C$. Therefore,

$$\liminf_{\substack{\varepsilon\to 0\\ y\to x}} \varepsilon \log P_y^\varepsilon[C] \geqq \liminf_{\substack{\varepsilon\to 0\\ y\to x}} \varepsilon \log P_y^\varepsilon[G] \geqq -\inf_{\omega\in G_x} I(\omega) .$$

Let $b \in B$ and N_b be a small neighborhood around b which is disconnected. Consider the trajectory $\omega(\tau)$, $0 \leqq \tau \leqq 1$, which is a geodesic from x to b, to be naturally parametrized. Then $I(\omega) = \frac{1}{2}d^2(x, b)$. The trajectory $\omega(\tau)$ can be extended slightly so that it penetrates into two different components of N_b and lies completely in N_b in between. This can always be achieved by choosing N_b small with only a slight increase in $I(\omega)$. The parametrization is to be adjusted by rescaling time so that the time interval is still $0 \leqq s \leqq 1$. The final conclusion is

$$\inf_{\omega \in G_x} I(\omega) \leqq \inf_{b \in B_0} \tfrac{1}{2} d^2(x, b) ,$$

where B_0 is the set of points in B for which the construction is possible. By assumption, B_0 is dense in B. Consequently,

$$\liminf_{\substack{\varepsilon \to 0 \\ y \to x}} \varepsilon \log F(y, \varepsilon) \geqq -\tfrac{1}{2} d^2(x, B) .$$

Along with (5.2) this implies

$$\lim_{\substack{\varepsilon \to 0 \\ y \to x}} \varepsilon \log F(y, \varepsilon) = -\tfrac{1}{2} d^2(x, B) ,$$

since

$$\phi(x, \lambda) = \int e^{-\lambda t} F(x, dt) .$$

Standard estimation yields

$$\lim_{\substack{\lambda \to \infty \\ y \to x}} \frac{1}{\sqrt{2\lambda}} \log \phi(y, \lambda) = -d(x, B) .$$

We still assume that we have a domain D with a boundary B satisfying the condition described early in the section. Let us define for any set $\Gamma \subset B$ (Γ measurable) the function $\psi(x, \lambda, \Gamma)$ as the solution of the equation

$$\begin{aligned} L\psi &= \lambda\psi \quad && \text{for} \quad x \in D , \\ \psi &= 1 && \text{for} \quad x \in \Gamma , \\ \psi &= 0 && \text{for} \quad x \in B - \Gamma . \end{aligned} \tag{5.3}$$

Then $\psi(x, \lambda, \Gamma)$ has a representation

$$\psi(x, \lambda, \Gamma) = \int e^{-\lambda t} F(x, \Gamma, dt) . \tag{5.4}$$

If τ is the first exit time from D, then

$$F(x, \Gamma, t) = P_x[\tau \leqq t, \omega(\tau) \in \Gamma] . \tag{5.5}$$

We assume that D is smooth in the sense of Section 4. Moreover, we assume that the property extends to the boundary B. The actual assumptions are as follows: If

$$G = \{\omega : \omega(\tau) \in D \quad \text{for} \quad 0 \leqq \tau < T\},$$

$$\bar{G} = \text{closure of } G,$$

$$\bar{d}_D(x, y) = \inf_{\substack{\omega \in G \\ \omega(0)=x \\ \omega(T)=y}} l(\omega),$$

$$\underline{d}_D(x, y) = \inf_{\substack{\omega \in \bar{G} \\ \omega(0)=x \\ \omega(T)=y}} l(\omega),$$

then $\bar{d}_D(x, y) = \underline{d}_D(x, y)$ for all $x \in D$ and $y \in D \cup B$. We denote the common value by $d_D(x, y)$ and by $d_D(x, b)$ if b is on the boundary.

LEMMA 5.2. *For $\Gamma \subset B$ open and $x \in D$,*

$$\liminf_{\substack{\lambda \to \infty \\ y \to x}} \frac{1}{\sqrt{2\lambda}} \log \psi(y, \lambda, \Gamma) \geqq -\inf_{b \in \Gamma} d_D(x, b).$$

For $\Gamma \subset B$ closed and $x \in D$,

$$\limsup_{\substack{\lambda \to \infty \\ y \to x}} \frac{1}{\sqrt{2\lambda}} \psi(y, \lambda, \Gamma) \leqq -\inf_{b \in \Gamma} d_D(x, b).$$

Proof: As in Theorem 5.1 we study

$$F(x, \Gamma, \varepsilon) = P_x[\tau \leqq \varepsilon, \omega(\tau) \in \Gamma] = P_x^\varepsilon[\tau \leqq 1, \omega(\tau) \in \Gamma].$$

Let U be the set

$$U = \{\omega : \tau \leqq 1, \omega(\tau) \in \Gamma\}.$$

In order to show that

$$\liminf_{\substack{\varepsilon \to 0 \\ y \to x}} \varepsilon \log F(y, \Gamma, \varepsilon) \geqq -\tfrac{1}{2} \inf_{b \in \Gamma} d_D^2(x, b), \tag{5.6}$$

it suffices to show that for any given $b \in \Gamma$ and $\delta > 0$ there exists an interior point ω_0 of U (U is not open in general) such that

$$I(\omega_0) \leqq \tfrac{1}{2} d_D^2(x, b) + \delta. \tag{5.7}$$

Let $b \in \Gamma$ be given. Then there is a b' near b for which the disconnectedness condition holds. Let us take a ω' in G with $\omega'(1) = b$ and $I(\omega')$ nearly equal to

$\frac{1}{2}d_D^2(x, b)$. We take a point near b on ω' inside D and connect the point with b' by a geodesic not necessarily inside D. Then we repeat the construction of Theorem 5.1 to get a ω_0 which penetrates into two different components of a small disconnected neighborhood $N_{b'}$ around b'. Any trajectory close to ω_0 will have to exit at a point near b or b'. Since Γ is open, things can be managed so that all these trajectories exit in Γ. $I(\omega_0)$ of course is not much larger than $\frac{1}{2}d_D^2(x, b)$. This proves (5.7) and hence (5.6). Now let Γ be closed. Then the set

$$C = \{\omega : \omega(s) \in \Gamma \text{ for some } 0 \leqq s \leqq 1, \omega \in \bar{G}\}$$

is closed and obviously

$$\limsup_{\substack{\varepsilon \to 0 \\ y \to x}} \varepsilon \log F(y, \Gamma, \varepsilon) \leqq \limsup_{\substack{\varepsilon \to 0 \\ y \to x}} \varepsilon \log P_y^\varepsilon[C] \tag{5.8}$$

$$\leqq -\inf_{\omega \in G_x} I(\omega) = -\tfrac{1}{2} \inf_{b \in \Gamma} d_D^2(x, b) .$$

The lemma is then proved using the relation (5.4) and the properties (5.6) and (5.8).

We now consider the solution $\phi(x, \lambda)$ to the equation

$$\begin{aligned} L\phi &= \lambda\phi && \text{for} \quad x \in D , \\ \phi &= h(b, \lambda) && \text{for} \quad b \in B , \end{aligned} \tag{5.9}$$

where

$$h(b, \lambda) = \exp\{\sqrt{2\lambda}\, h(b)\} \tag{5.10}$$

and $h(b)$ is a bounded continuous function on B.

THEOREM 5.3.

$$\lim_{\lambda \to \infty} \frac{1}{\sqrt{2\lambda}} \log \phi(x, \lambda) = \sup_{b \in B} [h(b) - d_D(x, b)]$$

uniformly over compact subsets $K \subset D$.

Proof: We use the formula

$$\phi(x, \lambda) = \int h(b, \lambda)\psi(x, \lambda, db) .$$

Lemma 5.2 is all we need to be able to use Theorem 3.4 of [6] and the theorem follows. It is elementary to see that the uniformity in Lemma 5.2 carries over to the uniformity asserted in the theorem.

Finally we turn to the following initial value problem: $u_\varepsilon(t, x)$ is the solution of the equation

$$
(5.11)\qquad
\begin{aligned}
\frac{\partial u_\varepsilon}{\partial t} &= \tfrac{1}{2}\varepsilon L u_\varepsilon + \tfrac{1}{2}\sum a_{ij}(x)\,\frac{\partial u_\varepsilon}{\partial x_i}\frac{\partial u_\varepsilon}{\partial x_j} + P(t, x)\,,\\
u_\varepsilon(0, x) &= V(x)\,.
\end{aligned}
$$

Let us suppose for simplicity that $V(x)$ and $P(t, x)$ are continuous and bounded. We are interested in the behavior of $u_\varepsilon(t, x)$ as $\varepsilon \to 0$.

THEOREM 5.4.

$$\lim_{\varepsilon\to 0} u_\varepsilon(t, x) = u(t, x)$$

exists and

$$u(t, x) = \sup_{\omega\in\Omega_x}\left[V(\omega(t)) + \int_0^t P(t-\tau, \omega-\tau)\,d\tau - I(\omega)\right].$$

Proof: Let us set

$$v_\varepsilon(t, x) = \exp\left\{\frac{1}{\varepsilon}\,u_\varepsilon(t, x)\right\}.$$

Then the equation satisfied by v_ε is

$$
(5.12)\qquad
\begin{aligned}
\frac{\partial v_\varepsilon}{\partial t} &= \tfrac{1}{2}\varepsilon\, L v_\varepsilon + \frac{1}{\varepsilon}\,v_\varepsilon P\,,\\
v_\varepsilon(0, x) &= \exp\left\{\frac{1}{\varepsilon}\,V(x)\right\};
\end{aligned}
$$

hence the solution is given by the function space integral

$$v_\varepsilon(t, x) = \int \exp\left\{\frac{1}{\varepsilon}\,V(\omega(t)) + \frac{1}{\varepsilon}\int_0^t P(t-\tau, \omega(\tau))\,d\tau\right\} P_x^\varepsilon(d\omega)\,.$$

We now appeal to Theorem 3.4 in [5] and conclude that

$$
\begin{aligned}
\lim_{\varepsilon\to 0} u_\varepsilon(t, x) &= \lim_{\varepsilon\to 0} \varepsilon \log v_\varepsilon(t, x)\\
&= \sup_{\omega\in\Omega_x}\left[V(\omega(t)) + \int_0^t P(t-\tau, \omega(\tau))\,d\tau - I(\omega)\right].
\end{aligned}
$$

Remark. With a little more effort one can show that the limit in Theorem 5.4 is actually uniform over compact subsets of t and x. One can also replace the boundedness of P and V by reasonable growth conditions. A similar approach for the case when L is the Laplacian was carried out in [2] using the results in [4].

6. The Case With a First Order Term

We assume that the operator (2.1) has an additional first order term:

$$L_0 f = \tfrac{1}{2}\sum a_{ij}(x)\frac{\partial^2 f}{\partial x_i\,\partial x_j} + \sum b_j(x)\frac{\partial f}{\partial x_j}\,.$$

The coefficients $\{a_{ij}(x)\}$ satisfy the assumptions postulated earlier. In addition we assume that

$$|b_j(x)| \leqq M < \infty\,,$$

$$|b_j(x) - b_j(y)| \leqq M\,|x-y|^{\alpha}\,.$$

Let $q(t, x, y)$ be the fundamental solution for the equation

$$\frac{\partial q}{\partial t} = L_0 q\,.$$

If we replace L_0 by εL_0, we get a family

$$q_\varepsilon(t, x, y) = q(\varepsilon t, x, y)\,.$$

Using the densities $q_\varepsilon(t, x, y)$ we obtain a family of measures Q_x^ε. The following Theorem is proved in [5].

THEOREM 6.1. *For every x and ε the measure Q_x^ε on Ω is absolutely continuous with respect to P_x^ε and*

$$\frac{dQ_x^\varepsilon}{dP_x^\varepsilon} = \exp\left\{\int_0^T b(\omega(\tau))a^{-1}(\omega(\tau))\,d\omega(\tau) - \tfrac{1}{2}\varepsilon\int_0^T [b(\omega(\tau))a^{-1}(\omega(\tau))b(\omega(\tau))]\,d\tau\right\}.$$

Remark. Although the assumptions in [5] are stronger (the Hölder exponent α is assumed there to be 1) the theorem can be proved under less restrictive conditions. For instance in [3], Volume II, Chapter XIV, Section 6, it is carried out by a different method when $a_{ij}(x) = \delta_{ij}$. The same method will work in the present case. It is elementary to obtain the following estimate.

LEMMA 6.2. *If we denote by $F(\varepsilon, \omega)$ the exponent in the Radon-Nikodym derivative of Theorem 6.1, then*

$$\int \exp\{\lambda F(\varepsilon, \omega)\}P_x^\varepsilon(d\omega) \leqq \exp\{C\varepsilon\tfrac{1}{2}\lambda^2\}\,.$$

Proof: One uses the fact that

$$\exp\left\{\lambda\int_0^t b(\omega(\tau))a^{-1}(\omega(\tau))\,d\omega(\tau) - \tfrac{1}{2}\lambda^2\varepsilon\int_0^t [b(\omega(\tau))a^{-1}(\omega(\tau))b(\omega(\tau))]\,d\tau\right\}$$

is a martingale relative to P_x^ε.

THEOREM 6.3. *For $G \subset \Omega$ open and $C \subset \Omega$ closed,*

$$\liminf_{\substack{\varepsilon\to 0\\ y\to x}} Q_y^\varepsilon[G] \geqq -\inf_{\omega\in G_x} I(\omega)\,,$$

$$\limsup_{\substack{\varepsilon\to 0\\ y\to x}} Q_y^\varepsilon[C] \leqq -\inf_{\omega\in C_x} I(\omega)\,.$$

Proof: We have

$$\begin{aligned} Q_y^\varepsilon[G] &= \int_G \exp\{F(\varepsilon,\omega)\}P_y^\varepsilon[d\omega] \\ &\geqq e^{-A}P_y^\varepsilon[G\cap\{F(\varepsilon,\omega)\geqq -A\}] \\ &\geqq e^{-A}P_y^\varepsilon[G] - e^{-A}P_y^\varepsilon[|F(\varepsilon,\omega)|\geqq A]\,. \end{aligned}$$

It therefore suffices to show that

$$\lim_{A\to\infty}\limsup_{\varepsilon\to 0}\varepsilon\log P_y^\varepsilon[|F(\varepsilon,\omega)|\geqq A] = -\infty \tag{6.1}$$

uniformly in y. Similarly,

$$\begin{aligned} Q_y^\varepsilon[C] &= \int_C \exp\{F(\varepsilon,\omega)\}P_y^\varepsilon(d\omega) \\ &\leqq e^{A}P_y^\varepsilon[C] + \int_{[F(\varepsilon,\omega)\geqq A]}\exp\{F(\varepsilon,\omega)\}P_y^\varepsilon(d\omega)\,. \end{aligned}$$

Hence we need only show that

$$\lim_{A\to\infty}\limsup_{\varepsilon\to 0}\varepsilon\log\int_{[F\geqq A]}\exp\{F(\varepsilon,\omega)\}P_y^\varepsilon(d\omega) = -\infty \tag{6.2}$$

uniformly in y. Relations (6.1) and (6.2) follow easily from Lemma 6.2.

THEOREM 6.4.

$$\lim_{t\to 0} -2t\log q(t,x,y) = d^2(x,y)$$

uniformly in x and y over compact sets.

Proof: We have the following estimates from Theorem 0.5, page 229, Volume II of [3]:

$$q(t, x, y) \leqq Mt^{-k/2} \exp\left\{-\frac{\alpha}{2t}|x-y|^2\right\}, \tag{6.3}$$

$$\begin{aligned} q(t, x, y) \geqq\ & M_1 t^{-k/2} \exp\left\{-\frac{\alpha_1}{2t}|x-y|^2\right\} \\ & - M_2 t^{-k/2+\lambda} \exp\left\{-\frac{\alpha_2}{2t}|x-y|^2\right\}. \end{aligned} \tag{6.4}$$

UPPER BOUND. Let x, y be two points in R_k. We choose a sphere S_δ around y of radius δ and a sphere $S_{2\delta}$ of radius δ around $S_{2\delta}$. Let B denote the boundary of $S_{2\delta}$. We assume that $y' \to y$ and hence $y' \in S_\delta$. Let τ be the first entrance time into the set B for a process starting at x'. Then,

$$\begin{aligned} q(t, x', y') &= \int_{\tau \leqq t} q(t-\tau, \omega(\tau), y') P_{x'}(d\omega) \\ &\leqq \sup_{\substack{0 \leqq u \leqq t \\ |z'-y'| \geqq \delta}} q(u, z', y') \cdot P_{x'}[\tau \leqq t]. \end{aligned}$$

From (6.3) we have

$$\sup_{\substack{0 \leqq u \leqq t \\ |z'-y'| \geqq \delta}} q(u, z', y') \leqq M < \infty.$$

Hence Theorem 6.3 yields

$$\begin{aligned} \limsup_{\substack{t \to 0 \\ x' \to x \\ y' \to y}} 2t \log q(t, x', y') &\leqq \limsup_{\substack{t \to 0 \\ x' \to x}} t \log P_{x'}[\tau \leqq t] \\ &\leqq -\inf_{z \in B} \tfrac{1}{2} d^2(x, z). \end{aligned}$$

Letting $\delta \to 0$, we find

$$\limsup_{\substack{t \to 0 \\ x' \to x \\ y' \to y}} 2t \log q(t, x', y') \leqq -\tfrac{1}{2} d^2(x, y).$$

LOWER BOUND. Equation (6.4) gives, for every $\Delta > 0$,

$$\operatorname*{Inf}_{0 \leqq t \leqq t_0} \inf_{|x-y| \leqq \Delta t} q(t, x, y) = \theta > 0 \quad \text{for} \quad t_0 \quad \text{small}. \tag{6.5}$$

Let x, y be points in R_k. Let $x' \to x$ and $y' \to y$. We define the sets

$$U_t = \{z : |z - y'| \leqq t\},$$

and

$$G_t = \{\omega : \omega(t) \in U_t\} .$$

Then,

$$\int_{G_t} p(t, x', v)\, dv = \int_{|z-y'| \leqq t} p(t, x', v)\, dv$$

$$\geqq ct^k \inf_{v \in G_t} p(t, x', v) .$$

Consequently, from Theorem 2.5, we obtain

$$\liminf_{\substack{t \to 0 \\ x' \to x \\ y' \to y}} t \log P_{x'}[G_t] \geqq -\tfrac{1}{2} d^2(x, y) .$$

Rewriting this inequality by introducing the measures P_x^ε, we have

$$\liminf_{\substack{\varepsilon \to 0 \\ x' \to x \\ y' \to y}} \varepsilon \log P_{x'}^\varepsilon[G_\varepsilon] \geqq -\tfrac{1}{2} d^2(x, y) ,$$

where we redefine

$$G_\varepsilon = \{\omega : \omega(1) \in U_\varepsilon\} .$$

From relation (6.1), it follows that

$$\liminf_{\substack{\varepsilon \to 0 \\ x' \to x \\ y' \to y}} \varepsilon \log Q_{x'}^\varepsilon[G_\varepsilon] \geqq -\tfrac{1}{2} d^2(x, y) ,$$

or equivalently that

$$\liminf_{\substack{t \to 0 \\ x' \to x}} t \log \int_{G_t} q(t, x', v)\, dv \geqq -\tfrac{1}{2} d^2(x, y) . \tag{6.6}$$

Now,

$$q(t, x', y') = \int q(t(1 - \Delta), x', v) q(\Delta t, v, y')\, dv$$

$$\geqq \int_{G_t} q(t(1 - \Delta), x', v) q(\Delta t, v, y')\, dv$$

$$\geqq \inf_{v \in G_t} q(\Delta t, v, y') \int_{G_t} q(t(1 - \Delta), x', v)\, dv .$$

Using (6.5) and (6.6) we arrive at

$$\liminf_{\substack{t \to 0 \\ x' \to x \\ y' \to y}} t \log q(t, x', y') \geqq -\frac{1}{2(1 - \Delta)} d^2(x, y) .$$

Allowing $\Delta \to 0$ we prove the theorem.

Bibliography

[1] Ciesielski, Z., *Heat conduction and the principle of not feeling the boundary*, Bull. Acad. Polon. Sci. Ser. Sci. Math. Astronom. Phys., Vol. 15, 1966, pp. 435–440.

[2] Donsker, M. D., *An application of function space integrals to a non-linear equation*, to appear.

[3] Dynkin, E. B., *Markov Processes*, Vols. I, II, Academic Press, New York, 1965.

[4] Schilder, M., *On a Laplace asymptotic formula for Wiener integrals*, Trans. Amer. Math. Soc., to appear.

[5] Skorokhod, A. V., *On the differentiability of measures corresponding to random processes, II, Markov processes*, Theor. Probability Appl., Vol. 5, 1960, pp. 40–49.

[6] Varadhan, S. R. S., *Asymptotic probabilities and differential equations*, Comm. Pure Appl. Math., Vol. 19, 1966, pp. 261–286.

[7] Varadhan, S. R. S., *On the behavior of the fundamental solution of the heat equation with variable coefficients*, Comm. Pure Appl. Math., Vol. 20, 1967, pp. 431–455.

Received January, 1967.

Proc. Nat. Acad. Sci. USA
Vol. 72, No. 3, pp. 780–783, March 1975

On a Variational Formula for the Principal Eigenvalue for Operators with Maximum Principle

(semigroups/Rayleigh-Ritz formula/Feynman-Kac formula)

MONROE D. DONSKER AND S. R. S. VARADHAN

Courant Institute of Mathematical Sciences, New York University, New York, N.Y. 10012

Communicated by Louis Nirenberg, December 11, 1974

ABSTRACT In this paper a variational formula is obtained for the principal eigenvalue for operators with maximum principle. This variational formula does not require the operators to be self-adjoint. But if they are self-adjoint this formula reduces to the classical Rayleigh-Ritz formula.

Let X be a compact metric space, and for $t \geq 0$ let T_t be a strongly continuous semigroup mapping $C(X) \rightarrow C(X)$ having the properties that if $f \geq 0$, then $T_t f \geq 0$ and $T_t 1 = 1$. Under these hypotheses the infinitesimal generator L of T_t will have domain $\mathfrak{D}$ dense in $C(X)$, and L will satisfy the maximum principle. Examples of operators L arising from such semigroups are

(*a*) X is a compact manifold and L is a second order elliptic operator with reasonable coefficients.

(*b*) $(Lf)(x) = \int_X [f(y) - f(x)]\pi(x,dy)$ where $\pi(x,dy)$ is a nonnegative measure on X for each $x \in X$ and is weakly continuous in x.

In fact, the most general L satisfying our assumptions is a limit of examples of type (*b*).

Let $V(x) \in C(X)$ and let λ_V be the principal eigenvalue of the operator $L+V$. If the operator L is self-adjoint with respect to a measure ν on X, then there is a classical variational formula (Rayleigh-Ritz) for λ_V, namely,

$$\lambda_V = \sup_{\substack{f \in L^2(\nu) \\ \|f\|_2 = 1}} \left[\int_X V(x) f^2(x) \nu(dx) + \langle Lf, f \rangle \right]. \quad [1]$$

In this note we obtain a variational formula for λ_V for all L considered above whether L is self-adjoint or not. In the self-adjoint case the new variational formula reduces to [1].

Let $\mathfrak{D}^+$ denote the functions $u \in \mathfrak{D}$ that are positive and let $\mathfrak{M}$ be the space of all probability measures on X. For each $\mu \in \mathfrak{M}$ we define

$$I(\mu) = - \inf_{u \in \mathfrak{D}^+} \int_X \left(\frac{Lu}{u}\right)(x)\mu(dx). \quad [2]$$

It is easy to see that $I(\mu)$ is nonnegative, lower semicontinuous, and convex.

The operator $L+V$ is the infinitesimal generator of a semigroup T_t^V given by a family of measures $p_V(t,x,dy)$, i.e.,

$$T_t^V f(x) = \int_X f(y) p_V(t,x,dy)$$

and we note

$$\|T_t^V\| = \sup_{x \in X} \int_X p_V(t,x,dy).$$

Moreover, if we let $\phi(t,V) = \log \|T_t^V\|$, we see that $\phi(t,V)$ is subadditive in t and therefore

$$\phi(V) = \lim_{t \to \infty} \frac{\phi(t,V)}{t} = \lim_{t \to \infty} \frac{1}{t} \log \sup_{x \in X} \int_X p_V(t,x,dy) = \lambda_V$$

exists. Because T_t is a positive semigroup, λ_V is in the spectrum of $L+V$ and is in fact the principal eigenvalue. We will prove

THEOREM. *The principal eigenvalue* λ_V of $L+V$ is given by*

$$\lambda_V = \sup_{\mu \in \mathfrak{M}} \left[\int_X V(x)\mu(dx) - I(\mu) \right]$$

where $I(\mu)$ is defined by [2]. *The theorem follows from Lemmas 1 and 2 below. Let us define*

$$\phi(V) = \lim_{t \to \infty} \frac{1}{t} \log \sup_{x \in X} \int_X p_V(t,x,dy),$$

$$\psi_1(V) = \sup_{\mu \in \mathfrak{M}} \left[\int_X V(x)\mu(dx) - I(\mu) \right],$$

$$\psi_2(V) = \inf_{u \in \mathfrak{D}^+} \sup_{x \in X} \left[V(x) + \left(\frac{Lu}{u}\right)(x) \right].$$

The theorem states that $\phi(V) = \psi_1(V)$. We will first prove in Lemma 1 that $\psi_1(V) \leq \phi(V) \leq \psi_2(V)$ and then in Lemma 2 that $\psi_2(V) = \psi_1(V)$.

---

* In general $\phi(V)$ is not an eigenvalue. However, it belongs to the spectrum of $L+V$ which is contained in $[z{:}\mathrm{Re}\ z \leq \phi(V)]$. We therefore call it the principal eigenvalue λ_V.

LEMMA 1.

$$\psi_1(V) \le \phi(V) \le \psi_2(V).$$

Proof: Let $\psi_2(V) = l$ and $\epsilon > 0$ be given. Then there exists $u \in \mathfrak{D}^+$ such that for all $x \in X$

$$V(x) + \left(\frac{Lu}{u}\right)(x) \le l+\epsilon.$$

For this u, let $v(t,x) = u(x)e^{(l+\epsilon)t}$ and we have

$$(L+V)v(t,x) = \left[\left(\frac{Lu}{u}\right)(x) + V(x)\right]v(t,x) \le (l+\epsilon)u(x)e^{(l+\epsilon)t}.$$

Thus, for all x and t,

$$\frac{\partial v(t,x)}{\partial t} \ge (L+V)v(t,x)$$

so that from the maximum principle we conclude

$$v(t,x) \ge \int_X v(0,y)p_V(t,x,dy) = \int_X u(y)p_V(t,x,dy).$$

Since X is compact and $u \in \mathfrak{D}^+$ we have $\inf_{x\in X} u(x) > 0$, and therefore for all x and t

$$\int_X p_V(t,x,dy) \le \frac{v(t,x)}{\inf_{x\in X} u(x)} \le \frac{\sup_{x\in X} u(x)}{\inf_{x\in X} u(x)} e^{(l+\epsilon)t}.$$

Thus,

$$\phi(V) = \lim_{t\to\infty} \frac{1}{t} \log \sup_{x\in X} \int_X p_V(t,x,dy) \le l+\epsilon,$$

and hence $\phi(V) \le \psi_2(V)$.

Now suppose $\phi(V) = h$. Let $u(t,x)$ be the solution for $t \ge 0$ of

$$\frac{\partial u}{\partial t} = Lu + Vu,$$
$$u(0,x) = 1,$$

i.e.,

$$u(t,x) = \int_X p_V(t,x,dy).$$

Since $\phi(V) = h$, corresponding to any $\epsilon > 0$ there is a t_0 such that $t \ge t_0$ implies

$$\sup_{x\in X} u(t,x) \le e^{(h+\epsilon)t}. \quad [3]$$

Let $V^\epsilon(x) = V(x) - h - \epsilon$ and $u_\epsilon(t,x)$ be the solution of

$$\frac{\partial u_\epsilon}{\partial t} = Lu_\epsilon + V^\epsilon u_\epsilon,$$
$$u_\epsilon(0,x) = 1.$$

We see that $u_\epsilon(t,x) = u(t,x)e^{-(h+\epsilon)t}$ and from [3] we have for all $t \ge t_0$

$$\sup_{x\in X} u_\epsilon(t,x) \le 1.$$

In particular, $u_\epsilon(t_0,x) \le 1$ for all $x \in X$. If we consider then $u_\epsilon(t,x)$ for $x \in X$ and $0 \le t \le t_0$, we have

$$\frac{\partial u_\epsilon}{\partial t} = Lu_\epsilon + V^\epsilon u_\epsilon,$$
$$u_\epsilon(0,x) = 1,$$
$$u_\epsilon(t_0,x) \le 1,$$

and $u_\epsilon(t,x)$ is bounded below. Thus,

$$\frac{\partial \log u_\epsilon}{\partial t} = \frac{Lu_\epsilon}{u_\epsilon} + V^\epsilon,$$

so that for any $\mu \in \mathfrak{M}$ and $0 \le t \le t_0$

$$\int_X \frac{\partial \log u_\epsilon(t,x)}{\partial t}\mu(dx) = \int_X \left(\frac{Lu_\epsilon}{u_\epsilon}\right)(t,x)\mu(dx) + \int_X V^\epsilon(x)\mu(dx).$$

Therefore,

$$\int_0^{t_0} dt\left[\int_X \left(\frac{Lu_\epsilon}{u_\epsilon}\right)(t,x)\mu(dx) + \int_X V^\epsilon(x)\mu(dx)\right] = \int_X \log u_\epsilon(t_0,x)\mu(dx) - \int_X \log u_\epsilon(0,x)\mu(dx) = \int_X \log u_\epsilon(t_0,x)\mu(dx) \le 0.$$

This implies that there is some point $0 \le t' \le t_0$ such that

$$\int_X \left(\frac{Lu_\epsilon}{u_\epsilon}\right)(t',x)\mu(dx) + \int_X V^\epsilon(x)\mu(dx) \le 0,$$

and hence, letting $u(x) = u_\epsilon(t',x)$, we see that for each $\mu \in \mathfrak{M}$ there is a $u \in \mathfrak{D}^+$ such that

$$\int_X \left(\frac{Lu}{u}\right)(x)\mu(dx) + \int_X V^\epsilon(x)\mu(dx) \le 0.$$

This means

$$\sup_{\mu\in\mathfrak{M}}\left[\inf_{u\in\mathfrak{D}^+}\int_X \left(\frac{Lu}{u}\right)(x)\mu(dx) + \int_X V^\epsilon(x)\mu(dx)\right] \le 0,$$

and thus, from the definition of $I(\mu)$, we have

$$\sup_{\mu\in\mathfrak{M}}\left[-I(\mu) + \int_X V^\epsilon(x)\mu(dx)\right] \le 0,$$

i.e., $\psi_1(V^\epsilon) \le 0$. But obviously $\psi_1(V^\epsilon) = \psi_1(V) - h - \epsilon$, so that $\psi_1(V) \le h+\epsilon$ which implies $\psi_1(V) \le \phi(V)$.

LEMMA 2.

$$\psi_1(V) = \psi_2(V).$$

Proof: We must show $\psi_2(V) \le \psi_1(V)$. Let $\psi_1(V) = l$ and $\epsilon > 0$ be given. We then want to find $u_\epsilon \in \mathfrak{D}^+$ such that for all $x \in X$

$$\left(\frac{Lu_\epsilon}{u_\epsilon}\right)(x) + V(x) \leq l+\epsilon.$$

Since $\psi_1(V) = l$, we have from the definition of $I(\mu)$

$$\sup_{\mu\in\mathfrak{M}} \inf_{u\in\mathfrak{D}^+} \left[\int_X \left(\frac{Lu}{u}\right)(x)\mu(dx) + \int_X V(x)\mu(dx)\right] = l.$$

Thus, for any $\mu \in \mathfrak{M}$ and our given $\epsilon > 0$, there exists $u_{\epsilon,\mu} \in \mathfrak{D}^+$ such that

$$\int_X \left(\frac{Lu_{\epsilon,\mu}}{u_{\epsilon,\mu}}\right)(x)\mu(dx) + \int_X V(x)\mu(dx) \leq l + \frac{\epsilon}{8}.$$

Since the left side of this last inequality is continuous in the weak topology on $\mathfrak{M}$, there exists a neighborhood N_μ of μ such that if $\lambda \in N_\mu$,

$$\int_X \left(\frac{Lu_{\epsilon,\mu}}{u_{\epsilon,\mu}}\right)(x)\lambda(dx) + \int_X V(x)\lambda(dx) \leq l + \frac{\epsilon}{4}.$$

Thus $\bigcup_{\mu\in\mathfrak{M}} N_\mu$ is an open covering of the compact space $\mathfrak{M}$. Hence, there exists a finite covering $N_{\mu_1}, N_{\mu_2}, \ldots, N_{\mu_k}$. Let $u_i = u_{\epsilon,\mu_i}$ for $i = 1,2,\ldots,k$ so that we now have

$$\sup_{\mu\in\mathfrak{M}} \inf_{1\leq i\leq k} \left[\int_X \frac{Lu_i}{u_i}(x)\mu(dx) + \int_X V(x)\mu(dx)\right] \leq l + \frac{\epsilon}{4}. \quad [4]$$

From the definition of the infinitesimal generator, $\frac{T_h u_i - u_i}{h} \to Lu_i$ uniformly in x as $h \to 0$. Since u_i has a positive lower bound, and

$$\int_X \left(\frac{T_h u_i - u_i}{hu_i}\right)(x)\mu(dx) \to \int_X \left(\frac{Lu_i}{u_i}\right)(x)\mu(dx)$$

uniformly in μ, we conclude from [4] that there exists an h_0 such that if $h \leq h_0$

$$\sup_{\mu\in\mathfrak{M}} \inf_{1\leq i\leq k} \left[\int_X \left(\frac{T_h u_i - u_i}{hu_i}\right)(x)\mu(dx) + \int_X V(x)\mu(dx)\right] \leq l + \frac{\epsilon}{3}. \quad [5]$$

Since u_i is continuous and is bounded above and bounded below by a positive constant, we can write $u_i = e^{g_i}$ where $g_i \in C(X)$. Let G be the convex hull of $\{g_1,g_2,\ldots,g_k\}$. From [5] we obtain for $h \leq h_0$

$$\sup_{\mu\in\mathfrak{M}} \inf_{g\in G} \left[\int_X \left(\frac{T_h e^g - e^g}{he^g}\right)(x)\mu(dx) + \int_X V(x)\mu(dx)\right] \leq l + \frac{\epsilon}{3}. \quad [6]$$

Since T_h is bounded and nonnegative it is easy to see that $\frac{T_h e^g - e^g}{he^g}$ is a convex functional of g for each $x \in X$. Hence, we have from one of the mini-max theorems (see e.g., ref. 1, Thm. 3.4) that

$$\inf_{g\in G} \sup_{\mu\in\mathfrak{M}} \left[\int_X \left(\frac{T_h e^g - e^g}{he^g}\right)(x)\mu(dx) + \int_X V(x)\mu(dx)\right] \leq l + \frac{\epsilon}{3}, \quad [7]$$

which implies

$$\inf_{g\in G} \sup_{x\in X} \left[\left(\frac{T_h e^g - e^g}{he^g}\right)(x) + V(x)\right] \leq l + \frac{\epsilon}{3}.$$

Since G is compact the infimum is attained at some point of G, call it g_h, i.e., for all $x \in X$

$$\left[\left(\frac{T_h e^{g_h} - e^{g_h}}{he^{g_h}}\right)(x) + V(x)\right] \leq l + \frac{\epsilon}{3}.$$

Letting $u_h = e^{g_h}$ we have for all $x \in X$ and $h \leq h_0$

$$\left[\left(\frac{T_h u_h - u_h}{hu_h}\right)(x) + V(x)\right] \leq l + \frac{\epsilon}{3}. \quad [8]$$

Since the family $\{u_h\}$ for $h \leq h_0$ belongs to a compact set, we can assume without loss of generality that $u_h \to u_0$ as $h \to 0$ where $u_0 \in C^+(X)$ (continuous functions on X bounded below by a positive number) and $u_0 = e^g$ where $g \in G$. Without loss of generality we can assume $u_h \in \mathfrak{D}^+$ because $\mathfrak{D}^+$ is dense in $C^+(X)$. Hence, we get from [8]

$$\left[\left(\frac{T_h u_h - u_h}{hu_h}\right)(x) + V(x)\right] \leq l + \frac{\epsilon}{2}$$

for all $x \in X$, $h \leq h_0$, and where $u_h \to u_0$ as $h \to 0$ with $u_h \in \mathfrak{D}^+$. Since $u_h \in \mathfrak{D}^+$ we can rewrite this last as

$$\left[\frac{L\left(\frac{1}{h}\int_0^h T_s u_h ds\right)}{u_h}\right](x) + V(x) \leq l + \frac{\epsilon}{2} \quad [9]$$

for all $x \in X$ and $h \leq h_0$. Let

$$v_h = \frac{1}{h}\int_0^h T_s u_h ds \in \mathfrak{D}^+.$$

Since $u_h \to u_0$ and T_s is a strongly continuous semigroup, we have $v_h \to u_0$ as $h \to 0$. Now from [9] we get for all $x \in X$ and $h \leq h_0$

$$\left(\frac{Lv_h}{v_h u_h} \cdot v_h\right)(x) + V(x) \leq l + \frac{\epsilon}{2}$$

or

$$\left(\frac{Lv_h}{v_h}\right)(x) + V(x) \leq \left(l + \frac{\epsilon}{2}\right) + \left(\frac{u_h(x)}{v_h(x)} - 1\right) \times \left(l + \frac{\epsilon}{2}\right) + V(x)\left(1 - \frac{u_h(x)}{v_h(x)}\right)$$

$$\leq \left(l + \frac{\epsilon}{2}\right) + \left[\left(l + \frac{\epsilon}{2}\right) + \|V\|\right]\left(\left\|\frac{u_h}{v_h} - 1\right\|\right).$$

But $u_h \to u_0$ uniformly and, as noted earlier, $v_h \to u_0$ so that $\frac{u_h}{v_h} \to 1$ uniformly, i.e., for h sufficiently small

$$\left\|\frac{u_h}{v_h} - 1\right\|\left[\left(l + \frac{\epsilon}{2}\right) + \|V\|\right] \leq \frac{\epsilon}{2}.$$

Thus, there is an h_1 such that for all $x \in X$

$$\left(\frac{Lv_{h_1}}{v_{h_1}}\right)(x) + V(x) \leq l + \epsilon$$

which gives $\psi_2(V) \leq \psi_1(V)$. This completes the proof of *Lemma 2* and the theorem.

Now in ref. 2 the authors proved (Section 4 of ref. 2) that if L is self-adjoint with respect to a reference measure ν on X and if $I(\mu) < \infty$ for some $\mu \in \mathfrak{M}$, then under mild assumptions $\mu \ll \nu$ and letting $g = \frac{d\mu}{d\nu}$ and $f = g^{1/2}$ we have

$$I(\mu) = -\langle Lf,f\rangle.$$

This shows that in the case where L is self-adjoint the variational expression in the theorem reduces to the classical formula [1].

The relation between $\phi(V)$ and $I(\mu)$ is clearly that they are conjugate convex functionals. The relation $\phi(V) = \psi_2(V)$ has been noticed before. See, for instance, ref. 3.

The work reported in this paper was supported by the Office of Naval Research, Contract N00014-67-A-L467-0031 and the National Science Foundation, Grant NSF-GP-41985.

1. Sion, M. (1958) "On general minimax theorem," *Pac. J. Math.* **8,** 171–176.
2. Donsker, M. D. & Varadhan, S. R. S. (1975) "Asymptotic evaluation of certain Markov process expectations for large time—I," *Comm. Pure Appl. Math.*, in press.
3. Protter, M. H. & Weinberger, H. F. (1966) "On the spectrum of general second order operators," *Bull. Amer. Math. Soc.* **72,** 251–255.

COMMUNICATIONS ON PURE AND APPLIED MATHEMATICS, VOL. XXVIII, 1–47 (1975)

Asymptotic Evaluation of Certain Markov Process Expectations for Large Time, I*

M. D. DONSKER AND S. R. S. VARADHAN

1. Introduction

Let Ω_x be the space of real-valued continuous functions $z(s)$, $0 \leqq s < \infty$, with $z(0) = x$. Let P_x denote Brownian motion measure (Wiener measure) on Ω_x and $E_x\{\ \}$ integration with respect to P_x (Wiener integral). For $t > 0$ and $\omega \in \Omega_x$, let

$$L(t, \omega, y) = \frac{1}{t} \lambda\{s : 0 \leqq s \leqq t, z(s) \leqq y\},$$

where $\lambda\{\ \}$ denotes Lebesgue measure, i.e., $L(t, \omega, y)$ is the proportion of time on $[0, t]$ for which a particular path $\omega = z(\cdot)$ is less than or equal to y. For each $t > 0$ and each $\omega \in \Omega_x$, $L(t, \omega, y)$ is a probability distribution function on $-\infty < y < \infty$ possessing a density function which is often called the "local time". Let $\mathcal{F}$ denote the space of all probability distribution functions F on R_1. In [1] the authors proved that, for a wide class of functionals Φ defined on $\mathcal{F}$,

$$\lim_{t \to \infty} \frac{1}{t} \log E_x\{\exp\{-t\,\Phi(L(t, \omega, \cdot))\}\} = -\inf_{f \in \mathcal{F}_0} \left[\frac{1}{8} \int_{-\infty}^{\infty} \frac{[f'(y)]^2}{f(y)}\, dy + \Phi(f)\right], \tag{1.1}$$

where $\mathcal{F}_0$ denotes the class of probability density functions f which are continuously differentiable on $-\infty < y < \infty$ and either $f(y) > 0$ for all $-\infty < y < \infty$ or f has compact support and is strictly positive on the interior of that support. In (1.1), $\Phi(f)$ means $\Phi(F)$, where F is the distribution function corresponding to the density f.

If, for example, in (1.1) we let $\Phi(F) = \int_{-\infty}^{\infty} V(y)\, dF(y)$, where $V(y) \geqq 0$ is continuous and $V(y) \to +\infty$ as $y \to \pm\infty$, then Φ satisfies the hypotheses

* The research for this paper was performed at the Courant Institute. The first author was supported by the Office of Naval Research under Contract No. N00014-67A-0467-0031 and the second by the National Science Foundation under Grant No. NSF-GP-41985. Reproduction in whole or in part is permitted for any purpose of the United States Government.

imposed in [1] and (1.1) becomes

$$\lim_{t\to\infty}\frac{1}{t}\log E_x\left\{\exp\left\{-\int_0^t V(z(s))\,ds\right\}\right\} \tag{1.2}$$
$$= -\inf_{f\in\mathcal{F}_0}\left[\frac{1}{8}\int_{-\infty}^{\infty}\frac{[f'(y)]^2}{f(y)}\,dy+\int_{-\infty}^{\infty}V(y)f(y)\,dy\right].$$

Now, on the right of (1.2) let $\psi(y)=\sqrt{f(y)}$ so that $\psi(y)\in L_2(R_1)$ and, indeed, $\int_{-\infty}^{\infty}\psi^2(y)\,dy=1$. Moreover,

$$\int_{-\infty}^{\infty}V(y)f(y)\,dy=\int_{-\infty}^{\infty}V(y)\psi^2(y)\,dy\,,$$

and

$$\frac{1}{8}\int_{-\infty}^{\infty}\frac{[f'(y)]^2}{f(y)}\,dy=\frac{1}{2}\int_{-\infty}^{\infty}[\psi'(y)]^2\,dy\,.$$

Thus, the right side of (1.2) becomes

$$-\inf_{\substack{\psi\in L_2(R_1)\\ \int_{-\infty}^{\infty}\psi^2(y)dy=1}}\left[\frac{1}{2}\int_{-\infty}^{\infty}[\psi'(y)]^2\,dy+\int_{-\infty}^{\infty}V(y)\psi^2(y)\,dy\right]$$

which is $-\lambda_1$, λ_1 being the lowest eigenvalue of

$$\tfrac{1}{2}\psi''(y)-V(y)\psi(y)=-\lambda\psi(y)\,.$$

Hence the special case of (1.1) when $\Phi(F)=\int_{-\infty}^{\infty}V(y)\,dF(y)$ is the old result of Kac [2] that, for $V(y)$ as described above,

$$\lim_{t\to\infty}\frac{1}{t}\log E_x\left\{\exp\left\{-\int_0^t V(z(s))\,ds\right\}\right\}=-\lambda_1\,.$$

The point of (1.1) is that it holds for functionals Φ of much greater generality—in particular for functionals Φ where the corresponding expectations on the left of (1.1) are not solutions of partial differential equations as obtains in the special case (1.2).

In the present paper the authors consider the analogous asymptotic behavior for expectations on Markov processes—both discrete parameter and continuous parameter. In that generality we shall, however, confine ourselves in this paper to Markov processes having a state space which is a compact metric space (whereas in [1] the Brownian motion state space was all of R_1).

It is clear that in the Brownian motion case (1.1) a special role is played by the functional

$$\frac{1}{8}\int_{-\infty}^{\infty}\frac{[f'(y)]^2}{f(y)}\,dy\,.$$

We shall make additional comments about this functional now and more so later in this introduction. Let $\mathcal{U}$ be the space of functions $u(y)$, $-\infty<y<\infty$, having two bounded continuous derivatives, each function having two positive constants α and β such that, for all y, $0<\alpha\leqq u(y)\leqq\beta<\infty$. For $F\in\mathcal{F}$ (defined above), let

$$I(F)=-\inf_{u\in\mathcal{U}}\int_{-\infty}^{\infty}\frac{u''(y)}{2u(y)}\,dF(y)\,. \tag{1.3}$$

It is easy to see that, as a functional on $\mathcal{F}$, $I(F)$ is non-negative, lower semicontinuous in the weak topology on $\mathcal{F}$, and convex. In [1] (Lemma 2) we proved that if f is a density function which is positive and continuously differentiable on $-\infty<y<\infty$, then for its corresponding distribution function F

$$I(F)=\frac{1}{8}\int_{-\infty}^{\infty}\frac{[f'(y)]^2}{f(y)}\,dy\,, \tag{1.4}$$

in the sense that if one side is finite so is the other, and they are equal.

We shall now define the corresponding I-functions for discrete and continuous parameter Markov processes.

Let $\mathbf{X}$ be a compact metric space and let $\lambda(dx)$ be a probability measure on $\mathbf{X}$. Let $X_0, X_1, X_2, \cdots$ be a stationary Markov process whose state space is $\mathbf{X}$, with $X_0=x$, having transition probability function $\pi(x, dy)$ about which we assume:

1. $\pi(x, dy)=\pi(x, y)\lambda(dy)$,

2. there exist constants a and A such that $0<a\leqq\pi(x, y)\leqq A<\infty$ for all $x\in\mathbf{X}$ and almost all (λ-measure) $y\in\mathbf{X}$,

3. for any function $u(y)\in L_1(\lambda)$,

$$\int_{\mathbf{X}}\pi(x, y)u(y)\lambda(dy)$$

is a continuous function of x.

Let $\mathcal{U}_1$ be the set of functions $u \in C(X)$ such that $u > 0$ on X. With

$$\pi u(x) = \int_X u(y)\pi(x, dy) ,$$

we define, for any probability measure μ on X,

$$I(\mu) = -\inf_{u \in \mathcal{U}_1} \int_X \log\left(\frac{\pi u}{u}\right)(x)\mu(dx) . \tag{1.5}$$

Let $\mathcal{M}$ be the space of all probability measures on X, and impose the weak topology on $\mathcal{M}$. Denote by Ω_x the space of all sequences $X_0, X_1, X_2, \cdots$ with $X_0 = x$ and $X_j \in X$. We have a probability measure on Ω_x induced by $\pi(x, dy)$ which we shall call P_x. For each $\omega \in \Omega_x$, each positive integer n, and A a Borel set in X, let

$$L_n(\omega, A) = \frac{1}{n}\sum_{j=0}^{n-1} \chi_A(X_j) . \tag{1.6}$$

We note that L_n maps Ω_x into $\mathcal{M}$. Let $Q_{n,x}$ be the measure induced on $\mathcal{M}$ by L_n from P_x, i.e., $Q_{n,x} = P_x L_n^{-1}$. Thus, if B is a set of measures in $\mathcal{M}$, we define for each positive integer n and each $x \in X$

$$Q_{n,x}(B) = P_x\{\omega : L_n(\omega, \cdot) \in B\} .$$

In Section 2 of this paper, with the I-function defined in (1.5), we shall prove

THEOREM 1. *For any closed set $C \subset \mathcal{M}$,*

$$\overline{\lim_{n\to\infty}} \frac{1}{n} \log Q_{n,x}(C) \leqq -\inf_{\mu \in C} I(\mu) . \tag{1.7}$$

For any open set $G \subset \mathcal{M}$,

$$\underline{\lim_{n\to\infty}} \frac{1}{n} \log Q_{n,x}(G) \geqq -\inf_{\mu \in G} I(\mu) . \tag{1.8}$$

A corollary of this theorem, which follows from it by the arguments in Section 3 of [3], is that if $\Phi(\mu)$ is a weakly continuous functional on $\mathcal{M}$, then

$$\lim_{n\to\infty} \frac{1}{n} \log E^{Q_{n,x}}\{e^{-n\Phi(\mu)}\} = -\inf_{\mu \in \mathcal{M}} [\Phi(\mu) + I(\mu)] . \tag{1.9}$$

Or, going back to the P_x-measure on Ω_x, we have equivalently

THEOREM 2. *If* Φ *is a real-valued weakly continuous functional on* $\mathcal{M}$, *then*

$$\lim_{n\to\infty}\frac{1}{n}\log E^{P_x}\{\exp\{-n\Phi(L_n(\omega,\cdot))\}\}=-\inf_{\mu\in\mathcal{M}}[\Phi(\mu)+I(\mu)], \tag{1.10}$$

where $I(\mu)$ *is defined in* (1.5).

We see that Theorem 2 is the analogue of (1.1) for discrete parameter Markov processes. Let us now examine the situation for continuous parameter Markov processes.

Let $\mathfrak{X}$ be a compact metric space with a probability measure $\lambda(dx)$. Let $C(\mathfrak{X})$ be continuous functions on $\mathfrak{X}$ and T_t a strongly continuous Markovian semigroup such that T_t maps $C(\mathfrak{X})\to C(\mathfrak{X})$ (Feller semigroup). We assume the associated transition probabilities $p(t, x, dy)$ have densities relative to $\lambda(dx)$, i.e.,

$$p(t, x, dy)=p(t, x, y)\lambda(dy),$$

where for each $t>0$

$$0<a(t)\leqq p(t, x, y)\leqq A(t)<\infty.$$

Let L be the infinitesimal generator of the semigroup T_t having domain $\mathcal{D}\subset C(\mathfrak{X})$. Our assumptions above imply that $\mathcal{D}$ is dense in $C(\mathfrak{X})$.

Let Ω_x be the space of functions $x(t)$, $0\leqq t<\infty$, which have values in $\mathfrak{X}$, $x(0)=x$, are right continuous and have left limits at every point t. The process described above induces a measure P_x on Ω_x.

For each $t>0$, $\omega\in\Omega_x$, and Borel set $A\subset\mathfrak{X}$, let

$$L_t(\omega, A)=\frac{1}{t}\int_0^t \chi_A(x(s))\,ds. \tag{1.11}$$

In other words, $L_t(\omega, A)$ is the proportion of time up to t that a particular sample path $\omega=x(\cdot)$ spends in the set A. We note that, for each $t>0$ and each ω, $L_t(\omega,\cdot)$ is a probability measure on $\mathfrak{X}$.

Let $\mathcal{M}$ be the space of all probability measures on $\mathfrak{X}$. For each $x\in\mathfrak{X}$ and each $t>0$, let $Q_{x,t}$ be the measure induced on $\mathcal{M}$ by L_t from P_x, i.e., $Q_{x,t}=P_xL_t^{-1}$ or, in other words, if $B\subset\mathcal{M}$,

$$Q_{x,t}(B)=P_x\{\omega\in\Omega_x : L_t(\omega,\cdot)\in B\}.$$

We endow $\mathcal{M}$ with the weak topology.

Now, for any $\mu \in \mathcal{M}$, we define

$$I(\mu) = -\inf_{\substack{u>0 \\ u\in\mathcal{D}}} \int_{\mathcal{X}} \left(\frac{Lu}{u}\right)(x)\mu(dx) . \tag{1.12}$$

In Section 3 of this paper we shall prove

THEOREM 3. *For any closed set* $C \subset \mathcal{M}$,

$$\overline{\lim_{t\to\infty}} \frac{1}{t} \log Q_{x,t}(C) \leqq -\inf_{\mu\in C} I(\mu) . \tag{1.13}$$

For any open set $G \subset \mathcal{M}$,

$$\underline{\lim_{t\to\infty}} \frac{1}{t} \log Q_{x,t}(G) \geqq -\inf_{\mu\in G} I(\mu) . \tag{1.14}$$

As before, a corollary of Theorem 3 is that if Φ is a real-valued weakly continuous functional on $\mathcal{M}$, then

$$\lim_{t\to\infty} \frac{1}{t} \log E^{Q_{x,t}}\{e^{-t\Phi(\mu)}\} = -\inf_{\mu\in\mathcal{M}} [\Phi(\mu) + I(\mu)] . \tag{1.15}$$

Going back to the P_x-measure on Ω_x we have equivalently

THEOREM 4. *If Φ is a real-valued weakly continuous functional on $\mathcal{M}$, then*

$$\lim_{t\to\infty} \frac{1}{t} \log E^{P_x}\{\exp\{-t\Phi(L_t(\omega, \cdot))\}\} = -\inf_{\mu\in\mathcal{M}} [\Phi(\mu) + I(\mu)] , \tag{1.16}$$

where $I(\mu)$ is defined by (1.12) *and $L_t(\omega, A)$ is defined by* (1.11).

Since, for Brownian motion in R_1, the infinitesimal generator L is $\frac{1}{2}\partial^2/\partial x^2$, we see the correspondence between (1.3) and (1.12). Now, we want to take a closer look at the I-function and in particular determine conditions for a general Markov process under which one has a simple expression for the I-function as happens for Brownian motion in (1.4). Section 4 of this paper will be devoted to this topic, and we now describe the assumptions and results.

Let X be an arbitrary space and $\mathcal{B}$ a σ-field of subsets of X. Let $p(t, x, dy)$ be the transition probabilities of a Markov process with $(X, \mathcal{B})$ as state space. Let T_t be the corresponding semigroup on $B(X)$, the space of bounded measurable functions on X. Let B_0 be the strongly continuous center of T_t. We assume the following:

1. There exists a σ-finite measure $\lambda(dx)$ such that $p(t, x, dy)$ has a density $p(t, x, y)$ with respect to λ for each $t>0$; the density $p(t, x, y)$ is jointly measurable in x and y and satisfies

$$p(t, x, y) = p(t, y, x) \quad \text{almost everywhere} \quad (\lambda \times \lambda).$$

2. The center B_0 is large enough in the sense that $B_0 \cap L_2(\lambda)$ is dense in $L_2(\lambda)$.

3. For any probability measure μ on X and any $f \in B(X)$ there is a sequence $\{f_n\} \in B_0$ such that $\|f_n\| \leqq \|f\|$ and $f_n \to f$ almost everywhere with respect to μ.

We note that if X is a locally compact separable metric space, if λ is a Borel measure which is finite on compact sets, and if B_0 contains continuous functions which have compact support, then assumptions 2 and 3 are automatically satisfied.

Under the assumptions above, T_t generates a strongly continuous contraction semigroup in $L_2(\lambda)$ which in addition is *selfadjoint* for each t. Thus we have the infinitesimal generators L on a domain $\mathfrak{D}$ dense in $B_0 \subset B(X)$ and $\tilde{L}$ on a domain $\tilde{\mathfrak{D}}$ dense in $L_2(\lambda)$. Since the T_t are contractions, $\tilde{L}$ is selfadjoint, negative semidefinite, and we denote by $\sqrt{-\tilde{L}}$ the canonical positive semidefinite square root of the selfadjoint positive semidefinite operator $-\tilde{L}$. Let $\tilde{\mathfrak{D}}_{1/2}$ be the natural domain of $\sqrt{-\tilde{L}}$.

Denote by $\mathfrak{D}^+$ those functions (bounded) in $\mathfrak{D} \subset B(X)$ which have a positive lower bound. Since constants belong to $\mathfrak{D}$, $\mathfrak{D}^+$ is quite rich. We define, as before, for any probability measure μ on X,

$$I(\mu) = -\inf_{u \in \mathfrak{D}^+} \int_X \left(\frac{Lu}{u}\right)(x)\mu(dx).$$

In Section 4 of this paper we shall prove

THEOREM 5. *$I(\mu) < \infty$ if and only if $\mu \ll \lambda$ and $(d\mu/d\lambda)^{1/2} \in \tilde{\mathfrak{D}}_{1/2}$. Moreover, letting $f = d\mu/d\lambda$ and $g = f^{1/2}$, we have*

$$I(\mu) = \|\sqrt{-\tilde{L}}\, g\|^2. \tag{1.17}$$

2. Discrete Parameter Markov Process

In this section we shall prove Theorem 1 of the introduction. We first prove (1.7).

Let $u \in \mathcal{U}_1$, $V = \pi u$, and $e^{-W} = u/V$. From our hypotheses on $\pi(x, dy)$ we see that W is a bounded continuous function on $\mathfrak{X}$. Denoting integration on Ω_x with respect to P_x measure by $E^{P_x}\{\ \}$, we show first that

$$(2.1) \qquad E^{P_x}\{\exp\{-[W(X_0) + W(X_1) + \cdots + W(X_{n-1})]\} V(X_{n-1})\} = u(x)\,,$$

for all $x \in \mathfrak{X}$ and any positive integer n. Now,

$$E^{P_x}\{\exp\{-[W(X_0) + \cdots + W(X_{n-1})]\} V(X_{n-1})\} = E^{P_x}\left\{\frac{u(X_0) \cdots u(X_{n-2})}{V(X_0) \cdots V(X_{n-2})} u(X_{n-1})\right\}.$$

Since the process is Markovian,

$$\begin{aligned} E^{P_x}\{u(X_{n-1}) \mid X_0, X_1, \cdots, X_{n-2}\} &= E^{P_x}\{u(X_{n-1}) \mid X_{n-2}\} \\ &= \int_{\mathfrak{X}} u(y)\pi(X_{n-2}, dy) = V(X_{n-2})\,, \end{aligned}$$

and hence

$$E^{P_x}\left\{\frac{u(X_0) \cdots u(X_{n-2})}{V(X_0) \cdots V(X_{n-2})} u(X_{n-1})\right\} = E^{P_x}\left\{\frac{u(X_0) \cdots u(X_{n-3})}{V(X_0) \cdots V(X_{n-3})} u(X_{n-2})\right\}.$$

Thus,

$$E^{P_x}\{\exp\{-[W(X_0) + \cdots + W(X_{n-1})]\} V(X_{n-1})\} = E^{P_x}\{u(X_0)\} = u(x)$$

which is (2.1).

Since u is bounded above and V is bounded below, it follows from (2.1) that there exists a constant $M > 0$ such that

$$E^{P_x}\{\exp\{-[W(X_0) + W(X_1) + \cdots + W(X_{n-1})]\}\} \leqq \frac{u(x)}{\inf_{x \in \mathfrak{X}} V(x)} \leqq M\,.$$

But,

$$\begin{aligned} \exp\{-[W(X_0) + W(X_1) + \cdots + W(X_{n-1})]\} &= \exp\{-n[[W(X_0) + \cdots + W(X_{n-1})]/n]\} \\ &= \exp\left\{-n\int_{\mathfrak{X}} W(y) L_n(\omega, dy)\right\}, \end{aligned}$$

and therefore, from the definition of $Q_{n,x}$ measure on $\mathcal{M}$,

$$(2.2) \quad E^{Q_{n,x}}\left\{\exp\left\{-n\int_{\mathfrak{X}} W(y) l(dy)\right\}\right\} = E^{P_x}\{\exp\{-[W(X_0) + \cdots + W(X_{n-1})]\}\} \leqq M\,.$$

On the left of (2.2), $l(dy)$ is a measure in $\mathcal{M}$ and is, of course, the variable of integration.

For the moment, let C be any measurable set in $\mathcal{M}$. Then

$$E^{Q_{n,x}}\left\{\exp\left\{-n\int_{\mathcal{X}} W(y)l(dy)\right\}\right\} \geqq Q_{n,x}(C)\exp\left\{-n\sup_{l\in C}\int_{\mathcal{X}} W(y)l(dy)\right\},$$

and from (2.2) we have

$$Q_{n,x}(C) \leqq M\exp\left\{n\sup_{l\in C}\int_{\mathcal{X}} W(y)l(dy)\right\}.$$

Thus,

$$\begin{aligned} \overline{\lim_{n\to\infty}}\frac{1}{n}\log Q_{n,x}(C) &\leqq \sup_{l\in C}\int_{\mathcal{X}} W(y)l(dy) \\ &= \sup_{l\in C}\int_{\mathcal{X}} \log\left(\frac{\pi u}{u}\right)(y)l(dy). \end{aligned} \tag{2.3}$$

Since, in (2.3), u was an arbitrary element of $\mathcal{U}_1$, we have

$$\overline{\lim_{n\to\infty}}\frac{1}{n}\log Q_{n,x}(C) \leqq \inf_{u\in\mathcal{U}_1}\sup_{l\in C}\int_{\mathcal{X}} \log\left(\frac{\pi u}{u}\right)(y)l(dy). \tag{2.4}$$

Now, if $C\subset\bigcup_{j=1}^{k} C_j$, then

$$\overline{\lim_{n\to\infty}}\frac{1}{n}\log Q_{n,x}(C) \leqq \sup_{1\leqq j\leqq k}\inf_{u\in\mathcal{U}_1}\sup_{l\in C_j}\int_{\mathcal{X}} \log\left(\frac{\pi u}{u}\right)(y)l(dy),$$

and hence, considering all possible finite unions covering C,

$$\overline{\lim_{n\to\infty}}\frac{1}{n}\log Q_{n,x}(C) \leqq \inf_{\substack{C_1,C_2,\cdots,C_k\\ C\subset\bigcup_{j=1}^{k}C_j}}\sup_{1\leqq j\leqq k}\inf\sup_{l\in C_j}\int_{\mathcal{X}} \log\left(\frac{\pi u}{u}\right)(y)l(dy). \tag{2.5}$$

In (2.5), C is an arbitrary measurable set in $\mathcal{M}$. If C is compact, then it is easy to show[1] that the right side of inequality (2.5) is, in fact, just

$$\sup_{l\in C}\inf_{u\in\mathcal{U}_1}\int_{\mathcal{X}} \log\left(\frac{\pi u}{u}\right)(y)l(dy) = -\inf_{l\in C}\left[-\inf_{u\in\mathcal{U}_1}\int_{\mathcal{X}} \log\left(\frac{\pi u}{u}\right)(y)l(dy)\right] = -\inf_{l\in C} I(l).$$

[1] The argument can be found in [1] (just after (2.6) of that paper) where a closely related situation was encountered.

Since X is compact, $\mathcal{M}$ is compact in the weak topology, and hence the hypothesis for (1.7) that C be closed implies it is compact. From (2.5) and the last remark we have (1.7).

The lower estimate (1.8) is much more delicate and we prove first a succession of nine lemmas.

LEMMA 2.1. *Let $(X, \mathcal{S})$ be a metric space and $\mathcal{S}$ its Borel sets. Let α and β be probability measures on X, and denote by $\mathcal{U}^*$ the set of functions $u(x)$ which are continuous on X and for each of which there exist constants c_1 and c_2 such that $0 < c_1 \leqq u(x) \leqq c_2 < \infty$ for all $x \in X$. Define*

$$I_\beta(\alpha) = -\inf_{u \in \mathcal{U}^*} \left[\log \int_X u(x)\beta(dx) - \int_X \log u(x)\alpha(dx) \right]. \tag{2.6}$$

Then, $I_\beta(\alpha) < \infty$ if and only if $\alpha \ll \beta$ with a density $a(x) = d\alpha/d\beta$ such that

$$\int_X a(x) \,|\log a(x)|\, \beta(dx) < \infty .$$

Moreover, in this case,

$$I_\beta(\alpha) = \int_X a(x) \log a(x)\beta(dx) = \int_X \log a(x)\alpha(dx) .$$

Proof: Assume first that $\alpha \ll \beta$ and the Radon-Nikodym derivative $a(x) = d\alpha/d\beta$ is such that

$$\int_X a(x) \,|\log a(x)|\, \beta(dx) < \infty .$$

Let $\int_X a(x) \log a(x)\beta(dx) = l$. If we show that, for all $u \in \mathcal{U}^*$,

$$\log \int_X u(x)\beta(dx) - \int_X \log u(x)\alpha(dx) \geqq -l, \tag{2.7}$$

then we can conclude that

$$I_\beta(\alpha) \leqq l = \int_X a(x) \log a(x)\beta(dx) . \tag{2.8}$$

So, let $u \in \mathcal{U}^*$, and we have

$$\int_X \log u(x)\alpha(dx) = \int_X (\log u(x))a(x)\beta(dx)$$
$$= \int_X \left(\log\left[\frac{u(x)}{a(x)}\, a(x)\right]\right)a(x)\beta(dx)$$
$$= \int_{a(x)>0} \left(\log \frac{u(x)}{a(x)}\right)a(x)\beta(dx) + l,$$

because $a(x)>0$ almost everywhere (α-measure).

From Jensen's inequality we then have

$$\int_X \log u(x)\alpha(dx) \leqq \log \int_{a(x)>0} \frac{u(x)}{a(x)} \cdot a(x)\beta(dx) + l$$
$$\leqq \log \int_X u(x)\beta(dx) + l,$$

and therefore (2.7), and hence (2.8), holds for any $u \in \mathcal{U}^*$.

Now let us assume $I_\beta(\alpha)<\infty$. We want to show that $d\alpha/d\beta = a(x)$ exists and satisfies

$$\int_X a(x)\,|\log a(x)|\,\beta(dx) < \infty\,;$$

moreover,

$$\int_X a(x) \log a(x)\beta(dx) \leqq I_\beta(\alpha)\,.$$

Let $I_\beta(\alpha)=l<\infty$ so that

(2.9) $$\log \int_X u(x)\beta(dx) - \int_X \log u(x)\alpha(dx) \geqq -l,$$

for all $u \in \mathcal{U}^*$. Let $B^+(X)$ be the space of all bounded measurable functions u on X which are bounded away from zero and infinity. From Lusin's theorem we know that (2.9) holds not just for all $u \in \mathcal{U}^*$ but for all $u \in B^+(X)$ as well.

Now assume $A \in \mathcal{S}$, and let, for any $k>0$, $u(x)=k$ on A and $u(x)=1$ on $\tilde{A}$ so that $u \in B^+(X)$ and hence, from (2.9),

(2.10) $$(\log k)\alpha(A) \leqq \log\,[k\beta(A)+\beta(\tilde{A})]+l\,.$$

Consequently, if $\beta(A)=0$, (2.10) implies that

$$\alpha(A) \leqq \frac{l}{\log k},$$

and hence, letting $k \to \infty$, we obtain $\alpha(A)=0$, i.e., $\alpha \ll \beta$. Let $a(x)=d\alpha/d\beta$ and, for any positive integer n, define

$$u_n(x)=\left(a(x)\vee\frac{1}{n}\right)\wedge n .$$

We see that $u_n \in B^+(X)$ and, moreover,

$$\lim_{n\to\infty}\int_X |u_n(x)-a(x)|\,\beta(dx)=0 . \tag{2.11}$$

From (2.9)

$$\int_X \log u_n(x)\alpha(dx) \leqq \log \int_X u_n(x)\beta(dx)+l ,$$

and (2.11) implies that the right side of this last inequality approaches l as $n\to\infty$. Hence,

$$\overline{\lim_{n\to\infty}}\int_X \log u_n(x)\alpha(dx) \leqq l . \tag{2.12}$$

Writing $\log u_n=(\log u_n)^+-(\log u_n)^-$, we note that $(\log u_n)^+ \uparrow (\log a)^+$, $(\log u_n)^- \uparrow (\log a)^-$, and (2.12) becomes

$$\overline{\lim_{n\to\infty}}\left[\int_X (\log u_n)^+\alpha(dx)-\int_X (\log u_n)^-\alpha(dx)\right] \leqq l . \tag{2.13}$$

Since $\int_X a(x)(\log a(x))^-\beta(dx)$ is clearly finite, we have by monotone convergence

$$\int_X (\log u_n(x))^-\alpha(dx)=\int_X a(x)(\log u_n(x))^-\beta(dx)\uparrow\int_X a(x)(\log a(x))^-\beta(dx) .$$

Thus, from (2.13)

$$\overline{\lim_{n\to\infty}}\int_X (\log u_n(x))^+\alpha(dx) \leqq l+\int_X a(x)(\log a(x))^-\beta(dx) .$$

Again, using the monotone convergence theorem, we have

$$\int_X (\log a(x))^+ \alpha(dx) \leqq l + \int_X a(x)(\log a(x))^- \beta(dx) .$$

We therefore conclude that

$$\int_X a(x) \, |\log a(x)| \, \beta(dx) < \infty ,$$

and moreover,

$$\int_X a(x) \log a(x) \beta(dx) \leqq l ,$$

which completes Lemma 2.1.

COROLLARY. *The function*

$$I_\beta(\alpha) = \int_X a(x) \log a(x) \beta(dx) = \int_X \log a(x) \alpha(dx)$$

is a lower semicontinuous function of α *for fixed* β *in the weak topology of measures on* X.

Proof: For fixed $u \in \mathcal{U}^*$,

$$\log \int_X u(x) \beta(dx) - \int_X \log u(x) \alpha(dx)$$

is a continuous functional of α for fixed β in the weak topology on X. Therefore,

$$\inf_{u \in \mathcal{U}^*} \left[\log \int_X u(x) \beta(dx) - \int_X \log u(x) \alpha(dx) \right]$$

is upper semicontinuous and hence $I_\beta(\alpha)$, from its definition (2.6), is lower semicontinuous.

Using the definitions and notation of the introduction just before the statement of Theorem 1, we now state

LEMMA 2.2. *Assume* $\mu \in \mathcal{M}$ *is such that* $d\mu/d\lambda = m(x)$ *exists and satisfies*

$$0 < \eta \leqq m(x) \leqq \xi < \infty .$$

Let $\mathcal{M}_\mu$ be the set of all probability measures $\alpha(dx_1, dx_2)$ on $\mathbf{X}\times\mathbf{X}$ such that both marginal measures are μ. Let $\beta(dx_1, dx_2)$ be the probability measure on $\mathbf{X}\times\mathbf{X}$ defined by

$$\begin{aligned}\beta(dx_1, dx_2) &= \pi(x_1, x_2)\mu(dx_1)\lambda(dx_2)\\ &= \pi(x_1, x_2)m(x_1)\lambda(dx_1)\lambda(dx_2)\,.\end{aligned}$$

Then, $\inf_{\alpha\in\mathcal{M}_\mu} I_\beta(\alpha)$ is attained at some point of $\mathcal{M}_\mu$ — call it α_0. Moreover, $\alpha_0 \ll \beta$ and $d\alpha_0/d\beta = a_0(x_1, x_2)$ is strictly positive almost everywhere in $\mathbf{X}\times\mathbf{X}$ ($\lambda\times\lambda$-measure).

Proof: In the corollary to Lemma 2.1 we showed that $I_\beta(\alpha)$ is lower semicontinuous as a function of α for fixed β. For $\beta(dx_1, dx_2)$ as defined above and $\alpha\in\mathcal{M}_\mu$ it then follows, from the fact that $\mathcal{M}_\mu$ is compact in the weak topology of measures on $\mathbf{X}\times\mathbf{X}$, that $\inf_{\alpha\in\mathcal{M}_\mu} I_\beta(\alpha)$ is attained at some point of $\mathcal{M}_\mu$ — call it α_0.

Now the product measure $\mu\times\mu$ is, of course, in $\mathcal{M}_\mu$ and

$$\begin{aligned}a(x_1, x_2) = a(x) = \frac{d(\mu\times\mu)}{d\beta} &= \frac{m(x_1)m(x_2)\lambda(dx_1)\lambda(dx_2)}{\pi(x_1, x_2)m(x_1)\lambda(dx_1)\lambda(dx_2)}\\ &= \frac{m(x_2)}{\pi(x_1, x_2)}\end{aligned}$$

is bounded above and below due to our assumptions on $m(x)$ and $\pi(x, y)$. Therefore,

$$\int_{\mathbf{X}\times\mathbf{X}} a(x)\,|\log a(x)|\,\beta(dx_1, dx_2) < \infty\,,$$

and from Lemma 2.1 this implies $I_\beta(\mu\times\mu)<\infty$. Thus, $I_\beta(\alpha_0)<\infty$. Again from Lemma 2.1 we now know that $\alpha_0 \ll \beta$. All that remains to prove is that $a_0(x_1, x_2) = d\alpha_0/d\beta$ is positive almost everywhere ($\lambda\times\lambda$-measure).

Let $a(x_1, x_2)$ be any density such that

$$a(x_1, x_2)\lambda(dx_1)\lambda(dx_2) = \alpha(dx_1, dx_2)\in\mathcal{M}_\mu\,.$$

The marginal measures of any convex linear combination of probability measures in $\mathcal{M}_\mu$ are still μ. Thus, for any $\varepsilon>0$, $(1-\varepsilon)\alpha_0+\varepsilon\alpha$ is a measure in $\mathcal{M}_\mu$ and by the definition of α_0 we have

$$I_\beta(\alpha_0) \leqq I_\beta((1-\varepsilon)\alpha_0+\varepsilon\alpha)\,.$$

Since $I_\beta(\alpha)$ is a convex functional of α, this implies that $I_\beta((1-\varepsilon)\alpha_0+\varepsilon\alpha)$ is a nondecreasing function of ε in $0 \leqq \varepsilon \leqq 1$ so that

$$\frac{d}{d\varepsilon}[I_\beta((1-\varepsilon)\alpha_0+\varepsilon\alpha)] \geqq 0\,, \qquad 0 \leqq \varepsilon \leqq 1\,, \tag{2.14}$$

provided it exists. We have

$$\frac{d[(1-\varepsilon)\alpha_0+\varepsilon\alpha]}{d\beta}=(1-\varepsilon)a_0+\frac{\varepsilon a(x_1, x_2)}{m(x_1)\pi(x_1, x_2)}\,;$$

we shall call it $a_\varepsilon(x_1, x_2)$. Therefore, from Lemma 2.1,

$$I_\beta[(1-\varepsilon)\alpha_0+\varepsilon\alpha]=\int_{\mathcal{X}\times\mathcal{X}} a_\varepsilon(x_1, x_2)\log a_\varepsilon(x_1, x_2)\beta(dx_1, dx_2)\,.$$

Hence, for $0<\varepsilon<1$, $(d/d\varepsilon)I_\beta[(1-\varepsilon)\alpha_0+\varepsilon\alpha]$ exists and is seen to equal

$$\begin{aligned}&\int_{\mathcal{X}\times\mathcal{X}}\left[\frac{a(x_1, x_2)}{m(x_1)\pi(x_1, x_2)}-a_0(x_1, x_2)\right][\log a_\varepsilon(x_1, x_2)]\\ &\qquad\times m(x_1)\pi(x_1, x_2)\lambda(dx_1)\lambda(dx_2)\\ &\qquad+\int_{\mathcal{X}\times\mathcal{X}} a_\varepsilon(x_1, x_2)\frac{[a(x_1, x_2)/m(x_1)\pi(x_1, x_2)-a_0(x_1, x_2)]}{a(x_1, x_2)}\\ &\qquad\times m(x_1)\pi(x_1, x_2)\lambda(dx_1)\lambda(dx_2)\,.\end{aligned} \tag{2.15}$$

The second integral on the right of (2.15) is just

$$\int_{\mathcal{X}\times\mathcal{X}} a(x_1, x_2)\lambda(dx_1)\lambda(dx_2)-\int_{\mathcal{X}\times\mathcal{X}} a_0(x_1, x_2)\beta(dx_1, dx_2)=1-1=0\,.$$

Thus from (2.14) and (2.15) we have, for $1>\varepsilon>0$,

$$\int_{\mathcal{X}\times\mathcal{X}} a(x_1, x_2)\log a_\varepsilon(x_1, x_2)\lambda(dx_1)\lambda(dx_2) \geqq \int_{\mathcal{X}\times\mathcal{X}} a_0(x_1, x_2)\log a_\varepsilon(x_1, x_2)\beta(dx_1, dx_2)\,.$$

As $\varepsilon\to 0$, $a_\varepsilon(x_1, x_2)\to a_0(x_1, x_2)$ and, by the dominated convergence theorem,

$$\begin{aligned}&\lim_{\varepsilon\to 0}\int_{\mathcal{X}\times\mathcal{X}} a_0(x_1, x_2)\log a_\varepsilon(x_1, x_2)\beta(dx_1, dx_2)\\ &\qquad=\int_{\mathcal{X}\times\mathcal{X}} a_0(x_1, x_2)\log a_0(x_1, x_2)\beta(dx_1, dx_2)=I_\beta(\alpha_0)\geqq 0\,.\end{aligned}$$

On the other hand, if $I_\beta(\alpha)<\infty$, then $I_\beta[\frac{1}{2}\alpha_0+\frac{1}{2}\alpha]<\infty$ by convexity and this implies that $|\log a_{1/2}(x_1, x_2)|$ is integrable with respect to $\frac{1}{2}\alpha_0+\frac{1}{2}\alpha$, and hence with respect to both α and α_0. In particular,

$$\log^+ a(x_1, x_2)=\log^+\left[(1-\varepsilon)a_0(x_1, x_2)+\varepsilon\frac{a(x_1, x_2)}{m(x_1)\pi(x_1, x_2)}\right]$$

$$\leqq\log^+\left[a_0(x_1, x_2)+\frac{a(x_1, x_2)}{m(x_1)\pi(x_1, x_2)}\right]$$

$$\leqq\log 2+\log^+ a_{1/2}(x_1, x_2)$$

is uniformly integrable with respect to α. It now follows from Fatou's lemma that

$$\int_{\mathcal{X}\times\mathcal{X}} a(x_1, x_2)\log^- a_0(x_1, x_2)\lambda(dx_1)\lambda(dx_2)>-\infty.$$

Therefore, $a_0(x_1, x_2)>0$ whenever $a(x_1, x_2)>0$. If we pick $\alpha(dx_1, dx_2)=\mu\times\mu$ so that $a(x_1, x_2)=m(x_1)m(x_2)$, then $a(x_1, x_2)$ never vanishes and moreover $I_\beta(\alpha)<\infty$. Hence $a_0(x_1, x_2)$ cannot vanish on a set of positive $\lambda\times\lambda$-measure which completes Lemma 2.2.

The following lemma is needed in the proof of Lemma 2.4.

LEMMA 2.3. *Let $(X, \mathcal{B}, \lambda)$ be a probability measure space. Assume there exist two sequences of functions $\{f_n(x_1)\}$, $\{g_n(x_2)\}$ such that, in the product space $X\times X$,*

$$\phi(x_1, x_2)=\lim_{n\to\infty}\text{ in meas. }[f_n(x_1)+g_n(x_2)],$$

where the measure indicated on the right is the product measure $\lambda\times\lambda$. Then, there exists a subsequence n_j and a sequence of constants $\{a_{n_j}\}$ such that

$$\lim_{j\to\infty}[f_{n_j}(x_1)-a_{n_j}]=f(x_1)$$

exists for almost all x_1,

$$\lim_{j\to\infty}[g_{n_j}(x_2)+a_{n_j}]=g(x_2)$$

exists for almost all x_2, and

$$\phi(x_1, x_2)=f(x_1)+g(x_2)$$

for almost all (x_1, x_2) $(\lambda\times\lambda$-measure).

Proof: Let $\phi_n(x_1, x_2)=f_n(x_1)+g_n(x_2)$ so that our assumption is that $\phi(x_1, x_2)=\lim \text{in meas}_{n\to\infty} \phi_n(x_1, x_2)$. There exists a subsequence, which we again denote by $\{\phi_n(x_1, x_2)\}$, which converges to $\phi(x_1, x_2)$ almost everywhere ($\lambda\times\lambda$-measure). Let

$$E=\{(x_1, x_2) : f_n(x_1)+g_n(x_2)\to \phi(x_1, x_2)\}$$

so that $(\lambda\times\lambda)(E)=1$. Let

$$E_{x_1}=\{x_2 : f_n(x_1)+g_n(x_2)\to \phi(x_1, x_2)\} .$$

By the Fubini theorem, $\lambda(E_{x_1})=1$ for almost all x_1 (λ-measure). Let $E^*=\{x_1 : \lambda(E_{x_1})=1$ so that $\lambda(E^*)=1$ and hence $E^*\neq\varnothing$. Choose $\bar{x}_1\in E^*$; then $f_n(\bar{x}_1)+g_n(x_2)\to\phi(\bar{x}_1, x_2)$ for all $x_2\in E_{\bar{x}_1}$. Thus, if we let $a_n=f_n(\bar{x}_1)$, we see that

$$\lim_{n\to\infty}[g_n(x_2)+a_n]$$

exists for all $x_2\in E_{\bar{x}_1}$ and $\lambda(E_{\bar{x}_1})=1$. But a_n is independent of x_2 and therefore there exists a function $g(x_2)$ such that

$$\lim_{n\to\infty}[g_n(x_2)+a_n]=g(x_2)$$

for almost all x_2 (λ-measure). Now

$$f_n(x_1)+g_n(x_2)=f_n(x_1)-a_n+g_n(x_2)+a_n$$

and so it follows from the fact that $g_n(x_2)+a_n\to g(x_2)$ for almost all (x_1, x_2) ($\lambda\times\lambda$-measure) that $f_n(x_1)-a_n$ converges for almost all (x_1, x_2) ($\lambda\times\lambda$-measure). But $f_n(x_1)-a_n$ is independent of x_2 and therefore $f_n(x_1)-a_n$ converges for almost all x_1 to a function of x_1—call it $f(x_1)$. Thus $f_n(x_1)+g_n(x_2)\to f(x_1)+g(x_2)$ for almost all (x_1, x_2) ($\lambda\times\lambda$-measure) which implies that $\phi(x_1, x_2)=f(x_1)+g(x_2)$ almost everywhere in $X\times X$ ($\lambda\times\lambda$-measure).

LEMMA 2.4. *Assume $\mu\in\mathcal{M}$ is such that $d\mu/d\lambda=m(x)$ exists and satisfies*

$$0<\eta\leqq m(x)\leqq\xi<\infty .$$

Then there exists a function $V(y)$ on $\mathfrak{X}$ which is non-negative, not identically zero, an element of $L_1(\lambda)$, such that the discrete parameter Markov process with transition function

$$\pi'(x, dy)=\frac{\pi(x, y)V(y)\lambda(dy)}{\int_{\mathfrak{X}}\pi(x, y)V(y)\lambda(dy)}$$

has μ *for its invariant probability measure, i.e., for every Borel set* $A \subset \mathcal{X}$

$$\int_{\mathcal{X}} \pi'(x, A)\mu(dx) = \mu(A) .$$

Proof: Let $\mathcal{M}_\mu$ and $\beta(dx_1, dx_2)$ be defined as in Lemma 2.2. In that lemma we proved that $\inf_{\alpha \in \mathcal{M}_\mu} I_\beta(\alpha)$ was attained in $\mathcal{M}_\mu$ at some point α_0 and that $\alpha_0 \ll \beta$. Moreover, we established in Lemma 2.2 that $d\alpha_0/d\beta = a_0(x_1, x_2)$ was strictly positive almost everywhere on $\mathcal{X} \times \mathcal{X}$ ($\lambda \times \lambda$-measure). We now show that $a_0(x_1, x_2)$ factors, i.e., we find functions $U(x_1)$ and $V(x_2)$ such that $a_0(x_1, x_2) = U(x_1)V(x_2)$ for almost all (x_1, x_2) ($\lambda \times \lambda$-measure). We shall then prove that this function V fulfills the requirements of the present lemma. To show first of all that $a_0(x_1, x_2)$ factors in this way we make use of the Euler equation characterizing the fact that α_0 minimizes $I_\beta(\alpha)$ for α varying in $\mathcal{M}_\mu$.

Let $k(x_1, x_2)$ be a bounded function on $\mathcal{X} \times \mathcal{X}$ having the property

$$\int_{\mathcal{X} \times \mathcal{X}} k(x_1, x_2) a_0(x_1, x_2)[f(x_1) + g(x_2)]\beta(dx_1, dx_2) = 0 \tag{2.16}$$

for all continuous (hence bounded) functions f and g. Choose ε such that $|\varepsilon| \leqq 1/C$, where $|k| \leqq C$. Consider the density $a_\varepsilon(x_1, x_2) = a_0(x_1, x_2)(1 + \varepsilon k(x_1, x_2))$ which is non-negative almost everywhere on $\mathcal{X} \times \mathcal{X}$ ($\lambda \times \lambda$-measure). We define a measure on $\mathcal{X} \times \mathcal{X}$ by $\alpha_\varepsilon(dx_1, dx_2) = a_\varepsilon(x_1, x_2)\beta(dx_1, dx_2)$. Now $\alpha_\varepsilon \in \mathcal{M}_\mu$ because $\alpha_0 \in \mathcal{M}_\mu$ and the defining property of $k(x_1, x_2)$ implies that for all continuous f and g

$$\int_{\mathcal{X} \times \mathcal{X}} a_\varepsilon(x_1, x_2) f(x_1)\beta(dx_1, dx_2) = \int_{\mathcal{X} \times \mathcal{X}} a_0(x_1, x_2) f(x_1)\beta(dx_1, dx_2) ,$$

$$\int_{\mathcal{X} \times \mathcal{X}} a_\varepsilon(x_1, x_2) g(x_2)\beta(dx_1, dx_2) = \int_{\mathcal{X} \times \mathcal{X}} a_0(x_1, x_2) g(x_2)\beta(dx_1, dx_2) .$$

To obtain the Euler equation for $I_\beta(\alpha)$ at $\alpha = \alpha_0$, we consider

$$\frac{\partial}{\partial \varepsilon} I_\beta(\alpha_\varepsilon)\Big|_{\varepsilon=0} = 0 . \tag{2.17}$$

From Lemma 2.1 and the definition of $a_\varepsilon(x_1, x_2)$ we have

$$\begin{aligned} I_\beta(\alpha_\varepsilon) &= \int_{\mathcal{X} \times \mathcal{X}} a_\varepsilon(x_1, x_2) \log a_\varepsilon(x_1, x_2)\beta(dx_1, dx_2) \\ &= \int_{\mathcal{X} \times \mathcal{X}} [a_0(x_1, x_2)(1 + \varepsilon k(x_1, x_2))] \log [a_0(x_1, x_2) \\ &\qquad \times (1 + \varepsilon k(x_1, x_2)]\beta(dx_1, dx_2) . \end{aligned}$$

Hence,

$$\left.\frac{\partial I_\beta(\alpha_\varepsilon)}{\partial \varepsilon}\right|_{\varepsilon=0}=\int_{\mathcal{X}\times\mathcal{X}} a_0(x_1, x_2)k(x_1, x_2)\log a_0(x_1, x_2)\beta(dx_1, dx_2)$$
$$+\int_{\mathcal{X}\times\mathcal{X}} a_0(x_1, x_2)k(x_1, x_2)\beta(dx_1, dx_2)\,.$$

From (2.16) with $f\equiv 1$ and $g\equiv 0$ we see that the second integral on the right in this last expression is zero. Thus equation (2.17) is just

$$\int_{\mathcal{X}\times\mathcal{X}} k(x_1, x_2)\log a_0(x_1, x_2)\alpha_0(dx_1, dx_2)=0\,. \tag{2.18}$$

Our hypothesis on $k(x_1, x_2)$, (2.16), can be rewritten as

$$\int_{\mathcal{X}\times\mathcal{X}} k(x_1, x_2)[f(x_1)+g(x_2)]\alpha_0(dx_1, dx_2)=0 \tag{2.19}$$

for all continuous f and g. Therefore, what we have shown is that if $k(x_1, x_2)$ is a bounded function satisfying (2.19), then (2.18) holds. It now follows from the Hahn-Banach theorem that there exist sequences of continuous functions $\{f_n(x_1)\}$, $\{g_n(x_2)\}$, such that

$$\log a_0(x_1, x_2)=\lim_{n\to\infty}[f_n(x_1)+g_n(x_2)]\,, \tag{2.20}$$

where the limit on the right is in the sense of $L_1(\alpha_0)$. Hence, there exist sequences of continuous functions $\{f_n(x_1)\}$, $\{g_n(x_2)\}$ such that

$$\log a_0(x_1, x_2)=\lim_{n\to\infty}\text{ in prob }[f_n(x_1)+g_n(x_2)]\,, \tag{2.21}$$

where the probability measure indicated on the right of (2.21) is, of course, $\alpha_0(dx_1, dx_2)$. However, from Lemma 2.2 and our hypotheses on $\pi(x_1, x_2)$ and $m(x)$ we see that $\alpha_0\sim\beta\sim\lambda\times\lambda$ so that in fact

$$\log a_0(x_1, x_2)=\lim_{n\to\infty}\text{ in prob }[f_n(x_1)+g_n(x_2)]\,, \tag{2.22}$$

where now the probability measure on the right of (2.22) is $\lambda\times\lambda$. Therefore, from Lemma 2.3 we conclude that there exist functions $f(x_1)$ and $g(x_2)$ such that

$$\log a_0(x_1, x_2)=f(x_1)+g(x_2) \tag{2.23}$$

for almost all (x_1, x_2) ($\lambda\times\lambda$-measure).

Thus, letting $U(x_1)=\exp\{f(x_1)\}$ and $V(x_2)=\exp\{g(x_2)\}$, we have

$$a_0(x_1, x_2)=U(x_1)V(x_2) \tag{2.24}$$

for almost all (x_1, x_2) ($\lambda\times\lambda$-measure).

Now since $\beta(dx_1, dx_2)=\pi(x_1, x_2)m(x_1)\lambda(dx_1)\lambda(dx_2)$ and $d\alpha_0/d\beta=a_0(x_1, x_2)$, (2.24) implies that

$$\alpha_0(dx_1, dx_2)=U(x_1)V(x_2)\pi(x_1, x_2)m(x_1)\lambda(dx_1)\lambda(dx_2)\,. \tag{2.25}$$

Since $\alpha_0\in\mathcal{M}_\mu$ and $d\mu/d\lambda=m(x)$, we get from (2.25)

$$U(x_1)m(x_1)\int_{\mathfrak{X}} V(x_2)\pi(x_1, x_2)\lambda(dx_2)=m(x_1) \tag{2.26}$$

for almost all x_1 (λ-measure).

In (2.26) alter U on a set of measure zero so that the equation is valid for all x_1 instead of almost all x_1. This revised U could just as well have been used instead of the original U in (2.24) since the set of measure zero on which we have changed U does not affect the Radon-Nikodym derivative. With this new U we then have

$$\int_{\mathfrak{X}} V(x_2)\pi(x_1, x_2)\lambda(dx_2)=\frac{1}{U(x_1)} \tag{2.27}$$

for all x_1. By its definition we know that $a_0\in L_1(\beta)$, and, since $d\beta/d(\lambda\times\lambda)=\pi(x_1, x_2)m(x_1)$ is bounded above and below, we have $a_0\in L_1(\lambda\times\lambda)$. Therefore, from the Fubini theorem and (2.24) we see that $V\in L_1(\lambda)$. It follows then from (2.27) and our hypothesis on $\pi(x_1, x_2)$ that U is continuous. We now define

$$\pi'(x, dy)=U(x)\pi(x, y)V(y)\lambda(dy) \tag{2.28}$$

and note from (2.27) that it is a transition probability.

Finally, we want to show that μ is an invariant probability measure for $\pi'(x, dy)$. Consider, for any Borel set $A\subset\mathfrak{X}$,

$$\begin{aligned}\int_{\mathfrak{X}} \pi'(x, A)\mu(dx) &= \int_{\mathfrak{X}}\int_A U(x)\pi(x, y)V(y)\lambda(dy)\mu(dx)\\ &= \int_{\mathfrak{X}}\int_A U(x)\pi(x, y)V(y)m(x)\lambda(dy)\lambda(dx)\\ &= \int_{\mathfrak{X}}\int_A \alpha_0(dx, dy)=\mu(A)\,,\end{aligned}$$

since $\alpha_0\in\mathcal{M}_\mu$. This completes Lemma 2.4.

COROLLARY. *The V of Lemma 2.4 is bounded above and below. In fact,*

$$\frac{\eta}{\left(\sup_{x\in\mathfrak{X}} U(x)\right)\xi A} \leqq V(y) \leqq \frac{\xi}{\left(\inf_{x\in\mathfrak{X}} U(x)\right)a\eta}.$$

Proof: From (2.27) and the fact that $V\in L_1(\lambda)$ we know that, for any $x\in\mathfrak{X}$ and $x'\in\mathfrak{X}$,

$$\frac{U(x')}{U(x)} = \frac{\int_{\mathfrak{X}} V(y)\pi(x, y)\lambda(dy)}{\int_{\mathfrak{X}} V(y)\pi(x', y)\lambda(dy)} \leqq \frac{A}{a}.$$

Thus $\sup_{x\in\mathfrak{X}} U(x)/\inf_{x\in\mathfrak{X}} U(x)\leqq A/a$. Since $\alpha_0\in\mathcal{M}_\mu$, we have for almost all y (λ-measure)

$$m(y) = \int_{\mathfrak{X}} U(x)V(y)\pi(x, y)m(x)\lambda(dx).$$

Moreover, since, by Lemma 2.2, $a_0(x_1, x_2)>0$ almost everywhere ($\lambda\times\lambda$-measure), $V(y)>0$ almost everywhere (λ-measure). Thus

$$\frac{m(y)}{V(y)} = \int_{\mathfrak{X}} U(x)\pi(x, y)m(x)\lambda(dx)$$
$$\leqq \left(\sup_{x\in\mathfrak{X}} U(x)\right)A\xi,$$

so that

$$V(y) \geqq \frac{\eta}{\left(\sup_{x\in\mathfrak{X}} U(x)\right)A\xi}.$$

Similarly,

$$\frac{m(y)}{V(y)} \geqq \left(\inf_{x\in\mathfrak{X}} U(x)\right)a\eta,$$

so that

$$V(y) \leqq \frac{\xi}{\left(\inf_{x\in\mathfrak{X}} U(x)\right)a\eta}.$$

The remaining lemmas are concerned with the function $I(\mu)$ defined in (1.5).

LEMMA 2.5. *$I(\mu)=0$ if and only if μ is the invariant measure of $\pi(x, dy)$.*

Proof: Since $u\equiv 1$ is in $\mathcal{U}_1$ and, for this u, $\log(\pi u/u)(x)=0$, it clearly suffices to show that

$$\int_{\mathcal{X}} \log\left(\frac{\pi u}{u}\right)(x)\mu(dx)\geqq 0 \tag{2.29}$$

for all $u\in\mathcal{U}_1$ if and only if for every Borel set $A\subset\mathcal{X}$

$$\int_{\mathcal{X}} \pi(x, A)\mu(dx)=\mu(A)\,. \tag{2.30}$$

First, assume (2.29); then as just noted the left side of (2.29) is zero if $u\equiv 1$. Therefore, if we consider

$$\int_{\mathcal{X}} \log\left(\frac{\pi(1+\varepsilon u)}{1+\varepsilon u}\right)(x)\mu(dx)=\int_{\mathcal{X}} \log\left(\frac{1+\varepsilon\pi u}{1+\varepsilon u}\right)(x)\mu(dx)$$

as a function of ε for continuous u such that $|u|\leqq c$ and $|\varepsilon|<1/c$, we see that it has a minimum at $\varepsilon=0$. Hence, setting the derivative with respect to ε equal to zero at $\varepsilon=0$, we obtain for all u as considered above

$$\int_{\mathcal{X}} (\pi u)(x)\mu(dx)=\int_{\mathcal{X}} u(x)\mu(dx)$$

which implies (2.30).

Now assume (2.30). Let $u\in\mathcal{U}_1$ and define h by $u=e^h$. From Jensen's inequality, $\pi e^h\geqq e^{\pi h}$ and therefore $\log \pi e^h\geqq \pi h$. This last and (2.30) imply that

$$\begin{aligned}\int_{\mathcal{X}} \log(\pi e^h)(x)\mu(dx)&\geqq\int_{\mathcal{X}} (\pi h)(x)\mu(dx)\\ &=\int_{\mathcal{X}} h(x)\mu(dx)\end{aligned}$$

which is (2.29).

LEMMA 2.6. *Let $\mathcal{U}_2$ be the set of measurable functions u on $\mathfrak{X}$ for each of which there exist constants c_1 and c_2 such that*

$$0 < c_1 \leqq u(x) \leqq c_2 < \infty$$

for all $x \in \mathfrak{X}$. Then, for any $\mu \in \mathcal{M}$,

$$I(\mu) = -\inf_{u \in \mathcal{U}_2} \int_{\mathfrak{X}} \log\left(\frac{\pi u}{u}\right)(x)\mu(dx)\,.$$

Proof: By definition,

$$I(\mu) = -\inf_{u \in \mathcal{U}_1} \int_{\mathfrak{X}} \log\left(\frac{\pi u}{u}\right)(x)\mu(dx)\,.$$

Let $I(\mu) = l$ and let

$$l' = -\inf_{u \in \mathcal{U}_2} \int_{\mathfrak{X}} \log\left(\frac{\pi u}{u}\right)(x)\mu(dx)\,.$$

Since $\mathcal{U}_1 \subset \mathcal{U}_2$, $l \leqq l'$ and we must show that $l' \leqq l$. Now, for all $u \in \mathcal{U}_1$,

$$\int_{\mathfrak{X}} \log(\pi u)(x)\mu(dx) - \int_{\mathfrak{X}} \log u(x)\mu(dx) \geqq -l\,.$$

Let

$$\mathcal{U}_3 = \left\{u \in \mathcal{U}_2 : \int_{\mathfrak{X}} \log(\pi u)(x)\mu(dx) - \int_{\mathfrak{X}} \log u(x)\mu(dx) \geqq -l\right\},$$

so that $\mathcal{U}_3 \supset \mathcal{U}_1$. Let $\{u_n(x)\}$ be a sequence in $\mathcal{U}_3$ converging for each $x \in \mathfrak{X}$ to a function $u(x)$. It is clear from the bounded convergence theorem that $u(x) \in \mathcal{U}_3$. Hence, $\mathcal{U}_2 = \mathcal{U}_3$ and $l' \leqq l$.

LEMMA 2.7. *Assume $\mu \in \mathcal{M}$ is such that $\mu \ll \lambda$ and $d\mu/d\lambda = m(x)$ satisfies $0 < \eta \leqq m(x) \leqq \xi < \infty$ for all $x \in \mathfrak{X}$. Then,*

$$\begin{aligned} I(\mu) &= -\int_{\mathfrak{X}} \log\left(\frac{\pi V}{V}\right)(x)\mu(dx) \\ &= \int_{\mathfrak{X}} \log[U(x)V(x)]\mu(dx) \end{aligned}$$

for the $U(x)$ and $V(x)$ determined in Lemma 2.4.

Proof: By definition,

$$I(\mu)=-\inf_{u\in\mathcal{U}_1}\int_{\mathcal{X}}\log\left(\frac{\pi u}{u}\right)(x)\mu(dx)\,.$$

In Lemma 2.6 we saw that

$$I(\mu)=-\inf_{u\in\mathcal{U}_2}\int_{\mathcal{X}}\log\left(\frac{\pi u}{u}\right)(x)\mu(dx)\,, \tag{2.31}$$

so that the present lemma states that in (2.31) the infimum is in fact attained at V which, by Lemma 2.4 and its corollary, we know to be an element of $\mathcal{U}_2$.

In Lemma 2.4 we defined $\pi'(x, y)=U(x)V(y)\pi(x, y)$, and therefore

$$\begin{aligned}(\pi u)(x)&=\int_{\mathcal{X}}u(y)\pi(x, y)\lambda(dy)\\ &=\int_{\mathcal{X}}\frac{\pi(x, y)}{\pi'(x, y)}\,\pi'(x, y)u(y)\lambda(dy)\\ &=\int_{\mathcal{X}}\frac{1}{U(x)V(y)}\,\pi'(x, y)u(y)\lambda(dy)\\ &=\frac{1}{U(x)}\left(\pi'\left(\frac{u}{V}\right)\right)(x)\\ &=\frac{1}{U(x)}\int_{\mathcal{X}}\frac{u(y)}{V(y)}\,\pi'(x, dy)\,.\end{aligned}$$

From this and Lemma 2.6 we have

$$\begin{aligned}I(\mu)&=-\inf_{u\in\mathcal{U}_2}\int_{\mathcal{X}}\log\left(\frac{\pi u}{u}\right)(x)\mu(dx)\\ &=-\inf_{u\in\mathcal{U}_2}\int_{\mathcal{X}}\log\left(\frac{\pi'(u/V)}{uU}\right)(x)\mu(dx)\\ &=-\inf_{u\in\mathcal{U}_2}\int_{\mathcal{X}}\log\left(\frac{\pi'(u/V)}{(u/V)\cdot VU}\right)(x)\mu(dx)\\ &=-\inf_{u\in\mathcal{U}_2}\left[\int_{\mathcal{X}}\log\left(\frac{\pi'(u/V)}{u/V}\right)(x)\mu(dx)-\int_{\mathcal{X}}\log\,[U(x)V(x)]\mu(dx)\right]\\ &=\int_{\mathcal{X}}\log\,[U(x)V(x)]\mu(dx)-\inf_{u\in\mathcal{U}_2}\int_{\mathcal{X}}\log\left(\frac{\pi'(u/V)}{u/V}\right)(x)\mu(dx)\,.\end{aligned}$$

Now if we can show that

$$\inf_{u\in\mathcal{U}_2}\int_{\mathcal{X}}\log\left(\frac{\pi'(u/V)}{u/V}\right)(x)\mu(dx)=0\,, \tag{2.32}$$

then indeed

$$I(\mu)=\int_{\mathcal{X}}\log\,[U(x)\,V(x)]\mu(dx)$$

and the lemma is completed by recalling from (2.27) that

$$\left(\frac{\pi V}{V}\right)(x)=\frac{1}{U(x)\,V(x)}$$

so that

$$-\int_{\mathcal{X}}\log\left(\frac{\pi V}{V}\right)(x)\mu(dx)=\int_{\mathcal{X}}\log\,[U(x)\,V(x)]\mu(dx)\,.$$

To see that (2.32) holds, consider first

$$I'(\mu)=-\inf_{u\in\mathcal{U}_1}\int_{\mathcal{X}}\log\left(\frac{\pi' u}{u}\right)(x)\mu(dx)\,,$$

i.e., the function $I(\mu)$ based on the transition function $\pi'(x, dy)$ instead of $\pi(x, dy)$. From Lemma 2.4 we know that μ is the invariant measure for $\pi'(x, dy)$ so that Lemma 2.5 implies $I'(\mu)=0$. From the argument of Lemma 2.6 we have therefore

$$\inf_{u\in\mathcal{U}_2}\int_{\mathcal{X}}\log\left(\frac{\pi' u}{u}\right)(x)\mu(dx)=0\,.$$

Since $V\in\mathcal{U}_2$, the infimum is not affected if we replace u by u/V, i.e.,

$$\inf_{u\in\mathcal{U}_2}\int_{\mathcal{X}}\log\left(\frac{\pi'(u/V)}{u/V}\right)(x)\mu(dx)=0$$

which is (2.32).

LEMMA 2.8. *Let* $\mu\in\mathcal{M}$. *Then,*

$$I_\lambda(\mu)-\log A\leqq I(\mu)\leqq I_\lambda(\mu)-\log a\,, \tag{2.33}$$

so that a necessary and sufficient condition for $I(\mu)<\infty$ *is that* $I_\lambda(\mu)<\infty$.

Proof: For $u\in\mathcal{U}_1$,

$$\int_{\mathcal{X}}\log\left(\frac{\pi u}{u}\right)(x)\mu(dx)=\int_{\mathcal{X}}\log \pi u(x)\mu(dx)-\int_{\mathcal{X}}\log u(x)\mu(dx)\,,$$

where, since $\pi u(x)=\int_{X}\pi(x,y)u(y)\lambda(dy)$, we have for all $x\in\mathcal{X}$

$$a\int_{\mathcal{X}}u(y)\lambda(dy)\leqq\pi u(x)\leqq A\int_{\mathcal{X}}u(y)\lambda(dy)\,.$$

Thus,

$$\begin{aligned}\log a+\log\int_{\mathcal{X}}u(y)\lambda(dy)-\int_{\mathcal{X}}\log u(x)\mu(dx)&\\ \leqq\int_{\mathcal{X}}\log\left(\frac{\pi u}{u}\right)(x)\mu(dx)&\\ \leqq\log A+\log\int_{\mathcal{X}}u(y)\lambda(dy)-\int_{\mathcal{X}}\log u(x)\mu(dx)\,.&\end{aligned}$$

Taking infimum over all $u\in\mathcal{U}_1$ we get (2.33).

COROLLARY. *Let* $\mu\in\mathcal{M}$. *Then* $I(\mu)<\infty$ *if and only if* $\mu\ll\lambda$ *and* $d\mu/d\lambda=m(x)$ *is such that*

$$\int_{\mathcal{X}}m(x)\,|\log m(x)|\,\lambda(dx)<\infty\,.$$

Proof: The corollary follows immediately from Lemma 2.1 and Lemma 2.8.

LEMMA 2.9. *Let* G *be an open set in* $\mathcal{M}$ *and let* $\mathcal{M}_0\subset\mathcal{M}$ *be the set of probability measures* μ *on* $\mathcal{X}$ *such that* $d\mu/d\lambda=m(x)$ *exists and satisfies* $0<\eta\leqq m(x)\leqq\xi<\infty$ (ξ *and* η *depend on* μ). *Then,*

$$\inf_{\mu\in G}I(\mu)=\inf_{\mu\in G\cap\mathcal{M}_0}I(\mu)\,.$$

Proof: It will suffice to show that for any $\mu\in\mathcal{M}$ such that $I(\mu)<\infty$ there exists a sequence $\{\mu_n\}\in\mathcal{M}_0$ with

$$\mu_n\Rightarrow\mu\,,$$

$$\lim_{n\to\infty}I(\mu_n)=I(\mu)\,.$$

So let $\mu \in \mathcal{M}$ be such that $I(\mu)<\infty$, and let $\mathcal{M}_1=\{\mu \in \mathcal{M} : \mu \ll \lambda$ and $d\mu/d\lambda = m(x)$ satisfies $0<\eta \leqq m(x) \leqq \infty\}$. We shall show first that there exists a sequence $\{\mu_n\} \in \mathcal{M}_1$ such that $\mu_n \Rightarrow \mu$ and $\lim_{n\to\infty} I(\mu_n)=I(\mu)$.

For any positive integer n, let $\mu_n=(1-1/n)\mu+(1/n)\lambda$. Since $I(\mu)<\infty$, we know from the corollary to Lemma 2.8 that $\mu \ll \lambda$ and therefore $\mu_n \ll \lambda$. Moreover, $d\mu_n/d\lambda \geqq 1/n$ so that the sequence $\{\mu_n\} \in \mathcal{M}_1$. Clearly, $\mu_n \Rightarrow \mu$ and since I is lower semicontinuous we have

$$\varliminf_{n\to\infty} I(\mu_n) \geqq I(\mu) .$$

But λ itself is in $\mathcal{M}_0$ so again by the corollary to Lemma 2.8 we have $I(\lambda)<\infty$. Therefore, since due to the convexity of I

$$I(\mu_n) \leqq \left(1-\frac{1}{n}\right) I(\mu)+\frac{1}{n} I(\lambda) ,$$

we have

$$\varlimsup_{n\to\infty} I(\mu_n) \leqq I(\mu)$$

so that indeed $\lim_{n\to\infty} I(\mu_n)=I(\mu)$.

Next we take a $\mu \in \mathcal{M}_1$ for which $I(\mu)<\infty$ and show there exists a sequence $\{\mu_n\} \in \mathcal{M}_0$ such that $\mu_n \Rightarrow \mu$ and $\lim_{n\to\infty} I(\mu_n)=I(\mu)$. Since $\mu \in \mathcal{M}_1$ and $I(\mu)<\infty$, we know from the corollary to Lemma 2.8 that $d\mu/d\lambda = m(x)$ satisfies

$$\int_{\mathcal{X}} m(x) \, |\log m(x)| \, \lambda(dx) < \infty . \tag{2.34}$$

Since $\mu \in \mathcal{M}_1$, there exists an $\eta>0$ such that $\eta \leqq m(x)$ for all $x \in \mathcal{X}$. Define

$$m_n(x)=\frac{m(x) \wedge n}{\int_{\mathcal{X}} [m(x) \wedge n] \lambda(dx)} ,$$

where n is chosen greater than or equal to η and so large that

$$\int_{\mathcal{X}} [m(x) \wedge n] \lambda(dx) \geqq \tfrac{1}{2} .$$

Thus $\eta \leqq m_n(x) \leqq 2n$; hence defining $\mu_n(dx)=m_n(x)\lambda(dx)$ we see that $\mu_n \Rightarrow \mu$ and each $\mu_n \in \mathcal{M}_0$. By lower semicontinuity,

$$\varliminf_{n\to\infty} I(\mu_n) \geqq I(\mu) ;$$

consequently it remains to show that

$$\overline{\lim_{n\to\infty}}\, I(\mu_n) \leqq I(\mu)\,. \tag{2.35}$$

Since $\mu_n \in \mathcal{M}_0$, it follows from Lemma 2.7 that there exists a $V_n \in \mathcal{U}_2$ (the existence of the V_n is guaranteed by Lemma 2.4) such that

$$\begin{aligned} I(\mu_n) &= -\int_{\mathfrak{X}} \log\left(\frac{\pi V_n}{V_n}\right)(x)\mu_n(dx) \\ &= -\int_{\mathfrak{X}} \log\left(\frac{\pi V_n}{V_n}\right)(x) m_n(x)\lambda(dx)\,. \end{aligned}$$

On the other hand, from the definition of $I(\mu)$ and from Lemma 2.6 we know that

$$\begin{aligned} I(\mu) &\geqq -\int_{\mathfrak{X}} \log\left(\frac{\pi V_n}{V_n}\right)(x)\mu(dx) \\ &= -\int_{\mathfrak{X}} \log\left(\frac{\pi V_n}{V_n}\right)(x) m(x)\lambda(dx)\,. \end{aligned}$$

Thus

$$\begin{aligned} I(\mu) &\geqq I(\mu_n) + \int_{\mathfrak{X}} \left[\log\left(\frac{\pi V_n}{V_n}\right)(x)\right][m_n(x) - m(x)]\lambda(dx) \\ &\geqq I(\mu_n) - \int_{\mathfrak{X}} \left|\log\left(\frac{\pi V_n}{V_n}\right)(x)\right| |m_n(x) - m(x)|\, \lambda(dx)\,, \end{aligned}$$

and (2.35) will follow if we show that

$$\lim_{n\to\infty} \int_{\mathfrak{X}} \left|\log\left(\frac{\pi V_n}{V_n}\right)(x)\right| |m_n(x) - m(x)|\, \lambda(dx) = 0\,. \tag{2.36}$$

The corollary to Lemma 2.4 and its proof yield

$$\begin{aligned} \frac{\sup_{x\in\mathfrak{X}} V(x)}{\inf_{x\in\mathfrak{X}} V(x)} &\leqq \frac{\xi}{\left(\inf_{x\in\mathfrak{X}} U(x)\right) a\eta} \times \frac{\left(\sup_{x\in\mathfrak{X}} U(x)\right)\xi A}{\eta} \\ &\leqq \frac{\xi^2 A^2}{a^2\eta^2}\,. \end{aligned}$$

Therefore, applying this last to V_n (where the corresponding $\xi=2n$), we have

$$\left|\log\left(\frac{\pi V_n}{V_n}\right)(x)\right|\leqq\log\frac{4n^2A^2}{a^2\eta^2}=2\log n+c\,.$$

Hence

$$\int_{\mathcal{X}}\left|\log\left(\frac{\pi V_n}{V_n}\right)(x)\right|\,|m_n(x)-m(x)|\,\lambda(dx) \tag{2.37}$$
$$\leqq[2\log n+c]\int_{\mathcal{X}}|m_n(x)-m(x)|\,\lambda(dx)\,.$$

But, remembering that $\int_{\mathcal{X}}[m(x)\wedge n]\lambda(dx)\geqq\frac{1}{2}$, we obtain

$$\int_{\mathcal{X}}|m_n(x)-m(x)|\,\lambda(dx)=\int_{\mathcal{X}}\left|\frac{[m(x)\wedge n]}{\int_{\mathcal{X}}[m(x)\wedge n]\lambda(dx)}-m(x)\right|\lambda(dx)$$

$$\leqq\int_{\mathcal{X}}\left|\frac{[m(x)\wedge n]}{\int_{\mathcal{X}}[m(x)\wedge n]\lambda(dx)}-[m(x)\wedge n]\right|\lambda(dx)+\int_{\mathcal{X}}|[m(x)\wedge n]-m(x)|\,\lambda(dx)$$

$$\leqq\int_{\mathcal{X}}[m(x)\wedge n]\left|\frac{1}{\int_{\mathcal{X}}[m(x)\wedge n]\lambda(dx)}-1\right|\lambda(dx)+\int_{\mathcal{X}}|[m(x)\wedge n]-m(x)|\,\lambda(dx)$$

$$\leqq\frac{1-\int_{\mathcal{X}}[m(x)\wedge n]\lambda(dx)}{\int_{\mathcal{X}}[m(x)\wedge n]\lambda(dx)}+\int_{\mathcal{X}}|[m(x)\wedge n]-m(x)|\,\lambda(dx)$$

$$\leqq2\int_{\mathcal{X}}(m(x)-[m(x)\wedge n])\lambda(dx)+\int_{\mathcal{X}}|[m(x)\wedge n]-m(x)|\,\lambda(dx)$$

$$\leqq3\int_{\mathcal{X}}|m(x)-[m(x)\wedge n]|\,\lambda(dx)$$

$$\leqq3\int_{m(x)\geqq n}m(x)\lambda(dx)\leqq\frac{3}{\log n}\int_{m(x)\geqq n}m(x)\,|\log m(x)|\,\lambda(dx)$$

$$=\frac{3}{\log n}\,o(1)\,,$$

because of (2.34). Thus from (2.37)

$$\int_{\mathfrak{X}} \left|\log\left(\frac{\pi V_n}{V_n}\right)(x)\right| |m_n(x)-m(x)|\,\lambda(dx) \leqq [2\log n+c]\left(\frac{3}{\log n}\,o(1)\right) \to 0$$

as $n\to\infty$ which gives (2.36) and hence (2.35) and the lemma.

Now we are ready to return to the proof of (1.8). Let $\mu\in\mathcal{M}_0$, i.e., $\mu\ll\lambda$ and $d\mu/d\lambda=m(x)$ satisfies the condition $0<\eta\leqq m(x)\leqq\xi<\infty$ for all $x\in\mathfrak{X}$. For any $\varepsilon>0$, let $S(\mu,\varepsilon)$ be the sphere with center at μ of radius ε in the weak topology on $\mathcal{M}$. Consider

$$Q_{n,x}[S(\mu,\varepsilon)]=P_x\{L_n(\omega,\cdot)\in S(\mu,\varepsilon)\}\,.$$

Let

$$E_{n,\mu,\varepsilon}=\{\omega : L_n(\omega,\cdot)\in S(\mu,\varepsilon)\}\,,$$

and, for any $\varepsilon'>0$, let

$$F_{n,\mu,\varepsilon'}=\left\{\omega : \left|\frac{W(X_0)+W(X_1)+\cdots+W(X_{n-1})}{n}-\int_{\mathfrak{X}} W(y)\mu(dy)\right|<\varepsilon'\right\}\,,$$

where $W(y)=\log V(y)+\log U(y)$ with the functions U and V as determined in Lemma 2.4. We note from Lemma 2.7 that

$$\int_{\mathfrak{X}} W(y)\mu(dy)=I(\mu)\,.$$

Now

$$\begin{aligned}Q_{n,x}[S(\mu,\varepsilon)]&=\int_{E_{n,\mu,\varepsilon}} dP_x\\&=\int_{E_{n,\mu,\varepsilon}} \pi(x,x_1)\pi(x_1,x_2)\cdots\pi(x_{n-2},x_{n-1})\lambda(dx_1)\cdots\lambda(dx_{n-1})\\&=\int_{E_{n,\mu,\varepsilon}} \pi'(x,x_1)\pi'(x_1,x_2)\cdots\pi'(x_{n-2},x_{n-1})\\&\qquad\times\left[\frac{\pi(x,x_1)\cdots\pi(x_{n-2},x_{n-1})}{\pi'(x,x_1)\cdots\pi'(x_{n-2},x_{n-1})}\right]\lambda(dx_1)\cdots\lambda(dx_{n-1})\,.\end{aligned}$$

From (2.28)

$$\frac{\pi(x,y)}{\pi'(x,y)}=\frac{1}{U(x)V(y)}\,,$$

and, therefore,

$$Q_{n,x}[S(\mu,\varepsilon)]=\int_{E_{n,\mu,\varepsilon}}\frac{1}{U(x)V(x_1)}\frac{1}{U(x_1)V(x_2)}\cdots\frac{1}{U(x_{n-2})V(x_{n-1})}$$
$$\times\pi'(x,x_1)\cdots\pi'(x_{n-2},x_{n-1})\lambda(dx_1)\cdots\lambda(dx_{n-1})$$
$$=\int_{E_{n,\mu,\varepsilon}}\frac{1}{U(x)V(x_1)}\frac{1}{U(x_1)V(x_2)}\cdots\frac{1}{U(x_{n-2})V(x_{n-1})}\,dP'_x\,,$$

where P'_x is the probability measure on Ω_X induced by the transition probability $\pi'(x,dy)$. Hence

$$Q_{n,x}[S(\mu,\varepsilon)]=\int_{E_{n,\mu,\varepsilon}}V(x)U(x_{n-1})\exp\{-[W(x_0)+W(x_1)+\cdots+W(x_{n-1})]\}\,dP'_x$$
$$=\int_{E_{n,\mu,\varepsilon}}V(x)U(x_{n-1})\exp\{-n[(W(x_0)+W(x_1)+\cdots$$
$$+W(x_{n-1}))/n]\}\,dP'_x$$
$$\geqq\int_{E_{n,\mu,\varepsilon}\cap F_{n,\mu,\varepsilon'}}V(x)U(x_{n-1})\exp\{-n[(W(x_0)+\cdots+W(x_{n-1}))/n]\}\,dP'_x\,.$$

Since, on $F_{n,\mu,\varepsilon'}$,

$$\left|\frac{W(x_0)+W(x_1)+\cdots+W(x_{n-1})}{n}-I(\mu)\right|<\varepsilon'\,,$$

we have

$$Q_{n,x}[S(\mu,\varepsilon)]\geqq\exp\{-n[I(\mu)+\varepsilon']\}\int_{E_{n,\mu,\varepsilon}\cap F_{n,\mu,\varepsilon'}}V(x)U(x_{n-1})\,dP'_x\,.$$

But both U and V are bounded below so that we get, for some constant C,

$$Q_{n,x}[S(\mu,\varepsilon)]\geqq C\exp\{-n[I(\mu)+\varepsilon']\}P'_x\{E_{n,\mu,\varepsilon}\cap F_{n,\mu,\varepsilon'}\}\,. \tag{2.38}$$

In Lemma 2.4 we showed that μ is the invariant measure for π', and therefore by the ergodic theorem

$$\lim_{n\to\infty}P'_x\{E_{n,\mu,\varepsilon}\}=1\quad\text{for all}\quad\varepsilon>0\,,$$

$$\lim_{n\to\infty}P'_x\{F_{n,\mu,\varepsilon'}\}=1\quad\text{for all}\quad\varepsilon'>0\,.$$

Thus from (2.38) we have

$$\underline{\lim}_{n\to\infty} \frac{1}{n} \log Q_{n,x}[S(\mu, \varepsilon)] \geqq -I(\mu) - \varepsilon' ,$$

and, since $\varepsilon' > 0$ was arbitrary, we obtain on letting $\varepsilon' \to 0$

$$\underline{\lim}_{n\to\infty} \frac{1}{n} \log Q_{n,x}[S(\mu, \varepsilon)] \geqq -I(\mu) . \tag{2.39}$$

Now let $\mu \in G \cap \mathcal{M}_0$, where G is an open set in $\mathcal{M}$. Since G is open, there exists an $\varepsilon > 0$ such that the sphere $S(\mu, \varepsilon) \subset G$. Thus, from (2.39),

$$\underline{\lim}_{n\to\infty} \frac{1}{n} \log Q_{n,x}(G) \geqq \underline{\lim}_{n\to\infty} \frac{1}{n} \log Q_{n,x}[S(\mu, \varepsilon)] \geqq -I(\mu) .$$

This last is true for every $\mu \in G \cap \mathcal{M}_0$ and therefore

$$\underline{\lim}_{n\to\infty} \frac{1}{n} \log Q_{n,x}(G) \geqq - \inf_{\mu \in G \cap \mathcal{M}_0} I(\mu) .$$

But in Lemma 2.9 we showed that

$$\inf_{\mu \in G \cap \mathcal{M}_0} I(\mu) = \inf_{\mu \in G} I(\mu) ,$$

and thus we obtain

$$\underline{\lim}_{n\to\infty} \frac{1}{n} \log Q_{n,x}(G) \geqq - \inf_{\mu \in G} I(\mu)$$

which is (1.8). This completes the proof of Theorem 1.

3. Continuous Parameter Markov Process

In this section we shall prove Theorem 3. The notation and definitions will be those of the introduction just prior to the statement of Theorem 3. We recall that $\mathcal{M}$, the space of probability measures on $\mathcal{X}$, was given the weak topology. This topology on $\mathcal{M}$ is induced by a metric, the Prokhorov metric, which we now recall:

Let $\rho(x, y)$ be the metric on $\mathcal{X}$. For any closed set $C \subset \mathcal{X}$ and any $\delta > 0$, let

$$C^{\delta} = \{y \in \mathcal{X}, \rho(x, y) < \delta \quad \text{for} \quad \text{some} \quad x \in C\} .$$

Given two measures μ_1 and μ_2 in $\mathcal{M}$ the Prokhorov metric $d(\mu_1, \mu_2)$ is defined by

$$d(\mu_1, \mu_2) = \inf_{\delta>0} \{\mu_1(c) \leqq \mu_2(c^\delta) + \delta,\ \mu_2(c) \leqq \mu_1(c^\delta) + \delta \text{ for all closed sets } C \subset \mathcal{X}\}.$$

$\mathcal{M}$ then becomes a compact metric space.

In addition to the definition of $I(\mu)$ given in (1.12) we find it useful to define, for any $h>0$ and any $\mu \in \mathcal{M}$,

$$(3.1)\qquad \begin{aligned} I_h(\mu) &= -\inf_{\substack{u>0\\ u\in C(\mathcal{X})}} \int_{\mathcal{X}} \log\left(\frac{T_h u}{u}\right)(x)\mu(dx) \\ &= -\inf_{\substack{u>0\\ u\in\mathcal{D}}} \int_{\mathcal{X}} \log\left(\frac{T_h u}{u}\right)(x)\mu(dx). \end{aligned}$$

We shall first prove (1.14) which will require four lemmas.

LEMMA 3.1. *For any* $h>0$ *and* $\mu \in \mathcal{M}$,

$$(3.2)\qquad I_h(\mu) \leqq hI(\mu).$$

Moreover, for every $\mu \in \mathcal{M}$,

$$(3.3)\qquad \lim_{h\to 0} \frac{1}{h} I_h(\mu) = I(\mu).$$

Proof: Let $u>0$ and $u \in \mathcal{D}$. Consider, for any $\mu \in \mathcal{M}$ and $0 \leqq t < \infty$,

$$\phi(t) = \int_{\mathcal{X}} \log\left(\frac{T_t u}{u}\right)(x)\mu(dx).$$

Since T_t maps positive elements of $\mathcal{D}$ into positive elements of $\mathcal{D}$, and since $dT_t u/dt = LT_t u$, we have

$$\frac{d\phi(t)}{dt} = \int_{\mathcal{X}} \left(\frac{LT_t u}{T_t u}\right)(x)\mu(dx) \geqq -I(\mu).$$

Now $T_0 u = u$, $\phi(0) = 0$, and hence for any $h>0$

$$\phi(h) = \int_0^h \frac{d\phi(s)}{ds}\, ds \geqq -\int_0^h I(\mu)\, ds = -hI(\mu).$$

Thus

$$\int_{\mathcal{X}} \log\left(\frac{T_h u}{u}\right)(x)\mu(dx) \geqq -hI(\mu)$$

holds for any $h>0$ and any positive element u of $\mathcal{D}$ which implies (3.2).

Let $\mu \in \mathcal{M}$ and consider, for any positive element u of $\mathcal{D}$,

$$\frac{1}{h} I_h(\mu) \geqq -\frac{1}{h}\int_{\mathcal{X}} \log\left(\frac{T_h u}{u}\right)(x)\mu(dx)\,. \tag{3.4}$$

Since $u \in \mathcal{D}$, $T_h u = u + hLu + o(h)$, where $o(h)$ is uniform in x. Since $\mathcal{X}$ is compact, u is actually bounded away from zero, and so

$$\frac{T_h u}{u} = 1 + h\frac{Lu}{u} + o(h)\,,$$

$$\log\left(\frac{T_h u}{u}\right) = \log\left(1 + h\frac{Lu}{u} + o(h)\right)$$

$$= h\frac{Lu}{u} + o(h)\,,$$

from which we obtain

$$\frac{1}{h}\log\left(\frac{T_h u}{u}\right) = \frac{Lu}{u} + o(1)\,,$$

where $o(1)$ is uniform in x. Thus, from (3.4),

$$\varliminf_{h\to 0}\frac{1}{h} I_h(\mu) \geqq -\int_{\mathcal{X}} \left(\frac{Lu}{u}\right)(x)\mu(dx)\,, \tag{3.5}$$

and since (3.5) is true for every positive element of $\mathcal{D}$, we have

$$\varliminf_{h\to 0}\frac{1}{h} I_h(\mu) \geqq -\inf_{\substack{u>0 \\ u\in\mathcal{D}}}\int_{\mathcal{X}} \left(\frac{Lu}{u}\right)(x)\mu(dx) = I(\mu)\,.$$

But (3.2) implies

$$\varlimsup_{h\to 0}\frac{1}{h} I_h(\mu) \leqq I(\mu)\,,$$

and hence we have (3.3).

LEMMA 3.2. *Let (Ω, Σ, P) be any probability space and let ξ and η be two maps of Ω into $\mathfrak{X}$. If we define $\alpha = P\xi^{-1}$, $\beta = P\eta^{-1}$ (so that α and β are in $\mathcal{M}$), then for any $\delta > 0$*

$$P\{\omega : \rho(\xi, \eta) \geqq \delta\} \leqq \delta \quad \textit{implies} \quad d(\alpha, \beta) \leqq \delta .$$

Proof: Let C be any closed set in $\mathfrak{X}$. By the definition of C^δ and the assumption that $P\{\omega : \rho(\xi, \eta) \geqq \delta\} \leqq \delta$, we have

$$\begin{aligned} \alpha(C) &= P\{\omega : \xi(\omega) \in C\} \\ &\leqq P\{\omega : \eta(\omega) \in C^\delta\} + P\{\omega : \rho(\xi, \eta) \geqq \delta\} \\ &\leqq \beta(C^\delta) + \delta . \end{aligned}$$

By a symmetric argument, $\beta(C) \leqq \alpha(C^\delta) + \delta$. Since these two inequalities hold for any closed set C, the definition of the Prokhorov metric implies that $d(\alpha, \beta) \leqq \delta$, which is the lemma.

For any $h > 0$, let $X_j^{(h)} = x(jh)$, $j = 0, 1, 2, \cdots$, so that $X_0^{(h)}, X_1^{(h)}, \cdots$ is a discrete parameter Markov process whose transition probability function is $\pi^{(h)}(x, dy) = p(h, x, dy)$. Note that, for a positive element u of $C(\mathfrak{X})$,

$$T_h u = \pi^{(h)} u = \int_{\mathfrak{X}} u(y) p(h, x, y) \lambda(dy) ,$$

and therefore the function $I_h(\mu)$ defined in (3.1) for any $\mu \in \mathcal{M}$ is just the $I(\mu)$ of Section 2 (as defined by (1.5)) for the discrete parameter Markov process $X_0^{(h)}, X_1^{(h)}, \cdots$.

For any positive integer n and any Borel set $A \subset \mathfrak{X}$, let

$$L_n^{(h)}(\omega, A) = \frac{1}{n} \sum_{j=0}^{n-1} \chi_A(X_j^{(h)}) .$$

Now, let $t > 0$ be given and write $t = nh$, where n is a positive integer and $h > 0$. We define the process $\{x^h(s), 0 \leqq s \leqq t\}$ as follows:

$$x^h(s) = \begin{cases} X_j^{(h)} & \text{if} \quad jh \leqq s < (j+1)h , \\ X_{n-1}^{(h)} & \text{if} \quad s = t = nh . \end{cases} \qquad j = 0, 1, 2, \cdots, n-1 ,$$

LEMMA 3.3. *Let $t > 0$ be given and write $t = nh$, where n is a positive integer and $h > 0$. For any $\delta > 0$ and any $\omega \in \Omega_x$, let*

$$N_t^\delta(\omega) = \left\{ \#j\text{'s } \textit{such that} \sup_{jh \leqq s \leqq (j+1)h} \rho(X_j^{(h)}, x(s)) > \delta, \ 0 \leqq j \leqq n-1 \right\} .$$

Then

$$P_x\{d(L_t(\omega,\cdot), L_n^{(h)}(\omega,\cdot))>\delta\}\leqq P_x\left\{\frac{N_t^\delta(\omega)}{n}>\delta\right\}. \tag{3.6}$$

Proof: Both $\{x(s), 0\leqq s\leqq t\}$ and $\{x^h(s), 0\leqq s\leqq t\}$ map $[0,t]$ into X. On $[0,t]$ impose the measure $\nu=ds/t$ so that

$$L_t(\omega,\cdot)=\nu[X(\cdot)]^{-1},$$
$$L_n^{(h)}(\omega,\cdot)=\nu[X^h(\cdot)]^{-1}.$$

Then, by Lemma 3.2, for any $\omega\in\Omega_x$, $\nu\{s:\rho(x(s),x^h(s))\geqq\delta\}\leqq\delta$ implies $d(L_t(\omega,\cdot), L_n^{(h)}(\omega,\cdot))\leqq\delta$.

But, for every $\omega\in\Omega_x$,

$$\nu\{s:\rho(x(s),x^h(s))\geqq\delta\}\leqq\frac{h}{t}N_t^\delta(\omega)=\frac{1}{n}N_t^\delta(\omega),$$

and hence

$$P_x\{d(L_t(\omega,\cdot), L_n^{(h)}(\omega,\cdot))>\delta\}\leqq P_x\{\nu\{s:\rho(x(s),x^h(s))\geqq\delta\}>\delta\}\leqq P_x\left\{\frac{N_t^\delta(\omega)}{n}>\delta\right\}.$$

LEMMA 3.4. *For every* $\varepsilon>0$,

$$\overline{\lim_{h\to 0}}\,\overline{\lim_{n\to\infty}}\,\frac{1}{nh}\log P_x\{d(L_{nh}(\omega,\cdot), L_n^{(h)}(\omega,\cdot))>\varepsilon\}=-\infty. \tag{3.7}$$

Proof: By Lemma 3.3 it suffices to show that

$$\overline{\lim_{h\to 0}}\,\overline{\lim_{n\to\infty}}\,\frac{1}{nh}\log P_x\left\{\frac{N_{nh}^\varepsilon(\omega)}{n}>\varepsilon\right\}=-\infty \tag{3.8}$$

for every $\varepsilon>0$. Let, for $h>0$ and $\varepsilon>0$,

$$\phi^\varepsilon(h)=\sup_{x\in X}P_x\left\{\sup_{0\leqq s\leqq h}\rho(x(s),x)\geqq\varepsilon\right\}.$$

Our hypotheses on the process $\{X(s), 0\leqq s<\infty\}$ imply that, for every $\varepsilon>0$, $\phi^\varepsilon(h)\to 0$ as $h\to 0$.

For $j=0,1,2,\cdots,n-1$, let

$$\chi_j^\varepsilon(\omega)=\begin{cases}1 & \text{if } \sup_{jh\leqq s<(j+1)h}\rho(x(s),x_j^{(h)})\geqq\varepsilon,\\ 0 & \text{otherwise}.\end{cases}$$

For $\theta>0$, consider

$$E^{P_x}\{\exp\{\theta N^{\varepsilon}_{nh}(\omega)\}\}=E^{P_x}\left\{\exp\left\{\theta\sum_{j=0}^{n-1}\chi^{\varepsilon}_j(\omega)\right\}\right\}.$$

Since the process is Markovian,

$$\begin{aligned}E^{P_x}\left\{\exp\left\{\theta\sum_{j=0}^{n-1}\chi^{\varepsilon}_j(\omega)\right\}\right\}&=E^{P_x}\left\{\exp\left\{\theta\sum_{j=0}^{n-2}\chi^{\varepsilon}_j(\omega)\right\}[E^{P_{x((n-2)h)}}\{\exp\{\theta\chi^{\varepsilon}_{n-1}(\omega)\}\}]\right\}\\ &\leqq E^{P_x}\left\{\exp\left\{\theta\sum_{j=0}^{n-2}\chi^{\varepsilon}_j(\omega)\right\}\sup_{y\in\mathcal{X}}E^{P_y}\{\exp\{\theta\chi^{\varepsilon}_{n-1}(\omega)\}\}\right\}\\ &\leqq[e^{\theta}\phi^{\varepsilon}(h)+(1-\phi^{\varepsilon}(h))]E^{P_x}\left\{\exp\left\{\theta\sum_{j=0}^{n-2}\chi^{\varepsilon}_j(\omega)\right\}\right\}\\ &\leqq[e^{\theta}\phi^{\varepsilon}(h)+(1-\phi^{\varepsilon}(h))]^n.\end{aligned}$$

Thus,

$$E^{P_x}\{\exp\{\theta N^{\varepsilon}_{nh}(\omega)\}\}\leqq[e^{\theta}\phi^{\varepsilon}(h)+(1-\phi^{\varepsilon}(h))]^n.$$

On the other hand,

$$\begin{aligned}E^{P_x}\{e^{\theta N^{\varepsilon}_{nh}(\omega)}\}&\geqq E^{P_x}_{N^{\varepsilon}_{nh}(\omega)>\varepsilon n}\{e^{\theta N^{\varepsilon}_{nh}(\omega)}\}\\ &\geqq e^{n\theta\varepsilon}P_x\left\{\frac{N^{\varepsilon}_{nh}(\omega)}{n}>\varepsilon\right\},\end{aligned}$$

so that

$$P_x\left\{\frac{N^{\varepsilon}_{nh}(\omega)}{n}>\varepsilon\right\}\leqq e^{-n\varepsilon\theta}[e^{\theta}\phi^{\varepsilon}(h)+(1-\phi^{\varepsilon}(h))]^n.$$

Therefore,

$$\overline{\lim_{n\to\infty}}\frac{1}{nh}\log P_x\left\{\frac{N^{\varepsilon}_{nh}(\omega)}{n}>\varepsilon\right\}\leqq-\frac{\varepsilon\theta}{h}+\frac{1}{h}\log[e^{\theta}\phi^{\varepsilon}(h)+(1-\phi^{\varepsilon}(h))].\tag{3.9}$$

As noted earlier, $\phi^{\varepsilon}(h)\to 0$ as $h\to 0$, and hence the right side of (3.9) behaves as follows as $h\to 0$:

$$\frac{1}{h}\{\log[e^{\theta}\phi^{\varepsilon}(h)+(1-\phi^{\varepsilon}(h))]-\varepsilon\theta\}\sim-\frac{\varepsilon\theta}{h}.$$

This gives us (3.8) and the lemma. We shall find it convenient to use the obvious

COROLLARY. *Assume* $\mu \in \mathcal{M}$ *is such that* $I(\mu)<\infty$. *There exists an* h_0 *such that for given* $\varepsilon>0$

$$\overline{\lim_{n\to\infty}} \frac{1}{nh_0} \log P_x\{d(L_{nh_0}(\omega, \cdot), L_n^{(h_0)}(\omega, \cdot))>\varepsilon\}<-I(\mu)-1 . \tag{3.10}$$

Now we are ready to prove (1.14). Let G be an open set in $\mathcal{M}$. Let $\mu \in G$ be such that $I(\mu)<\infty$, and let $S(\mu, \varepsilon)$ be a sphere with center at μ and radius ε such that $S(\mu, \varepsilon)\subset G$. It will suffice to show

$$\underline{\lim_{t\to\infty}} \frac{1}{t} \log Q_{x,t}[S(\mu, \varepsilon)]=\underline{\lim_{t\to\infty}} \frac{1}{t} \log P_x\{L_t(\omega, \cdot)\in S(\mu, \varepsilon)\}\geqq -I(\mu) . \tag{3.11}$$

Let h_0 be as in the corollary to Lemma 3.4 so that (3.10) holds. For any $\varepsilon>0$, it is clear that

$$\begin{aligned} &P_x\{L_{nh_0}(\omega, \cdot)\in S(\mu, \tfrac{1}{2}\varepsilon)\} \\ &\qquad \geqq P_x\{L_n^{(h_0)}(\omega, \cdot)\in S(\mu, \tfrac{1}{4}\varepsilon)\}-P_x\{d(L_{nh_0}(\omega, \cdot), L_n^{(h_0)}(\omega, \cdot))\geqq \tfrac{1}{4}\varepsilon\} \end{aligned} \tag{3.12}$$

for all positive integers n. By (2.39) and Lemma 3.1,

$$\underline{\lim_{n\to\infty}} \frac{1}{nh_0} \log P_x\{L_n^{(h_0)}(\omega, \cdot)\in S(\mu, \tfrac{1}{4}\varepsilon)\}\geqq -\frac{1}{h_0} I_{h_0}(\mu)\geqq -I(\mu) . \tag{3.13}$$

From (3.13) and (3.10) we see that (3.12) implies that, for any $\varepsilon>0$,

$$\underline{\lim_{n\to\infty}} \frac{1}{nh_0} \log P_x\{L_{nh_0}(\omega, \cdot)\in S(\mu, \tfrac{1}{2}\varepsilon)\}\geqq -I(\mu) . \tag{3.14}$$

Now, let $n=[t/h_0]$. For any Borel set $A\subset X$ and any $\omega\in\Omega_x$,

$$\begin{aligned} |L_t(\omega, A)-L_{nh_0}(\omega, A)| &= \left|\frac{1}{t}\int_0^t \chi_A(x(s))\,ds-\frac{1}{nh_0}\int_0^{nh_0}\chi_A(x(s))\,ds\right| \\ &= \left|\frac{1}{nh_0}\int_0^t \chi_A(x(s))\,ds+\left(\frac{1}{t}-\frac{1}{nh_0}\right)\int_0^t \chi_A(x(s))\,ds\right. \\ &\qquad \left. -\frac{1}{nh_0}\int_0^{nh_0}\chi_A(x(s))\,ds\right| \\ &= \left|\frac{1}{nh_0}\int_{nh_0}^t \chi_A(x(s))\,ds+\frac{nh_0-t}{nh_0 t}\int_0^t \chi_A(x(s))\,ds\right| \\ &\leqq \frac{2(t-nh_0)}{nh_0}\leqq\frac{2}{n}=\frac{2}{[t/h_0]} . \end{aligned}$$

Thus, there exists a t_0 such that if $t > t_0$ we have, for all $\omega \in \Omega_x$,

$$d(L_t(\omega, \cdot), L_{nh_0}(\omega, \cdot)) < \tfrac{1}{2}\varepsilon .$$

Hence since, for any $\varepsilon > 0$,

$$P_x\{L_t(\omega, \cdot) \in S(\mu, \varepsilon)\} \geqq P_x\{L_{nh_0}(\omega, \cdot) \in S(\mu, \tfrac{1}{2}\varepsilon)\} - P_x\{d(L_t(\omega, \cdot), L_{nh_0}(\omega, \cdot)) \geqq \tfrac{1}{2}\varepsilon\},$$

we see that if $t > t_0$ and $n = [t/h_0]$

$$P_x\{L_t(\omega, \cdot) \in S(\mu, \varepsilon)\} \geqq P_x\{L_{nh_0}(\omega, \cdot) \in S(\mu, \tfrac{1}{2}\varepsilon)\}$$

or

$$\frac{1}{t} \log P_x\{L_t(\omega, \cdot) \in S(\mu, \varepsilon)\} \geqq \frac{nh_0}{t} \cdot \frac{1}{nh_0} \log P_x\{L_{nh_0}(\omega, \cdot) \in S(\mu, \tfrac{1}{2}\varepsilon)\} .$$

Therefore, from (3.14) and the fact that $nh_0/t \to 1$ for $n = [t/h_0]$ as $t \to \infty$, we have for all $\varepsilon > 0$

$$\varliminf_{t \to \infty} \frac{1}{t} \log P_x\{L_t(\omega, \cdot) \in S(\mu, \varepsilon)\} \geqq -I(\mu)$$

which is (3.11) and hence (1.14).

Now we prove (1.13). For $u > 0$ and $u \in \mathcal{D}$, let

$$\psi(x, t) = E^{P_x}\left\{u(x(t)) \exp\left\{-\int_0^t (Lu/u)(x(s))\, ds\right\}\right.$$

for $x \in \mathbf{X}$ and $t > 0$. From the Feynmann-Kac formula we know that $\psi(x, t)$ satisfies

$$\frac{\partial \psi}{\partial t} = L\psi - \left(\frac{Lu}{u}\right)(x)\psi ,$$

$$\psi(x, 0) = u(x) ,$$

but since the solution of this last is clearly $\psi(x, t) = u(x)$, we have

$$E^{P_x}\left\{u(x(t)) \exp\left\{-\int_0^t (Lu/u)(x(s))\, ds\right\}\right\} = u(x) . \tag{3.15}$$

Since $u > 0$ and $\mathbf{X}$ is compact, there is a constant α such that $u(y) \geqq \alpha > 0$ for all $y \in \mathbf{X}$. Thus

$$E^{P_x}\left\{\exp\left\{-\int_0^t (Lu/u)(x(s))\, ds\right\}\right\} \leqq \frac{u(x)}{\alpha} . \tag{3.16}$$

From the definition of $L_t(\omega, \cdot)$ and of $Q_{x,t}$-measure on $\mathcal{M}$,

$$E^{P_x}\left\{\exp\left\{-\int_0^t (Lu/u)(x(s))\,ds\right\}\right\}=E^{P_x}\left\{\exp\left\{-t\int_{\mathcal{X}} (Lu/u)(y)L_t(\omega, dy)\right\}\right\}$$
$$=E^{Q_{x,t}}\left\{\exp\left\{-t\int_{\mathcal{X}} (Lu/u)(y)\mu(dy)\right\}\right\}.$$

Thus, (3.16) implies that, for any measurable set $C\subset\mathcal{M}$,

$$Q_{x,t}(C)\leqq\frac{u(x)}{\alpha}\exp\left\{t\sup_{\mu\in C}\int_{\mathcal{X}} (Lu/u)(y)\mu(dy)\right\},$$

and hence

$$\overline{\lim_{t\to\infty}}\frac{1}{t}\log Q_{x,t}(C)\leqq\sup_{\mu\in C}\int_{\mathcal{X}}\left(\frac{Lu}{u}\right)(y)\mu(dy)\,. \tag{3.17}$$

Since (3.17) holds for any positive $u\in\mathcal{D}$,

$$\overline{\lim_{t\to\infty}}\frac{1}{t}\log Q_{x,t}(C)\leqq\inf_{\substack{u>0\\ u\in\mathcal{D}}}\sup_{\mu\in C}\int_{\mathcal{X}}\left(\frac{Lu}{u}\right)(y)\mu(dy)\,. \tag{3.18}$$

If $C\subset\bigcup_{j=1}^{k} C_j$, then this last implies that

$$\overline{\lim_{t\to\infty}}\frac{1}{t}\log Q_{x,t}(C)\leqq\sup_{1\leqq j\leqq k}\inf_{\substack{u>0\\ u\in\mathcal{D}}}\sup_{\mu\in C_j}\int_{\mathcal{X}}\left(\frac{Lu}{u}\right)(y)\mu(dy)\,.$$

Taking the infimum over all such finite unions which cover C, we have

$$\overline{\lim_{t\to\infty}}\frac{1}{t}\log Q_{x,t}(C)\leqq\inf_{\substack{C_1,C_2\cdots C_k\\ C\subset\bigcup_{j=1}^{k} C_j}}\sup_{1\leqq j\leqq k}\inf_{\substack{u>0\\ u\in\mathcal{D}}}\sup_{\mu\in C_j}\int_{\mathcal{X}}\left(\frac{Lu}{u}\right)(y)\mu(dy)\,. \tag{3.19}$$

Now by an elementary argument (see footnote 1) one can show that if C is compact, then (3.19) implies that

$$\overline{\lim_{t\to\infty}}\frac{1}{t}\log Q_{x,t}(C)\leqq\sup_{\mu\in C}\inf_{\substack{u>0\\ u\in\mathcal{D}}}\int_{\mathcal{X}}\left(\frac{Lu}{u}\right)(y)\mu(dy)$$
$$=-\inf_{\mu\in C} I(\mu)$$

which is (1.13). The hypothesis of (1.13) is that C is a closed set in $\mathcal{M}$ but $\mathcal{M}$ is compact and therefore C is compact. This completes the proof of Theorem 3.

4. The *I*-Function for Selfadjoint Generators

Let $\pi(x, dy)$ be the transition probability of a discrete parameter Markov process on a state space X. For any probability measure μ on X we define, as in (1.5),

$$I(\mu) = -\inf_{u \in \mathcal{U}} \int_X \log\left(\frac{\pi u}{u}\right)(x)\mu(dx)\,, \tag{4.1}$$

where $\mathcal{U}$ consists of all positive measurable functions which have an upper bound and a positive lower bound. In (4.1), $\pi u(x)$ stands, as before, for

$$(\pi u)(x) = \int_X u(y)\pi(x, dy)\,.$$

Let $\mathcal{M}$ be the space of all probability measures on X and consider the norm topology on $\mathcal{M}$. Thus, if μ_1 and μ_2 are two measures in $\mathcal{M}$, $\|\mu_1 - \mu_2\| = 2\sup_A |\mu_1(A) - \mu_2(A)|$. By $\mu\pi(A)$ we mean the measure

$$(\mu\pi)(A) = \int_X \pi(x, A)\mu(dx)\,.$$

Now in Lemma 2.5 we proved that a necessary and sufficient condition for $I(\mu) = 0$ is that $\mu\pi(A) = \mu(A)$ for every Borel set A, i.e., that μ is the invariant measure for $\pi(x, dy)$. The following lemma sharpens the result that if $I(\mu) = 0$, then $\mu\pi = \mu$, by giving an estimate on $\|\mu\pi - \mu\|$ in terms of a function $\phi(I(\mu))$ (which indeed vanishes when its argument does). Note that the function ϕ in the lemma is universal—same for all π and μ.

LEMMA 4.1. *There exists a function $\phi(l)$, such that $0 \leqq \phi(l) \leqq 2$ for $l \geqq 0$ and $\phi(l) \to 0$ as $l \to 0$, having the property that if $\pi(x, dy)$ is any transition probability and μ is any probability measure on X with the corresponding $I(\mu) = l$, then*

$$\|\mu\pi - \mu\| \leqq \phi(l)\,.$$

Proof: Since $I(\mu) = l$, we have

$$\int_X \log\left(\frac{\pi u}{u}\right)(x)\mu(dx) \geqq -l \tag{4.2}$$

for every bounded measurable u which is bounded below by a positive number. Let A be a Borel set contained in X and consider the function $u=a+1$ on A and $u=1$ on $\tilde{A}$, where $a>0$. From (4.2) we have, for all $a>0$,

$$\int_X \log\left[\frac{a\pi(x, A)+1}{a\chi_A(x)+1}\right]\mu(dx)\geqq -l$$

or

$$\int_X \log(1+a\pi(x, A))\mu(dx)\geqq -l+\mu(A)\log(1+a)\,. \tag{4.3}$$

Since $\log x$ is a concave function of x,

$$\log\int_X (1+a\pi(x, A))\mu(dx)\geqq\int_X \log(1+a\pi(x, A))\mu(dx)\,. \tag{4.4}$$

Combining (4.3) and (4.4), we get

$$\log[1+a\mu\pi(A)]\geqq -l+\mu(A)\log(1+a)\,.$$

Since $\log(1+\xi)\leqq\xi$ for all ξ, we obtain

$$a\mu\pi(A)\geqq -l+\mu(A)\log(1+a)\,; \tag{4.5}$$

this inequality holds for all Borel sets A and all $a>0$. Subtracting $\mu(A)$ on both sides of (4.5) after dividing by a, we get

$$\mu\pi(A)-\mu(A)\geqq\frac{-l+\mu(A)[\log(1+a)-a]}{a}\,.$$

Since $\mu(A)\leqq 1$ for all A, again using the fact that $\log(1+a)\leqq a$ for all $a>0$, we have

$$\mu\pi(A)-\mu(A)\geqq -\frac{l+a-\log(1+a)}{a}\,. \tag{4.6}$$

Since (4.6) is true for every Borel set A, and since $\mu\pi(\tilde{A})-\mu(\tilde{A})=-[\mu\pi(A)-\mu(A)]$, we obtain

$$\sup_A|\mu\pi(A)-\mu(A)|\leqq\frac{l+a-\log(1+a)}{a}\,.$$

Finally, since $\|\mu\pi-\mu\|=2\sup_A|\mu\pi(A)-\mu(A)|$, we get

$$\|\mu\pi-\mu\|\leqq 2\left[\frac{l+a-\log(1+a)}{a}\right],$$

which implies, since $a>0$ is arbitrary, that

$$\|\pi\mu-\mu\|\leqq 2\inf_{a>0}\left[\frac{l+a-\log(1+a)}{a}\right].$$

Thus, if we define

$$\phi(l)=2\inf_{a>0}\left[\frac{l+a-\log(1+a)}{a}\right],$$

then

$$\|\mu\pi-\mu\|\leqq\phi(l)$$

and $\phi(l)$ can easily be seen to have the properties stated in the lemma.

Now suppose $p(t, x, dy)$ is the transition probability of a continuous parameter Markov process whose state space is X. Then we can define, for each $h>0$, $\pi^{(h)}(x, dy)$ to be $p(h, x, dy)$ and use it to define $I_h(\mu)$ with $\pi^{(h)}$ playing the role of π in (4.1).

We now prove

LEMMA 4.2. *Let B_0 be the strongly continuous center of the semigroup T_t associated with the given Markov process acting on the space $B(X)$ of bounded measurable functions on X. We assume that for any measure μ on X and any $u\in B(X)$ there exists a sequence $\{u_n\}\in B_0$ which converges almost everywhere (μ) to u (assumption 3 in the introduction just preceding Theorem 5). Let L be the infinitesimal generator of the semigroup T_t having domain $\mathfrak{D}$ (which is dense in B_0), and let $\mathfrak{D}^+$ be the subset of $\mathfrak{D}$ consisting of those functions which have a positive lower bound. Let*

$$I(\mu)=-\inf_{u\in\mathfrak{D}^+}\int_X\left(\frac{Lu}{u}\right)(x)\mu(dx).$$

Then, for any $h>0$,

(4.7) $$I_h(\mu)\leqq hI(\mu).$$

Proof: As in Lemma 3.1 we can obtain

$$\int_X\log\pi^{(h)}u\mu(dx)-\int_X\log u\mu(dx)\geqq -hI(\mu)$$

for every u in B_0^+, where $B_0^+ \subset B_0$ consists of those functions in B_0 which have a positive lower bound.

Consider the measure $\mu + \pi^{(h)}\mu$, where

$$(\pi^{(h)}\mu)(A) = \int_X \pi^{(h)}(x, A)\mu(dx) .$$

Let $u \in B^+$, where $B^+ \subset B$ consists of those bounded measurable functions which have a positive lower bound. By our hypothesis there exists a sequence $\{u_n\} \in B_0$ which converges almost everywhere $(\mu + \pi^{(h)}\mu)$ to u. Without loss of generality we can assume that $\{u_n\} \in B_0^+$ and that $\{u_n\}$ is uniformly bounded above and below by the bounds on u. Let E be the exceptional set where $u_n \not\to u$ so that $(\mu + \pi^{(h)}\mu)(E) = 0$, i.e., $\mu(E) = 0$ and $(\pi^{(h)}\mu)(E) = 0$. Consider

$$(\pi^{(h)}u_n)(x) = \int_X u_n(y)\pi^{(h)}(x, dy) ,$$

and let $F = \{x : \pi^{(h)}(x, E) = 0\}$. Then, for $x \in F$, $\pi^{(h)}u_n \to \pi^{(h)}u$. On the other hand, $0 = (\pi^{(h)}\mu)(E) = \int_X \pi^{(h)}(x, E)\mu(dx)$ and therefore $\mu(F) = 1$, i.e., $\pi^{(h)}u_n \to \pi^{(h)}u$ almost everywhere (μ). Thus we can let $n \to \infty$ in

$$\int_X \log \pi^{(h)}u_n\mu(dx) - \int_X \log u_n\mu(dx) \geqq -hI(\mu) ,$$

and get for every $u \in B^+$

$$\int_X \log \pi^{(h)}u\mu(dx) - \int_X \log u\mu(dx) \geqq -hI(\mu) ,$$

which implies the lemma.

Lemmas 4.1 and 4.2 give us the

COROLLARY. *Let T_t be any Markov semigroup on any state space X satisfying the assumption of Lemma 4.2. If $I(\mu) = l < \infty$, then*

$$\|\mu T_t - \mu\| \leqq \phi(tl)$$

for the ϕ defined in Lemma 4.1. Hence

$$\|\mu T_t - \mu\| \to 0 \quad as \quad t \to 0 .$$

We now proceed to the proof of Theorem 5 in the introduction. The notation, definitions, and hypotheses are stated in the introduction just before the statement of that theorem.

Suppose, first of all, that $I(\mu)=l<\infty$. Then, from the corollary to Lemma 4.2, we know that $\|\mu T_t-\mu\|\to 0$ as $t\to 0$. Since $\mu T_t \ll \lambda$ for $t>0$, by assumption 1 on the process, we conclude that $\mu\ll\lambda$.

Let B^+ denote those elements u in $B(x)$ which have positive lower bounds. For $h>0$, let

$$I_h(\mu)=-\inf_{u\in B^+}\int_X \log\left(\frac{\pi^{(h)}u}{u}\right)(x)\mu(dx)\,.$$

Then, from Lemma 4.2, $I_h(\mu)\leqq hI(\mu)$. Hence, for any $u\in B^+$,

$$\int_X \log\left(\frac{\pi^{(h)}u}{u}\right)(x)\mu(dx)\geqq -hI(\mu)\,. \tag{4.8}$$

But

$$\log\left(\frac{\pi^{(h)}u}{u}\right)=\log\left(1-\frac{u-\pi^{(h)}u}{u}\right)$$
$$=\log\left(1-\frac{(I-\pi^{(h)})u}{u}\right)$$

and, since $u>0$ implies $(I-\pi^{(h)})u/u\leqq 1$, and $\log(1-x)\leqq -x$ for $-\infty<x\leqq 1$, we have from (4.8) that, for any $u\in B^+$,

$$-\int_X \left[\frac{(I-\pi^{(h)})u}{u}\right](x)\mu(dx)\geqq -hI(\mu)\,. \tag{4.9}$$

With $f=d\mu/d\lambda$ we get from (4.9), for any $u\in B^+$,

$$\int_X \left[\frac{(I-\pi^{(h)})u}{u}\right](x)f(x)\lambda(dx)\leqq hI(\mu)\,. \tag{4.10}$$

Now, let $\{u_n\}$ be a sequence such that $0\leqq u_n\leqq\sqrt{f}$ and $u_n\to\sqrt{f}$ with u_n bounded (e.g. we could take $u_n=\sqrt{f}\wedge n$). Let $\varepsilon>0$. Since $u_n+\varepsilon\in B^+$, we have from (4.10) that for all n

$$\int_X \left[\frac{(I-\pi^{(h)})u_n}{\varepsilon+u_n}\right](x)f(x)\lambda(dx)\leqq hI(\mu)\,.$$

Letting $n\to\infty$, this last yields

$$\int_X (I-\pi^{(h)})\sqrt{f(x)}\left(\frac{f(x)}{\varepsilon+\sqrt{f(x)}}\right)\lambda(dx)\leqq hI(\mu),$$

and then, letting $\varepsilon\to 0$, we get

$$\int_X \sqrt{f(x)}(I-\pi^{(h)})\sqrt{f(x)}\lambda(dx)\leqq hI(\mu), \tag{4.11}$$

i.e., if $E(\Lambda)$ is the spectral measure corresponding to $\sqrt{f}$ relative to the operator $-\tilde{L}$, then

$$\int_0^\infty (1-e^{-h\Lambda})E(d\Lambda)\leqq hI(\mu). \tag{4.12}$$

If in (4.12) we divide by h and then let $h\to 0$, we get, using Fatou's lemma,

$$\int_0^\infty \Lambda E(d\Lambda)\leqq I(\mu).$$

This proves that $g=\sqrt{f}\in\tilde{\mathscr{D}}_{1/2}$ and

$$\|\sqrt{-\tilde{L}}\,g\|^2\leqq I(\mu).$$

To prove the converse assume $g=\sqrt{f}\in\tilde{\mathscr{D}}_{1/2}$ where $f=d\mu/d\lambda$, and that $\|\sqrt{-\tilde{L}}\sqrt{f}\|^2=l$. We wish to show $I(\mu)\leqq l$.

Consider the semigroup

$$S_t^\phi f=E_x\left\{f(x(t))\exp\left\{-\int_0^t (L\phi/\phi)(x(s))\,ds\right\}\right\},$$

where $\phi\in\mathscr{D}^+$, and $E_x\{\ \}$ denotes expectation on the process $\{x(s), 0\leqq s<\infty\}$ (with $x(0)=x$) corresponding to our given transition probability $p(t,x,dy)$. Now $V=L\phi/\phi$ is bounded and S_t^ϕ is a selfadjoint semigroup in $L_2(\lambda)$ with a generator $\tilde{L}_\phi$ given by

$$\tilde{L}_\phi u=\tilde{L}u-Vu.$$

Since

$$\sup_t\sup_x E_x\left\{\exp\left\{-\int_0^t (L\phi/\phi)(x(s))\,ds\right\}\right\}\leqq\frac{\sup\phi}{\inf\phi}=C<\infty,$$

and S_t^ϕ is selfadjoint, it follows from Schwartz' inequality that $\|S_t^\phi f\|_2 \leqq C\,\|f\|_2$, and hence $\|S_t^\phi\| \leqq C$. Since S_t^ϕ is selfadjoint, $\|S_t^\phi\| \leqq 1$ so that it is a contraction. Thus, $\tilde{L}_\phi$ is negative semidefinite, i.e., $\langle \tilde{L}u - Vu, u\rangle \leqq 0$. In other words, $(\tilde{L}u, u) \leqq \langle Vu, u\rangle$ or $(Vu, u) \geqq (\tilde{L}u, u)$ for $u \in \tilde{\mathscr{D}}$. Thus $(Vu, u) \geqq -(-\tilde{L}u, u) = -\|\sqrt{-\tilde{L}}\, u\|^2$ for $u \in \tilde{\mathscr{D}}_{1/2}$. In particular, let $u = \sqrt{f}$ so that $(V\sqrt{f}, \sqrt{f}) \geqq -l$, i.e.,

$$\int_X \left(\frac{L\phi}{\phi}\right)(x) f(x) \lambda(dx) = \int_X \left(\frac{L\phi}{\phi}\right)(x) \mu(dx) \geqq -l$$

for all $\phi \in \mathscr{D}^+$, which implies $I(\mu) \leqq l$; hence we have completed the proof of Theorem 5.

Bibliography

[1] Donsker, M. D., and Varadhan, S. R. S., *Asymptotic evaluation of certain Wiener integrals for large time*, to appear in Proceedings of International Conference on Function Space Integration, Oxford, 1974.

[2] Kac, M., *On some connections between probability theory and differential and integral equations*, Proceedings of the Second Berkeley Symposium, 1950, pp. 189–215.

[3] Varadhan, S. R. S., *Asymptotic probabilities and differential equations*, Comm. Pure Appl. Math., Vol. XIX, No. 3, 1966, pp. 261–286.

Received September, 1974.

COMMUNICATIONS ON PURE AND APPLIED MATHEMATICS, VOL. XXVIII, 279–301 (1975)

Asymptotic Evaluation of Certain Markov Process Expectations for Large Time, II*

M. D. DONSKER AND S. R. S. VARADHAN

1. Introduction

Let $\mathcal{X}$ be a compact metric space. Let $p(t, x, dy)$ be the transition probability function of a homogeneous Markov Process having $\mathcal{X}$ as its state space. Denote by T_t the corresponding semigroup which we shall assume to be a strongly continuous Feller semigroup on $C(\mathcal{X})$. We denote by L the infinitesimal generator of the semigroup and by $\mathcal{D}$ the domain of L. Without loss of generality we can assume that the trajectories of the process are right-continuous and have left limits at every point of time $0 \leqq t < \infty$. For each starting point $x \in \mathcal{X}$ we get a measure P_x on the space Ω_x consisting of those trajectories $\omega = x(\cdot)$ satisfying $x(0) = x$.

Let $\mathcal{M}$ be the space of all probability measures on $\mathcal{X}$. We shall regard $\mathcal{M}$ as a compact metric space endowed with the weak topology. For each $t > 0$ there is a canonical map $\Omega_x \to \mathcal{M}$ provided by the occupation distribution. For each $\omega \in \Omega_x$ and $t > 0$, we define the occupation distribution $L_t(\omega, A)$ by

$$L_t(\omega, A) = \frac{1}{t} \text{ Lebesgue measure } \{\sigma : x(\sigma) \in A, 0 \leqq \sigma \leqq t\},$$

where $A \subset \mathcal{X}$ is a Borel subset of $\mathcal{X}$. Clearly $L_t(\omega, A)$ is the proportion of time up to t that a particular trajectory $\omega = x(\cdot)$ spends in the set A, and $L_t(\omega, \cdot) \in \mathcal{M}$. It is easy to see that, for each $\omega \in \Omega_x$ and $t > 0$,

$$\int_{\mathcal{X}} V(y) L_t(\omega, dy) = \frac{1}{t} \int_0^t V(x(\sigma))\, d\sigma ,$$

where V is any bounded continuous function on $\mathcal{X}$.

If $\mathcal{X}$ is one-dimensional and $p(t, x, dy)$ is a suitable diffusion, then $L_t(\omega, dy)$ is almost surely absolutely continuous with respect to the Lebesgue measure on $\mathcal{X}$, and the density is a multiple of the local time.

* Results obtained at the Courant Institute of Mathematical Sciences, New York University, with the Office of Naval Research, Contract N00014-67-A-0467-0031 and the National Science Foundation, Grant NSF-GP-41985. Reproduction in whole or in part is permitted for any purpose of the United States Government.

For each $x \in \mathcal{X}$ and $t>0$, let $Q_{x,t}$ be the measure on $\mathcal{M}$ induced by the map L_t from $\Omega_x \to \mathcal{M}$, i.e., $Q_{x,t}=P_x L_t^{-1}$. Thus, if B is a set of measures in $\mathcal{M}$,

$$Q_{x,t}(B)=P_x\{\omega : L_t(\omega, \cdot)\in B\}.$$

Let $\mathcal{U}$ be those elements of the domain $\mathcal{D}$ which are strictly positive. Since every element $u \in \mathcal{U}$ is continuous and $\mathcal{X}$ is compact, u satisfies bounds of the form $0<a\leqq u(x)\leqq b<\infty$.

For each $\mu \in \mathcal{M}$, we define

$$I(\mu)=-\inf_{u\in\mathcal{U}}\int_{\mathcal{X}}\left(\frac{Lu}{u}\right)(x)\mu(dx). \tag{1.1}$$

It is easy to see that $I(\mu)$ is a convex, lower semicontinuous, non-negative functional on $\mathcal{M}$.

We shall now assume that the transition probability $p(t, x, dy)$ satisfies the following:

ASSUMPTION A. There is a finite reference measure λ on $\mathcal{X}$ such that $p(t, x, dy)$ is absolutely continuous with respect to λ for each $x \in \mathcal{X}$ and $t>0$. Moreover, the density $p(t, x, y)$ satisfies bounds of the form

$$0<a(t)\leqq p(t, x, y)\leqq A(t)<\infty$$

for each $t>0$.

Under Assumption A, the following theorem relating the asymptotic behavior of $Q_{x,t}$ for large t to the functional $I(\mu)$ was proved in [1].

THEOREM 1.1. *If $C\subset\mathcal{M}$ is closed and $G\subset\mathcal{M}$ is open (in the weak topology on $\mathcal{M}$), then for each $x\in\mathcal{X}$*

$$\overline{\lim_{t\to\infty}}\frac{1}{t}\log Q_{x,t}(C)\leqq-\inf_{\mu\in C} I(\mu),$$

$$\underline{\lim_{t\to\infty}}\frac{1}{t}\log Q_{x,t}(G)\geqq-\inf_{\mu\in G} I(\mu).$$

Such a theorem has a corollary as was proved in [2] (Section 3).

COROLLARY. *If $\Phi(\mu)$ is a weakly continuous functional on $\mathcal{M}$, then*

$$\lim_{t\to\infty}\frac{1}{t}\log E^{Q_{x,t}}\{e^{t\Phi(\mu)}\}=\sup_{\mu}[\Phi(\mu)-I(\mu)],$$

or equivalently

$$\lim_{t\to\infty} \frac{1}{t} \log E^{P_x}\{e^{t\Phi(L_t(\omega,\cdot))}\} = \sup_{\mu} [\Phi(\mu) - I(\mu)] .$$

In some interesting applications (see e.g. [3] for the sausage problem arising in statistical mechanics) the functional Φ of interest is not continuous in the weak topology on $\mathcal{M}$ but perhaps has continuity properties in the norm topology on $\mathcal{M}$. This means that one needs a theorem like Theorem 1.1 but where C and G are, respectively, closed and open in the norm topology rather than in the weak topology. There are difficulties in this situation because the occupation distribution $L_t(\omega, dy)$ can have a density with respect to a reference measure only in the one-dimensional case. In higher dimensions, although $Q_{x,t}$ will be concentrated mostly on singular measures, in the limit as $t\to\infty$ the variational formula $\sup_{\mu} [\Phi(\mu)-I(\mu)]$ involves only the values of $\Phi(\mu)$ for measures μ such that $I(\mu)$ is finite and these in general are absolutely continuous with respect to a reference measure. Such a transition in the limit from singular to absolutely continuous measures can easily occur in the weak topology but cannot occur in the norm topology. To derive results that are valid in the norm topology a certain amount of mollification is necessary. We shall state shortly Theorem 1.2 which is the analogue of Theorem 1.1 but for the norm topology. First we must consider the mollification of the occupation distribution.

We start with a family $k_\varepsilon(x, y)$ of functions on $\mathfrak{X}\times\mathfrak{X}$ satisfying Assumption B which follows. The λ measure which appears in Assumption B is the same finite reference measure as in Assumption A.

ASSUMPTION B.

(i) The family $k_\varepsilon(x, y)$ is non-negative and jointly measurable in x and y for each $\varepsilon>0$.

(ii) $\int_{\mathfrak{X}} k_\varepsilon(x, y)\lambda(dy)=1$ for each $\varepsilon>0$ and $x\in\mathfrak{X}$.

(iii) For each $\varepsilon>0$, the map $x\to k_\varepsilon(x, \cdot)$ is a continuous map of $\mathfrak{X}$ into $L_1(\lambda)$.

(iv) For each $\phi\in C(\mathfrak{X})$,

$$\lim_{\varepsilon\to 0} \int_{\mathfrak{X}} k_\varepsilon(x, y)\phi(y)\lambda(dy) = \phi(x)$$

uniformly for $x\in\mathfrak{X}$.

(v) For each $f \in L_1(\lambda)$,

$$\lim_{\varepsilon \to 0} \int_{\mathcal{X}} k_\varepsilon(x, y) f(x) \lambda(dx) = f(y)$$

in $L_1(\lambda)$.

We then use $k_\varepsilon(x, y)$ as a mollifier and define

$$l_t^\varepsilon(\omega, y) = \int_{\mathcal{X}} k_\varepsilon(x, y) L_t(\omega, dx) .$$

We see that $l_t^\varepsilon(\omega, y)$ is a probability density on $\mathcal{X}$ relative to λ.

Let γ be the subset of $\mathcal{M}$ consisting of those probability measures on $\mathcal{X}$ which are absolutely continuous with respect to λ. We can think of each element of γ in terms of its density f relative to λ and hence identify γ with the subset of $L_1(\lambda)$ consisting of non-negative elements having total integral 1. Thus for each ω, $\varepsilon > 0$, and $t > 0$, $l_t^\varepsilon(\omega, y) \in \gamma$.

Now, we make a choice of ε for each t and denote the choice by $\varepsilon(t)$. Assume that $\varepsilon(t) \to 0$ as $t \to \infty$ in an appropriate manner. Assume also that the measure $R_{x,t}$ on γ is defined by $R_{x,t} = Q_{x,t} k_{\varepsilon(t)}^{-1}$, that is, if A is a Borel subset of γ,

$$\begin{aligned} R_{x,t}(A) &= Q_{x,t}\left\{\mu : \int_{\mathcal{X}} k_{\varepsilon(t)}(x, \cdot)\mu(dx) \in A\right\} \\ &= P_x\{\omega : l_t^{\varepsilon(t)}(\omega, \cdot) \in A\} . \end{aligned}$$

We would like to prove that $R_{x,t}$ obeys the following

THEOREM 1.2. *If $C \subset \gamma$ is closed and $G \subset \gamma$ is open (in the norm topology on γ), then*

$$\overline{\lim_{t \to \infty}} \frac{1}{t} \log R_{x,t}(C) \leqq -\inf_{f \in C} I(f) ,$$

$$\underline{\lim_{t \to \infty}} \frac{1}{t} \log R_{x,t}(G) \geqq -\inf_{f \in G} I(f) .$$

In Theorem 1.2,

$$I(f) = -\inf_{u \in \mathcal{U}} \int_{\mathcal{X}} \left(\frac{Lu}{u}\right)(x) f(x) \lambda(dx)$$

which is just $I(\mu)$ as previously defined but for $\mu \in \gamma$.

Just as we had a corollary to Theorem 1.1 we have a corollary to Theorem 1.2.

COROLLARY. *If $\Phi(f)$ is a continuous functional on γ which is bounded above, then*

$$\lim_{t\to\infty}\frac{1}{t}\log E^{R_{x,t}}\{e^{t\Phi(f)}\}=\sup_{f\in\gamma}[\Phi(f)-I(f)],$$

or equivalently

$$\lim_{t\to\infty}\frac{1}{t}\log E^{P_x}\{\exp\{t\Phi(l_t^{\varepsilon(t)}(\omega,\cdot))\}\}=\sup_{f\in\gamma}[\Phi(f)-I(f)].$$

Remark. If μ_0 is the unique invariant measure for the process, then the ergodic theorem asserts that $L_t(\omega, dy)$ converges weakly to μ_0 as $t\to\infty$ for almost all ω. Theorem 1.2 will imply that $l_t^{\varepsilon(t)}(\omega, y)$ converges to $f_0=d\mu_0/d\lambda$ in $L_1(\lambda)$ for almost all ω. Therefore, Theorem 1.2 is a non-trivial strengthening of Theorem 1.1 and should require among its assumptions that $\varepsilon(t)$ does not go to zero too rapidly as $t\to\infty$. We shall state below Assumption C which is of this nature.

Assumption B on $k_\varepsilon(x, y)$ implies that the operator K_ε defined by

$$(K_\varepsilon\phi)(x)=\int_{\mathfrak{X}} k_\varepsilon(x, y)\phi(y)\lambda(dy)$$

is a compact operator mapping the unit ball of $C(\mathfrak{X})$ into an equicontinuous and bounded family of functions on $\mathfrak{X}$. Let S_ε be the image of the unit ball under K_ε. Since S_ε is totally bounded, it can be covered by a finite number of spheres in $C(\mathfrak{X})$ of any given positive radius δ. Let $N_\delta(\varepsilon)$ be the smallest number of such spheres required to cover S_ε. Our assumption on the rate of convergence of $\varepsilon(t)$ to 0 is

ASSUMPTION C. $\varepsilon(t)\to 0$ as $t\to\infty$ in such a manner that

$$\overline{\lim_{t\to\infty}}\frac{1}{t}\log N_\delta(\varepsilon(t))<\infty$$

for each $\delta>0$.

Finally, we need a regularity assumption on the semigroup T_t which we state as

ASSUMPTION D. For each $t>0$, the transition probability density $p(t, x, y)$ gives a continuous map $x\to p(t, x, \cdot)$ of $\mathfrak{X}$ into $\gamma\subset L_1(\lambda)$.

Sections 2, 3, and 4 of this paper contain the proof of Theorem 1.2 under Assumptions A, B, C, and D. In Section 5 we treat a class of examples which are used in [3] for the sausage problem.

2. Some Preliminary Estimates

Let $\phi(x)\in C(\mathscr{X})$ and consider, for $\varepsilon>0$ and $\delta>0$, the function

$$\phi_{\varepsilon,\delta}(x)=(K_\varepsilon(T_\delta-I)\phi)(x).$$

Define for each ϕ

$$\lambda_{\varepsilon,\delta}(\phi)=\sup_{\mu\in\mathscr{M}}\left[\int_{\mathscr{X}}\phi_{\varepsilon,\delta}(x)\mu(dx)-I(\mu)\right].$$

We have

LEMMA 2.1. *For each* $0<\rho<\infty$,

$$\overline{\lim_{\substack{\varepsilon\to0\\ \delta\to0}}}\sup_{\|\phi\|\leqq\rho}\lambda_{\varepsilon,\delta}(\phi)\leqq0.$$

Proof: For any $\phi(x)$ which is bounded and continuous,

$$\lambda(\phi)=\sup_{\mu\in\mathscr{M}}\left[\int_{\mathscr{X}}\phi(x)\mu(dx)-I(\mu)\right]\leqq\sup_x\phi(x)\leqq\|\phi\|.$$

For each $\phi\in C(\mathscr{X})$, $\|K_\varepsilon(T_\delta-I)\phi\|\to0$ as $\varepsilon,\delta\to0$ and therefore

$$\overline{\lim_{\substack{\varepsilon\to0\\ \delta\to0}}}\lambda_{\varepsilon,\delta}(\phi)\leqq0.$$

However, the convergence of $K_\varepsilon(T_\delta-I)\phi$ is not uniform over bounded sets because $\|K_\varepsilon(T_\delta-I)\|$ does not tend to zero. So we must estimate more carefully.

Since $|\phi_{\varepsilon,\delta}(x)|\leqq2\rho$ for all x if $\|\phi\|\leqq\rho$, we have

$$\int_{\mathscr{X}}\phi_{\varepsilon,\delta}(x)\mu(dx)-I(\mu)\leqq0$$

if $I(\mu)\geqq2\rho$. To prove the lemma it is therefore sufficient to show that

$$\lim_{\substack{\varepsilon\to0\\ \delta\to0}}\sup_{\mu:I(\mu)\leqq2\rho}\left|\int_{\mathscr{X}}\phi_{\varepsilon,\delta}(x)\mu(dx)\right|=0. \tag{2.1}$$

Since

$$\int_{X} \phi_{\varepsilon,\delta}(x)\mu(dx) = \int_{X} (K_\varepsilon(T_\delta - I)\phi)(x)\mu(dx)$$
$$= \int_{X} \phi(x)[\mu K_\varepsilon(T_\delta - I)](dx),$$

it is enough to show that

$$\sup_{\mu: I(\mu)\leqq 2\rho} \|\mu K_\varepsilon(T_\delta - I)\| \to 0 \tag{2.2}$$

as ε and $\delta\to 0$. From Lemma 4.1 of [1] and its corollary we have

$$\lim_{\delta\to 0} \sup_{\mu: I(\mu)\leqq 2\rho} \|\mu T_\delta - \mu\| = 0 . \tag{2.3}$$

By Assumption D, for each $\delta>0$ the range of μT_δ as μ varies over $\mathcal{M}$ is a totally bounded subset of γ in the norm topology. Combining this with (2.3) the set of probability measures μ satisfying $I(\mu)\leqq 2\rho$ is a totally bounded subset of γ. Since K_ε is a contraction and $\|\mu K_\varepsilon - \mu\|\to 0$ as $\varepsilon\to 0$ for each $\mu\in\gamma$, we conclude from the fact that $\{\mu : I(\mu)\leqq 2\rho\}$ is totally bounded in γ that

$$\sup_{\mu: I(\mu)\leqq 2\rho} \|\mu K_\varepsilon - \mu\| \to 0 \quad \text{as} \quad \varepsilon\to 0 . \tag{2.4}$$

Now, we can write

$$\|\mu K_\varepsilon(T_\delta - I)\| = \|\mu K_\varepsilon T_\delta - \mu K_\varepsilon\|$$
$$\leqq \|(\mu K_\varepsilon - \mu)T_\delta\| + \|\mu T_\delta - \mu\| + \|\mu - \mu K_\varepsilon\|$$
$$\leqq 2\,\|\mu K_\varepsilon - \mu\| + \|\mu T_\delta - \mu\| ,$$

and therefore (2.3) and (2.4) imply (2.2) and the lemma.

LEMMA 2.2. *For each* $0<\rho<\infty$, *there is a constant* C_ρ *such that if* $V(x)$ *is any continuous function on* X *bounded in absolute value by* ρ *and if*

$$\lambda_V = \sup_{\mu\in\mathcal{M}} \left[\int_{X} V(x)\mu(dx) - I(\mu)\right],$$

then, for each $x \in \mathbf{X}$ *and* $t \geqq 0$,

$$E_x\left\{\exp\left\{\int_0^t V(x(s))\,ds\right\}\right\}=E^{P_x}\left\{\exp\left\{\int_0^t V(x(s))\,ds\right\}\right\}\leqq C_p e^{\lambda_V t}\,.$$

Proof: It was proved in [1] (see the corollary to Theorem 1.1 in the introduction of the present paper) that for any $V(x)$ which is bounded and continuous

$$\begin{aligned}\lambda_V&=\lim_{t\to\infty}\frac{1}{t}\log E_x\left\{\exp\left\{\int_0^t V(x(s))\,ds\right\}\right\}\\ &=\sup_{\mu\in\mathcal{M}}\left[\int V(x)\mu(dx)-I(\mu)\right].\end{aligned}\tag{2.5}$$

Let $v(t, x)=E_x\left\{\exp\left\{\int_0^t V(x(s))\,ds\right\}\right\}$, and $v^+(t)=\sup_x v(t, x)$, $v^-(t)=\inf_x v(t, x)$, so that

$$v^-(t)\leqq v(t, x)\leqq v^+(t)\,.\tag{2.6}$$

By the Markov property,

$$v^+(t+s)\leqq v^+(t)v^+(s)$$

and

$$v^-(t+s)\geqq v^-(t)v^-(s)\,.$$

Therefore, from the property of subadditive functions,

$$\lim_{t\to\infty}\frac{\log v^+(t)}{t}=\inf_{t>0}\frac{\log v^+(t)}{t}$$

and

$$\lim_{t\to\infty}\frac{\log v^-(t)}{t}=\sup_{t>0}\frac{\log v^-(t)}{t}\,.$$

From (2.5) and (2.6) it follows that, for every $t\geqq 0$,

$$v^-(t)\leqq e^{\lambda_V t}\leqq v^+(t)\,.\tag{2.7}$$

Now, let $t \geqq 1$. Then for any pair x, y of points in $\mathfrak{X}$, we have using Assumption A

$$
\begin{aligned}
E_x\left\{\exp\left\{\int_0^t V(x(s))\,ds\right\}\right\} &= E_x\left\{\exp\left\{\int_0^1 V(x(s))\,ds\right\}\exp\left\{\int_1^t V(x(s))\,ds\right\}\right\} \\
&= E_x\left\{\exp\left\{\int_0^1 V(x(s))\,ds\right\}E_{x(1)}\left\{\exp\left\{\int_0^{t-1} V(x(s))\,ds\right\}\right\}\right. \\
&= E_x\left\{\exp\left\{\int_0^1 V(x(s))\,ds\right\}v(t-1, x(1))\right\} \\
&\leqq e^{\rho} E_x\{v(t-1, x(1))\} \\
&= e^{\rho}\int_{\mathfrak{X}} v(t-1, z)p(1, x, dz) \\
&= e^{\rho}\int_{\mathfrak{X}} v(t-1, z)p(1, x, z)\lambda(dz) \\
&= e^{\rho}\int_{\mathfrak{X}} v(t-1, z)\frac{p(1, x, z)}{p(1, y, z)}p(1, y, z)\lambda(dz) \\
&\leqq e^{\rho}\frac{A(1)}{a(1)}\int_{\mathfrak{X}} v(t-1, z)p(1, y, z)\lambda(dz) \\
&\leqq e^{2\rho}\frac{A(1)}{a(1)} E_y\left\{\exp\left\{\int_0^t V(x(s))\right\}\right\}.
\end{aligned}
$$

Since x and y are arbitrary points in $\mathfrak{X}$, we obtain for $t \geqq 1$

$$
v^+(t) \leqq e^{2\rho}\frac{A(1)}{a(1)}v^-(t).
$$

On the other hand, for $0 \leqq t \leqq 1$,

$$
\begin{aligned}
v^+(t) &\leqq e^{t\rho} \leqq e^{\rho}, \\
v^-(t) &\geqq e^{-t\rho} \geqq e^{-\rho}.
\end{aligned}
$$

Thus, for $0 \leqq t \leqq 1$, $v^+(t) \leqq e^{2\rho} v^-(t)$ and, since $A(1)/a(1) \geqq 1$, we can take $C_\rho = e^{2\rho} A(1)/a(1)$ and have

$$v^+(t) \leqq C_\rho v^-(t) \quad \text{for all} \quad t \geqq 0 .$$

Using (2.7), we obtain from this last inequality

$$v(t, x) \leqq v^+(t) \leqq C_\rho v^-(t) \leqq C_\rho e^{\lambda_V t} ,$$

which is the lemma.

3. The Main Estimate

Let us suppose that a choice of $\varepsilon(t)$ has been made for each t in such a manner that Assumption C is satisfied. We recall the definitions:

$$L_t(\omega, A) = \frac{1}{t} \text{ Lebesgue measure } \{\sigma : x(\sigma) \in A,\ 0 \leqq \sigma \leqq t\} ,$$

$$l_t^\varepsilon(\omega, y) = \int_{\mathfrak{X}} k_\varepsilon(x, y) L_t(\omega, dx) .$$

For convenience we denote $l_t^{\varepsilon(t)}(\omega, y)$ by $g_t(\omega, y)$ and define, for any $\delta > 0$,

$$g_t^\delta(\omega, y) = \int_{\mathfrak{X}} p(\delta, x, y) g_t(\omega, x) \lambda(dx)$$

and

$$\Delta_t^\delta(\omega) = \int_{\mathfrak{X}} |g_t^\delta(\omega, y) - g_t(\omega, y)|\, \lambda(dy) .$$

We now prove

THEOREM 3.1. *For any $\theta > 0$ and $x \in \mathfrak{X}$,*

$$\overline{\lim_{\delta \to 0}}\ \overline{\lim_{t \to \infty}} \frac{1}{t} \log P_x\{\Delta_t^\delta(\omega) \geqq \theta\} = -\infty .$$

Proof: We note that

$$\Delta_t^\delta(\omega) = \sup_{\|\phi\| \leq 1} \int_{\mathscr{X}} \phi(y)[g_t^\delta(\omega, y) - g_t(\omega, y)]\lambda(dy)$$

$$= 2 \sup_{\|\phi\| \leq 1/2} \int_{\mathscr{X}} \phi(y)[g_t^\delta(\omega, y) - g_t(\omega, y)]\lambda(dy)$$

$$= 2 \sup_{\|\phi\| \leq 1/2} \int_{\mathscr{X}} \phi(y)[L_t(\omega, \cdot)K_{\varepsilon(t)}(T_\delta - I)](dy)$$

$$= 2 \sup_{\|\phi\| \leq 1/2} \int_{\mathscr{X}} (K_{\varepsilon(t)}(T_\delta - I)\phi)(y)L_t(\omega, dy)$$

$$= 2 \sup_{\|\phi\| \leq 1/2} \frac{1}{t}\int_0^t (K_{\varepsilon(t)}(T_\delta - I)\phi)(x(s))\, ds$$

$$= 2 \sup_{\psi \in A_{\varepsilon(t),\delta}} \frac{1}{t}\int_0^t \psi(x(s))\, ds\,,$$

where $A_{\varepsilon(t),\delta}$ is the set of functions of the form

$$\psi = \phi_{\varepsilon,\delta} = K_\varepsilon(T_\delta - I)\phi\,,$$

where ϕ is continuous and $\|\phi\| \leq \frac{1}{2}$.

If A is any set in a metric space and if $A \subset B$, where B can be covered by a finite number N of spheres of radius θ, then A can be covered by the same number N of spheres of radius 2θ with centers in A. To see this, let $s(\xi_1, \theta), \cdots, s(\xi_N, \theta)$ be the covering of B. For each $i = 1, 2, \cdots, N$, we pick a_i from $s(\xi_i, \theta) \cap A$ if it is non-empty. Otherwise, we pick any a_i from A. Then $s(a_i, 2\theta)$, $1 \leq i \leq N$, is a covering of A.

The set $A_{\varepsilon,\delta}$ defined above is clearly contained in the set B_ε which is the image of $\{\phi : \|\phi\| \leq 1\}$ under K_ε. Since B_ε can be covered by a finite number $N_{\theta/8}(\varepsilon)$ of spheres of radius $\frac{1}{8}\theta$, $A_{\varepsilon,\delta}$ can be covered by the same number of spheres of radius $\frac{1}{4}\theta$ with centers $\psi_1, \psi_2, \cdots, \psi_N$ in $A_{\varepsilon,\delta}$. Here N stands for $N_{\theta/8}(\varepsilon(t))$. We can now write

$$\sup_{\psi \in A_{\varepsilon(t),\delta}} \frac{1}{t}\int_0^t \psi(x(s))\, ds \leq \tfrac{1}{4}\theta + \sup_{1 \leq i \leq N} \frac{1}{t}\int_0^t \psi_i(x(s))\, ds\,.$$

Therefore,

$$P_x\{\Delta_t^\delta(\omega)\geqq\theta\}=P_x\left\{\sup_{\psi\in A_{\varepsilon,\delta}}\frac{1}{t}\int_0^t\psi(x(s))\,ds\geqq\tfrac{1}{2}\theta\right\}$$

$$\leqq P_x\left\{\sup_{1\leqq i\leqq N}\frac{1}{t}\int_0^t\psi_i(x(s))\,ds\geqq\tfrac{1}{4}\theta\right\}$$

(3.1)

$$\leqq\sum_{i=1}^N P_x\left\{\frac{1}{t}\int_0^t\psi_i(x(s))\,ds\geqq\tfrac{1}{4}\theta\right\}$$

$$\leqq N\sup_{1\leqq i\leqq N}P_x\left\{\frac{1}{t}\int_0^t\psi_i(x(s))\,ds\geqq\tfrac{1}{4}\theta\right\}.$$

But

(3.2) $$P_x\left\{\frac{1}{t}\int_0^t\psi_i(x(s))\,ds\geqq\tfrac{1}{4}\theta\right\}\leqq e^{-\rho t\theta/4}E_x\left\{\exp\left\{\int_0^t\rho\psi_i(x(s))\,ds\right\}\right\},$$

where $\rho>0$ is an arbitrary constant. Moreover, since $\|\psi_i\|\leqq 1$, we have $\|\rho\psi_i\|\leqq\rho$ and therefore from Lemma 2.2

$$E_x\left\{\exp\left\{\int_0^t\rho\psi_i(x(s))\,ds\right\}\right\}\leqq C_\rho e^{t\lambda_{\rho\psi_i}},$$

where

$$\lambda_{\rho\psi_i}=\sup_{\mu\in\mathcal{M}}\left[\int_{\mathcal{X}}\rho\psi_i(x)\mu(dx)-I(\mu)\right].$$

Let

$$\lambda_{\varepsilon,\delta,\rho}=\sup_{\psi\in A_{\varepsilon,\delta}}\lambda_{\rho\psi}$$

so that

$$\sup_{1\leqq i\leqq N}E_x\left\{\exp\left\{\int_0^t\rho\psi_i(x(s))\,ds\right\}\right\}\leqq C_\rho e^{t\lambda_{\varepsilon(t),\delta,\rho}}.$$

Combining this with (3.1) and (3.2) we have

$$\overline{\lim_{t\to\rho}}\,\frac{1}{t}\log P_x\{\Delta_t^\delta(\omega)\geqq\theta\}\leqq\overline{\lim_{t\to\infty}}\,\frac{1}{t}\log N_{\theta/8}(\varepsilon(t))+\overline{\lim_{t\to\rho}}\,\lambda_{\varepsilon(t),\delta,\rho}-\tfrac{1}{4}\theta\rho.$$

Recalling that $\psi \in A_{\varepsilon,\delta}$ if it is of the form $\phi_{\varepsilon,\delta} = K_\varepsilon(T_\delta - I)\phi$, where ϕ is continuous and $\|\phi\| \leqq \frac{1}{2}$, we see from Lemma 2.1 that

$$\overline{\lim_{\delta\to 0}}\ \overline{\lim_{t\to\infty}}\ \lambda_{\varepsilon(t),\delta,\rho} \leqq 0$$

for each fixed $0<\rho<\infty$. Thus

$$\overline{\lim_{\delta\to 0}}\ \overline{\lim_{t\to\infty}}\ \frac{1}{t}\log P_x\{\Delta_t^\delta(\omega) \geqq \theta\} \leqq \overline{\lim_{t\to\infty}}\ \frac{1}{t}\log N_{\theta/8}(\varepsilon(t)) - \tfrac{1}{4}\theta\rho$$

for each $0<\rho<\infty$. The first term on the right of this inequality is independent of ρ and by Assumption C is finite for each $\theta>0$. Hence keeping $\theta>0$ fixed and letting $\rho\to\infty$ we conclude that

$$\overline{\lim_{\delta\to 0}}\ \overline{\lim_{t\to\infty}}\ \frac{1}{t}\log P_x\{\Delta_t^\delta(\omega) \geqq \theta\} = -\infty$$

which is the theorem.

4. Proof of Theorem 1.2

We now turn to the proof of Theorem 1.2. First we recall our notation. For each $x \in \mathfrak{X}$, P_x is the basic measure on Ω_x corresponding to the Markov process starting at time 0 from the point x. We denoted by $L_t(\omega,\cdot)$ the map from $\Omega_x \to \mathcal{M}$ defined by the occupation distribution, where $\mathcal{M}$ is the space of all probability measures on $\mathfrak{X}$. We defined a mollifying kernel $k_\varepsilon(x, y)$ which led to a map K_ε of $C(\mathfrak{X}) \to C(\mathfrak{X})$ and the dual map also denoted by K_ε from $\mathcal{M} \to \gamma$ given by

$$(\mu K_\varepsilon)(y) = \int_{\mathfrak{X}} k_\varepsilon(x, y)\mu(dx)\,,$$

where γ is the subset of $\mathcal{M}$ consisting of those measures which are absolutely continuous with respect to the reference measure λ. We identified γ with the set of probability densities relative to λ. We note that, for any $\mu \in \mathcal{M}$, if $I(\mu)<\infty$, then $\mu \in \gamma$ (cf. Lemma 2.8 and its corollary in [1]). Moreover, $\{\mu \in \mathcal{M} : I(\mu) \leqq l\}$ is a compact subset of γ in the norm or L_1 topology for each finite l. We also consider the maps T_δ induced by the Markov semigroup

$$(\mu T_\delta)(y) = \int_{\mathfrak{X}} p(\delta, x, y)\mu(dx)$$

which also map $\mathcal{M} \to \gamma$.

Now $\mathcal{M}$ comes with two topologies. Weak convergence makes $\mathcal{M}$ a compact metric space, and $\mathcal{M}$ can also be considered as a metric space under the norm topology. In the norm topology, $\gamma\subset\mathcal{M}$ is closed.

We denote by $Q_{x,t}$ the measure on $\mathcal{M}$ induced by P_x via the occupation distribution map $L_t(\omega,\cdot)$ from $\Omega_x\to\mathcal{M}$. The map $K_{\varepsilon(t)}$ from $\mathcal{M}\to\gamma$ induces a measure $R_{x,t}$ on γ—namely $R_{x,t}=Q_{x,t}K_{\varepsilon(t)}^{-1}$. Similarly we consider the measure $R_{x,t}^{\delta}$ on γ given by $R_{x,t}^{\delta}=R_{x,t}T_{\delta}^{-1}$.

We want to prove Theorem 1.2 which is the analogue of Theorem 1.1 in the norm topology for the measures $R_{x,t}$. First we prove the theorem in the weak topology.

THEOREM 4.1. *Let $C\subset\mathcal{M}$ be closed and $G\subset\mathcal{M}$ be open (in the weak topology on $\mathcal{M}$). Then*

$$\overline{\lim_{t\to\infty}}\frac{1}{t}\log R_{x,t}(C)\leqq-\inf_{\mu\in C}I(\mu)\,,$$

$$\underline{\lim_{t\to\infty}}\frac{1}{t}\log R_{x,t}(G)\geqq-\inf_{\mu\in G}I(\mu)\,.$$

Proof: It is easy to see that $\mu K_\varepsilon\Rightarrow\mu$ (weakly) as $\varepsilon\to0$ uniformly over all $\mu\in\mathcal{M}$. Let C_1 and G_1 be, respectively, closed and open sets in $\mathcal{M}$ in the weak topology such that $C\subset C_1^0$ and $\bar{G}_1\subset G$, where C_1^0 is the interior of C_1 and $\bar{G}_1$ is the closure of G_1. Since $\mu K_\varepsilon\Rightarrow\mu$ uniformly over $\mathcal{M}$ as $\varepsilon\to0$, there exists $\varepsilon_0>0$ such that, for all $\varepsilon<\varepsilon_0$, $\mu K_\varepsilon\in C$ implies $\mu\in C_1$ and $\mu\in G_1$ implies $\mu K_\varepsilon\in G$. Thus, since $\varepsilon(t)\to0$ as $t\to\infty$, for $t\geqq t_0(\varepsilon_0)$, $R_{x,t}(C)\leqq Q_{x,t}(C_1)$ and $R_{x,t}(G)\geqq Q_{x,t}(G_1)$. From Theorem 1.1 we have then

$$\overline{\lim_{t\to\infty}}\frac{1}{t}\log R_{x,t}(C)\leqq-\inf_{\mu\in C_1}I(\mu)$$

and

$$\underline{\lim_{t\to\infty}}\frac{1}{t}\log R_{x,t}(G)\geqq-\inf_{\mu\in G_1}I(\mu)\,.$$

Since C_1 is arbitrary except that $C_1^0\supset C$ and G_1 is arbitrary except that $\bar{G}_1\subset G$, we conclude that

$$\overline{\lim_{t\to\infty}}\frac{1}{t}\log R_{x,t}(C)\leqq-\sup_{C_1:C_1^0\supset C}\inf_{\mu\in C_1}I(\mu)$$

and

$$\underline{\lim_{t\to\infty}}\frac{1}{t}\log R_{x,t}(G)\geqq-\inf_{G_1:\bar{G}_1\subset G}\inf_{\mu\in G_1}I(\mu)\,.$$

It is easy to see, using the lower semicontinuity of $I(\mu)$ and the compactness of $\mathcal{M}$, that

$$\sup_{C_1:C_1^0\supset C}\ \inf_{\mu\in C_1} I(\mu)=\inf_{\mu\in C} I(\mu)\,.$$

On the other hand, it is obvious that

$$\inf_{G_1:\bar{G}_1\subset G}\ \inf_{\mu\in G_1} I(\mu)=\inf_{\mu\in G} I(\mu)\,.$$

This proves the theorem.

We now prove a version of Theorem 1.2 for $R^{\delta}_{x,t}$ in the norm topology.

THEOREM 4.2. *Let C be closed in γ and G be open in γ in the norm topology. Then, for each $\delta>0$,*

$$\overline{\lim_{t\to\infty}}\frac{1}{t}\log R^{\delta}_{x,t}(C)\leqq -\inf_{\mu:\mu T_\delta\in C} I(\mu)\,,$$

$$\underline{\lim_{t\to\infty}}\frac{1}{t}\log R^{\delta}_{x,t}(G)\geqq -\inf_{\mu:\mu T_\delta\in G} I(\mu)\,.$$

Proof: We note that, for any set A, $R^{\delta}_{x,t}(A)=R_{x,t}(\tilde{A}_\delta)$, where $\tilde{A}_\delta=\{\mu:\mu T_\delta\in A\}$. Moreover, for each $\delta>0$, the map $\mu\to\mu T_\delta$ maps $\mathcal{M}$ with the weak topology continuously into γ with the norm topology. Therefore, $\{\mu:\mu T_\delta\in C\}$ and $\{\mu:\mu T_\delta\in G\}$ are, respectively, weakly closed and weakly open in $\mathcal{M}$ so long as C and G are norm closed and norm open in γ. We can now deduce Theorem 4.2 from Theorem 4.1.

We turn to the proof of Theorem 1.2. The transition from $R^{\delta}_{x,t}$ to $R_{x,t}$ is carried out by letting $\delta\to 0$ and using the estimates provided by Theorem 3.1.

Proof of Theorem 1.2: Let C be norm closed in γ. Let $C_\theta=\{\beta\in\gamma:\|\beta-\alpha\|<\theta$ for some $\alpha\in C\}$ and denote its closure by $\bar{C}_\theta$. We have, for any $t>0$, $\delta>0$, $\theta>0$,

$$\{\omega:g_t(\omega,\cdot)\in C\}\subset\{\omega:g^{\delta}_t(\omega,\cdot)\in\bar{C}_\theta\}\cup\{\omega:\|g^{\delta}_t(\omega,y)-g_t(\omega,y)\|\geqq\theta\}$$

so that

$$P_x\{g_t(\omega,\cdot)\in C\}\leqq P_x\{g^{\delta}_t(\omega,\cdot)\in\bar{C}_\theta\}+P_x\{\Delta^{\delta}_t(\omega)\geqq\theta\}\,,$$

or

$$R_{x,t}(C)\leqq R^{\delta}_{x,t}(\bar{C}_\theta)+P_x\{\Delta^{\delta}_t(\omega)\geqq\theta\}\,. \tag{4.1}$$

Hence,

$$\overline{\lim_{t\to\infty}} \frac{1}{t} \log R_{x,t}(C) \leqq \max\left[\overline{\lim_{t\to\infty}} \frac{1}{t} \log R^{\delta}_{x,t}(\bar{C}_\theta), \overline{\lim_{t\to\infty}} \frac{1}{t} \log P_x\{\Delta^{\delta}_t(\omega) \geqq \theta\}\right].$$

From Theorems 3.1 and 4.2, we obtain on letting $\delta \to 0$

$$\overline{\lim_{t\to\infty}} \frac{1}{t} \log R_{x,t}(C) \leqq -\overline{\lim_{\delta\to 0}} \inf_{\mu:\mu T_\delta \in \bar{C}_\theta} I(\mu). \tag{4.2}$$

Since the left side of (4.2) is independent of $\theta > 0$, we can let $\theta \to 0$ and conclude that

$$\overline{\lim_{t\to\infty}} \frac{1}{t} \log R_{x,t}(C) \leqq -\overline{\lim_{\theta\to 0}}\, \overline{\lim_{\delta\to 0}} \inf_{\mu:\mu T_\delta \in \bar{C}_\theta} I(\mu). \tag{4.3}$$

We have to show that

$$\overline{\lim_{\theta\to 0}}\, \overline{\lim_{\delta\to 0}} \inf_{\mu:\mu T_\delta \in \bar{C}_\theta} I(\mu) \geqq \inf_{\mu \in C} I(\mu). \tag{4.4}$$

In fact, we shall show that

$$\underline{\lim}_{\substack{\theta\to 0\\ \delta\to 0}} \inf_{\mu:\mu T_\delta \in \bar{C}_\theta} I(\mu) \geqq \inf_{\mu \in C} I(\mu). \tag{4.5}$$

Suppose there is a sequence $\{\mu_n\}$ such that $I(\mu_n) \to l$ and $\mu_n T_{\delta_n} \in \bar{C}_{\theta_n}$ for some sequence δ_n, θ_n going to zero. Since $I(\mu_n)$ is bounded, we know that $\{\mu_n\} \in \gamma$ is a totally bounded sequence in γ in the norm topology. We also know from the corollary to Lemma 4.1 of [1] that

$$\sup_n \|\mu_n T_\delta - \mu_n\| \to 0$$

as $\delta \to 0$. Therefore, $\{\mu_n\}$ has a subsequence converging in norm to a limit μ and, since $\mu_n T_{\delta_n} \to \mu$ in norm, we conclude that $\mu \in C$. By lower semicontinuity, $I(\mu) \leqq l$. This proves (4.5) and from (4.3) we have

$$\overline{\lim_{t\to\infty}} \frac{1}{t} \log R_{x,t}(C) \leqq -\inf_{\mu \in C} I(\mu).$$

To derive the lower bound of Theorem 1.2, we let $G \subset \gamma$ be open in the norm topology, and we must prove

$$\underline{\lim}_{t\to\infty} \frac{1}{t} \log R_{x,t}(G) \geqq -\inf_{\mu \in G} I(\mu). \tag{4.6}$$

It is clearly sufficient to prove that, for any $\mu \in G$,

$$\varliminf_{t\to\infty} \frac{1}{t} \log R_{x,t}(G) \geqq -I(\mu) . \tag{4.7}$$

To prove (4.7) we note that, since G is open and $\mu \in G$, there exists a sphere $S(\mu, \theta)$ of some positive radius θ around μ such that $S(\mu, \theta) \subset G$. Clearly,

$$R_{x,t}(G) \geqq R_{x,t}(S(\mu, \theta)) \geqq R^{\delta}_{x,t}(S(\mu, \tfrac{1}{2}\theta)) - P_x(\Delta^{\delta}_t(\omega) \geqq \tfrac{1}{2}\theta) .$$

We again use Theorems 3.1 and 4.2 and obtain

$$\varliminf_{t\to\infty} \frac{1}{t} \log R_{x,t}(G) \geqq -\varlimsup_{\delta\to 0} \inf_{\alpha : \alpha T_\delta \in S(\mu, \theta/2)} I(\alpha) . \tag{4.8}$$

Since $I(\mu)$ can be assumed to be finite, $\|\mu T_\delta - \mu\| \to 0$ as $\delta \to 0$ by the corollary to Lemma 4.1 of [1]. Therefore, for sufficiently small δ,

$$\inf_{\alpha : \alpha T_\delta \in S(\mu, \theta/2)} I(\alpha) \leqq I(\mu) .$$

This combined with (4.8) gives us (4.7) and hence (4.6) which completes the proof of Theorem 1.2.

5. A Class of Examples

We shall now consider in detail a class of examples for Theorem 1.2 which we use later in [3] to study the sausage problem.

Let R^d be d-dimensional Euclidean space. Let $M>0$ and let G^M_d be the discrete subgroup of R^d consisting of vectors having for each coordinate an integral multiple of M. We denote by T^M_d the quotient of R^d modulo G^M_d. With reference to Theorem 1.2, we shall take as our state space X the space T^M_d which is a compact torus.

For the Markov process with transition probability $p(t, x, dy)$ and state space X, we take a process with independent increments, i.e., we have a one-parameter convolution semigroup π_t of probability measures on X and

$$p(t, x, A) = \pi_t(A - x) .$$

If, for $t>0$, $\pi_t(A)$ has a density $\nu_t(y)$ with respect to the Haar measure (Lebesgue measure) on X and if the density $\nu_t(y)$ is continuous and strictly positive on X, then Assumptions A and D are clearly satisfied with the Haar measure as the reference measure λ.

Let us take a family $\tilde{k}_\varepsilon(x)$ of probability densities on R^d such that $\tilde{k}_\varepsilon(x)$ converges weakly as $\varepsilon \to 0$ to the degenerate distribution at the origin. We can form the induced density on $\mathcal{X}$ by

$$k_\varepsilon(x) = \sum_{g \in G_d^M} \tilde{k}_\varepsilon(x+g)$$

and now $k_\varepsilon(x)$ can be used as a mollifier on $\mathcal{X}$, i.e., $k_\varepsilon(x-y)$ satisfies Assumption B. We shall further assume that $\tilde{k}_\varepsilon(x)$ depends on ε by a scale change. That is,

$$\tilde{k}_\varepsilon(x) = \varepsilon^{-d}\tilde{k}\left(\frac{x}{\varepsilon}\right),$$

where $\tilde{k}(x)$ is a fixed probability density on R^d.

Let us find the impact of Assumption C so that we can use Theorem 1.2.

LEMMA 5.1. *Assumption C holds if $\varepsilon(t) \to 0$ as $t \to \infty$ in such a manner that $t^{1/d}\varepsilon(t)$ is bounded away from zero.*

Proof: We must compute, for fixed $\delta > 0$ and $\varepsilon > 0$, the number of δ spheres required to cover the image of the unit ball of $C(\mathcal{X})$ by the map

$$(K_\varepsilon\phi)(x) = \int_{\mathcal{X}} k_\varepsilon(x-y)\phi(y)\, dy.$$

Since $\|\phi\| \leqq 1$, denoting by ψ the function $\psi = K_\varepsilon\phi$, we have

$$|\psi(x) - \psi(x')| \leqq \int_{\mathcal{X}} |k_\varepsilon(x-y) - k_\varepsilon(x'-y)|\, dy$$

$$\leqq \int_{R^d} |\tilde{k}_\varepsilon(x-y) - \tilde{k}_\varepsilon(x'-y)|\, dy.$$

If we introduce the L_1-modulus of continuity

$$W_\varepsilon(h) = \sup_{\|z\| \leqq h} \int_{R^d} |\tilde{k}_\varepsilon(x+z) - \tilde{k}_\varepsilon(x)|\, dx,$$

then the image S_ε of the unit ball by K_ε is contained in the set of functions on $\mathcal{X}$ bounded by 1 and having a modulus of continuity dominated by $W_\varepsilon(h)$. Given a $\delta > 0$ and $\varepsilon > 0$ we pick $h_0 = h_0(\varepsilon, \delta)$ such that

$$W_\varepsilon(h_0) < \tfrac{1}{2}\delta.$$

Then the torus X can be covered by a finite number $N(h_0)$ of sets each of which has diameters less than h_0. We can assume these sets to be disjoint and call them $T_1, T_2, \cdots, T_N$. By the modulus of continuity estimate, any $\psi \in S_\varepsilon$ can be approximated by a simple function which is constant on each T_i uniformly to within $\frac{1}{2}\delta$. Each such simple function can be approximated to within $\frac{1}{2}\delta$ by a simple function which takes its values only from a finite grid on $[-1, 1]$ with gap size at most $\frac{1}{2}\delta$. The number of points on the grid can be taken to be $[\frac{1}{4}\delta]+1$. We have, therefore, an estimate on the number of spheres of radius δ needed to cover S_ε. It is an estimate of the form

$$N \leqq ([\tfrac{1}{4}\delta]+1)^{N(h_0)} .$$

It is easy to see that Assumption C means that $\varepsilon(t) \to 0$ as $t \to \infty$ in such a way that $N(h_0)$ grows at most linearly with t. We know that, as $h \to 0$,

$$N(h) \leqq Ch^{-d},$$

where C is a constant depending only on the size M of the torus. Therefore, we have to be sure that

$$\overline{\lim_{t\to\infty}} \frac{1}{[h(\varepsilon(t), \delta)]^d} \cdot \frac{1}{t} < \infty ,$$

or equivalently that

$$\underline{\lim_{t\to\infty}}\, t^{1/d} h(\varepsilon(t), \delta) > 0 . \tag{5.1}$$

If $\tilde{k}_\varepsilon$ is obtained from $\tilde{k}$ by a scale change, then

$$\begin{aligned} W_\varepsilon(h) &= \sup_{\|z\|\leq h} \int_{R^d} \varepsilon^{-d} \left| \tilde{k}\left(\frac{x+z}{\varepsilon}\right) - \tilde{k}\left(\frac{x}{\varepsilon}\right) \right| dx \\ &= \sup_{\|z\|\leq h} \int_{R^d} \left| \tilde{k}\left(x+\frac{z}{\varepsilon}\right) - \tilde{k}(x) \right| dx \\ &= \sup_{\|z\|\leq h/\varepsilon} \int_{R^d} |\tilde{k}(x+z) - \tilde{k}(x)|\, dx \\ &= W_1\left(\frac{h}{\varepsilon}\right) . \end{aligned}$$

So, for fixed $\delta>0$, since $W_1(h) \to 0$ as $h \to 0$, we can pick $h(\delta)$ such that

$$W_1(h(\delta)) < \tfrac{1}{2}\delta ,$$

so that

$$h_0(\varepsilon, \delta) = \varepsilon h(\delta) .$$

Therefore, condition (5.1) reduces to

$$\varliminf_{t\to\infty} t^{1/d} \varepsilon(t) > 0$$

which proves the lemma.

One way of generating processes with independent increments on $\mathfrak{X}$ is to start from a process with independent increments on R^d and project it onto $\mathfrak{X}$. We shall now look at a class of such processes and see how our theorem applies to them.

Let $0<\alpha\leqq 2$. By a symmetric stable process of order α, we mean a process with homogeneous independent increments $\tilde{x}(t)$ such that

$$E\{e^{i\langle\xi, \tilde{x}(t)\rangle}\} = e^{-tQ(\xi)} ,$$

where

$$\begin{aligned} Q(\xi) &= \int_0^\infty \int_{S^{d-1}} (1-\cos\langle\xi, rs\rangle) \frac{\Omega(ds)}{r^{1+\alpha}}\, dr \qquad \text{if} \qquad 0<\alpha<2 , \\ &= \tfrac{1}{2} \sum a_{ij} \xi_i \xi_j \qquad\qquad\qquad\qquad\qquad \text{if} \qquad \alpha=2 . \end{aligned}$$

If $0<\alpha<2$, then $\Omega(ds)$ is a symmetric finite mass distribution on the unit sphere S^{d-1}, $r>0$ is a positive scalar, and rs is interpreted as a vector in R^d. If $\alpha=2$, then $\{a_{ij}\}$ is a symmetric positive semidefinite matrix. An elementary computation yields

$$Q(\xi) = Q(-\xi)$$

and

$$Q(\lambda\xi) = |\lambda|^\alpha Q(\xi)$$

for real λ.

We make the non-degeneracy assumption: for $0<\alpha<2$, the support S_0 of $\Omega(ds)$ spans R^d as a vector space; for $\alpha=2$, we assume $\{a_{ij}\}$ is positive definite. Under these assumptions, it follows that

$$\inf_{\|\xi\|=1} Q(\xi) > 0 .$$

Therefore,

$$e^{-tQ(\xi)} \leqq e^{-tc\|\xi\|^{\alpha}}$$

and is rapidly decreasing. It follows from this that

$$\tilde{\nu}_t(y) = \left(\frac{1}{2\pi}\right)^d \int_{R^d} e^{-tQ(\xi)} e^{-i\langle\xi, y\rangle}\, d\xi$$

is bounded on all of R^d and is smooth. The same is true of

$$\begin{aligned}\nu_t(y) &= \sum_{g\in G_d^M} \tilde{\nu}_t(y+g)\\ &= c \sum_{\xi\in\tilde{G}_d^M} e^{-tQ(\xi)} e^{-i\langle\xi, y\rangle}\,.\end{aligned}$$

Here $\tilde{G}_d^M$ is the lattice which is dual to G_d^M. One can also check that, under the non-degeneracy assumption, $\tilde{\nu}_t(x)>0$ for each $t>0$ and x. Therefore the same is true of $\nu_t(x)$. The space X being compact, $\nu_t(x)$ has a positive lower bound for each $t>0$.

If we take a probability density $\tilde{k}(x)$ on R^d and project it onto X by

$$k(x) = \sum_{g\in G_d^M} \tilde{k}(x+g)\,,$$

then the family $k_\varepsilon(x)$ obtained from

$$\tilde{k}_\varepsilon(x) = \varepsilon^{-d}\tilde{k}\left(\frac{x}{\varepsilon}\right)$$

can be used as a mollifier. If we take $\varepsilon(t)=\varepsilon/t^{-1/d}$ for some $\varepsilon>0$, then Theorem 1.2 is applicable. In order to have a convenient reference for future use in the sausage problem (cf. [3]) we shall restate Theorem 1.2 as it applies to this case.

Let $\tilde{x}(t)$ be a symmetric stable process on R^d of order α, $0<\alpha\leqq 2$. It is characterized either by an Ω if $0<\alpha<2$, or by $\{a_{ij}\}$ if $\alpha=2$ (Brownian motion in R^d). We shall assume that it is non-degenerate. Let $x(t)$ be the process induced on the torus X by projection. Then $x(t)$ is again a process with independent increments on X. Let $L_t(\omega, A)$ be the occupation distribution of the $x(\cdot)$ process. We take a mollifier $k_\varepsilon(x)$ on X by projecting down a mollifier $\tilde{k}_\varepsilon(x)$ from R^d. The dependence of $\tilde{k}_\varepsilon(x)$ on ε is by a scale change

$$\tilde{k}_\varepsilon(x) = \varepsilon^{-d}\tilde{k}\left(\frac{x}{\varepsilon}\right).$$

We define

$$l_t^{\varepsilon}(\omega, y) = \int k_{\varepsilon}(y-x) L_t(\omega, dx)$$

and denote

$$g_t(\omega, y) = l_t^{\varepsilon(t)}(\omega, y),$$

where $\varepsilon(t) = \varepsilon t^{-1/d}$ with $\varepsilon > 0$ fixed. We denote by $R_{x,t}$ the distribution of $g_t(\omega, y)$ on γ and by $I(\mu)$ the I-functional corresponding to the given process. We then have the following theorem and its corollary.

THEOREM 5.1. *Let $C \subset \gamma$ be closed and $G \subset \gamma$ be open in the norm topology. Then*

$$\overline{\lim_{t\to\infty}} \frac{1}{t} \log R_{x,t}(C) \leqq -\inf_{\mu \in C} I(\mu),$$

$$\underline{\lim_{t\to\infty}} \frac{1}{t} \log R_{x,t}(G) \geqq -\inf_{\mu \in G} I(\mu).$$

COROLLARY. *Let $\Phi_t(f)$ be a family of functionals defined on γ such that, for each $f \in \gamma$ with $I(f) < \infty$ and each family $f_t \in \gamma$ converging to f in norm,*

$$\underline{\lim_{t\to\infty}} \Phi_t(f) \geqq \Phi(f),$$

where $\Phi(f)$ is a lower semicontinuous functional on γ. We assume further that $\Phi_t(f) \geqq 0$ for all t. Then,

$$\overline{\lim_{t\to\infty}} \frac{1}{t} \log E^{P_x}\{e^{-t\Phi_t(g_t(\omega,\cdot))}\} = \overline{\lim_{t\to\infty}} \frac{1}{t} \log E^{R_{x,t}}\{e^{-t\Phi_t(f)}\}$$

$$\leqq -\inf_{f \in \gamma} [\Phi(f) + I(f)].$$

Remark. If $f \in \gamma$, then f can be thought of as a measure μ on $\boldsymbol{X}$ and we use $I(\mu)$ and $I(f)$ interchangeably. The corollary follows from Theorem 5.1 by using Section 3 of [2].

Bibliography

[1] Donsker, M. D., and Varadhan, S. R. S., *Asymptotic evaluation of certain Markov process expectations for large time, I*, Comm. Pure App. Math. Vol. 28, 1975, pp. 1–47.

[2] Varadhan, S. R. S., *Asymptotic probabilities and differential equations*, Comm. Pure App. Math. Vol. 19, 1966, pp. 261–286.

[3] Donsker, M. D., and Varadhan, S. R. S., *Asymptotics for the Wiener sausage*, to appear in Comm. Pure App. Math., Vol. 28, issue 4, 1975.

Received October, 1974.

Functional Integration and its Applications, Clarendon Press, 1975, 15-33

Asymptotic evaluation of certain Wiener integrals for large time

M. D. DONSKER AND S. R. S. VARADHAN

2.1. Introduction

LET $V(y) \geqslant 0$ and continuous on $-\infty < y < \infty$. If $V(y) \to +\infty$ as $y \to \pm\infty$, the eigenvalue problem

$$\tfrac{1}{2}\psi''(y) - V(y)\psi(y) = -\lambda\psi(y) \tag{2.1}$$

has a discrete spectrum, and an old result of Kac [1] is that, for the least eigenvalue λ_1,

$$\lim_{t\to\infty} \frac{1}{t} \ln E_x\left\{\exp\left[-\int_0^t V(z(\tau))\,d\tau\right]\right\} = -\lambda_1 \tag{2.2}$$

where $E_x\{\ \}$ denotes expectation on the Brownian-motion process $\{z_\tau, 0 \leqslant \tau < \infty\}$ with $z(0) = x$.

The formal idea of his proof was that if, for any real y, we let $u(t, x, y)$ be the solution of

$$\frac{\partial u}{\partial t} = \frac{1}{2}\frac{\partial^2 u}{\partial x^2} - V(x)u,$$

$$u(x, 0) = \delta(x - y),$$

then, on the one hand, by the Feynman–Kac formula,

$$u(t, x, y) = E_x\left\{\exp\left[-\int_0^t V(z(\tau))\,d\tau\right]\delta(z(t) - y)\right\}$$

and, on the other hand,

$$u(t, x, y) = \sum_{j=1}^{\infty} \exp(-\lambda_j t)\psi_j(x)\psi_j(y)$$

where $\{\lambda_j\}$ and $\{\psi_j\}$ are respectively the eigenvalues and normalized eigenfunctions of (2.1). Thus, integrating on y

$$E_x\left\{\exp\left[-\int_0^t V(z(\tau))\,d\tau\right]\right\} = \sum_{j=1}^{\infty} \exp(-\lambda_j t)\psi_j(x) \int_{-\infty}^{\infty} \psi_j(y)\,dy$$

which gives (2.2).

A long-standing question has been the relation between (2.2) and the fact that we also know

$$\lambda_1 = \inf\left\{\int_{-\infty}^{\infty} V(y)\psi^2(y)\,dy + \tfrac{1}{2}\int_{-\infty}^{\infty} [\psi'(y)]^2\,dy\right\} \tag{2.3}$$

where the infimum is taken over all $\psi \in L^2$ such that $\int_{-\infty}^{\infty} \psi^2(y)\,dy = 1$. It is reasonable to expect that an expression like the right side of (2.3) should come from a direct asymptotic evaluation of $E_x\{\exp[-\int_0^t V(z(\tau))\,d\tau]\}$. Moreover, if such a direct connection can be made, one should be able to find the asymptotic behaviour of Brownian-motion expectations of more general functionals—including those for which there is no associated differential equation at all. Such a theorem is proved in this chapter.

Let $\mathcal{F}$ be the space of probability distribution functions $F(y)$, $-\infty < y < \infty$, with the Lévy metric:

$$\rho(F_1, F_2) = \inf_{h>0} [F_1(y-h) - h \leq F_2(y) \leq F_1(y+h) + h \text{ for all } y].$$

Let $\Phi(F)$ be a function on $\mathcal{F}$ satisfying:

(1) $0 \leq \Phi(F) \leq \infty$.

(2) $\Phi(F)$ is lower semicontinuous on $\mathcal{F}$.

(3) $\{F \in \mathcal{F} : \Phi(F) \leq K\}$ is a compact subset of $\mathcal{F}$ for every finite $K > 0$.

(4) If $F_n \Rightarrow F$ and the support of F_n is contained in an interval $[a, b]$ for all n, then $\Phi(F_n) \to \Phi(F)$.

(5) Let $F \in \mathcal{F}$ be such that $\Phi(F) < \infty$ and let $\{g_n(y)\}$ be a sequence of continuous functions such that $0 \leq g_n(y) \leq 1$, $-\infty < y < \infty$, with $g_n(y) = 1$ for $|y| \leq n$. If we define $F_n(y) = \int_{-\infty}^{y} g_n(\xi)\,dF(\xi) \Big/ \int_{-\infty}^{\infty} g_n(\xi)\,dF(\xi)$, then $\Phi(F_n) \to \Phi(F)$.

If $F \in \mathcal{F}$ has a density function f, we will sometimes for convenience write $\Phi(f)$ for $\Phi(F)$.

Let Ω_x be the space of continuous functions $z(\tau)$, $0 \leq \tau < \infty$, with $z(0) = x$. For $t > 0$ and $\omega \in \Omega_x$, we define

$$L(t, \omega, y) = \frac{1}{t}\lambda\{s : 0 \leq s \leq t,\ z(s) \leq y\},$$

where $\lambda\{\ \}$ denotes Lebesgue measure, i.e. $L(t, \omega, y)$ is the proportion of time on $[0, t]$ that a particular path $\omega = z(\cdot)$ is less than or equal to y. We note that for each ω and each $t > 0$, $L(t, \omega, .) \in \mathcal{F}$. Since each ω is bounded on $[0, t]$, $L(t, \omega, y)$ is a distribution function whose support is $[\inf_{0\leq\tau\leq t} z(\tau), \sup_{0\leq\tau\leq t} z(\tau)]$. $L(t, \omega, y)$ has a density function, $l(t, \omega, y)$, often called the 'local time', and conveniently expressed as

$$\frac{1}{t}\int_0^t \delta(z(\tau) - y)\,d\tau.$$

Let $\not{f}$ denote the space of probability density functions f which are continuously differentiable on $-\infty < y < \infty$ and either $f(y) > 0$ for all $-\infty < y < \infty$ or f has compact support and is strictly positive on the interior of that support. The main result of this chapter is the following.

THEOREM 2.1. *Let $\Phi(F)$ satisfy hypotheses* (1)–(5) *above. Then*

$$\lim_{t\to\infty}\frac{1}{t}\ln E_x\{e^{-t\Phi[L(t,\omega,.)]}\} = -\inf_{f\in\mathcal{f}}\left\{\frac{1}{8}\int_{-\infty}^{\infty}\frac{[f'(y)]^2}{f(y)}\,dy + \Phi(f)\right\}, \tag{2.4}$$

where $E_x\{\ \}$ denotes integration on Ω_x with respect to Brownian-motion measure.

Let us look at some examples of this theorem. In the particular case, $\Phi(F)=\int_{-\infty}^{\infty} V(y)\,dF(y)$, where $V(y)\geq 0$ and is continuous, and $V(y)\to+\infty$ as $y\to\pm\infty$, it is easy to see that $\Phi(F)$ satisfies hypotheses (1)–(5).† In this case,

$$\Phi(\dot{L}(t,\omega,.)) = \int_{-\infty}^{\infty} V(y)\,dL(t,\omega,y) = \frac{1}{t}\int_0^t V(z(\tau))\,d\tau,$$

and if, for any $f\in\mathcal{f}$, we let $\psi(y)=\sqrt{(f(y))}$, then $\psi\in L^2$ and, indeed, $\int_{-\infty}^{\infty}\psi^2(y)\,dy=1$. Moreover,

$$\int_{-\infty}^{\infty} V(y)\,dF(y) = \int_{-\infty}^{\infty} V(y)f(y)\,dy = \int_{-\infty}^{\infty} V(y)\psi^2(y)\,dy$$

and

$$\frac{1}{8}\int_{-\infty}^{\infty}\frac{[f'(y)]^2}{f(y)}\,dy = \frac{1}{2}\int_{-\infty}^{\infty}[\psi'(y)]^2\,dy.$$

Thus, in the special case $\Phi(F)=\int_{-\infty}^{\infty} V(y)dF(y)$, (2.4) becomes

$$\lim_{t\to\infty}\frac{1}{t}\ln E_x\left\{\exp\left[-\int_0^t V(z(\tau))\,d\tau\right]\right\}$$

$$= -\inf\left\{\int_{-\infty}^{\infty} V(y)\psi^2(y)\,dy + \frac{1}{2}\int_{-\infty}^{\infty}[\psi'(y)]^2\,dy\right\} \tag{2.5}$$

where the infimum is taken over all $\psi\in L^2$ such that $\int_{-\infty}^{\infty}\psi^2(y)\,dy=1$. This solves the problem mentioned earlier.

If we let

$$\Phi(F) = \int_{-\infty}^{\infty}\int_{-\infty}^{\infty}\rho(y_1, y_2,)\,dF(y_1)\,dF(y_2),$$

where $\rho(y_1,y_2)$ is continuous and $\rho(y_1,y_2)\to+\infty$ if $|y_1|+|y_2|\to\infty$, then $\Phi(F)$ satisfies hypotheses (1)–(5) and (2.4) becomes

$$\lim_{t\to\infty}\frac{1}{t}\ln E_x\left\{\exp\left[-\frac{1}{t}\int_0^t\int_0^t\rho(z(s),z(\tau))\,ds\,d\tau\right]\right\}$$

$$= -\inf_{f\in\mathcal{f}}\left\{\frac{1}{8}\int_{-\infty}^{\infty}\frac{[f'(y)]^2}{f(y)}\,dy + \int_{-\infty}^{\infty}\int_{-\infty}^{\infty}\rho(y_1,y_2)f(y_1)f(y_2)\,dy_1\,dy_2\right\}. \tag{2.6}$$

† The condition $V(y)\to+\infty$ as $y\to\pm\infty$ is used only for hypothesis (3).

If $\Phi(F)=\exp[\int_{-\infty}^{\infty} y^2\,\mathrm{d}F(y)]$, one obtains readily from (2.4) that

$$\lim_{t\to\infty}\frac{1}{t}\ln E_x\left(\exp\left\{-t\exp\left[\frac{1}{t}\int_0^t z^2(\tau)\,\mathrm{d}\tau\right]\right\}\right)=-\frac{\alpha+1}{8\alpha^2},$$

where $\alpha>0$ is the root of $\mathrm{e}^u=1/8u^2$.

Before we proceed with the proof we shall try to explain the idea behind the proof. By the ergodic theorem, since Brownian motion has no finite invariant measure on the line, as time increases, most of the Brownian paths will spend only a negligible proportion of time in a finite region. However, there will be a few Brownian paths, with very low probability, which hang around too long. Let us fix a nice probability density $f(y)$ and ask the following question. What is the probability that the occupation distribution $L(t,\omega,y)$ of a Brownian path is close to the distribution $F(y)$ with density $f(y)$? The probability will of course go to zero with t: the question is, at what rate? It turns out that, roughly,

$$\operatorname{Prob}\{L(t,\omega,.)\sim F(.)\}\sim\exp\left\{-\frac{t}{8}\int_{-\infty}^{\infty}\frac{[f'(y)]^2}{f(y)}\,\mathrm{d}y\right\}.$$

For practical purposes one can view the measure Q_x^t on the space of distribution functions as behaving like

$$Q_x^t\sim\exp\left\{-\frac{t}{8}\int_{-\infty}^{\infty}\frac{[f'(y)]^2}{f(y)}\,\mathrm{d}y\right\}\mathrm{d}f.$$

The theorem is then a Laplace asymptotic formula.

This type of reasoning generalizes to a wide class of Markov processes both with discrete and continuous time and will be dealt with in several forthcoming articles.

2.2. The upper estimate

Let P_x denote Brownian-motion measure on Ω_x. For each x and $t>0$ we define a probability measure $Q_{x,t}$ on $\mathscr{F}$ by

$$Q_{x,t}(C)=P_x\{\omega: L(t,\omega,.)\in C\}$$

for measurable sets $C\subset\mathscr{F}$, i.e. for which the right side is defined.

Denoting expectation with respect to the measure $Q_{x,t}$ by $E^{Q_{x,t}}\{\ \}$, we have

$$E_x\{\mathrm{e}^{-t\Phi[L(t,\omega,.)]}\}=E^{Q_{x,t}}(\mathrm{e}^{-t\Phi(F)}).\tag{2.7}$$

In order to get an upper estimate on the left side of (2.7) we first get an upper estimate on $Q_{x,t}(C)$ for appropriate sets C. Let $\mathscr{U}$ be the space of functions $u(y)$, $-\infty<y<\infty$, having two bounded continuous derivatives and for each of which there are two positive numbers α and β such that for all y, $0<\alpha\leqslant u(y)\leqslant\beta<\infty$.

For $u \in \mathcal{U}$, define

$$\psi(x,t) = E_x\left\{u(z(t)) \exp\left[-\int_0^t \left(\frac{u''}{2u}\right)(z(s))\, ds\right]\right\}.$$

From the Feynman–Kac formula we see that $\psi(x, t)$ satisfies

$$\frac{\partial\psi}{\partial t} = \frac{1}{2}\frac{\partial^2\psi}{\partial x^2} - \frac{u''(x)}{2u(x)}\ \psi,$$

but, since the solution of this last is clearly $\psi(x, t) = u(x)$, we have

$$E_x\left\{u(z(t)) \exp\left[-\int_0^t \left(\frac{u''}{2u}\right)(z(s))\, ds\right]\right\} = u(x). \tag{2.8}$$

Since $u(y) \geqslant \alpha > 0$ for all y,

$$E_x\left\{\exp\left[-\int_0^t \left(\frac{u''}{2u}\right)(z(s))\, ds\right]\right\} \leqslant \frac{u(x)}{\alpha}.$$

But

$$\begin{aligned} E_x\left\{\exp\left[-\int_0^t \left(\frac{u''}{2u}\right)(z(s))\, ds\right]\right\} &= E_x\left\{\exp\left[-t\int_{-\infty}^{\infty} \left(\frac{u''}{2u}\right)(y)\, dL(t, \omega, y)\right]\right\} \\ &= E^{Q_{x,t}}\left\{\exp\left[-t\int_{-\infty}^{\infty} \left(\frac{u''}{2u}\right)(y)\, dF(y)\right]\right\}, \end{aligned}$$

and therefore for any measurable $C \subset \mathcal{F}$,

$$Q_{x,t}(C) \leqslant \frac{u(x)}{\alpha} \exp\left\{t \sup_{F\in C} \int_{-\infty}^{\infty} \left(\frac{u''}{2u}\right)(y)\, dF(y)\right\}. \tag{2.9}$$

Since (2.9) holds for any $u \in \mathcal{U}$, we obtain for any measurable $C \subset \mathcal{F}$,

$$\overline{\lim_{t\to\infty}} \frac{1}{t} \ln Q_{x,t}(C) \leqslant \inf_{u\in\mathcal{U}} \sup_{F\in C} \int_{-\infty}^{\infty} \left(\frac{u''}{2u}\right)(y)\, dF(y). \tag{2.10}$$

Let

$$I(F) = -\inf_{u\in\mathcal{U}} \int_{-\infty}^{\infty} \frac{u''(y)}{2u(y)}\, dF(y).$$

We see that $I(F)$ is lower semicontinuous on $\mathcal{F}$ and $0 \leqslant I(F) \leqslant \infty$. We will show now that for compact sets $C \subset \mathcal{F}$ it follows from (2.10) that

$$\begin{aligned} \overline{\lim_{t\to\infty}} \frac{1}{t} \ln Q_{x,t}(C) &\leqslant \sup_{F\in C} \inf_{u\in\mathcal{U}} \int_{-\infty}^{\infty} \frac{u''(y)}{2u(y)}\, dF(y) \\ &= -\inf_{f\in C} I(F) \end{aligned}$$

First of all, for any measurable set $C \subset \mathscr{F}$ which is contained in a finite union $\bigcup_{j=1}^{k} C_j$ of measurable sets, we have from (2.10)

$$\overline{\lim_{t\to\infty}} \frac{1}{t} \ln Q_{x,t}(C) \leq \max_{1\leq j\leq k} \inf_{u\in\mathscr{U}} \sup_{F\in C_j} \int_{-\infty}^{\infty} \frac{u''(y)}{2u(y)} \, dF(y),$$

and for any measurable set $C \subset \mathscr{F}$

$$\overline{\lim_{t\to\infty}} \frac{1}{t} \ln Q_{x,t}(C) \leq \inf_{\substack{C_1, C_2, \ldots, C_k \\ C \subset \bigcup_{j=1}^{k} C_j}} \max_{1\leq j\leq k} \inf_{u\in\mathscr{U}} \sup_{F\in C_j} \int_{-\infty}^{\infty} \frac{u''(y)}{2u(y)} \, dF(y).$$

Let C be a compact subset of $\mathscr{F}$ and let

$$l = \sup_{F\in C} \inf_{u\in\mathscr{U}} \int_{-\infty}^{\infty} \frac{u''(y)}{2u(y)} \, dF(y),$$

so that for every $F \in C$,

$$\inf_{u\in\mathscr{U}} \int_{-\infty}^{\infty} \frac{u''(y)}{2u(y)} \, dF(y) \leq l.$$

Thus, given $\varepsilon > 0$, for every $F \in C$ there exists $u_F \in \mathscr{U}$ such that

$$\int_{-\infty}^{\infty} \frac{u_F''(y)}{2u_F(y)} \, dF(y) \leq l + \varepsilon.$$

Since $u''(y)/2u(y)$ is bounded and continuous on $-\infty < y < \infty$ for every $u \in \mathscr{U}$, $\int_{-\infty}^{\infty} u_F''(y)/2u_F(y) \, dF(y)$ is a continuous function on $\mathscr{F}$. Thus for every $F \in C$ there is a Levy neighbourhood N_F such that for every $G \in N_F$,

$$\int_{-\infty}^{\infty} \frac{u_F''(y)}{2u_F(y)} \, dG(y) \leq l + \varepsilon.$$

Now $\bigcup_{F\in C} N_F$ is an open covering of C and therefore there exists $F_1, F_2, \ldots, F_k$ in C such that $C \subset \bigcup_{j=1}^{k} N_{F_j}$. Let $C_j = N_{F_j}$ and we have

$$\sup_{G\in C_j} \int_{-\infty}^{\infty} \frac{u_{F_j}''(y)}{2u_{F_j}(y)} \, dG(y) \leq l + 2\varepsilon,$$

$$\inf_{u\in\mathscr{U}} \sup_{G\in C_j} \int_{-\infty}^{\infty} \frac{u''(y)}{2u(y)} \, dG(y) \leq l + 2\varepsilon,$$

$$\max_{1\leq j\leq k} \inf_{u\in\mathscr{U}} \sup_{G\in C_j} \int_{-\infty}^{\infty} \frac{u''(y)}{2u(y)} \, dG(y) \leq l + 2\varepsilon,$$

and finally

$$\inf_{\substack{C_1,C_2,\ldots,C_k \\ \bigcup_{j=1}^{k} C_j \supset C}} \max_{l\leqslant j\leqslant k} \inf_{u\in\mathcal{U}} \sup_{G\in C_j} \int_{-\infty}^{\infty} \frac{u''(y)}{2u(y)}\,\mathrm{d}G(y) \leqslant l+2\varepsilon.$$

Since ε is arbitrary, we conclude that for compact sets C†

$$\inf_{\substack{C_1,C_2,\ldots,C_k \\ \bigcup_{j=1}^{k} C_j \supset C}} \max_{l\leqslant j\leqslant k} \inf_{u\in\mathcal{U}} \sup_{F\in C_j} \int_{-\infty}^{\infty} \frac{u''(y)}{2u(y)}\,\mathrm{d}F(y) \leqslant \sup_{F\in C} \inf_{u\in\mathcal{U}} \int_{-\infty}^{\infty} \frac{u''(y)}{2u(y)}\,\mathrm{d}F(y)$$

$$= -\inf_{F\in C} I(F). \tag{2.13}$$

Thus (2.13) and (2.12) imply (2.11).

For $0<K<\infty$, let $A_K=\{F\in\mathscr{F}:\Phi(F)\leqslant K\}$ which is compact from hypothesis (3) on Φ. From (2.1) we have

$$\begin{aligned} E_x\{e^{-t\Phi[L(t,\omega,.)]}\} &= E^{Q_{x,t}}_{A_K}\{e^{-t\Phi(F)}\} + E^{Q_{x,t}}_{\bar{A}_K}\{e^{-t\Phi(F)}\} \\ &\leqslant E^{Q_{x,t}}_{A_K}\{e^{-t\Phi(F)}\} + e^{-tK}. \end{aligned} \tag{2.14}$$

Let $F^*\in A_K$ and let $\varepsilon>0$ be given. Assume first of all that $I(F^*)>\infty$. Then, since both $\Phi(F)$ and $I(F)$ are lower semicontinuous at F^*, there exists a neighbourhood N_{F^*} such that for all $F\in N_{F^*}$,

$$\Phi(F^*)\leqslant\Phi(F)+\varepsilon,$$

$$I(F^*)\leqslant I(F)+\varepsilon.$$

Let G_{F^*} be another neighbourhood of F^* such that $\bar{G}_{F^*}\subset N_{F^*}$. We have

$$E^{Q_{x,t}}_{G_{F^*}\cap A_K}(e^{-t\Phi(F)})\leqslant e^{-t[\Phi(F^*)-\varepsilon]}Q_{x,t}(\bar{G}_{F^*}\cap A_K).$$

Since $\bar{G}_{F^*}\cap A_K$ is a closed subset of a compact set it is itself compact, and therefore from (2.11)

$$\begin{aligned} \overline{\lim_{t\to\infty}}\frac{1}{t}\ln E^{Q_{x,t}}_{G_{F^*}\cap A_K}(e^{-t\Phi(F)}) &\leqslant -\Phi(F^*)+\varepsilon-\inf_{F\in G_{F^*}\cap A_K} I(F) \\ &\leqslant -\Phi(F^*)+\varepsilon-[I(F^*)-\varepsilon] \\ &\leqslant -\inf_{F\in\mathscr{F}}[\Phi(F)+I(F)]+2\varepsilon. \end{aligned} \tag{2.15}$$

On the other hand, if $I(F^*)=\infty$ we have from the lower semicontinuity of $I(F)$ that there exists a neighbourhood M_{F^*} such that for all $F\in M_{F^*}$, $I(F)\geqslant K$. Let G_{F^*} be another neighbourhood of F^* such that $\bar{G}_{F^*}\subset M_{F^*}$.

† In fact it is easy to show that for compact sets C equality holds in (2.13).

Then

$$E^{Q_{x,t}}_{G_{F^*}\cap A_K}(e^{-t\Phi(F)}) \leq Q_{x,t}(\bar{G}_{F^*}\cap A_K),$$

and again from (2.11) we have

$$\overline{\lim_{t\to\infty}} \frac{1}{t} \ln E^{Q_{x,t}}_{G_{F^*}\cap A_K}(e^{-t\Phi(F)}) \leq - \inf_{F\in\bar{G}_{F^*}\cap A_K} I(F) \leq -K. \tag{2.16}$$

Thus each point F^* of A_K is covered by a neighbourhood G_{F^*} and since A_K is compact there exists $F_1^*, F_2^*, \ldots, F_n^*$ such that $A_K \subset \bigcup_{j=1}^n G_{F^*_j}$. From (2.15) and (2.16) we have, therefore,

$$\overline{\lim_{t\to\infty}} \frac{1}{t} \ln E^{Q_{x,t}}_{A_K}(e^{-t\Phi(F)}) \leq -\min\{K, \inf_{F\in\mathscr{F}} [\Phi(F)+I(F)]-2\varepsilon\}. \tag{2.17}$$

From (2.14) and (2.17) we obtain

$$\overline{\lim_{t\to\infty}} \frac{1}{t} \ln E_x(e^{-t\Phi[L(t,\omega,.)]}) \leq -\min\{K, \inf_{F\in\mathscr{F}} [\Phi(F)+I(F)]-2\varepsilon\},$$

and letting $\varepsilon \to 0$ we get

$$\overline{\lim_{t\to\infty}} \frac{1}{t} \ln E_x(e^{-t\Phi[L(t,\omega,.)]}) \leq -\min\{K, \inf_{F\in\mathscr{F}} [\Phi(F)+I(F)]\}.$$

Finally, letting $K \to \infty$ we find

$$\overline{\lim_{t\to\infty}} \frac{1}{t} \ln E_x(e^{-t\Phi[L(t,\omega,.)]}) \leq - \inf_{F\in\mathscr{F}} [\Phi(F)+I(F)]. \tag{2.18}$$

In the next section we obtain a lower estimate on

$$\underline{\lim_{t\to\infty}} \frac{1}{t} \ln E_x(e^{-t\Phi[L(t,\omega,.)]}),$$

and in § 2.4 we show that this lower estimate and the upper estimate on the right of (2.18) are equal and in fact are the same as the right side of (2.4). This will prove the theorem.

2.3. The lower estimate

Let $[a, b]$ be a finite closed interval such that $x\in(a, b)$. Let $f(y)$ be a density function having two continuous derivatives on $-\infty<y<\infty$, $f(y)>0$ for $y\in(a, b)$, $f(y)=0$ for y outside of $[a, b]$, and

$$\int_a^b \frac{[f'(y)]^2}{f(y)}\,dy<\infty.$$

Let $F(y)$ be the distribution function corresponding to $f(y)$.

For each $\omega = z(\,.\,) \in \Omega_x$ such that $a \leq z(s) \leq b$ for all $0 \leq s \leq t$, $L(t, \omega, y)$ is a distribution function whose support is contained in $[a, b]$. Thus, by hypothesis (4) (p.16) on Φ, for any $\varepsilon > 0$ there exists a Lévy neighbourhood of F, say N_F, so that if $L(t, \omega, .) \in N_F$ for any ω such that $a \leq z(s) \leq b$ for all $0 \leq s \leq t$, then

$$|\Phi(L(t, \omega, .)) - \Phi(F)| \leq \varepsilon.$$

Thus, if we let $C^t_{a,b} = \{z \in \Omega_x : a \leq z(s) \leq b \text{ for all } 0 \leq s \leq t\}$,

$$\begin{aligned} E_x(\mathrm{e}^{-t\Phi[L(t,\omega,.)]}) &\geq E_x\{\mathrm{e}^{-t\Phi[L(t,\omega,.)]};\ L(t, \omega, .) \in N_F,\ z \in C^t_{a,b}\} \\ &\geq \mathrm{e}^{-t[\Phi(F)+\varepsilon]} P_x\{L(t, \omega, .) \in N_F,\ z \in C^t_{a,b}\}. \qquad (2.19) \end{aligned}$$

In Lemma 2.1, which follows, we will show

$$\varliminf_{t\to\infty} \frac{1}{t} \ln P_x\{L(t, \omega, .) \in N_F,\ z \in C^t_{a,b}\} \geq -\frac{1}{8}\int_a^b \frac{[f'(y)]^2}{f(y)}\,\mathrm{d}y. \qquad (2.20)$$

From (2.19) we obtain that

$$\varliminf_{t\to\infty} \frac{1}{t} \ln E_x\{\mathrm{e}^{-t\Phi[L(t,\omega,.)]}\} \geq -\Phi(F) - \varepsilon - \frac{1}{8}\int_a^b \frac{[f'(y)]^2}{f(y)}\,\mathrm{d}y,$$

and since ε is arbitrary we obtain

$$\begin{aligned} \varliminf_{t\to\infty} \frac{1}{t} \ln E_x(\mathrm{e}^{-t\Phi[L(t,\omega,.)]}) &\geq -\left\{\Phi(F) + \frac{1}{8}\int_a^b \frac{[f'(y)]^2}{f(y)}\,\mathrm{d}y\right\} \\ &\geq -\inf_{f\in\mathcal{f}_x}\left\{\Phi(f) + \frac{1}{8}\int_a^b \frac{[f'(y)]^2}{f(y)}\,\mathrm{d}y\right\}, \qquad (2.21) \end{aligned}$$

where $\mathcal{f}_x$ is the set of all density functions f with two continuous derivatives for which there is a closed interval $[a, b]$ containing x in its interior, outside of which $f = 0$, $f > 0$ on (a, b), and

$$\int_a^b \frac{[f'(y)]^2}{f(y)}\,\mathrm{d}y < \infty.$$

Since (2.20) is of independent interest we prove it as a lemma.

LEMMA 2.1. *Let $f \in \mathcal{f}_x$ and let N_F be any Lévy neighbourhood of its corresponding distribution function. Then*

$$\varliminf_{t\to\infty} \frac{1}{t} \ln P_x\{L(t, \omega, .) \in N_F,\ a \leq z(s) \leq b \quad \text{for all} \quad 0 \leq s \leq t\} \geq -\frac{1}{8}\int_a^b \frac{[f'(y)]^2}{f(y)}\,\mathrm{d}y.$$

Proof. Let $b(y)=\frac{1}{2}f'(y)/f(y)$ and consider Brownian motion with drift b, i.e. the process whose generator is

$$\frac{1}{2}\frac{\partial^2}{\partial x^2}+b(x)\frac{\partial}{\partial x}.$$

Now, the solution of

$$\frac{1}{2}\frac{\partial^2 u}{\partial x^2}+b(x)\frac{\partial u}{\partial x}=0$$

being, in the case $b(x)=\frac{1}{2}f'(x)/f(x)$, simply $u'=1/f$, we have

$$u(y)=\int_\eta^y \frac{\mathrm{d}\xi}{f(\xi)}$$

for $y\in(a,b)$, $\eta\in(a,b)$. Hence $u(y)\to-\infty$ as $y\to a$ and $u(y)\to+\infty$ as $y\to b$, which means according to Feller's test [2] that with probability 1 Brownian-motion paths with this drift never leave (a,b).

If we let $\tilde{P}_x$ denote the measure on Ω_x induced by this drift process, we have the Cameron–Martin formula [3]

$$P_x\{A\cap[z(s)\in[a,b]\quad\text{for all}\quad 0\leqslant s\leqslant t]\}$$
$$=E_A^{\tilde{P}_x}\left\{\exp\left[-\int_0^t b(z(s))\,\mathrm{d}z(s)+\frac{1}{2}\int_0^t b^2(z(s))\,\mathrm{d}s\right]\right\}.$$

In our case, this becomes

$$P_x\{L(t,\omega,.)\in N_F,\ z(s)\in[a,b]\quad\text{for all}\quad 0\leqslant s\leqslant t\}$$
$$=E^{\tilde{P}_x}\left\{\exp\left[-\frac{1}{2}\int_0^t\left(\frac{f'}{f}\right)(z(s))\,\mathrm{d}z(s)+\frac{1}{8}\int_0^t\left(\frac{f'}{f}\right)^2(z(s))\,\mathrm{d}s\right];\quad L(t,\omega,.)\in N_F\right\}. \tag{2.22}$$

Since f is twice continuously differentiable, we have by Ito's formula

$$\mathrm{d}\ln f(z(s))=(\ln f)'(z(s))\,\mathrm{d}z(s)+\tfrac{1}{2}(\ln f)''(z(s))\,\mathrm{d}s,$$

and, integrating this from 0 to t using $z(0)=x$, we obtain

$$\ln f(z(t))-\ln f(x)$$
$$=\int_0^t\left(\frac{f'}{f}\right)(z(s))\,\mathrm{d}z(s)+\frac{1}{2}\int_0^t\left(\frac{f''}{f}\right)(z(s))\,\mathrm{d}s-\frac{1}{2}\int_0^t\left(\frac{f'}{f}\right)^2(z(s))\,\mathrm{d}s. \tag{2.23}$$

Using (2.23) in (2.22) we obtain

$$P_x\{L(t,\omega,.)\in N_F,\ z(s)\in[a,b]\quad \text{for all}\quad 0\leqslant s\leqslant t\}$$
$$=E^{\tilde{P}_x}\left(\sqrt{\left[\frac{f(x)}{f(z(t))}\right]}\exp\left\{-\left[\frac{1}{8}\int_0^t\left(\frac{f'}{f}\right)^2(z(s))\,ds-\frac{1}{4}\int_0^t\left(\frac{f''}{f}\right)(z(s))\,ds\right]\right\};\right.$$
$$\left. L(t,\omega,.)\in N_F\right). \tag{2.24}$$

Let

$$J(y)=\frac{1}{8}\left(\frac{f'}{f}\right)^2(y)-\frac{1}{4}\left(\frac{f''}{f}\right)(y),$$

and let $\varepsilon>0$ be given. Define

$$S(t,\varepsilon)=\left\{z\in\Omega_x:\left|\int_a^b J(y)\,dL(t,\omega,y)-\int_a^b J(y)f(y)\,dy\right|<\varepsilon,\right.$$

and

$$S'(t,\varepsilon)=\{z\in\Omega_x:L(t,\omega,.)\in N_F\}\cap S(t,\varepsilon).$$

Since, for almost all $z\in\Omega_x$ ($\tilde{P}_x$-measure),

$$\int_a^b J(y)\,dL(t,\omega,y)=\int_{-\infty}^{\infty} J(y)L(t,\omega,y)=\frac{1}{t}\int_0^t J(z(s))\,ds,$$

we have from (2.24)

$$P_x\{L(t,\omega,.)\in N_F,\ z(s)\in[a,b]\quad \text{for all}\quad 0\leqslant s\leqslant t\}$$
$$=E^{\tilde{P}_x}\left(\sqrt{\left[\frac{f(x)}{f(z(t))}\right]}\exp\left\{-t\int_a^b J(y)L(t,\omega,y)\right];\quad L(t,\omega,.)\in N_F\right)$$
$$\geqslant\exp\left\{-t\left[\varepsilon+\int_a^b J(y)f(y)\,dy\right]\right\}E^{\tilde{P}_x}\left\{\sqrt{\left[\frac{f(x)}{f(z(t))}\right]};\quad S'(t,\varepsilon)\right\}$$
$$=\exp\left\{-\varepsilon t-\frac{t}{8}\int_a^b\frac{[f'(y)]^2}{f(y)}\,dy\right\}E^{\tilde{P}_x}\left\{\sqrt{\left[\frac{f(x)}{f(z(t))}\right]};\quad S'(t,\varepsilon)\right\}. \tag{2.25}$$

Because of the hypotheses on f there exists a constant γ such that

$$\sqrt{\left[\frac{f(x)}{f(z(t))}\right]}\geqslant\gamma>0$$

for almost all $z \in \Omega_x$ ($\tilde{P}_x$-measure). Therefore

$$E^{\tilde{P}_x}\left\{\sqrt{\left[\frac{f(x)}{f(z(t))}\right]};\quad S'(t,\varepsilon)\right\} \geqslant \gamma\tilde{P}_x(S'(t,\varepsilon))$$
$$= \gamma - \gamma\tilde{P}_x(\Omega_x - S'(t,\varepsilon)). \quad (2.26)$$

Now the reason the drift $b(y) = \frac{1}{2}f'(y)/f(y)$ was selected is because then f itself is the invariant measure for the drift process. This is easy to see since if the drift is $\frac{1}{2}f'/f$ then $u = f$ is the solution of

$$\tfrac{1}{2}u'' - [b(x)u(x)]' = 0,$$

the adjoint equation. From the ergodic properties of Markov processes (see e.g. [2])

$$\lim_{t\to\infty} \tilde{P}_x(\Omega_x - S(t,\varepsilon))$$
$$= \lim_{t\to\infty} \tilde{P}_x\left(z \in \Omega_x : \left|\int_a^b J(y)\,\mathrm{d}L(t,\omega,y) - \int_a^b J(y)f(y)\,\mathrm{d}y\right| \geqslant \varepsilon\right)$$
$$= 0$$

and

$$\lim_{t\to\infty} \tilde{P}_x(z \in \Omega_x : L(t,\omega,.) \notin N_F) = 0.$$

These last two statements imply

$$\lim_{t\to\infty} \tilde{P}_x(\Omega_x - S'(t,\varepsilon)) = 0,$$

and, therefore, from (2.25) and (2.26) we obtain

$$\varliminf_{t\to\infty} \frac{1}{t} \ln P_x\{L(t,\omega,.) \in N_F,\ z(s) \in [a,b] \quad \text{for all} \quad 0 \leqslant s \leqslant t\}$$
$$\geqslant -\varepsilon - \frac{1}{8}\int_a^b \frac{[f'(y)]^2}{f(y)}\,\mathrm{d}y.$$

Finally, letting $\varepsilon \to 0$, we obtain (2.20).

2.4. The equality of the estimates

In this section we want to show that the expressions on the right of (2.18) and (2.21) are equal and the same as the expression on the right of (2.4). We shall first show Lemma 2.2.

LEMMA 2.2. *Let f be a density function such that $f>0$ on (a, b) where $-\infty\leqslant a<b\leqslant+\infty$, $f=0$ outside of $[a, b]$ and f is continuously differentiable on $-\infty<y<\infty$. Then, letting F be the corresponding distribution function,*

$$I(F)=\frac{1}{8}\int_a^b \frac{[f'(y)]^2}{f(y)}\,dy, \tag{2.27}$$

in the sense that if one side is finite so is the other and they are equal.

Proof. Assume first that

$$\int_a^b \frac{[f'(y)]^2}{f(y)}\,dy<\infty.$$

We show now that for every $u\in\mathcal{U}$

$$\int_{-\infty}^{\infty} \frac{u''(y)}{2u(y)}\,dF(y)\geqslant -\frac{1}{8}\int_a^b \frac{[f'(y)]^2}{f(y)}\,dy, \tag{2.28}$$

which will imply

$$I(F)\leqslant\frac{1}{8}\int_a^b \frac{[f'(y)]^2}{f(y)}\,dy<\infty. \tag{2.29}$$

So, for any $u\in\mathcal{U}$, let $h=\ln u$ and we have

$$\begin{aligned}\int_{-\infty}^{\infty} \frac{u''(y)}{2u(y)}\,dF(y)&=\frac{1}{2}\int_{-\infty}^{\infty} \{h''(y)+[h'(y)]^2\}f(y)\,dy\\ &=\frac{1}{2}\int_a^b \{h''(y)+[h'(y)]^2\}f(y)\,dy.\end{aligned} \tag{2.30}$$

Using elementary inequalities

$$\begin{aligned}\left|\int_a^b h'(y)f'(y)\,dy\right| &= \left|\int_a^b \frac{h'(y)f'(y)}{\sqrt{[f(y)]}}\sqrt{[f(y)]}dy\right|\\ &\leqslant \sqrt{\left[\int_a^b [h'(y)]^2 f(y)\,dy\right]}\sqrt{\left[\int_a^b \frac{[f'(y)]^2}{f(y)}\,dy\right]}\\ &=2\sqrt{\left[\int_a^b [h'(y)]^2 f(y)\,dy\right]}\sqrt{\left[\frac{1}{4}\int_a^b \frac{[f'(y)]^2}{f(y)}\,dy\right]}\\ &\leqslant \int_a^b [h'(y)]^2 f(y)\,dy+\frac{1}{4}\int_a^b \frac{[f'(y)]^2}{f(y)}\,dy,\end{aligned}$$

and hence

$$\frac{1}{2}\left\{\int_a^b [h'(y)]^2 f(y)\,\mathrm{d}y - \int_a^b h'(y)f'(y)\,\mathrm{d}y\right\} \geq -\frac{1}{8}\int_a^b \frac{[f'(y)]^2}{f(y)}\,\mathrm{d}y.$$

Integrating by parts we obtain (2.28) and hence (2.29).

To complete the proof it suffices to show that if F is such that $I(F) < \infty$, then

$$\frac{1}{8}\int_a^b \frac{[f'(y)]^2}{f(y)}\,\mathrm{d}y \leq I(F). \tag{2.31}$$

From the definition of $I(F)$, we have for every $u \in \mathcal{U}$ and every $h = \ln u$

$$\int_a^b \frac{u''(y)}{2u(y)}\,\mathrm{d}F(y) = \frac{1}{2}\int_a^b [h'(y)]^2 f(y)\,\mathrm{d}y + \frac{1}{2}\int_a^b h''(y)f(y)\,\mathrm{d}y \geq -I(F). \tag{2.32}$$

Now (2.32) holds in particular for these $u \in \mathcal{U}$ such that $h = \ln u$ are constant outside of a closed bounded interval contained in (a, b). If we let $\mathcal{H}$ denote this latter class of functions h, we have from (2.32) for any $h \in \mathcal{H}$

$$\frac{1}{2}\int_a^b [h'(y)]^2 f(y)\,\mathrm{d}y - \frac{1}{2}\int_a^b h'(y)f'(y)\,\mathrm{d}y \geq -I(F). \tag{2.33}$$

If $h \in \mathcal{H}$ then $\lambda h \in \mathcal{H}$ for any real λ, so from (2.33) we have for all $h \in \mathcal{H}$

$$\frac{\lambda^2}{2}\int_a^b [h'(y)]^2 f(y)\,\mathrm{d}y - \frac{\lambda}{2}\int_a^b h'(y)f'(y)\,\mathrm{d}y + I(F) \geq 0.$$

Thus, the discriminant of this quadratic in λ must be less than or equal to zero, i.e. for all $h \in \mathcal{H}$,

$$\left|\int_a^b h'(y)f'(y)\,\mathrm{d}y\right| \leq \sqrt{(8I(F))}\sqrt{\left\{\int_a^b [h'(y)]^2 f(y)\,\mathrm{d}y\right\}}. \tag{2.34}$$

If we let $g(y) = h'(y)$, then we have

$$\left|\int_a^b g(y)f'(y)\,\mathrm{d}y\right| \leq \sqrt{[8I(F)]}\sqrt{\left[\int_a^b g^2(y)f(y)\,\mathrm{d}y\right]} \tag{2.35}$$

is valid for all functions $g(y)$ which have compact support contained in (a, b) and which have a continuous derivative. By approximation we then have (2.35) valid for all functions $g(y)$ which are continuous and have compact support contained in (a, b). Let $\phi(y) \in C^\infty$ with compact support contained in (a, b) and $0 \leq \phi(y) \leq 1$ for all $-\infty < y < \infty$. We now apply (2.35) to

$g(y)=[f'(y)/f(y)]\phi(y)$, obtaining

$$\left|\int_a^b \frac{[f'(y)]^2}{f(y)}\phi(y)\,dy\right| \leqslant \sqrt{(8I(F))}\sqrt{\left\{\int_a^b \frac{[f'(y)]^2}{f(y)}\phi^2(y)\,dy\right\}}$$

$$\leqslant \sqrt{(8I(F))}\sqrt{\left\{\int_a^b \frac{[f'(y)]^2}{f(y)}\phi(y)\,dy\right\}}.$$

Dividing and squaring we obtain

$$\frac{1}{8}\int_a^b \frac{[f'(y)]^2}{f(y)}\phi(y)\,dy \leqslant I(F). \tag{2.36}$$

Since (2.36) holds for all $\phi(y)$ such that $0\leqslant\phi(y)\leqslant 1$, $\phi(y)\in C^\infty$, $\phi(y)$ has compact support contained in (a, b) we conclude that in fact

$$\frac{1}{8}\int_a^b \frac{[f'(y)]^2}{f(y)}\,dy \leqslant I(F).$$

We now prove Lemma 2.3.

LEMMA 2.3.

$$\inf_{f\in f_x}\left\{\frac{1}{8}\int_a^b \frac{[f'(y)]^2}{f(y)}dy+\Phi(f)\right\}=\inf_{F\in\mathscr{F}}[I(F)+\Phi(F)]. \tag{2.37}$$

Proof. Let $f\in f_x$, then from Lemma 2.2 we have

$$\frac{1}{8}\int_a^b \frac{[f'(y)]^2}{f(y)}\,dy = I(F)$$

where F is the distribution function with f. Thus, the left side of (2.37) is clearly greater than or equal to the right side. We prove the reverse inequality in successive stages.

Let $\mathscr{F}_c$ be the class of distribution functions having compact support. We show now

$$\inf_{F\in\mathscr{F}_c}[I(F)+\Phi(F)]=\inf_{F\in\mathscr{F}}[I(F)+\Phi(F)]. \tag{2.38}$$

It suffices to show the left side of (2.38) is less than or equal to the right side, and for this let $F\in\mathscr{F}$ such that $\Phi(F)=a_1<\infty$ and $I(F)=a_2<\infty$. We then show that for any $\varepsilon>0$ there exists a distribution function $G\in\mathscr{F}_c$ such that $\Phi(G)<a_1+\varepsilon$ and $I(G)<a_2+\varepsilon$.

Define a sequence of functions $\{h_n(y)\}$ such that $0 \leq h_n(y) \leq 1$ for $-\infty < y < \infty$, $h_n(h) = 1$ if $|y| \leq n$, $h_n(y) = 0$ if $|y| > 2n$, and $|h_n'(y)| < 1/n$ for $n < |y| \leq 3n$. For the given $F \in \mathscr{F}$ and $g_n(y) = h_n^2(y)$, define

$$F_n(y) = \frac{\int_{-\infty}^{y} g_n(\xi)\,\mathrm{d}F(\xi)}{\int_{-\infty}^{\infty} g_n(\xi)\,\mathrm{d}F(\xi)}.$$

For each n, $F_n \in \mathscr{F}_c$ and from hypothesis (5) (p. 16) on $\Phi(F)$, there exists an n_0 so that $n \geq n_0$ implies $\Phi(F_n) \leq \Phi(F) + \varepsilon = a_1 + \varepsilon$.

We show now that $\lim_{n\to\infty} I(F_n) = I(F)$. First, since I is lower semicontinuous on $\mathscr{F}$, $\underline{\lim}_{n\to\infty} I(F_n) \geq I(F)$, and so we want to show $\overline{\lim}_{n\to\infty} I(F_n) \leq I(F)$. For any $\eta > 0$, let ψ_η be the Gaussian distribution with mean 0 and variance η. For our given $F \in \mathscr{F}$, define $F_\eta = F * \psi_\eta$. By convexity†, $I(F_\eta) \leq I(F)$, and moreover, since F_η has a strictly positive density f_η with continuous derivatives, we have from Lemma 2.2,

$$I(F_\eta) = \frac{1}{8}\int_{-\infty}^{\infty} \frac{[f_\eta'(y)]^2}{f_\eta(y)}\,\mathrm{d}y. \tag{2.39}$$

Let

$$F_{n,\eta}(y) = \frac{\int_{-\infty}^{y} g_n(\xi)\,\mathrm{d}F_\eta(\xi)}{\int_{-\infty}^{\infty} g_n(\xi)\,\mathrm{d}F_\eta(\xi)},$$

and note $F_{n,\eta}(y)$ has a strictly positive density with continuous derivative, namely,

$$\frac{g_n f_\eta}{\int_{-\infty}^{\infty} g_n(\xi) f_\eta(\xi)\,\mathrm{d}\xi}.$$

Thus again from Lemma 2.2, and the definition of g_n,

$$\begin{aligned}
I(F_{n,\eta}) &= \frac{1}{8\int_{-\infty}^{\infty} g_n(\xi) f_\eta(\xi)\,\mathrm{d}\xi}\int_{-\infty}^{\infty} \frac{[f_\eta(y) g_n(y)']^2}{f_\eta(y) g_n(y)}\,\mathrm{d}y \\
&= \frac{1}{8\int_{-\infty}^{\infty} g_n(\xi) f_\eta(\xi)\,\mathrm{d}\xi}\left\{\int_{-\infty}^{\infty} \frac{[f_\eta'(y)]^2}{f_\eta(y)} g_n(y)\,\mathrm{d}y \; + \right. \\
&\qquad \left. + \int_{-\infty}^{\infty} f_\eta(y)\frac{[g_n'(y)]^2}{g_n(y)}\,\mathrm{d}y + 2\int_{-\infty}^{\infty} f_\eta'(y) g_n'(y)\,\mathrm{d}y\right\} \\
&\leq \frac{1}{8\int_{-\infty}^{\infty} g_n(\xi) f_\eta(\xi)\mathrm{d}\xi}\left\{\int_{-\infty}^{\infty} \frac{[f_\eta'(y)]^2}{f_\eta(y)}\,\mathrm{d}y + \frac{4}{n^2}\int_{-\infty}^{\infty} f_\eta(y)\,\mathrm{d}y \right. \\
&\qquad \left. + \frac{2}{n}\int_{-\infty}^{\infty} |f_\eta'(y)|\mathrm{d}y\right\}.
\end{aligned}$$

† From the definition of $I(F)$ it follows that it is translation invariant, is lower semicontinuous, and is convex. Therefore it decreases under convolution.

From (2.39) and the fact, noted earlier, that $I(F_\eta) \leqslant I(F)$,

$$\int_{-\infty}^{\infty} |f'_\eta(y)|\,dy = \int_{-\infty}^{\infty} \frac{|f'_\eta(y)|}{\sqrt{[f_\eta(y)]}} \sqrt{[f_\eta(y)]}\,dy$$

$$\leqslant \sqrt{\left\{\int_{-\infty}^{\infty} \frac{[f'_\eta(y)]^2}{f_\eta(y)}\right\}dy} \sqrt{\left\{\int_{-\infty}^{\infty} f_\eta(y)\,dy\right\}}$$

$$= \sqrt{\left\{\int_{-\infty}^{\infty} \frac{[f'_\eta(y)]^2}{f_\eta(y)}\right\}dy} = \sqrt{[8I(F_\eta)]} \leqslant \sqrt{[8I(F)]}.$$

Therefore, again using (2.39), and $I(F_\eta) \leqslant I(F)$,

$$I(F_{n,\eta}) \leqslant \frac{1}{8\int_{-\infty}^{\infty} g_n(\xi) f_\eta(\xi)\,d\xi}\left\{8I(F) + \frac{4}{n^2} + \frac{2}{n}\sqrt{[8I(F)]}\right\}. \tag{2.40}$$

Holding n fixed and letting $\eta \to 0$ we see that

$$\lim_{\eta \to 0} I(F_{n,\eta}) = I(F_n),$$

$$\lim_{\eta \to 0} \int_{-\infty}^{\infty} g_n(\xi) f_\eta(\xi)\,d\xi = \int_{-\infty}^{\infty} g_n(\xi)\,d\xi,$$

and therefore from (2.40), letting $\eta \to 0$,

$$I(F_n) \leqslant \frac{1}{8\int_{-\infty}^{\infty} g_n(\xi)\,d\xi}\left\{8I(F) + \frac{4}{n^2} + \frac{2}{n}\sqrt{[8I(F)]}\right\}.$$

Thus $\overline{\lim}_{n\to\infty} I(F_n) \leqslant I(F)$, and so there exists an n_1 such that $n > n_1$ implies $I(F_n) \leqslant I(F) + \varepsilon = a_2 + \varepsilon$. For any $n > \max(n_0, n_1)$ choose $G(y) = F_n(y) \in \mathscr{F}_c$ and $\Phi(G) < a_1 + \varepsilon$ and $I(G) < a_2 + \varepsilon$ so that we have (2.38).

Now let $\mathscr{F}_{c'}$ be the space of distribution functions having compact support and possessing twice continuously differentiable density functions. We show now

$$\inf_{F \in \mathscr{F}_{c'}} [I(F) + \Phi(F)] = \inf_{F \in \mathscr{F}_c} [I(F) + \Phi(F)]. \tag{2.41}$$

Again, of course, it suffices to show the left side of (2.41) is less than or equal to the right side. Let $F \in \mathscr{F}_c$. For any $\varepsilon > 0$ we want to find $G \in \mathscr{F}_{c'}$ such that $I(G) < I(F) + \varepsilon$ and $\Phi(G) < \Phi(F) + \varepsilon$. For any $\eta > 0$, let $\eta(y)$ be a density function which is twice continuously differentiable with compact support $\subset (-\eta, \eta)$. For our $F \in \mathscr{F}_c$, define

$$g_\eta(x) = \int \eta(x-y)\,dF(y).$$

Thus $g_\eta(x)$ is a density function which is twice continuously differentiable and has compact support, since both $\eta(y)$ and $F(y)$ have compact support. If we let G_η be the distribution function corresponding to the density g_η we have that for all $\eta>0$, $G_\eta \in \mathcal{F}_{c'}$. Moreover, by convexity, for all $\eta>0$,

$$I(G_\eta) \leq I(F) \leq I(F)+\varepsilon.$$

On the other hand, as $\eta \to 0$, $G_\eta \Rightarrow F$ and the support of G_η is for every $\eta>0$ contained in a fixed interval independent of η. Thus by hypothesis (4) (p. 16) on Φ we can choose η so small that $\Phi(G_\eta) \leq \Phi(F)+\varepsilon$. Hence, for any η so small, let $G=G_\eta$ and we have $I(G) \leq I(F)+\varepsilon$ and $\Phi(G) \leq \Phi(F)+\varepsilon$, establishing (2.41).

Finally, we show

$$\inf_{f \in f_x} \left\{\frac{1}{8}\int_a^b \frac{[f'(y)]^2}{f(y)}\,dy + \Phi(f)\right\} = \inf_{F \in \mathcal{F}_{c'}} [I(F)+\Phi(F)]. \tag{2.42}$$

Since $f_x \subset \mathcal{F}_{c'}$ it again suffices to show that for $F^* \in \mathcal{F}_{c'}$ and given $\varepsilon>0$ there exists a density $f \in f_x$ such that

$$\frac{1}{8}\int_a^b \frac{[f'(y)]^2}{f(y)}\,dy \leq I(F^*)+\varepsilon$$

and

$$\Phi(f) \leq \Phi(F^*)+\varepsilon.$$

Let $h \in f_x$ such that the interval $[a, b] \ni x$ and contains the support of F^* on the interior of which $h>0$. Let for any $\eta>0$,

$$g_\eta = (1-\eta)f^* + \eta h$$

where f^* is the density of F^*. Thus g_η has two continuous derivatives and $g_\eta>0$ on (a, b). Again by convexity, for any $\eta>0$

$$I(g_\eta) \leq (1-\eta)I(F^*)+\eta I(h)$$

so by letting $\eta \to 0$ and using hypothesis (4) (p. 16) on Φ we have for small enough η,

$$I(g_\eta) \leq I(F^*)+\varepsilon$$

and

$$\Phi(g_\eta) \leq \Phi(F^*)+\varepsilon.$$

Also from Lemma 2.2, since $I(g_\eta)<\infty$, we have

$$\int_a^b \frac{[g'_\eta(y)]^2}{g_\eta(y)}\,dy < \infty.$$

This gives (2.42).

Now, since $\not{f}_x \subset \not{f} \subset \mathcal{F}$, we have finally that

$$\inf_{f\in \not{f}_x}\left\{\frac{1}{8}\int_a^b \frac{[f'(y)]^2}{f(y)}\mathrm{d}y+\Phi(f)\right\}=\inf_{f\in \not{f}}\left\{\frac{1}{8}\int_{-\infty}^{\infty} \frac{[f'(y)]^2}{f(y)}\mathrm{d}y+\Phi(f)\right\}$$

$$=\inf_{F\in\mathcal{F}}[I(F)+\Phi(F)].$$

This completes the proof of the theorem.

Acknowledgements

This research was supported by the Office of Naval Research, Contract N00014-67-A-0467-0031 and the National Science Foundation, Grant NSF-GP-41985.

References

1. KAC, M. On some connections between probability theory and differential and integral equations, *Proceedings of the 2nd Berkeley Symposium*, pp. 189–215 (1950).
2. ITO, K. and MCKEAN JR., H. P. *Diffusion processes and their sample paths.* Springer–Verlag, New York (1965).
3. CAMERON, R. H. and MARTIN, W. T. The transformation of Wiener integrals by non-linear transformations. *Trans. Am. math. Soc.* **75**, 552–75 (1953).

COMMUNICATIONS ON PURE AND APPLIED MATHEMATICS, VOL. XXVIII, 525–565 (1975)

Asymptotics for the Wiener Sausage*

M. D. DONSKER AND S. R. S. VARADHAN

(Dedicated to Mark Kac on the occasion of his sixtieth birthday.)

1. Introduction

Let $\{\beta(s), 0 \leqq s < \infty\}$ be the Brownian motion in d dimensions with $\beta(0) = x \in R^d$. For any $t > 0$, let $C_t = C_t(\beta(\cdot))$ be the image set of the particular path $\beta(s)$ for $0 \leqq s \leqq t$ so that $C_t \subset R^d$. For any set $A \subset R^d$ and any $\epsilon > 0$, let $A^\epsilon = \cup_{y \in A} S(y,\epsilon)$, where $S(y,\epsilon)$ is the sphere in R^d of radius ϵ with center at y. We shall denote the d-dimensional Lebesgue measure of a set A by $|A|$. In particular then we shall be interested in the set $C_t^\epsilon = C_t^\epsilon(\beta(\cdot))$ which has been called the Wiener sausage (see e.g. [1]). Its d-dimensional volume $|C_t^\epsilon(\beta(\cdot))|$ is clearly a functional of the Brownian motion path whose increase with t depends on how much the sausage intersects itself for that particular path as t grows.

We shall be principally concerned in this paper with the asymptotic behavior as $t \to \infty$ of

$$E_x\{\exp\{-\nu|C_t^\epsilon(\beta(\cdot))|\}\}, \tag{1.1}$$

where $\nu > 0$, $x \in R^d$, and $E_x\{\ \}$ denotes expectation on the Brownian motion paths in R^d starting at x. We shall prove

THEOREM 1. *For any $x \in R^d$ and any $\epsilon > 0$,*

$$\lim_{t\to\infty} \frac{1}{t^{d/(d+2)}} \log E_x\{\exp\{-\nu|C_t^\epsilon(\beta(\cdot))|\}\} = -k(\nu,d), \tag{1.2}$$

where

$$k(\nu,d) = \nu^{2/(d+2)}\left(\frac{d+2}{2}\right)\left(\frac{2\gamma_d}{d}\right)^{d/(d+2)},$$

and γ_d is the lowest eigenvalue of $-\frac{1}{2}\Delta$ for the sphere of unit volume in d dimensions with zero boundary values.

*This research was performed at the Courant Institute and supported by the U. S. Office of Naval Research, Contract N00014-67A-0467-0031 and by the National Science Foundation, Grant NSF-GP-41985. Reproduction in whole or in part is permitted for any purpose of the United States Government.

It is quite easy to show that

$$\varliminf_{t\to\infty} \frac{1}{t^{d/(d+2)}} \log E_x\{\exp\{-\nu|C_t^\epsilon(\beta(\cdot))|\}\} \geqq -k(\nu,d) \tag{1.3}$$

and we do this in Section 2. This lower estimate has already been obtained by others (c.f. [1]), and, in fact, the result (1.2) has been conjectured on physical grounds (cf. [2]). However, to prove

$$\varlimsup_{t\to\infty} \frac{1}{t^{d/(d+2)}} \log E_x\{\exp\{-\nu|C_t^\epsilon(\beta(\cdot))|\}\} \leqq -k(\nu,d) \tag{1.4}$$

is a much more delicate matter. We prove (1.4) in Section 3 of this paper making use of results recently obtained by the authors for the asymptotic evaluation of Markov process expectations as the time becomes infinite (cf. [3], [4], [5]). To see how these latter results apply to (1.1), it will be convenient to make some changes of variable. Since the expectation in (1.4) is independent of the starting point, we can take x equal to zero.

For any $\theta>0$, let, as usual, θA denote the set of points $\{\theta y\}$ where $y\in A$. Then, for any $\lambda>0$, we have, using the scaling properties of Brownian motion,

$$\begin{aligned} E_0\{\exp\{-\nu|C_t^\epsilon(\beta(\cdot))|\}\} &= E_0\left\{\exp\left\{-\nu|C^\epsilon_{t/\lambda}(\sqrt{\lambda}\ \beta(\cdot))|\right\}\right\} \\ &= E_0\left\{\exp\left\{-\nu|\sqrt{\lambda}\, C^{\epsilon/\sqrt{\lambda}}_{t/\lambda}(\beta(\cdot))|\right\}\right\} \\ &= E_0\left\{\exp\left\{-\nu\lambda^{d/2}|C^{\epsilon\circ\sqrt{\lambda}}_{t/\lambda}(\beta(\cdot))|\right\}\right\}. \end{aligned}$$

Thus, if we let $\lambda=t^{2/(d+2)}$, we have

$$E_0\{\exp\{-\nu|C_t^\epsilon(\beta(\cdot))|\}\} = E_0\left\{\exp\left\{-\nu t^{d/(d+2)}|C^{\epsilon/t^{1/(d+2)}}_{t^{d/(d+2)}}(\beta(\cdot))|\right\}\right\}, \tag{1.5}$$

which means that to examine the asymptotic behavior of (1.1) as $t\to\infty$ we can just as well examine the asymptotic behavior of

$$E_0\left\{\exp\left\{-\nu t|C_t^{\epsilon/t^{1/d}}(\beta(\cdot))|\right\}\right\} \tag{1.6}$$

as $t\to\infty$, i.e., replace $t^{d/(d+2)}$ by t. Therefore, what we actually prove is

$$\varliminf_{t\to\infty} \frac{1}{t}\log E_0\left\{\exp\left\{-\nu t|C_t^{\epsilon/t^{1/d}}(\beta(\cdot))|\right\}\right\} \geqq -k(\nu,d), \tag{1.7}$$

and

$$\overline{\lim_{t\to\infty}} \frac{1}{t} \log E_0\left\{\exp\left\{-\nu t\left|C_t^{\epsilon/t^{1/d}}(\beta(\cdot))\right|\right\}\right\} \leqq -k(\nu,d), \tag{1.8}$$

which then imply Theorem 1.

We now describe how we handle the left side of (1.8). For any $t>0$, Borel set $A\subset R^d$, and Brownian path $\omega=\beta(\cdot)$, let

$$L_t(\omega,A)=\frac{1}{t}\lambda\left\{s:0\leqq s\leqq t,\beta(s)\in A\right\},$$

where $\lambda\{\ \}$ denotes one-dimensional Lebesgue measure. $L_t(\omega,A)$ is then the proportion of time up to t that a particular Brownian path spends in the set A. For each $t>0$, and each Brownian path ω, $L_t(\omega,\cdot)$ is a probability measure on R^d the support of which is $C_t(\beta(\cdot))$.

For any $\epsilon>0$, let $\phi^{(\epsilon)}(y)$ be non-negative, an element of C^∞ on R^d, and such that $\{y:\phi^{(\epsilon)}(y)>0\}=\{y:|y|<\epsilon\}$ and $\int_{R^d}\phi^{(\epsilon)}(y)\,dy=1$. Let $\phi_t^{(\epsilon)}(y)=t\phi^{(\epsilon)}(t^{1/d}y)$ so that

$$\int_{R^d}\phi_t^{(\epsilon)}(y)dy=1 \quad\text{and}\quad \left\{y:\phi_t^{(\epsilon)}(y)>0\right\}=\left\{y:|y|<\epsilon/t^{1/d}\right\}.$$

For any $\epsilon>0$, $t>0$, and Brownian path $\omega=\beta(\cdot)$, let

$$f_t^{(\epsilon)}(\omega,y)=\int_{R^d}\phi_t^{(\epsilon)}(y-x)L_t(\omega,dx). \tag{1.9}$$

We see that $f_t^{(\epsilon)}(\omega,\cdot)$ is a probability density on R^d, and, what is important, since the support of the convolution of $\phi_t^{(\epsilon)}$ and L_t is the closure of the sum of their supports,

$$\left|\left\{y:f_t^{(\epsilon)}(\omega,y)>0\right\}\right|\leqq\left|C_t^{\epsilon/t^{1/d}}(\beta(\cdot))\right|.$$

Thus, to show (1.8) it suffices to show that

$$\overline{\lim_{t\to\infty}} \frac{1}{t} \log E_0\left\{\exp\left\{-\nu t\left|\left\{y:f_t^{(\epsilon)}(\omega,y)>0\right\}\right|\right\}\right\} \leqq -k(\nu,d). \tag{1.10}$$

The earlier results we wish to use on (1.8) apply to Markov processes where the state space is a compact metric space while here the d-dimensional Brownian motion has all of R^d as its state space. We find it convenient, therefore, to consider Brownian motion on a d-dimensional torus and, as we shall see shortly, this suffices to obtain (1.8). For $M>0$, let T_d^M be the torus in R^d of size M, i.e., the set of equivalence classes defined by $x\sim y$ if $x\in R^d, y\in R^d$, and $x-y$ has coordinates which are integral multiples of M. We can, of course, think of T_d^M as $0\leqq x_i\leqq M$, $i=1,2,3,\cdots,d$, with the sides identified. Let g_M be the canonical mapping of $R^d\to T_d^M$, i.e., any element of R^d maps into the equivalence class to which it belongs. If we have Brownian motion $\{\beta(s), 0\leqq s<\infty\}$ on R^d, then by Brownian motion on the torus T_d^M we mean $\{g_M(\beta(s)), 0\leqq s<\infty\}$. Now, as will be shown in Lemma 3.3, for any $M>0$,

$$(1.11)\quad E_0\left\{\exp\{-\nu t|C_t^{\epsilon/t^{1/d}}(\beta(\cdot))|\}\right\}\leqq E_0\left\{\exp\{-\nu t|C_t^{\epsilon/t^{1/d}}(g_M(\beta(\cdot)))|\}\right\},$$

so that, for any $M>0$,

$$(1.12)\quad \begin{aligned}&\overline{\lim_{t\to\infty}}\frac{1}{t}\log E_0\left\{\exp\{-\nu t|C_t^{\epsilon/t^{1/d}}(\beta(\cdot))|\}\right\}\\ &\leqq \overline{\lim_{t\to\infty}}\frac{1}{t}\log E_0\left\{\exp\{-\nu t|C_t^{\epsilon/t^{1/d}}(g_M(\beta(\cdot)))|\}\right\}.\end{aligned}$$

If we let

$$\tilde{L}_t(\omega,A)=\frac{1}{t}\lambda\left\{s:0\leqq s\leqq t, g_M(\beta(s))\in A\right\}$$

and

$$\psi_t^{(\epsilon)}(y)=\sum_{g\in G_d^M}\phi_t^{(\epsilon)}(y+g),$$

where G_d^M is the set of vectors in R^d each of the coordinates of which is an integral multiple of M, then

$$\tilde{f}_t^{(\epsilon)}(\omega,y)=\int_{T_d^M}\psi_t^{(\epsilon)}(y-x)\tilde{L}_t(\omega,dx)$$

is the projection of $f_t^{(\epsilon)}(\omega,y)$ (defined by (1.9)) on the torus T_d^M. Moreover, as before,

$$|\{y: \tilde{f}_t^{(\epsilon)}(\omega,y)>0\}| \leqq |C_t^{\epsilon/t^{1/d}}(g_M(\beta(\cdot)))|,$$

and therefore from (1.12) we have

$$\begin{aligned}&\overline{\lim_{t\to\infty}}\frac{1}{t}\log E_0\{\exp\{-\nu t|C_t^{\epsilon/t^{1/d}}(\beta(\cdot))|\}\}\\ (1.13)\qquad &\leqq \overline{\lim_{t\to\infty}}\frac{1}{t}\log E_0^M\{\exp\{-\nu t|\{y:\tilde{f}_t^{(\epsilon)}(\omega,y)>0\}|\}\},\end{aligned}$$

where $E_0^M\{\ \}$ denotes integration over Brownian motion paths on T_d^M starting at the origin.

We note that $\tilde{f}_t^{(\epsilon)}(\omega,\cdot)$ is a probability density on T_d^M. If we let $\mathfrak{F}_M$ be the space of all probability densities on T_d^M, then the functional $|\{y: f(y)>0\}|$ for $f\in\mathfrak{F}_M$ is lower semicontinuous in the L_1 topology on $\mathfrak{F}_M$.

But in [5] (corollary of Theorem 5.1) we proved that, for any non-negative functional $\Phi(f)$ which is lower semicontinuous in the L_1 topology on $\mathfrak{F}_M$,

$$(1.14)\qquad \overline{\lim_{t\to\infty}}\frac{1}{t}\log E_0^M\{\exp\{-t\Phi(\tilde{f}_t^{(\epsilon)}(\omega,\cdot))\}\} \leqq -\inf_{f\in\mathfrak{F}_M}[\Phi(f)+I_M(f)],$$

where

$$I_M(f)=\frac{1}{8}\int_{T_d^M}\frac{|\nabla f|^2}{f}dy.$$

Thus, in particular, for $\Phi(f)=\nu|\{y: f(y)>0\}|$ we get

$$\begin{aligned}&\overline{\lim_{t\to\infty}}\frac{1}{t}\log E_0^M\{\exp\{-\nu t|\{y:\tilde{f}_t^{(\epsilon)}(\omega,y)>0\}|\}\}\\ (1.15)\qquad &\leqq -\inf_{f\in\mathfrak{F}_M}[\nu|\{y: f(y)>0\}|+I_M(f)].\end{aligned}$$

Hence, from (1.13) and (1.15), we have, for any $M>0$,

$$\begin{aligned}&\overline{\lim_{t\to\infty}}\frac{1}{t}\log E_0\{\exp\{-\nu t|C_t^{\epsilon/t^{1/d}}(\beta(\cdot))|\}\}\\ (1.16)\qquad &\leqq -\inf_{f\in\mathfrak{F}_M}[\nu|\{y: f(y)>0\}|+I_M(f)].\end{aligned}$$

Since (1.16) holds for all $M>0$, we get

$$\overline{\lim_{t\to\infty}}\frac{1}{t}\log E_0\{-\exp\{-\nu t|C_t^{\epsilon/t^{1/d}}(\beta(\cdot))|\}\}$$

(1.17)
$$\leqq -\sup_M \inf_{f\in\mathcal{F}_M}[\nu|\{y: f(y)>0\}|+I_M(f)].$$

Let $\mathcal{G}$ be the space of all probability density functions on R^d, i.e., $g\in\mathcal{G}$ if $g\geqq 0$ on R^d and $\int_{R^d} g(y)dy=1$. Let

$$I(g)=\frac{1}{8}\int_{R^d}\frac{|\nabla g|^2}{g}dy.$$

In Lemma 3.5 we show that

$$-\sup_M \inf_{f\in\mathcal{F}_M}[\nu|\{y: f(y)>0\}|+I_M(f)]$$

(1.18)
$$\leqq -\inf_{g\in\mathcal{G}}[\nu|\{y: g(y)>0\}|+I(g)],$$

and in Lemma 3.9 we show that

(1.19)
$$\inf_{g\in\mathcal{G}}[\nu|\{y: g(y)>0\}|+I(g)]=k(\nu,d),$$

where $k(\nu,d)$ is defined in Theorem 1. Therefore, from (1.17), (1.18), and (1.19) we obtain (1.8).

The function space integral (1.1), the asymptotic behavior of which as $t\to\infty$ is determined by Theorem 1, is actually a special case of what arises in statistical mechanics. Let $V(y)$ be non-negative on R^d and consider for $\nu>0$

(1.20)
$$E_x\left\{\exp\left\{-\nu\int_{R_d}\left[1-\exp\left\{-\int_0^t V(\beta(s)-y)ds\right\}\right]dy\right\}\right\},$$

where $E_x\{\ \}$ denotes expectation on the Brownian motion paths $\beta(\cdot)$ in R^d starting at x. If, for $\epsilon>0$, we let $V(y)=\infty$ for $|y|<\epsilon$ and $V(y)=0$ for $|y|\geqq\epsilon$ (i.e., the potential for "hard spheres"), then

$$1-\exp\left\{-\int_0^t V(\beta(s)-y)\,ds\right\}$$

is just the indicator function of the Wiener sausage of size ϵ. Hence, for this particular V, (1.20) reduces to (1.1).

In Section 4 of this paper we shall show that, under very broad assumptions on $V(y)$, the asymptotic behavior of (1.20) as $t\to\infty$ is exactly the same as it is in the special case of the sausage. We shall prove

THEOREM 2. *Let V be non-negative but not identically zero. Then,*

$$\overline{\lim_{t\to\infty}}\ \frac{1}{t^{d/(d+2)}}\log E_x\left\{\exp\left\{-\nu\int_{R_d}\left[1-\exp\left\{-\int_0^t V(\beta(s)-y)\,ds\right\}\right]dy\right\}\right\} \tag{1.21}$$
$$\leqq -k(\nu,d).$$

Moreover, if $V(y)=o(1/|y|^{d+2})$ as $|y|\to\infty$, then

$$\underline{\lim_{t\to\infty}}\ \frac{1}{t^{d/(d+2)}}\log E_x\left\{\exp\left\{-\nu\int_{R_d}\left[1-\exp\left\{-\int_0^t V(\beta(s)-y)\,ds\right\}\right]dy\right\}\right\} \tag{1.22}$$
$$\geqq -k(\nu,d),$$

where $k(\nu,d)$ is the same as in Theorem 1.

The proof of Theorem 2 follows the same lines as Theorem 1, i.e., for (1.21) we go to the torus in R^d and use the general result in [5]. The proof of (1.22) is easier and analogous to the proof of (1.3) (or equivalently (1.7)), but the asymptotic condition on V seems to be necessary for (1.22).

Since Brownian motion is a particular symmetric stable process, it is interesting to ask how the asymptotic rate in Theorem 1 would be affected if instead of the Wiener sausage treated there we consider the general symmetric stable sausage. Only in the special case of the Wiener sausage will the paths be almost surely continuous which means that all other symmetric stable sausages come in disconnected links. However our methods still apply because of the scaling properties of symmetric stable processes in R^d and the generality of the results in [5].

Thus, if $\{x(s), 0 \leqq s < \infty\}$ is a symmetric stable process[1] in R^d with exponent $0 < \alpha \leqq 2$, and now $E_0\{\ \}$ denotes expectation on the paths of this process starting at 0, we have, for any $\lambda > 0$,

$$\begin{aligned} E_0\{\exp\{-\nu|C_t^\epsilon(x(\cdot))|\}\} &= E_0\left\{\exp\left\{-\nu|C_{t/\lambda}^\epsilon(\lambda^{1/\alpha}x(\cdot))|\right\}\right\} \\ &= E_0\left\{\exp\left\{-\nu|\lambda^{1/\alpha}C_{t/\lambda}^{\epsilon/(\lambda^{1/\alpha})}(x(\cdot))|\right\}\right\} \\ &= E_0\left\{\exp\left\{-\nu\lambda^{1/\alpha}|C_{t/\lambda}^{\epsilon/\lambda^{1/\alpha}}(x(\cdot))|\right\}\right\}. \end{aligned}$$

Hence, if we let $\lambda = t^{\alpha/(d+\alpha)}$, we have

$$(1.23)\quad E_0\{\exp\{-\nu|C_t^\epsilon(x(\cdot))|\}\} = E_0\left\{\exp\left\{-\nu t^{d/(d+\alpha)}|C_{t^{d/(d+\alpha)}}^{\epsilon/t^{1/(d+\alpha)}}(x(\cdot))|\right\}\right\},$$

which means that to examine the asymptotic behavior of the expectation on the left of (1.23) we can just as well examine the asymptotic behavior of

$$(1.24)\qquad E_0\left\{\exp\left\{-\nu t|C_t^{\epsilon/t^{1/d}}(x(\cdot))|\right\}\right\}$$

as $t \to \infty$, i.e., replace $t^{d/(d+\alpha)}$ by t. But (1.24) is exactly of the form we discussed earlier in the special case of Brownian motion and is susceptible to the same treatment since the asymptotic results in [5] apply to Markov processes in general.

In order to state precisely the asymptotic behavior of (1.24) (and hence of the left side of (1.23)) as $t \to \infty$ we must introduce the I-function for symmetric stable processes when $0 < \alpha < 2$, i.e., the function which plays the same role in the asymptotics as

$$(1.25)\qquad I(g) = \frac{1}{8}\int_{R^d} \frac{|\nabla g|^2}{g}\, dy$$

did in the special case of Brownian motion.

Let $Q(\xi)$, defined for $\xi \in R^d$, be such that $Q(-\xi) = Q(\xi)$, $Q(\lambda\xi) = \lambda^\alpha Q(\xi)$ for some α with $0 < \alpha \leqq 2$ and all $\lambda > 0$, and $\phi(t,\xi) = e^{-tQ(\xi)}$ is positive definite for $t > 0$. If we then let

$$p(t,x) = \frac{1}{(2\pi)^d}\int_{R^d} e^{-i\langle\xi,x\rangle}\phi(t,\xi)\, d\xi,$$

[1]For a discussion of these processes in the present context see [5], Section 5.

then $p(t,x)$ is the transition probability for a symmetric stable process in R^d of exponent α, $0<\alpha\leqq 2$. Now, if $\alpha=2$, then $Q(\xi)$ is of the form $\frac{1}{2}\Sigma a_{ij}\xi_i\xi_j$, where (a_{ij}) is a symmetric positive definite matrix. Brownian motion is the special case when $\alpha=2$ and the matrix (a_{ij}) is the identity, but, of course, this last can be accomplished by a suitable change of coordinates.

When $0<\alpha<2$, the infinitesimal generator L of the semigroup corresponding to $p(t,x)$ can be represented by the Lévy-Khintchine representation as

$$(Lf)(x)=\int_{R^d}\left[f(x+y)-f(x)-\frac{(y\cdot\nabla f(x))}{1+|y|^2}\right]\Omega(dy), \tag{1.26}$$

for suitably smooth functions f, where Ω is a σ-finite measure which is finite outside of any neighborhood of the origin such that

$$\int_{R^d}\frac{|y|^2}{1+|y|^2}\Omega(dy)<\infty.$$

In particular, $\Omega\{y:|y|\geqq M\}\to 0$ as $M\to\infty$. Thus a symmetric stable process in R^d of order α, where $0<\alpha<2$, is characterized by the measure Ω.

In our earlier work on the asymptotic evaluation of Markov process expectations, the essential role is played by the I-function (defined for general Markov processes in [4]). For Brownian motion in R^d the I-function is given by (1.25) for any $g\in\mathcal{G}$, the space of all probability densities on R^d. For any Markov process with a symmetric transition probability density it was proved in [4] (Theorem 5) that the I-function takes the form

$$I(\mu)=\|(-L)^{1/2}\sqrt{g}\,\|^2,$$

where L is the infinitesimal generator of the process viewed as a selfadjoint negative definite operator on square integrable functions. The formula above says that, in order that $I(\mu)$ be finite, μ must have a density g with respect to the reference measure (Lebesgue measure in the present case) and $\sqrt{g}$ must be in the domain of $(-L)^{1/2}$. In such a case $I(\mu)$, which will be denoted by $I(g)$, is given by $\|(-L)^{1/2}\sqrt{g}\,\|^2$. In the case of symmetric stable processes it is easy to see that

$$I(g)=\frac{1}{8}\int_{R^d}\frac{|\nabla g|^2}{g}\,dx$$

for Brownian motion, and

$$I(g)=\frac{1}{4}\int_{R^d}\int_{R^d}\left|\sqrt{g(x+y)}-\sqrt{g(x)}\right|^2\Omega(dy)\,dx \tag{1.27}$$

when $0<\alpha<2$. Here, $g\in\mathcal{G}$ and Ω is defined by (1.26).

Following along the same discussion as in the Brownian motion case we arrive at

THEOREM 3. *Let $\{x(s),\ 0\leqq s<\infty\}$ be a symmetric stable process in R^d of order $0<\alpha\leqq 2$, with infinitesimal generator L. For any $x\in R^d$, $\nu>0$, and $\epsilon>0$,*

$$\lim_{t\to\infty}\frac{1}{t^{d/(d+\alpha)}}\log E_x\{\exp\{-\nu|C_t^\epsilon(x(\cdot))|\}\}=-k(\nu,L), \tag{1.28}$$

where

$$k(\nu,L)=\nu^{d/(d+\alpha)}\left(\frac{d+\alpha}{\alpha}\right)\left(\frac{\alpha\lambda_L}{d}\right)^{d/(d+\alpha)}$$

and

$$\lambda_L=\inf_{G:|G|=1}\lambda(G),$$

the infimum being over all open sets G in R^d of unit volume with $\lambda(G)$ the smallest eigenvalue of $-L$ with zero boundary conditions on G.

In Lemma 3.10 we show that $\lambda_L>0$ for all L under consideration.

Theorem 1 is a special case of Theorem 3 because, in the case of Brownian motion, $L=-\frac{1}{2}\Delta$ and $\inf_{G:|G|=1}\lambda(G)$ is attained for the sphere of unit volume (see Lemma 3.13). Hence $\lambda_L=\gamma_d$, as defined in Theorem 1, and $k(\nu,L)$ is just the $k(\nu,d)$ of that theorem.

It will also be proved in Lemma 3.13 that if $0<\alpha<2$ and Ω is spherically symmetric, then again $\inf_{G:|G|=1}\lambda(G)$ is attained for the sphere of unit volume.

It is interesting to observe that the 2 that appears in $t^{d/(d+2)}$ in Theorem 1 is the 2 of the Gaussian to be replaced by α for a general symmetric stable process as in (1.28).

To prove Theorem 3 we use exactly the same arguments as outlined above for Brownian motion. In Lemma 3.9 we shall show that

$$\inf_{g\in\mathcal{G}}\left[\nu|\{y:\ g(y)>0\}|+I(g)\right]=k(\nu,L), \tag{1.29}$$

where

$$I(g)=\frac{1}{8}\int_{R^d}\frac{|\nabla g|^2}{g}\,dy$$

in the Brownian motion case and

$$I(g)=\frac{1}{4}\int_{R^d}\int_{R^d}\left|\sqrt{g(x+y)}-\sqrt{g(x)}\right|^2\Omega(dy)\,dx,$$

for the symmetric stable process with $0<\alpha<2$.

In what follows, all proofs are framed so as to include symmetric stable processes of order $0<\alpha\leqq 2$ unless the Brownian motion case is sufficiently simpler to justify a separate proof.

2. Lower Bounds

Let E be a bounded closed set in R^d and let $\mathcal{E}$ be the collection of such sets. Let $\{x(s), 0\leqq s<\infty\}$ be a symmetric stable process in R^d of order $0<\alpha\leqq 2$, and let $C_t^\epsilon(x(\cdot))$ be the symmetric stable sausage up to time t. Let $I(\mu)$ be the I-function for this symmetric stable process (given by (1.25) in the case of Brownian motion and by (1.27) if $0<\alpha<2$). In this section we prove

THEOREM 2.1. *For any* $\epsilon>0$, $x\in R^d$, *and* $\nu>0$,

$$\begin{aligned}
&\varliminf_{t\to\infty}\frac{1}{t}\log E_x\left\{\exp\{-\nu t|C_t^{\epsilon/t^{1/d}}(x(\cdot))|\}\right\}\\
(2.1)\qquad &\geqq -\inf_{E\in\mathcal{E}}\left[\nu|E|+\inf_{\mu:\operatorname{supp}\mu\subset E}I(\mu)\right].
\end{aligned}$$

Before proving this theorem we make a few comments. Let, as before, $\mathcal{G}$ be the set of all probability densities on R^d. In Section 3 we shall show that

$$(2.2)\qquad \inf_{E\in\mathcal{E}}\left[\nu|E|+\inf_{\mu:\operatorname{supp}\mu\subset E}I(\mu)\right]=\inf_{g\in\mathcal{G}}\left[\nu|\{y: g(y)>0\}|+I(g)\right],$$

and also in Section 3 we shall show that

$$(2.3)\qquad \inf_{g\in\mathcal{G}}\left[\nu|\{y: g(y)>0\}|+I(g)\right]=k(\nu,L),$$

where $k(\nu,L)$ is defined just after (1.28). Hence Theorem 2.1, together with (2.2) and (2.3), implies

(2.4) $$\varliminf_{t\to\infty} \frac{1}{t}\log E_x\left\{\exp\{-\nu t|C_t^{\epsilon/t^{1/d}}(x(\cdot))|\}\right\} \geqq -k(\nu,L).$$

Proof of Theorem 2.1: Let $p(t,x,y)$ be the transition density of the symmetric stable process $\{x(s), 0\leqq s<\infty\}$. Let G be a bounded open set in R^d and A, a Borel set such that $A\subset G$. We define

(2.5) $$q_G(t,x,A)=P_x\left\{x(t)\in A,\ x(s)\in G \quad \text{for} \quad \text{all} \quad 0\leqq s\leqq t\right\},$$

where, of course, $P_x\{\ \}$ denotes the probability measure on the paths of the process starting at x. Now $q_G(t,x,\cdot)$ has a density $q_G(t,x,y)\leqq p(t,x,y)$ which satisfies the Chapman-Kolmogorov equation and is symmetric and bounded for $t>0$. Thus, $q_G(t,x,y)$ is a compact symmetric operator for each $t>0$ on $L^2(G)$ so that

(2.6) $$q_G(t,x,y)=\sum_{j=1}^{\infty} e^{-\lambda_j^\delta t}\ \psi_j(x)\psi_j(y)$$

on $L^2(G\times G)$, where $\{\lambda_j\}$ and $\{\psi_j\}$ are, respectively, the eigenvalues and normalized eigenfunctions of L_G, the infinitesimal generator of the semigroup associated with $q_G(t,x,y)$. Since $p(t,x,y)$ is a contraction, so is $q_G(t,x,y)$, and therefore all the λ_j are positive.

Now let $E\in\mathfrak{E}$ and let $\delta>0$ be given. Since E^δ is a bounded open set, we have from the preceding discussion that

(2.7) $$q_{E^\delta}(t,x,y)=\sum_{j=1}^{\infty} e^{-\lambda_j^\delta t}\,\psi_j^\delta(x)\psi_j^\delta(y)$$

on $L^2(E^\delta\times E^\delta)$, where now $\{\lambda_j^\delta\}$ and $\{\psi_j^\delta\}$ are the eigenvalues and eigenfunctions of L_{E^δ}, the infinitesimal generator of the semigroup associated with $q_{E^\delta}(t,x,y)$. Since the series on the right of (2.7) converges in $L^2(E^\delta\times E^\delta)$,

(2.8) $$\int_{R^d} q_{E^\delta}(t,x,y)\psi_1^\delta(y)\,dy=e^{-\lambda_1^\delta t}\,\psi_1^\delta(x)$$

for almost all $x\in E^\delta$. Therefore we can pick an x, call it x_δ, such that

$$\int_{R^d} q_{E^\delta}(t,x_\delta,y)\psi_1^\delta(y)\,dy=e^{-\lambda_1^\delta t}\,\psi_1^\delta(x_\delta)$$

and $|\psi_1^\delta(x_\delta)| = c \neq 0$. Since $q_{E^\delta}(t,x,y)$ has its support on E^δ and is bounded there,

$$\left\| \int_{R^d} q_{E^\delta}(t,x,y)\psi_1^\delta(y)dy \right\| \leqq \text{const.}\|\psi_1^\delta\| \leqq \text{const.},$$

which means from (2.8) that $\psi_1^\delta(x)e^{-\lambda_1^\delta t}$ is bounded, i.e., $\sup_{x\in E^\delta}|\psi_1^\delta(x)| < \infty$. Now, since $q_{E^\delta}(t,x,y) = 0$ for $y \notin E^\delta$,

$$ce^{-\lambda_1^\delta t} = \left| \int_{R^d} q_{E^\delta}(t,x_\delta,y)\psi_1^\delta(y)dy \right|$$
$$\leqq \sup_{y\in E^\delta} |\psi_1^\delta(y)| \int_{E^\delta} q_{E^\delta}(t,x_\delta,y)dy.$$

Therefore,

$$P_{x_\delta}\left\{x(s)\in E^\delta \quad \text{for all} \quad 0 \leqq s \leqq t\right\} \geqq \frac{ce^{-\lambda_1^\delta t}}{\sup_{y\in E^\delta}|\psi_1^\delta(y)|}. \tag{2.9}$$

Since,

$$E_{x_\delta}\left\{\exp\left\{-\nu t|C_t^{\epsilon/t^{1/d}}(x(\cdot))|\right\}\right\}$$
$$\geqq E_{x_\delta}\left\{\exp\left\{-\nu t|C_t^{\epsilon/t^{1/d}}(x(\cdot))|\right\}; x(s)\in E^\delta \quad \text{for all} \quad 0\leqq s\leqq t\right\}$$
$$\geqq \exp\left\{-\nu t|E^{\delta+\epsilon/t^{1/d}}|\right\}P_{x_\delta}\left\{x(s)\in E^\delta \quad \text{for all} \quad 0\leqq s\leqq t\right\},$$

we have from (2.9) and the fact that the asymptotic behavior of

$$E_x\left\{\exp\left\{-\nu t|C_t^{\epsilon/t^{1/d}}(x(\cdot))|\right\}\right\}$$

is independent of the starting point x,

$$\varliminf_{t\to\infty} \frac{1}{t}\log E_x\left\{\exp\left\{-\nu t|C_t^{\epsilon/t^{1/d}}(x(\cdot))|\right\}\right\} \geqq -\nu|E^{2\delta}| - \lambda_1^\delta. \tag{2.10}$$

Since $E \in \mathfrak{S}$, $\lim_{\delta\to 0}|E^{2\delta}| = |E|$. Moreover, we prove below in Lemma 2.3, that

$$\varlimsup_{\delta\to 0} \lambda_1^\delta \leqq \inf_{\mu:\operatorname{supp}\mu\subset E} I(\mu). \tag{2.11}$$

Thus, from (2.10) and (2.11), we obtain on letting $\delta \to 0$

$$\overline{\lim_{t\to\infty}} \frac{1}{t} \log E_x \left\{ \exp\{ -\nu t |C_t^{\epsilon/t^{1/d}}(x(\cdot))| \} \right\} \tag{2.12}$$
$$\geqq -\left[\nu|E| + \inf_{\mu : \operatorname{supp}\mu \subset E} I(\mu) \right].$$

But (2.12) holds for any set $E \in \mathcal{E}$ and hence we have (2.1).

Let G be a bounded open set in R^d and define $q_G(t,x,y)$ by (2.5). Let L_G be the infinitesimal generator associated with $q_G(t,x,y)$ and L be the infinitesimal generator associated with $p(t,x,y)$.

LEMMA 2.2. *Let $h \in C^\infty$ on R^d and $h = 0$ off of G. Then $h \in \mathcal{D}(L_G)$, the domain of L_G, and moreover*

$$L_G h = Lh|_G .$$

Proof: We have to show that

$$\int_G h(y) q_G(t,x,y)\,dy - h(x) = \int_0^t ds \int_G (Lh)(y) q_G(s,x,y)\,dy, \tag{2.13}$$

i.e., $T_t^G h - h = \int_0^t T_s^G W\, ds$, where $W = Lh$. Let $\{x(s),\ 0 \leqq s < \infty\}$ be the symmetric stable process and consider

$$h(x(t)) - \int_0^t W(x(s))\,ds .$$

This is a martingale and the proof of the lemma is just an application of Doob's stopping theorem. By definition,

$$E_x\left\{ h(x(t)) - \int_0^t W(x(s))\,ds \right\}\Big|_{t=0} = h(x),$$

and by Doob's theorem,

$$E_x\left\{ h(x(\tau\wedge t)) - \int_0^{\tau\wedge t} W(x(s))\,ds \right\} = h(x) \tag{2.14}$$

for any stopping time τ. Thus, let, in particular, $\tau = \tau_G$ be the first time the process exits from G. From (2.14), with this τ,

$$E_x\{h(x(\tau\wedge t))\} - h(x) = E_x\left\{\int_0^{\tau\wedge t} W(x(s))ds\right\}. \tag{2.15}$$

First of all,

$$E_x\{h(x(\tau\wedge t))\} = E_x\{h(x(t)),\tau>t\} + E_x\left\{h(x(\tau)),\tau\leqq t\right\}.$$

But, from the definition of τ, $x(\tau)\in\tilde{G}$ (the complement of G) and since $h=0$ on $\tilde{G}$ we have $h(x(\tau))=0$. Thus,

$$\begin{aligned} E_x\{h(x(\tau\wedge t))\} &= E_x\{h(x(t)),\tau>t\} \\ &= \int_G h(y)q_G(t,x,y)dy\,. \end{aligned} \tag{2.16}$$

Secondly, if we let $\overline{W}= W=Lh$ on G and $\overline{W}=0$ on $\tilde{G}$, then

$$\begin{aligned} E_x\left\{\int_0^{\tau\wedge t} W(x(s))\,ds\right\} &= E_x\left\{\int_0^{t}\overline{W}(x(\tau\wedge s))\,ds\right\} \\ &= \int_0^t E_x\{\overline{W}(x(\tau\wedge s))\}\,ds\,. \end{aligned}$$

But, as before,

$$\begin{aligned} E_x\{\overline{W}(x(\tau\wedge s))\} &= E_x\{\overline{W}(x(s)),\tau>s\} + E_x\left\{\overline{W}(x(\tau)),\tau\leqq s\right\} \\ &= E_x\{\overline{W}(x(s)),\tau>s\} \\ &= \int_G \overline{W}(y)q_G(s,x,y)\,dy \\ &= \int_G W(y)q_G(s,x,y)dy, \end{aligned}$$

so that

$$E_x\left\{\int_0^{\tau\wedge t} W(x(s))ds\right\} = \int_0^t ds\int_G W(y)q_G(s,x,y)dy\,. \tag{2.17}$$

Now from (2.15), (2.16), and (2.17) we have

$$\int_G h(y)q_G(t,x,y)dy - h(x) = \int_0^t ds \int_G W(y)q_G(s,x,y)dy$$

which is (2.13).

LEMMA 2.3. *Let $E \in \mathcal{E}$. Then,*

$$\overline{\lim_{\delta \to 0}}\, \lambda_1^\delta \leqq \inf_{\mu : \operatorname{supp} \mu \subset E} I(\mu).$$

Proof: Since λ_1^δ is the principal eigenvalue of L_{E^δ}, the infinitesimal generator of the semigroup associated with $q_{E^\delta}(t,x,y)$, we have, for any $\delta > 0$,

$$\lambda_1^\delta = \inf_{\substack{\|\psi\|=1 \\ \operatorname{supp} \psi \subset E^\delta}} \|(-L_{E^\delta})^{1/2}\psi\|^2. \tag{2.18}$$

Now, let μ be a probability measure on R^d such that $\mu(E)=1$ and such that $I(\mu)=l<\infty$. As noted in the introduction, $I(\mu)<\infty$ implies that the measure μ has a density f which now, moreover, is zero almost everywhere off of E. If we let $g=\sqrt{f}$ and again let L be the infinitesimal generator of the semigroup associated with $p(t,x,y)$, then, as was also noted in the introduction,

$$I(\mu) = \|(-L)^{1/2} g\|^2 = l. \tag{2.19}$$

For any $\delta > 0$, let $\phi_\delta(x) \in C^\infty$ on R^d, $\phi_\delta(x) > 0$ for $|x| < \delta$, $\phi_\delta(x) = 0$ if $|x| > \delta$ and $\int_{R^d} \phi_\delta(x)\, dx = 1$. We define the function $g_\delta(x)$ to be the convolution of g with this mollifier, i.e., $g_\delta(x) = (\phi_\delta * g)(x)$. We note that $g_\delta \in C^\infty$ on R^d and supp $g_\delta \subset E^\delta$. It is easy to see that $\|(-L)^{1/2} g_\delta\|^2 \leqq l$ and $\|(-L)^{1/2} g_\delta\|^2 \to l$ as $\delta \to 0$. Thus, if we let $h_\delta(x) = g_\delta(x)/\|g_\delta\|$, we observe that supp $h_\delta \subset E^\delta$, $\|h_\delta\| = 1$, $h_\delta \in C^\infty$ on R^d, and

$$\|(-L)^{1/2} h_\delta\|^2 = -\langle L h_\delta, h_\delta \rangle \to l \quad \text{as} \quad \delta \to 0. \tag{2.20}$$

From Lemma 2.2 we have, therefore,

$$-\langle L_{E^\delta} h_\delta, h_\delta \rangle = -\langle L h_\delta, h_\delta \rangle \to l \quad \text{as} \quad \delta \to 0. \tag{2.21}$$

But from (2.18), since $\|h_\delta\|=1$ and supp $h_\delta \subset E^\delta$, we have

$$\lambda_1^\delta \leqq \|(-L_{E^\delta})^{1/2} h_\delta\|^2 = -\langle L_{E^\delta} h_\delta, h_\delta \rangle,$$

so that (2.21) implies

$$\overline{\lim_{\delta\to 0}}\, \lambda_1^\delta \leqq l = I(\mu). \tag{2.22}$$

Since (2.22) holds for any probability measure μ on R^d such that $\mu(E)=1$, we have finally

$$\overline{\lim_{\delta\to 0}}\, \lambda_1^\delta \leqq \inf_{\mu : \operatorname{supp} \mu \subset E} I(\mu),$$

which is the lemma.

3. Proof of Theorems 3 and 1

As pointed out in the introduction, Theorem 1 is a special case of Theorem 3. So again let $\{x(s),\ 0 \leqq s < \infty\}$ be a symmetric stable process in R^d of order $0 < \alpha \leqq 2$, and let $C_t^\epsilon(x(\cdot))$ be the symmetric stable sausage up to time t. Let $I(\mu)$ be the I-function for this symmetric stable process (given by (1.25) in the case of Brownian motion and by (1.27) if $0<\alpha<2$). In this section we prove

THEOREM 3.1. *For any* $\epsilon > 0$, $x \in R^d$, *and* $\nu > 0$,

$$\begin{aligned} &\overline{\lim_{t\to\infty}}\, \frac{1}{t} \log E_x \left\{ \exp\{ -\nu t |C_t^{\epsilon/t^{1/d}}(x(\cdot))| \} \right\} \\ &\qquad \leqq - \inf_{g \in \mathcal{G}} \left[\nu |\{y : g(y) > 0\}| + I(g) \right]. \end{aligned} \tag{3.1}$$

We shall then prove, with $\mathfrak{S}$ defined just before Theorem 2.1,

THEOREM 3.2.

$$\begin{aligned} &\inf_{E \in \mathfrak{S}} \left[\nu |E| + \inf_{\mu : \operatorname{supp} \mu \subset E} I(\mu) \right] \\ &\qquad = \inf_{g \in \mathcal{G}} \left[\nu |\{y : g(y) > 0\}| + I(g) \right]. \end{aligned} \tag{3.2}$$

From Theorems 2.1, 3.1, and 3.2 we deduce

$$\lim_{t\to\infty} \frac{1}{t} \log E_x \left\{ \exp\left\{ -\nu t | C_t^{\epsilon/t^{1/d}}(x(\cdot))| \right\} \right\} \tag{3.3}$$
$$= - \inf_{g \in \mathcal{G}} \left[\nu |\{y : g(y) > 0\}| + I(g) \right].$$

Finally in Lemma 3.9 we show that

$$\inf_{g\in\mathcal{G}} \left[\nu|\{y : g(y)>0\}| + I(g) \right] = k(\nu, L),$$

as defined in Theorem 3, which gives us Theorems 3 and 1 as stated in the introduction.

We first prove

LEMMA 3.3. *Let $\{x(s),\ 0 \leqq s < \infty\}$ be the symmetric stable process in R^d of order $0 < \alpha \leqq 2$. Let $M > 0$ and g_M be the canonical mapping of $R^d \to T_d^M$, the torus in R^d of size M. Then*

$$E_0\left\{ \exp\left\{ -\nu t|C_t^{\epsilon/t^{1/d}}(x(\cdot))| \right\} \right\} \leqq E_0\left\{ \exp\left\{ -\nu t|C_t^{\epsilon/t^{1/d}}(g_M(x(\cdot)))| \right\} \right\}.$$

Proof: This is really quite obvious, since if B is any set in R^d, we have clearly $|g_M(B)| \leqq |B|$ and hence, in particular, $|C_t^{\epsilon/t^{1/d}}(x(\cdot))| \geqq |C_t^{\epsilon/t^{1/d}}(g_M(x(\cdot)))|$ which gives the lemma.

LEMMA 3.4. *Let $f \in \mathcal{F}_M$ and $0 < \lambda < M$. Define*

$$E_M^\lambda = \cup_{i=1}^d \left\{ \left(0 \leqq x_i \leqq \lambda\right) \cup \left(M - \lambda \leqq x_i \leqq M\right) \right\}.$$

Then there exists an $a \in R^d$ such that

$$\int_{E_M^\lambda} f(x-a)\,dx \leqq \frac{2\lambda d}{M}.$$

Proof: Let $M_d = \cap_{i=1}^d \{0 \leqq x_i \leqq M\}$. We note that $E_M^\lambda \subset M_d$ and $|E_M^\lambda| \leqq 2\lambda d M^{d-1}$. Let $E_M^\lambda + a$ be the translation modulo M of E_M^λ by any $a \in R^d$, and consider

$$\int_{E_M^\lambda + a} f(x)\,dx$$

as a function of a. We obtain, for its mean value over M_d,

$$\frac{1}{|M_d|}\int_{M_d} da \int_{E_M^\lambda + a} f(x)\,dx = \frac{1}{|M_d|}\int_{M_d} da \int_{M_d} \chi_{E_M^\lambda}(x-a) f(x)\,dx$$

$$= \frac{1}{|M_d|}|E_M^\lambda|\int_{M_d} f(x)\,dx$$

$$\leq \frac{2\lambda d}{M},$$

and hence there exists an a such that

$$\int_{E_M^\lambda + a} f(x)\,dx \leqq \frac{2\lambda d}{M} \quad \text{or} \quad \int_{E_M^\lambda} f(x-a)\,dx \leqq \frac{2\lambda d}{M}.$$

LEMMA 3.5. *Let $M > 0$, $\nu > 0$, and let $\mathfrak{F}_M$ be the space of probability density functions on the torus T_d^M. Let $\mathcal{G}$ be the space of probability density functions on R^d. Let*

$$I_M(f) = \frac{1}{8}\int_{T_d^M} \frac{|\nabla f|^2}{f}\,dx,$$

and

$$I(g) = \frac{1}{8}\int_{R^d} \frac{|\nabla g|^2}{g}\,dx.$$

Then

$$\sup_M \inf_{f\in\mathfrak{F}_M} \left[\nu|\{y: f(y) > 0\}| + I_M(f)\right]$$

$$\geqq \inf_{g\in\mathcal{G}} \left[\nu|\{y: g(y) > 0\}| + I(g)\right]. \tag{3.4}$$

Proof: Let $J = \sup_M \inf_{f\in\mathfrak{F}_M} [\nu|\{y:\ f(y) > 0\}| + I_M(f)]$, and let $\epsilon > 0$ be given. We must find $g \in \mathcal{G}$ such that $\nu|\{y: g(y) > 0\}| + I(g) \leqq J + \epsilon$. Let

$l = J + \frac{1}{2}\epsilon$ and choose M so large that

$$(3.5)\qquad \frac{1}{8\left(1-2d/\sqrt{M}\right)}\left[\frac{16ld}{\sqrt{M}}+\frac{4d}{M}+\frac{4\sqrt{8dl}}{\sqrt{M}}\right]<\frac{1}{2}\epsilon.$$

For this fixed M, we know there exists an $f \in \mathfrak{F}_M$ such that

$$(3.6)\qquad \nu|\{y: f(y)>0\}| + I_M(f) \leqq l.$$

For this f which we just picked and any $a \in R^d$, let $f^*(x) = f(x-a)$. Then, $\nu|\{y: f^*(y)>0\}| + I_M(f^*) \leqq l$, and thus, taking $\lambda = \sqrt{M}$ in Lemma 3.4, we can without loss of generality assume that the $f \in \mathfrak{F}_M$ which satisfies (3.6) also satisfies

$$(3.7)\qquad \int_{E_M^{\sqrt{M}}} f(x)\,dx \leqq \frac{2d}{\sqrt{M}},$$

where $E_M^{\sqrt{M}}$ is defined in Lemma 3.4. Now, let

$$\phi_M(x) = \begin{cases} 0 & \text{for } x \leqq 0, \\ \dfrac{x}{\sqrt{M}} & \text{for } 0 \leqq x \leqq \sqrt{M}, \\ 1 & \text{for } \sqrt{M} \leqq x \leqq M-\sqrt{M}, \\ -\dfrac{x}{\sqrt{M}}+\sqrt{M} & \text{for } M-\sqrt{M} \leqq x \leqq M, \\ 0 & \text{for } x \geqq M, \end{cases}$$

and $\hat{\phi}_M(x_1, x_2, \cdots, x_d) = \phi_M(x_1)\phi_M(x_2)\cdots\phi_M(x_d)$, so that

$$\left|\frac{\partial\hat{\phi}_M}{\partial x_i}\right| \leqq \frac{1}{\sqrt{M}} \quad \text{and hence} \quad |\nabla\hat{\phi}_M| \leqq \frac{\sqrt{d}}{\sqrt{M}}.$$

Let $\psi(x_1, x_2, \cdots, x_d) = [\hat{\phi}_M]^2 = \phi_M^2(x_1)\cdots\phi_M^2(x_d)$, and define $g \in \mathcal{G}$ by

$$g(x) = \frac{f(x)\psi(x)}{\int_{R^d} f(x)\psi(x)\,dx}$$

for the f which satisfies (3.6) and (3.7). We note from the definition of g that $\{y: g(y)>0\} \subset \{y: f(y)>0\}$ and hence

$$|\{y: g(y)>0\}| \leqq |\{y: f(y)>0\}|. \tag{3.8}$$

Let $\alpha = \int_{R^d} f(x)\psi(x)dx$. From the definition of g we have

$$\frac{1}{8}\int_{T_d^M} \frac{|\nabla g|^2}{g} dx = \frac{1}{8\alpha}\left[\int_{T_d^M} \frac{|\nabla f|^2}{f}\psi\, dx + \int_{T_d^M} \frac{|\nabla \psi|^2}{\psi} f dx + 2\int_{T_d^M} \langle \nabla \psi, f\rangle dx\right].$$

Since $0 \leqq \psi(x) \leqq 1$, $f \in \mathcal{F}_M$, $|\nabla \psi| \leqq 2\sqrt{d}\,/\sqrt{M}$, and

$$\frac{|\nabla \psi|^2}{\psi} = \frac{4\hat{\phi}_M^2 |\nabla \hat{\phi}_M|^2}{\hat{\phi}_M^2} = 4|\nabla \hat{\phi}_M|^2 \leqq \frac{4d}{M},$$

we have

$$\frac{1}{8}\int_{T_d^M} \frac{|\nabla g|^2}{g} dx \leqq \frac{1}{8\alpha}\left[\int_{T_d^M} \frac{|\nabla f|^2}{f} dx + \frac{4d}{M} + \frac{4\sqrt{d}}{\sqrt{M}} \int_{T_d^M} |\nabla f| dx\right]. \tag{3.9}$$

Now, from (3.6),

$$\int_{T_d^M} |\nabla f| dx = \int_{T_d^M} \frac{|\nabla f|}{\sqrt{f}} \sqrt{f}\, dx \leqq \sqrt{\int_{T_d^M} \frac{|\nabla f|^2}{f}\, dx} = \sqrt{8I_M(f)} \leqq \sqrt{8l},$$

and hence

$$\begin{aligned} I(g) &= \frac{1}{8}\int_{R^d} \frac{|\nabla g|^2}{g} dx = \frac{1}{8}\int_{T_d^M} \frac{|\nabla g|^2}{g} dx \\ &\leqq \frac{1}{8\alpha}\left[8I_M(f) + \frac{4d}{M} + \frac{4\sqrt{8dl}}{\sqrt{M}}\right]. \end{aligned} \tag{3.10}$$

We note from the definition of ψ and (3.7) that

$$\alpha=\int_{R^d} f(x)\psi(x)\,dx=\int_{T_d^M} f(x)\psi(x)\,dx \geqq \int_{\substack{\sqrt{M}\leqq x_i\leqq M-\sqrt{M}\\ i=1,2,\cdots,d}} f(x)\psi(x)\,dx,$$

$$=\int_{\substack{\sqrt{M}\leqq x_i\leqq M-\sqrt{M}\\ i=1,2,,\cdots,d}} f(x)\,dx=1-\int_{E_M^{\sqrt{M}}} f(x)\,dx \geqq 1-\frac{2d}{\sqrt{M}}.$$

Thus from (3.8), (3.10), and this last inequality, we have

$$\nu|\{y:g(y)>0\}|+I(g)$$

$$\leqq \nu|\{y:f(y)>0\}|+\frac{1}{8(1-2d/\sqrt{M})}\left[8I_M(f)+\frac{4d}{M}+\frac{4\sqrt{8dl}}{\sqrt{M}}\right]$$

$$=\nu|\{y:f(y)>0\}|+I_M(f)+\left[\frac{I_M(f)}{(1-2d/\sqrt{M})}-I_M(f)\right]$$

$$+\frac{1}{8(1-2d/\sqrt{M})}\left[\frac{4d}{M}+\frac{4\sqrt{8dl}}{\sqrt{M}}\right].$$

Using (3.6) and (3.5), we obtain finally

$$\nu|\{y:g(y)>0\}|+I(g)\leqq l+\frac{1}{8(1-2d/\sqrt{M})}\left[\frac{16dl}{\sqrt{M}}+\frac{4d}{M}+\frac{4\sqrt{8dl}}{\sqrt{M}}\right]$$

$$\leqq l+\tfrac{1}{2}\epsilon=J+\epsilon,$$

which completes the proof of this lemma.

We now prove the analogue of Lemma 3.5 for the symmetric stable process in R^d with $0<\alpha<2$. Let $I(f)$ be the I-function for the symmetric stable process in R^d and let $I_M(f)$ be the corresponding I-function for the symmetric stable process in T_d^M. We want to show (3.4) with these I-functions instead of the Brownian motion I-functions which appear there. It is just as easy to prove

$$\text{(3.11)}\qquad \overline{\lim_{M\to\infty}}\ \inf_{f\in\mathcal{F}_M}\left[\nu|\{y:f(y)>0\}|+I_M(f)\right] \geqq \inf_{g\in\mathcal{G}}\left[\nu|\{y:g(y)>0\}|+I(g)\right],$$

and, as a matter of fact, (3.11) is implied by

LEMMA 3.6. *For each $l<\infty$ there exists $\epsilon_M \to 0$ as $M \to \infty$ with the property that if $f \in \mathcal{F}_M$ with $I_M(f) \leqq l$, then there is a $g \in \mathcal{G}$ such that*

$$|\{y: g(y)>0\}| \leqq |\{y: f(y)>0\}| \quad \textit{and} \quad I(g) \leqq I_M(f)+\epsilon_M.$$

Proof: Let $M>0$ and let $\hat{\phi}_M(x)$ be as defined in Lemma 3.5. With f as in the statement of the present lemma and satisfying (3.7), define

$$g(x)=\frac{f(x)[\hat{\phi}_M(x)]^2}{\int_{R^d} f(x)[\hat{\phi}_M(x)]^2 dx},$$

so that $g \in \mathcal{G}$, and $|\{y: g(y)>0\}| \leqq |\{y: f(y)>0\}|$. Now,

$$I(g)=\frac{1}{4}\int_{R^d}\int_{R^d}\left|\sqrt{g(x+y)}-\sqrt{g(x)}\right|^2 \Omega(dy)\,dx$$

and

$$I_M(f)=\frac{1}{4}\int_{T_d^M}\int_{T_d^M}\left|\sqrt{f(x\dotplus y)}-\sqrt{f(x)}\right|^2 \Omega_M(dy)\,dx,$$

where in this last expression $x\dotplus y$ means addition modulo M and

$$\Omega_M(A)=\sum_{\xi\in G_d^M}\Omega(A+\xi),$$

where, as before, G_d^M is the set of vectors in R^d each of the coordinates of which is an integral multiple of M.

To estimate $I(g)$ we decompose $R^d \times R^d$ into the four sets:

$$S_1=\{(x,y): x+y\in T_d^M,\ x\in T_d^M\},$$

$$S_2=\{(x,y): x+y\in T_d^M,\ x\notin T_d^M\},$$

$$S_3=\{(x,y): x+y\notin T_d^M,\ x\in T_d^M\},$$

$$S_4=\{(x,y): x+y\notin T_d^M,\ x\notin T_d^M\}.$$

Since the support of g is contained in T_d^M,

$$I(g)=\frac{1}{4}\int_{S_1}\int\left|\sqrt{g(x+y)}-\sqrt{g(x)}\right|^2\Omega(dy)\,dx$$
$$+\frac{1}{4}\int_{S_2}\int g(x+y)\Omega(dy)\,dx+\frac{1}{4}\int_{S_3}\int g(x)\Omega(dy)\,dx. \tag{3.12}$$

Letting $\alpha_M=\int_{R^d}f(x)[\hat{\phi}_M(x)]^2\,dx$, we estimate the first term on the right of (3.12),

$$\frac{1}{4}\int_{S_1}\int\left|\sqrt{g(x+y)}-\sqrt{g(x)}\right|^2\Omega(dy)\,dx$$
$$\leqq\frac{1}{4\alpha_M}\int_{S_1}\int\left|\sqrt{f(x+y)}\,\hat{\phi}_M(x+y)-\sqrt{f(x)}\,\hat{\phi}_M(x)\right|^2\Omega(dy)\,dx$$
$$\leqq\frac{1}{4\alpha_M}\left[\sqrt{\int_{S_1}\int[\hat{\phi}_M(x+y)]^2\left|\sqrt{f(x+y)}-\sqrt{f(x)}\right|^2\Omega(dy)\,dx}\right.$$
$$\left.+\sqrt{\int_{S_1}\int f(x)|\hat{\phi}_M(x+y)-\hat{\phi}_M(x)|^2\,\Omega(dy)\,dx}\,\right]^2. \tag{3.13}$$

We have

$$\int_{S_1}\int|\hat{\phi}_M(x+y)|^2\left|\sqrt{f(x+y)}-\sqrt{f(x)}\right|^2\Omega(dy)dx$$
$$\leqq\int_{x+y\in T_d^M}\int_{x\in T_d^M}\left|\sqrt{f(x+y)}-\sqrt{f(x)}\right|^2\Omega(dy)\,dx,$$

and, if we extend f periodically off of T_d^M, the last integral is at most

$$\int_{x\in T_d^M}\int_{x+y\in R^d}\left|\sqrt{f(x+y)}-\sqrt{f(x)}\right|^2\Omega(dy)\,dx$$
$$=\int_{x\in T_d^M}\int_{y\in T_d^M}\left|\sqrt{f(x+y)}-\sqrt{f(x)}\right|^2\Omega(dy)dx$$
$$=\int_{x\in T_d^M}\int_{y\in T_d^M}\left|\sqrt{f(x+y)}-\sqrt{f(x)}\right|^2\Omega_M(dy)\,dx=4I_M(f).$$

Also

$$\int_{S_1}\int f(x)|\hat{\phi}_M(x+y)-\hat{\phi}_M(x)|^2\,\Omega(dy)\,dx$$

$$\leqq \int_{x\in T_d^M} f(x)\int_{y\in R^d}|\hat{\phi}_M(x+y)-\hat{\phi}_M(x)|^2\,\Omega(dy)\,dx$$

$$\leqq \rho\int_{T_d^M} f(x)\,dx=\rho,$$

where

$$\rho=\sup_{x\in R^d}\int_{R^d}|\hat{\phi}_M(x+y)-\hat{\phi}_M(x)|^2\,\Omega(dy).$$

Thus, from (3.13)

$$\frac{1}{4}\int_{S_1}\int\left|\sqrt{g(x+y)}-\sqrt{g(x)}\right|^2\Omega(dy)dx\leqq\frac{1}{4\alpha_M}\left[2\sqrt{I_M(f)}+\sqrt{\rho}\,\right]^2. \tag{3.14}$$

Taking the third integral on the right of (3.12),

$$\frac{1}{4}\int_{S_3}\int g(x)\,\Omega(dy)\,dx$$

$$\begin{aligned}
&=\frac{1}{4\alpha_M}\iint_{\substack{x+y\notin T_d^M\\ x\in T_d^M}} f(x)[\hat{\phi}_M(x)]^2\,\Omega(dy)\,dx\\
&=\frac{1}{4\alpha_M}\iint_{\substack{x+y\notin T_d^M\\ x\in T_d^M}} f(x)|\hat{\phi}_M(x+y)-\hat{\phi}_M(x)|^2\,\Omega(dy)\,dx\\
&\leqq\frac{1}{4\alpha_M}\int_{x\in T_d^M}\int_{R^d} f(x)|\hat{\phi}_M(x+y)-\hat{\phi}_M(x)|^2\,\Omega(dy)dx\leqq\frac{\rho}{4\alpha_M}.
\end{aligned} \tag{3.15}$$

Finally the second integral on the right of (3.12) satisfies,

$$\frac{1}{4}\int_{S_2}\int g(x+y)\,\Omega(dy)\,dx=\frac{1}{4\alpha_M}\iint_{\substack{x+y\in T_d^M\\ x\notin T_d^M}} f(x+y)[\hat{\phi}_M(x+y)]^2\,\Omega(dy)\,dx$$

$$=\frac{1}{4\alpha_M}\iint_{\substack{x+y\in T_d^M\\ x\notin T_d^M}} f(x+y)|\hat{\phi}_M(x+y)-\hat{\phi}_M(x)|^2\,\Omega(dy)\,dx.$$

In the last integral we keep y fixed and let $x'=x+y, y'=y$. Thus,

$$\frac{1}{4}\int_{S_2}\int g(x+y)\,\Omega(dy)\,dx$$
$$=\frac{1}{4\alpha_M}\iint_{\substack{x'-y'\notin T_d^M\\ x'\in T_d^M}} f(x')|\hat{\phi}_M(x')-\hat{\phi}_M(x'-y')|^2\,\Omega(dy')\,dx.$$

Change y' to $-y'$ and recall that $\Omega(A)=\Omega(-A)$, so that

$$\frac{1}{4}\int_{S_2}\int g(x+y)\Omega(dy)dx=\frac{1}{4\alpha_M}\iint_{\substack{x+y\notin T_d^M\\ x\in T_d^M}} f(x)|\hat{\phi}_M(x+y)-\hat{\phi}_M(x)|^2\,\Omega(dy)dx,$$

which is an integral we estimated in handling the third term above. Hence,

$$\frac{1}{4}\int_{S_2}\int g(x+y)\Omega(dy)dx \leqq \frac{\rho}{4\alpha_M}, \tag{3.16}$$

and finally we have from (3.12), (3.13), (3.14) and (3.15)

$$I(g)\leqq\frac{1}{4\alpha_M}\left[2\sqrt{I_M(f)}+\sqrt{\rho}\,\right]^2+\frac{2\rho}{4\alpha_M}. \tag{3.17}$$

We have

$$\rho=\rho_M=\sup_{x\in R^d}\int_{R^d}|\hat{\phi}_M(x+y)-\hat{\phi}_M(x)|^2\,\Omega(dy)$$
$$\leqq\sup_{x\in R^d}\left[\int_{|y|\leqq N}|\hat{\phi}_M(x+y)-\hat{\phi}_M(x)|^2\,\Omega(dy)+2\Omega\{|y|>N\}\right].$$

We recall that $|\nabla\hat{\phi}_M|\leqq\sqrt{d}\,/\sqrt{M}$, and hence we have

$$\rho_M\leqq\frac{d}{M}\int_{|y|\leqq N}|y|^2\Omega(dy)+2\Omega\{|y|>N\}.$$

But, as noted earlier, the measure Ω is such that $\int_{|y|\leqq N}|y|^2\,\Omega(dy)<\infty$ for all $N>0$ and $\Omega\{|y|>N\}\to0$ as $N\to\infty$. Thus, first choosing N large and then letting

$M \to \infty$, we see that $\rho_M \to 0$. Hence, from (3.17)

$$I(g) \leqq I_M(f) + \left(\frac{I_M(f)}{\alpha_M} - I_M(f) \right) + \frac{\sqrt{I_M(f)}\sqrt{\rho}}{\alpha_M} + \frac{3\rho}{4\alpha_M} \tag{3.18}$$

implies there is a sequence $\epsilon_M \to 0$ as $M \to \infty$ such that $I(g) \leqq I_M(f) + \epsilon_M$. In this last observation we make use of the fact that $I_M(f) \leqq l$, $\rho_M \to 0$, and, as in the preceding lemma, $\alpha_M \geqq 1 - 2d/\sqrt{M}$ because without loss of generality we picked $f \in \mathfrak{F}_M$ so that (3.7) would be satisfied. This completes the lemma.

As noted in the introduction, Theorem 3.1 follows from Lemmas 3.3, 3.5, 3.6 and the corollary to Theorem 5.1 in [5].

We now wish to prove Theorem 3.2. First of all, it is clear that, if $\mathfrak{B}$ is the class of Borel sets in R^d,

$$\begin{aligned} &\inf_{g \in \mathcal{G}} [\nu|\{y : g(y) > 0\}| + I(g)] \\ &\quad = \inf_{A \in \mathfrak{B}} \left[\nu|A| + \inf_{\mu : \mu(A) = 1} I(\mu) \right]. \end{aligned} \tag{3.19}$$

In Lemma 3.7 we show for $\mathcal{C}$, the class of closed sets in R^d,

$$\inf_{E \in \mathcal{C}} \left[\nu|E| + \inf_{\mu : \mu(E) = 1} I(\mu) \right] \leq \inf_{A \in \mathfrak{B}} \left[\nu|A| + \inf_{\mu : \mu(A) = 1} I(\mu) \right], \tag{3.20}$$

and then in Lemma 3.8 that

$$\inf_{E \in \mathcal{E}} \left[\nu|E| + \inf_{\mu : \operatorname{supp} \mu \subset E} I(\mu) \right] \leqq \inf_{E \in \mathcal{C}} \left[\nu|E| + \inf_{\mu : \mu(E) = 1} I(\mu) \right]. \tag{3.21}$$

Thus (3.19), (3.20), and (3.21) will give Theorem 3.2. To verify (3.20) it suffices to prove

LEMMA 3.7. *Let A be a Borel set, $|A| = a > 0$, and let f be a probability density in R^d which equals 0 off of A and is such that $I(f) = l < \infty$. Let $\epsilon > 0$ be given. There exists a closed set C_ϵ and a density f_ϵ supported on C_ϵ such that $|C_\epsilon| \leqq a + \epsilon$, and $I(f_\epsilon) \leqq l + \epsilon$.*

Proof: Let $\theta > 0$ and $\delta > 0$ be given and let $g(x) = \sqrt{f(x)}$. Let $g_\theta = g * \phi_\theta$ where ϕ_θ is some mollifier. Let $h_\theta^\delta = (g_\theta - \delta)^+$ and $E_\theta^\delta(x) = g(x) - h_\theta^\delta(x)$. We shall show first that

$$\lim_{\substack{\delta\to 0\\ \theta\to 0}} \int_{R^d} |E_\theta^\delta(x)|^2\,dx = 0, \tag{3.22}$$

$$\lim_{\substack{\delta\to 0\\ \theta\to 0}} \int_{R^d}\int_{R^d} |E_\theta^\delta(x+y) - E_\theta^\delta(x)|^2\,\Omega(dy)\,dx = 0. \tag{3.23}$$

If we write $E_\theta^\delta(x) = g(x) - g_\theta(x) + g_\theta(x) - h_\theta^\delta(x)$, we note that as $\theta \to 0$

$$\begin{gathered}\int_{R^d} |g_\theta(x) - g(x)|^2\,dx \to 0,\\ \int_{R^d} |(g_\theta - g)(x+y) - (g_\theta - g)(x)|^2\,\Omega(dy)\,dx \to 0;\end{gathered} \tag{3.24}$$

thus to verify (3.22) and (3.23) we must estimate $g_\theta(x) - h_\theta^\delta(x)$. For $\delta > 0$ and $x \in R^1$, let $\psi_\delta(x) = x \wedge \delta$ so that

$$\begin{aligned} g_\theta(x) - h_\theta^\delta(x) &= g_\theta(x) - (g_\theta(x) - \delta)^+ \\ &= g_\theta(x) \wedge \delta = \psi_\delta(g_\theta(x)). \end{aligned}$$

Let Φ be the class of functions ϕ on R^d such that

$$\int_{R^d} |\phi(x)|^2\,dx < \infty \quad\text{and}\quad \int_{R^d}\int_{R^d} |\phi(x+y) - \phi(x)|^2\,\Omega(dy)\,dx < \infty. \tag{3.25}$$

We show now that

$$\lim_{\delta\to 0} \int_{R^d} |\psi_\delta(\phi(x))|^2\,dx = 0, \tag{3.26}$$

$$\lim_{\delta\to 0} \int_{R^d}\int_{R^d} |\psi_\delta(\phi(x+y)) - \psi_\delta(\phi(x))|^2\,\Omega(dy)\,dx = 0, \tag{3.27}$$

both limits being uniform for ϕ in a compact set $K \subset \Phi$, where compactness is with respect to both norms in (3.25).

Take (3.26) first; if K is a compact set of ϕ's, then there exists an R independent of ϕ such that $\int_{|x|>R} \phi^2(x)\,dx < \epsilon$. Thus

$$\begin{aligned} \int_{R^d} [\phi(x)\wedge\delta]^2\,dx &= \int_{|x|\leq R} [\phi(x)\wedge\delta]^2\,dx + \int_{|x|>R} [\phi(x)\wedge\delta]^2\,dx \\ &< M_R \cdot \delta + \epsilon, \end{aligned}$$

and hence we have (3.26).

Note that $|\psi_\delta(x)-\psi_\delta(y)| \leqq |x-y|$ and, of course, $|\psi_\delta(x)| \leqq \delta$ so that $\psi_\delta(x)\to 0$ as $\delta\to 0$ for each x. Since K is a compact family of ϕ's and the measure Ω is σ-finite, we can find ρ and R independent of ϕ such that

$$\int_{\tilde{E}}\int |\phi(x+y)-\phi(x)|^2\,\Omega(dy)dx<\epsilon,$$

E being the set in $R^d\times R^d$ where $|y| \geqq \rho$ and $|x| \leqq R$. Thus

$$\int_{R^d}\int_{R^d}|\psi_\delta(\phi(x+y))-\psi_\delta(\phi(x))|^2\,\Omega(dy)\,dx$$

$$\leqq 4\delta^2\int_E\int\Omega(dy)dx+\int_{\tilde{E}}\int|\phi(x+y)-\phi(x)|^2\,\Omega(dy)\,dx$$

$$\leqq 4\delta^2\int_E\int\Omega(dy)\,dx+\epsilon,$$

which establishes (3.27).

Now if we let $\phi(x)=g_\theta(x)$, we see that (3.24), (3.26), and (3.27) imply the double limits (3.22) and (3.23).

Let $k_\theta^\delta=h_\theta^\delta/\|h_\theta^\delta\|_2$ and let $f_\theta^\delta=(k_\theta^\delta)^2$ which is then a probability density. Moreover, (3.22) and (3.23) imply that $\lim_{\substack{\delta\to 0\\ \theta\to 0}} I(f_\theta^\delta)=l$. So let $\epsilon>0$ be given and choose $\theta_0>0$, $\delta_0>0$ such that

$$I\left(f_\theta^{\delta_0}\right)\leqq l+\epsilon \quad\text{for all } \theta\leqq\theta_0.$$

The support of $f_\theta^{\delta_0}$ is the same as the support of $h_\theta^{\delta_0}$ which is $\{x:g_\theta(x)\geqq\delta_0\}$. But $g_\theta\in C^\infty$ and therefore $\{x:g_\theta(x)\geqq\delta_0\}$ is closed—call it $C_\theta^{\delta_0}$. To complete the proof of this lemma we want to show that, for sufficiently small θ,

$$|C_\theta^{\delta_0}|\leqq A+\epsilon. \tag{3.28}$$

But, since $g=\sqrt{f}$, we have from Tchebycheff's inequality,

$$\left|\left\{x:\,g_\theta(x)\geqq\delta_0\right\}\right|\leqq\left|\left\{x:\,g(x)\geqq\frac{1}{2}\delta_0\right\}\right|+\left|\left\{x:\,|g(x)-g_\theta(x)|\geqq\frac{1}{2}\delta_0\right\}\right|$$

$$\leqq|A|+\frac{4\|g-g_\theta\|^2}{\delta_0^2}.$$

Since $\| g-g_\theta\|^2 \to 0$ as $\theta \to 0$, for θ small enough, $|C_\theta^{\delta_0}| \leqq |A|+\epsilon$ which completes the proof of the lemma.

Now to see that (3.21) holds it suffices to prove

LEMMA 3.8. *Let C be a closed set in R^d and $g \in \mathcal{G}$ such that* supp $g \subset C$ *and $I(g)=l<\infty$. For any $\epsilon>0$ there exists a probability density f on R^d with $I(f) \leqq l+\epsilon$ and* supp $f \subset K$ *(compact)* $\subset C$.

Proof: Let $\phi(x)=1$ for $|x| \leqq 1$, $\phi(x)=0$ for $|x|>2$, $\phi(x) \in C^\infty$ on R^d and $0 \leqq \phi(x) \leqq 1$ for all $x \in R^d$. For $R>0$, let $\phi_R(x)=\phi(x/R)$ and define

$$C_R = \frac{1}{\int_{R^d} \phi_R^2(x)\, g(x)\, dx}.$$

We note that $C_R \to 1$ as $R \to \infty$. Let $f_R(x)=C_R \phi_R^2(x) g(x)$ so that $f_R(x)$ is a probability density on R^d with supp $f_R \subset C \cap B_{2R}$, where B_{2R} is the closed ball in R^d of radius $2R$. Note that $C \cap B_{2R}$ is a compact set contained in C. Thus, if we can show that

$$\overline{\lim_{R\to\infty}}\, I(f_R) \leqq l, \tag{3.29}$$

the proof is completed by merely choosing $f=f_R$ for large enough R. To establish (3.29) we use the explicit form of the I-function and therefore we shall consider the Brownian motion case and the symmetric stable case $(0<\alpha<2)$ separately.

CASE 1. Symmetric stable process $(0<\alpha<2)$.
We have

$$\begin{aligned} I(f_R) &= \frac{1}{4}\int_{R^d}\int_{R^d} \left|\sqrt{f_R(x+y)} - \sqrt{f_R(x)}\right|^2 \Omega(dy)\,dx \\ &= \frac{1}{4} C_R \int_{R^d}\int_{R^d} \left|(\phi_R\sqrt{g})(x+y) - (\phi_R\sqrt{g})(x)\right|^2 \Omega(dy)\,dx \\ &\leqq \frac{1}{4} C_R \Bigg[\sqrt{\int_{R^d}\int_{R^d} \phi_R^2(x+y)\left|\sqrt{g(x+y)} - \sqrt{g(x)}\right|^2 \Omega(dy)\,dx} \\ &\qquad + \sqrt{\int_{R^d}\int_{R^d} g(x)\left|\phi_R(x+y) - \phi_R(x)\right|^2 \Omega(dy)\,dx}\,\Bigg]^2 \\ &\leqq \frac{1}{4} C_R \left[\sqrt{\int_{R^d}\int_{R^d} \left|\sqrt{g(x+y)} - \sqrt{g(x)}\right|^2 \Omega(dy)\,dx} + \sqrt{\rho}\,\right]^2, \end{aligned}$$

where

$$\rho = \sup_{x \in R^d} \int_{R^d} |\phi_R(x+y) - \phi_R(x)|^2 \, \Omega(dy).$$

Now, for any $M > 0$,

$$\rho \leqq \frac{\text{const.}}{R^2} \int_{|y| \leqq M} |y|^2 \, \Omega(dy) + 2\Omega\{y : |y| > M\}.$$

But, as pointed out in the introduction, the measure Ω is such that, for every $M > 0$, $\int_{|y| \leqq M} |y|^2 \Omega(dy) < \infty$ and also $\Omega\{y : |y| > M\} \to 0$ as $M \to \infty$. Therefore, if we first choose M large enough and then R large, we can make ρ as small as we please. This together with the fact that $C_R \to 1$ as $R \to \infty$ gives us

$$\begin{aligned}\overline{\lim_{R\to\infty}}\, I(f_R) &\leqq \frac{1}{4} \int_{R^d} \int_{R^d} \left| \sqrt{g(x+y)} - \sqrt{g(x)} \right|^2 \Omega(dy)\, dx \\ &= I(g) = l\end{aligned}$$

which is (3.29).

CASE 2. Brownian motion.

Again we wish to establish (3.29), where now

$$\begin{aligned} I(f_R) &= \frac{1}{8} \int_{R^d} \frac{|\nabla f_R|^2}{f_R}\, dx \\ &= \frac{1}{8} C_R \int_{R^d} \frac{\left[\phi_R^2(x) \nabla g(x) + 2g(x)\phi_R(x) \nabla \phi_R(x)\right]^2}{\phi_R^2(x) g(x)}\, dx \\ &= \frac{1}{8} C_R \int_{R^d} \phi_R^2(x) \frac{|\nabla g|^2}{g}\, dx + \frac{1}{4} C_R \int_{R^d} \phi_R(x) \langle \nabla g, \nabla \phi_R \rangle\, dx \\ &\quad + \frac{1}{2} C_R \int_{R^d} g(x) |\nabla \phi_R(x)|^2\, dx. \end{aligned}$$

Since $|\nabla \phi_R| \leqq \text{const.}/R$, the third term goes to zero as $R \to \infty$. The second term is at most $\frac{1}{4} C_R$ (const./R)$\int_{R^d} |\nabla g| dx$, and since

$$\int_{R^d} |\nabla g| dx = \int_{R^d} (|\nabla g| / \sqrt{g}) \cdot \sqrt{g}\, dx$$

$$\leqq \sqrt{\int_{R^d} (|\nabla g|^2 / g) dx} \sqrt{\int_{R^d} g(x) dx} < \infty,$$

we see that the second term goes to zero as $R \to \infty$. Since $C_R \to 1$ as $R \to \infty$ and $\phi_R(x) \to 1$ as $R \to \infty$, the first term converges to $I(g)$. This completes Lemma 3.8.

Finally we end this section by carrying out the computation of

$$\inf_{g \in \mathcal{G}} \left[\nu |\{ y : g(y) > 0 \}| + I(g) \right]$$

as explicitly as possible. Let L be the generator of a symmetric stable process characterized either by the positive definite covariance matrix (a_{ij}) when $\alpha = 2$ or by a suitable measure Ω when $0 < \alpha < 2$. Let G be an open set in R^d and let $\lambda(G)$ be the smallest eigenvalue of $-L$ with zero boundary conditions on G. We denote

$$\lambda_L = \inf_{G : |G| = 1} \lambda(G).$$

LEMMA 3.9. *We have*

$$\inf_{g \in \mathcal{G}} \left[\nu |\{ y : g(y) > 0 \}| + I(g) \right] = k(\nu, L),$$

where

$$k(\nu, L) = \nu^{\alpha/(d+\alpha)} \left(\frac{d+\alpha}{\alpha} \right) \left(\frac{\alpha \lambda_L}{d} \right)^{d/(d+\alpha)}$$

Proof: As we saw earlier, if we let $\mathcal{B}$ be the Borel sets in R^d, then

$$\inf_{g \in \mathcal{G}} \left[\nu |\{ y : g(y) > 0 \}| + I(g) \right]$$

$$= \inf_{B \in \mathcal{B}} \left[\nu |B| + \inf_{\operatorname{supp} g \subset B} I(g) \right].$$

It follows from Lemma 2.3 that the above infimum is equal to

$$\inf_{G \text{ open}} \left[\nu|G|+\lambda(G)\right] = \inf_{a>0} \inf_{\substack{G \text{ open}\\ |G|=a}} \left[\nu|G|+\lambda(G)\right]$$

$$= \inf_{a>0} \left[\nu a + \inf_{\substack{G \text{ open}\\ |G|=a}} \lambda(G)\right].$$

From the scaling properties we have

$$\inf_{\substack{G \text{ open}\\ |G|=a}} \lambda(G) = \frac{1}{a^{\alpha/d}} \inf_{\substack{G \text{ open}\\ |G|=1}} \lambda(G) = \frac{1}{a^{\alpha/d}} \lambda_L,$$

so that

$$\inf_{g \in \mathcal{G}} \left[\nu|\{y: g(y)>0\}| + I(g)\right]$$

$$= \inf_{a>0} \left[\nu a + \frac{\lambda_L}{a^{\alpha/d}}\right],$$

and it is elementary to show that this equals

$$\nu^{\alpha/(d+\alpha)} \left(\frac{d+\alpha}{\alpha}\right) \left(\frac{\alpha\lambda_L}{d}\right)^{d/(d+\alpha)} = k(\nu, L).$$

LEMMA 3.10. *For any generator L under consideration*

$$\lambda_L > 0.$$

Proof: As noted in the introduction, the transition probability density is given by

$$p(t,x) = \frac{1}{(2\pi)^d} \int_{R^d} e^{-i\langle \xi, x\rangle_{\mathcal{g}} - tQ(\xi)} \, d\xi,$$

and, since $\sup_x p(t,x) \to 0$ as $t \to \infty$, we can choose $t_0 < \infty$ such that

$$\sup_x p(t_0, x) \leq \tfrac{1}{2}.$$

Let G be an open set with $|G|=1$. Let $q_G(t,x,y)$ be as defined in (2.5), so that

$$\int_G q_G(t_0,x,y)\,dy \leqq \int_G p(t_0,y-x)\,dy \leqq \frac{1}{2}.$$

Thus, if τ_G is the exit time from G,

$$P_x\left\{\tau_G \geqq nt_0\right\} \leqq \frac{1}{2^n} \quad \text{for every} \quad n.$$

In view of (2.9), we have

$$\lambda(G) \geqq \frac{\log 2}{t_0}.$$

Since this is true for every G with $|G|=1$,

$$\lambda_L \geqq \frac{\log 2}{t_0},$$

which completes the lemma.

We conclude Section 3 by proving that if L is spherically symmetric, i.e., if L is either $C\Delta$ when $\alpha=2$, or if Ω is spherically symmetric for $0<\alpha<2$, then the infimum in the definition of λ_L is attained when G is the sphere of unit volume.

Let $f(x)$ be a non-negative function on R^d such that the d-dimensional Lebesgue measure of $\{x:\ f(x)>\lambda\}$ is finite for every $\lambda>0$. By the spherical symmetrization of f, denoted by f^*, we mean the unique function f^* such that for each $\lambda>0$ the set $\{x:\ f^*(x)>\lambda\}$ is the sphere around the origin of the same measure as $\{x:\ f(x)>\lambda\}$.

LEMMA 3.11. *Let f, g, h be three non-negative functions on R^d and f^*, g^*, h^* their symmetrizations; then*

$$\int_{R^d}\int_{R^d} f(x)g(y)h(x-y)\,dx\,dy \leqq \int_{R^d}\int_{R^d} f^*(x)g^*(x)h^*(x-y)\,dx\,dy.$$

Proof: This lemma, a generalization of a result of F. Riesz in one dimension, is due to Sobolev and can be found in [6].

LEMMA 3.12. *Let f be any density such that, for some spherically symmetric stable process, $I(f)=l<\infty$. If f^* is the spherical symmetrization of f, then $I(f^*) \leqq l$.*

Proof: This is classical for $\alpha=2$ and therefore we shall prove it for $0<\alpha<2$.

In this case,

$$I(f)=\frac{1}{4}\int_{R^d}\int_{R^d}\left|\sqrt{f(x+y)}-\sqrt{f(x)}\right|^2 \Omega(dy)\, dx.$$

Since $(f^*)^{1/2}=(f^{1/2})^*$, it is enough to show that

$$\int_{R^d}\int_{R^d}|f(x+y)-f(x)|^2\,\Omega(dy)\,dx \geqq \int_{R^d}\int_{R^d}|f^*(x+y)-f^*(x)|^2\,\Omega(dy)\,dx.$$

By hypothesis, $\Omega(dy)$ is of the form

$$\Omega(dy)=\Phi(r)dy,$$

where $\Phi(r)$ is a nonincreasing function of r. We can then consider

$$\Phi_\lambda(r)=\min[\Phi(r),\lambda]$$

which is an integrable function on R^d having the same nonincreasing properties. We shall show that, for every $\lambda>0$,

$$\begin{aligned}&\int_{R^d}\int_{R^d}|f(x+y)-f(x)|^2\,\Phi_\lambda(|y|)\,dy\,dx\\ &\qquad\geqq \int_{R^d}\int_{R^d}|f^*(x+y)-f^*(x)|^2\,\Phi_\lambda(|y|)\,dy\,dx\end{aligned}\tag{3.30}$$

and then obtain the lemma by letting $\lambda\to\infty$. But, to prove (3.30) means showing that

$$\begin{aligned}&\int_{R^d}\int_{R^d}f(x+y)f(x)\Phi_\lambda(|y|)\,dx\,dy\\ &\qquad\leqq \int_{R^d}\int_{R^d}f^*(x+y)f^*(x)\Phi_\lambda(|y|)\,dx\,dy,\end{aligned}$$

or equivalently that

$$\begin{aligned}&\int_{R^d}\int_{R^d}f(x)f(y)\Phi_\lambda(|x-y|)\,dx\,dy\\ &\qquad\leqq \int_{R^d}\int_{R^d}f^*(x)f^*(y)\Phi_\lambda(|x-y|)\,dx\,dy,\end{aligned}$$

and this last inequality follows from Lemma 3.11 noting that $\Phi_\lambda^*=\Phi_\lambda$.

LEMMA 3.13. *If L is spherically symmetric, then*

$$\lambda_L = \inf_{G:|G|=1} \lambda(G) = \lambda(S),$$

where S is the sphere of unit volume.

Proof: Let G be any set with $|G|=1$ and $\lambda(G)=\lambda$. Then there is an f supported in G such that $I(f) \leqq \lambda+\epsilon$, where $\epsilon>0$ is given. Let f^* be the symmetrization of f. Then, from Lemma 3.12, $I(f^*) \leqq \lambda+\epsilon$ and supp f^* is contained in S, the sphere of unit volume. Therefore, $\lambda(S) \leqq \lambda+\epsilon$ and by letting $\epsilon \to 0$ we get $\lambda(S) \leqq \lambda$, which gives the lemma.

4. Proof of Theorem 2

We shall first prove the upper bound (1.21). If $V_1 \leqq V_2$ on R^d, then

$$E_x\left\{\exp\left\{-\nu\int_{R^d}\left[1-\exp\left\{-\int_0^t V_1(\beta(s)-y)\,ds\right\}\right]dy\right\}\right\}$$
$$\geqq E_x\left\{\exp\left\{-\nu\int_{R^d}\left[1-\exp\left\{-\int_0^t V_2(\beta(s)-y)\,ds\right\}\right]dy\right\}\right\}. \tag{4.1}$$

Now, if V is non-negative but not identically zero, there exists a V_1 such that $0 \leqq V_1 \leqq V$ and $\int_{R^d} V_1(x)\,dx = a \neq 0$. Thus in proving (1.21) we can, without loss of generality, assume because of (4.1) that V is of the form aV, where V is a probability density on R^d and $a>0$. We find it convenient to make the same sort of scale changes that we made on (1.1). Also, since the asymptotic behavior of (1.20) as $t\to\infty$ is independent of the starting point, we set $x=0$. Thus, for any $\lambda>0$,

$$E_0\left\{\exp\left\{-\nu\int_{R^d}\left[1-\exp\left\{-a\int_0^t V(\beta(s)-y)\,ds\right\}\right]dy\right\}\right\}$$
$$= E_0\left\{\exp\left\{-\nu\int_{R^d}\left[1-\exp\left\{\lambda a\int_0^{t/\lambda} V(\beta(\lambda\sigma)-y)\,d\sigma\right\}\right]dy\right\}\right\}$$
$$= E_0\left\{\exp\left\{-\nu\lambda^{d/2}\int_{R^d}\left[1-\exp\left\{-\lambda a\int_0^{t/\lambda} V\big(\sqrt{\lambda}\,(\beta(\sigma)-x)\big)\,d\sigma\right\}\right]dx\right\}\right\}.$$

In particular, letting $\lambda = t^{2/(d+2)}$ we get

$$E_0\left\{\exp\left\{-\nu\int_{R^d}\left[1-\exp\left\{-a\int_0^t V(\beta(s)-y)ds\right\}\right]dy\right\}\right\}$$

$$= E_0\left\{\exp\left\{-\nu t^{d/(d+2)}\int_{R^d}\left[1-\exp\left\{-at^{2/(d+2)}Q(x)\right\}\right]dx\right\}\right\}.$$

where

$$Q(x)=\int_0^{t^{d/(d+2)}} V\left(t^{1/(d+2)}(\beta(\sigma)-x)\right)d\sigma.$$

Thus, to examine the asymptotic behavior of (1.20) as $t\to\infty$, it suffices to examine the asymptotic behavior of

$$E_0\left\{\exp\left\{-\nu t\int_{R^d}\left[1-\exp\left\{-at^{2/d}\int_0^t V\left(t^{1/d}(\beta(s)-y)\right)ds\right\}\right]dy\right\}\right\}$$

as $t\to\infty$, i.e., replace $t^{d/(d+2)}$ by t. To prove (1.21) we shall, therefore, prove

$$\overline{\lim_{t\to\infty}}\frac{1}{t}\log E_0\left\{\exp\left\{-\nu t\int_{R^d}\left[1-\exp\left\{-at^{2/d}\int_0^t V\left(t^{1/d}(\beta(s)-y)\right)ds\right\}\right]dy\right\}\right\} \tag{4.2}$$
$$\leq -k(\nu,d).$$

With $L_t(\omega,A)$ defined as before, i.e., the proportion of time up to t that a particular Brownian motion path $\omega=\beta(\cdot)$ spends in the Borel set A, we have

$$\int_0^t V\left(t^{1/d}(\beta(s)-y)\right)ds = t\int_{R^d} V\left(t^{1/d}(\xi-y)\right)L_t(\omega,d\xi). \tag{4.3}$$

Let $V_t(y)=tV(t^{1/d}y)$ and

$$f_t(\omega,y)=\int_{R^d} V_t(\xi-y)L_t(\omega,d\xi),$$

so that the left side of (4.2) becomes

$$\overline{\lim_{t\to\infty}}\frac{1}{t}\log E_0\left\{\exp\left\{-\nu t\int_{R^d}\left[1-\exp\left\{-at^{2/d}f_t(\omega,y)\right\}\right]dy\right\}\right\}. \tag{4.4}$$

As in the proof of Theorem 1, let T_d^M be the torus in R^d of size M and let $\tilde{f}_t(\omega,y)$ be the projection of $f_t(\omega,y)$ on T_d^M. We show now that, for any $M>0$,

$$E_0\left\{\exp\left\{-\nu t\int_{R^d}\left[1-\exp\{-at^{2/d}f_t(\omega,y)\}\right]dy\right\}\right\}$$
$$\leqq E_0^M\left\{\exp\left\{-\nu t\int_{T_d^M}\left[1-\exp\{-at^{2/d}\tilde{f}_t(\omega,y)\}\right]dy\right\}\right\}, \tag{4.5}$$

where $E_0^M\{\ \}$ denotes integration with respect to Brownian motion on the torus T_d^M. To see (4.5), first note that if $\xi_i \geqq 0$, then it is elementary that

$$1-\exp\left\{-\sum_i \xi_i\right\} \leqq \sum_i (1-\exp\{-\xi_i\}).$$

Thus, if $h(y)$ is an L_1 function on R^d and if $\tilde{h}(y)$ is its projection on T_d^M, i.e.,

$$\tilde{h}(y)=\sum_{g\in G_d^M} h(y+g),$$

we have, for any constant c,

$$\int_{T_d^M}[1-\exp\{-c\tilde{h}(y)\}]dy=\int_{T_d^M}\left[1-\exp\left\{-c\sum_{g\in G_d^M} h(y+g)\right\}\right]dy$$
$$\leqq \int_{T_d^M}\sum_{g\in G_d^M}[1-\exp\{-ch(y+g)\}]dy \tag{4.6}$$
$$=\int_{R^d}[1-\exp\{-ch(y)\}]dy.$$

If in (4.6) we let $c=at^{2/d}$ and $h(y)=f_t(\omega,y)$, we obtain (4.5). This means that

$$\varlimsup_{t\to\infty}\frac{1}{t}\log E_0\left\{\exp\left\{-\nu t\int_{R^d}\left[1-\exp\{-at^{2/d}f_t(\omega,y)\}\right]dy\right\}\right\}$$
$$\leqq \varlimsup_{t\to\infty}\frac{1}{t}\log E_0^M\left\{\exp\left\{-\nu t\int_{T_d^M}\left[1-\exp\{-at^{2/d}\tilde{f}_t(\omega,y)\}\right]dy\right\}\right\}. \tag{4.7}$$

Now, for any $f \in \mathcal{F}_M$, let

$$\Phi_t(f) = \int_{T_d^M} \left[1 - \exp\left\{-at^{2/d} f(y)\right\}\right] dy.$$

We note that this family of functionals, $\Phi_t(f)$, on $\mathcal{F}_M$ has the property that if $f_t \in \mathcal{F}_M$ and f_t converges to f in L_1, then

$$\varliminf_{t \to \infty} \Phi_t(f_t) \geqq |\{y : f(y) > 0\}|.$$

Therefore, from the corollary to Theorem 5.1 in [5] we have

$$\begin{aligned} &\varlimsup_{t\to\infty} \frac{1}{t} \log E_0^M\left\{\exp\left\{-\nu t \int_{T_d^M} \left[1 - \exp\left\{-at^{2/d} \tilde{f}_t(\omega, y)\right\}\right] dy\right\}\right\} \\ (4.8) \qquad &\leqq -\inf_{f \in \mathcal{F}_M} \left[\nu |\{y : f(y) > 0\}| + I_M(f)\right], \end{aligned}$$

where $I_M(f) = \frac{1}{8} \int_{T_d^M} (|\nabla f|^2 / f) dx$. The estimate in (4.8) is now exactly the same as we obtained in the special case of the sausage and we can proceed in the same way to obtain from (4.7) and (4.8)

$$\varlimsup_{t\to\infty} \frac{1}{t} \log E_0\left\{\exp\left\{-\nu t \int_{R^d} \left[1 - \exp\left\{-at^{2/d} f_t(\omega, y)\right\}\right] dy\right\}\right\} \leqq -k(\nu, d)$$

which is (1.21).

Now we want to obtain (1.22). From the previous discussion it suffices to show that

$$\begin{aligned} &\varliminf_{t\to\infty} \frac{1}{t} \log E_0\left\{\exp\left\{-\nu t \int_{R^d} \left[1 - \exp\left\{-t^{2/d} \int_0^t V\left(t^{1/d}(\beta(s) - y)\right) ds\right\}\right] dy\right\}\right\} \\ (4.9) \qquad &\geqq -k(\nu, d), \end{aligned}$$

where we are assuming that V is non-negative on R^d and satisfies

$$V(y) = o\left(\frac{1}{|y|^{d+2}}\right) \quad \text{as} \quad |y| \to \infty.$$

Let G be a bounded open set in R^d containing the origin. Then

$$E_0\left\{\exp\left\{-\nu t\int_{R^d}\left[1-\exp\left\{-t^{2/d}\int_0^t V\left(t^{1/d}(\beta(s)-y)\right)ds\right\}\right]dy\right\}\right\}$$

$$\geqq E_0\left\{\exp\left\{-\nu t\int_{R^d}\left[1-\exp\left\{-t^{2/d}\int_0^t V\left(t^{1/d}(\beta(s)-y)\right)ds\right\}\right]dy\right\};\right. \tag{4.10}$$

$$\left.\beta(s)\in G \quad \text{for all } s, 0\leqq s\leqq t\right\}.$$

If $\beta(s)\in G$ for $0\leqq s\leqq t$, then

$$\int_{R^d}\left[1-\exp\left\{-t^{2/d}\int_0^t V\left(t^{1/d}(\beta(s)-y)\right)ds\right\}\right]dy$$

$$\leqq \int_{R^d}\left[1-\exp\left\{-t^{2/d}\sup_{x\in G} tV\left(t^{1/d}(x-y)\right)\right\}\right]dy$$

$$=\int_{R^d}\left[1-\exp\left\{-t^{(d+2)/d}\sup_{x\in G} V\left(t^{1/d}(x-y)\right)\right\}\right]dy.$$

Therefore, from (4.10) we get

$$E_0\left\{\exp\left\{-\nu t\int_{R^d}\left[1-\exp\left\{-t^{2/d}\int_0^t V\left(t^{1/d}(\beta(s)-y)\right)ds\right\}\right]dy\right\}\right\}$$

$$\geqq \exp\left\{-\nu t\int_{R^d}\left[1-\exp\left\{-t^{(d+2)/d}\sup_{x\in G} V\left(t^{1/d}(x-y)\right)\right\}\right]dy\right\} \tag{4.11}$$

$$\times P_0\left\{\beta(s)\in G, 0\leqq s\leqq t\right\},$$

where $P_0\{\ \}$ indicates Brownian motion measure for paths starting at the origin in R^d.

Our hypothesis on V implies that, for any $\epsilon>0$, there exists T such that $t\geqq T$ implies

$$V\left(t^{1/d}(x-y)\right)\leqq \frac{\epsilon}{|t^{1/d}(x-y)|^{d+2}},$$

and hence it is easy to see that

$$\int_{R^d}\left[1-\exp\left\{-t^{(d+2)/d}\sup_{x\in G} V\left(t^{1/d}(x-y)\right)\right\}\right]dy \leqq |G|+o(1).$$

Using this in (4.11) we deduce

$$\lim_{t\to\infty}\frac{1}{t}\log E_0\left\{\exp\left\{-\nu t\int_{R^d}\left[1-\exp\left\{-t^{2/d}\int_0^t V\left(t^{1/d}(\beta(s)-y)\right)ds\right\}\right]dy\right\}\right\}$$

$$\geqq -\nu|G|-\lambda_1(g), \tag{4.12}$$

where $\lambda_1(G)$ is the least eigenvalue for the Laplacian, $-\frac{1}{2}\Delta$, on G. Since (4.12) holds for all bounded open sets G ,we can replace the right side of (4.12) by $-\inf_G\ [\nu|G|+\lambda_1(G)]$ and by our discussion in Section 3 we know that this is $k(\nu,d)$. This completes the proof of Theorem 2.

Bibliography

[1] Kac, M., and Luttinger, J. M., *Bose-Einstein condensation in the presence of impurities*, J. Math. Phys., Vol. 15, No. 2, February 1974, pp. 183–186.
[2] Lifschitz, I. M., *Theory of fluctuation levels in disordered systems*, Soviet Physics JETP, Vol. 26, pp. 462–469, (English translation).
[3] Donsker, M. D., and Varadhan, S. R. S., *Asymptotic evaluation of certain Wiener integrals for large time*, to appear in *Integration in Function Space,* Oxford Univ. Press.
[4] Donsker, M. D., and Varadhan, S. R. S., *Asymptotic evaluation of certain Markov process expectations for large time, I*, Comm. Pure Appl. Math., Vol. 28, 1975, pp. 1–47.
[5] Donsker, M. D., and Varadhan, S. R. S., *Asymptotic evaluation of certain Markov process expectations for large time*, II, Comm. Pure Appl. Math., Vol. 28, 1975, pp. 279–301.
[6] Sobolev, S., *On a theorem of functional analysis*, Math. Sbornik, Vol. 4 (46), 1938, pp. 471–497.

Received February, 1975.

ERRATA

Communications on Pure and Applied Mathematics. Issue 4, 1975

The following misprints which distort the meaning have crept in after the last proofs.

Reflection of Singularities at the Boundary
by Andrew Majda and Stanley Osher

p. 487 – Equation (1.15), second line should read:

$$\operatorname{supp} \rho_k \cap \Gamma_j = \varnothing$$

p. 493 – Line 5 from above should read:

$(\xi, \tau), Q_0(x_0, \xi, \tau) = Q_0^+(x_0(x_0, \xi, \tau) Q_0^-(x_0, \xi, \tau)$, where $Q_0^+(x_0, \xi, \tau)(Q_0^-(x_0, \xi, \tau))$ is a ...

Decay of Solutions to Nondissipative Hyperbolic Systems on Compact Manifolds
by Jeffrey Rauch and Michael Taylor

p. 514, line 5 – **R** should be $\mathbb{R}$.

Asymptotics for the Wiener Sausage
by M. D. Donsker and S. R. S. Varadhan:

p. 526 – Last line of displayed equation above equation (1.5) should read:

$$= E_0\left\{ \exp\left\{ -\nu \lambda^{d/2} |C_{t/\lambda}^{\epsilon/\sqrt{\lambda}}(\beta(\cdot))| \right\} \right\}.$$

p. 531 – In inequality (1.22)

$\overline{\lim}_{t \to \infty}$ should be replaced by $\underline{\lim}_{t \to \infty}$

p. 538–In inequality (2.12)

$$\overline{\lim_{t\to\infty}} \quad \text{should be replaced by} \quad \underline{\lim_{t\to\infty}}$$

p. 560–Middle line of last equation should read:

$$= E_0\left\{\exp\left\{-\nu\int_{R^d}\left[1-\exp\left\{-\lambda\alpha\int_0^{t/\lambda}V(\beta(\lambda\sigma)-y)\,d\sigma\right\}\right]dy\right\}\right\}$$

p. 562–The right side of the first line of relation (4.6) should read:

$$= \int_{T_d^M}\left[1-\exp\left\{-c\sum_{g\in G_d^M}h(y+g)\right\}\right]dy$$

p. 562–In relation (4.7)

$$\underline{\lim_{t\to\infty}} \quad \text{should be replaced by} \quad \overline{\lim_{t\to\infty}}$$

COMMUNICATIONS ON PURE AND APPLIED MATHEMATICS, VOL. XXIX, 389–461 (1976)

Asymptotic Evaluation of Certain Markov Process Expectations for Large Time—III*

M. D. DONSKER AND S. R. S. VARADHAN

1. Introduction

Before stating hypotheses and theorems precisely we give a brief description of the kind of results we are considering.

Let $X_0, X_1, X_2, \cdots$ be a discrete parameter Markov process with transition probabilities $\pi(x, dy)$ on some state space X. Each realization ω of the process is a sequence $x_0, x_1, x_2, \cdots$ where $x_i \in X$. If we denote by Ω the space of all such sequences, then, for each starting point $x \in X$, $\pi(x, dy)$ induces a probability measure P_x on Ω. For each $\omega \in \Omega$, positive integer n, and set $A \subset X$, let

$$L_n(\omega, A) = \frac{1}{n} \sum_{j=0}^{n-1} \chi_A(x_j) \tag{1.1}$$

be the proportion of time the process spends in the set A during the first n steps. We note that $L_n(\omega, \cdot)$ maps Ω into $\mathcal{M}$, the space of probability measures on X. For each $x \in X$ and each positive integer n we define a probability measure $Q_{n,x}$ on $\mathcal{M}$ by $Q_{n,x} = P_x L_n^{-1}$, i.e., if B is a set of measures in $\mathcal{M}$, then

$$Q_{n,x}(B) = P_x\{\omega \in \Omega : L_n(\omega, \cdot) \in B\}. \tag{1.2}$$

The basic problem is to analyze the asymptotic behavior of $Q_{n,x}$ for large n.

In [1] it was assumed that X is a compact metric space and $\pi(x, dy)$ is a Feller transition probability. It was further assumed that $\pi(x, dy)$ has a density $\pi(x, y)$ relative to a reference probability measure $\beta(dy)$ and that $\pi(x, y)$ is uniformly bounded above and below. Under these assumptions it was proved that the asymptotic behavior of $Q_{n,x}$ for large n is determined by the functional $I(\mu)$ defined for $\mu \in \mathcal{M}$ by

$$I(\mu) = -\inf_{u>0} \int_X \log \frac{(\pi u)(x)}{u(x)} \, \mu(dx), \tag{1.3}$$

* This research was performed at the Courant Institute and was partially supported by the National Science Foundation, Grant NSF-MPS-74-01921-A01.

where $(\pi u)(x) = \int_X u(y)\pi(x, dy)$ and the infimum is taken over all positive u in $C(X)$. More precisely we proved in [1] that for any closed set $C \subset \mathcal{M}$ (weak topology on $\mathcal{M}$) and any open set $G \subset \mathcal{M}$,

$$\overline{\lim_{n\to\infty}} \frac{1}{n} \log Q_{n,x}(C) \leqq -\inf_{\mu\in C} I(\mu)\,, \tag{1.4}$$

$$\underline{\lim_{n\to\infty}} \frac{1}{n} \log Q_{n,x}(G) \geqq -\inf_{\mu\in G} I(\mu)\,. \tag{1.5}$$

It follows from (1.4) and (1.5) that if $\Phi(\mu)$ is a real-valued weakly continuous functional on $\mathcal{M}$ (cf. Theorem 2 in [1]), then

$$\lim_{n\to\infty} \frac{1}{n} \log E_x\{\exp\{-n\Phi(L_n(\omega, \cdot))\}\} = -\inf_{\mu\in\mathcal{M}} [\Phi(\mu) + I(\mu)]. \tag{1.6}$$

Corresponding results were obtained in [1] in the case of continuous time Markov processes where analogous assumptions are made on the transition probability $p(t, x, dy)$ and the state space X is again a compact metric space. Now (1.1) is replaced by

$$L_t(\omega, A) = \frac{1}{t} \int_0^t \chi_A(x(s))\, ds\,,$$

and (1.3) is replaced by

$$I(\mu) = -\inf_{u\in\mathcal{D}^+} \int_X \left(\frac{Lu}{u}\right)(x)\mu(dx)\,,$$

where L is the infinitesimal generator of the semigroup T_t associated with $p(t, x, dy)$ and $\mathcal{D}^+$ is the set of positive functions u in the domain of L. With $Q_{t,x} = P_x L_t^{-1}$, theorems corresponding to (1.4), (1.5), and (1.6), where now $t \to \infty$, were obtained in [1] (Theorems 3 and 4).

The fact that in both the discrete and continuous parameter versions (1.6) holds for such a large class of functionals Φ is due to the assumption that the state space X is compact. If the state space X is *not* compact, then the class of functionals Φ for which a result like (1.6) holds depends on the properties imposed on $\pi(x, dy)$ or $p(t, x, dy)$. For example, in the special case of Brownian motion, which we treated in [2], where the state space is R_1, we obtained a result like (1.6) only under fairly stringent hypotheses on the functional Φ. In Section 9 of this paper we shall exhibit some examples to show that such conditions are necessary.

The present paper is concerned with the general case where the state space X is a complete separable metric space. In this generality, to obtain asymptotic results like (1.6) one must choose to prove the result either for a large class of functionals Φ by imposing special hypotheses on the transition probabilities or for a restricted class of functionals Φ but assuming much less about the transition probabilities. To examine this point further we consider separately the two basic results (1.4) and (1.5).

Under the assumption of this paper that X is a complete separable metric space, to get lower bounds, in both the discrete and continuous parameter cases, like (1.5), holding for each starting point x, requires some strong transitivity assumptions. In the discrete case, the transitivity assumption we make is that, for any two points x_1 and x_2, the resolvents

$$R(x_1, dy) = \sum_{n=1}^{\infty} \frac{1}{2^n} \Pi^{(n)}(x_1, dy) \qquad \text{and} \qquad R(x_2, dy) = \sum_{n=1}^{\infty} \frac{1}{2^n} \Pi^{(n)}(x_2, dy)$$

are mutually absolutely continuous. Under this assumption, called Hypothesis H in Section 3, the lower bounds are obtained. In Section 8 the lower bounds are proved in the case of continuous parameter Markov processes where we have to assume the continuous version of Hypothesis H above. In addition some mild assumptions involving the strong continuity of the semigroup of transition operators as $t \to 0$ are made.

In the case of upper bounds like (1.4), in both the discrete and continuous parameter cases, the problem is to obtain the upper estimate for *closed* sets, that is, if one adopts the point of view which we do here—to have (1.6) holding for as wide a class of functionals as possible and pay the price of imposing stronger conditions on the transition probabilities. The point is that (1.4) holds for *compact* sets just under the hypothesis that the transition probability satisfies the Feller property. Thus we see that if X is a compact space, (1.4) holds for closed sets also, just under that single hypothesis. But if X is not compact, to have (1.4) hold for closed sets in $\mathcal{M}$ requires a strong positive recurrence condition on the transition probability. This is carried out in Section 4 under Hypothesis H* of that section. In Section 7 we obtain upper bounds for continuous time Markov processes under the continuous analogue of Hypothesis H* which again allows us to obtain the upper estimate for closed, not just compact, sets.

In the special case of independent identically distributed random variables, the Hypothesis H* is always fulfilled. However since our state space is so general, this leads to interesting applications to identically distributed random variables with values in a Banach space. This is carried out in Sections 5 and 6 and yields new derivations of some known estimates on the tail probabilities of Gaussian processes.

To summarize some of the results in this paper, let X be a complete separable metric space and $\pi(x, dy)$ a transition probability satisfying the Feller property. If $\pi(x, dy)$ has a density $\pi(x, y)$ with respect to a reference measure $\beta(dy)$, and if, for each $x \in X$ and each $E \subset X$ such that $\beta(E) > 0$, there exists a $k = k(x, E)$ for which the k-step transition $\pi^{(k)}(x, E) > 0$, then, for any open set $G \subset \mathcal{M}$,

$$(1.7) \qquad \varliminf_{n\to\infty} \frac{1}{n} \log Q_{n,x}(G) \geqq -\inf_{\mu \in G} I(\mu).$$

Inequality (1.7) follows from Corollary 3.4 in Section 3.

To get the proper upper bound we assume in addition to the fact that $\pi(x, dy)$ is a Feller transition probability that there exists a function $u_0(x)$ on X such that $u_0(x) \geqq 1$ for all x, that $v(x) = (\pi u_0)(x) = \int_X u_0(dy)\pi(x, dy)$ is bounded on every compact set and that, for every l, $\{x \in X : u_0(x)/v(x) \leqq l\}$ is a compact set. Then for C, a closed set in $\mathcal{M}$,

$$(1.8) \qquad \varlimsup_{n\to\infty} \frac{1}{n} \log Q_{n,x}(C) \leqq -\inf_{\mu \in C} I(\mu).$$

Inequality (1.8) is a consequence of Theorem 4.4 in Section 4.

Using (1.7) and (1.8) it follows from Section 3 of [9] that, for real-valued weakly continuous functionals $\Phi(\mu)$ on $\mathcal{M}$,

$$(1.9) \qquad \lim_{n\to\infty} \frac{1}{n} \log E_x\{\exp\{-n\Phi(L_n(\omega, \cdot))\}\} = -\inf_{\mu \in \mathcal{M}} [\Phi(\mu) + I(\mu)].$$

For the continuous parameter case assume $p(t, x, dy)$ is a Feller transition probability on a complete separable metric space X. Let L be the infinitesimal generator of the semigroup associated with $p(t, x, dy)$ having domain $\mathcal{D}$. We assume there exists a $v(x)$ on X such that $\{x \in X : v(x) \geqq l\}$ is a compact set for each $l > -\infty$ and that there exists a sequence $\{u_n\}$ in $\mathcal{D}$ satisfying:

(i) $u_n(x) \geqq 1$ for all n and all $x \in X$;

(ii) for each compact set $W \subset X$,

$$\sup_{x \in W} \sup_n u_n(x) < \infty;$$

(iii) for each $x \in X$,

$$\lim_{n\to\infty} \left(\frac{Lu_n}{u_n}\right)(x) = v(x);$$

(iv) for some $A<\infty$,

$$\sup_{n,x}\left(\frac{Lu_n}{u_n}\right)(x)\leqq A\,.$$

Then, for any closed set $C\subset \mathcal{M}$,

$$\overline{\lim_{t\to\infty}}\frac{1}{t}\log Q_{x,t}(C)\leqq -\inf_{\mu\in C} I(\mu)\,. \tag{1.10}$$

Inequality (1.10) follows from Theorem 7.2 in Section 7.

To get the proper lower bound in the continuous parameter case, we assume $p(t, x, dy)$ is a Feller transition probability having a density $p(t, x, y)$ with respect to a reference measure $\beta(dy)$. Let B_0 be the center of the semigroup T_t associated with $p(t, x, dy)$, i.e., $B_0=\{f\in C(X):\lim_{t\to 0}\|T_tf-f\|=0\}$, where the norm is the usual norm for $C(X)$. Let

$$B_{00}=\left\{f\in C(X):\lim_{t\to 0}\sup_{x}\int_X|f(y)-f(x)|\,p(t,x,dy)=0\right\},$$

so that $B_{00}\subset B_0\subset C(X)$. Let $\alpha\in\mathcal{M}$ be such that $I(\alpha)=l<\infty$ and let N_α be a neighborhood of α in $\mathcal{M}$ of the form

$$N_\alpha=\left\{\mu\in\mathcal{M}:\left|\int_X f_j(x)[\mu(dx)-\alpha(dx)]\right|<\varepsilon,\quad 1\leqq j\leqq k,\right.\\ \left.\text{where}\quad f_j\in B_{00},\quad 1\leqq j\leqq k\right\}. \tag{1.11}$$

We assume that $p(t, x, dy)$ is such that every neighborhood of $\alpha\in\mathcal{M}$ contains a neighborhood of the form (1.11). We also assume that, for any $\mu\in\mathcal{M}$ and any bounded measurable function $f(x)$, there exists a sequence $\{f_n\}\in B_0$ such that $\|f_n\|\leqq\|f\|$ and $f_n\to f$ almost everywhere with respect to μ. Finally we impose on $p(t, x, dy)$ the condition that, for any $x\in X$, $\sigma>0$, and $E\subset X$ such that $\beta(E)>0$, we have $\int_0^\infty e^{-\sigma t}p(t,x,E)\,dt>0$.

Under these hypotheses we prove that, for any open set $G\subset\mathcal{M}$,

$$\underline{\lim_{t\to\infty}}\frac{1}{t}\log Q_{x,t}(G)\geqq -\inf_{\mu\in G} I(\mu)\,. \tag{1.12}$$

Inequality (1.12) follows from Theorem 8.1 in Section 8. From (1.10) and

(1.12) we obtain the result that, for weakly continuous functionals $\Phi(\mu)$ on $\mathcal{M}$,

$$\lim_{t\to\infty}\frac{1}{t}\log E_x\{\exp\{-t\Phi(L_t(\omega,\cdot))\}\} = -\inf_{\mu\in\mathcal{M}}[\Phi(\mu)+I(\mu)]. \tag{1.13}$$

In some of the theorems in this paper we actually prove somewhat more than is necessary for the results quoted above. This is because we need such estimates in a forthcoming paper on the laws of the iterated logarithm for the local time.

2. Some Preliminary Results

Let X be a complete separable metric space and let $\pi(x, dy)$ be the transition probability function of a discrete parameter Markov process with X as state space. We denote by $\mathcal{M}$ the space of probability measures on X.

Let $\mathcal{U}$ be the space of Borel measurable functions u on X for each of which there exist constants a and b such that $0 < a \leqq u(x) \leqq b < \infty$ for all $x \in X$. With

$$(\pi u)(x) = \int_X u(y)\pi(x, dy),$$

we define, for each $\mu \in \mathcal{M}$,

$$I(\mu) = -\inf_{u\in\mathcal{U}}\int_X \log\left(\frac{\pi u}{u}\right)(x)\mu(dx). \tag{2.1}$$

If we let $\mathcal{U}_1$ be the subset of $\mathcal{U}$ consisting of functions u which are also continuous on X, then it is easy to see that

$$-\inf_{u\in\mathcal{U}_1}\int_X \log\left(\frac{\pi u}{u}\right)(x)\mu(dx) = I(\mu). \tag{2.2}$$

Let $\alpha \in \mathcal{M}$, $\beta \in \mathcal{M}$ and define

$$I(\alpha, \beta) = -\inf_{u\in\mathcal{U}_1}\left[\int_X \log(\pi u)(x)\alpha(dx) - \int_X \log u(x)\beta(dx)\right] \tag{2.3}$$

so that, in particular, $I(\alpha, \alpha) = I(\alpha)$.

Let $\mathcal{M}_{X\times X}$ be the space of all probability measures on $X \times X$, and let $\mathcal{M}_{\alpha,\beta}$ be the set of all $\lambda \in \mathcal{M}_{X\times X}$ with first marginal α and second marginal β. We denote by $\mathcal{M}_{\alpha,\cdot}$ the union of $\mathcal{M}_{\alpha,\beta}$ for all β. For any $\lambda \in \mathcal{M}_{X\times X}$, we define

$$(2.4)\ \bar{I}(\lambda) = -\inf_{u\in\mathcal{U}_2}\left[\log \iint_{X\times X} u(x, y)\lambda_0(dx, dy) - \iint_{X\times X} \log u(x, y)\lambda(dx, dy)\right],$$

where $\lambda_0(dx, dy) = \gamma(dx)\pi(x, dy)$, γ being the first marginal of λ. In (2.4), $\mathcal{U}_2$ is the set of functions $u(x, y)$ on $X \times X$ which are continuous and for each of which there are constants a and b such that $0 < a \leqq u(x, y) \leqq b < \infty$ for all $(x, y) \in X \times X$. Finally, we define, for any $\alpha \in \mathcal{M}$, $\beta \in \mathcal{M}$,

$$(2.5) \qquad J(\alpha, \beta) = \inf_{\lambda\in\mathcal{M}_{\alpha,\beta}} \bar{I}(\lambda) .$$

We now prove

THEOREM 2.1. *For all $\alpha \in \mathcal{M}$, $\beta \in \mathcal{M}$,*

$$I(\alpha, \beta) = J(\alpha, \beta) .$$

Proof: Since *this* theorem involves X only as a Borel structure and since by our assumption on X (complete separable metric) it is a standard space, by a theorem of Kuratowski [4] one can assume without loss of generality that X is a compact metric space.

We show first that $I(\alpha, \beta) \leqq J(\alpha, \beta)$. Assume $J(\alpha, \beta) = J_0 < \infty$, and let $\varepsilon > 0$ be given. There exists $\lambda \in \mathcal{M}_{\alpha,\beta}$ such that $\bar{I}(\lambda) < J_0 + \varepsilon$, i.e., such that, for every $u \in \mathcal{U}_2$,

$$\iint_{X\times X} \log u(x, y)\lambda(dx, dy) - \log \iint_{X\times X} u(x, y)\lambda_0(dx, dy) \leqq J_0 + \varepsilon .$$

In particular, if we set $u(x, y) = g(y)/(\pi g)(x)$ where $g \in \mathcal{U}_1$, we have

$$\iint_{X\times X} [\log g(y) - \log (\pi g)(x)]\lambda(dx, dy)$$

$$\leqq \log \iint_{X\times X} \frac{g(y)}{(\pi g)(x)}\lambda_0(dx, dy) + J_0 + \varepsilon .$$

Since $\lambda \in \mathcal{M}_{\alpha,\beta}$ and hence $\lambda_0(dx, dy) = \alpha(dx)\pi(x, dy)$, we obtain

$$\int_X \log g(y)\beta(dy) - \int_X \log(\pi g)(x)\alpha(dx) \leqq \log \int_X \frac{(\pi g)(x)}{(\pi g)(x)} \alpha(dx) + J_0 + \varepsilon .$$

Thus,

$$-\left[\int_X \log(\pi g)(x)\alpha(dx) - \int_X \log g(y)\beta(dy)\right] \leqq J_0 + \varepsilon ,$$

and this inequality holds for all $g \in \mathcal{U}_1$ which implies

$$-\inf_{g \in \mathcal{U}_1} \left[\int_X \log(\pi g)(x)\alpha(dx) - \int_X \log g(x)\beta(dx)\right] \leqq J_0 + \varepsilon .$$

Consequently, $I(\alpha, \beta) \leqq J(\alpha, \beta)$.

To show that $J(\alpha, \beta) \leqq I(\alpha, \beta)$ for all $\alpha \in \mathcal{M}$, $\beta \in \mathcal{M}$, assume that, for some fixed $\alpha \in \mathcal{M}$ and $\beta \in \mathcal{M}$, $J(\alpha, \beta) > J_0$; then it suffices to show $I(\alpha, \beta) \geqq J_0$ for that α and β. But $J(\alpha, \beta) > J_0$ means $\mathcal{M}_{\alpha,\beta} \cap \{\lambda \in \mathcal{M}_{\alpha,\cdot} : \bar{I}(\lambda) \leqq J_0\} = \emptyset$. Thus, $\mathcal{M}_{\alpha,\beta}$ and $\{\lambda \in \mathcal{M}_{\alpha,\cdot} : \bar{I}(\lambda) \leqq J_0\}$ are disjoint, non-empty, convex, compact subsets of $\mathcal{M}_{X \times X}$. By the separation theorem there exists a $v(x, y) \in C(X \times X)$ such that

$$\iint_{X \times X} v(x, y)\lambda(dx, dy) \leqq 0 \quad \text{for all} \quad \lambda \in \mathcal{M}_{\alpha,\beta} , \tag{2.6}$$

and

$$\iint_{X \times X} v(x, y)\lambda(dx, dy) \geqq 1 \quad \text{for all} \quad \lambda \in \{\lambda \in \mathcal{M}_{\alpha,\cdot} : \bar{I}(\lambda) \leqq J_0\} . \tag{2.7}$$

From (2.6) and Lemma 2.2 below we can choose, for any $\varepsilon > 0$, $f_\varepsilon(x) \in C(X)$ and $g_\varepsilon(x) \in C(X)$ such that $v(x, y) \leqq f_\varepsilon(x) - g_\varepsilon(y) + \varepsilon$ for all $(x, y) \in X \times X$ which moreover satisfy

$$\int_X f_\varepsilon(x)\alpha(dx) = \int_X g_\varepsilon(y)\beta(dy) .$$

Therefore, from (2.7) we see that for any $\varepsilon > 0$ there exist $f_\varepsilon(x) \in C(X)$, $g_\varepsilon(x) \in C(X)$ such that

$$\iint_{X \times X} [f_\varepsilon(x) - g_\varepsilon(y) + \varepsilon]\lambda(dx, dy) \geqq 1 \quad \text{for all} \quad \lambda \in \{\lambda \in \mathcal{M}_{\alpha,\cdot} : \bar{I}(\lambda) \leqq J_0\} \tag{2.8}$$

and also

$$\int_X f_\varepsilon(x)\alpha(dx) = \int_X g_\varepsilon(y)\beta(dy)\,.$$

Let $0<\varepsilon<1$. Relation (2.8) shows that if $\lambda \in \mathcal{M}_{\alpha,\cdot}$ is such that

$$\iint\limits_{X\times X} [f_\varepsilon(x) - g_\varepsilon(y)]\lambda(dx, dy) = 0\,, \tag{2.9}$$

then certainly, for that λ, $\bar{I}(\lambda) > J_0$. Hence, for our given $0<\varepsilon<1$,

$$\inf_{\substack{\lambda\in\mathcal{M}_{\alpha,\cdot} \\ \iint_{X\times X}[f_\varepsilon(x)-g_\varepsilon(y)]\lambda(dx,dy)=0}} \bar{I}(\lambda) \geqq J_0\,. \tag{2.10}$$

Using Lemma 2.3 below (with $\Omega = X\times X$, $W(x, y) = f_\varepsilon(x) - g_\varepsilon(y)$, $P = \alpha(dx)\pi(x, dy)$, and $a=0$) we get from (2.10)

$$-\inf_\theta \log M(\theta) \geqq J_0\,, \tag{2.11}$$

where

$$M(\theta) = \iint\limits_{X\times X} \exp\{\theta[f_\varepsilon(x) - g_\varepsilon(x)]\}\alpha(dx)\pi(x, dy)\,.$$

Thus, for any $\eta>0$, there exists θ_η such that $\log M(\theta_\eta) \leqq -J_0 + \eta$, i.e.,

$$\iint\limits_{X\times X} \exp\{\theta_\eta[f_\varepsilon(x) - g_\varepsilon(y)]\}\alpha(dx)\pi(x, dy) \leqq e^{-J_0+\eta}\,. \tag{2.12}$$

In (2.12) let $V(y) = \exp\{-\theta_\eta g_\varepsilon(y)\}$, $W(x) = \exp\{-\theta_\eta f_\varepsilon(x)\}$, then we have

$$\iint\limits_{X\times X} \frac{V(y)}{W(x)}\alpha(dx)\pi(x, dy) = \int\limits_X \left(\frac{\pi V}{W}\right)(x)\alpha(dx) \leqq e^{-J_0+\eta}\,.$$

Hence, from Jensen's inequality,

$$\int_X \log\left(\frac{\pi V}{W}\right)(x)\alpha(dx) \leqq \log\int_X \left(\frac{\pi V}{W}\right)(x)\alpha(dx) \leqq -J_0 + \eta\,. \tag{2.13}$$

Since we were able to choose $f_\varepsilon(x)$ and $g_\varepsilon(y)$ such that

$$\int_X f_\varepsilon(x)\alpha(dx)=\int_X g_\varepsilon(y)\beta(dy),$$

we see that

$$\int_X \log W(x)\alpha(dx)=\int_X \log V(y)\beta(dy),$$

and thus from (2.13) we get

$$\int_X \log(\pi V)(x)\alpha(dx)-\int_X \log V(x)\beta(dx)\leqq -J_0+\eta. \tag{2.14}$$

But (2.14) implies that, for any $\eta>0$,

$$\inf_{u\in\mathcal{U}_1}\left[\int_X \log(\pi u)(x)\alpha(dx)-\int_X \log u(x)\beta(dx)\right]\leqq -J_0+\eta,$$

so that

$$-\inf_{u\in\mathcal{U}_1}\left[\int_X \log(\pi u)(x)\alpha(dx)-\int_X \log u(x)\beta(dx)\right]\geqq J_0,$$

and hence $I(\alpha,\beta)\geqq J_0$ which completes Theorem 2.1.

LEMMA 2.2. *Suppose X is compact and $v(x,y)\in C(X\times X)$ is such that*

$$\iint_{X\times X} v(x,y)\lambda(dx,dy)\leqq 0 \quad \text{for all} \quad \lambda\in\mathcal{M}_{\alpha,\beta}. \tag{2.15}$$

Then, for any $\varepsilon>0$, there exist $f_\varepsilon(x)\in C(X)$, $g_\varepsilon(x)\in C(X)$ such that $v(x,y)\leqq f_\varepsilon(x)-g_\varepsilon(y)+\varepsilon$ for all $(x,y)\in X\times X$ and

$$\int_X f_\varepsilon(x)\alpha(dx)=\int_X g_\varepsilon(y)\beta(dy).$$

Proof: Before proving this lemma we note that if for any $\varepsilon>0$ there exist functions $f_\varepsilon(x)$ and $g_\varepsilon(y)$ such that $v(x,y)\leqq f_\varepsilon(x)-g_\varepsilon(y)+\varepsilon$ for all $(x,y)\in X\times X$ and also $\int_X f_\varepsilon(x)\alpha(dx)=\int_X g_\varepsilon(y)\beta(dy)$, then surely, for every

$\lambda \in \mathcal{M}_{\alpha,\beta}$, $\iint_{X\times X} v(x, y)\lambda(dx, dy) = 0$. Therefore, this lemma characterizes functions $v(x, y)$ on $X \times X$ satisfying (2.15).

Let $\phi(x) \in C(X)$, $\psi(x) \in C(X)$ and consider, for $\lambda \in \mathcal{M}_{X\times X}$ and $v(x, y)$ satisfying (2.15),

$$H[\lambda, \phi, \psi] = \iint_{X\times X} v(x, y)\lambda(dx, dy) - \iint_{X\times X} [\phi(x) - \psi(y)]\lambda(dx, dy) + \int_X \phi(x)\alpha(dx) - \int_X \psi(y)\beta(dy).$$

If $\lambda \in \mathcal{M}_{\alpha,\beta}$ then, clearly, $H[\lambda, \phi, \psi] \leqq 0$ for all $\phi \in C(X)$ and $\psi \in C(X)$. Moreover, if $\lambda \in \mathcal{M}_{X\times X}$ but $\lambda \notin \mathcal{M}_{\alpha,\beta}$, then

$$\inf_{\substack{\phi \in C(X)\\ \psi \in C(X)}} H[\lambda, \phi, \psi] = -\infty .$$

Thus, we have

$$\sup_{\lambda \in \mathcal{M}_{X\times X}} \inf_{\substack{\phi \in C(X)\\ \psi \in C(X)}} H[\lambda, \phi, \psi] \leqq 0 . \tag{2.16}$$

Since X is compact, $\mathcal{M}_{X\times X}$ is compact, and hence by one of the mini-max theorems (cf. [7]) we have

$$\sup_{\lambda \in \mathcal{M}_{X\times X}} \inf_{\substack{\phi \in C(X)\\ \psi \in C(X)}} H[\lambda, \phi, \psi] = \inf_{\substack{\phi \in C(X)\\ \psi \in C(X)}} \sup_{\lambda \in \mathcal{M}_{X\times X}} H[\lambda, \phi, \psi],$$

so that

$$\inf_{\substack{\phi \in C(X)\\ \psi \in C(X)}} \sup_{\lambda \in \mathcal{M}_{X\times X}} H[\lambda, \phi, \psi] \leqq 0 . \tag{2.17}$$

But

$$\begin{aligned}
&\sup_{\lambda \in \mathcal{M}_{X\times X}} H[\lambda, \phi, \psi] \\
&= \sup_{\lambda \in \mathcal{M}_{X\times X}} \left[\iint_{X\times X} \{v(x, y) - \phi(x) + \psi(y)\}\lambda(dx, dy) + \int_X \phi(x)\alpha(dx) - \int_X \psi(y)\beta(dy) \right] \\
&= \sup_{(x,y)\in X\times X} [v(x, y) - \phi(x) + \psi(y)] + \int_X \phi(x)\alpha(dx) - \int_X \psi(y)\beta(dy),
\end{aligned}$$

and (2.17) becomes

$$(2.18)\quad \inf_{\substack{\phi\in C(X)\\ \psi\in C(X)}}\left\{\sup_{(x,y)\in X\times X}[v(x,y)-\phi(x)+\psi(y)] + \int_X \phi(x)\alpha(dx)-\int_X \psi(y)\beta(dy)\right\}\leqq 0\,.$$

Consequently, for any $\varepsilon>0$, there exist $\phi_\varepsilon\in C(X)$ and $\psi_\varepsilon\in C(X)$ such that

$$(2.19)\quad \sup_{(x,y)\in X\times X}[v(x,y)-\phi_\varepsilon(x)+\psi_\varepsilon(y)]+\int_X \phi_\varepsilon(x)\alpha(dx)-\int_X \psi_\varepsilon(y)\beta(dy)\leqq \varepsilon\,.$$

If we let

$$l=\int_X \phi_\varepsilon(x)\alpha(dx)-\int_X \psi_\varepsilon(y)\beta(dy)\,,$$

$f_\varepsilon(x)=\phi_\varepsilon(x)-l$, and $g_\varepsilon(y)=\psi_\varepsilon(y)$, then (2.19) becomes

$$(2.20)\qquad \sup_{(x,y)\in X\times X}[v(x,y)-f_\varepsilon(x)+g_\varepsilon(y)]\leqq \varepsilon$$

and, moreover,

$$\int_X f_\varepsilon(x)\alpha(dx)=\int_X g_\varepsilon(y)\beta(dy)\,.$$

This completes the proof of the lemma.

The next lemma is a special case of Theorem 5.2 (of Section 5) and we refer to that theorem for the proof.

LEMMA 2.3. *Let (Ω,Σ,P) be a probability space and let W be a bounded random variable. Let $\mathcal{M}$ be the space of probability measures γ on Ω and $\Gamma_a\subset\mathcal{M}$ the set of probability measures γ on Ω such that $\int_\Omega W(x)\gamma(dx)=a$. For any $\gamma\in\mathcal{M}$, let*

$$I_P(\gamma)=-\inf_{u\in\mathcal{U}_1}\left[\log\int_\Omega u(x)P(dx)-\int_\Omega \log u(x)\gamma(dx)\right],$$

and

$$h(a)=\inf_{\gamma\in\Gamma_a} I_P(\gamma)\,.$$

Then,

$$h(a)=\sup_\theta[\theta a-\log M(\theta)]\,,$$

where

$$M(\theta)=\int_{\Omega} e^{\theta W(x)}P(dx)\,,$$

i.e., $h(a)$ is the conjugate convex function of $\log M(\theta)$.

In proving Theorem 2.1 we made minimal assumptions on the Markov process transition probability $\pi(x, dy)$. We now assume that $\pi(x, dy)$ is a Feller transition probability and that there exists a reference measure β on X such that $\pi(x, dy)=\pi(x, y)\beta(dy)$. Under this condition, if $\mu\in\mathcal{M}$ is such that $I(\mu)<\infty$, then μ is absolutely continuous with respect to the reference measure β.

Let $\mu\in\mathcal{M}$ and for convenience denote $\mathcal{M}_{\mu,\mu}$ by $\mathcal{M}_\mu$. From Theorem 2.1 we know that

$$I(\mu)=I(\mu,\mu)=J(\mu,\mu)=\inf_{\lambda\in\mathcal{M}_\mu}\bar{I}(\lambda)\,. \tag{2.21}$$

Under the hypotheses made on $\pi(x, dy)$ above, $\bar{I}(\lambda)$ is lower semicontinuous and, since $\mathcal{M}_\mu$ is compact, the infimum in (2.21) is actually attained. Denote the element of $\mathcal{M}_\mu$ where this infimum is attained by $\bar{\lambda}$, i.e., $\bar{I}(\bar{\lambda})=I(\mu)$. We now write $\bar{\lambda}(dx, dy)=\mu(dx)\bar{\pi}(x, dy)$, where $\bar{\pi}(x, dy)$ is defined for almost all x relative to μ. However we can extend $\bar{\pi}$ arbitrarily so that we have a transition probability defined for all x. We shall continue to call this extension $\bar{\pi}(x, dy)$, and consider it as a transition probability for a new discrete parameter Markov process on X. We note that, for $\bar{\pi}$, μ is an invariant measure since μ is not only a first marginal for $\bar{\lambda}$ but also a second marginal. Denote by $\bar{P}$ the probabilities for the stationary discrete parameter Markov process having $\bar{\pi}(x, dy)$ as its transition probability and the invariant measure μ as initial distribution. We denote by $\bar{P}_x$ the corresponding probabilities for this process starting at x. In Section 3 we shall need the fact that the stationary process $\bar{P}$ is ergodic. The next two lemmas lead up to Theorem 2.6 where the ergodicity of $\bar{P}$ is established under the assumption that $\pi(x, y)>0$ almost everywhere with respect to the product measure $\mu\times\mu$.

LEMMA 2.4. *Let $\mu\in\mathcal{M}$ be such that $I(\mu)<\infty$, and let $\{u_n\}$ be a sequence in $\mathcal{U}_1$ such that*

$$\lim_{n\to\infty}\int_X \log\left(\frac{\pi u_n}{u_n}\right)(x)\mu(dx)=-I(\mu)\,.$$

Let $\lambda_n(dx, dy)=\mu(dx)(u_n(y)/(\pi u_n)(x))\pi(x, dy)$, and let $\bar{\lambda}$ be that element of $\mathcal{M}_\mu$ for which $\bar{I}(\bar{\lambda})=I(\mu)$. Then, $\lim_{n\to\infty}\|\lambda_n-\bar{\lambda}\|=0$, where the norm is the variation norm.

Proof: It follows from [1] (Lemma 2.1) that $\bar{\lambda} \ll \mu(dx)\pi(x, dy)$, and since u_n and πu_n are bounded above and below, we have $\bar{\lambda} \ll \lambda_n$. Let

$$I_{\lambda_n}(\bar{\lambda}) = -\inf_{u \in \mathcal{U}_2} \left[\log \iint_{X \times X} u(x, y)\lambda_n(dx, dy) - \iint_{X \times X} \log u(x, y)\bar{\lambda}(dx, dy)\right].$$

Again, from [1] (Lemma 2.1) and recalling that $\lambda_0(dx, dy) = \mu(dx)\pi(x, dy)$, we have

$$\begin{aligned} I_{\lambda_n}(\bar{\lambda}) &= \iint_{X \times X} \log \frac{d\bar{\lambda}}{d\lambda_n} \bar{\lambda}(dx, dy) \\ &= \iint_{X \times X} \log \left[\left(\frac{d\bar{\lambda}}{d\lambda_0}\right)\left(\frac{d\lambda_0}{d\lambda_n}\right)\right] \bar{\lambda}(dx, dy) \\ &= \iint_{X \times X} \log \frac{d\bar{\lambda}}{d\lambda_0} \bar{\lambda}(dx, dy) + \iint_{X \times X} \log \frac{d\lambda_0}{d\lambda_n} \bar{\lambda}(dx, dy) \\ &= \bar{I}(\bar{\lambda}) + \iint_{X \times X} \log \frac{(\pi u_n)(x)}{u_n(y)} \bar{\lambda}(dx, dy) \\ &= I(\mu) + \iint_{X \times X} \log (\pi u_n)(x)\bar{\lambda}(dx, dy) - \iint_{X \times X} \log u_n(y)\bar{\lambda}(dx, dy) \\ &= I(\mu) + \int_X \log (\pi u_n)(x)\mu(dx) - \int_X \log u_n(y)\mu(dy) \\ &= I(\mu) + \int_X \log \left(\frac{\pi u_n}{u_n}\right)(x)\mu(dx)\,. \end{aligned}$$

By hypothesis the second term on the right in this last expression converges to $-I(\mu)$; hence we conclude that

$$\lim_{n \to \infty} I_{\lambda_n}(\bar{\lambda}) = 0\,. \tag{2.22}$$

In [1] (Lemma 4.1) we showed the existence of a universal function $\phi(l)$ such that $\phi(l) \geqq 0$ for $l \geqq 0$ and $\phi(l) \to 0$ as $l \to 0$, having the property that if

$\pi(x, dy)$ is any transition probability and μ is any probability measure on X with the corresponding $I(\mu)=l$, then, in the variation norm, $\|\mu\pi-\mu\|\leqq\phi(l)$. Here $\mu\pi$ means the measure $(\mu\pi)(A)=\int_X \pi(x, A)\mu(dx)$. Consider a very special case of this, namely, the case where $\pi(x, dy)=\alpha(dy)$. Note first that, in this case, $(\mu\pi)(A)=\alpha(A)$ and second that the I-function, $I(\mu)$, becomes

$$I_\alpha(\mu)=-\inf_{u\in\mathcal{U}_1}\left[\log\int_X u(x)\alpha(dx)-\int_X \log u(x)\mu(dx)\right].$$

Thus Lemma 4.1 of [1] says that, whenever $I_\alpha(\mu)=l$ for any two measures α and μ, we have $\|\alpha-\mu\|\leqq\phi(l)$. Hence (2.22) and the fact that $\phi(l)\to 0$ as $l\to 0$ implies that $\lim_{n\to\infty}\|\lambda_n-\bar{\lambda}\|=0$.

LEMMA 2.5. *Let (X, Σ, α) and (Y, S, β) be two probability measure spaces, and let (Z, Γ, λ) be the product space with product measure. Let $\{h_n(x, y)\}$ be a sequence of non-negative functions in Z of the form $h_n(x, y)=f_n(x)/g_n(y)$ where, for each n, $f_n(x)$ and $g_n(y)$ are bounded strictly positive functions on X and Y, respectively. If $\lim_{n\to\infty} h_n(x, y)=h(x, y)$ exists almost everywhere in Z and $0\leqq h(x, y)<\infty$ almost everywhere in Z, then $h(x, y)=f(x)/g(y)$ where $0\leqq f(x)<\infty$ almost everywhere in X and $0<g(y)\leqq\infty$ almost everywhere in Y.*

Proof: Let $E^*=\{(x, y): f_n(x)/g_n(y)\to h(x, y)\}$ so that $\lambda(E^*)=1$. Let $E_x=\{y:(x, y)\in E^*\}$ so that $\alpha\{x:\beta(E_x)=1\}=1$. Without loss of generality we can assume $h(x, y)\not\equiv 0$. Therefore, there exists an x such that $h(x, y)\not\equiv 0$ in y. In fact, $\alpha\{x:\beta\{y: h(x, y)=0\}<1\}>0$. Now pick $\bar{x}$ such that $\beta(E_{\bar{x}})=1$ and $\beta\{y: h(\bar{x}, y)=0\}<1$.

For any $(x, y)\in E^*$,

$$\lim_{n\to\infty} h_n(x, y)=\lim_{n\to\infty}\frac{f_n(x)f_n(\bar{x})}{f_n(\bar{x})g_n(y)}=h(x, y),$$

and

$$\lim_{n\to\infty}\frac{f_n(\bar{x})}{g_n(y)}=h(\bar{x}, y)\quad\text{for all}\quad y\in E_{\bar{x}}.$$

Let $E_1=\{x:\beta(E_x)=1\}$. For any $x\in E_1$ we have $\lim_{n\to\infty} h_n(x, y)=h(x, y)$ almost everywhere in y (β-measure). On the other hand, for $x\in E_1$ and any y, $h_n(x, y)=(f_n(x)/f_n(\bar{x}))(f_n(\bar{x})/g_n(y))$; thus letting $l_n(y)=f_n(\bar{x})/g_n(y)$ and $a_n(x)=f_n(x)/f_n(\bar{x})$ we get that $\lim_{n\to\infty} l_n(y)=h(\bar{x}, y)$ almost everywhere in y (β-measure) and $\lim_{n\to\infty} a_n(x)l_n(y)=h(x, y)$ almost everywhere in y (β-measure). Moreover, $h(\bar{x}, y)\not\equiv 0$ in y. Therefore, $\lim_{n\to\infty} a_n(x)$ exists—call it

$f(x)$, and $h(x, y) = f(x)h(\bar{x}, y)$. Letting $g(y) = 1/h(\bar{x}, y)$ we have $h(x, y) = f(x)/g(y)$. Clearly $0 \leqq f(x) < \infty$ almost everywhere in X and $0 < g(y) \leqq \infty$ almost everywhere in Y.

THEOREM 2.6. *Let μ be such that $I(\mu) < \infty$. Let $\bar{P}$ be the stationary process that was constructed earlier. If $\pi(x, y) > 0$ a.e. with respect to the product measure $\mu \times \mu$, then $\bar{P}$ is ergodic.*

Proof: As in Lemma 2.4, let $\{u_n\}$ be a sequence in $\mathcal{U}_1$ such that

$$\lim_{n\to\infty} \int_X \log\left(\frac{\pi u_n}{u_n}\right)(x)\mu(dx) = -I(\mu)$$

and let

$$\lambda_n(dx, dy) = \mu(dx)\frac{u_n(y)}{(\pi u_n)(x)}\pi(x, dy) = \frac{\mu(dx)u_n(y)}{(\pi u_n)(x)}\pi(x, y)\beta(dy).$$

From that lemma, $\lim_{n\to\infty} \lambda_n = \bar{\lambda}$ in variation norm, where $\bar{\lambda}$ is the element of $\mathcal{M}_\mu$ for which $\bar{I}(\bar{\lambda}) = I(\mu)$. Hence

$$\lim_{n\to\infty} \frac{u_n(y)}{(\pi u_n)(x)}\pi(x, y) = \theta(x, y)$$

exists in $L_1(\mu \times \beta)$ and $\bar{\lambda}(dx, dy) = \theta(x, y)\mu(dx)\beta(dy)$. We can choose a subsequence if necessary and conclude, from the assumption $\pi(x, y) > 0$ a.e. $\mu \times \mu$, that

$$\lim_{n\to\infty} \frac{u_n(y)}{(\pi u_n)(x)} = \frac{\theta(x, y)}{\pi(x, y)} = h(x, y) \quad \text{a.e.} \quad \mu \times \mu.$$

Lemma 2.5 implies that $h(x, y) = a(x)/b(y)$ a.e. $\mu \times \mu$, where $0 \leqq a(x) < \infty$ a.e. μ and $0 < b(y) \leqq \infty$ a.e. μ. So $\theta(x, y) = (a(x)/b(y))\pi(x, y)$ a.e. $\mu \times \mu$. Since $\mu \ll \beta$, let B be the set in X where $d\mu/d\beta > 0$. Then $\mu(\tilde{B}) = 0$ and since $\bar{\lambda}$ has both marginals equal to μ, $\theta(x, y)$ must be equal to zero a.e. $\mu \times \beta$ on the set $X \times \tilde{B}$. Therefore we can extend $b(y)$ defined on B to be equal to $+\infty$ on $\tilde{B}$ and, with this extension,

$$\theta(x, y) = \frac{a(x)}{b(y)}\pi(x, y) \quad \text{a.e.} \quad \mu \times \beta.$$

Let $E = \{x : a(x) = 0\}$ and $F = \{y : b(y) = \infty\}$. Since

$$\int_{E\times X} \bar{\lambda}(dx, dy) = 0 = \int_{X\times F} \bar{\lambda}(dx, dy),$$

we have $\bar{\lambda}(\tilde{E}\times\tilde{F})=1$, where $\tilde{E}$ denotes the complement of E. We shall now show that if $A\subset X$ with the property that $\bar{\lambda}(A\times\tilde{A})=\bar{\lambda}(\tilde{A}\times A)=0$, then either $\mu(A)=0$ or $\mu(\tilde{A})=0$. To see this, we first note that since

$$\iint_{A\times\tilde{A}} \frac{a(x)}{b(y)}\,\pi(x, y)\mu(dx)\beta(dy)=0\,,$$

and $\pi(x, y)>0$ almost everywhere (β) for $x\in X$, we must have $a(x)/b(y)=0$ almost everywhere $(\mu\times\beta)$ on $A\times\tilde{A}$. This implies

$$(2.23)\quad (\mu\times\beta)\{(A\times\tilde{A})\cap(\tilde{E}\times\tilde{F})\}=0 \quad\text{and}\quad (\mu\times\beta)\{(\tilde{A}\times A)\cap(\tilde{E}\times\tilde{F})\}=0\,.$$

The first equation in (2.23) is the same as $\mu(A\cap\tilde{E})\beta(\tilde{A}\cap\tilde{F})=0$, and the second is the same as $\mu(\tilde{A}\cap\tilde{E})\beta(A\cap\tilde{F})=0$. Hence, we have four possibilities; either

$$\mu(A\cap\tilde{E})=0 \quad\text{and}\quad \mu(\tilde{A}\cap\tilde{E})=0\,,$$

or

$$\mu(A\cap\tilde{E})=0 \quad\text{and}\quad \beta(A\cap\tilde{F})=0\,,$$

or

$$\beta(\tilde{A}\cap\tilde{F})=0 \quad\text{and}\quad \mu(\tilde{A}\cap\tilde{E})=0\,,$$

or

$$\beta(\tilde{A}\cap\tilde{F})=0 \quad\text{and}\quad \beta(A\cap\tilde{F})=0\,.$$

The first and fourth of these possibilities are impossible since then either $\beta(\tilde{F})=0$ or $\mu(\tilde{E})=0$, and, since $\bar{\lambda}\ll(\mu\times\beta)$, this contradicts $\bar{\lambda}(\tilde{E}\times\tilde{F})=1$. Thus either the second or third of these possibilities holds. Since $\bar{\lambda}\in\mathcal{M}_\mu$ and $\bar{\lambda}(\tilde{E}\times\tilde{F})=1$, we see that $\mu(\tilde{E})=1$ and $\mu(\tilde{F})=1$. Therefore, if the second possibility holds, we conclude that $\mu(A)=0$, and if the third possibility holds we conclude that $\mu(\tilde{A})=0$. That this implies the ergodicity of the $\bar{P}$ process is now easy to see in the following way.

Since the $\bar{P}$ process is a Markov process with stationary distribution μ, if it were not ergodic, there would be a subset $A\subset X$ such that $0<\mu(A)<1$ and the process never leaves A and never leaves $\tilde{A}$, i.e., $\bar{\lambda}(A\times\tilde{A})=0=\bar{\lambda}(\tilde{A}\times A)$. But we just proved that this last equation implies either $\mu(A)=0$ or $\mu(\tilde{A})=0$ which contradicts $0<\mu(A)<1$. Hence the $\bar{P}$ process is ergodic.

3. Lower Bounds—Discrete Case

Let X be a complete separable metric space and $\pi(x, dy)$ the transition probability of a discrete parameter Markov process with X as state space. We assume $\pi(x, dy)$ is a Feller transition probability and that there exists a reference measure β on X such that $\pi(x, dy) = \pi(x, y)\beta(dy)$. Denote by Ω the space of all sequences $x_0, x_1, x_2, \cdots$ with $x_i \in X$. For each starting point $x \in X$, the transition probability $\pi(x, dy)$ induces a probability measure on Ω which we denote by P_x. For each $\omega \in \Omega$, positive integer n, and A a Borel set in X, let

$$L_n(\omega, A) = \frac{1}{n} \sum_{j=0}^{n-1} \chi_A(x_j) .$$

We note that L_n maps Ω into $\mathcal{M}$, the space of probability measures on X. Let $Q_{n,x}$ be the measure induced on $\mathcal{M}$ by L_n from P_x, i.e., if B is a set of measures in $\mathcal{M}$, we define, for each $x \in X$ and each positive integer n,

$$Q_{n,x}(B) = P_x\{\omega \in \Omega : L_n(\omega, \cdot) \in B\} .$$

In this section we obtain asymptotic lower bounds on $Q_{n,x}$ as $n \to \infty$. In doing so we use the ergodicity of the $\bar{P}$ process proved in the preceding section. We begin with

THEOREM 3.1. *Let $\mu \in \mathcal{M}$ be such that $I(\mu) < \infty$ and let $A \subset X$ be any subset of the state space with $\mu(A) = 1$. Let N be a weak neighborhood of μ in $\mathcal{M}$, and denote by $\mathcal{M}(A)$ the set of all probability measures ν on X with $\nu(A) = 1$. If*

$$\pi(x, y) > 0 \quad \textit{almost everywhere} \quad (\mu \times \mu) , \tag{3.1}$$

and if $x \in A$ is such that

$$\pi(x, dy) \quad \textit{is not orthogonal to} \quad \mu(dy) , \tag{3.2}$$

then

$$\lim_{n \to \infty} \frac{1}{n} \log Q_{n,x}(N \cap \mathcal{M}(A)) \geqq -I(\mu) . \tag{3.3}$$

If, in addition, $\pi(x, E)$ is a continuous function of x for each Borel set E, and K is a compact set of points in A for each of which (3.2) *holds, then* (3.3) *holds uniformly over K.*

Proof: Let $\mu \in \mathcal{M}$ be given such that $I(\mu) < \infty$ and $A \subset X$ be such that $\mu(A) = 1$. We note that

$$Q_{n,x}(N \cap \mathcal{M}(A)) = P_x\{\omega \in \Omega : L_n(\omega, \cdot) \in N; x_0, x_1, \cdots, x_{n-1} \in A\} .$$

As in the preceding section, we denote by $\bar{\lambda} = \bar{\lambda}(dx, dy)$ that element of $\mathcal{M}_\mu$ for which $\bar{I}(\bar{\lambda}) = I(\mu)$ and define $\bar{\pi}(x, dy)$ by $\bar{\lambda}(dx, dy) = \mu(dx)\bar{\pi}(x, dy)$. Although $\bar{\pi}(x, dy)$ is defined only for almost all x with respect to μ, we can extend it arbitrarily and treat $\bar{\pi}(x, dy)$ as a transition probability. As noted in the proof of Lemma 2.4, $\bar{\lambda} \ll \mu(dx)\pi(x, dy)$, and therefore $\bar{\pi}(x, dy)$ is absolutely continuous with respect to $\pi(x, dy)$ for almost all x with respect to μ. If we write $\bar{\pi}(x, dy) = h(x, y)\pi(x, dy)$, then we have from [1] (Lemma 2.1) that

$$I(\mu) = \iint_{X \times X} \log h(x, y)\bar{\lambda}(dx, dy) . \tag{3.4}$$

Let N_1 be a weak neighborhood of μ contained in N and let $\delta > 0$ be so chosen that if $\nu \in N_1$ and $\|\nu' - \nu\| < \delta$, then $\nu' \in N$ where $\|\cdot\|$ denotes the variation norm. Set $E_n = \{\omega : L_n(\omega, \cdot) \in N_1\}$ and $D_n = \{\omega : x_j \in A, 0 \leqq j \leqq n-1\}$. For each $\varepsilon > 0$, let

$$F_{n,\varepsilon} = \left\{ \omega : \frac{1}{n-1} \sum_{j=1}^{n-1} \log h(x_{j-1}, x_j) \leqq I(\mu) + \varepsilon \right\} .$$

We denote by $\bar{P}_x$ probabilities for the Markov process with $\bar{\pi}(x, dy)$ as transition function and starting from the point x. We denote by $\bar{P}$ probabilities for the Markov process with transition function $\bar{\pi}(x, dy)$ and initial distribution μ. We note that the $\bar{P}$ process is a stationary process and, by (3.1) and Theorem 2.6, is an ergodic process.

Let $L'_n(\omega, A) = \frac{1}{n} \sum_{i=1}^{n} \chi_A(x_i)$. If n is so large that $2/n < \delta$, then

$$\|L_n(\omega, \cdot) - L'_n(\omega, \cdot)\| < \delta ,$$

and therefore, for large n,

$$\{\omega : L_n(\omega, \cdot) \in N\} \supset \{\omega : L'_n(\omega, \cdot) \in N_1\} .$$

Thus

$$\begin{aligned} P_x\{\omega &: L_n(\omega, \cdot) \in N; x_i \in A, 0 \leqq i \leqq n-1\} \\ &\geqq P_x\{\omega : L'_n(\omega, \cdot) \in N_1 ; x_i \in A, 0 \leqq i \leqq n\} \\ &= \int_A P_y\{E_n \cap D_n\}\pi(x, dy) . \end{aligned}$$

But

$$P_y\{E_n\cap D_n\}=\int_{E_n\cap D_n}\pi(y,dx_1)\cdots\pi(x_{n-2},dx_{n-1})$$

$$\geqq\int_{E_n\cap D_n}\frac{\pi(y,dx_1)}{\bar\pi(y,dx_1)}\cdots\frac{\pi(x_{n-2},dx_{n-1})}{\bar\pi(x_{n-2},dx_{n-1})}\bar\pi(y,dx_1)\cdots\bar\pi(x_{n-2},dx_{n-1})$$

$$=\int_{E_n\cap D_n}\exp\left\{-\sum_{j=1}^{n-1}\log h(x_{j-1},x_j)\right\}d\bar P_y$$

$$\geqq\int_{E_n\cap D_n\cap F_{n,\varepsilon}}\exp\{-(n-1)[I(\mu)+\varepsilon]\}\,d\bar P_y$$

$$=\exp\{-(n-1)[I(\mu)+\varepsilon]\}\bar P_y\{E_n\cap D_n\cap F_{n,\varepsilon}\}.$$

Since μ is an invariant measure for $\bar\pi(x,dy)$ and $\mu(A)=1$, we have $\bar P_y(D_n)=1$ for almost all y with respect to μ. Therefore, if we denote by $\alpha(x,dy)$ the part of $\pi(x,dy)$ that is absolutely continuous with respect to μ, then

$$(3.5)\quad Q_{n,x}(N\cap\mathcal{M}(A))\geqq\exp\{-(n-1)[I(\mu)+\varepsilon]\}\int_A\bar P_y\{E_n\cap F_{n,\varepsilon}\}\alpha(x,dy).$$

By the ergodic theorem, $\bar P_y\{E_n\cap F_{n,\varepsilon}\}\to1$ as $n\to\infty$ for almost all y with respect to μ. Since α is absolutely continuous with respect to μ and $\alpha(x,A)>0$ by (3.2), we conclude that

$$\lim_{n\to\infty}\frac{1}{n}\log Q_{n,x}(N\cap\mathcal{M}(A))\geqq-[I(\mu)+\varepsilon].$$

Since $\varepsilon>0$ is arbitrary, we have (3.3). To prove uniformity over K we see that, by the properties of Lebesgue decomposition, $\alpha(x,E)$ is also continuous as a function of x for each E. Moreover, for $x\in K$, $\alpha(x,dy)$ is uniformly absolutely continuous with respect to μ. It is now routine to see that

$$\lim_{n\to\infty}\int_A\bar P_y\{E_n\cap F_{n,\varepsilon}\}\alpha(x,dy)=\alpha(x,A)$$

uniformly over K, and therefore (3.3) holds uniformly over K.

It will be useful for us to have the analogue of Theorem 3.1 for substochastic transition probabilities. This becomes important if one wants to estimate in the continuous case the probabilities relative to a process that is killed outside of a set. It is also advantageous to replace the strong transitivity assumptions (3.1) and (3.2) by slightly weaker versions.

Let $p(x, dy)$ be a substochastic transition probability on a state space X which we assume to be a complete separable metric space. We can think of $1-p(x, X)$ as the probability of extinction for a chain starting from x in a single step. We can adjoin a point ∞ to our state space (isolated point) and call $X'=X\cup\{\infty\}$. We can then redefine a stochastic transition probability on X' by

$$\begin{aligned}
&\pi(x, A)=p(x, A) \quad \text{for} \quad x\in X \quad \text{and} \quad A\subset X,\\
&\pi(x, \{\infty\})=1-p(x, X),\\
&\pi(\infty, \{\infty\})=1.
\end{aligned}$$

Even though $p(x, dy)$ is substochastic, we can define an I-function for it in the usual way:

$$I(\mu)=-\inf_{u\in\mathcal{U}_1}\int_X \log\frac{\int u(y)p(x, dy)}{u(x)}\,\mu(dx),$$

for any probability measure μ on X. For the stochastic transition function $\pi(x, dy)$ on X' we have the usual I-function defined for probability measures on X', i.e.,

$$I'(\mu)=-\inf_{u'\in\mathcal{U}_1'}\int \log\left(\frac{\pi u'}{u'}\right)(x)\mu(dx).$$

We can state

LEMMA 3.2. *For any measure μ on X' which is concentrated on X,*

$$I'(\mu)=I(\mu).$$

Proof: A function u' on X' can be thought of as a function u on X and a constant k which is the value of u' at ∞. Clearly,

$$I'(\mu)=-\inf_{u\in\mathcal{U}_1,k>0}\int_{X'} \log\left(\frac{\pi u'}{u'}\right)(x)\mu(dx).$$

Since $\mu(\{\infty\})=0$, the integration above is carried out over X only and for $x\in X$

$$(\pi u')(x)=\int_X u(y)p(x, dy)+k(1-p(x, X)).$$

Therefore,

$$I'(\mu)=-\inf_{u\in\mathcal{U}_1, k>0}\int_X \log\left(\frac{\pi u'}{u}\right)(x)\mu(dx),$$

but clearly $k=0$ produces the infimum for any $u\in\mathcal{U}_1$ so that

$$I'(\mu)=-\inf_{u\in\mathcal{U}_1}\int_X \log\frac{\int p(x, dy)u(y)}{u(x)}\mu(dx)=I(\mu).$$

We make the following assumptions on $p(x, dy)$.

HYPOTHESIS H. There is a reference measure $\beta(dy)$ such that $p(x, dy)=p(x, y)\beta(dy)$ for each $x\in X$. Moreover, β can be so chosen that, for each $x\in X$ and $E\subset X$ with $\beta(E)>0$, there is an integer $k=k(x, E)$ such that $p^{(k)}(x, E)>0$. Here, $p^{(k)}(x, E)$ denotes the k-step transition probability from x into E.

Just as before, for each integer n we have the map $L_n(\omega, \cdot)$ from Ω into $\mathcal{M}$ and the measure $\rho_{n,x}$ induced on $\mathcal{M}$ from the $p(x, dy)$ chain. The only difference now is that the total mass of $\rho_{n,x}$ is less than or equal to 1 and may vary with n and x. Our aim is to prove

THEOREM 3.3. *Under Hypothesis* H, *if μ is any measure with $I(\mu)<\infty$ and N is any weak neighborhood of μ in $\mathcal{M}$, then*

$$\varliminf_{n\to\infty}\frac{1}{n}\log\rho_{n,x}(N)\geqq -I(\mu).$$

Moreover, if $p(x, E)$ is continuous as a function of x for each $E\subset X$, then the conclusion above is uniform over every compact subset of X.

We shall reduce the proof of the theorem to an application of Theorem 3.1. First of all we carry out the construction just preceding Lemma 3.2 to obtain a stochastic transition probability $\pi(x, dy)$ on the augmented space $X'=X\cup\{\infty\}$. We can take for reference measure β' which is β on X and δ-measure at ∞. We denote by $\pi(x, y)$ the density of $\pi(x, dy)$ relative to β'.

For each ξ in $0<\xi<1$ we define a transition probability density (relative to β') on X' by

$$\begin{aligned}\pi_\xi(x, y) &= (\pi+\xi\pi^2+\xi^2\pi^3+\cdots)(1-\xi) \\ &= (1-\xi)\pi(1-\xi\pi)^{-1} = \pi R_\xi,\end{aligned} \tag{3.6}$$

where $R_\xi = (1-\xi)(1-\xi\pi)^{-1}$. If $p(x, dy)$ satisfies Hypothesis H, then we note that $\pi_\xi(x, y)$ satisfies

(3.7) $\pi_\xi(x, y)>0$ almost everywhere (β') on X, for each $x\in X$.

If $p(x, E)$ is a continuous function of x for each $E\subset X$, then

$$\pi_\xi(x, E) = \int_E \pi_\xi(x, y)\beta'(dy)$$

is again a continuous function of x for each $E\subset X'$. Let $I^{(\xi)}(\mu)$ be the I-function associated with π_ξ. We now have

LEMMA 3.4. *For each μ on X' with $I'(\mu)<\infty$,*

$$\lim_{\xi\to 0} I^{(\xi)}(\mu) = I'(\mu).$$

In particular, if μ is concentrated on X, we have

$$\lim_{\xi\to 0} I^{(\xi)}(\mu) = I'(\mu) = I(\mu).$$

Proof: Consider

$$\begin{aligned} I^{(\xi)}(\mu) &= -\inf_{u\in\mathcal{U}_1'} \int_{X'} \log\left(\frac{\pi_\xi u}{u}\right)(x)\mu(dx) \\ &= -\inf_{u\in\mathcal{U}_1'} \left[\int_{X'} \log\left(\frac{\pi R_\xi u}{R_\xi u}\right)(x)\mu(dx) + \int_{X'} \log\left(\frac{R_\xi u}{u}\right)(x)\mu(dx)\right] \\ &\leqq -\inf_{u\in\mathcal{U}_1'} \int_{X'} \log\left(\frac{\pi R_\xi u}{R_\xi u}\right)(x)\mu(dx) - \inf_{u\in\mathcal{U}_1'} \int_{X'} \log\left(\frac{R_\xi u}{u}\right)(x)\mu(dx).\end{aligned}$$

Since R_ξ is itself a transition probability and $R_\xi u \geqq (1-\xi)u$, we conclude from this last inequality that, for each $0<\xi<1$, $I^{(\xi)}(\mu)\leqq I'(\mu)-\log(1-\xi)$. Thus, if $I'(\mu)<\infty$, we see that, for each $0<\xi<1$, $I^{(\xi)}(\mu)<\infty$. Moreover,

$\overline{\lim}_{\xi\to 0} I^{(\xi)}(\mu) \leqq I'(\mu)$. On the other hand,

$$\begin{aligned}\underline{\lim}_{\xi\to 0} I^{(\xi)}(\mu) &= \underline{\lim}_{\xi\to 0}\left[-\inf_{u\in\mathcal{U}_1'}\int_{X'}\log\left(\frac{\pi_\xi u}{u}\right)(x)\mu(dx)\right]\\ &= -\overline{\lim}_{\xi\to 0}\inf_{u\in\mathcal{U}_1'}\int_{X'}\log\left(\frac{\pi_\xi u}{u}\right)(x)\mu(dx)\\ &\geqq -\inf_{u\in\mathcal{U}_1'}\int_{X'}\log\left(\frac{\pi u}{u}\right)(x)\mu(dx)\\ &= I'(\mu),\end{aligned}$$

so that we have $\lim_{\xi\to 0} I^{(\xi)}(\mu) = I'(\mu)$. If $\mu(X)=1$, then the second conclusion of the lemma follows from Lemma 3.2.

For $0<\xi<1$ consider an integer-valued random variable τ haveing distribution

$$P\{\tau=i\}=(1-\xi)\xi^{i-1}, \qquad i=1,2,3,\cdots. \tag{3.8}$$

Let $\tau_1, \tau_2, \cdots, \tau_n, \cdots$ be a sequence of independent random variables each having the same distribution (3.8) and, moreover, independent of our Markov process with transition probability $\pi(x, dy)$. Let $s_n=\tau_1+\tau_2+\cdots+\tau_n$ and $s_0=0$. The chain with transition probabilities $\pi_\xi(x, dy)$ can be thought of as the chain with transition probabilities $\pi(x, dy)$ sampled at times $s_0, s_1, \cdots, s_n, \cdots$. Since $\tau_i \geqq 1$ with probability 1, $s_0, s_1, \cdots$ are all distinct. If $s_{n-1}=l$, the sets $[s_0, s_1, \cdots, s_{n-1}]$ and $[0, 1, 2, \cdots, n-1]$ as subsets of the integers have at most $2(l-n+1)$ points which are in one set and not in the other. We denote by P_x and P_x^ξ the processes corresponding to $\pi(x, dy)$ and $\pi_\xi(x, dy)$, respectively. We denote by $Q_{n,x}$ and $Q_{n,x}^\xi$ the induced measures on $\mathcal{M}'$ by the function $L_n(\omega, \cdot)$. Now, if D is any subset of $\mathcal{M}'$ and $D^\varepsilon = \{\gamma\in\mathcal{M}' : \|\gamma-\alpha\|<\varepsilon$ for some $\alpha\in D\}$, then the above considerations yield the inequality

$$\begin{aligned}Q_{n,x}(D^\varepsilon) &\geqq Q_{n,x}^\xi(D) - P\left\{\frac{\tau_1+\tau_2+\cdots+\tau_{n-1}-n+1}{n}\geqq\tfrac{1}{2}\varepsilon\right\}\\ &\geqq Q_{n,x}^\xi(D) - P\left\{\frac{\tau_1+\cdots+\tau_n-n}{n}\geqq\tfrac{1}{2}\varepsilon\right\}.\end{aligned} \tag{3.9}$$

We first estimate the second term on the right which is independent of the starting point x of the chain.

LEMMA 3.5. *For any* $\varepsilon > 0$,

$$\lim_{\xi\to 0}\overline{\lim_{n\to\infty}}\frac{1}{n}\log P\left\{\frac{\tau_1+\tau_2+\cdots+\tau_n}{n}\geqq 1+\varepsilon\right\}=-\infty .$$

Proof: Let $\sigma > 0$ be arbitrary and choose $0<\xi<1$ such that $\xi e^{\sigma}<1$. Clearly,

$$P\{\tau_1+\tau_2+\cdots+\tau_n\geqq n(1+\varepsilon)\}\leqq \exp\{-n(1+\varepsilon)\sigma\}$$
$$\times E\{\exp\{\sigma(\tau_1+\tau_2+\cdots+\tau_n)\}\},$$

and since

$$E\{e^{\sigma\tau}\}=(1-\xi)\sum_{i=1}^{\infty}\xi^{i-1}e^{i\sigma}=\frac{(1-\xi)e^{\sigma}}{1-\xi e^{\sigma}},$$

we have

$$E\{\exp\{\sigma(\tau_1+\tau_2+\cdots+\tau_n)\}\}=\frac{(1-\xi)^n e^{n\sigma}}{(1-\xi e^{\sigma})^n}.$$

Therefore,

$$\overline{\lim_{n\to\infty}}\frac{1}{n}\log P\left\{\frac{\tau_1+\tau_2+\cdots+\tau_n}{n}\geqq 1+\varepsilon\right\}\leqq -\varepsilon\sigma+\log(1-\xi)-\log(1-\xi e^{\sigma}),$$

and hence

$$\lim_{\xi\to 0}\overline{\lim_{n\to\infty}}\frac{1}{n}\log P\left\{\frac{\tau_1+\tau_2+\cdots+\tau_n}{n}\geqq 1+\varepsilon\right\}\leqq -\varepsilon\sigma .$$

But $\varepsilon>0$ and $\sigma>0$ is arbitrary so that the lemma follows.

We now are ready for the

Proof of Theorem 3.3: We need first a slightly modified version of inequality (3.9). Let $\mathcal{M}'(X)$ be the set of measures in $\mathcal{M}'$ concentrated on X. If $D\subset\mathcal{M}'(X)$, then we actually have

$$Q_{n,x}(D^{\varepsilon}\cap\mathcal{M}'(X))\geqq Q^{\xi}_{n,x}(D)-P\left\{\frac{\tau_1+\tau_2+\cdots+\tau_n-n}{n}\geqq\tfrac{1}{2}\varepsilon\right\}. \tag{3.10}$$

This follows from the fact that ∞ is an absorbing point of the π chain, and if the chain remains in X at times $s_0, s_1, s_2, \cdots, s_{n-1}$, then, s_{n-1} being at least $n-1$, the chain remains in X for times up to $n-1$. If in addition $(\tau_1+\tau_2+\cdots+\tau_n-1)/n<\frac{1}{2}\varepsilon$, then of course $L_n(\omega,\cdot)$ is in D^{ε} as well as in $\mathcal{M}'(X)$.

So let $\mu \in \mathcal{M}$ be such that $I(\mu) < \infty$ and let N be a neighborhood of μ in $\mathcal{M}$. We seek a lower estimate for $\rho_{n,x}(N)$. Let $\varepsilon > 0$ be chosen so that we can find a smaller neighborhood N_1 of μ such that, if $\gamma \in N_1$ and $\|\alpha - \gamma\| < \varepsilon$, then $\alpha \in N$. We can think of N_1 as coming from a neighborhood U of μ in $\mathcal{M}'$, i.e., $N_1 = U \cap \mathcal{M}'(X)$. Then

$$\rho_{n,x}(N) = Q_{n,x}(N \cap \mathcal{M}'(X)), \tag{3.11}$$

and

$$N \supset N_1^\varepsilon, \tag{3.12}$$

where $N_1^\varepsilon = \{\gamma : \|\gamma - \alpha\| < \varepsilon \text{ for some } \alpha \in N_1\}$. Taking $D = U \cap \mathcal{M}'(X)$ in (3.10) we obtain

$$\rho_{n,x}(N) \geqq Q_{n,x}^\xi(U \cap \mathcal{M}'(X)) - P\left\{\frac{\tau_1 + \tau_2 + \cdots + \tau_n - n}{n} \geqq \tfrac{1}{2}\varepsilon\right\}. \tag{3.13}$$

We now pick $0 < \xi < 1$ small enough so that

$$I^{(\xi)}(\mu) \leqq I(\mu) + \varepsilon \tag{3.14}$$

and

$$\overline{\lim_{n\to\infty}} \frac{1}{n} \log P\left\{\frac{\tau_1 + \cdots + \tau_n - n}{n} \geqq \tfrac{1}{2}\varepsilon\right\} \leqq -2(I(\mu) + \varepsilon), \tag{3.15}$$

which we can do because of Lemmas 3.4 and 3.5. Since μ is supported in X, we see that, for each $\xi > 0$, $\pi_\xi(x, y)$ satisfies assumptions (3.1) and (3.2) of Theorem 3.1 (see (3.6) and (3.7)). Therefore, Theorem 3.1 can be applied and, if in that theorem we choose X for the set A, then we obtain from Theorem 3.1 and (3.14)

$$\underline{\lim_{n\to\infty}} \frac{1}{n} \log Q_{n,x}^\xi(U \cap \mathcal{M}'(X)) \geqq -[I(\mu) + \varepsilon]. \tag{3.16}$$

In (3.13) we now use (3.15) and (3.16) to see that, since ε is arbitrary,

$$\underline{\lim_{n\to\infty}} \frac{1}{n} \log \rho_{n,x}(N) \geqq -I(\mu),$$

which is Theorem 3.3.

If we assume that $p(x, E)$ is continuous in x for each E, then as noted earlier $\pi_\xi(x, E)$ is also continuous in x for each E. This gives the required uniformity in (3.16) which carries over to give the uniformity asserted in Theorem 3.3.

COROLLARY 3.4. *Let G be an open set in $\mathcal{M}$. If the substochastic transition function $p(x, dy)$ satisfies Hypothesis* H, *then*

$$\varliminf_{n\to\infty} \frac{1}{n} \log \rho_{n,x}(G) \geqq -\inf_{\mu\in G} I(\mu)\,. \tag{3.17}$$

Moreover, if $p(x, E)$ is continuous in x for each E, then (3.17) *holds uniformly over compact subsets of X.*

Proof: G can be thought of as a neighborhood of each point μ in G and Theorem 3.3 can be applied.

4. Upper Bounds—Discrete Case

The state space X will in general not be compact, and in order to obtain asymptotic upper bounds on the $Q_{n,x}$ measure appropriate to the asymptotic evaluation of expectations with respect to the Markov process with transition probability $\pi(x, dy)$, we would have to impose further hypotheses on $\pi(x, dy)$. However, we note that if K is a compact subset of $\mathcal{M}$, then we can prove without any further hypotheses, in exactly the same way as we did in [2], that

$$\varlimsup_{n\to\infty} \frac{1}{n} \log \sup_x Q_{n,x}(K) \leqq -\inf_{\mu\in K} I(\mu)\,. \tag{4.1}$$

In order to obtain the asymptotic theorems described in the introduction we need a result like (4.1) but for closed sets—not just compact sets. To obtain such a result in general we impose a new Hypothesis H* stated below, and subsequently we observe that in the special applications described in the introduction this hypothesis is automatically satisfied.

We now make the following hypothesis:

HYPOTHESIS H*. There exists a function $u_0(x)$ on X such that $u_0(x) \geqq 1$ for all $x \in X$, $(\pi u_0)(x) = V(x)$ is bounded on every compact set, and, for every l, $\{x \in X : u_0(x)/V(x) \leqq l\}$ is a compact set.

Hypothesis H* implies that π has an invariant measure—it forces the discrete parameter Markov process with transition probability $\pi(x, dy)$ to be positive recurrent. Before proving the analogue of (4.1) for closed sets we need three lemmas. First of all we note that H* implies that $\inf_{x\in X} u_0(x)/V(x)$ is positive. Let $\inf_{x\in X} u_0(x)/V(x) \geqq \rho > 0$ for some $\rho < 1$.

LEMMA 4.1. *For any compact set* $W \subset X$ *let* $K_l = W \cup \{x \in X : u_0(x)/V(x) \leqq l\}$. *For any* $\varepsilon > 0$,

$$\overline{\lim_{n\to\infty}} \frac{1}{n} \log \sup_{x\in W} Q_{n,x}\{\mu \in \mathcal{M} : \mu(K_l) \leqq 1-\varepsilon\} \leqq -\log \rho - \varepsilon \log \frac{l}{\rho}.$$

Proof: Since the process $X_0, X_1, X_2, \cdots$ is Markovian, it is easy to see that (cf. (2.1) of [1])

$$E_x\left\{\frac{u_0(X_1)u_0(X_2)\cdots u_0(X_n)}{V(X_0)V(X_1)\cdots V(X_{n-1})}\right\} = 1,$$

and because $X_0 = x$, this is the same as

$$E_x\left\{\exp\left\{\sum_{i=1}^{n-1} \log (u_0/V)(X_i)\right\} u_0(X_n)\right\} = V(x).$$

But $u_0(x) \geqq 1$ for all $x \in X$, so

$$E_x\left\{\exp\left\{\sum_{i=1}^{n-1} \log (u_0/V)(X_i)\right\}\right\} \leqq V(x). \tag{4.2}$$

By the definition of $L_n(\omega, \cdot)$, for $X_0 = x \in W \subset K_l$,

$$\begin{aligned}\sum_{i=1}^{n-1} \log\left(\frac{u_0}{V}\right)(X_i) &= \sum_{i=1}^{n-1} \log\left(\frac{u_0}{V}\right)(X_i)[\chi_{\tilde{K}_l}(X_i) + (1-\chi_{\tilde{K}_l}(X_i))] \\ &\geqq (\log \rho) \sum_{i=1}^{n-1} (1-\chi_{\tilde{K}_l}(X_i)) + \log l \sum_{i=1}^{n-1} \chi_{\tilde{K}_l}(X_i) \\ &= (n-1)\log \rho + \log \frac{l}{\rho} \sum_{i=1}^{n-1} \chi_{\tilde{K}_l}(X_i).\end{aligned}$$

Since $X_0 \in K_l$ and $\rho < 1$, we have

$$\begin{aligned}\sum_{i=1}^{n-1} \log\left(\frac{u_0}{V}\right)(X_i) &\geqq n \log \rho + \log \frac{l}{\rho} \sum_{i=0}^{n-1} \chi_{\tilde{K}_l}(X_i) \\ &= n \log \rho + n \log \frac{l}{\rho} L_n(\omega, \tilde{K}_l).\end{aligned} \tag{4.3}$$

From (4.2), (4.3) and the definition of $Q_{n,x}$,

$$E^{Q_{n,x}}[\exp\{n \log (l/\rho)\mu(\tilde{K}_l)\}] \leqq V(x) \exp\{-n \log \rho\}. \tag{4.4}$$

From (4.4) we get for $x \in W$

$$Q_{n,x}\{\mu \in \mathcal{M} : \mu(\tilde{K}_l) > \varepsilon\} \leqq V(x) \exp\{-n \log \rho - n\varepsilon \log (l/\rho)\}, \tag{4.5}$$

and this inequality gives us, since $V(x)$ is bounded on W,

$$\overline{\lim_{n\to\infty}} \frac{1}{n} \log \sup_{x \in W} Q_{n,x}\{\mu \in \mathcal{M} : \mu(K_l) \leqq 1-\varepsilon\} \leqq -\log \rho - \varepsilon \log \left(\frac{l}{\rho}\right).$$

LEMMA 4.2. *For each $l<\infty$, the set of measures $A_l = \{\mu \in \mathcal{M} : I(\mu) \leqq l\}$ is a tight family, i.e., for any $\varepsilon > 0$, there exists a compact set $K \subset X$ such that $\mu(\tilde{K}) < \varepsilon$ for all $\mu \in A_l$.*

Proof: If $\mu \in A_l$, then by the definition of $I(\mu)$

$$\int_X \log\left(\frac{\pi u}{u}\right)(x)\mu(dx) \geqq -l \tag{4.6}$$

for all $u \in \mathcal{U}$. By hypothesis H* there exists a function u_0 such that $u_0(x) \geqq 1$ for all $x \in X$, and, for any σ, $K_\sigma = \{x \in X : u_0(x)/(\pi u_0)(x) \leqq \sigma\}$ is compact. Let $u_n(x) = \min[u_0(x), n]$ for $n = 1, 2, 3, \cdots$, and consider $u_n(x)/(\pi u_n)(x)$. Clearly,

$$\lim_{n\to\infty} \frac{u_n(x)}{(\pi u_n)(x)} = \frac{u_0(x)}{(\pi u_0)(x)} \quad \text{for all} \quad x \in X. \tag{4.7}$$

Moreover, from (4.6) and the definition of u_n we get, for any $\mu \in A_l$,

$$\int_X \log\left(\frac{u_n}{\pi u_n}\right)(x)\mu(dx) \leqq l \quad \text{for all} \quad n. \tag{4.8}$$

We easily observe that $u_n(x)/(\pi u_n)(x)$ is uniformly bounded below because for each $x \in X$ either $1 \leqq u_n(x) < n$ or $u_n(x) = n$. In the first case, $u_n(x)/(\pi u_n)(x) = u_0(x)/(\pi u_n)(x) \geqq u_0(x)/(\pi u_0)(x) \geqq \rho$, and in the second case, $u_n(x)/(\pi u_n)(x) \geqq n/n = 1$. Therefore, from (4.7), (4.8) and Fatou's lemma, we get

$$\int_X \log\left(\frac{u_0}{\pi u_0}\right)(x)\mu(dx) \leqq l, \tag{4.9}$$

for all $\mu \in A_l$. Writing this last inequality as

$$\int_{K_\sigma} \log\left(\frac{u_0}{\pi u_0}\right)(x)\mu(dx) + \int_{\tilde{K}_\sigma} \log\left(\frac{u_0}{\pi u_0}\right)(x)\mu(dx) \leqq l,$$

we obtain, for all $\mu \in A_l$ and all σ,

$$(\log \rho)\mu(K_\sigma) + (\log \sigma)\mu(\tilde{K}_l) \leqq l. \tag{4.10}$$

If we now choose $\sigma > \rho$ we have

$$\mu(\tilde{K}_\sigma) \leqq \frac{l - \log \rho}{\log \sigma - \log \rho},$$

for all $\mu \in A_l$. Indeed, for any $\varepsilon > 0$ we can choose σ so large that $\mu(\tilde{K}_\sigma) < \varepsilon$ for all $\mu \in A_l$ which means, since K_σ is compact, that A_l is a tight family.

LEMMA 4.3. *For any $A < \infty$ and for any compact set $W \subset X$, there exists a compact set (a tight family of measures) $\bar{K}_A \subset \mathcal{M}$ such that*

$$\overline{\lim_{n\to\infty}} \frac{1}{n} \log \sup_{x\in W} Q_{n,x}(\tilde{\bar{K}}_A) \leqq -A.$$

Proof: Let $\{\sigma_i\}$, $i = 1, 2, 3, \cdots$, be an increasing sequence with $\lim_{i\to\infty} \sigma_i = \infty$ and let $\{\varepsilon_i\}$, $i = 1, 2, 3, \cdots$, be a decreasing sequence with $\lim_{i\to\infty} \varepsilon_i = 0$. With $K_{\sigma_i} = \{x \in X : u_0(x)/(\pi u_0)(x) \leqq \sigma_i\} \cup W$ consider

$$\bar{K}_{\sigma,\varepsilon} = \{\mu \in \mathcal{M} : \mu(K_{\sigma_i}) \geqq 1 - \varepsilon_i \text{ for all } i\}.$$

Now $\bar{K}_{\sigma,\varepsilon}$ is obviously a tight family of measures in $\mathcal{M}$ for every $\sigma = (\sigma_1, \sigma_2, \cdots)$ and $\varepsilon = (\varepsilon_1, \varepsilon_2, \cdots)$ with the described properties.

From (4.5) we have for $x \in W$

$$\begin{aligned} Q_{n,x}(\tilde{\bar{K}}_{\sigma,\varepsilon}) &\leqq \sum_{i=1}^{\infty} Q_{n,x}\{\mu \in \mathcal{M} : \mu(K_{\sigma_i}) < 1 - \varepsilon_i\} \\ &\leqq \sum_{i=1}^{\infty} V(x) \exp\{-n \log \rho - n\varepsilon_i \log(\sigma_i/\rho)\} \\ &= V(x) \exp\{-n \log \rho\} \sum_{i=1}^{\infty} \exp\{-n\varepsilon_i \log(\sigma_i/\rho)\} \\ &= \frac{V(x)}{\rho^n} \sum_{i=1}^{\infty} \left(\frac{\rho}{\sigma_i}\right)^{n\varepsilon_i}. \end{aligned} \tag{4.11}$$

We choose $\varepsilon_i = 1/i$ and without loss of generality we can assume that $\rho < 1$. From (4.11) we then have

$$Q_{n,x}(\bar{\bar{K}}_{\sigma,\varepsilon}) \leqq \frac{V(x)}{\rho^n} \sum_{i=1}^{\infty} \frac{1}{(\sigma_i^{1/i})^n} \tag{4.12}$$

in which we now take $\sigma_i = (A/\rho)^{i^2}$ noting that again without loss of generality we can choose $A > 1$. Thus

$$Q_{n,x}(\bar{\bar{K}}_{\sigma,\varepsilon}) \leqq \frac{V(x)}{\rho^n} \sum_{i=1}^{\infty} \left(\frac{\rho}{A}\right)^{ni} = \frac{V(x)}{A^n - \rho^n}. \tag{4.13}$$

For the above choices of $\sigma = (\sigma_1, \sigma_2, \cdots)$ and $\varepsilon = (\varepsilon_1, \varepsilon_2, \cdots)$, we let $\bar{K}_{\sigma,\varepsilon} = \bar{K}_A$ and we see from (4.13) that, since V is bounded on W,

$$\overline{\lim_{n\to\infty}} \frac{1}{n} \log \sup_{x\in W} Q_{n,x}(\bar{\bar{K}}_A) \leqq -A.$$

THEOREM 4.4. *Let C be a closed set in $\mathcal{M}$, and assume Hypothesis* H*. *Then, for any compact set $W \subset X$,*

$$\overline{\lim_{n\to\infty}} \frac{1}{n} \log \sup_{x\in W} Q_{n,x}(C) \leqq -\inf_{\mu\in C} I(\mu).$$

Proof: Let $A < \infty$ and let $\bar{K}_A$ be selected as in Lemma 4.3. Since

$$\begin{aligned} Q_{n,x}(C) &= Q_{n,x}(C \cap \bar{K}_A) + Q_{n,x}(C \cap \bar{\bar{K}}_A) \\ &\leqq Q_{n,x}(C \cap \bar{K}_A) + Q_{n,x}(\bar{\bar{K}}_A), \end{aligned}$$

we have

$$\overline{\lim_{n\to\infty}} \frac{1}{n} \log Q_{n,x}(C) \leqq \max\left[\overline{\lim_{n\to\infty}} \frac{1}{n} \log Q_{n,x}(C \cap \bar{K}_A), \overline{\lim_{n\to\infty}} \frac{1}{n} \log Q_{n,x}(\bar{\bar{K}}_A)\right].$$

Since $C \cap \bar{K}_A$ is compact in $\mathcal{M}$, we obtain from (4.1)

$$\overline{\lim_{n\to\infty}} \frac{1}{n} \log \sup_{x\in W} Q_{n,x}(C \cap \bar{K}_A) \leqq -\inf_{\mu\in C\cap\bar{K}_A} I(\mu),$$

and from Lemma 4.3,

$$\overline{\lim_{n\to\infty}} \frac{1}{n} \log \sup_{x\in W} Q_{n,x}(\tilde{\tilde{K}}_A) \leqq -A,$$

so that

$$\overline{\lim_{n\to\infty}} \frac{1}{n} \log \sup_{x\in W} Q_{n,x}(C) \leqq \max\left[-\inf_{\mu\in C\cap\bar{K}_A} I(\mu), -A\right]$$
$$\leqq \max\left[-\inf_{\mu\in C} I(\mu), -A\right].$$

But this last inequality is true for every A, and therefore letting $A\to\infty$ we obtain the theorem.

Now we want to observe, and use in the sequel, the fact that in the special case of independent random variables with values in X, i.e., the transition probability $\pi(x, dy) = \beta(dy)$, the Hypothesis H*, imposed at the beginning of this section in order to obtain the upper bound, is automatically satisfied. To see this, we choose first of all a sequence of compact sets $K_j \subset X$ with $K_1 \subset K_2 \subset \cdots$ and $\beta(\tilde{K}_j) \leqq 1/2^j$. Then, for any $x \in X$ define $u_0(x) \equiv \inf\{j : x \in K_j\}$. Observe that $u_0(x) \geqq 1$ for all $x \in X$ and that

$$V(x) = (\pi u_0)(x) = \int_X u_0(x)\beta(dx) = m < \infty$$

so that $V(x)$ is trivially bounded on every compact set. Moreover, for every l, $\{x \in X : u_0(x)/V(x) \leqq l\} = \{x \in X : u_0(x) \leqq ml\}$ is a compact set and hence H* is satisfied.

We also note that, in the case of independent random variables with values in X, the Hypothesis H of Section 3 (on the lower bound) is also automatically satisfied. Thus for independent random variables with values in X both Theorems 3.4 and 4.4 hold without imposing Hypothesis H* for the upper bound or H for the lower bound.

In the case of independent random variables it is more natural to consider

$$L'_n(\omega, A) = \sum_{j=1}^{n} \chi_A(X_j). \tag{4.14}$$

In such a case, the distribution of L'_n is independent of the starting point $X_0 = x$ and we shall denote it by Q_n. Since $\|L_n - L'_n\| \leqq 2/n$ asymptotically, there is no difference between L_n and L'_n. Theorems 3.4 and 4.4 hold just as well with L'_n instead of L_n.

Moreover, in this case

$$\begin{aligned} I(\mu) &= -\inf_{u\in\mathcal{U}_1}\left[\int_X \log(\pi u)(x)\mu(dx) - \int_X \log u(x)\mu(dx)\right] \\ &= -\inf_{u\in\mathcal{U}_1}\left[\log\int_X u(x)\beta(dx) - \int_X \log u(x)\mu(dx)\right] \\ &= I_\beta(\mu). \end{aligned}$$

We then have the following corollary to Theorems 3.4 and 4.4.

THEOREM 4.5. *Let X be a complete separable metric space and let $X_1, X_2, \cdots$ be independent random variables with values in X and having common distribution $\beta(dy)$. For any closed set $C\subset\mathcal{M}$,*

$$\overline{\lim_{n\to\infty}}\frac{1}{n}\log Q_n(C) \leqq -\inf_{\mu\in C} I_\beta(\mu),$$

and, for any open set $G\subset\mathcal{M}$,

$$\underline{\lim_{n\to\infty}}\frac{1}{n}\log Q_n(G) \geqq -\inf_{\mu\in G} I_\beta(\mu).$$

5. X a Banach Space

For the purposes of the applications in the next section, we want to consider independent random variables with values in a Banach space.

Let X be a Banach space and λ a probability measure on X such that

$$\int_X e^{\sigma\|x\|}\lambda(dx) < \infty \quad \text{for all} \quad \sigma > 0. \tag{5.1}$$

As usual, let $\mathcal{M}$ be the space of all probability measures on X. Now, however, instead of considering general discrete parameter Markov processes with X as state space, we want to focus on the special case of independent random variables with values in X having common distribution $\lambda(dy)$. In other words, the transition probability function $\pi(x, dy)$ is $\lambda(dy)$. In this special case, the

I-function becomes, for any $\mu \in \mathcal{M}$,

$$I(\mu) = -\inf_{u\in\mathcal{U}}\left[\int_X \log(\pi u)(x)\mu(dx) - \int_X \log u(x)\mu(dx)\right]$$

(5.2)
$$= -\inf_{u\in\mathcal{U}}\left[\log\int_X u(x)\lambda(dx) - \int_X \log u(x)\mu(dx)\right]$$

$$= I_\lambda(\mu)\,.$$

In the sequel we shall use several times results about $I_\lambda(\mu)$ obtained in [1] (Lemma 2.1), namely, $I_\lambda(\mu) < \infty$ if and only if $\mu \ll \lambda$ with a density $f(x) = d\mu/d\lambda$ such that

$$\int_X f(x)\,|\log f(x)|\,\lambda(dx) < \infty\,.$$

Moreover, in this case,

$$I_\lambda(\mu) = \int_X f(x)\log f(x)\lambda(dx) = \int_X \log f(x)\mu(dx)\,. \tag{5.3}$$

We first prove

LEMMA 5.1. *Let $\{\lambda_n\}$ and $\{\mu_n\}$ be sequences in $\mathcal{M}$. If $\{\lambda_n\}$ is a tight sequence and $I_{\lambda_n}(\mu_n) \leqq l < \infty$ for all n, then $\{\mu_n\}$ is a tight sequence. If, moreover,*

$$\sup_n \int_X e^{\sigma\|x\|}\lambda_n(dx) < \infty \quad \textit{for all} \quad \sigma > 0\,, \tag{5.4}$$

then $\int_X \|x\|\,\mu_n(dx)$ is uniformly bounded in n, and

$$\lim_{A\to\infty}\sup_n \int_{\|x\|\geqq A} \|x\|\,\mu_n(dx) = 0\,. \tag{5.5}$$

Proof: The main content of this lemma is that if $\int_X e^{\sigma\|x\|}\lambda_n(dx)$ is uniformly bounded in n for every $\sigma>0$ and $I_{\lambda_n}(\mu_n)$ is uniformly bounded, then $\int_X \|x\|\,\mu_n(dx)$ is uniformly integrable.

First of all, if $\{\lambda_n\}$ is a tight sequence, by definition, for any $\varepsilon>0$ there exists a compact set $K\subset X$ such that $\lambda_n(\tilde{K})<\varepsilon$ for all n. By the remark just preceding this lemma and the hypothesis that $I_{\lambda_n}(\mu)\leqq l<\infty$, we see that $\mu_n\ll\lambda_n$ for every n. We show now that $\{\mu_n\}$ is a tight sequence. We have

$$l\geqq I_{\lambda_n}(\mu_n)=-\inf_{u\in\mathcal{U}}\left[\log\int_X u(x)\lambda_n(dx)-\int_X \log u(x)\mu_n(dx)\right],$$

and, therefore, for all $u\in\mathcal{U}$,

$$\log\int_X u(x)\lambda_n(dx)-\int_X \log u(x)\mu_n(dx)\geqq -l\,.$$

Assume $A\subset X$; if, in particular, we let $u(x)=1$ on A and $M>0$ on $\tilde{A}$, then this last inequality becomes

$$\mu_n(\tilde{A})\leqq\frac{1}{\log M}[l+\log\{1+\lambda_n(\tilde{A})(M-1)\}]\,.$$

By hypothesis, $\{\lambda_n\}$ is a tight sequence and therefore, for any $\varepsilon>0$, there exists a compact set K such that $\lambda_n(\tilde{K})<\varepsilon$ for all n. If we set $A=K$ and $M=1+1/\varepsilon$ in the last displayed inequality, we get

$$\mu_n(\tilde{K})\leqq\frac{1}{\log(1+1/\varepsilon)}[l+\log 2]$$

which shows that $\{\mu_n\}$ is also a tight sequence.

For $\sigma>0$, $x\geqq 0$, $y\geqq 0$,

$$\inf_x\,[e^{\sigma x}-xy]=\frac{y}{\sigma}\left(1-\log\frac{y}{\sigma}\right),$$

and therefore

$$xy\leqq e^{\sigma x}+\frac{y}{\sigma}\log\frac{y}{\sigma}\,. \tag{5.6}$$

Using (5.6) and letting $d\mu_n/d\lambda_n = f_n(x)$, we have

$$
\begin{aligned}
\int_{\|x\|\geqq A} \|x\|\, \mu_n(dx) &= \int_{\|x\|\geqq A} \|x\|\, f_n(x)\lambda_n(dx) \\
&\leqq \int_{\|x\|\geqq A} \left(e^{\sigma\|x\|} + \frac{f_n(x)}{\sigma}\log\frac{f_n(x)}{\sigma}\right)\lambda_n(dx) \\
&\leqq \int_{\|x\|\geqq A} e^{\sigma\|x\|}\lambda_n(dx) + \frac{1}{\sigma}\int_{\|x\|\geqq A} f_n(x)\,|\log f_n(x)|\,\lambda_n(dx) \\
&\quad - \frac{\log\sigma}{\sigma}\int_{\|x\|\geqq A} \lambda_n(dx)\,.
\end{aligned}
\tag{5.7}
$$

If we take $\sigma \geqq 1$ in (5.7) so that $(\log\sigma)/\sigma \geqq 0$, we obtain

$$
\begin{aligned}
\sup_n \int_{\|x\|\geqq A} \|x\|\, \mu_n(dx) \leqq \sup_n \int_{\|x\|\geqq A} e^{\sigma\|x\|}\lambda_n(dx) \\
+\frac{1}{\sigma}\sup_n \int_X f_n(x)\,|\log f_n(x)|\,\lambda_n(dx)\,.
\end{aligned}
\tag{5.8}
$$

The assumption (5.4) implies

$$
\lim_{A\to\infty} \sup_n \int_{\|x\|\geqq A} e^{\sigma\|x\|}\lambda_n(dx) = 0 \quad \text{for all} \quad \sigma > 0\,,
$$

and therefore to show (5.5) it suffices to prove

$$
\sup_n \int_X f_n(x)\,|\log f_n(x)|\,\lambda_n(dx) < \infty\,,
\tag{5.9}
$$

since we can take σ as large as we please. Before doing this we note that inequalities (5.7) and (5.8) also hold if the region of integration is all of X instead of $\|x\| \geqq A$, and therefore the uniform boundedness of $\int_X \|x\|\, \mu_n(dx)$ will also follow from (5.9).

But

$$
\begin{aligned}
&\int_X f_n(x)\,|\log f_n(x)|\,\lambda_n(dx) \\
&= \int_{f_n\geqq 1} f_n(x)\log f_n(x)\lambda_n(dx) - \int_{f_n\leqq 1} f_n(x)\log f_n(x)\lambda_n(dx) \\
&= \int_X f_n(x)\log f_n(x)\lambda_n(dx) - 2\int_{f_n\leqq 1} f_n(x)\log f_n(x)\lambda_n(dx)\,,
\end{aligned}
\tag{5.10}
$$

and by (5.3) the first term on the right is $I_{\lambda_n}(\mu_n) \leqq l$. Thus, since $|\log x| \leqq c/\sqrt{x}$ for $|x| \leqq 1$,

$$\begin{aligned}\int_X f_n(x)\,|\log f_n(x)|\,\lambda_n(dx) &\leqq l + 2\int_{f_n \leqq 1} f_n(x)\,|\log f_n(x)|\,\lambda_n(dx)\\ &\leqq l + 2c\int_{f_n \leqq 1} \sqrt{f_n(x)}\,\lambda_n(dx)\\ &\leqq l + 2c\sqrt{\int_X f_n(x)\lambda_n(dx)}\\ &= l + 2c,\end{aligned}$$

which gives (5.9) and the lemma.

For $a \in X$ we define

$$h(a) = \inf_{\mu : \int x\mu(dx) = a} I(\mu),$$

and let X^* be the dual of X.

THEOREM 5.2.

(i) *$h(a)$ is non-negative and lower semicontinuous on X;*

(ii) *$h(a) = 0$ if and only if $a = \int x\lambda(dx)$;*

(iii) *for any l, $\{a \in X : h(a) \leqq l\}$ is a compact subset of X;*

(iv) *$h(a) = \sup_{\theta \in X^*}[\langle \theta, a\rangle - \log M(\theta)]$, where*

$$M(\theta) = \int_X e^{\langle \theta, x\rangle}\lambda(dx).$$

Proof: (i) $I(\mu)$ itself is non-negative so that $h(a)$ is non-negative. Let $\{a_n\}$ be a sequence in X converging to a. We want to show that $h(a) \leqq \underline{\lim}_{n\to\infty} h(a_n)$. Let $l = \underline{\lim}_{n\to\infty} h(a_n)$. By definition there exists a sequence of measures $\{\mu_n\}$ such that $\int x\mu_n(dx) = a_n$ and $I(\mu_n) \to l$. Since $I(\mu_n) = I_\lambda(\mu_n)$ is bounded, we have from the preceding lemma that $\{\mu_n\}$ is a tight sequence. Thus without loss of generality we can assume μ_n converges weakly to a measure μ. Moreover, the uniform integrability proved in Lemma 5.1 implies that $\int x\mu_n(dx) \to \int x\mu(dx) = a$. But $I(\mu) \leqq l$ by the lower semicontinuity of $I(\mu)$ so that indeed $h(a) \leqq l$.

(ii) If $a=\int x\lambda(dx)$, then $h(a)=0$ since $I(\lambda)=I_\lambda(\lambda)=0$. Conversely, if $h(a)=0$ there exists a sequence $\{\mu_n\}$ in $\mathcal{M}$ such that $\int x\mu_n(dx)=a$ and $I(\mu_n)\to 0$. By the preceding lemma, there then exists a subsequence $\{\mu_{n_j}\}$ such that $\{\mu_{n_j}\}$ converges weakly to some measure $\mu\in\mathcal{M}$ and $I(\mu)=0$. By the uniform integrability part of the preceding lemma, this implies $\int x\mu(dx)=a$. But $I(\mu)=0$ if and only if $\mu=\lambda$ so that $a=\int x\lambda(dx)$.

(iii) Let $\{a_n\}$ be a sequence in X such that, for each n, $h(a_n)\leqq l$. Corresponding to the $\{a_n\}$ sequence there exists a sequence $\{\mu_n\}$ in $\mathcal{M}$ such that $\int x\mu_n(dx)=a_n$ and $I(\mu_n)\leqq l+1$. Since $I(\mu_n)=I_\lambda(\mu_n)$ is bounded we know from the preceding lemma that $\{\mu_n\}$ is tight, i.e., compact. Thus there exists a subsequence $\{\mu_{n_j}\}$ converging weakly to some measure $\mu\in\mathcal{M}$. Again because of the uniform integrability shown in the preceding lemma, we have $\int x\mu_{n_j}(dx)\to\int x\mu(dx)$ which implies $\{a_{n_j}\}$ is a convergent subsequence.

(iv) Let

$$\phi(a)=\sup_{\theta\in X^*}\left[\langle\theta, a\rangle-\log M(\theta)\right].$$

We prove first that $h(a)\geqq\phi(a)$. Let μ be such that $\int x\mu(dx)=a$ and $I(\mu)=l<\infty$. We must show that, for all $\theta\in X^*$, $I(\mu)\geqq\langle\theta, a\rangle-\log M(\theta)$. By definition,

$$I(\mu)=-\inf_{u\in\mathcal{U}}\left[\int_X\log\frac{\int_X u(x)\lambda(dx)}{u(x)}\mu(dx)\right]$$

$$=\sup_{u\in\mathcal{U}}\left[\int_X\log u(x)\mu(dx)-\log\int_X u(x)\lambda(dx)\right].$$

Thus, for all $u\in\mathcal{U}$,

$$I(\mu)\geqq\int_X\log u(x)\mu(dx)-\log\int_X u(x)\lambda(dx).$$

In particular, let $u(x)=e^{\langle\theta,x\rangle}$ ($e^{\langle\theta,x\rangle}$ is not in $\mathcal{U}$ but we can truncate and pass to the limit inside the integrals), and we get

$$I(\mu) \geqq \int_X \langle \theta, x \rangle \mu(dx) - \log M(\theta).$$

Since $\int x\mu(dx) = a$, this last inequality becomes $I(\mu) \geqq \langle \theta, a \rangle - \log M(\theta)$.

To show that $h(a) \leqq \phi(a)$ is more delicate, and we prove it first for $X = R_1$ and then use this result to show that $h(a) \leqq \phi(a)$ for X a Banach space. So consider first $X = R_1$ where now

$$\phi(a) = \sup_\theta [\theta a - \log M(\theta)] \quad \text{and} \quad M(\theta) = \int_X e^{\theta x} \lambda(dx).$$

Suppose, as a first case, that the supremum in $\phi(a)$ is actually attained. Then it is easy to see that $h(a) \leqq \phi(a)$. This is so because if θ_0 is the point where the supremum is attained, i.e., $\phi(a) = \theta_0 a - \log M(\theta_0)$, then $a = M'(\theta_0)/M(\theta_0)$ or

$$\frac{1}{M(\theta_0)} \int_X x e^{\theta_0 x} \lambda(dx) = a.$$

Hence, if we let μ be a probability measure on R_1 such that $d\mu/d\lambda = 1/M(\theta_0)e^{\theta_0 x}$, we have $\int_X x\mu(dx) = a$ and, what is important, from (5.3),

$$I(\mu) = \int_X \log\left(\frac{d\mu}{d\lambda}\right)(x)\mu(dx) = \int_X \log \frac{e^{\theta_0 x}}{M(\theta_0)} \mu(dx)$$
$$= \theta_0 a - \log M(\theta_0) = \phi(a)$$

so that clearly $h(a) \leqq \phi(a)$.

Now, as a second case, still with $X = R_1$, suppose the supremum in $\phi(a)$ is not attained. For any $0 < \varepsilon < 1$ let $\lambda_\varepsilon = (1-\varepsilon)\lambda + \varepsilon\nu$, where ν is Gaussian measure on R_1 with mean 0 and variance 1. Since

$$\log M_\varepsilon(\theta) = \log \int_X e^{\theta x} \lambda_\varepsilon(dx)$$
$$= \log [(1-\varepsilon)M(\theta) + \varepsilon e^{\theta^2/2}]$$
$$\geqq \log \varepsilon + \tfrac{1}{2}\theta^2,$$

we see that $\sup_\theta [a\theta - \log M_\varepsilon(\theta)]$ is actually attained and therefore, by the preceding case,

$$\inf_{\mu:\int x\mu(dx)=a} I_{\lambda_\varepsilon}(\mu) = \sup_\theta [a\theta - \log M_\varepsilon(\theta)]. \tag{5.11}$$

But

$$\begin{aligned}\sup_\theta [a\theta - \log M_\varepsilon(\theta)] &= \sup_\theta [a\theta - \log\{(1-\varepsilon)M(\theta) + \varepsilon e^{\theta^2/2}\}] \\ &\leqq \sup_\theta [a\theta - \log(1-\varepsilon) - \log M(\theta)] \\ &= \sup_\theta [a\theta - \log M(\theta)] - \log(1-\varepsilon),\end{aligned}$$

and so the argument that $h(a) \leqq \phi(a)$ in R_1 will be completed if we show that

$$\inf_{\mu:\int x\mu(dx)=a} I_\lambda(\mu) \leqq \overline{\lim_{\varepsilon\to 0}} \inf_{\mu:\int x\mu(dx)=a} I_{\lambda_\varepsilon}(\mu). \tag{5.12}$$

To do this, let μ_ε be such that

$$\inf_{\mu:\int x\mu(dx)=a} I_{\lambda_\varepsilon}(\mu) = l_\varepsilon = I_{\lambda_\varepsilon}(\mu_\varepsilon),$$

and let $l = \overline{\lim}_{\varepsilon\to 0} l_\varepsilon$. If $l = \infty$, there is nothing to prove; so let $l < \infty$. Thus, for all ε, $I_{\lambda_\varepsilon}(\mu_\varepsilon) \leqq l$ and, since $\{\lambda_\varepsilon\}$ is a tight family, we have by Lemma 5.1 that $\{\mu_\varepsilon\}$ is a tight family. Moreover, the uniform integrability part of that same lemma implies that there exists a subsequence μ_{ε_j} with $\varepsilon_j \to 0$ such that $\lim_{j\to\infty} \mu_{\varepsilon_j} = \bar{\mu}$ exists and

$$\int x\mu_{\varepsilon_j}(dx) \to \int x\bar{\mu}(dx).$$

Since, for each $\varepsilon > 0$, $\int x\mu_\varepsilon(dx) = a$, we have $\int x\bar{\mu}(dx) = a$. $I_\lambda(\mu)$ is jointly lower semicontinuous in λ and μ and therefore

$$I_\lambda(\bar{\mu}) \leqq \underline{\lim}_{\varepsilon\to 0} I_{\lambda_\varepsilon}(\mu_\varepsilon) \leqq \overline{\lim_{\varepsilon\to 0}} I_{\lambda_\varepsilon}(\mu_\varepsilon) = l,$$

which implies that

$$\inf_{\mu:\int x\mu(dx)=a} I_\lambda(\mu) \leqq l$$

and (5.12).

Having shown that $h(a) \leqq \phi(a)$ when $X = R_1$, we are prepared to show the same inequality for X a Banach space. To recapitulate, we have a probability measure λ on a Banach space X such that

$$\int_X e^{\sigma\|x\|}\lambda(dx)<\infty$$

for all $\sigma>0$. For any $a\in X$, we have

$$h(a)=\inf_{\mu:\int x\mu(dx)=a} I(\mu)$$

and

$$\phi(a)=\sup_{\theta\in X^*}[\langle a,\theta\rangle-\log M(\theta)],$$

where

$$M(\theta)=\int_X e^{\langle\theta,x\rangle}\lambda(dx).$$

Thus we assume $h(a)>l>\phi(a)$ and we show that this leads to a contradiction.

Let

$$k_l=\left\{y\in X:y=\int x\mu(dx)\text{ for some }\mu\text{ with }I_\lambda(\mu)\leqq l\right\}.$$

We note that k_l is a compact, convex set and $a\notin k_l$. Therefore by the separation theorem there exists $\theta_0\in X^*$ such that $\langle\theta,a\rangle=a_0$ and $\langle\theta_0,y\rangle\geqq a_1$, for all $y\in k_l$ with $a_1>a_0$. For this θ_0 let λ_0, a probability measure on R_1, be the distribution of the linear functional $\langle\theta_0,x\rangle$, i.e., if $A\subset R_1$, then $\lambda_0(A)=\lambda\{x\in X:\langle\theta_0,x\rangle\in A\}$. Moreover, let μ_0 be a probability measure on R_1 such that $I_{\lambda_0}(\mu_0)\leqq l$. Then we know that $\mu_0\ll\lambda_0$; so let $(d\mu_0/d\lambda_0)(\xi)=f_0(\xi)$. Define μ on X by $d\mu/d\lambda=f_0(\langle\theta_0,x\rangle)$. Thus by (5.3),

$$\begin{aligned}I_\lambda(\mu)&=\int_X[\log f_0(\langle\theta_0,x\rangle)]f_0(\langle\theta_0,x\rangle)\lambda(dx)\\&=\int_{R_1}[\log f_0(\xi)]f_0(\xi)\lambda_0(d\xi)=I_{\lambda_0}(\mu_0)\leqq l.\end{aligned}\tag{5.13}$$

Therefore for this particular μ we have $\int x\mu(dx)\in k_l$ and so

$$\left\langle\theta_0,\int x\mu(dx)\right\rangle\geqq a_1,$$

i.e.,

$$\int_X\langle\theta_0,x\rangle\mu(dx)\geqq a_1.$$

But this means that

$$\int_X \langle\theta_0, x\rangle f_0(\langle\theta_0, x\rangle)\lambda(dx) = \int_{R_1} \xi f_0(\xi)\lambda_0(d\xi) = \int_{R_1} \xi\mu_0(d\xi) \geqq a_1 .$$

To summarize, we have shown that if $I_{\lambda_0}(\mu_0) \leqq l$, then $\int_{R_1} \xi\mu_0(d\xi) \geqq a_1 > a_0$. We conclude that

$$\inf_{\mu_0 : \int_{R_1}^{\xi} \mu_0(d\xi) = a_0} I_{\lambda_0}(\mu_0) \geqq l . \tag{5.14}$$

But we have shown that, in one dimension, $h(a) = \phi(a)$ so that (5.14) implies

$$\sup_\sigma [a_0\sigma - \log M_0(\sigma)] \geqq l , \tag{5.15}$$

where

$$M_0(\sigma) = \int_{R_1} e^{\sigma\xi}\lambda_0(d\xi) .$$

Therefore, since $\langle\theta_0, a\rangle = a_0$ and

$$\begin{aligned} M(\sigma\theta_0) &= \int_X e^{\sigma\langle\theta_0, x\rangle}\lambda(dx) \\ &= \int_{R_1} e^{\sigma\xi}\lambda_0(d\xi) = M_0(\sigma) , \end{aligned}$$

we get from (5.15)

$$\sup_\sigma [\sigma\langle\theta_0, a\rangle - \log M(\sigma\theta_0)] \geqq l ,$$

or

$$\sup_\sigma [\langle\sigma\theta_0, a\rangle - \log M(\sigma\theta_0)] \geqq l , \tag{5.16}$$

which implies

$$\sup_{\theta \in X^*} [\langle\theta, a\rangle - \log M(\theta)] \geqq l . \tag{5.17}$$

This last inequality states that $\phi(a) \geqq l$ which contradicts our assumption that $h(a) > l > \phi(a)$; we have completed part (iv) of the theorem.

Let X be a Banach space and λ a probability measure on X such that $\int_X e^{\sigma\|x\|}\lambda(dx)<\infty$ for all $\sigma>0$. Let $\tilde{X}$ be a random variable with values in X having distribution λ, i.e., if $A\subset X$, $P\{\omega:\tilde{X}(\omega)\in A\}=\lambda(A)$. Let $\tilde{X}_1,\tilde{X}_2,\cdots,\tilde{X}_n$ be independent copies of $\tilde{X}$ and let ν_n be the distribution of $(\tilde{X}_1+\tilde{X}_2+\cdots+\tilde{X}_n)/n$. For any $a\in X$ let, as before,

$$\phi(a)=\sup_{\theta\in X^*}[\langle\theta,a\rangle-\log M(\theta)].$$

We now prove

THEOREM 5.3. *For any closed set $C\subset X$,*

$$\overline{\lim_{n\to\infty}}\frac{1}{n}\log\nu_n(C)\leqq-\inf_{a\in C}\phi(a), \tag{5.18}$$

and, for any open set $G\subset X$,

$$\underline{\lim_{n\to\infty}}\frac{1}{n}\log\nu_n(G)\geqq-\inf_{a\in G}\phi(a). \tag{5.19}$$

Proof: We prove this theorem in two parts—first we assume there exists an l such that $P\{\omega:\|\tilde{X}(\omega)\|\leqq l\}=\lambda\{x\in X:\|x\|\leqq l\}=1$, i.e., that the random variables are bounded, and then in the second part we remove this hypothesis by truncation.

Assuming therefore that $\lambda\{\|x\|\leqq l\}=1$ we let $\mathcal{M}_l=\{\mu\in\mathcal{M}:\mu(\|x\|\leqq l\}=1\}$ and note that $L'_n\in\mathcal{M}_l$ for each n. We denote by Q_n, which is a measure on $\mathcal{M}_l$, the distribution of L'_n. Consider the map $\Phi:\mathcal{M}_l\to X$ given by $\Phi(\mu)=\int x\mu(dx)$. It is clear that Φ is continuous with $\mathcal{M}_l$ given the weak topology and X its usual topology, and moreover $Q_n\Phi^{-1}=\nu_n$. If we let C be a closed set in X, we have

$$\overline{\lim_{n\to\infty}}\frac{1}{n}\log\nu_n(C)=\overline{\lim_{n\to\infty}}\frac{1}{n}\log Q_n[\Phi^{-1}C]. \tag{5.20}$$

But by Theorem 4.5 and Theorem 5.2 we then get

$$\begin{aligned}\overline{\lim_{n\to\infty}}\frac{1}{n}\log Q_n[\Phi^{-1}C]&\leqq-\inf_{\mu:\int x\mu(dx)\in C}I_\lambda(\mu)\\&=-\inf_{a\in C}\inf_{\mu:\int x\mu(dx)=a}I_\lambda(\mu)\\&=-\inf_{a\in C}h(a)=-\inf_{a\in C}\phi(a).\end{aligned} \tag{5.21}$$

Thus (5.20) and (5.21) give us (5.18) under the hypothesis that the random variables are bounded.

Similarly, to prove (5.19) under the hypothesis of boundedness, we again use Theorem 4.5 and Theorem 5.2 to see that

$$
\begin{aligned}
\underline{\lim}_{n\to\infty} \frac{1}{n} \log \nu_n(G) &= \underline{\lim}_{n\to\infty} \frac{1}{n} \log Q_n[\Phi^{-1} G] \\
&\geqq - \inf_{\mu : \int x\mu(x) \in G} I_\lambda(\mu) \\
&= - \inf_{a\in G} \inf_{\mu : \int x\mu(dx) = a} I_\lambda(\mu) \\
&= - \inf_{a\in G} h(a) = - \inf_{a\in G} \phi(a) .
\end{aligned} \tag{5.22}
$$

Now we wish to remove the hypothesis that the random variables are bounded. For any $l>0$, define

$$
\tilde{Y}_j^l = \begin{cases} \tilde{X}_j & \text{if } \|\tilde{X}_j\| \leqq l, \\ 0 & \text{if } \|\tilde{X}_j\| > l, \end{cases}
$$

and

$$
\tilde{Z}_j^l = \begin{cases} \tilde{X}_j & \text{if } \|\tilde{X}_j\| > l, \\ 0 & \text{if } \|\tilde{X}_j\| \leqq l, \end{cases}
$$

so that $\tilde{X}_j = \tilde{Y}_j^l + \tilde{Z}_j^l$. Then, for C a closed set in X and any $\varepsilon>0$, $l>0$,

$$
\begin{aligned}
\nu_n(C) &= P\left\{\frac{\tilde{X}_1+\tilde{X}_2+\cdots+\tilde{X}_n}{n} \in C\right\} \\
&\leqq P\left\{\frac{\tilde{Y}_1^l+\tilde{Y}_2^l+\cdots+\tilde{Y}_n^l}{n} \in \bar{C}^\varepsilon\right\} + P\left\{\frac{\|\tilde{Z}_1^l+\cdots+\tilde{Z}_n^l\|}{n} > \varepsilon\right\},
\end{aligned} \tag{5.23}
$$

where C^ε is the set of all points in X the distance of which from C is less than ε and $\bar{C}^\varepsilon$ is its closure. From (5.23) and what we just proved for bounded random variables, we have

$$
\begin{aligned}
&\overline{\lim}_{n\to\infty} \frac{1}{n} \log \nu_n(C) \\
&\qquad \leqq \max\left[\overline{\lim}_{n\to\infty} \frac{1}{n} \log P\left\{\frac{\tilde{Y}_1^l+\tilde{Y}_2^l+\cdots+\tilde{Y}_n^l}{n} \in \bar{C}^\varepsilon\right\},\right. \\
&\qquad\qquad \left.\overline{\lim}_{n\to\infty} \frac{1}{n} \log P\left\{\frac{\|\tilde{Z}_1^l+\cdots+\tilde{Z}_n^l\|}{n} > \varepsilon\right\}\right] \\
&\qquad \leqq \max\left[-\inf_{a\in\bar{C}^\varepsilon} \phi^l(a),\ \overline{\lim}_{n\to\infty} \frac{1}{n} \log P\left\{\frac{\|\tilde{Z}_1^l+\cdots+\tilde{Z}_n^l\|}{n} > \varepsilon\right\}\right],
\end{aligned} \tag{5.24}
$$

where, if we let λ^l be the distribution of $\tilde{Y}^l$ and $M^l(\theta)=\int_X e^{\langle\theta,x\rangle}\lambda^l(dx)$,

$$\phi^l(a)=\sup_{\theta\in X^*}[\langle\theta, a\rangle-\log M^l(\theta)].$$

In Lemma 5.4 below we show that, for every $\varepsilon>0$,

$$\overline{\lim_{l\to\infty}}\ \overline{\lim_{n\to\infty}}\ \frac{1}{n}\log P\left\{\frac{\|\tilde{Z}_1^l+\tilde{Z}_2^l+\cdots+\tilde{Z}_n^l\|}{n}>\varepsilon\right\}=-\infty. \tag{5.25}$$

If, in (5.24), we first let $l\to\infty$ and then $\varepsilon\to 0$, we get

$$\overline{\lim_{n\to\infty}}\ \frac{1}{n}\log \nu_n(C)\leqq-\underline{\lim}_{\varepsilon\to 0}\ \underline{\lim}_{l\to\infty}\ \inf_{a\in\overline{C^\varepsilon}}\phi^l(a). \tag{5.26}$$

It follows from Lemma 5.5 below that, in fact,

$$\underline{\lim}_{\varepsilon\to 0}\ \underline{\lim}_{l\to\infty}\ \inf_{a\in\overline{C^\varepsilon}}\phi^l(a)\geqq\inf_{a\in C}\phi(a), \tag{5.27}$$

and hence (5.26) and (5.27) give us (5.18) in the general case.

For the lower bound (5.19) in the general case, we truncate exactly as for the upper bound and proceed as follows. Let G be an open set in X. Pick $a\in G$ such that $\phi(a)<\infty$ and $\delta>0$ such that $S(a,\delta)\subset G$. From Lemma 5.4 below we can pick l_0 so large that, for $l>l_0$,

$$\overline{\lim_{n\to\infty}}\ \frac{1}{n}\log P\left\{\frac{\|\tilde{Z}_1^l+\tilde{Z}_2^l+\cdots+\tilde{Z}_n^l\|}{n}\geqq\tfrac{1}{4}\delta\right\}\leqq-(\phi(a)+1). \tag{5.28}$$

In Lemma 5.6 below we shall show that if $\phi(a)<\infty$, there exist a_l such that $a_l\to a$ as $l\to\infty$ and $\phi^l(a_l)\leqq\phi(a)$ for all l. Therefore, we can assume that the l_0 we just picked for (5.28) is also so large that $\|a_l-a\|<\frac{1}{4}\delta$ for all $l>l_0$. Hence, for a fixed $l>l_0$,

$$\begin{aligned}\nu_n(G)&\geqq\nu_n(S(a,\delta))\geqq\nu_n(S(a_l,\tfrac{1}{2}\delta))\\ &=P\left\{\frac{\tilde{X}_1+\tilde{X}_2+\cdots+\tilde{X}_n}{n}\in S(a_l,\tfrac{1}{2}\delta)\right\}\\ &\geqq P\left\{\frac{\tilde{Y}_1^l+\tilde{Y}_2^l+\cdots+\tilde{Y}_n^l}{n}\in S(a_l,\tfrac{1}{4}\delta)\right\}-P\left\{\frac{\|\tilde{Z}_1^l+\cdots+\tilde{Z}_n^l\|}{n}\geqq\tfrac{1}{4}\delta\right\}.\end{aligned} \tag{5.29}$$

Let $0<\eta<\frac{1}{2}$. By (5.28) we can take n so large that

$$P\left\{\frac{\|\tilde{Z}_1^l+\tilde{Z}_2^l+\cdots+\tilde{Z}_n^l\|}{n}\geqq\tfrac{1}{4}\delta\right\}\leqq\exp\{-n[\phi(a)+1-\eta]\}, \tag{5.30}$$

and since the $\tilde{Y}^l$ variables are bounded, we have from the first part of the proof that, for n large enough,

$$P\left\{\frac{\tilde{Y}_1^l+\tilde{Y}_2^l+\cdots+\tilde{Y}_n^l}{n}\in S(a_l,\tfrac{1}{4}\delta)\right\}\geqq \exp\{-n[\phi^l(a_l)+\eta]\}\,. \tag{5.31}$$

But, for all l, $\phi^l(a_l)\leqq\phi(a)$ (Lemma 5.6) and therefore by (5.31), for n large enough,

$$P\left\{\frac{\tilde{Y}_1^l+\tilde{Y}_2^l+\cdots+\tilde{Y}_n^l}{n}\in S(a_l,\tfrac{1}{4}\delta)\right\}\geqq \exp\{-n[\phi(a)+\eta]\}\,. \tag{5.32}$$

Using (5.30) and (5.32) in (5.29) we have, for n large enough,

$$\nu_n(G)\geqq \exp\{-n[\phi(a)+\eta]\}-\exp\{-n[\phi(a)+1-\eta]\}\,,$$

and therefore, since $0<\eta<\frac{1}{2}$,

$$\varliminf_{n\to\infty}\frac{1}{n}\log \nu_n(G)\geqq -(\phi(a)+\eta)\,. \tag{5.33}$$

But η is arbitrary and a is any point in G so that (5.33) implies (5.19). This completes Theorem 5.3 except for three lemmas.

LEMMA 5.4. *For every* $\varepsilon>0$,

$$\varlimsup_{l\to\infty}\varlimsup_{n\to\infty}\frac{1}{n}\log P\left\{\frac{\|\tilde{Z}_1^l+\tilde{Z}_2^l+\cdots+\tilde{Z}_n^l\|}{n}>\varepsilon\right\}=-\infty\,.$$

Proof: The random variables $\tilde{Z}_1^l, \tilde{Z}_2^l, \cdots, \tilde{Z}_n^l$ are independent, identically distributed, and using obvious inequalities we have, for any $\varepsilon>0$ and any $\sigma>0$,

$$\begin{aligned} P\left\{\frac{\|\tilde{Z}_1^l+\tilde{Z}_2^l+\cdots+\tilde{Z}_n^l\|}{n}>\varepsilon\right\} &\leqq P\left\{\frac{\|\tilde{Z}_1^l\|+\|\tilde{Z}_2^l\|+\cdots+\|\tilde{Z}_n^l\|}{n}>\varepsilon\right\}\\ &\leqq e^{-\sigma\varepsilon}E\left\{\exp\left\{\frac{\sigma}{n}\sum_{j=1}^{n}\|\tilde{Z}_j^l\|\right\}\right\}\\ &= e^{-\sigma\varepsilon}\left[E\left\{\exp\left\{\frac{\sigma}{n}\|\tilde{Z}_1^l\|\right\}\right\}\right]^n\\ &= e^{-n\sigma\varepsilon}[E\{\exp\{\sigma\|\tilde{Z}_1^l\|\}\}]^n\,. \end{aligned} \tag{5.34}$$

Thus,

$$\overline{\lim_{n\to\infty}} \frac{1}{n} \log P\left\{\frac{\|\tilde{Z}_1^l + \tilde{Z}_2^l + \cdots + \tilde{Z}_n^l\|}{n} > \varepsilon\right\} \leqq -\sigma\varepsilon + \log E\{\exp\{\sigma \|\tilde{Z}_1^l\|\}\},$$

and from the definition of $\tilde{Z}_1^l$ we conclude that

$$\overline{\lim_{l\to\infty}}\, \overline{\lim_{n\to\infty}} \frac{1}{n} \log P\left\{\frac{\|\tilde{Z}_1^l + \cdots + \tilde{Z}_n^l\|}{n} > \varepsilon\right\} \leqq -\sigma\varepsilon .$$

Since the last inequality is true for all $\sigma > 0$ and since $\varepsilon > 0$, we have the lemma.

LEMMA 5.5. *Let C be a closed subset of X and C^ε the set of points in X all of which are a distance less than ε from C. Then,*

$$\underline{\lim}_{\substack{\varepsilon\to 0\\ l\to\infty}} \inf_{a\in \overline{C^\varepsilon}} \phi^l(a) \geqq \inf_{a\in C} \phi(a) . \tag{5.35}$$

Proof: Denote the left side of (5.35) by q. Then there exist sequences $\{l_n\} \to \infty$, $\{\varepsilon_n\} \to 0$, $\{a_n\} \in \overline{C^{\varepsilon_n}}$ such that $\phi^{l_n}(a_n) \to q$. By Theorem 5.2 and the fact that the infima are attained, we have a sequence $\{\mu^{l_n}\}$ such that $I_{\lambda^{l_n}}(\mu^{l_n}) \to q$ and $\int x\mu^{l_n}(dx) = a_n$. The family of measures $\{\lambda^{l_n}\}$ is tight and since $I_{\lambda^{l_n}}(\mu^{l_n})$ is uniformly bounded, we know by Lemma 5.1 that $\{\mu^{l_n}\}$ is tight. Thus there exists a subsequence $\{\mu^{l_{n_j}}\}$ which converges to some $\mu \in \mathcal{M}$ and $\int x\mu(dx)$ exists. Moreover, the uniform integrability part of Lemma 5.1 implies that $\int x\mu^{l_{n_j}}(dx) \to \int x\mu(dx)$. Thus $\{a_n\}$ is compact. Let a be any limit point of $\{a_n\}$. By Theorem 5.2, $\phi(a) \leqq I_\lambda(\mu)$ and since $I_\lambda(\mu)$ is lower semicontinuous, we have

$$\phi(a) \leqq I_\lambda(\mu) \leqq \underline{\lim}_{j\to\infty} I_{\lambda^{l_{n_j}}}(\mu^{l_{n_j}}) = q ,$$

which completes the lemma.

LEMMA 5.6. *Let $\phi(a) < \infty$. There exist a_l such that $a_l \to a$ as $l \to \infty$ and $\phi^l(a_l) \leqq \phi(a)$ for all l.*

Proof: Let $p=\phi(a)<\infty$. From Theorem 5.2,

$$p=\inf_{\mu:\int x\mu(dx)=a} I(\mu)$$

and this infimum is attained. Thus, there exists μ such that $I(\mu)=I_\lambda(\mu)=p$ and $\int x\mu(dx)=a$. Let μ^l be the distribution of $\tilde{y}^l$ with respect to μ measure and λ^l be the distribution of $\tilde{y}^l$ with respect to λ measure. Since $I_{\lambda^l}(\mu^l)\leqq I_\lambda(\mu)$, if we define $a_l=\int x\mu^l(dx)$, we have $a_l\to a$ as $l\to\infty$ and $\phi^l(a_l)\leqq I_{\lambda^l}(\mu^l)\leqq I_\lambda(\mu)=\phi(a)$ for all l.

6. Some Applications of Section 5

Let $C(0,1)$ be the space of continuous functions $z(\tau)$, $0\leqq\tau\leqq 1$, with $z(0)=0$. Let $F[z]$ be a real-valued function defined on $C(0,1)$. For a large class of F, Schilder [6] showed[1]

$$\lim_{n\to\infty}\frac{1}{n}\log E\{\exp\{-nF[z/\sqrt{n}]\}\}=-\inf_{z\in C(0,1)}\left[F[z]+\frac{1}{2}\int_0^1[z'(\tau)]^2\,d\tau\right], \tag{6.1}$$

where $E\{\ \}$ denotes expectation (Wiener integral) with respect to Brownian motion measure on $C(0,1)$, i.e., Gaussian measure on $C(0,1)$ with mean function 0 and covariance function $\rho(s,\tau)=\min(s,\tau)$. Later Pincus [5] (in an application of the asymptotics to Hammerstein nonlinear equations) showed that if in (6.1) instead of Brownian motion one has a suitable Gaussian process $\{y_\tau, 0\leqq\tau\leqq 1\}$ with mean function 0, covariance function $\rho(s,\tau)$, having continuous trajectories, then

$$\lim_{n\to\infty}\frac{1}{n}\log E\{\exp\{-nF[y/\sqrt{n}]\}\}=-\inf_{y\in C(0,1)}[F[y]+\tfrac{1}{2}\langle y,\rho^{-1}y\rangle]. \tag{6.2}$$

Obviously,

$$\langle y,\rho^{-1}y\rangle=\int_0^1[y'(\tau)]^2\,d\tau$$

when $\rho(s,\tau)=\min(s,\tau)$.

---

[1] Actually he obtained the whole asymptotic expansion with which we are concerned here only with the leading term.

For any Gaussian process $\{y_\tau, 0 \leqq \tau \leqq 1\}$ with continuous trajectories, Marcus and Shepp [3] proved

$$\lim_{\alpha\to\infty} \frac{1}{\alpha^2} \log P\left\{\sup_{0\leqq s\leqq 1} |y(s)| \geqq \alpha\right\} = -\frac{1}{2a}, \tag{6.3}$$

where $a = \sup_{0\leqq s\leqq 1} \rho(s, s)$.

We shall show how the results (6.1)–(6.3) can be obtained from the considerations of the preceding section. Our starting point is a lemma due to Skorohod [8].

Lemma 6.1. *Let λ be a Gaussian measure on a separable Banach space X. Then*

$$\int_X e^{\sigma\|x\|}\lambda(dx) < \infty$$

for some $\sigma > 0$, and therefore, a fortiori for all $\sigma > 0$.

Proof: If ξ is a symmetric Gaussian random variable with values in a Banach space X and if $\xi_1, \xi_2, \cdots, \xi_n$ are independent copies of ξ, then the distribution of $(\xi_1 + \cdots + \xi_n)/\sqrt{n}$ is the same as that of ξ. Hence

$$\begin{aligned} E\{e^{\sigma\|\xi\|}\} &= E\{\exp\{(\sigma/\sqrt{n})\,\|\xi_1 + \cdots + \xi_n\|\}\} \\ &\leqq E\{\exp\{(\sigma/\sqrt{n})[\|\xi_1\| + \cdots + \|\xi_n\|]\}\} \\ &= (E\{\exp\{(\sigma/\sqrt{n})\,\|\xi\|\}\})^n . \end{aligned}$$

Therefore if $E\{e^{\sigma\|\xi\|}\} < \infty$ for some $\sigma > 0$, then the same is true for all $\sigma < \infty$.

For a symmetric Gaussian measure λ on a separable Banach space X we define λ_n as the distribution of $x/\sqrt{n}$. Estimating $\lambda_n(A)$ for various sets A is the same as estimating the tail of λ. We can also interpret λ_n as the distribution of $(\xi_1 + \cdots + \xi_n)/n$, where $\xi_1, \xi_2, \cdots, \xi_n$ are independent and have the common distribution λ. The results of Section 5 are now applicable. In particular we can deduce from Theorem 5.3 the following

Theorem 6.2. *Let λ be a symmetric Gaussian distribution on a separable Banach space X with dual X^*. Let $M(\theta)$ be the moment generating function defined for $\theta \in X^*$ by*

$$M(\theta) = \int \exp\{\langle\theta, x\rangle\}\lambda(dx) = \exp\{\tfrac{1}{2}Q(\theta)\},$$

where

$$Q(\theta)=\int\langle\theta, x\rangle^2\lambda(dx)\,.$$

Let $f(x)$ be defined by

$$f(x)=\sup_\theta\,[\langle\theta, x\rangle-\tfrac{1}{2}Q(\theta)]\,.$$

Then, for any closed set $C\subset X$,

$$\limsup_{n\to\infty}\frac{1}{n}\log\lambda\left[x:\frac{x}{\sqrt{n}}\in C\right]\leqq-\inf_{x\in C}f(x) \tag{6.4}$$

and, for any open set $G\subset X$,

$$\liminf_{n\to\infty}\frac{1}{n}\log\lambda\left[x:\frac{x}{\sqrt{n}}\in G\right]\geqq-\inf_{x\in G}f(x). \tag{6.5}$$

Combining this with Section 3 of [9] we have

COROLLARY 6.3. *Let $F(x)$ be a bounded continuous function on X. Then*

$$\lim_{n\to\infty}\frac{1}{n}\log\int\exp\left\{nF\left(\frac{x}{\sqrt{n}}\right)\right\}\lambda(dx)=\sup_x\,[F(x)-f(x)]\,.$$

This is clearly the result of Schilder [6] and Pincus [5]. In order to connect Theorem 6.2 with the result of Marcus and Shepp [3] we have to perform the following calculation.

LEMMA 6.4. *We have the relation*

$$\inf_{\|x\|\geqq 1}f(x)=\frac{1}{2a}\,,$$

where

$$a=\sup_{\|\theta\|\leqq 1}Q(\theta)\,.$$

Proof: Let us look at $\lambda[\|x\|\geqq\sqrt{n}]$. Clearly, by (6.4),

$$\limsup_{n\to\infty}\frac{1}{n}\log\lambda[\|x\|\geqq\sqrt{n}]\leqq-\inf_{\|x\|\geqq 1}f(x)\,. \tag{6.6}$$

On the other hand, for any θ with $\|\theta\| \leqq 1$,

$$\lambda[\|x\| \geqq \sqrt{n}] \geqq \lambda[|\langle x, \theta\rangle| \geqq \sqrt{n}],$$

and, by the tail behavior of the Gaussian random variables on the real line,

$$\limsup_{n\to\infty} \frac{1}{n} \log \lambda[|\langle \theta, x\rangle| \geqq \sqrt{n}] \geqq -\frac{1}{2Q(\theta)}. \tag{6.7}$$

Therefore, combining (6.6) and (6.7), we see that

$$\inf_{\|x\|\geqq 1} f(x) \leqq \frac{1}{2Q(\theta)},$$

for any θ with $\|\theta\| \leqq 1$. Since θ with $\|\theta\| \leqq 1$ is arbitrary,

$$\inf_{\|x\|\geqq 1} f(x) \leqq \inf_{\|\theta\|\leqq 1} \frac{1}{2Q(\theta)} = \frac{1}{2a},$$

where

$$a = \sup_{\|\theta\|\leqq 1} Q(\theta).$$

Now with a defined as above, $Q(\theta) \leqq a\,\|\theta\|^2$ for all θ. Therefore,

$$\begin{aligned} f(x) &= \sup_{\theta} [\langle \theta, x\rangle - \tfrac{1}{2}Q(\theta)] \\ &\geqq \sup_{\theta} [\langle \theta, x\rangle - \tfrac{1}{2}a\,\|\theta\|^2] \\ &= \sup_{k\geqq 0} \sup_{\|\theta\|=k} [\langle \theta, x\rangle - \tfrac{1}{2}ak^2] \\ &= \sup_{k\geqq 0} [k\,\|x\| - \tfrac{1}{2}ak^2] \\ &= \frac{1}{2a}\|x\|^2. \end{aligned}$$

Hence

$$\inf_{\|x\|\geqq 1} f(x) \geqq \frac{1}{2a}.$$

This completes the proof of the lemma.

7. Upper Bounds—Continuous Case

We return to the setting of Sections 3 and 4 except that now we want to consider a continuous parameter Markov process. Let X be a complete separable metric space, and let $p(t, x, dy)$ be the transition probability of a Markov process with X as state space. We shall assume $p(t, x, dy)$ is a Feller transition probability. Let T_t be the semigroup associated with $p(t, x, dy)$, i.e., for $f \in C(X)$, $(T_t f)(x) = \int_X f(y)p(t, x, dy)$. Let L be the infinitesimal generator of this semigroup and $\mathcal{D} = \mathcal{D}(L)$ its domain. Again let $\mathcal{M}$ be the space of probability measures on X.

We define the I-function in the continuous case for any $\mu \in \mathcal{M}$ by

$$I(\mu) = -\inf_{\substack{u>0 \\ u\in\mathcal{D}}} \int_X \left(\frac{Lu}{u}\right)(x)\mu(dx) .$$

Let Ω be the space of right continuous functions $x(t)$, $0 \leqq t < \infty$, which have values in X. We assume that for each $x \in X$, $p(t, x, dy)$ induces a measure P_x on Ω. For each $t > 0$, $\omega \in \Omega$, and Borel set $A \subset X$, let

$$L_t(\omega, A) = \frac{1}{t}\int_0^t \chi_A(x(s))\, ds .$$

In other words, $L_t(\omega, A)$ is the proportion of time up to t that a particular sample path $\omega = x(\cdot)$ spends in the set A. For each $t > 0$ and each ω, $L_t(\omega, \cdot) \in \mathcal{M}$. For each $x \in X$ and each $t > 0$, let $Q_{x,t}$ be the measure induced on $\mathcal{M}$ by L_t from P_x, i.e., $Q_{x,t} = P_x L_t^{-1}$ or, in other words, if $B \subset \mathcal{M}$,

$$Q_{x,t}(B) = P_x\{\omega \in \Omega : L_t(\omega, \cdot) \in B\} .$$

In this section we shall prove, under hypotheses stated below, that for any closed set $C \subset \mathcal{M}$ and any compact set $W \subset X$

$$\overline{\lim_{t\to\infty}} \frac{1}{t} \log \sup_{x\in W} Q_{x,t}(C) \leqq -\inf_{\mu\in C} I(\mu) , \tag{7.1}$$

i.e., the continuous analogue of Theorem 4.4. If we are content with having (7.1) for just compact sets K instead of closed sets C, then (7.1) can be proved without further hypotheses using exactly the method of [2]. The reason that we want (7.1) for closed sets is that then, in conjunction with the corresponding lower bound for open sets, one obtains a useful asymptotic theorem for expectations on the Markov process.

To prove (7.1), we now make the hypothesis on the semigroup T_t that there exists a function $V(x)$ such that $\{x \in X : V(x) \geqq l\}$ is a compact set for each $l > -\infty$ and that there exists a sequence $\{u_n\} \in \mathcal{D}$ such that

(i) $u_n \geqq 1$ for all n and all $x \in X$,
(ii) for each compact set $W \subset X$, $\sup_{x \in W} \sup_n u_n(x) < \infty$,
(iii) for each $x \in X$,

$$\lim_{n\to\infty} \left(\frac{Lu_n}{u_n}\right)(x) = V(x)\,,$$

(iv) for some $A < \infty$,

$$\sup_{n,x} \left(\frac{Lu_n}{u_n}\right)(x) \leqq A\,.$$

We prove

LEMMA 7.1. *Under the hypotheses above, $\{\mu \in \mathcal{M} : I(\mu) \leqq l\}$ is compact for each l, and for any $N < \infty$ there exists a compact set $K \subset \mathcal{M}$ such that, for any compact set $W \subset X$,*

$$\overline{\lim_{t\to\infty}}\, \frac{1}{t} \log \sup_{x \in W} Q_{x,t}(\tilde{K}) \leqq -N\,. \tag{7.2}$$

Proof: Let $\mu \in \mathcal{M}$ be such that $I(\mu) \leqq l$. Then $\int_X (Lu/u)(x)\mu(dx) \geqq -l$ for all $u \in \mathcal{D}$ with $u > 0$. In particular, using the sequence $\{u_n\}$ of the hypothesis, we have $\int_X (Lu_n/u_n)(x)\mu(dx) \geqq -l$ for all n, and, by Fatou's lemma, this implies that $\int_X V(x)\mu(dx) \geqq -l$. Since $\lim_{n\to\infty} (Lu_n/u_n)(x) = V(x)$ for each $x \in X$ and $(Lu_n/u_n)(x) \leqq A$ uniformly in n and x, we have $V(x) \leqq A$ for all x. Thus, for any $h < 0$,

$$\int_X V(x)\mu(dx) = \int_{V<h} V(x)\mu(dx) + \int_{V \geqq h} V(x)\mu(dx)$$
$$\leqq h\mu\{x : V(x) < h\} + A\,.$$

Hence we have $h\mu\{x : V(x) < h\} \geqq -(l + A)$ or $\mu\{x : V(x) < h\} \leqq (l + A)/|h|$. Let $K_h = \{x \in X : V(x) \geqq h\}$ which by hypothesis is compact. We have just seen that for any l

$$\{\mu \in \mathcal{M} : I(\mu) \leqq l\} \subset \{\mu \in \mathcal{M} : \mu(\tilde{K}_h) \leqq (l + A)/|h|\}\,. \tag{7.3}$$

For any $\varepsilon > 0$, we can choose $|h|$ so large that $\mu(\tilde{K}_h) \leqq \varepsilon$ which means that the family of measures on the right of (7.3) is tight and therefore a compact subset of $\mathcal{M}$. But, since $I(\mu)$ is lower semicontinuous, the family of measures on the left of (7.3) is closed and being a subset of a compact set is itself compact.

Let, for $x \in X$ and $t > 0$,

$$\psi_n(x, t) = E^{P_x}\left\{u_n(x(t)) \exp\left\{-\int_0^t (Lu_n/u_n)(x(s))\, ds\right\}\right\}. \tag{7.4}$$

From the Feynmann–Kac formula, $\psi_n(x, t)$ satisfies

$$\begin{gathered}\frac{\partial \psi_n}{\partial t} = L\psi_n - \left(\frac{Lu_n}{u_n}\right)(x)\psi_n, \\ \psi_n(x, 0) = u_n(x),\end{gathered} \tag{7.5}$$

but, since the solution of (7.5) is clearly $\psi_n(x, t) \equiv u_n(x)$, we have

$$u_n(x) = E^{P_x}\left\{u_n(x(t)) \exp\left\{-\int_0^t (Lu_n/u_n)(x(s))\, ds\right\}\right\}. \tag{7.6}$$

By hypothesis, $u_n \geqq 1$ so that, for all $x \in X$,

$$E^{P_x}\left\{\exp\left\{-\int_0^t (Lu_n/u_n)(x(s))\, ds\right\}\right\} \leqq u_n(x). \tag{7.7}$$

From the hypothesis and Fatou's lemma we see that

$$\varliminf_{n\to\infty} \left(-\int_0^t \left(\frac{Lu_n}{u_n}\right)(x(s))\, ds\right) \geqq -\int_0^t V(x(s))\, ds,$$

or

$$\varliminf_{n\to\infty} \exp\left\{-\int_0^t (Lu_n/u_n)(x(s))\, ds\right\} \geqq \exp\left\{-\int_0^t V(x(s))\, ds\right\},$$

which implies, using Fatou's lemma again, that

$$\varliminf_{n\to\infty} E^{P_x}\left\{\exp\left\{-\int_0^t (Lu_n/u_n)(x(s))\, ds\right\}\right\} \geqq E^{P_x}\left\{\exp\left\{-\int_0^t V(x(s))\, ds\right\}\right\}. \tag{7.8}$$

By the hypothesis, for each compact set $W \subset X$, $\{u_n(x)\}$ is uniformly bounded on W so that from (7.7) and (7.8) we get

$$E^{P_x}\left\{\exp\left\{-\int_0^t V(x(s))\,ds\right\}\right\} \leqq \varliminf_{n\to\infty} u_n(x) \leqq \sup_{x\in W}\sup_n u_n(x) \leqq c, \tag{7.9}$$

for each $x \in W$, where c is a constant independent of t. Recalling the definition of $Q_{x,t}$ measure and noting that

$$\int_0^t V(x(s))\,ds = t\int_X V(y)L_t(\omega, dy),$$

we see that (7.9) is the same as

$$E^{Q_{x,t}}\left\{\exp\left\{-t\int_X V(x)\mu(dx)\right\}\right\} \leqq c \quad \text{for} \quad x \in W. \tag{7.10}$$

For $\sigma > 0$ and any positive integer n, let $K_n^\sigma = \{x \in X : V(x) \geqq -n\sigma\}$. By hypothesis this is a compact set. For any $\varepsilon > 0$, let $M_{n,\sigma}^\varepsilon = \{\mu \in \mathcal{M} : \mu(\tilde{K}_n^\sigma) > \varepsilon\}$. We noted earlier that $V(x) \leqq A$ for all $x \in X$, and therefore, if $\mu \in M_{n,\sigma}^\varepsilon$, we have

$$\begin{aligned} \int_X V(x)\mu(dx) &= \int_{K_n^\sigma} V(x)\mu(dx) + \int_{\tilde{K}_n^\sigma} V(x)\mu(dx) \\ &\leqq A(1-\mu(\tilde{K}_n^\sigma)) - n\sigma\mu(\tilde{K}_n^\sigma) \leqq A - \varepsilon n\sigma. \end{aligned} \tag{7.11}$$

From (7.10) and (7.11) we see that, for $x \in W$,

$$Q_{x,t}(M_{n,\sigma}^\varepsilon) \leqq ce^{tA}e^{-t\varepsilon n\sigma},$$

and, in particular, $Q_{x,t}(M_{n^2,\sigma}^{1/n}) \leqq ce^{tA}e^{-n\sigma t}$, so that

$$Q_{x,t}\left(\bigcup_{n=1}^\infty M_{n^2,\sigma}^{1/n}\right) \leqq ce^{tA}\frac{e^{-\sigma t}}{1-e^{-\sigma t}}. \tag{7.12}$$

Let $J_\sigma = \widetilde{\bigcup_{n=1}^\infty M_{n^2,\sigma}^{1/n}}$; then (7.12) yields

$$\varlimsup_{t\to\infty}\frac{1}{t}\log\sup_{x\in W} Q_{x,t}(\tilde{J}_\sigma) \leqq A - \sigma. \tag{7.13}$$

But

$$J_\sigma = \bigcap_{n=1}^\infty \widetilde{M_{n^2,\sigma}^{1/n}} = \{\mu \in \mathcal{M} : \mu(\tilde{K}_{n^2}^\sigma) \leqq 1/n \text{ for all } n\},$$

and this family of measures is obviously tight and hence compact. Since we can choose σ as large as we please, the proof of the lemma is complete.

With this lemma we can now prove the upper bound in the continuous case completely, using a techsnique which is analogous to that for the discrete case.

THEOREM 7.2. *Let C be a closed set in $\mathcal{M}$, and assume the hypothesis stated above. Then, for any compact set $W \subset X$,*

$$\overline{\lim_{t\to\infty}} \frac{1}{t} \log \sup_{x\in W} Q_{x,t}(C) \leqq - \inf_{\mu\in C} I(\mu) .$$

Proof: Let $N<\infty$ and choose a compact set $K \subset \mathcal{M}$, as in Lemma 7.1, so that

$$\overline{\lim_{t\to\infty}} \frac{1}{t} \log \sup_{x\in W} Q_{x,t}(\tilde{K}) \leqq -N . \tag{7.14}$$

Since

$$\begin{aligned} Q_{x,t}(C) &= Q_{x,t}(C\cap K) + Q_{x,t}(C\cap \tilde{K}) \\ &\leqq Q_{x,t}(C\cap K) + Q_{x,t}(\tilde{K}) , \end{aligned}$$

we have

(7.15)

$$\overline{\lim_{t\to\infty}} \frac{1}{t} \log \sup_{x\in W} Q_{x,t}(C) \leqq \max \left[\overline{\lim_{t\to\infty}} \frac{1}{t} \log \sup_{x\in W} Q_{x,t}(C\cap K), \ \overline{\lim_{t\to\infty}} \frac{1}{t} \log \sup_{x\in W} Q_{x,t}(\tilde{K}) \right] .$$

But $C\cap K$ is compact and therefore by our earlier remark (just after (7.1)) we know that

$$\overline{\lim_{t\to\infty}} \frac{1}{t} \log \sup_{x\in W} Q_{x,t}(C\cap K) \leqq - \inf_{\mu\in C\cap K} I(\mu) . \tag{7.16}$$

Thus, from (7.14) and (7.15),

$$\begin{aligned} \overline{\lim_{t\to\infty}} \frac{1}{t} \log \sup_{x\in W} Q_{x,t}(C) &\leqq \max \left[- \inf_{\mu\in C\cap K} U(\mu), -N \right] \\ &\leqq \max \left[- \inf_{\mu\in C} I(\mu), -N \right] . \end{aligned}$$

The last inequality holding for all N, we obtain the theorem by letting $N\to\infty$.

8. Lower Bounds—Continuous Case

In order to obtain lower bounds in the continuous case, the state space X being a complete separable metric space, we consider several hypotheses on the transition probability $p(t, x, dy)$.

HYPOTHESIS H_1. We assume there exists a reference measure β on X such that, for all $x \in X$, $p(t, x, dy) = p(t, x, y)\beta(dy)$.

Let B_0 be the center of the semigroup T_t associated with $p(t, x, dy)$, i.e., $B_0 = \{f \in C(X) : \lim_{t\to 0} \|T_t f - f\| = 0\}$, where the norm is the usual norm for $C(X)$ and, of course,

$$(T_t f)(x) = \int_X f(y) p(t, x, dy) .$$

Further, let

$$B_{00} = \left\{ f \in C(X) : \lim_{t\to 0} \sup_x \int_X |f(y) - f(x)|\, p(t, x, dy) = 0 \right\},$$

so that $B_{00} \subset B_0 \subset C(X)$. We note that if both f and f^2 are in B_0, then $f \in B_{00}$. Indeed, if B_0 contains any algebra of functions, then the same algebra is contained in B_{00}. Let $\alpha \in \mathcal{M}$ be such that $I(\alpha) = l < \infty$ and let N_α be a neighborhood of α in $\mathcal{M}$ of the form

$$N_\alpha = \left\{ \mu \in \mathcal{M} : \left| \int_X f_j(x)[\mu(dx) - \alpha(dx)] \right| < \varepsilon, \quad 1 \leqq j \leqq k, \right. \tag{8.1}$$
$$\left. \text{where} \quad f_j \in B_{00}, \quad i \leqq j \leqq k \right\} .$$

We make the second hypothesis on $p(t, x, dy)$:

HYPOTHESIS H_2. B_{00} is such that every neighborhood of $\alpha \in \mathcal{M}$ contains a neighborhood of the form (8.1).

We assume (cf. (3) just after the statement of Theorem 4 in the introduction of [1]):

HYPOTHESIS H_3. For any $\mu \in \mathcal{M}$ and any bounded measurable function $f(x)$ there is a sequence $\{f_n\} \in B_0$ such that $\|f_n\| \leqq \|f\|$ and $f_n \to f$ almost everywhere with respect to μ.

Let $\mu \in \mathcal{M}$ be such that $I(\mu) < \infty$ and let C be a closed set in X such that $\mu(C) = 1$. Let $G \supset C$ be an open set and define for $A \subset G$

$$p_G(t, x, A) = P_x\{x(s) \in G, 0 \leqq s \leqq t \text{ and } x(t) \in A\}.$$

For $\sigma > 0$, let

$$R_G^\sigma(x, A) = \int_0^\infty e^{-\sigma t} p_G(t, x, A)\, dt.$$

From Hypothesis H_1 we see that $R_G^\sigma(x, dy) \ll \beta(dy)$. We now make

HYPOTHESIS H_4. If $x \in G$ and $E \subset G$ with $\beta(E) > 0$, then $R_G^\sigma(x, E) > 0$.

It is clear that H_4 means that, if the process starts from $x \in G$ and E is a subset of G having positive β measure, then with positive probability the process will spend positive time in E before exiting from G.

We shall also consider

HYPOTHESIS H_5. For each $E \subset G$ and $\sigma > 0$, $R_G^\sigma(x, E)$ is a continuous function of x for $x \in G$.

We now state

THEOREM 8.1. *Let $\mu \in \mathcal{M}$ be such that $I(\mu) < \infty$ and let N be a neighborhood of μ in $\mathcal{M}$. Let* $\operatorname{supp} \mu \subset G$ *for some open $G \subset X$. $\mathcal{M}(G)$ is the set of probability measures in $\mathcal{M}$ whose support is contained in G. Under Hypotheses* H_1–H_4,

$$\underline{\lim}_{t \to \infty} \frac{1}{t} \log Q_{x,t}(N \cap \mathcal{M}(G)) \geqq -I(\mu). \tag{8.2}$$

Moreover, if in addition H_5 *holds, then* (8.2) *holds uniformly for x in compact subsets of G.*

Proof: Because of Hypothesis H_2 we can assume N is of the form

$$N_\varepsilon = \left\{ \alpha \in \mathcal{M} : \left| \int_X f_j(x)[\mu(dx) - \alpha(dx)] \right| < \varepsilon,\ 1 \leqq j \leqq k, \right.$$
$$\left. \text{where } \quad f_j \in B_{00},\ 1 \leqq j \leqq k \right\}.$$

Consider, then,

$$Q_{x,t}(N_\varepsilon \cap \mathcal{M}(G)) = P_x\{L_t(\omega, \cdot) \in N_\varepsilon;\ x(s) \in G, 0 \leqq s \leqq t\}. \tag{8.3}$$

If we choose $\Delta>0$ so that $2\Delta \max_{1\leqq j\leqq k}|f_j|<\frac{1}{2}\varepsilon$, then

$$(8.4)\quad \begin{aligned} &P_x\{L_t(\omega,\cdot)\in N_\varepsilon\,;\,x(s)\in G, 0\leqq s\leqq t\} \\ &\qquad \geqq P_x\{L_{t(1+\Delta)}(\omega,\cdot)\in N_{\varepsilon/2}\,;\,x(s)\in G, 0\leqq s\leqq t\}\,. \end{aligned}$$

For each $\sigma>0$, let $\tau_1, \tau_2, \cdots$ be a sequence of independent random variables which are independent of the process and have as common distribution $P\{\tau\in dt\}=\sigma e^{-\sigma t}\,dt$. Let $s_0=0$ and $s_n=\tau_1+\tau_2+\cdots+\tau_n$. For any Borel set $A\subset X$, we have our usual

$$L_t(\omega, A)=\frac{1}{t}\int_0^t \chi_A(x(s))\,ds\,,$$

but now we consider also

$$L_{n,\sigma}(\omega, A)=\frac{1}{n}\sum_{j=0}^{n-1}\chi_A(x(s_j))\,.$$

It is important to note that if we let $x_0=x(s_0)$, $x_1=x(s_1), \cdots, x_j=x(s_j), \cdots$, then $x_0, x_1, x_2, \cdots$ form a discrete parameter Markov process with transition probability

$$\pi_\sigma(x, dy)=\sigma\int_0^\infty e^{-\sigma t}p(t, x, dy)\,dt\,.$$

We see that, for all n and $\sigma>0$,

$$(8.5)\quad \begin{aligned} &P_x\{L_{t(1+\Delta)}(\omega,\cdot)\in N_{\varepsilon/2}\,;\,x(s)\in G, 0\leqq s\leqq t\} \\ &\qquad \geqq P_x\{L_{n,\sigma}(\omega,\cdot)\in N_{\varepsilon/4}\,;\,x(s)\in G, 0\leqq s\leqq t\} \\ &\qquad\quad -P_x\Big\{\sup_{1\leqq j\leqq k}\Big|\int_X f_j(x)L_{n,\sigma}(\omega, dx)-\int_X f_j(x)L_{t(1+\Delta)}(\omega, dx)\Big|>\tfrac{1}{4}\varepsilon\Big\} \\ &\qquad \geqq P_x\{L_{n,\sigma}(\omega,\cdot)\in N_{\varepsilon/4}\,;\,x(s)\in G, 0\leqq s\leqq \tau_1+\tau_2+\cdots+\tau_n\} \\ &\qquad\quad -P_x\Big\{\sup_{1\leqq j\leqq k}\Big|\int_X f_j(x)L_{n,\sigma}(\omega, dx)-\int_X f_j(x)L_{t(1+\Delta)}(\omega, dx)\Big|>\tfrac{1}{4}\varepsilon\Big\} \\ &\qquad\quad -P\{\tau_1+\tau_2+\cdots+\tau_n\leqq t\}\,. \end{aligned}$$

Thus from (8.3), (8.4), and (8.5) we have, for any n and $\sigma>0$,

$$
\begin{aligned}
&Q_{x,t}(N_\varepsilon \cap \mathcal{M}(G)) \\
&\qquad \geqq P_x\{L_{n,\sigma}(\omega,\cdot)\in N_{\varepsilon/4}\,;\, x(s)\in G,\, 0\leqq s\leqq \tau_1+\tau_2+\cdots+\tau_n\} \\
(8.6)\qquad &\qquad -P_x\left\{\sup_{1\leqq j\leqq k}\left|\int_X f_j(x)L_{n,\sigma}(\omega,dx)-\int_X f_j(x)L_{t(1+\Delta)}(\omega,dx)\right|>\tfrac{1}{4}\varepsilon\right\} \\
&\qquad -P\{\tau_1+\tau_2+\cdots+\tau_n\leqq t\}\,.
\end{aligned}
$$

In particular, choose $n=\sigma t(1+\Delta)$. In Lemma 8.3 below we show that, for $f\in B_{00}$, $n=\sigma t$, and any $\varepsilon>0$,

$$\lim_{\sigma\to\infty}\overline{\lim_{t\to\infty}}\frac{1}{t}\log\sup_x P_x\left\{\left|\int_X f(y)L_t(\omega,dy)-\int_X f(y)L_{n,\sigma}(\omega,dy)\right|\geqq\varepsilon\right\}=-\infty\,.$$

This implies that, for $n=\sigma t(1+\Delta)$,

$$
\begin{aligned}
(8.7)\qquad &P_x\left\{\sup_{1\leqq j\leqq k}\left|\int_X f_j(x)L_{n,\sigma}(\omega,dx)-\int_X f_j(x)L_{t(1+\Delta)}(\omega,dx)\right|>\tfrac{1}{4}\varepsilon\right\} \\
&\qquad\qquad \leqq \exp\{-t[h_{\varepsilon/4}(\sigma)+o(1)]\}\,,
\end{aligned}
$$

where $h_{\varepsilon/4}(\sigma)\to\infty$ as $\sigma\to\infty$ and $o(1)$ refers to $t\to\infty$.

With

$$R_G^\sigma(x,dy)=\int_0^\infty e^{-\sigma t}p_G(t,x,dy)\,dt\,,$$

we let $q_\sigma(x,dy)=\sigma R_G^\sigma(x,dy)$ and note that $q_\sigma(x,dy)$ is a substochastic transition probability on X. Let I_G^σ be the I-function coming from $q_\sigma(x,dy)$. If we let $\bar{Q}_{n,x}$ be the Q-measure coming from $q_\sigma(x,dy)$, then the important observation is that

$$P_x\{L_{n,\sigma}(\omega,\cdot)\in N_{\varepsilon/4}\,;\,x(s)\in G,\,0\leqq s\leqq\tau_1+\tau_2+\cdots+\tau_n\}=\bar{Q}_{n,x}(N_{\varepsilon/4})\,.$$

Since, by H_1 and H_4, $q_\sigma(x,dy)$ satisfies the hypotheses imposed in Theorem 3.3, that theorem implies that

$$(8.8)\quad \underline{\lim}_{n\to\infty}\frac{1}{n}\log P_x\{L_{n,\sigma}(\omega,\cdot)\in N_{\varepsilon/4}\,;\,x(s)\in G,\,0\leqq s\leqq\tau_1+\cdots+\tau_n\}\geqq -I_G^\sigma(\mu)\,,$$

and indeed that (8.8) is uniform in x if H_5 obtains. Now (8.8) means that

$$P_x\{L_{n,\sigma}(\omega,\cdot)\in N_{\varepsilon/4}\,;\,x(s)\in G,\,0\leqq s\leqq \tau_1+\cdots+\tau_n\}\geqq \exp\{-n[I_G^\sigma(\mu)+o(1)]\},$$

where here the $o(1)$ refers to $n\to\infty$. In Lemma 8.2 below we prove that $\sigma I_G^\sigma(\mu)\leqq I(\mu)$. In Lemma 8.4 we prove that, for $\Delta>0$ and $n=\sigma t(1+\Delta)$,

$$\lim_{\sigma\to\infty}\overline{\lim_{t\to\infty}}\frac{1}{t}\log P\{\tau_1+\cdots+\tau_n\leqq t\}=-\infty$$

which means, for $n=\sigma t(1+\Delta)$, that

$$P\{\tau_1+\cdots+\tau_n\leqq t\}\leqq \exp\{-t[r(\sigma,\Delta)+o(1)]\},$$

where $r(\sigma,\Delta)\to\infty$ for each $\Delta>0$ as $\sigma\to\infty$ and $o(1)$ refers to $t\to\infty$.

Putting all these estimates together into (8.6), we obtain, for $n=\sigma t(1+\Delta)$,

$$Q_{x,t}(N_\varepsilon\cap\mathcal{M}(G))\geqq \exp\{-\sigma t(1+\Delta)[I_G^\sigma+o(1)]\}$$
$$-\exp\{-t[h_{\varepsilon/4}(\sigma)+o(1)]\}-\exp\{-t[r(\sigma,\Delta)+o(1)]\}.$$

For any $\eta>0$, pick Δ_0 so small (still satisfying the condition imposed on Δ so that (8.4) holds) that $\Delta_0 I(\mu)<\eta$. Now choose σ_0 so large that $h_{\varepsilon/4}(\sigma_0)>2(I(\mu)+\eta+1)$ and $r(\sigma_0,\Delta_0)>2(I(\mu)+\eta+1)$. Then since $\sigma_0 I_G^{\sigma_0}(\mu)\leqq I(\mu)$, for large t,

$$\begin{aligned}Q_{x,t}(N_\varepsilon\cap\mathcal{M}(G))&\geqq \exp\{-t(1+\Delta_0)[I(\mu)+o(1)]\}\\&\quad-\exp\{-t[2I(\mu)+2\eta+2+o(1)]\}\\&\quad-\exp\{-t[2I(\mu)+2\eta+2+o(1)]\}\\&\geqq \exp\{-t[(I(\mu)+\eta+o(1)]\}\\&\quad-2\exp\{-t[2I[\mu)+2\eta+2+o(1)]\},\end{aligned}\tag{8.9}$$

where now all $o(1)$ refer to $t\to\infty$. Thus,

$$\underline{\lim}_{t\to\infty}\frac{1}{t}\log Q_{x,t}(N_\varepsilon\cap\mathcal{M}(G))\geqq -I(\mu)-\eta,$$

and since $\eta>0$ is arbitrary we have the inequality (8.2). Also, since under H_5

the estimates for all the terms on the right of (8.9) are uniform for x in a compact subset of G, we have (8.2) uniformly for x in such sets.

LEMMA 8.2. *Let $\mu \in \mathcal{M}$ be such that $I(\mu)<\infty$ and let C be a closed set in X such that $\mu(C)=1$. Let $G \supset C$ be an open set and define*

$$p_G(t, x, A) = P_x\{x(s) \in G, 0 \leqq s \leqq t, x(t) \in A\}.$$

Let

$$R_G^\sigma(x, A) = \int_0^\infty e^{-\sigma t} p_G(t, x, A)\, dt$$

and $q_\sigma(x, dy) = \sigma R_G^\sigma(x, dy)$. The I-function, I_G^σ, corresponding to $q_\sigma(x, dy)$ satisfies

$$\sigma I_G^\sigma(\mu) \leqq I(\mu).$$

Proof: The transition probability $p_G(t, x, dy)$ is substochastic as is $q_\sigma(x, dy)$. Denote the I-function corresponding to $p_G(t, x, dy)$ by $I_{G,t}$. We show first that

$$I_{G,t}(\mu) \leqq tI(\mu). \tag{8.10}$$

Let $I(\mu)=l$. From Hypothesis H_3 and [1] (Lemma 3.1) we have, for every u and $h>0$,

$$\int \log\left(\frac{T_h u}{u}\right)(x)\mu(dx) \geqq -lh, \tag{8.11}$$

where $(T_h u)(x) = \int_X u(y)p(h, x, dy)$. Let W and v be bounded measurable functions on X with strict positive lower bound and $W=1$ on C. Then if we let

$$u(x) = WT_h WT_h W \cdots T_h Wv,$$

we have from (8.11)

$$\begin{aligned} &\int_X \log(T_h WT_h W \cdots T_h Wv)(x)\mu(dx) \\ &\qquad - \int_X \log(WT_h WT_h W \cdots T_h Wv)(x)\mu(dx) \geqq -lh. \end{aligned} \tag{8.12}$$

Set $S_h = T_h W$; noting that $W=1$ on C and $\mu(C)=1$, we rewrite (8.12) as

$$\int_X \log (S_h^k v)(x)\mu(dx) - \int_X \log (S_h^{k-1} v)(x)\mu(dx) \geqq -lh .$$

Summing over k from 1 to n, we get

$$\int_X \log (S_h^n v)(x)\mu(dx) - \int_X \log v(x)\mu(dx) \geqq -nlh . \tag{8.13}$$

But

$$(S_h^n v)(x) = E_x\{v(x(nh))W(x(nh))W(x(n-1)h)) \cdots W(x(h))\},$$

and if we let W decrease to the indicator function of C, we have from (8.13) and monotone convergence that

$$\int_X \log E_x\{v(x(nh))\chi_C(x(h))\chi_C(x(2h)) \cdots \chi_C(x(nh))\}\mu(dx)$$
$$- \int_X \log v(x)\mu(dx) \geqq -nlh .$$

If in this last inequality we let $n \to \infty$ and $h \to 0$ so that $nh = t$ is fixed we get, by Fatou's lemma,

$$\int_X \log E_x\{v(x(t)); x(s) \in C, 0 \leqq s \leqq t\}\mu(dx) - \int_X \log v(x)\mu(dx) \geqq -lt,$$

i.e.,

$$\int_X \log \left(\int_X v(y) p_C(t, x, dy) \right) \mu(dx) - \int_X \log v(x)\mu(dx) \geqq -lt .$$

Since $G \supset C$ and v is positive, we conclude that

$$\int_X \log \left(\int_X v(y) p_G(t, x, dy) \right) \mu(dx) - \int_X \log v(x)\mu(dx) \geqq -lt$$

which, holding for all v in the class, implies $I_{G,t}(\mu) \leqq tI(\mu)$.

Consider

$$\begin{aligned}\int_X \log\left(\frac{q_\sigma u}{u}\right)(x)\mu(dx) &= \int_X \log(q_\sigma u)(x)\mu(dx) - \int_X \log u(x)\mu(dx)\\ &= \int_X \log\left[\int_X u(y)\int_0^\infty \sigma e^{-\sigma t} p_G(t,x,dy)\,dt\right]\mu(dx)\\ &\quad - \int_X \log u(x)\mu(dx)\\ &= \int_X \log\left[\int_0^\infty (T_{t,G}u)(x)\sigma e^{-\sigma t}\,dt\right]\mu(dx)\\ &\quad - \int_X \log u(x)\mu(dx),\end{aligned} \tag{8.14}$$

where $(T_{t,G}u)(x) = \int_X u(y)p_G(t,x,dy)$. By Jensen's inequality,

$$\log \int_0^\infty (T_{t,G}u)(x)\sigma e^{-\sigma t}\,dt \geqq \int_0^\infty [\log(T_{t,G}u)(x)]\sigma e^{-\sigma t}\,dt,$$

for each $x \in X$, and therefore

$$\int_X \log(q_\sigma u)(x)\mu(dx) \geqq \int_X \int_0^\infty [\log(T_{t,G}u)(x)]\sigma e^{-\sigma t}\,dt\,\mu(dx). \tag{8.15}$$

Using (8.15) in (8.14) we get

$$\begin{aligned}\int_X \log\left(\frac{q_\sigma u}{u}\right)(x)\mu(dx) &\geqq \int_X \int_0^\infty \left[\log\left(\frac{T_{t,G}u}{u}\right)(x)\right]\sigma e^{-\sigma t}\,dt\,\mu(dx)\\ &= \int_0^\infty \sigma e^{-\sigma t}\int_X \log\left(\frac{T_{t,G}u}{u}\right)(x)\mu(dx)\,dt\\ &\geqq -\int_0^\infty \sigma e^{-\sigma t} I_{G,t}(\mu)\,dt.\end{aligned}$$

Applying (8.10) to this last inequality, we obtain

$$\int_X \log\left(\frac{q_\sigma u}{u}\right)(x)\mu(dx) \geqq -\int_0^\infty \sigma t e^{-\sigma t}\, dt I(\mu)$$
$$= -\frac{1}{\sigma} I(\mu)\,,$$

and since this holds for all eligible u, we have $\sigma I_G^\sigma(\mu) \leqq I(\mu)$.

LEMMA 8.3. *Let $f \in B_{00}$. For any $\varepsilon > 0$,*

$$\lim_{\sigma\to\infty} \overline{\lim_{t\to\infty}} \frac{1}{t} \log \sup_x P_x\left\{\left|\int_X f(y)L_t(\omega, dy) - \int_X f(y)L_{n,\sigma}(\omega, dy)\right| \geqq \varepsilon\right\} = -\infty\,,$$

where $n = \sigma t$.

Proof: Let $\{x_t, 0 \leqq t < \infty\}$ again denote the Markov process with transition probability $p(t, x, dy)$. Let f be a given element of B_{00}, and define a stochastic process $\{z_t, 0 \leqq t < \infty\}$ by $z(t) = f(x(t))$. Then

$$\int_X f(y)L_t(\omega, dy) = \frac{1}{t}\int_0^t f(x(s))\, ds = \frac{1}{t}\int_0^t z(s)\, ds\,,$$

(8.16)

$$\int_X f(y)L_{n,\sigma}(\omega, dy) = \frac{1}{n}\sum_{j=0}^{n-1} f(x(s_j)) = \frac{z(s_0) + z(s_1) + \cdots + z(s_{n-1})}{n}\,.$$

Since $t = n/\sigma$, (8.16) yields

$$\int_X f(y)L_t(\omega, dy) - \int_X f(y)L_{n,\sigma}(\omega, dy)$$

(8.17)

$$= \frac{\sigma}{n}\int_0^{n/\sigma} z(s)\, ds - \frac{1}{n}[z(s_0) + z(s_1) + \cdots + z(s_{n-1})]$$
$$= \frac{\sigma}{n}\int_0^{s_n} z(s)\, ds - \frac{1}{n}[z(s_0) + z(s_1) + \cdots + z(s_{n-1})] + E\,,$$

where $|E| \leqq (\sigma/n)\,|s_n - n/\sigma|\, A = (1/n)\,|\sigma s_n - n|\, A$ and $|f| \leqq A$. With $n = rl$, we can rewrite (8.17) as

$$\int_X f(y)L_t(\omega, dy) - \int_X f(y)L_{n,\sigma}(\omega, dy)$$

$$= \frac{1}{n}\sum_{j=1}^{r}\left(\sigma\int_{s_{l(j-1)}}^{s_{lj}} z(s)\,ds - [z(s_{l(j-1)}) + \cdots + z(s_{lj-1})]\right) + E$$

$$(8.18) \qquad = \frac{1}{n}\sum_{j=1}^{r}\left(\sigma\int_{s_{l(j-1)}}^{s_{lj}} [z(s) - z(s_{l(j-1)})]\,ds + \sigma(s_{lj} - s_{l(j-1)})z(s_{l(j-1)}) - lz(s_{l(j-1)})\right.$$

$$\left. + [z(s_{l(j-1)}) - z(s_{l(j-1)})] + \cdots + [z(s_{l(j-1)}) - z(s_{lj-1})]\right) + E,$$

so that

$$\left|\int_X f(y)L_t(\omega, dy) - \int_X f(y)L_{n,\sigma}(\omega, dy)\right|$$

$$(8.19) \qquad \leqq \frac{1}{n}\sum_{j=1}^{r}\left([\sigma\,|s_{lj} - s_{l(j-1)}| + l]\sup_{s_{l(j-1)}\leqq s\leqq s_{lj}} |z(s) - z(s_{l(j-1)})|\right.$$

$$\left. + A\,|\sigma(s_{lj} - s_{l(j-1)}) - l|\right) + \frac{A}{n}|\sigma s_n - n|.$$

But

$$|\sigma s_n - n| = \left|\sum_{j=1}^{r} [\sigma(s_{lj} - s_{l(j-1)}) - l]\right| \leqq \sum_{j=1}^{r} |\sigma(s_{lj} - s_{l(j-1)}) - l|,$$

or

$$\frac{A}{n}|\sigma s_n - n| \leqq \frac{1}{n}\sum_{j=1}^{r} A\,|\sigma(s_{lj} - s_{l(j-1)}) - l|.$$

Thus, from (8.19) we get

$$\left|\int_X f(y)L_t(\omega, dy) - \int_X f(y)L_{n,\sigma}(\omega, dy)\right|$$

$$(8.20) \qquad \leqq \frac{1}{n}\sum_{j=1}^{r}\left([\sigma\,|s_{lj} - s_{l(j-1)}| + l]\sup_{s_{l(j-1)}\leqq s\leqq s_{lj}} |z(s) - z(s_{l(j-1)})|\right.$$

$$\left. + 2A\,|\sigma(s_{lj} - s_{l(j-1)}) - l|\right).$$

Moreover,

$$\begin{aligned}[\sigma\,|s_{lj}-s_{l(j-1)}|+l] &\sup_{s_{l(j-1)}\leqq s\leqq s_{lj}} |z(s)-z(s_{l(j-1)})| \\ &\leqq 2l \sup_{s_{l(j-1)}\leqq s\leqq s_{lj}} |z(s)-z(s_{l(j-1)})| + 2A[\sigma\,|s_{lj}-s_{l(j-1)}|-l] \\ &\leqq 2l \sup_{s_{l(j-1)}\leqq s\leqq s_{lj}} |z(s)-z(s_{l(j-1)})| + 2A\,|\sigma(s_{lj}-s_{l(j-1)})-l|\,,\end{aligned}$$

and hence, from (8.20), using $n=lr$, we have

$$\begin{aligned}&\left|\int_X f(y)L_t(\omega,dy)-\int_X f(y)L_{n,\sigma}(\omega,dy)\right| \\ &\leqq \frac{2l}{n}\sum_{j=1}^{r} \sup_{s_{l(j-1)}\leqq s\leqq s_{lj}} |z(s)-z(s_{l(j-1)})| + \frac{4A\sigma}{n}\sum_{j=1}^{r}\left|(s_{lj}-s_{l(j-1)})-\frac{l}{\sigma}\right| \qquad (8.21)\\ &= \frac{2}{r}\sum_{j=1}^{r} \sup_{s_{l(j-1)}\leqq s\leqq s_{lj}} |z(s)-z(s_{l(j-1)})| + \frac{4A}{r}\sum_{j=1}^{r}\left|\left(\frac{\sigma}{l}\right)(s_{lj}-s_{l(j-1)})\,1\right|.\end{aligned}$$

From (8.21) we get, for each $x\in X$,

$$\begin{aligned}&P_x\left\{\left|\int_X f(y)L_t(\omega,dy)-\int_X f(y)L_{n,\sigma}(\omega,dy)\right|>\varepsilon\right\} \\ &= P_x\left\{\left|\frac{1}{t}\int_0^t z(s)\,ds-\frac{1}{n}[z(s_0)+z(s_1)+\cdots+z(s_{n-1})]\right|>\varepsilon\right\} \\ &\leqq P_x\left\{\frac{2}{r}\sum_{j=1}^{r} \sup_{s_{l(j-1)}\leqq s\leqq s_{lj}} |z(s)-z(s_{l(j-1)})|\geqq \tfrac{1}{2}\varepsilon\right\} \qquad (8.22)\\ &\quad + P\left\{\frac{4A}{r}\sum_{j=1}^{r}\left|\left(\frac{\sigma}{l}\right)(s_{lj}-s_{l(j-1)})-1\right|\geqq\tfrac{1}{2}\varepsilon\right\} \\ &= F_1+F_2\,.\end{aligned}$$

We now estimate F_1 and F_2. For any $\theta>0$, by the obvious estimate,

$$F_2 \leqq e^{-\theta\varepsilon/2} E\left\{\exp\left\{\frac{\theta 4A}{r}\sum_{j=1}^{r}\left|\frac{\sigma}{l}(s_{l_j}-s_{l(j-1)})-1\right|\right\}\right\}.$$

Replacing θ by $r\theta$ and then using the independence of $\tau_1, \tau_2, \cdots$, we obtain

$$F_2 \leqq e^{-r\theta\varepsilon/2}\left[E\left\{\exp\left\{\theta 4A\left|\frac{\sigma}{l}s_l-1\right|\right\}\right\}\right]^r. \tag{8.23}$$

As for F_1, let E_j, for $j=1,2,\cdots,r$, be the event

$$\left\{\sup_{s_{l(j-1)}\leqq s\leqq s_{lj}}|z(s)-z(s_{l(j-1)})|>\tfrac{1}{8}\varepsilon\right\},$$

and consider $(1/r)\sum_{j=1}^{r}\chi_{E_j}$. Clearly, the event $\left\{(1/r)\sum_{j=1}^{r}\chi_{E_j}>\varepsilon/16A\right\}$ contains the event

$$\left\{\frac{2}{r}\sum_{j=1}^{r}\sup_{s_{l(j-1)}\leqq s\leqq s_{l_j}}|z(s)-z(s_{l(j-1)})|\geqq\tfrac{1}{2}\varepsilon\right\}$$

so that

$$F_1 \leqq P_x\left\{\frac{1}{r}\sum_{j=1}^{r}\chi_{E_j}\geqq\frac{\varepsilon}{16A}\right\}. \tag{8.24}$$

For any $\nu>0$, by the usual estimate,

$$P_x\left\{\frac{1}{r}\sum_{j=1}^{r}\chi_{E_j}\geqq\frac{\varepsilon}{16A}\right\}\leqq e^{-\nu\varepsilon/16A}E_x\left\{\exp\left\{\frac{\nu}{r}\sum_{j=1}^{r}\chi_{E_j}\right\}\right\},$$

and therefore replacing ν by $r\nu$ we get

$$F_1 \leqq e^{-r\varepsilon\nu/16A}E_x\left\{\exp\left\{\nu\sum_{j=1}^{r}\chi_{E_j}\right\}\right\}. \tag{8.25}$$

In handling F_2 at the corresponding stage we were able to exploit the independence of the τ_j's so that the expectation of the exponential factored (cf. (8.23)). Now, however, the events $E_1, E_2, \cdots$ are not independent and so we must be more delicate. Let $\sigma(E_1, E_2, \cdots, E_{j-1})$ be the σ-field generated by $E_1, E_2, \cdots, E_{j-1}$. Let

$$p=\sup_{1\leqq j\leqq r}\sup_{\omega}P\{E_j\,|\sigma(E_1, E_2, \cdots, E_{j-1})\},$$

where $\sup_\omega$ means the supremum over all possible conditionings. Then (8.25) yields

$$F_1 \leqq e^{-r\varepsilon\nu/16A}(pe^\nu + q)^r, \tag{8.26}$$

where $q = 1 - p$. To get an estimate on p, we note that

$$p = \sup_{1 \leqq j \leqq r} \sup_\omega P\left\{ \sup_{s_{l(j-1)} \leqq s \leqq s_{lj}} |z(s) - z(s_{l(j-1)})| > \tfrac{1}{8}\varepsilon \mid \sigma(E_1, E_2, \cdots, E_{j-1}) \right\}$$

$$\leqq \sup_x P_x\left\{ \sup_{0 \leqq s \leqq s_l} |z(s) - z(0)| > \tfrac{1}{8}\varepsilon \right\} = g(l, \sigma, \varepsilon).$$

We shall show below that, for any $\varepsilon > 0$ and fixed l,

$$\lim_{\sigma \to \infty} g(l, \sigma, \varepsilon) = 0. \tag{8.27}$$

Now we fix l and σ and recall that $n = rl = \sigma t$, so that, by (8.26), $r \to \infty$, $t \to \infty$ but r/t is fixed at σ/l and

$$\overline{\lim_{t \to \infty}} \frac{1}{t} \log F_1 \leqq -\frac{\sigma}{l} \frac{\varepsilon\nu}{16A} + \frac{\sigma}{l} \log(pe^\nu + q). \tag{8.28}$$

Similarly from (8.23) we get

$$\overline{\lim_{t \to \infty}} \frac{1}{t} \log F_2 \leqq -\frac{\sigma}{l} \frac{\varepsilon\theta}{2} + \frac{\sigma}{l} \log E\left\{ \exp\left\{ \theta 4A \left| \frac{\sigma}{l} S_l - 1 \right| \right\} \right\}. \tag{8.29}$$

To complete the proof of the lemma, except for (8.27) which we show below, we must, in view of (8.22), (8.28) and (8.29), pick l as a function of σ and ε, θ as a function of σ and ε, and ν as a function of σ and ε so that the expressions on the right of (8.28) and (8.29) both go to $-\infty$ as $\sigma \to \infty$.

Let J_1 and J_2 be, respectively, the expressions on the right of inequalities (8.28) and (8.29). Recall that $s_l = \tau_1 + \tau_2 + \cdots + \tau_l$, where each τ_i has as its density $\sigma e^{-\sigma t}$. Let $\bar{s}_l = \mu_1 + \mu_2 + \cdots + \mu_l$, where the μ_i are independent random variables each having density e^{-t}. We see that

$$E\left\{ \exp\left\{ 4\theta A \left| \frac{\sigma}{l} s_l - 1 \right| \right\} \right\} = E\{\exp\{4\theta A\, |\bar{s}_l/l - 1|\}\},$$

and if we choose $\theta = \nu = 1$ we have

$$J_2 = -\frac{\sigma}{l}[\tfrac{1}{2}\varepsilon - \log E\{\exp\{4A\,|\bar{s}_l/l - 1|\}\}],$$

(8.30)

$$J_1 = -\frac{\sigma}{l}\left[\frac{\varepsilon}{16A} - \log(pe+q)\right].$$

For our given $\varepsilon > 0$, we can choose l_0 so large that

$$[\tfrac{1}{2}\varepsilon - \log E\{\exp\{4A\,|\bar{s}_{l_0}/l_0 - 1|\}\}] > 0 .$$

By (8.27) we have, for this l_0, $\lim_{\sigma\to\infty} g(l_0, \sigma, \varepsilon) = 0$. Since $p \leqq g(l_0, \sigma, \varepsilon)$, we see that $q \to 1$ as $\sigma \to \infty$, and therefore $[\varepsilon/16A - \log(pe+q)]$ becomes positive as $\sigma \to \infty$. Thus, with $l = l_0$, both J_1 and J_2 approach $-\infty$ as $\sigma \to \infty$. This completes the proof of the lemma except for (8.27).

To estimate $g(l, \sigma, \varepsilon)$ we first observe that, for fixed l and Δ,

$$g(l, \sigma, \varepsilon) = \sup_x P_x\left\{\sup_{0 \leqq s \leqq s_l} |z(s) - z(0)| > \tfrac{1}{8}\varepsilon\right\}$$

$$\leqq \sup_x P_x\left\{\sup_{0 \leqq s \leqq \Delta} |z(s) - z(0)| > \tfrac{1}{8}\varepsilon\right\} + P\{s_l > \Delta\}.$$

By Tchebycheff's inequality, $P\{s_l > \Delta\} \to 0$ as $\sigma \to \infty$ and hence, for fixed l and Δ,

(8.31)
$$\overline{\lim_{\sigma\to\infty}}\, g(l, \sigma, \varepsilon) \leqq \sup_x P_x\left\{\sup_{0 \leqq s \leqq \Delta} |z(s) - z(0)| > \tfrac{1}{8}\varepsilon\right\}.$$

If (Ω, Σ, P) is a probability space and Σ_t an increasing family of σ-fields with $z(t)$ a stochastic process with right continuous paths such that $z(t)$ is Σ_t-measurable for every t, then

$$P\left\{\sup_{0 \leqq t \leqq \Delta} |z(t) - z(0)| \geqq \varepsilon\right\} \leqq \frac{\rho_{\varepsilon/2}(\Delta)}{1 - \rho_{\varepsilon/2}(\Delta)},$$

where

$$\rho_\varepsilon(\delta) = \sup_t \sup_{0 \leqq h \leqq \delta} \operatorname*{ess\,sup}_\omega P\{|z(t+h) - z(t)| > \varepsilon \mid \Sigma_t\}.$$

If $z(t)$ were a process with independent increments this would be just Lévy's inequality. Since we have uniform conditional estimates, the proof of the above inequality is no different from that of Lévy's inequality. Thus, by (8.31), it will suffice to show that for our process $z(t)$ and with $f \in B_{00}$ we have $\rho_\varepsilon(\Delta) \to 0$ as $\Delta \to 0$.

First of all

$$E\{|f(x(t+h))-f(x(t))| \,|\, \Sigma_t\} = \int |f(y)-f(x)|\, p(h, x, dy)\big|_{x=x(t)},$$

and therefore,

$$P\{|z(t+h)-z(t)| > \varepsilon \,|\, \Sigma_t\} \leqq \frac{1}{\varepsilon} \sup_x \int |f(y)-f(x)|\, p(h, x, dy)\,. \tag{8.32}$$

The right side of (8.32) $\to 0$ as $h \to 0$ since $f \in B_{00}$. Thus $\rho_\varepsilon(\Delta) \to 0$ as $\Delta \to 0$ and we have (8.27) and the lemma.

LEMMA 8.4. *For $\Delta > 0$ and $n = \sigma t(1+\Delta)$,*

$$\lim_{\sigma \to \infty} \overline{\lim_{t \to \infty}} \frac{1}{t} \log P\{\tau_1 + \tau_2 + \cdots + \tau_n \leqq t\} = -\infty\,.$$

Proof: Since the τ_j are independent and $P\{\tau \in dt\} = \sigma e^{-\sigma t}\, dt$ we have

$$E\{\exp\{\theta(\tau_1+\tau_2+\cdots+\tau_n)\}\} = \left(1 - \frac{\theta}{\sigma}\right)^{-n},$$

and therefore, for $\theta > 0$,

$$P\{\tau_1+\tau_2+\cdots+\tau_n \leqq t\} \leqq e^{\theta t}\left(1+\frac{\theta}{\sigma}\right)^{-n} = e^{\theta t}\left(1+\frac{\theta}{\sigma}\right)^{-\sigma t(1 \geqq \Delta)}.$$

Consequently,

$$\overline{\lim_{t \to \infty}} \frac{1}{t} \log P\{\tau_1+\tau_2+\cdots+\tau_n \leqq t\} \leqq \theta - \sigma(1+\Delta)\log\left(1+\frac{\theta}{\sigma}\right),$$

and since this is true for all $\theta > 0$,

$$\overline{\lim_{t \to \infty}} \frac{1}{t} \log P\{\tau_1+\tau_2+\cdots+\tau_n \leqq t\} \leqq \inf_{\theta > 0} \left[\theta - \sigma(1+\Delta)\log\left(1+\frac{\theta}{\sigma}\right)\right].$$

But

$$\lim_{\sigma\to\infty}\inf_{\theta>0}\left[\theta-(1+\Delta)\sigma\log\left(1+\frac{\theta}{\sigma}\right)\right]\leqq\inf_{\theta>0}\lim_{\sigma\to\infty}\left[\theta-(1+\Delta)\sigma\log\left(1+\frac{\theta}{\sigma}\right)\right]$$
$$=\inf_{\theta>0}[-\Delta\theta]=-\infty,$$

and hence the lemma.

9. Examples and Counterexamples

In this section we exhibit some examples and counterexamples to illustrate the scope of our hypotheses. First of all without some sort of hypothesis like H* we cannot have an upper bound for all closed sets. To see this consider the Markov chain on the integers $\{1,2,\cdots\}$ with transition probability $\pi_{ij}=\delta_{i+1,j}$. The chain is purely deterministic and moves one step to the right each time. One can verify that $I(\mu)=\infty$ for all μ. If we take for the closed set C the whole space, i.e., all probability measures on the positive integers, then

$$\overline{\lim_{n\to\infty}}\frac{1}{n}\log Q_{n,x}(C)=0.$$

On the other hand, $\inf_{\mu\in C}I(\mu)=\infty$, so that (1.4) cannot be valid in this case for all closed sets.

In the case of continuous time, if we take the process with generator

$$\frac{1}{2}\frac{\partial^2}{\partial x^2}+a\frac{\partial}{\partial x},$$

i.e., Brownian motion with a constant drift a, then

$$I(\mu)=\frac{1}{8}\int_{-\infty}^{\infty}\frac{[f'(y)]^2}{f(y)}\,dy+\tfrac{1}{2}a^2,$$

where $\mu(dx)=f(x)\,dx$. Clearly, $\inf_{\mu\in\mathcal{M}}I(\mu)=\frac{1}{2}a^2>0$. We see the continuous analogue of (1.4) cannot hold when C is the whole space.

These two examples are also counterexamples to (1.6) and its continuous version with $\Phi(\mu)\equiv 0$.

To see that some sort of transitivity condition is needed, we consider the case of a finite state space Markov chain with transition probability $\pi_{ij}=\delta_{i,j}$. Again the chain is deterministic and never moves. In this case, $I(\mu)=0$ for all μ and so if we take for the open set G the set of all nondegenerate probability distributions on the state space, then $Q_{n,x}(G)=0$ for all n and x. On the other hand, if the state space is at least two points, $\inf_{\mu\in G}I(\mu)=0$ so that (1.5) cannot hold.

Finally, we give an example to illustrate when Condition H* is valid. If we take the one-dimensional Ornstein–Uhlenbeck process, i.e., the process whose generator is

$$\frac{1}{2}\frac{\partial^2}{\partial x^2} - x\frac{\partial}{\partial x},$$

we see that Condition H* is satisfied with

$$u_n(x) = \cosh n\theta\left(\frac{x}{n}\right),$$

$$V(x) = \tfrac{1}{2} - x \sinh x,$$

where

$$\theta(x) = \begin{cases} x & \text{for } \; 0 \leqq x \leqq 1, \\ 2 & \text{for } \; x \geqq 2, \\ \text{smooth and increasing} & \text{for } \; 1 \leqq x \leqq 2, \end{cases}$$

and $\theta(-x) = -\theta(x)$.

Bibliography

[1] Donsker, M. D., and Varadhan, S. R. S., *Asymptotic evaluation of certain Markov process expectations for large time—I*, Comm. Pure Appl. Math., Vol. 27, 1975, pp. 1–47.

[2] Donsker, M. D., and Varadhan, S. R. S., *Asymptotic evaluation of certain Wiener integrals for large time*, Functional Integration and Its Applications, Proceedings of the International Conference held at the Cumberland Lodge, Windsor Great Park, London in April 1974. A. M. Arthurs (editor), Clarendon Press, Oxford, 1975.

[3] Marcus, M. B., and Shepp, L. A., *Sample behavior of Gaussian processes*, Sicth Berkeley Symposium on Math. Stat. and Probability, Vol. II, 1972, pp. 423–439.

[4] Kuratowski, C., *Topologie I*, Warszawa-Lwow, 1933.

[5] Pincus, M., *Gaussian processes and Hammerstein integral equations*, Trans. Amer. Math. Soc., Vol. 134, 1968, pp. 193–216.

[6] Schilder, M., *Some asymptotic formulae for Wiener integrals*, Trans. Amer. Math. Soc., Vol. 125, 1966, pp. 63–85.

[7] Sion, M., *On general minimax theorem*, Pacific Journal of Math., Vol. VIII, 1958, pp. 171–176.

[8] Skorohod, A. V., *A note on Gaussian measures in a Bãnach space*, Theory of Prob. and Its Appl., Vol. XV, 1970, p. 508.

[9] Varadhan, S. R. S., *Asymptotic probabilities and differential equations*, Comm. Pure Appl. Math., Vol. 19, 1966, pp. 261–286.

Received April, 1976.

COMMUNICATIONS ON PURE AND APPLIED MATHEMATICS, VOL. XXIX, 595-621 (1976)

On the Principal Eigenvalue of Second-Order Elliptic Differential Operators*

M. D. DONSKER AND S. R. S. VARADHAN

1. Introduction

Let X be a compact metric space and T_t a strongly continuous semigroup of operators on $C(X)$ such that $T_t f$ is non-negative for each non-negative f and $T_t 1 = 1$ for all t. Let L be the infinitesimal generator with domain $\mathfrak{D}$ and let $\mathfrak{D}_+$ denote the strictly positive functions in $\mathfrak{D}$. We define

$$I(\mu) = -\inf_{u \in \mathfrak{D}_+} \int_X \left(\frac{Lu}{u}\right)(x)\mu(dx), \tag{1.1}$$

where μ is a probability measure on X. The function $I(\mu)$ is lower semicontinuous on the space $\mathcal{M}$ of all probability measures on X.

We proved in [1] that if $V \in C(X)$, then the principal eigenvalue λ_V of the operator $L + V$ is given by

$$\lambda_V = \sup_{\mu \in \mathcal{M}} \left[\int_X V(x)\mu(dx) - I(\mu) \right]. \tag{1.2}$$

If the operator L is selfadjoint with respect to some reference measure on X, then $I(\mu)$ can be readily computed and formula (1.2) reduces to the classical variational formula for the principal eigenvalue. If L is not selfadjoint, then in general $I(\mu)$ is difficult to evaluate explicitly.

In this paper we investigate the situation when X is R^d (it can be replaced by any connected manifold without boundary) and L is an elliptic second-order differential operator. $I(\mu)$ can be defined as the exact analogue of (1.1) but we need some compactness for (1.2) and therefore we consider the operator $L + V$ only on bounded regions with Dirichlet boundary conditions. After establishing the analogue of (1.2) in such cases, we look at the problem of when two operators L_1 and L_2 of the type we are dealing with can have identical I-functions. Equivalently, we seek the relation between L_1 and L_2 if

* This research was performed at the Courant Institute and partially supported by the National Science Foundation, Grant NSF-MPS-74-01921-A01. Reproduction in whole or in part is permitted for any purpose of the United States Government.

L_1+V and L_2+V have the same principal eigenvalue for all regions G and all potentials V. The answer to this question is the content of Theorem 3.1.

In order to prove that theorem we need to evaluate the I-function as explicitly as we can for operators of the type described. It turns out that, for elliptic second-order differential operators, $I(\mu)$ can be calculated quite explicitly (cf. Lemma 3.3). For example, if X is the circle parametrized by real numbers x (mod 1) and L is the operator

$$L=\frac{1}{2}\frac{\partial^2}{\partial x^2}+b(x)\frac{\partial}{\partial x}, \tag{1.3}$$

then L is selfadjoint with respect to some measure if and only if $\int_0^1 b(x)\,dx=0$. In any case, the I-function for (1.3) on the circle is given by

$$I(f)=\frac{1}{8}\int_0^1\frac{[f'(x)]^2}{f(x)}\,dx+\frac{1}{2}\int_0^1[b^2(x)+b_x(x)]f(x)\,dx-\frac{1}{2}\frac{\left(\int_0^1 b(x)\,dx\right)^2}{\int_0^1\frac{dx}{f(x)}}, \tag{1.4}$$

where $f(x)$ is a probability density on the circle.

2. The Principal Eigenvalue

Consider the operator

$$L=\frac{1}{2}\sum_{i,j=1}^d a_{ij}(x)\frac{\partial^2}{\partial x_i\,\partial x_j}+\sum_{j=1}^d b_j(x)\frac{\partial}{\partial x_j}, \tag{2.1}$$

where $a_{ij}(x)$ and $b_j(x)$ are in $C^\infty(R^d)$ for $1\leqq i,\ j\leqq d$, and the matrix $(a_{ij}(x))$ is positive definite for each $x\in R^d$. We shall assume, for Lemma 2.1 and Theorems 2.2 and 2.3 below, that, for large x, $a_{ij}(x)=\delta_{i,j}$ and $b_j(x)=0$, $1\leqq i$, $j\leqq d$. It will then be shown that for our purposes this hypothesis on the behavior of the coefficients for large x can be discarded.

Under the hypotheses above, L is the infinitesimal generator of the semigroup T_t associated with the fundamental solution of $\partial u/\partial t=Lu$. Let us denote the domain of L by $\mathcal{D}$ and by $\mathcal{D}_+$ those elements u in $\mathcal{D}$ for each of which there exist constants c_1 and c_2 such that $0<c_1\leqq u(x)\leqq c_2<\infty$ for all $x\in R^d$. Let $\mathcal{M}$ be the space of probability measures on R^d. For any $\mu\in\mathcal{M}$ we define

$$I(\mu)=-\inf_{u\in\mathcal{D}_+}\int_{R^d}\left(\frac{Lu}{u}\right)(x)\mu(dx). \tag{2.2}$$

For any $\mu \in \mathcal{M}$ we also define

$$\bar{I}(\mu) = -\inf_{u \in C_+^\infty(R^d)} \int_{R^d} \left(\frac{Lu}{u}\right)(x)\mu(dx), \tag{2.3}$$

where $C_+^\infty(R^d)$, analogous to $\mathcal{D}_+$, is the set of positive functions $u \in C^\infty(R^d)$ which are bounded above and below. We now prove

LEMMA 2.1. *For all* $\mu \in \mathcal{M}$,

$$I(\mu) = \bar{I}(\mu).$$

Proof: Clearly $\bar{I}(\mu) \leqq I(\mu)$; thus we must show the reverse inequality. Let $f \in C_+^\infty(R^d)$. Since $T_t f \in C_+^\infty(R^d)$, we have

$$\frac{\partial}{\partial t} \int_{R^d} \log\left(\frac{T_t f}{f}\right)(x)\mu(dx) = \int_{R^d} \left(\frac{LT_t f}{T_t f}\right)(x)\mu(dx) \geqq -\bar{I}(\mu). \tag{2.4}$$

When $t = 0$, $\log(T_t f/f) = 0$ so that integrating both sides of (2.4) from 0 to t we get

$$\int_{R^d} \log\left(\frac{T_t f}{f}\right)(x)\mu(dx) \geqq -t\bar{I}(\mu) \tag{2.5}$$

which holds for all $t \geqq 0$ and all $f \in C_+^\infty(R^d)$. Actually, if $f \in C_+(R^d)$, then (2.5) holds for this f also because we can find a sequence $\{f_n\} \in C_+^\infty(R^d)$ (even satisfying the same bounds) which converges to f uniformly on compact sets. Thus, if $u \in \mathcal{D}_+$, then $u \in C_+(R^d)$, and from (2.5) we have, for all $t \geqq 0$,

$$\int_{R^d} \log\left(\frac{T_t u}{u}\right)(x)\mu(dx) \geqq -t\bar{I}(\mu).$$

Both sides of this inequality vanish at $t = 0$, and hence the same inequality holds for the derivatives at $t = 0$, i.e.,

$$\int_{R^d} \left(\frac{Lu}{u}\right)(x)\mu(dx) \geqq -\bar{I}(\mu).$$

Since this last inequality holds for any $u \in \mathcal{D}_+$, we conclude that $I(\mu) \leqq \bar{I}(\mu)$ which proves the lemma.

Let G be an open set in R^d with compact closure $\bar{G}$ and having a smooth boundary ∂G. Let $V(x)$ be continuous on R^d and denote by $T_t^{V,G}$ the semigroup defined by

$$\begin{aligned} &\frac{\partial u}{\partial t} = Lu + Vu && \text{on} \quad [0, \infty) \times G, \\ &u(t, x) = 0 && \text{on} \quad [0, \infty) \times \partial G, \\ &u(0, x) = f(x) && \text{on} \quad G. \end{aligned} \tag{2.6}$$

Then, $T_t^{V,G}$ defines a semigroup on $C(G)$ and it leaves $C_0(G)$, the space of continuous functions on $\bar{G}$ which vanish on ∂G, invariant; $T_t^{V,G}$ is strongly continuous on $C_0(G)$.

Since $\|T_t^{V,G}\| = \sup_{x \in \bar{G}} (T_t^{V,G} \cdot 1)(x)$ is submultiplicative in t, the following limit exists:

$$\lambda_{V,G} = \lim_{t \to \infty} \frac{1}{t} \log \|T_t^{V,G}\|. \tag{2.7}$$

Let $L_{V,G}$ be the infinitesimal generator of $T_t^{V,G}$ on $C_0(G)$, and denote the spectrum of $L_{V,G}$ by $\sigma_{V,G}$.

THEOREM 2.2. *We always have* $\lambda_{V,G} \in \sigma_{V,G}$ *and if* $\mathcal{R}e\, z > \lambda_{V,G}$, *then* $z \notin \sigma_{V,G}$. *Moreover,*

$$\lambda_{V,G} = \sup_{\mu : \mu(\bar{G}) = 1} \left[\int_{R^d} V(x) \mu(dx) - I(\mu) \right]. \tag{2.8}$$

Proof: If $\mathcal{R}e\, z > \lambda_{V,G}$, then

$$\int_0^\infty e^{-zt} T_t^{V,G}\, dt$$

is a convergent integral and defines explicitly the resolvent so that $z \notin \sigma_{V,G}$.

Let

$$l_{V,G} = \sup_{\mu : \mu(\bar{G}) = 1} \left[\int_{R^d} V(x) \mu(dx) - I(\mu) \right]. \tag{2.9}$$

We show first that $\lambda_{V,G} \leqq l_{V,G}$. From Lemma 2.1,

$$
\begin{aligned}
l_{V,G} &= \sup_{\mu:\mu(\bar{G})=1} \left[\int_{R^d} V(x)\mu(dx) + \inf_{u \in C_+^\infty(R^d)} \int_{R^d} \left(\frac{Lu}{u}\right)(x)\mu(dx) \right] \\
&= \sup_{\mu:\mu(\bar{G})=1} \inf_{u \in C_+^\infty(R^d)} \left[\int_{R^d} V(x)\mu(dx) + \int_{R^d} \left(\frac{Lu}{u}\right)(x)\mu(dx) \right].
\end{aligned} \tag{2.10}
$$

For $u \in C_+^\infty(R^d)$ we can write $u = e^h$, where $h \in C^\infty(R^d)$. Moreover, from (2.1) we see that $(Le^h/e^h)(x) = [Lh + \frac{1}{2}\langle a\, \nabla h, \nabla h\rangle](x)$, where $a = a(x) = (a_{ij}(x))$. Thus, from (2.10) we get

$$
l_{V,G} = \sup_{\mu:\mu(\bar{G})=1} \inf_{h \in C^\infty(R^d)} \left[\int_{R^d} V(x)\mu(dx) + \int_{R^d} [Lh + \tfrac{1}{2}\langle a\, \nabla h, \nabla h\rangle](x)\mu(dx) \right].
$$

If we impose the C^2-topology on $C^\infty(R^d)$, then

$$
\int_{R^d} V(x)\mu(dx) + \int_{R^d} [Lh + \tfrac{1}{2}\langle a\, \nabla h, \nabla h\rangle](x)\mu(dx)
$$

is a continuous convex function of h for each fixed μ, and for each fixed h, it is a continuous linear function of μ. Since μ varies over a compact set, the mini-max theorem of Sion ([2], Corollary 3.3) applies, and we have

$$
\begin{aligned}
l_{V,G} &= \inf_{h \in C^\infty(R^d)} \sup_{\mu:\mu(\bar{G})=1} \left[\int_{R^d} V(x)\mu(dx) + \int_{R^d} [Lh + \tfrac{1}{2}\langle a\, \nabla h, \nabla h\rangle](x)\mu(dx) \right] \\
&= \inf_{u \in C_+^\infty(R^d)} \sup_{\mu:\mu(\bar{G})=1} \left[\int_{R^d} V(x)\mu(dx) + \int_{R^d} \left(\frac{Lu}{u}\right)(x)\mu(dx) \right] \\
&= \inf_{u \in C_+^\infty(R^d)} \sup_{x \in \bar{G}} \left[V(x) + \left(\frac{Lu}{u}\right)(x) \right].
\end{aligned}
$$

Therefore, given $\varepsilon > 0$, there exists $u \in C_+^\infty(R^d)$ such that

$$
V(x) + \left(\frac{Lu}{u}\right)(x) \leqq l_{V,G} + \varepsilon, \tag{2.11}
$$

for all $x \in \bar{G}$. For this ε and this u, let $W(t, x) = u(x) \exp\{(\varepsilon + l_{V,G})t\}$. Then, (2.11) implies that, for all $x \in \bar{G}$,

$$\begin{aligned} \frac{\partial W}{\partial t} - LW - VW &= (\varepsilon + l_{V,G})W - \exp\{(\varepsilon + l_{V,G})t\}Lu - \exp\{(\varepsilon + l_{V,G})t\}Vu \\ &= W\left[\varepsilon + l_{V,G} - \frac{Lu}{u} - V\right] \geqq 0 . \end{aligned} \tag{2.12}$$

For the function u in (2.12), let $c_1 = \inf_{x \in \bar{G}} u(x)$ and $c_2 = \sup_{x \in \bar{G}} u(x)$. For a given $f(x)$ defined on $\bar{G}$, let $U(t, x)$ be the solution of (2.6), and let $\|f\| = \sup_{x \in \bar{G}} |f(x)|$. From (2.12) we see that the function $(\|f\|/c_1)W$ satisfies

$$\frac{\partial}{\partial t}\left(\frac{\|f\|}{c_1} W\right) - L\left(\frac{\|f\|}{c_1} W\right) - V\left(\frac{\|f\|}{c_1} W\right) \geqq 0 . \tag{2.13}$$

Also, at $t = 0$, $\|f\|\, W/c_1 \geqq f$ on $\bar{G}$. Moreover, on $[0, \infty) \times \partial G$ we have $\|f\|\, W/c_1 \geqq 0$. Thus, from the maximum principle,

$$\frac{\|f\|\, W}{c_1} \geqq U(t, x) . \tag{2.14}$$

On the other hand, $\|f\|\, W/c_1 \leqq (\|f\|/c_1)c_2 \exp\{(\varepsilon + l_{V,G})t\}$, so that

$$U(t, x) \leqq \frac{c_2}{c_1} \|f\| \exp\{(\varepsilon + l_{V,G})t\} . \tag{2.15}$$

Since $U(t, x) = (T_t^{V,G} f)(x)$, we get from (2.15), by taking the supremum over all $x \in \bar{G}$,

$$\|T_t^{V,G} f\| \leqq \frac{c_2}{c_1} \|f\| \exp\{(\varepsilon + l_{V,G})t\} ,$$

and therefore

$$\lim_{t \to \infty} \frac{1}{t} \log \|T_t^{V,G}\| \leqq \varepsilon + l_{V,G} . \tag{2.16}$$

In (2.16), ε is arbitrary and we conclude that $\lambda_{V,G} \leqq l_{V,G}$.

To show that $l_{V,G} \leqq \lambda_{V,G}$ we first note that the supremum in (2.9) is actually attained because $\int_{R^d} V(x)\mu(dx)$ is continuous in μ, $I(\mu)$ is lower semicontinuous in μ, and $\{\mu : \mu(\bar{G})=1\}$ is a compact set. So let μ_0 be the μ, where the supremum is attained, i.e.,

$$l_{V,G}=\int_{R^d} V(x)\mu_0(dx)-I(\mu_0)\,. \tag{2.17}$$

For $h>0$ consider the operator $S_h=\exp\{hV(x)\}T_h$. With $u\in C_+^\infty(R^d)$ and k a positive integer, let $u_k=S_h^k u$. Then,

$$\begin{aligned}\int_{R^d}\log\left(\frac{S_h u_k}{u_k}\right)(x)\mu_0(dx)&=\int_{R^d}\log\left[\exp\{hV(x)\}\left(\frac{T_h u_k}{u_k}\right)(x)\right]\mu_0(dx)\\&=h\int_{R^d}V(x)\mu_0(dx)+\int_{R^d}\log\left(\frac{T_h u_k}{u_k}\right)(x)\mu_0(dx)\,.\end{aligned} \tag{2.18}$$

From Lemma 2.1 and its proof, we know that

$$\int_{R^d}\log\left(\frac{T_h u_k}{u_k}\right)(x)\mu_0(dx)\geqq -hI(\mu_0)\,,$$

and thus from (2.18) and (2.17) we get

$$\begin{aligned}\int_{R^d}\log\left(\frac{S_h u_k}{u_k}\right)(x)\mu_0(dx)&\geqq h\left[\int_{R^d}V(x)\mu_0(dx)-I(\mu_0)\right]\\&=hl_{V,G}\,.\end{aligned} \tag{2.19}$$

Since $S_h u_k=u_{k+1}$, we rewrite (2.19) as

$$\int_{R^d}\log\left(\frac{u_{k+1}}{u_k}\right)(x)\mu_0(dx)\geqq hl_{V,G}\,,$$

and summing on k from 0 to $n-1$, we obtain

$$\int_{R^d}\log\left(\frac{u_n}{u}\right)(x)\mu_0(dx)\geqq nhl_{V,G}\,. \tag{2.20}$$

In (2.20) we let $n \to \infty$ and $h \to 0$ so that $nh \to t$. First of all, we have by Trotter's formula

$$\lim_{\substack{n\to\infty\\ h\to 0\\ nh\to t}} u_n = \lim_{\substack{n\to\infty\\ h\to 0\\ nh\to t}} S_h^n u = \lim_{\substack{n\to\infty\\ h\to 0\\ nh\to t}} [e^{hV(x)}T_h]^n u = T_t^V u,$$

where $T_t^V = T_t^{V,G}$, as defined by (2.6), but with $G = R^d$. Thus from (2.20) we obtain

$$\int_{R^d} \log\left(\frac{T_t^V u}{u}\right)(x)\mu_0(dx) \geqq tl_{V,G},$$

or

$$\int_{R^d} \log (T_t^V u)(x)\mu_0(dx) \geqq tl_{V,G} + \int_{R^d} \log u(x)\mu_0(dx). \tag{2.21}$$

Inequality (2.21) holds for any $u \in C_+^\infty(R^d)$ and we choose $u(x) \equiv 1$ on $\bar{G}$. Since $\mu_0(\bar{G}) = 1$, we get from (2.21), for such u,

$$\int_{R^d} \log (T_t^V u)(x)\mu_0(dx) \geqq tl_{V,G}. \tag{2.22}$$

From Jensen's inequality this gives us

$$\log \int_{R^d} (T_t^V u)(x)\mu_0(dx) \geqq tl_{V,G},$$

or

$$\int_{R^d} (T_t^V u)(x)\mu_0(dx) \geqq \exp\{tl_{V,G}\}. \tag{2.23}$$

In (2.23), $\mu_0(\bar{G}) = 1$ so that $l_{V,G} = \int_{R^d} V(x)\mu_0(dx) - I(\mu_0)$ is insensitive to the values of $V(x)$ for $x \notin \bar{G}$; we also recall that $u(x) \equiv 1$ on $\bar{G}$. Thus, since (2.23) holds for all $V(x)$ continuous on R^d, we can choose a sequence $\{V_n(x)\}$ of such functions converging to a V which is continuous on $\bar{G}$ and equal to $-\infty$ off $\bar{G}$. In this way we get

$$\int_{R^d} (T_t^{V,G} \cdot 1)(x)\mu_0(dx) \geqq \exp\{tl_{V,G}\}. \tag{2.24}$$

But (2.24) implies that

$$\|T_t^{V,G}\| = \sup_{x\in\bar{G}} (T_t^{V,G}\cdot 1)(x) \geqq \exp\{tl_{V,G}\},$$

and hence that $\lambda_{V,G} \geqq l_{V,G}$ which completes the proof that $\lambda_{V,G} = l_{V,G}$.

Finally, we want to show that $\lambda_{V,G} \in \sigma_{V,G}$. Assume $\lambda_{V,G} \notin \sigma_{V,G}$. Then, from the first part of the proof of this theorem we know that, for $\lambda > \lambda_{V,G}$,

$$(R_\lambda f)(x) = \int_0^\infty e^{-\lambda t}(T_t^{V,G} f)(x)\,dt$$

exists, and therefore assuming $\lambda_{V,G} \notin \sigma_{V,G}$ implies that

$$(R_{\lambda_{V,G}} f)(x) = \lim_{\lambda \downarrow \lambda_{V,G}} \int_0^\infty e^{-\lambda t}(T_t^{V,G} f)(x)\,dt$$

exists. If we take $f \geqq 0$, then by monotone convergence

$$(R_{\lambda_{V,G}} f)(x) = \int_0^\infty \exp\{-\lambda_{V,G} t\}(T_t^{V,G} f)(x)\,dt,$$

so that, in particular,

$$\|R_{\lambda_{V,G}}\| = \sup_{x\in\bar{G}} \int_0^\infty \exp\{-\lambda_{V,G} t\}(T_t^{V,G}\cdot 1)(x)\,dt < c < \infty.$$

But this implies

$$\int_{R^d} \mu_0(dx) \int_0^\infty \exp\{-\lambda_{V,G} t\}(T_t^{V,G}\cdot 1)(x)\,dt < c,$$

and, by Fubini's theorem,

$$\int_0^\infty \exp\{-\lambda_{V,G} t\}\,dt \int_{R^d} (T_t^{V,G}\cdot 1)(x)\mu_0(dx) < c. \tag{2.25}$$

Now (2.24) and (2.25) yield

$$\int_0^\infty \exp\{-\lambda_{V,G} t\}\exp\{tl_{V,G}\}\,dt < c < \infty,$$

and, having just proved that $\lambda_{V,G} = l_{V,G}$, we obtain a contradiction so that $l_{V,G} \in \sigma_{V,G}$. This completes the proof of Theorem 2.2.

With $T_t^{V,G}$ defined by (2.6) and $\lambda_{V,G}$ defined by (2.7) we now prove

THEOREM 2.3. *For any* μ *with* supp $\mu \subset \bar{G}$,

$$I(\mu) = -\inf_{V \in C^\infty(R^d)} \left[\lambda_{V,G} - \int_{R^d} V(x)\mu(dx)\right]. \tag{2.26}$$

Proof: From (2.8) of Theorem 2.2 we know that the right side of (2.26) is

$$-\inf_{V \in C^\infty(R^d)} \left[\sup_{\alpha:\alpha(\bar{G})=1} \left\{\int_{R^d} V(x)\alpha(dx) - I(\alpha)\right\} - \int_{R^d} V(x)\mu(dx)\right]$$

$$= -\inf_{V \in C^\infty(R^d)} \sup_{\alpha:\alpha(\bar{G})=1} \left[-I(\alpha) + \int_{R^d} V(x)\alpha(dx) - \int_{R^d} V(x)\mu(dx)\right].$$

We use again the mini-max theorem of Sion ([2], Corollary 3.3) to see that the right side of (2.26) is

$$-\sup_{\alpha:\alpha(\bar{G})=1} \inf_{V \in C^\infty(R^d)} \left[-I(\alpha) + \int_{R^d} V(x)\alpha(dx) - \int_{R^d} V(x)\mu(dx)\right]. \tag{2.27}$$

But since supp $\mu \subset \bar{G}$, we note that

$$\inf_{V \in C^\infty(R^d)} \left[-I(\alpha) + \int_{R^d} V(x)\alpha(dx) - \int_{R^d} V(x)\mu(dx)\right]$$

is clearly equal to $-\infty$ if $\alpha \neq \mu$ and is equal to $-I(\mu)$ if $\alpha = \mu$. Thus (2.27) is just $I(\mu)$ as claimed.

As noted at the beginning of this section, Theorems 2.2 and 2.3 have been proved under the hypothesis that the coefficients in the operator L behave suitably at infinity; namely, $a_{ij}(x) = \delta_{ij}$ and $b_j(x) = 0$ if $|x|$ is large, for $1 \leqq i$, $j \leqq d$. We needed this assumption in order to define the semigroup T_t which we used in the proofs.

Consider

$$L = \frac{1}{2} \sum_{i,j=1}^{d} a_{ij}(x) \frac{\partial^2}{\partial x_i \, \partial x_j} + \sum_{j=1}^{d} b_j(x) \frac{\partial}{\partial x_j}, \tag{2.28}$$

where $a_{ij}(x)$ and $b_j(x)$ are in $C^\infty(R^d)$ for $1 \leqq i, j \leqq d$ and the matrix $(a_{ij}(x))$ is positive definite for each $x \in R^d$ but where now we make no hypotheses on the behavior of the coefficients at infinity. If μ is a probability measure on R^d with compact support, we can still define

$$(2.29)\qquad I(\mu) = -\inf_{\substack{u>0\\ u\in C^\infty(R^d)}} \int_{R^d} \left(\frac{Lu}{u}\right)(x)\mu(dx).$$

The following lemma is then obvious.

LEMMA 2.4. *Suppose L and L' are both operators satisfying the hypotheses imposed on* (2.28) *and that, for some $R, L = L'$ on $|x| \leqq R$. If $\mu \in \mathcal{M}$ and $\mu(|x| \leqq R) = 1$, then we have*

$$I'(\mu) = I(\mu).$$

Thus if L and L' are two such operators and they agree on $\bar{G}$, it follows that $\lambda_{V,G} = \lambda'_{V,G}$ and $I'(\mu) = I(\mu)$. This means Theorems 2.2 and 2.3 are true for operators (2.28) without imposing any conditions on the behavior of the coefficients as $|x| \to \infty$.

3. The Inverse Problem

As in the last section, let

$$(3.1)\qquad L = \frac{1}{2}\sum_{i,j=1}^{d} a_{ij}(x)\frac{\partial^2}{\partial x_i\,\partial x_j} + \sum_{j=1}^{d} b_j(x)\frac{\partial}{\partial x_j},$$

where $a_{ij}(x)$ and $b_j(x)$ are in $C^\infty(R^d)$ for $1 \leqq i, j \leqq d$, and the matrix $(a_{ij}(x))$ is positive definite for each $x \in R^d$. If G is an open set in R^d with compact closure $\bar{G}$ and smooth boundary ∂G, and if $V(x)$ is a continuous function in R^d, then we saw in the last section that the principal eigenvalue $\lambda_{V,G}$ of the region G and potential V is given by a variational formula

$$(3.2)\qquad \lambda_{V,G} = \sup_{\mu:\mu(\bar{G})=1} \left[\int_{R^d} V(x)\mu(dx) - I(\mu)\right].$$

Here, $I(\mu)$ is defined for μ with compact support by

$$(3.3)\qquad I(\mu) = -\inf_{\substack{u>0\\ u\in C^\infty(R^d)}} \int_{R^d} \left(\frac{Lu}{u}\right)(x)\mu(dx).$$

We also saw in the last section that for any μ with supp $\mu \subset \bar{G}$

$$I(\mu) = -\inf_{V \in C^{\infty}(R^d)} \left[\lambda_{V,G} - \int_{R^d} V(x)\mu(dx) \right], \tag{3.4}$$

so that $\lambda_{V,G}$ and $I(\mu)$ are conjugate convex functions.

Suppose now we have two operators L_1 and L_2 of the type above. For each continuous potential $V(x)$ and each admissible open set G we have then the principal eigenvalue $\lambda_{V,G}^{(1)}$ corresponding to L_1 and the principal eigenvalue $\lambda_{V,G}^{(2)}$ corresponding to L_2. In this section we consider the question: If $\lambda_{V,G}^{(1)} = \lambda_{V,G}^{(2)}$ for all such regions G and all such potentials V, then what can be said about the operators L_1 and L_2? The main result is the following

THEOREM 3.1. *Let L_1 and L_2 be two diffusion operators satisfying the hypotheses imposed on* (3.1). *Suppose that for every open set $G \subset R^d$ with compact closure $\bar{G}$ and smooth boundary ∂G and for every function $V(x) \in C^{\infty}(R^d)$ we have*

$$\lambda_{V,G}^{(1)} = \lambda_{V,G}^{(2)}. \tag{3.5}$$

Then, either there exists a positive harmonic function u for L_2 (i.e., $L_2 u = 0$ and $u > 0$) such that

$$L_1 f = \frac{L_2(uf)}{u},$$

or there exists a positive invariant density for L_2 (i.e., $L_2^* \rho = 0$ and $\rho > 0$) such that

$$L_1 f = \frac{L_2^*(\rho f)}{\rho}.$$

To prove this theorem we make use of the relation between $\lambda_{V,G}$ and $I(\mu)$ as exemplified by (3.4) and the special nature of $I(\mu)$ corresponding to an operator of the form (3.1). To the two operators L_1 and L_2 correspond $I_1(\mu)$ and $I_2(\mu)$ defined by (3.3) for μ with compact support. It is clear from (3.4) that the hypothesis $\lambda_{V,G}^{(1)} = \lambda_{V,G}^{(2)}$ of Theorem 3.1 implies that $I_1(\mu) = I_2(\mu)$ for all μ with compact support. Indeed, to conclude this, all we really need to assume is that $\lambda_{V,G}^{(1)} = \lambda_{V,G}^{(2)}$ for all $V \in C^{\infty}(R^d)$ and for an increasing sequence of bounded open sets G whose union is R^d. In any event, since (3.5) implies $I_1(\mu) = I_2(\mu)$ for all μ with compact support, we have in particular $I_1(\mu) = I_2(\mu)$ for all μ with compact support and having a density $\phi \in C^{\infty}(R^d)$. Hence, Theorem 3.1 will follow from

THEOREM 3.2. *If L_1 and L_2 are two operators satisfying the hypotheses imposed on* (3.1) *and if $I_1(\mu) = I_2(\mu)$ for all μ with compact support having a* *density* $\phi \in C^\infty(R^d)$, *then either there exists a positive harmonic function u for* L_2 *such that* $L_1 f = L_2(uf)/u$ *or there exists a positive invariant density* ρ *for* L_2 *such that* $L_1 f = L_2^*(\rho f)/\rho$.

To prove Theorem 3.2 we first prove a succession of lemmas. We shall find it convenient to express the operator (3.1) in divergence form, i.e.,

$$L = \tfrac{1}{2}\nabla \cdot a\nabla + \tilde{b} \cdot \nabla . \tag{3.6}$$

We first prove

LEMMA 3.3. *Let L be given by* (3.6) *and let* $I(\mu)$ *be its I-function defined by* (3.3). *If* μ *is a probability measure on* R^d *with compact support such that* $\mu(dx) = \phi(x)\,dx$, *where* $\phi(x) \in C^\infty(R^d)$, *then*

$$\begin{aligned} I(\mu) = \frac{1}{8}\int_{R^d} \frac{\langle a\,\nabla\phi, \nabla\phi\rangle}{\phi}\,dx + \frac{1}{2}\int_{R^d} [\nabla \cdot \tilde{b} + \langle \tilde{b}, a^{-1}\tilde{b}\rangle]\phi\,dx \\ - \frac{1}{2}\inf_{W \in C^\infty(R^d)} \int_{R^d} (a^{-1}\tilde{b} - \nabla W)^T a (a^{-1}\tilde{b} - \nabla W)\phi\,dx . \end{aligned} \tag{3.7}$$

Remark. The last term on the right-hand side of (3.7) is clearly zero if $d = 1$.

Proof: First of all,

$$\begin{aligned} I(\mu) &= -\inf_{\substack{u>0 \\ u \in C^\infty(R^d)}} \int_{R^d} \left(\frac{Lu}{u}\right)(x)\phi(x)\,dx \\ &= -\inf_{h \in C^\infty(R^d)} \int_{R^d} \left(\frac{Le^h}{e^h}\right)(x)\phi(x)\,dx \\ &= -\inf_{h \in C^\infty(R^d)} \int_{R^d} [Lh + \tfrac{1}{2}\langle a\,\nabla h, \nabla h\rangle]\phi\,dx . \end{aligned}$$

For $\varepsilon > 0$ and $h \in C^\infty(R^d)$, define $W(x) = \frac{1}{2}\log[\phi(x) + \varepsilon] - h(x)$ so that $W(x)$ is also in $C^\infty(R^d)$. An elementary calculation shows that

$$\begin{aligned} Lh + \tfrac{1}{2}\langle a\,\nabla h, \nabla h\rangle = \frac{1}{4}\frac{\nabla \cdot a\,\nabla\phi}{(\phi+\varepsilon)} - \frac{1}{8}\frac{\langle a\,\nabla\phi, \nabla\phi\rangle}{(\phi+\varepsilon)^2} + \frac{\tilde{b}\cdot\nabla\phi}{2(\phi+\varepsilon)} - \tilde{b}\cdot\nabla W \\ - \tfrac{1}{2}\nabla \cdot a\,\nabla W + \tfrac{1}{2}\langle a\,\nabla W, \nabla W\rangle - \frac{\langle a\,\nabla\phi, \nabla W\rangle}{2(\phi+\varepsilon)} . \end{aligned}$$

Therefore,

$$I(\mu) = -\inf_{W \in C^\infty(R^d)} \Bigg[\int_{R^d} \Bigg\{\frac{\nabla \cdot a\,\nabla\phi}{4(\phi+\varepsilon)} - \frac{\langle a\,\nabla\phi, \nabla\phi\rangle}{8(\phi+\varepsilon)^2} + \frac{\tilde b \cdot \nabla\phi}{2(\phi+\varepsilon)} - \tilde b \cdot \nabla W$$

$$-\tfrac{1}{2}\nabla \cdot a\,\nabla W + \tfrac{1}{2}\langle a\,\nabla W, \nabla W\rangle - \frac{\langle a\,\nabla\phi, \nabla W\rangle}{2(\phi+\varepsilon)}\Bigg\}\phi\,dx\Bigg]$$

$$(3.8) \quad = -\frac{1}{4}\int_{R^d} \frac{\nabla \cdot a\,\nabla\phi}{(\phi+\varepsilon)}\phi\,dx + \frac{1}{8}\int_{R^d} \frac{\langle a\,\nabla\phi, \nabla\phi\rangle}{(\phi+\varepsilon)^2}\phi\,dx - \frac{1}{2}\int_{R^d} \frac{\tilde b \cdot \nabla\phi}{(\phi+\varepsilon)}\phi\,dx$$

$$-\inf_{W \in C^\infty(R^d)} \Bigg[\int_{R^d} \Bigg\{\tfrac{1}{2}\langle a\,\nabla W, \nabla W\rangle - \tilde b \cdot \nabla W - \tfrac{1}{2}\nabla \cdot a\,\nabla W$$

$$- \frac{\langle a\,\nabla\phi, \nabla W\rangle}{2(\phi+\varepsilon)}\Bigg\}\phi\,dx\Bigg].$$

Integrating by parts, we see that

$$\frac{1}{2}\int_{R^d} (\nabla \cdot a\,\nabla W)\phi\,dx = -\frac{1}{2}\int_{R^d} \langle a\,\nabla\phi, \nabla W\rangle\,dx,$$

and hence

$$-\frac{1}{2}\int_{R^d} (\nabla \cdot a\,\nabla W)\phi\,dx - \frac{1}{2}\int_{R^d} \frac{\langle a\,\nabla\phi, \nabla W\rangle}{\phi+\varepsilon}\phi\,dx = \tfrac{1}{2}\varepsilon\int_{R^d} \frac{\langle a\,\nabla\phi, \nabla W\rangle}{\phi+\varepsilon}\,dx.$$

Thus

$$I(\mu) = -\frac{1}{4}\int_{R^d} \frac{\nabla \cdot a\,\nabla\phi}{\phi+\varepsilon}\phi\,dx + \frac{1}{8}\int_{R^d} \frac{\langle a\,\nabla\phi, \nabla\phi\rangle}{\phi+\varepsilon^{\,2}}\phi\,dx$$

$$(3.9) \quad -\frac{1}{2}\int_{R^d} \frac{\tilde b \cdot \nabla\phi}{\phi+\varepsilon}\phi\,dx - \inf_{W \in C^\infty(R^d)} \Bigg[\int_{R^d} \{\tfrac{1}{2}\langle a\,\nabla W, \nabla W\rangle - \tilde b \cdot \nabla W\}\phi\,dx$$

$$+ \tfrac{1}{2}\varepsilon\int_{R^d} \frac{\langle a\,\nabla\phi, \nabla W\rangle}{\phi+\varepsilon}\,dx\Bigg].$$

Since $\phi \in C^\infty(R^d)$ with compact support, $\int_{R^d} \nabla \cdot a\,\nabla\phi\,dx = 0$, and therefore

$$\lim_{\varepsilon \to 0}\Bigg[-\frac{1}{4}\int_{R^d} \frac{\nabla \cdot a\,\nabla\phi}{\phi+\varepsilon}\phi\,dx + \frac{1}{8}\int_{R^d} \frac{\langle a\,\nabla\phi, \nabla\phi\rangle}{(\phi+\varepsilon)^2}\phi\,dx - \frac{1}{2}\int_{R^d} \frac{\tilde b \cdot \nabla\phi}{\phi+\varepsilon}\phi\,dx\Bigg]$$

$$(3.10) \quad = \frac{1}{8}\int_{R^d} \frac{\langle a\,\nabla\phi, \nabla\phi\rangle}{\phi}\,dx + \frac{1}{2}\int_{R^d} (\nabla \cdot \tilde b)\phi\,dx.$$

We shall show below that

$$\lim_{\varepsilon\to 0}\inf_{W\in C^\infty(R^d)}\left[\int_{R^d}\{\tfrac{1}{2}\langle a\,\nabla W,\nabla W\rangle-\tilde{b}\cdot\nabla W\}\phi\,dx+\tfrac{1}{2}\varepsilon\int_{R^d}\frac{\langle a\,\nabla\phi,\nabla W\rangle}{\phi+\varepsilon}\,dx\right]$$

(3.11)

$$=\inf_{W\in C^\infty(R^d)}\lim_{\varepsilon\to 0}\left[\int_{R^d}\{\tfrac{1}{2}\langle a\,\nabla W,\nabla W\rangle-\tilde{b}\cdot\nabla W\}\phi\,dx+\tfrac{1}{2}\varepsilon\int_{R^d}\frac{\langle a\,\nabla\phi,\nabla W\rangle}{\phi+\varepsilon}\,dx\right]$$

$$=\inf_{W\in C^\infty(R^d)}\int_{R^d}\{\tfrac{1}{2}\langle a\,\nabla W,\nabla W\rangle-\tilde{b}\cdot\nabla W\}\phi\,dx.$$

Thus, if we use (3.10) and (3.11) in (3.9), we get

$$I(\mu)=\frac{1}{8}\int_{R^d}\frac{\langle a\,\nabla\phi,\nabla\phi\rangle}{\phi}\,dx+\frac{1}{2}\int_{R^d}(\nabla\cdot\tilde{b})\phi\,dx$$

$$-\inf_{W\in C^\infty(R^d)}\int_{R^d}\{\tfrac{1}{2}\langle a\,\nabla W,\nabla W\rangle-\tilde{b}\cdot\nabla W\}\phi\,dx,$$

and by adding and subtracting the term $\dfrac{1}{2}\displaystyle\int_{R^d}\langle\tilde{b},a^{-1}\tilde{b}\rangle\phi\,dx$, we obtain

$$I(\mu)=\frac{1}{8}\int_{R^d}\frac{\langle a\,\nabla\phi,\nabla\phi\rangle}{\phi}\,dx+\frac{1}{2}\int_{R^d}[\nabla\cdot\tilde{b}+\langle\tilde{b},a^{-1}\tilde{b}\rangle]\phi\,dx$$

$$-\frac{1}{2}\inf_{W\in C^\infty(R^d)}\int_{R^d}(a^{-1}\tilde{b}-\nabla W)^T a(a^{-1}\tilde{b}-\nabla W)\phi\,dx$$

which is what the lemma states. It remains to prove (3.11). Let

$$H[W,\varepsilon]=\int_{R^d}\{\tfrac{1}{2}\langle a\,\nabla W,\nabla W\rangle-\tilde{b}\cdot\nabla W\}\phi\,dx+\tfrac{1}{2}\varepsilon\int_{R^d}\frac{\langle a\,\nabla\phi,\nabla W\rangle}{\phi+\varepsilon}\,dx,$$

and since always $\overline{\lim}_{\varepsilon\to 0}\inf_{W\in C^\infty(R^d)}H[W,\varepsilon]\leqq\inf_{W\in C^\infty(R^d)}\lim_{\varepsilon\to 0}H[W,\varepsilon]$, it suffices to show that

(3.12) $$\underline{\lim}_{\varepsilon\to 0}\inf_{W\in C^\infty(R^d)}H[W,\varepsilon]\geqq\inf_{W\in C^\infty(R^d)}\lim_{\varepsilon\to 0}H[W,\varepsilon].$$

We note that, for $W\equiv 0$, $H[W,\varepsilon]=0$ so that $\inf_{W\in C^\infty(R^d)}H[W,\varepsilon]\leqq 0$. Let $\mathscr{W}$ be the set of $W\in C^\infty(R^d)$ such that $H[W,\varepsilon]\leqq 0$. We show now that

(3.13) $$\lim_{\varepsilon\to 0}\tfrac{1}{2}\varepsilon\int_{R^d}\frac{\langle a\,\nabla\phi,\nabla W\rangle}{\phi+\varepsilon}\,dx=0$$

uniformly for $W \in \mathcal{W}$ which clearly implies (3.12). If we denote the vectors ∇W and $\tilde{b} - \varepsilon a \nabla\phi/(2\phi(\phi+\varepsilon))$ by ξ and η, respectively, then

$$H[W, \varepsilon] = \int_{R^d} \{\tfrac{1}{2}\langle a\xi, \xi\rangle - \langle \eta, \xi\rangle\}\phi\, dx$$

$$= \int_{R^d} \{\tfrac{1}{4}\langle a\xi, \xi\rangle + \tfrac{1}{4}\langle a\xi, \xi\rangle - \langle \eta, \xi\rangle + \langle \eta, a^{-1}\eta\rangle - \langle \eta, a^{-1}\eta\rangle\}\phi\, dx$$

$$= \int_{R^d} \{\tfrac{1}{4}\langle a\xi, \xi\rangle + \langle a(\tfrac{1}{2}\xi - a^{-1}\eta), (\tfrac{1}{2}\xi - a^{-1}\eta)\rangle - \langle \eta, a^{-1}\eta\rangle\}\phi\, dx$$

$$\geqq \int_{R^d} \{\tfrac{1}{4}\langle a\xi, \xi\rangle - \langle \eta, a^{-1}\eta\rangle\}\phi\, dx.$$

Thus, if $W \in \mathcal{W}$,

$$\int_{R^d} \tfrac{1}{4}\langle a\xi, \xi\rangle\phi\, dx \leqq \int_{R^d} \langle \eta, a^{-1}\eta\rangle\phi\, dx,$$

i.e.,

$$\int_{R^d} \langle a \nabla W, \nabla W\rangle\phi\, dx$$

(3.14)

$$\leqq 4\left[\int_{R^d} \langle \tilde{b}, a^{-1}\tilde{b}\rangle\phi\, dx - \varepsilon \int_{R^d} \frac{\langle \tilde{b}, \nabla\phi\rangle}{\phi+\varepsilon}\, dx + \tfrac{1}{4}\varepsilon^2 \int_{R^d} \frac{\langle a\nabla\phi, \nabla\phi\rangle}{\phi(\phi+\varepsilon)^2}\, dx\right].$$

The right side of (3.14) is bounded in ε, and hence there exists a constant C such that, for all $W \in \mathcal{W}$,

$$\int_{R^d} \langle a \nabla W, \nabla W\rangle\phi\, dx \leqq C. \tag{3.15}$$

From (3.15) and obvious inequalities we get

$$\tfrac{1}{2}\varepsilon \int_{R^d} \frac{\langle a\nabla\phi, \nabla W\rangle}{\phi+\varepsilon}\, dx \leqq \tfrac{1}{2}\varepsilon \int_{R^d} \frac{\langle a\nabla\phi, \nabla\phi\rangle^{1/2}\langle a\nabla W, \nabla W\rangle^{1/2}}{\sqrt{\phi}\,(\phi+\varepsilon)}\sqrt{\phi}\, dx$$

$$\leqq \tfrac{1}{2}\varepsilon \sqrt{\int_{R^d} \frac{\langle a\nabla\phi, \nabla\phi\rangle}{\phi(\phi+\varepsilon)^2}\, dx}\sqrt{\int_{R^d} \langle a\nabla W, \nabla W\rangle\phi\, dx}$$

$$\leqq \tfrac{1}{2}\varepsilon\sqrt{C}\sqrt{\int_{R^d} \frac{\langle a\nabla\phi, \nabla\phi\rangle}{\phi(\phi+\varepsilon)^2}\, dx}$$

$$= \tfrac{1}{2}\sqrt{C}\sqrt{\int_{R^d} \frac{\langle a\nabla\phi, \nabla\phi\rangle\varepsilon^2}{\phi(\phi+\varepsilon)^2}\, dx}.$$

But, as $\varepsilon \to 0$, $\varepsilon/(\phi+\varepsilon)\to 0$ and $\langle a\,\nabla\phi, \nabla\phi\rangle\varepsilon^2/\phi(\phi+\varepsilon)^2 \leqq \langle a\,\nabla\phi, \nabla\phi\rangle/\phi$ so that, by dominated convergence,

$$\lim_{\varepsilon\to 0} \tfrac{1}{2}\varepsilon \int_{R^d} \frac{\langle a\,\nabla\phi, \nabla W\rangle}{\phi+\varepsilon}\,dx = 0$$

uniformly for $W \in \mathscr{W}$ which is (3.13). This completes the proof of Lemma 3.3.

Now let L_1 and L_2 be two operators satisfying the hypotheses imposed on (3.1). We express L_1 and L_2 in divergence form:

$$L_1 = \tfrac{1}{2}\nabla \cdot a\nabla + \tilde{b}\cdot\nabla\,,$$

$$L_2 = \tfrac{1}{2}\nabla \cdot \alpha\nabla + \tilde{\beta}\cdot\nabla\,,$$

and denote their I-functions as defined by (3.3) with $I_1(\mu)$ and $I_2(\mu)$.

LEMMA 3.4. *If $I_1(\mu) = I_2(\mu)$ for all μ with compact support having a density $\phi \in C^\infty(R^d)$, then the matrices $a = (a_{ij}(x))$ and $\alpha = (\alpha_{ij}(x))$ are identical.*

Proof: From Lemma 3.3 we have for any such μ

$$\begin{aligned}I_1(\mu) = {}&\frac{1}{8}\int_{R^d}\frac{\langle a\,\nabla\phi, \nabla\phi\rangle}{\phi}\,dx + \frac{1}{2}\int_{R^d}[\nabla\cdot\tilde{b} + \langle\tilde{b}, a^{-1}\tilde{b}\rangle]\phi\,dx\\ &-\frac{1}{2}\inf_{W\in C^\infty(R^d)}\int_{R^d}(a^{-1}\tilde{b}-\nabla W)^T a(a^{-1}\tilde{b}-\nabla W)\phi\,dx\,,\end{aligned}\tag{3.16}$$

and

$$\begin{aligned}I_2(\mu) = {}&\frac{1}{8}\int_{R^d}\frac{\langle \alpha\,\nabla\phi, \nabla\phi\rangle}{\phi}\,dx + \frac{1}{2}\int_{R^d}[\nabla\cdot\tilde{\beta} + \langle\tilde{\beta}, \alpha^{-1}\tilde{\beta}\rangle]\phi\,dx\\ &-\frac{1}{2}\inf_{W\in C^\infty(R^d)}\int_{R^d}(\alpha^{-1}\tilde{\beta}-\nabla W)^T \alpha(\alpha^{-1}\tilde{\beta}-\nabla W)\phi\,dx\,.\end{aligned}\tag{3.17}$$

Let $\phi(x)$ be a probability density in $C^\infty(R^d)$ with compact support. For any $\lambda > 0$, $\phi_\lambda(x) = (1/\lambda^d)\phi(x/\lambda)$ is also a probability density in $C^\infty(R^d)$ with compact support. By hypothesis, $I_1(\phi_\lambda) = I_2(\phi_\lambda)$. Since

$$\frac{1}{8}\int_{R^d}\frac{\langle a(x)\,\nabla\phi_\lambda, \nabla\phi_\lambda\rangle}{\phi_\lambda}\,dx = \frac{1}{8\lambda^2}\int_{R^d}\frac{\langle a(\lambda x)\,\nabla\phi, \nabla\phi\rangle}{\phi}\,dx\,,$$

it is clear from (3.16) that

$$\lim_{\lambda\to 0}\lambda^2 I_1(\phi_\lambda) = \frac{1}{8}\int_{R^d}\frac{\langle a(0)\,\nabla\phi, \nabla\phi\rangle}{\phi}\,dx\,.$$

Similarly, from (3.17) we see that

$$\lim_{\lambda\to 0} \lambda^2 I_2(\phi_\lambda) = \frac{1}{8}\int_{R^d} \frac{\langle \alpha(0)\,\nabla\phi, \nabla\phi\rangle}{\phi}\,dx\,,$$

and hence that, for all ϕ,

$$\frac{1}{8}\int_{R^d} \frac{\langle a(0)\,\nabla\phi, \nabla\phi\rangle}{\phi}\,dx = \frac{1}{8}\int_{R^d} \frac{\langle \alpha(0)\,\nabla\phi, \nabla\phi\rangle}{\phi}\,dx\,.$$

This implies $a(0)=\alpha(0)$, i.e., $(a_{ij}(0))=(\alpha_{ij}(0))$. If we take $\phi_\lambda(x) = (1/\lambda^d)\phi((x-x_0)/\lambda)$, where x_0 is any point in R^d, the same argument leads to $(a_{ij}(x_0))=(\alpha_{ij}(x_0))$.

LEMMA 3.5. *If $I_1(\mu)=I_2(\mu)$ for all μ with compact support having a density $\phi\in C^\infty(R^d)$, then*

$$[\nabla\cdot\tilde{b}+\langle\tilde{b}, a^{-1}\tilde{b}\rangle](x) \equiv [\nabla\cdot\tilde{\beta}+\langle\tilde{\beta}, \alpha^{-1}\tilde{\beta}\rangle](x)\,. \tag{3.18}$$

Proof: From Lemmas 3.3 and 3.4, the hypothesis $I_1(\mu)=I_2(\mu)$ implies

$$\begin{aligned}\frac{1}{2}\int_{R^d}[\nabla\cdot\tilde{b}+\langle\tilde{b}, a^{-1}\tilde{b}\rangle]\phi\,dx - \frac{1}{2}\inf_{W\in C^\infty(R^d)}\int_{R^d}(a^{-1}\tilde{b}-\nabla W)^T a(a^{-1}\tilde{b}-\nabla W)\phi\,dx \\ = \frac{1}{2}\int_{R^d}[\nabla\cdot\tilde{\beta}+\langle\tilde{\beta}, \alpha^{-1}\tilde{\beta}\rangle]\phi\,dx \\ -\frac{1}{2}\inf_{W\in C^\infty(R^d)}\int_{R^d}(\alpha^{-1}\tilde{\beta}-\nabla W)^T\alpha(\alpha^{-1}\tilde{\beta}-\nabla W)\phi\,dx\,.\end{aligned} \tag{3.19}$$

If we take a sequence of probability densities $\phi_k\in C^\infty(R^d)$ having compact support and approaching δ_{x_0}, then (3.18) follows from (3.19) providing we can show

$$\lim_{k\to\infty}\frac{1}{2}\inf_{W\in C^\infty(R^d)}\int_{R^d}(a^{-1}\tilde{b}-\nabla W)^T a(a^{-1}\tilde{b}-\nabla W)\phi_k\,dx = 0\,. \tag{3.20}$$

Now $(a^{-1}\tilde{b})(x_0)$ is a vector θ, and if we take $W_0=\langle\theta, x\rangle$ for x near x_0 and extend it arbitrarily but so that $W_0\in C^\infty(R^d)$, then clearly $(a^{-1}\tilde{b}-\nabla W_0)(x_0)=0$. Thus

$$\lim_{k\to\infty}\int_{R^d}(a^{-1}\tilde{b}-\nabla W_0)a(a^{-1}\tilde{b}-\nabla W_0)\phi_k(x)\,dx = 0$$

which implies (3.20) and the lemma.

LEMMA 3.6. *If $I_1(\mu)=I_2(\mu)$ for all μ with compact support having a density $\phi\in C^\infty(R^d)$, then*

$$\left|\int_C \langle a^{-1}\tilde{b}, dx\rangle\right|^2=\left|\int_C \langle \alpha^{-1}\tilde{\beta}, dx\rangle\right|^2 \tag{3.21}$$

for any smooth simple closed contour C in R^d.

Remark. The lemma is vacuous if $d=1$. We assume that $d\geqq 2$.

Proof: Let

$$K_1(\phi)=\frac{1}{2}\inf_{W\in C^\infty(R^d)}\int_{R^d}(a^{-1}\tilde{b}-\nabla W)^T a(a^{-1}\tilde{b}-\nabla W)\phi\,dx$$

and

$$K_2(\phi)=\frac{1}{2}\inf_{W\in C^\infty(R^d)}\int_{R^d}(\alpha^{-1}\tilde{\beta}-\nabla W)^T \alpha(\alpha^{-1}\tilde{\beta}-\nabla W)\phi\,dx\,.$$

From Lemmas 3.3, 3.4, and 3.5 we see that the hypothesis $I_1(\mu)=I_2(\mu)$ implies $K_1(\phi)=K_2(\phi)$ for all probability densities $\phi\in C^\infty(R^d)$ with compact support.

Let C be a smooth simple closed contour in R^d. Consider a tubular neighborhood of C and foliate it as $D\times S^1$, where D is an open neighborhood of 0 in R^{d-1} and S^1 is the circle. For each $y\in D$ we have a simple closed contour C_y in the tubular neighborhood of C with C itself being $\{0\}\times S^1$. For any $y\in D$, consider the non-negative measure $\bar{\sigma}_y(dx)$ equal to the arc length on C_y in the metric $a^{ij}\,dx_i\,dx_j$, where $(a^{ij}(x))$ is the matrix inverse to $(a_{ij}(x))$. Let $l(y)$ be the total length of C_y in this metric, i.e., the total mass of $\bar{\sigma}_y(dx)$. In particular, $l(0)$ is the length of C in this metric. Since $I_1(\mu)=I_2(\mu)$ implies $(a_{ij}(x))=(\alpha_{ij}(x))$ for all x (Lemma 3.4), we note that $l(y)$ is the same whether based on the metric coming from a or α. We define $\sigma_y(dx)=\bar{\sigma}_y(dx)/l(y)$. Let $\psi_\varepsilon(y)$ be a family of smooth probability densities supported in D and such that $\psi_\varepsilon(y)\to\delta_0$ as $\varepsilon\to 0$. We define a probability measure μ_ε on R^d by setting, for $A\subset R^d$,

$$\mu_\varepsilon(A)=\int_D\frac{\bar{\sigma}_y(A)}{l(y)}\psi_\varepsilon(y)\,dy\,. \tag{3.22}$$

We note that μ_ε has compact support and has a density $\phi_\varepsilon\in C^\infty(R^d)$. We shall show that

$$\lim_{\varepsilon\to 0}K_1(\phi_\varepsilon)=\frac{1}{2}\frac{\left|\int_C\langle a^{-1}\tilde{b}, dx\rangle\right|^2}{[l(0)]^2}\,; \tag{3.23}$$

the same proof shows that

$$\lim_{\varepsilon\to 0} K_2(\phi_\varepsilon)=\frac{1}{2}\frac{\left|\int_C \langle \alpha^{-1}\tilde{\beta}, dx\rangle\right|^2}{[l(0)]^2}.$$

Since $K_1(\phi_\varepsilon)=K_2(\phi_\varepsilon)$ for all $\varepsilon>0$ and, as noted above, $l(0)$ is the same whether based on a or α, we see that (3.21) follows. So we must show that (3.23) is true.

Fix an orientation on the contours C_y and let $T(x)$ be the oriented tangent vector at x to the contour through x normalized so that $\langle a^{-1}(x)T(x), T(x)\rangle = 1$. Consider this vector field $T(x)$. Choose a constant λ so that

$$\int_C \langle a^{-1}\tilde{b}-\lambda a^{-1}T(x), dx\rangle = 0\,, \tag{3.24}$$

i.e.,

$$\lambda=\frac{\int_C \langle a^{-1}\tilde{b}, dx\rangle}{\int_C \langle a^{-1}T(x), dx\rangle}=\frac{\int_C \langle a^{-1}\tilde{b}, dx\rangle}{l(0)}. \tag{3.25}$$

The last equality in (3.25) follows from the normalization of $T(x)$ because $dx = T(x)\,d\sigma$ and hence

$$\int_C \langle a^{-1}T(x), dx\rangle = \int_C \langle a^{-1}T(x), T(x)\rangle\, d\sigma$$
$$=\int_C d\sigma = l(0)\,.$$

Now

$$\begin{aligned}&\left|\sqrt{\frac{1}{2}\inf_{W\in C^\infty(R^d)}\int_{R^d}(a^{-1}\tilde{b}-\nabla W)^T a(a^{-1}\tilde{b}-\nabla W)\phi\,dx}\right.\\ &\left.\quad-\sqrt{\frac{1}{2}\inf_{W\in C^\infty(R^d)}\int_{R^d}(\lambda a^{-1}T(x)-\nabla W)^T a(\lambda a^{-1}T(x)-\nabla W)\phi\,dx}\right|\\ &\leqq\sqrt{\frac{1}{2}\inf_{W\in C^\infty(R^d)}\int_{R^d}[(a^{-1}\tilde{b}-\lambda a^{-1}T(x))-\nabla W]^T a[(a^{-1}\tilde{b}-\lambda a^{-1}T(x))-\nabla W]\phi\,dx}\,.\end{aligned} \tag{3.26}$$

To verify inequality (3.26) notice that $f(y)=\sqrt{\int_{R^d} y^T a y\phi\,dx}$ is a norm; if we denote by M the linear manifold of functions on R^d which are gradients, then $g(y)=\inf_{z\in M} f(y-z)$ is again a norm and (3.26) is just the triangle inequality for this norm.

In (3.26) replace ϕ by the ϕ_ε defined just after (3.22). In Lemma 3.7 below we shall show that if $f=(f_1, f_2, \cdots, f_d)$ is a vector of smooth functions defined on C, then a necessary and sufficient condition for the existence of a function W on R^d such that $\nabla W=f$ on C is that $\int_C \langle f, dx\rangle=0$. Now we chose λ so that $\int_C \langle a^{-1}\tilde{b}-\lambda a^{-1}T(x), dx\rangle=0$ and hence, by Lemma 3.7, there exists W_0 on R^d such that, on C, $\nabla W_0=a^{-1}\tilde{b}-\lambda a^{-1}T(x)$. Moreover, with ϕ replaced by ϕ_ε, the integral over R^d on the right of (3.26) becomes the integral over C as $\varepsilon\to 0$. We conclude that

$$\overline{\lim_{\varepsilon\to 0}}\left|\sqrt{\frac{1}{2}\inf_{W\in C^\infty(R^d)}\int_{R^d}(a^{-1}\tilde{b}-\nabla W)^T a(a^{-1}\tilde{b}-\nabla W)\phi_\varepsilon\,dx}\right.$$
$$\left.-\sqrt{\frac{1}{2}\inf_{W\in C^\infty(R^d)}\int_{R^d}(\lambda a^{-1}T(x)-\nabla W)^T a(\lambda a^{-1}T(x)-\nabla W)\phi_\varepsilon\,dx}\right|=0. \tag{3.27}$$

Each radical is bounded because the integral under each radical is bounded for $W=0$. Thus we can remove the square roots and obtain

$$\overline{\lim_{\varepsilon\to 0}}\left|K_1(\phi_\varepsilon)-\frac{1}{2}\inf_{W\in C^\infty(R^d)}\int_{R^d}(\lambda a^{-1}T(x)-\nabla W)^T a(\lambda a^{-1}T(x)-\nabla W)\phi_\varepsilon\,dx\right|=0. \tag{3.28}$$

We show now that, for each $\varepsilon>0$,

$$\frac{1}{2}\inf_{W\in C^\infty(R^d)}\int_{R^d}(\lambda a^{-1}T(x)-\nabla W)^T a(\lambda a^{-1}T(x)-\nabla W)\phi_\varepsilon\,dx=\tfrac{1}{2}\lambda^2 = \frac{\left|\int_C\langle a^{-1}\tilde{b}, dx\rangle\right|^2}{2l^2(0)}, \tag{3.29}$$

which, because of (3.28), proves (3.23).

To show (3.29) consider first of all

$$\int_{R^d}\langle T(x), \nabla W\rangle\phi_\varepsilon\,dx.$$

We note that

$$\int_{R^d} \langle T(x), \nabla W\rangle \phi_\varepsilon \, dx = \int_D \int_{R^d} \langle T(x), \nabla W\rangle \frac{\bar{\sigma}_y(dx)}{l(y)} \psi_\varepsilon(y)\, dy$$

$$= \int_D \frac{1}{l(y)} \left[\int_{R^d} \langle T(x), \nabla W\rangle \bar{\sigma}_y(dx) \right] \psi_\varepsilon(y)\, dy$$

$$= \int_D \frac{1}{l(y)} \left[\int_{C_y} \nabla W \, dx \right] \psi_\varepsilon(y)\, dy,$$

because $T(x)\bar{\sigma}_y(dx)$ is the element dx on C_y. But, for any W, $\int_{C_y} \nabla W\, dx = 0$ and hence

$$\int_{R^d} \langle T(x), \nabla W\rangle \phi_\varepsilon \, dx = 0 . \tag{3.30}$$

Returning to the left side of (3.29) we have then

$$\frac{1}{2} \inf_{W \in C^\infty(R^d)} \int_{R^d} (\lambda a^{-1} T(x) - \nabla W) a (\lambda a^{-1} T(x) - \nabla W) \phi_\varepsilon \, dx$$

$$= \tfrac{1}{2}\lambda^2 \inf_{W \in C^\infty(R^d)} \int_{R^d} (a^{-1} T(x) - \nabla W) a (a^{-1} T(x) - \nabla W) \phi_\varepsilon \, dx$$

$$= \tfrac{1}{2}\lambda^2 \inf_{W \in C^\infty(R^d)} \left[\int_{R^d} \langle a^{-1} T(x), T(x)\rangle \phi_\varepsilon \, dx \right.$$

$$\left. + \int_{R^d} \langle a \nabla W, \nabla W\rangle \phi_\varepsilon \, dx - 2 \int_{R^d} \langle T(x), \nabla W\rangle \phi_\varepsilon \, dx \right]$$

$$= \tfrac{1}{2}\lambda^2 \inf_{W \in C^\infty(R^d)} \left[\int_{R^d} \langle a^{-1} T(x), T(x)\rangle \phi_\varepsilon \, dx + \int_{R^d} \langle a \nabla W, \nabla W\rangle \phi_\varepsilon \, dx \right].$$

But both terms inside the brackets are non-negative for each $\varepsilon > 0$ so the infimum is taken on when $W = 0$. Thus, from the normalization of $T(x)$,

$$\frac{1}{2} \inf_{W \in C^\infty(R^d)} \int_{R^d} (\lambda a^{-1} T(x) - \nabla W) a (\lambda a^{-1} T(x) - \nabla W) \phi_\varepsilon \, dx$$

$$= \tfrac{1}{2}\lambda^2 \int_{R^d} \langle a^{-1} T(x), T(x)\rangle \phi_\varepsilon \, dx = \tfrac{1}{2}\lambda^2 ,$$

for each $\varepsilon > 0$. This proves (3.29) and the lemma except for

LEMMA 3.7. *Let $f = (f_1, f_2, \cdots, f_d)$ be a vector of smooth functions defined on C. In order that there exists a function W on R^d such that $\nabla W = f$ on C, it is necessary and sufficient that*

$$\int_C \langle f, dx \rangle = 0 .$$

Proof: The necessity is obvious and for the sufficiency it is clearly enough to construct W in a tubular neighborhood of C. We can change the coordinates in such a way that the tubular neighborhood is $G \times S^1$, where G is an open subset of R^{d-1} containing the origin and S^1 is the circle. The contour C itself is $\{0\} \times S^1$. Then, our hypothesis becomes

$$\int_{S^1} f_d(0, 0, 0, \cdots, \theta) \, d\theta = 0 .$$

There exists a function $F(\theta)$ such that $dF(\theta)/d\theta = f_d(0, 0, 0, \cdots, \theta)$ and so if we define $W(y, \theta)$ for $y \in G$ and $\theta \in S^1$ by

$$W(y, \theta) = F(\theta) + \sum_{j=1}^{d-1} y_j f_j(0, 0, 0, \cdots, \theta) ,$$

then $W(y, \theta)$ satisfies $\nabla W = f$ on C.

LEMMA 3.8. *Let f and g be two smooth R^d-valued functions on R^d such that*

$$\left| \int_C \langle f(x), dx \rangle \right|^2 = \left| \int_C \langle g(x), dx \rangle \right|^2$$

for all simple smooth closed contours C in R^d. Then either

$$\int_C \langle f(x), dx \rangle = \int_C \langle g(x), dx \rangle$$

for all closed rectifiable contours C in R^d or

$$\int_C \langle f(x), dx \rangle = -\int_C \langle g(x), dx \rangle$$

for all closed rectifiable contours C in R^d.

Remark. Again we can assume that $d \geqq 2$. The case $d = 2$ is the difficult one. The following proof covers all cases $d \geqq 2$.

Proof: First we show that, for any two nonintersecting simple smooth closed contours C_1 and C_2,

$$\int_{C_1} \langle f(x), dx\rangle \int_{C_2} \langle f(x), dx\rangle = \int_{C_1} \langle g(x), dx\rangle \int_{C_2} \langle g(x), dx\rangle . \tag{3.31}$$

To see this, let C_1 and C_2 be two such contours and connect them by an arc γ which has no additional intersections with C_1 and C_2. Then, with a suitable orientation of C_1 and C_2, the closed contour $C = C_1 \gamma C_2 \gamma^{-1}$ can be approximated by simple smooth closed contours. Therefore, from our hypothesis,

$$\left| \int_C \langle f(x), dx\rangle \right|^2 = \left| \int_C \langle g(x), dx\rangle \right|^2 ,$$

i.e.,

$$\left| \int_{C_1} \langle f(x), dx\rangle + \int_{C_2} \langle f(x), dx\rangle \right|^2 = \left| \int_{C_1} \langle g(x), dx\rangle + \int_{C_2} \langle g(x), dx\rangle \right|^2$$

which again using the hypothesis implies (3.31).

Now let C be any closed rectifiable contour in R^d. We can approximate C by contours C_n each of which is a finite union of mutually disjoint smooth simple closed contours. Therefore, from (3.31) and the hypothesis, it follows that

$$\left| \int_C \langle f(x), dx\rangle \right|^2 = \left| \int_C \langle g(x), dx\rangle \right|^2 \tag{3.32}$$

for all closed rectifiable contours C in R^d. If we now let C_1 and C_2 be any two closed rectifiable contours and let γ be an arc connecting them, then $C = C_1 \gamma C_2 \gamma^{-1}$ is again a closed rectifiable contour and from (3.32) we have

$$\int_{C_1} \langle f(x), dx\rangle \int_{C_2} \langle f(x), dx\rangle = \int_{C_1} \langle g(x), dx\rangle \int_{C_2} \langle g(x), dx\rangle .$$

This last equation clearly implies the lemma.

We are now ready for the

Proof of Theorem 3.2: We have two operators L_1 and L_2 satisfying the hypotheses imposed on (3.1) and for which $I_1(\mu) = I_2(\mu)$ for all μ with compact support having a density $\phi \in C^\infty(R^d)$. If we write L_1 and L_2 in divergence form:

$$L_1 = \tfrac{1}{2}\nabla \cdot a\,\nabla + \tilde{b} \cdot \nabla ,$$

$$L_2 = \tfrac{1}{2}\nabla \cdot \alpha\,\nabla + \tilde{\beta} \cdot \nabla ,$$

then the preceding lemmas showed that the hypothesis $I_1(\mu)=I_2(\mu)$ implies

(3.33) $$a(x)=(a_{ij}(x))=(\alpha_{ij}(x))=\alpha(x) \quad \text{for all} \quad x\in R^d,$$

(3.34) $$\nabla\cdot\tilde{b}+\tilde{b}^T a^{-1}\tilde{b}=\nabla\cdot\tilde{\beta}+\tilde{\beta}^T\alpha^{-1}\tilde{\beta} \quad \text{for all} \quad x\in R^d,$$

and either

(3.35a) $$\int_C \langle a^{-1}\tilde{b}, dx\rangle=\int_C\langle a^{-1}\tilde{\beta}, dx\rangle \text{ for all closed rectifiable contours } C \text{ in } R^d,$$

or

(3.35b) $$\int_C \langle a^{-1}\tilde{b}, dx\rangle=-\int_C\langle a^{-1}\tilde{\beta}, dx\rangle \text{ for all closed rectifiable contours } C \text{ in } R^d.$$

If $d=1$, (3.35) is a vacuous statement and we work only with (3.33) and (3.34). Equations (3.35) are used only to show that $a^{-1}(\tilde{b}-\tilde{\beta})$ is a gradient of a function. In case $d=1$ this of course is always true. Another feature of the one-dimensional situation is that the two alternatives $L_1f=L_2(uf)/u$ and $L_1f=L_2^*(\rho f)/\rho$ lead to the same class of operators L_1. This is because any L is selfadjoint with respect to some measure on the line.

Suppose the first alternative in (3.35) holds so that

$$\int_C\langle a^{-1}(\tilde{b}-\tilde{\beta}), dx\rangle=0$$

for all closed rectifiable contours C in R^d. This implies that there exists a function V on R^d such that $a^{-1}(\tilde{b}-\tilde{\beta})=\nabla V$ or $\tilde{b}-\tilde{\beta}=a\,\nabla V$. Thus from (3.33) and (3.34) we get

$$\begin{aligned}\nabla\cdot\tilde{\beta}+\tilde{\beta}^T\alpha^{-1}\tilde{\beta}&=\nabla\cdot\tilde{\beta}+\tilde{\beta}^T a^{-1}\tilde{\beta}\\&=\nabla\cdot\tilde{b}+\tilde{b}^T a^{-1}\tilde{b}\\&=\nabla\cdot(\tilde{\beta}+a\,\nabla V)+(\tilde{\beta}+a\,\nabla V)^T a^{-1}(\tilde{\beta}+a\,\nabla V)\\&=\nabla\cdot\tilde{\beta}+\nabla\cdot a\,\nabla V+\tilde{\beta}^T a^{-1}\tilde{\beta}+(a\,\nabla V,\nabla V)+2(\tilde{\beta},\nabla V),\end{aligned}$$

from which we conclude that

(3.36) $$\nabla\cdot a\,\nabla V+(a\,\nabla V,\nabla V)+2(\tilde{\beta},\nabla V)=0.$$

Let $V=\log H$, where $H>0$; then we get from (3.36)

$$\nabla\cdot a\frac{\nabla H}{H}+\left(a\frac{\nabla H}{H},\frac{\nabla H}{H}\right)+2\left(\tilde{\beta},\frac{\nabla H}{H}\right)=0$$

or

$$\frac{\nabla\cdot a\,\nabla H}{H}-\frac{(a\,\nabla H,\nabla H)}{H^2}+\frac{(a\,\nabla H,\nabla H)}{H^2}+\frac{2(\tilde{\beta},\nabla H)}{H}=0\,.$$

Since $H>0$, this last equation gives us

$$\tfrac{1}{2}\nabla\cdot a\,\nabla H+\tilde{\beta}\cdot\nabla H=\tfrac{1}{2}\nabla\cdot \alpha\,\nabla H+\tilde{\beta}\cdot\nabla H=0\,,$$

which says that $L_2H=0$. Thus H is a positive harmonic function for L_2. We noted earlier that choosing the first alternative in (3.35) implies $\tilde{b}-\tilde{\beta}=a\,\nabla H/H$. If we use this fact together with (3.33), (3.34), and $L_2H=0$, an easy calculation shows that $L_1f=L_2(Hf)/H$.

If the second alternative in (3.35) holds, then

$$\int_C\langle a^{-1}(\tilde{b}+\tilde{\beta}),dx\rangle=0$$

for all closed rectifiable contours C in R^d. Thus there exists a function V on R^d such that $\tilde{b}+\tilde{\beta}=a\,\nabla V$, and again using (3.33) and (3.34) we have

$$\begin{aligned}\nabla\cdot\tilde{\beta}+\tilde{\beta}^T\alpha^{-1}\tilde{\beta}&=\nabla\cdot\tilde{b}+\tilde{b}^Ta^{-1}\tilde{b}\\&=\nabla\cdot(-\tilde{\beta}+a\,\nabla V)+(-\tilde{\beta}+a\,\nabla V)^Ta^{-1}(-\tilde{\beta}+a\,\nabla V)\\&=-\nabla\tilde{\beta}+\nabla\cdot a\,\nabla V+\tilde{\beta}^Ta^{-1}\tilde{\beta}+(a\,\nabla V,\nabla V)-2(\tilde{\beta},\nabla V)\,.\end{aligned}$$

Hence

$$\nabla\cdot a\,\nabla V+(a\,\nabla V,\nabla V)-2(\tilde{\beta},\nabla V)-2\nabla\cdot\tilde{\beta}=0\,. \tag{3.37}$$

Let $V=\log\rho$, where $\rho>0$; then from (3.37) we get

$$\nabla\cdot a\frac{\nabla\rho}{\rho}+\left(a\frac{\nabla\rho}{\rho},\frac{\nabla\rho}{\rho}\right)-2\left(\tilde{\beta},\frac{\nabla\rho}{\rho}\right)-2\nabla\cdot\tilde{\beta}=0$$

or

$$\frac{\nabla\cdot a\,\nabla\rho}{\rho}-\frac{(a\,\nabla\rho,\nabla\rho)}{\rho^2}+\frac{(a\,\nabla\rho,\nabla\rho)}{\rho^2}-2\frac{(\tilde{\beta},\nabla\rho)}{\rho}-2\nabla\cdot\tilde{\beta}=0\,.$$

Since $\rho>0$, this last equation implies

$$\tfrac{1}{2}\nabla\cdot a\,\nabla\rho-(\tilde{\beta},\nabla\rho)-(\nabla\cdot\tilde{\beta})\rho=0\,,$$

i.e., $L_2^* \rho = 0$. Thus there exists a positive invariant density ρ for L_2. We saw above that the second alternative in (3.35) implies $\tilde{b} + \tilde{\beta} = a\, \nabla\rho/\rho$. If we use this fact together with (3.33), (3.34), and $L_2^* \rho = 0$, an elementary calculation shows that $L_1 f = L_2^*(\rho f)/\rho$. This completes the proof of Theorem 3.2.

Bibliography

[1] Donsker, M. D., and Varadhan, S. R. S., *On a variational formula for the principal eigenvalue for operators with maximum principle*, Proc. Nat. Acad. Sci. U.S.A., Vol. 72, No. 3, March 1975, pp. 780–783.

[2] Sion, M., *On general minimax theorem*, Pac. J. Math., Vol. 8, 1958, pp. 171–176.

Received May, 1976.

COMMUNICATIONS ON PURE AND APPLIED MATHEMATICS, VOL. XXX, 707–753 (1977)

On Laws of the Iterated Logarithm for Local Times*

M. D. DONSKER AND S. R. S. VARADHAN

1. Introduction

Let $\{\beta(s),\ s \geqq 0\}$ be the standard Brownian motion in one dimension. There are various types of laws of the iterated logarithm that are known in this case. For example, (see Chung [1])

$$\overline{\lim_{t\to\infty}} \frac{1}{(2t \log\log t)^{1/2}} \sup_{0\leqq s\leqq t} |\beta(s)| = 1 \quad \text{a.e.}, \tag{1.1}$$

and (see Chung [1], Jain and Pruitt [2])

$$\underline{\lim_{t\to\infty}} \left(\frac{\log\log t}{t}\right)^{1/2} \sup_{0\leqq s\leqq t} |\beta(s)| = \frac{\pi}{\sqrt{8}} \quad \text{a.e.} \tag{1.2}$$

These are really two different types of laws of the iterated logarithm. The typical Brownian path wanders a distance $\sqrt{t}$ in the time interval $[0, t]$. The first law, (1.1), concerns the possibility of the path wandering farther than $\sqrt{t}$ whereas the second law, (1.2), is concerned with the possibility of wandering much less.

Laws of the first type, of which (1.1) is only a special case, are all essentially contained in Strassen's version of the law of the iterated logarithm for Brownian motion. For each $t>0$, we define

$$\beta_t(s) = \frac{1}{(t \log\log t)^{1/2}} \beta(st) \quad \text{for} \quad 0 \leqq s \leqq 1,$$

so that $\beta_t(s)$ is again a Brownian motion but with a variance at time s equal to $s/\log\log t$). As $t\to\infty$, although $\beta_t(\cdot)$ is nearly zero for most ω, the family $\beta_t(\cdot)$ is compact and has as limit points the set C of continuous trajectories

---

* The research for this paper was done at the Courant Institute of Mathematical Sciences and supported by the National Science Foundation under NSF Grant No. MPS 74-01921. Reproduction in whole or in part is permitted for any purpose of the United States Government.

$f(\cdot)$ such that $\frac{1}{2}\int_0^1 [f'(s)]^2\, ds \leqq 1$. In particular, this implies that, for any functional $\Phi(f)$ continuous on $C[0, 1]$,

$$\overline{\lim_{t\to\infty}}\, \Phi(\beta_t(\cdot)) = \sup_{f\in C} \Phi(f) \quad \text{a.e.} \tag{1.3}$$

This well-known result of Strassen [3] contains, of course, (1.1). In this paper we obtain a theorem like (1.3) but which covers the other type of law of the iterated logarithm of which (1.2) is a special case. Before stating this theorem we give another special case of it which is perhaps more revealing of the second type of law of the iterated logarithm. Let $\xi(t, x)$ be the local time at x of $\beta(\cdot)$ during $[0, t]$, i.e., $\xi(t, x) = \int_0^t \delta(\beta(\sigma) - x)\, d\sigma$. Then, Kesten [4] proved

$$\overline{\lim_{t\to\infty}}\, \frac{\xi(t)}{(2t \log\log t)^{1/2}} = 1 \quad \text{a.e.}, \tag{1.4}$$

where $\xi(t) = \sup_x \xi(t, x)$. The reason (1.4) should be considered as a law of the second type is that in order for the local time to build up to more than the normal $\sqrt{t}$, the path itself must remain around zero much more than usual.

In order to obtain a Strassen type law encompassing (1.2), (1.4) and other examples of the second type, we look at

$$\beta_t(s) = \left(\frac{\log\log t}{t}\right)^{1/2} \beta(st) \quad \text{for} \quad 0 \leqq s \leqq 1\,.$$

Now $\beta_t(s)$ is a Brownian motion on $[0, 1]$ with variance $s \log\log t$ and should dissipate as $t \to \infty$. To make it precise, we look at the occupation distribution

$$L_t(\omega, A) = \int_0^1 \chi_A(\beta_t(s))\, ds\,,$$

or its density $\hat{L}_t(\omega, y)$ determined by

$$L_t(\omega, A) = \int_A \hat{L}_t(\omega, y)\, dy\,.$$

One can view $\hat{L}_t$ as an element of all continuous subprobability densities on R_1 with the topology of uniform convergence on bounded intervals. One can show that, for almost all ω, $\{\hat{L}_t\}$ is compact and the set of limit points of $\hat{L}_t$

can be characterized as the set C of subprobability densities f such that

$$I(f) = \tfrac{1}{8} \int_{R_1} \frac{[f'(y)]^2}{f(y)}\, dy \leqq 1 .$$

It then follows that, for suitable functionals $\Phi(\cdot)$,

$$\varlimsup_{t \to \infty} \Phi(\hat{L}_t(\omega, \cdot)) = \sup_{f \in C} \Phi(f) \quad \text{a.e.} \tag{1.5}$$

By taking $\Phi(f) = f(0)$ we obtain (1.4) (cf. Corollary 3.12 below), and by taking

$$\Phi(f) = \inf \left\{ l : \int_{-l}^{l} f(y)\, dy = 1 \right\}$$

we obtain (1.2). The I-function entering into (1.5), i.e.,

$$I(f) = \tfrac{1}{8} \int_{R_1} \frac{[f'(y)]^2}{f(y)}\, dy$$

is the special case for Brownian motion of the I-function introduced by the authors in a recent series of papers, [5]–[8], on asymptotics for Markov processes. In proving (1.5) essential use is made of these asymptotics.

Another feature of the second type of law of the iterated logarithm is that there is an analogous result for symmetric stable processes as well. If the stable law index α is larger than one, then only the normalization and the I-function have to be changed appropriately in (1.5). If $\alpha \leqq 1$, then, of course, local times do not exist and we have to work with L_t instead of $\hat{L}_t$. In this latter case we view L_t as an element of the space of subprobability measures on R_1, and the topology will be the vague topology. One then obtains the exact analogue of (1.5).

In Section 2, we treat this latter case, i.e., symmetric stable processes of index α, $0 < \alpha \leqq 2$. In Section 3, we prove (1.5) in its more general form, i.e., not just for Brownian motion ($\alpha = 2$), but for any symmetric stable process with index $\alpha > 1$. In Section 4, we present a few examples to illustrate the general theorems.

2. The Weak Topology

Let $\{x(s),\ 0 \leqq s < \infty\}$ with $x(0) = x$ be a symmetric stable process of index α, i.e., the process has stationary independent increments and

$$E_x\{\exp\{i\xi[x(t+h) - x(h)]\}\} = \exp\{-t\,|\xi|^{\alpha}\},$$

where $0<\alpha\leqq 2$. We first prove

LEMMA 2.1. *For a fixed interval* $[a, b]$, *let*

$$f_\alpha^T(x)=P_x\{x(t)\in[a,b] \text{ for some } t, 0\leqq t\leqq T\}. \tag{2.1}$$

Then,

$$\int_{-\infty}^{\infty} f_\alpha^T(x)\,dx\leqq \gamma_1(T+1), \tag{2.2}$$

where γ_1 *depends only on* a, b, *and* α. *Furthermore, for any* $d>0$ *and* $0\leqq c\leqq d$,

$$\sum_{n=-\infty}^{\infty} f_\alpha^T(c+nd)\leqq \gamma_2(T+1), \tag{2.3}$$

where γ_2 *depends only on* a, b, d, *and* α.

Proof: Let

$$g_\alpha^T(x)=E_x\left\{\int_0^{T+1}\chi_{[a-1,b+1]}(x(s))\,ds\right\}$$

and let $\tau=\inf[t : x(t)\in[a,b]]$. Then, clearly,

$$g_\alpha^T(x)\geqq E_x\left\{\int_\tau^{\tau+1}\chi_{[x(\tau)-1,x(\tau)+1]}(x(s))\,ds;\quad \tau\leqq T\right\}. \tag{2.4}$$

By the strong Markov property,

$$\begin{aligned} &E_x\left\{\int_\tau^{\tau+1}\chi_{[x(\tau)-1,x(\tau)+1]}(x(s))\,ds;\tau\leqq T\right\}\\ &\qquad=E_x\left\{E_{x(\tau)}\left\{\int_0^1\chi_{[x(\tau)-1,x(\tau)+1]}(x(s))\,ds\right\};\tau\leqq T\right\}, \end{aligned} \tag{2.5}$$

but, since the process has independent increments,

$$\begin{aligned} E_{x(\tau)}\left\{\int_0^1\chi_{(x(\tau)-1,x(\tau)+1]}(x(s))\,ds\right\}&=E_0\left\{\int_0^1\chi_{[-1,1]}(x(s))\,ds\right\}\\ &=p>0, \end{aligned} \tag{2.6}$$

where p depends only on α. Thus, from (2.4)–(2.6) we see that

$$g_\alpha^T(x) \geqq pP_x\{\tau \leqq T\} = pf_\alpha^T(x). \tag{2.7}$$

Now

$$\begin{aligned} g_\alpha^T(x) &= E_x\left\{\int_0^{T+1} \chi_{[a-1,b+1]}(x(s))\,ds\right\} \\ &= E_0\left\{\int_0^{T+1} \chi_{[a-1-x,b+1-x]}(x(s))\,ds\right\}, \end{aligned}$$

and

$$\int_{-\infty}^{\infty} \chi_{[a-1-x,b+1-x]}(y)\,dx = b-a+2,$$

so that

$$\int_{-\infty}^{\infty} g_\alpha^T(x)\,dx = (b-a+2)(T+1). \tag{2.8}$$

Thus, from (2.7) and (2.8) with $\gamma_1 = (b-a+2)/p$, we see that

$$\int_{-\infty}^{\infty} f_\alpha^T(x)\,dx \leqq \gamma_1(T+1).$$

Similarly,

$$\sum_{n=-\infty}^{\infty} \chi_{[a-1-nd-c,b+1-nd-c]} \leqq \frac{b-a+2}{d}+1,$$

and therefore

$$\sum_{n=-\infty}^{\infty} g_\alpha^T(c+nd) \leqq \left(\frac{b-a+2}{d}\right)(T+1). \tag{2.9}$$

Hence, (2.7) and (2.9) with $\gamma_2 = (1/p)((b-a+2)/d+1)$ imply

$$\sum_{n=-\infty}^{\infty} f_\alpha^T(c+nd) \leqq \gamma_2(T+1).$$

Corresponding to the symmetric stable process under consideration, there is an infinitesimal generator $L = L_\alpha$. If we let U denote the space of functions $u \in C^\infty(R_1)$ which are strictly positive and are equal to some positive constant for large $|x|$, then U is contained in the domain of L.

Let M be the space of measures λ on R_1 such that $\lambda(R_1) \leqq 1$. The space M will be provided with the vague topology, i.e., the topology induced by defining $\lambda_n \to \lambda$ if $\int f\lambda_n(dx) \to \int f\lambda(dx)$ for all f continuous on R_1 with $f \to 0$ as $|x| \to \infty$. In the vague topology M is compact.

If L is the infinitesimal generator of the stable process, then it is easy to show that

$$I(\lambda) = -\inf_{u \in U} \int_{R_1} \left(\frac{Lu}{u}\right)(x)\lambda(dx) = -\inf_{u \in \mathcal{D}^+} \int_{R_1} \left(\frac{Lu}{u}\right)(x)\lambda(dx), \tag{2.10}$$

where $\mathcal{D}^+$ consists of those functions in the domain of L which have a positive lower bound. Moreover, $I(\lambda)$ is a lower semicontinuous, convex functional on M which is homogeneous of degree 1.

For $t > 0$, E a Borel set in R_1, and any sample path $\omega = x(\cdot)$, we define

$$L_t(\omega, E) = \frac{1}{t} \int_0^t \chi_E(x(s))\, ds,$$

i.e., the proportion of time up to t that the particular path $\omega = x(\cdot)$ spends in the set E. For each fixed t and ω, $L_t(\omega, \cdot)$ is a probability measure on R_1 so that, for fixed t and x, $L_t(\omega, \cdot)$ maps the sample paths starting at x into M. For each $t > 0$, and $x \in R_1$, define for sets $A \subset M$

$$Q_{t,x}(A) = P_x\{L_t(\omega, \cdot) \in A\}.$$

If we compactify R_1 by adding a point at infinity, then M can be identified naturally as the space of probability measures on the compactified space by placing any missing mass at infinity. The vague topology we imposed on M is nothing but the weak topology in the compactified space. It is therefore easy to deduce from our earlier work

LEMMA 2.2. *For any closed set $A \subset M$,*

$$\overline{\lim_{t \to \infty}} \frac{1}{t} \log \sup_x Q_{t,x}(A) \leqq -\inf_{\lambda \subset A} I(\lambda). \tag{2.11}$$

Let $\beta \in M$, $x \in R_1$, and let E be a Borel set in R_1. We define $\beta_x(E) = \beta(E - x)$ so that $\beta_x \in M$. If $A \subset M$ and $x \in R_1$, we define $A_x = \{\beta_x\}$ for $\beta \in A$.

We shall use the notation $D_A = \bigcup_x A_x$. We now prove

LEMMA 2.3. *For any set $A \subset M$, let $C_0(A) = \bigcup_{|x| \leqq 1/2} A_x$. Let $0 < a \leqq 1$ be given, and define*

$$C_0^a(A) = \left\{ \beta' \in M : \beta'(E) = \theta\beta\left(\frac{1}{\theta^{1/\alpha}} E\right) \text{ for all Borel sets } E \right.$$

$$\left. \text{and all } 1 \leqq \theta \leqq 2/a, \text{ where } \beta \in C_0(A) \right\}.$$

Then, for any $\lambda \in M$ and any $\varepsilon > 0$, there exists a neighborhood N of λ such that

$$\inf_{\beta \in C_0^a(N)} I(\beta) \geqq I(\lambda) - \varepsilon. \tag{2.12}$$

Proof: We note from the definition of $C_0^a(A)$ that if $A \subset B$, then $A \subset C_0^a(A) \subset C_0^a(B)$. From the lower semicontinuity of I, there exists a neighborhood N_1 of λ such that $\inf_{\beta \in N_1} I(\beta) \geqq I(\lambda) - \varepsilon$. Let N be a neighborhood of λ such that $N \subseteq \bar{N} \subset N_1$, so that $C_0^a(N) \subset C_0^a(\bar{N})$. Since M is compact, it is clear that if A is closed, then $C_0(A)$ and $C_0^a(A)$ are closed. Thus $C_0^a(\bar{N})$ is closed and $\overline{C_0^a(N)} \subseteq C_0^a(\bar{N})$. This implies that

$$\inf_{\beta \in \overline{C_0^a(N)}} I(\beta) \geqq \inf_{\beta \in C_0^a(\bar{N})} I(\beta).$$

Thus, since

$$\inf_{\beta \in \bar{N}} I(\beta) \geqq \inf_{\beta \in N_1} I(\beta) \geqq I(\lambda) - \varepsilon,$$

it suffices to show that

$$\inf_{\beta \in C_0^a(\bar{N})} I(\beta) = \inf_{\beta \in \bar{N}} I(\beta).$$

But this follows from the fact that $I(\beta)$ is translation invariant and that for symmetric stable processes the I-function has the property that

$$I\left(\theta\beta\left(\frac{1}{\theta^{1/\alpha}}\cdot\right)\right) = \theta I\left(\beta\left(\frac{1}{\theta^{1/\alpha}}\cdot\right)\right) = I(\beta).$$

This last identity for the I-function of a symmetric stable process of index α is the reflection in the I-function of the fundamental scaling property of such a process—namely, that $x(cs)$ has the same distribution as $c^{1/\alpha}x(s)$ for any constant c. This scaling property is shared by the semigroup, the infinitesimal generator, and hence also the I-function.

We state the following elementary property of the vague topology as

LEMMA 2.4. *Let $\lambda \in M$. For any neighborhood N of λ there exists a neighborhood V of λ and a $K<\infty$ such that if $\beta \in V$ and $\beta(A)=\beta'(A)$ for all Borel sets $A \subset [-K, K]$, then $\beta' \in N$.*

We recall the notation that for any set $A \subset M$, $D_A = \bigcup_x A_x$ and prove next

LEMMA 2.5. *Let $\lambda \in M$ and $\varepsilon > 0$ be given. There exists a neighborhood $V = V_\varepsilon$ of λ such that*

$$\overline{\lim_{t\to\infty}} \frac{1}{t} \log Q_{t,0}(D_V) \leqq -I(\lambda) + \varepsilon . \tag{2.13}$$

Proof: First of all, if $\lambda(R_1)=0$ there is nothing to prove since in that case $I(\lambda)=0$. So assume $\lambda(R_1)=a>0$. For the given $\lambda \in M$, the given $\varepsilon>0$, and this $a=\lambda(R_1)$, we have from Lemma 2.3 that there exists a neighborhood N of λ such that, for the corresponding $C_0^a(N)$,

$$\inf_{\beta \in \overline{C_0^a(N)}} I(\beta) \geqq I(\lambda) - \varepsilon . \tag{2.14}$$

Corresponding to this neighborhood N of λ we have from Lemma 2.4 that there exists a neighborhood V of λ and a $K<\infty$ such that if $\beta \in V$ and $\beta(E)=\beta'(E)$ for all $E \subset [-K, K]$, then $\beta' \in N$. Without loss of generality we can assume that V and K have been selected so that if $\beta \in V$, then $\beta([-K, K]) \geqq \frac{1}{2}a$. It is this neighborhood V of λ which we shall show satisfies (2.13).

Let $B_0 = \bigcup_{|x|\leqq 1/2} V_x$ and, for any integer n, let $B_n = \bigcup_{|x-n|\leqq 1/2} V_x$. Thus, $D_V = \bigcup_x V_x = \bigcup_{n=-\infty}^{\infty} B_n$, and

$$Q_{t,0}(D_V) \leqq \sum_{n=-\infty}^{\infty} Q_{t,0}(B_n) = \sum_{n=-\infty}^{\infty} P_0\{L_t(\omega, \cdot) \in B_n\} . \tag{2.15}$$

If $\beta \in B_n$, then for some x with $|x-n| \leqq \frac{1}{2}$, $\beta \in V_x$ and hence $\beta_{-x} \in V$. From the way we selected V and K this means that for some x with $|x-n| \leqq \frac{1}{2}$ we have $\beta_{-x}([-K, K]) \geqq \frac{1}{2}a$, i.e., $\beta([-K+x, K+x]) \geqq \frac{1}{2}a$. If we let $K' = K + \frac{1}{2}$, this last inequality implies $\beta([-K'+n, K'+n]) \geqq \frac{1}{2}a$. Thus we see that if $\beta \in B_n$, then $\beta([-K'+n, K'+n]) \geqq \frac{1}{2}a$.

Let $\tau_n = \inf[t : x(t) \in [-K'+n, K'+n]]$, and the conclusion of the last paragraph implies

$$\{\omega : L_t(\omega, \cdot) \in B_n\} \subset \{\omega : \tau_n \leqq (1-\tfrac{1}{2}a)t\}. \tag{2.16}$$

Let

$$\beta(E) = L_t(\omega, E) = \frac{1}{t} \int_0^t \chi_E(x(s))\, ds,$$

and let

$$\beta'(E) = \frac{1}{t} \int_{\tau_n}^t \chi_E(x(s))\, ds.$$

Obviously, $\beta'(E) = \beta(E)$ for all $E \subset [-K'+n, K'+n]$. This means, with our choice of V and K and since $K' = K + \frac{1}{2}$, that

$$\{\omega : L_t(\omega, \cdot) \in B_n\} \subset \left\{\omega : \frac{1}{t} \int_{\tau_n}^t \chi_{\cdot}(x(s))\, ds \in C_n\right\}, \tag{2.17}$$

where $C_n = \bigcup_{|x-n| \leqq 1/2} N_x$. Hence, from (2.15)–(2.17), we see that

$$Q_{t,0}(D_V) \leqq \sum_{n=-\infty}^{\infty} P_0\left\{\tau_n \leqq (1-\tfrac{1}{2}a)t, \frac{1}{t} \int_{\tau_n}^t \chi_{\cdot}(x(s))\, ds \in C_n\right\}. \tag{2.18}$$

From the strong Markov property and the fact that the process has stationary independent increments,

$$\begin{aligned} &P_0\left\{\tau_n \leqq (1-\tfrac{1}{2}a)t, \frac{1}{t} \int_{\tau_n}^t \chi_{\cdot}(x(s))\, ds \in C_n\right\} \\ &\quad = E_0\left\{\chi_{\tau_n \leq (1-a/2)t}(x(\cdot)) P_{x(\tau_n)}\left[\frac{1}{t} \int_0^{t-\tau_n} \chi_{\cdot}(x(s))\, ds \in C_n\right]\right\} \\ &\quad \leqq \left(\sup_{\substack{x \\ 0 \leq \tau \leq t(1-a/2)}} P_x\left\{\frac{1}{t} \int_0^{t-\tau} \chi_{\cdot}(x(s))\, ds \in C_n\right\}\right) P_0\{\tau_n \leq t(1-\tfrac{1}{2}a)\} \\ &\quad = \left(\sup_{\substack{x \\ 0 \leq \tau \leq t(1-a/2)}} P_x\left\{\frac{1}{t} \int_0^{t-\tau} \chi_{\cdot}(x(s))\, ds \in C_0\right\}\right) P_{-n}\{\tau_0 \leq t(1-\tfrac{1}{2}a)\}. \end{aligned} \tag{2.19}$$

Using the notation of Lemma 2.1, we have from (2.3),

$$\sum_{n=-\infty}^{\infty} P_{-n}\{\tau_0 \leqq t(1-\tfrac{1}{2}a)\} = \sum_{n=-\infty}^{\infty} P_n\{\tau_0 \leqq t(1-\tfrac{1}{2}a)\}$$

$$= \sum_{n=-\infty}^{\infty} P_n\{x(s) \in [-K', K'] \text{ for some } s, 0 \leqq s \leqq t(1-\tfrac{1}{2}a)\} \tag{2.20}$$

$$= \sum_{n=-\infty}^{\infty} f_\alpha^{t(1-a/2)}(n) \leqq \gamma_2(t-\tfrac{1}{2}at+1) \leqq \gamma_2(t+1),$$

where γ_2 depends on K' and α but not on t. Thus, from (2.18)–(2.20), we obtain

$$\overline{\lim_{t\to\infty}} \frac{1}{t} \log Q_{t,0}(D_V) \leqq \overline{\lim_{t\to\infty}} \frac{1}{t} \log \sup_{\substack{x \\ 0 \leqq \tau \leqq t(1-a/2)}} P_x\left\{\frac{1}{t}\int_0^{t-\tau} \chi_{\cdot}(x(s))\, ds \in C_0\right\}. \tag{2.21}$$

The scaling property of a symmetric stable process of index α, i.e., that the distribution of $x(ks)$ is the same as the distribution of $k^{1/\alpha}x(s)$ for any positive constant k, implies that $L_{kt}(\omega, \cdot)$ has the same distribution as $L_t(\omega, (1/k^{1/\alpha})\cdot)$. This, together with the fact that

$$\left\{\omega : \frac{1}{t}\int_0^{t-\tau} \chi_{\cdot}(x(s))\, ds \in C_0\right\}$$

$$= \left\{\omega : \frac{1}{t-\tau}\int_0^{t-\tau} \chi_{\cdot}(x(s))\, ds \in \left(\frac{t}{t-\tau} C_0 \cap M\right)\right\}$$

$$= \left\{\omega : L_{t-\tau}(\omega, \cdot) \in \left(\frac{t}{t-\tau} C_0 \cap M\right)\right\},$$

gives us

$$\sup_{\substack{x \\ 0 \leqq \tau \leqq t(1-a/2)}} P_x\left\{\frac{1}{t}\int_0^{t-\tau} \chi_{\cdot}(x(s))\, ds \in C_0\right\}$$

$$= \sup_{\substack{x \\ 0 \leqq \tau \leqq t(1-a/2)}} P_x\left\{L_{t-\tau}(\omega, \cdot) \in \left(\frac{t}{t-\tau} C_0 \cap M\right)\right\}$$

$$= \sup_{\substack{x \\ 0 \leqq \tau \leqq t(1-a/2)}} P_x\left\{L_t\left(\omega, \left(\frac{t}{t-\tau}\right)^{1/\alpha} \cdot\right) \in \left(\frac{t}{t-\tau} C_0 \cap M\right)\right\} \tag{2.22}$$

$$= \sup_{\substack{x \\ 0 \leqq \tau \leqq t(1-a/2)}} P_x\{L_t(\omega, \cdot) \in F_{t,\tau}\},$$

where $F_{t,\tau}$ consists of all measures $\beta' \in M$ of the form $\beta'(A) = (t/(t-\tau))\beta(((t-\tau)/t)^{1/\alpha}A)$ with $\beta \in C_0$.

If now we let $F_{t,a} = \bigcup_{0 \leqq \tau \leqq t(1-a/2)} F_{t,\tau}$, then it follows from (2.22) that

$$\sup_{\substack{x \\ 0 \leqq \tau \leqq (1-a/2)t}} P_x\left\{\frac{1}{t}\int_0^{t-\tau} \chi_\cdot(x(s))\, ds \in C_0\right\}$$

$$(2.23) \qquad \leqq \sup_x P_x\{L_t(\omega, \cdot) \in (F_{t,a})\}$$

$$\leqq \sup_c P_x\{L_t(\omega, \cdot) \in \overline{F_{t,a}}\}$$

$$= \sup_x Q_{t,x}(\overline{F_{t,a}})\,.$$

From Lemma 2.2 we see that

$$(2.24) \qquad \overline{\lim_{t\to\infty}} \frac{1}{t} \log \sup_x Q_{t,x}(\overline{F_{t,a}}) \leqq - \inf_{\beta \in F_{t,a}} I(\beta)\,.$$

Putting together (2.21), (2.23), and (2.24) we obtain

$$(2.25) \qquad \overline{\lim_{t\to\infty}} \frac{1}{t} \log Q_{t,0}(D_V) \leqq - \inf_{\beta \in \overline{F_{t,a}}} I(\beta)\,.$$

But $F_{t,a} = C_0^a(N)$ and therefore by Lemma 2.3 and the choice of N,

$$\inf_{\beta \in \overline{F_{t,a}}} I(\beta) = \inf_{\beta \in \overline{C_0^a(N)}} I(\beta) \geqq I(\lambda) - \varepsilon\,.$$

This completes the proof of Lemma 2.5.

We can now prove

THEOREM 2.6. *Let $A \subset M$ be closed and let $D_A = \bigcup_x A_x$. Then,*

$$(2.26) \qquad \overline{\lim_{t\to\infty}} \frac{1}{t} \log Q_{t,0}(D_A) \leqq - \inf_{\beta \in A} I(\beta)\,.$$

Proof: Let $\inf_{\beta\in\lambda} I(\beta)=l$. Let $\lambda\in A$ and $\varepsilon>0$ be given. From Lemma 2.5 there exists a neighborhood V_λ of λ such that

$$\begin{aligned}\overline{\lim_{t\to\infty}}\frac{1}{t}\log Q_{t,0}(D_{V_\lambda}) &\leqq -I(\lambda)+\varepsilon\\ &\leqq -l+\varepsilon.\end{aligned}$$

The neighborhoods V_λ cover A, and since A is compact, there exists $\lambda_1, \lambda_2,\ldots,\lambda_n$ such that $V_{\lambda_1}, V_{\lambda_2},\ldots,V_{\lambda_n}$ cover A. Hence, $Q_{t,0}(D_A)\leqq \sum_{j=1}^n Q_{t,0}(D_{V\lambda_j})$, so that

$$\begin{aligned}\overline{\lim_{t\to\infty}}\frac{1}{t}\log Q_{t,0}(D_A) &\leqq \max_{1\leqq j\leqq n}\left(\overline{\lim_{t\to\infty}}\frac{1}{t}\log Q_{t,0}(D_{V_{\lambda_j}})\right)\\ &\leqq -l+\varepsilon,\end{aligned}$$

and since $\varepsilon>0$ is arbitrary, the theorem follows.

For the next theorem we need the following lemma. We denote the complement of a set A by $\tilde{A}$.

LEMMA 2.7. *Let G be a neighborhood in M and N another neighborhood such that $N\subset\bar{N}\subset G$. Suppose $\{\alpha_n\}$ is a sequence of measures in M such that, for sufficiently large n, $\alpha_n\notin D_{\tilde{N}}$. Let $\{\gamma_n\}$ be a sequence of measure in M such that*

$$\lim_{n\to\infty}\|\alpha_n-\gamma_n\|=0,$$

where the norm is the variation norm. If $\{\lambda_n\}$ is a sequence of constants such that $\lim_{n\to\infty}\lambda_n=1$ and we define, for any Borel set A, $\beta_n(A)=\gamma_n(\lambda_n A)$, then, for sufficiently large n, $\beta_n\notin D_{\tilde{G}}$.

Proof: Assume the contrary, i.e., $\beta_{n_j}\in D_{\tilde{G}}$ for some sequence $\{n_j\}$ with $\lim_{j\to\infty} n_j=\infty$. Relabeling, our assumption means that $\beta_n\in D_{\tilde{G}}$ for all n. This implies that for each n there is an x_n such that $\beta_n'=\beta_n(\cdot+x_n)\in\tilde{G}$. For any Borel set A, define

$$\gamma_n'(A)=\beta_n'\left(\frac{1}{\lambda_n}A\right)=\beta_n\left(\frac{1}{\lambda_n}A+x_n\right)=\gamma_n(a+\lambda_n x_n),$$

and also define

$$\alpha'_n(A) = \alpha_n(A + \lambda_n x_n).$$

Obviously, $\|\gamma'_n - \alpha'_n\| = \|\gamma_n - \alpha_n\|$ and hence, $\lim_{n\to\infty} \|\gamma'_n - \alpha'_n\| = 0$. Since $\beta'_n \in \tilde{G}$ for all n, and $\gamma'_n(A) = \beta'_n((1/\lambda_n)A)$ with $\lambda_n \to 1$ as $n \to \infty$, we see that any limit point of γ'_n is also in $\tilde{G}$. Thus, any limit point of α'_n is also in $\tilde{G}$. Now, $\tilde{N}$ is a neighborhood of $\tilde{G}$ so that $\alpha'_n \in \bar{\tilde{N}}$ for sufficiently large n, i.e., $\alpha'_n \in \tilde{N}$ for sufficiently large n. Since α'_n is a translate of α_n, this last statement implies that $\alpha_n \in D_{\tilde{N}}$ for sufficiently large n which contradicts the hypothesis that $\alpha_n \notin D_{\tilde{N}}$ for sufficiently large n.

We defined earlier $L_t(\omega, A) = (1/t)\int_0^t \chi_A(x(s))\, ds$, and now we define

$$\hat{L}_t(\omega, A) = L_t(\omega, \rho_t A), \tag{2.27}$$

where $\rho_t = (t/\log\log t))^{1/\alpha}$. Thus,

$$\hat{L}_t(\omega, A) = \frac{1}{t}\int_0^t \chi_A\left(\left(\frac{\log\log t}{t}\right)^{1/\alpha} x(s)\right) ds. \tag{2.28}$$

We now prove

THEOREM 2.8. *Let $C = \{\beta \in M : I(\beta) \leqq 1\}$. Then, for almost all ω, the set of limit points as $t \to \infty$ of any translate of $\hat{L}_t(\omega, \cdot)$ is in C, i.e., if we let*

$$C_t^\omega = \{\hat{L}_t(\omega, \cdot + x), x \in R_1\},$$

then, for almost all ω,

$$\bigcap_T \overline{\bigcup_{t \geqq T} C_t^\omega} \subset C. \tag{2.29}$$

Proof: Let G be a neighborhood of C in M, and let N be a neighborhood of C such that $C \subset N \subset \bar{N} \subset G$. Using the same notation introduced earlier, let $D_{\tilde{G}} = \bigcup_x \tilde{G}_x$ and $D_{\tilde{N}} = \bigcup_x \tilde{N}_x$. If we set $\theta = \inf_{\beta \in \tilde{N}} I(\beta)$, then since I is lower semicontinuous and $\tilde{N}$ is compact in M, we have $\theta > 1$. Choose $0 < k < 1$ but such that $k\theta' = k(\frac{1}{2}(1+\theta)) > 1$. Define, for each integer n, $t_n = e^{n^k}$.

As noted earlier, for any positive constant λ, $L_{\lambda t}(\omega, A)$ has the same distribution as $L_t(\omega, (1/\lambda^{1/\alpha})^A)$. Thus

$$\begin{aligned} P_0\{\hat{L}_t(\omega,\cdot)\in D_{\tilde N}\} &= P_0\{L_{\log\log t}(\omega,\cdot)\in D_{\tilde N}\} \\ &= Q_{\log\log t,0}(D_{\tilde N}), \end{aligned} \tag{2.30}$$

and since $\tilde N$ is closed in M, we have from Theorem 2.6, with $\theta'=\frac{1}{2}(1+\theta)$ and for sufficiently large t,

$$Q_{\log\log t,0}(D_{\tilde N}) \leqq \exp\{-(\log\log t)\theta'\}. \tag{2.31}$$

In particular, for large enough n, we see from (2.30) and (2.31) that

$$P_0\{\hat{L}_{t_n}(\omega,\cdot)\in D_{\tilde N}\} \leqq \exp\{-(\log\log t_n)\theta'\} = \frac{1}{n^{k\theta'}}. \tag{2.32}$$

Since k was chosen so that $k\theta'>1$, we conclude that $\sum_{n=1}^{\infty} P_0\{\hat{L}_{t_n}(\omega,\cdot)\in D_{\tilde N}\}<\infty$, and hence from the Borel–Cantelli lemma we have:

$$\begin{gathered} \text{for almost all } \omega, \quad \hat{L}_{t_n}(\omega,\cdot)\in D_{\tilde N} \\ \text{for only finitely many } n. \end{gathered} \tag{2.33}$$

Consider then a fixed ω for which $\hat{L}_{t_n}(\omega,\cdot)\in D_{\tilde N}$ for only finitely many n. For this ω we shall show below that, for sufficiently large t,

$$\hat{L}_t(\omega,\cdot)\notin D_{\tilde G}. \tag{2.34}$$

But (2.34) implies that no translate of $\hat{L}_t(\omega,\cdot)\in \tilde G$ for sufficiently large t. Thus all translates of $\hat{L}_t(\omega,\cdot)$ are in G for sufficiently large t, and since G is any neighborhood of C and M is compact metric, we conclude that all translates of $\hat{L}_t(\omega,\cdot)$ have limit points in C as $t\to\infty$. Hence we have (2.29) for almost all ω completing the proof of the theorem except for (2.34) which we show now.

Assume (2.34) *is false, then for a sequence* $\{\tau_j\}$ with $\lim_{j\to\infty}\tau_j=\infty$ we have $\hat{L}_{\tau_j}(\omega,\cdot)\in D_{\tilde G}$ for all $j=1,2,3,\ldots$. We defined above the sequence $t_n=e^{n^k}$ with $k<1$. Let t_{n_j} be such that $t_{n_j-1}\leqq \tau_j\leqq t_{n_j}$. Define

$$\alpha_j = L_{t_{n_j}}(\omega,\cdot),\quad \beta_j=\hat{L}_{\tau_j}(\omega,\cdot),$$

$$\gamma_j = L_{\tau_j}\left(\omega, \left(\frac{t_{n_j}}{\log\log t_{n_j}}\right)^{1/\alpha}\cdot\right),$$

and

$$\lambda_j = \left(\frac{t_{n_j}}{\log\log t_{n_j}} \frac{\log\log \tau_j}{\tau_j}\right)^{-1/\alpha}.$$

We observe first of all, that by our choice of ω, $\alpha_j \notin D_{\bar{N}}$ for sufficiently large j. Since $k<1$, $t_{n-1}/t_n = \exp\{(n-1)^k - n^k\} \to 1$ as $n \to \infty$, and therefore $t_{n_j-1}/t_{n_j} \to 1$ as $j \to \infty$. This implies, since $t_{n_j-1} \leqq \tau_j \leqq t_{n_j}$, that $\|L_{\tau_j}(\omega, \cdot) - L_{t_{n_j}}(\omega, \cdot)\| \to 0$ as $j \to \infty$, where the norm is the variation norm. Since

$$\alpha_j(A) = \hat{L}_{t_{n_j}}(\omega, A) = L_{t_{n_j}}\left(\omega, \left(\frac{t_{n_j}}{\log\log t_{n_j}}\right)^{1/\alpha} A\right),$$

we see that $\|\alpha_j - \gamma_j\| \to 0$ as $j \to \infty$. Next, we note that, for any Borel set A,

$$\beta_j(A) = \hat{L}_{\tau_j}(\omega, A) = L_{\tau_j}\left(\omega, \left(\frac{\tau_j}{\log\log \tau_j}\right)^{1/\alpha} A\right) = \gamma_j(\lambda_j A).$$

Moreover, since $t_{n_j-1}/t_{n_j} \to 1$ as $j \to \infty$ and $t_{n_j-1} \leqq \tau_j \leqq t_{n_j}$,

$$\lim_{j\to\infty} \lambda_j = \lim_{j\to\infty} \left(\frac{t_{n_j}}{\log\log t_{n_j}} \frac{\log\log \tau_j}{\tau_j}\right)^{-1/\alpha} = 1.$$

Hence, our assumption that (2.34) is false, i.e., $\beta_j \in D_{\bar{G}}$ for all $j = 1, 2, 3, \ldots$ contradicts Lemma 2.7. Thus (2.34) is true and the proof of Theorem 2.8 is complete.

In Theorem 2.8 we showed that for almost all ω, the set of limit points as $t \to \infty$ of any translate of $\hat{L}_t(\omega, \cdot)$ is in $C = \{\beta \in M : I(\beta) \leqq 1\}$, or in the notation of that theorem, $\bigcap_T \overline{\bigcup_{t \geqq T} C_t^\omega} \subset C$. Now we wish to prove

THEOREM 2.9. *Let $C = \{\beta \in M : I(\beta) \leqq 1\}$. Then, for almost all ω,*

$$\bigcap_T \overline{\bigcup_{t\geqq T} \{\hat{L}_t(\omega, \cdot)\}} \supset C. \tag{2.35}$$

Remark. Clearly, $\bigcap_T \overline{\bigcup_{t \leqq T} \{\hat{L}_t(\omega, \cdot)\}} \subset \bigcap_T \overline{\bigcup_{t \geqq T} C_t^\omega}$, and therefore from Theorems 2.8 and 2.9 we can conclude that, for almost all ω,

$$C \subset \bigcap_T \overline{\bigcup_{t \geqq T} \{\hat{L}_t(\omega, \cdot)} \subset \bigcap_T \overline{\bigcup_{t \leqq T} C_t^\omega} \subset C. \tag{2.36}$$

To prove Theorem 2.9 we shall first prove several lemmas and another theorem. Let $a>0$ and denote by $\tau_{(-a,a)}$ the first exit time for the process from $(-a, a)$, i.e.,

$$\tau_{(-a,a)}=\tau_{(-a,a)}(\omega)=\inf\{t : x(t)\notin(-a, a)\}.$$

We define, for any Borel set $E\subset(-a, a)$,

$$\hat{p}(t, x, E)=P_x\{x(t)\in E, \tau_{(-a,a)}>t\}. \tag{2.37}$$

We write, as usual, $p(t, x, E)=P_x\{x(t)\in E\}$ and begin with

LEMMA 2.10. *Let $f(y)$ be a bounded measurable function on $(-a, a)$ and define*

$$u(t, x)=E_x\{f(x(t)); \tau_{(-a,a)}>t\}.$$

Then, $u(t, x)$ is jointly continuous in t and x for $t>0$ and $|x|<a$. In particular, $\hat{p}(t, x, E)$ is jointly continuous in t and x for $t>0$ and $|x|<a$ for any Borel set $E\subset(-a, a)$.

Proof: The last assertion about $\hat{p}(t, x, E)$ follows, of course from the first statement of the lemma by letting $f(y)=\chi_E(y)$.

We can assume without loss of generality that $|f|\leqq 1$, and we write

$$u(t, x)=\int f(y)\hat{p}(t, x, dy).$$

Fix a $t>0$ and choose $h>0$ such that $t-h>0$. Defining $u(t, y)=0$ if $|y|>a$, we have from the semigroup property,

$$\begin{aligned} u(t, x)&=\int u(t-h, y)\hat{p}(h, x, dy)\\ &=\int u(t-h, y)p(h, x, dy)-\int u(t-h, y)[p(h, x, dy)-\hat{p}(h, x, dy)]. \end{aligned}$$

Thus, since $|u(t-h, y)|\leqq 1$,

$$\begin{aligned} \left|u(t, x)-\int u(t-h, y)p(h, x, dy)\right|&\leqq\int[p(h, x, dy)-\hat{p}(h, x, dy)]\\ &\leqq P_x\{\tau_{(-a,a)}\leqq h\}. \end{aligned} \tag{2.38}$$

Let $t\geqq t_0>0$ and $0<h\leqq t_0$. Define $V(t, h, x)=\int u(t-h, y)p(h, x, dy)$, and let K

be a closed set with $K \subset (-a, a)$. From (2.38) we have

$$\sup_{\substack{x_1, x_2 \in K \\ t \geqq t_0 \\ |x_1 - x_2| \leqq \delta}} |u(t, x_1) - u(t, x_2)| \tag{2.39}$$

$$\leqq \sup_{\substack{t \geqq t_0 \\ |x_1 - x_2| \leqq \delta}} |V(t, h, x_1) - V(t, h, x_2)| + 2 \sup_{x \in K} P_x\{\tau_{(-a,a)} \leqq h\}.$$

Since the process has stationary independent increments and $p(t, x, dy)$ has a density, we can write $V(t, h, x) = \int u(t-h, y) p(h, y-x)\, dy$ so that

$$|V(t, h, x_1) - V(t, h, x_2)| \leqq \int |p(h, y - x_1) - p(h, y - x_2)|\, dy.$$

Hence, from (2.39),

$$\sup_{\substack{x_1, x_2 \in K \\ t \geqq t_0 \\ |x_1 - x_2| \leqq \delta}} |u(t, x_1) - u(t, x_2)| \tag{2.40}$$

$$\leqq \sup_{|z| \leqq \delta} \int |p(h, y) - p(h, y - z)|\, dy + 2 \sup_{x \in K} \mathrm{P}_x\{\tau_{(-a,a)} \leqq h\}.$$

Because $p(h, y)$ is a density, the first term on the right of inequality (2.40) approaches zero as $\delta \to 0$. Thus

$$\overline{\lim_{\delta \to 0}} \sup_{\substack{x_1, x_2 \in K \\ t \geqq t_0 \\ |x_1 - x_2| \leqq \delta}} |u(t, x_1) - u(t, x_2)| \leqq 2 \sup_{x \in K} P_x\{\tau_{(-a,a)} \leqq h\}. \tag{2.41}$$

The left side of inequality (2.41) does not involve h, and since $0 < h \leqq t_0$ is arbitrary and the right side of (2.41) approaches zero as $h \to 0$, we conclude that

$$\overline{\lim_{\delta \to 0}} \sup_{\substack{x_1, x_2 \in K \\ t \geqq t_0 \\ |x_1 - x_2| \leqq \delta}} |u(t, x_1) - u(t, x_2)| = 0. \tag{2.42}$$

Therefore, $u(t, x)$ is continuous in x and uniformly so for $t \geqq t_0 > 0$ and x in a compact subset of $(-a, a)$.

To show the continuity in t, let $t_2 > t_1 > t_0 > 0$ and $t_2 - t_1 = h$. Consider

$$\begin{aligned} u(t_2, x) - u(t_1, x) &= \int u(t_1, y)\hat{p}(h, x, dy) - u(t_1, x) \\ &= \int u(t_1, y)p(h, x, dy) - u(t_1, x) \\ &\quad - \int u(t_1, y)[p(h, x, dy) - \hat{p}(h, x, dy)] \\ &= \int [u(t_1, y) - u(t_1, x)]p(h, x, dy) \\ &\quad - \int u(t_1, y)[p(h, x, dy) - \hat{p}(h, x, dy)] . \end{aligned}$$

Since $|u(t_1, y)| \leqq 1$,

$$\begin{aligned} |u(t_2, x) - u(t_1, x)| \leqq \left| \int [u(t_1, y) - u(t_1, x)]p(h, x, dy) \right| \\ + P_x\{\tau_{(-a,a)} \leqq h\} . \end{aligned} \tag{2.43}$$

From the first part of this lemma, the first term on the right of (2.43) approaches zero as $h \to 0$, and the second term also approaches 0 as $h \to 0$. Hence $u(t, x)$ is continuous in t and uniformly so for x in compact subsets of $(-a, a)$.

LEMMA 2.11. *If $E \subset (-a, a)$ and has positive Lebesgue measure, then*

$$\int_0^\infty \hat{p}(t, x, E)e^{-t}\, dt > 0 \tag{2.44}$$

for every $x \in (-a, a)$.

Proof: If $\alpha = 2$, i.e., the symmetric stable process is Brownian motion, then this lemma is obvious because in that case $\hat{p}(t, x, y)$ is strictly positive. If, however, $0 < \alpha < 2$, then (2.44) requires proof.

Let $Q(x, E) = \int_0^\infty \hat{p}(t, x, E)e^{-t}\, dt$. Without loss of generality we can assume $E \subset (-a, a)$ is a compact set. We show first that $Q(x, E)$ is not identically zero in x. Suppose it were, then $\hat{p}(t, x, E)$ must be 0 for almost all t. In the preceding lemma we showed that $\hat{p}(t, x, E)$ is jointly continuous in t and x so that we infer $\hat{p}(t, x, E)=0$ for all $t>0$ and all $x \in(-a, a)$. But, since $|p(t, x, E)-\hat{p}(t, x, E)| \leqq P_x\{\tau_{(-a,a)} \leqq t\}$, the fact that $\hat{p}(t, x, E)=0$ for all $t>0$ and all $x \in(-a, a)$ implies

$$\sup_{x \in E} p(t, x, E) \leqq \sup_{x \in E} P_x\{\tau_{(-a,a)} \leqq t\}. \tag{2.45}$$

Since E is compact, the right side of inequality (2.45) approaches zero as $t \rightarrow 0$ and therefore $\lim_{t \rightarrow 0} \sup_{x \in E} p(t, x, E)=0$. However, it is a basic property of $p(t, x, E)$ that $\lim_{t \rightarrow 0} p(t, x, E)=\chi_E(x)$ in L_1. Thus, we have a contradiction and conclude that $Q(x, E)$ is not identically zero in x.

We show now that if $Q(x, E)=0$ for some x, then $Q(x, E) \equiv 0$ in x which will finish the lemma. To show this we prove that if $Q(x, E)>0$ for some x, then $Q(x, E)>0$ for all x. So let x' be such that $Q(x', E)>0$. Since $\hat{p}(t, x, E)$ is continuous in x, $Q(x, E)$ is also continuous in x, and therefore there exists a neighborhood U of x' such that $Q(\xi, E) \geqq \delta>0$ for all $\xi \in U$. It will suffice to show that $Q(x, U)>0$ for all $x \in(-a, a)$ since, by the choice of U and the semigroup property, we have

$$\begin{aligned} 0<\int_U Q(y, E) Q(x, d y) & \leqq \int Q(y, E) Q(x, d y) \\ & =\int_0^{\infty} t \hat{p}(t, x, E) e^{-t} d t, \end{aligned} \tag{2.46}$$

and (2.46) implies $\hat{p}(t, x, E)>0$ on some set of t having positive Lebesgue measure which in turn implies $Q(x, E)>0$. So we must show $Q(x, U)>0$ for all $x \in(-a, a)$. This means proving that

$$P_x\{x(t) \text{ spends positive time in } U \text{ before exiting } (-a, a)\}>0.$$

Since U is open and almost all paths of the process are right continuous, if the process enters U before exiting $(-a, a)$ it will spend positive time in U. Thus it suffices to show that the probability of entering U before exiting from $(-a, a)$, starting from x, is positive. Consider two open sets N_1 and N_2 such that $x \in N_1 \subset(-a, a)$ and $N_1+N_2 \subset U$. Now the probability of entering U starting from x is at least the probability that the stable process remains in N_1 up to a certain time and then has a jump whose size is in N_2. That this last probability is positive is obvious since the probability that the process remains in N_1 for some length of time is positive because N_1 is open and, on the other hand, the Lévy measure for stable processes gives support for all possible jumps including those whose size is in N_2. This completes the proof of Lemma 2.11.

We now prove

LEMMA 2.12. *If β is a probability measure such that $\beta(x : |x| \leqq a) = 1$ and G is a weak neighborhood of β, then, for $a' > a$ and $\tau_{(-a',a')}$ denoting the first exit time for the process from $(-a', a')$, we have*

$$\lim_{t\to\infty} \frac{1}{t} \log \inf_{|x|\leqq a} P_x\{L_t(\omega,\cdot) \in G, \tau_{(-a',a')} > t\} \geqq -I(\beta) . \tag{2.47}$$

Proof: This lemma is a special case of Theorem 8.1 in [8], and Lemmas 2.10 and 2.11 are proved here to verify the hypotheses necessary for that theorem. In particular, Lemma 2.11 verifies hypothesis H_4 of Theorem 8.1. Hypotheses H_1—H_3 for Theorem 8.1 are obvious properties of any process with stationary independent increments and therefore, in particular, are satisfied by the symmetric stable processes considered here.

LEMMA 2.13. *Let (Ω, Σ, μ) be a probability space and $\mathcal{F}_n$ an increasing sequence of σ-fields. Let $\{f_n(\omega)\}$ be a sequence of functions which forms a martingale relative to $(\Omega, \mathcal{F}_n, \mu)$ such that $|f_n(\omega) - f_{n-1}(\omega)| \leqq 1$ for all n and ω. Then*

$$\mu\{\omega : \lim_{n\to\infty} f_n(\omega) \text{ exists and equals } +\infty\} = 0 .$$

Proof: Normalize the sequence so that $f_0(\omega) = 0$ and consider an interval $(-a, b)$. Let $\tau = \tau(\omega) = \inf[n : f_n(\omega) \notin (-a, b)]$ and note that τ might be $+\infty$. However, $\tau \wedge n$ is a finite stopping time and so, by Doob's theorem, $\int f_{\tau\wedge n}(\omega)\, d\mu = 0$, i.e.,

$$\int_{\tau\leqq n} f_\tau(\omega)\, d\mu + \int_{\tau>n} f_n(\omega)\, d\mu = 0 . \tag{2.48}$$

Let

$$J_1 = \int_{\substack{\tau\leqq n\\ f_\tau(\omega)\leqq -a}} f_\tau(\omega)\, d\mu, \qquad J_2 = \int_{\substack{\tau\leqq n\\ f_\tau(\omega)\geqq b}} f_\tau(\omega)\, d\mu ,$$

and

$$J_3 = \int_{\tau > n} f_n(\omega)\, d\mu\,,$$

so that (2.48) implies $J_1 + J_2 + J_3 = 0$. Since $|f_n(\omega) - f_{n-1}(\omega)| \leqq 1$ for all n and ω, we have

$$J_1 \geqq -(a+1)\mu\{\tau \leqq n\,,\ f_\tau(\omega) \leqq -a\}\,,$$

$$J_2 \geqq b\mu\{\tau \leqq n\,,\ f_\tau(\omega) \geqq b\}\,,$$

and

$$J_3 \geqq -a\mu\{\tau > n\}\,.$$

Since $J_1 + J_2 + J_3 = 0$ and

$$\mu\{\tau \leqq n, f_\tau(\omega) \leqq -a\} + \mu\{\tau \leqq n, f_\tau(\omega) \geqq b\} + \mu\{\tau > n\} = 1\,,$$

we get

$$\begin{aligned}(a+1)\mu\{\tau \leqq n, f_\tau(\omega) \leqq -a\} &\geqq b\mu\{\tau \leqq n, f_\tau(\omega) \geqq b\} - a\mu\{\tau > n\}\\ &= b\mu\{\tau \leqq n, f_\tau(\omega) \geqq b\}\\ &\quad -a[1 - \mu\{\tau \leqq n, f_\tau(\omega) \leqq -a\} - \mu\{\tau \leqq n, f_\tau(\omega) \geqq b\}]\,.\end{aligned}$$

Hence,

$$a + \mu\{\tau \leqq n, f_\tau(\omega) \leqq -a\} \geqq (a+b)\mu\{\tau \leqq n, f_\tau(\omega) \geqq b\}\,,$$

and therefore, for every n,

$$\mu\{\tau \leqq n, f_\tau(\omega) \geqq b\} \leqq (a+1)/(a+b)\,,$$

which means that

$$\lim_{a\to\infty} \lim_{b\to\infty} \lim_{n\to\infty} \mu\{\tau \leqq n, f_\tau(\omega) \geqq b\} = 0\,.$$

This last equation implies

$$\mu\{\lim_{n\to\infty} f_n(\omega) = +\infty\} = 0$$

which is the lemma.

LEMMA 2.14. *Let (Ω, Σ, μ) be a probability space and let $\mathscr{F}_n$ be an increasing sequence of σ-fields. Let $E_n \in \mathscr{F}_n$, and define $p_{n-1}(\omega) = \mu\{E_n \mid \mathscr{F}_{n-1}\}$. Then, if for almost all ω,*

$$\sum p_{n-1}(\omega) = \infty ,$$

we have

$$\sum \chi_{E_n}(\omega) = \infty ,$$

for almost all ω.

Proof: Let $g_n(\omega) = -\chi_{E_n}(\omega) + p_{n-1}(\omega)$, and note that $E\{g_n \mid \mathscr{F}_{n-1}\} = 0$. Defining $f_n(\omega) = g_1(\omega) + \cdots + g_n(\omega)$, we see that $\{f_n\}$ is a martingale because

$$\begin{aligned} E\{f_n \mid \mathscr{F}_{n-1}\} &= g_1(\omega) + \cdots + g_{n-1}(\omega) + E\{g_n \mid \mathscr{F}_{n-1}\} \\ &= g_1(\omega) + \cdots + g_{n-1}(\omega) = f_{n-1}(\omega) . \end{aligned}$$

Now, $f_n(\omega) - f_{n-1}(\omega) = g_n(\omega)$ and $|g_n(\omega)| \leqq 1$ for all n and all ω. Therefore from Lemma 2.13 we have

$$\mu\{\lim_{n\to\infty} f_n(\omega) = \infty\} = 0 ,$$

or

$$\mu\left\{\sum g_n(\omega) = \infty\right\} = 0 .$$

Since $g_n(\omega) = -\chi_{E_n}(\omega) + p_{n-1}(\omega)$, and by hypothesis $\sum p_{n-1}(\omega) = \infty$ for almost all ω, we conclude that $\sum \chi_{E_n}(\omega) = \infty$ for almost all ω.

We are now ready to prove

THEOREM 2.15. *Let β be a probability measure supported on $(-a, a)$ such* *that* $I(\beta)<1$. *Let* $a'>a$ *and* G *be a weak neighborhood of* β. *With* $\rho_t=(t/\log\log t))^{1/\alpha}$, *define*

$$E_t=\{\omega : \hat{L}_t(\omega,\cdot)\in G,\ \tau_{(-a'\rho_t,a'\rho_t)}>t\}.$$

Then, for almost all ω, $\omega\in E_t$ *for a sequence of times increasing to* ∞.

Proof: Let G_1 be a weak neighborhood of β such that $G_1\subset\bar{G}_1\subset G$. By hypothesis, $I(\beta)=\theta<1$. Choose $k>1$ but such that $(2k-1)(\frac{1}{2}(1+\theta))<1$. For this k, we let $t_n=e^{n^k}$ and for convenience we write $\rho_n=\rho_{t_n}=(t_n/(\log\log t_n))^{1/\alpha}$. By definition,

$$\hat{L}_{t_n}(\omega,A)=\frac{1}{t_n}\int_0^{t_n}\chi_{\rho_n A}(x(s))\,ds.$$

We introduce the notation

$$\tilde{L}_{t_{n-1},t_n}(\omega,A)=\frac{1}{t_n-t_{n-1}}\int_{t_{n-1}}^{t_n}\chi_{\rho_n A}(x(s))\,ds.$$

Since $k>1$, $t_{n-1}/t_n\to 0$ as $n\to\infty$, and since, in variation norm,

$$\|\tilde{L}_{t_{n-1},t_n}(\omega,\cdot)-\hat{L}_{t_n}(\omega,\cdot)\|\leqq\frac{2t_{n-1}}{t_n},$$

we see that, for sufficiently large n,

$$\tilde{L}_{t_{n-1},t_n}(\omega,\cdot)\in G_1\Rightarrow\hat{L}_{t_n}(\omega,\cdot)\in G.$$

Thus, defining

$$F_n=\{\omega : \tilde{L}_{t_{n-1},t_n}(\omega,\cdot)\in G_1,\ \tau_{(-a\rho_n,a\rho_n)}>t_{n-1},\ \tau_{(-a'\rho_n,a'\rho_n)}>t_n\},$$

it is clear that if F_n occurs infinitely often so does E_{t_n} occur infinitely often. This means that to prove the theorem it will suffice to show that, for almost all ω, F_n occurs infinitely often.

Let $\mathscr{F}_n=\sigma(x(s),0\leqq s\leqq t_n)$, i.e., let $\mathscr{F}_n$ be the σ-field generated by the process up to time t_n. We see that

$$\begin{aligned}&P_0\{F_n\mid\mathscr{F}_{n-1}\}\\ (2.49)\qquad&=\chi_{[\tau_{(-a\rho_n,a\rho_n)}>t_{n-1}]}(\omega)\,P_{x(t_{n-1})}\{\tau_{(-a'\rho_n,a'\rho_n)}>(t_n-t_{n-1}),\ L'_{t_n-t_{n-1},t_n}(\omega,\cdot)\in G_1\},\end{aligned}$$

where

$$L'_{t,s}(\omega, A) = \frac{1}{t}\int_0^t \chi_{\rho_s A}(x(\sigma))\, d\sigma = \frac{1}{t}\int_0^t \chi_A\left(\frac{x(\sigma)}{\rho_s}\right) d\sigma .$$

Let $B_{n-1} = \{\omega : \tau_{(-a\rho_n, a\rho_n)} > t_{n-1}\}$. Then, from (2.49), we have

$$(2.50)\quad P_0\{F_n | \mathscr{F}_{n-1} \geqq \chi_{B_{n-1}}(\omega) \inf_{|x|\leqq a\rho_n} P_x\{\tau_{(-a'\rho_n, a'\rho_n)} > (t_n - t_{n-1}),$$

$$L'_{t_n - t_{n-1}, t_n}(\omega, \cdot) \geqslant G_1\} .$$

The process being symmetric stable of index α, the distribution of $x(s/\rho_n^\alpha)$ is the same as $(1/\rho_n)x(s)$. Therefore, letting $s_n = (t_n - t_{n-1})/\rho_n^\alpha$, we see that

$$\inf_{|x|\leqq a\rho_n} P_x\{\tau_{(-a'\rho_n, a'\rho_n)} > (t_n - t_{n-1}), L'_{t_n - t_{n-1}, t_n}(\omega, \cdot) \in G_1\}$$

$$(2.51)\quad = \inf_{|x|\leqq a\rho_n} P_x\left\{ \sup_{0\leqq s\leqq t_n - t_{n-1}} |x(s)| \leqq a'\rho_n, \frac{1}{t_n - t_{n-1}} \int_0^{t_n - t_{n-1}} \chi_\cdot\left(\frac{x(s)}{\rho_n}\right) ds \in G_1\right\}$$

$$= \inf_{|x|\leqq a} P_x\left\{ \sup_{0\leqq s\leqq s_n} |x(s)| \leqq a', L_{s_n}(\omega, \cdot) \in G_1\right\}.$$

Now

$$s_n = \frac{t_n - t_{n-1}}{\rho_n^\alpha} = \left(\frac{t_n - t_{n-1}}{t_n}\right)(\log\log t_n)$$

$$= \left(1 - \frac{t_{n-1}}{t_n}\right)(\log\log t_n),$$

and, as noted earlier, $t_{n-1}/t_n \to 0$ as $n \to \infty$. Thus, $s_n \to \infty$ as $n \to \infty$ and indeed $s_n \sim \log\log t_n$, i.e., $s_n \sim k \log n$ as $n \to \infty$. Hence, from Lemma 2.12 we have, for sufficiently large n,

$$\inf_{|x|\leqq a} P_x\left\{ \sup_{0\leqq s\leqq s_n} |x(s)| \leqq a', L_{s_n}(\omega, \cdot) \in G_1\right\}$$

$$\geqq \exp\{-s_n(\tfrac{1}{2}(1+\theta))\}$$

$$(2.52)\quad \geqq \exp\{-(2k-1)\tfrac{1}{2}(1+\theta)\log n\}$$

$$= \exp\{-\theta' \log n\} = 1/n^{\theta'},$$

where we recall that $k>1$ was picked so that $\theta' = (2k-1)(\frac{1}{2}(1+\theta))<1$.

From (2.50)–(2.52) we conclude that there exists an N such that, for all ω,

$$\sum_{n=N}^{\infty} P_0\{F_n \mid \mathfrak{F}_{n-1}\} \geqq \sum_{n=N}^{\infty} \frac{\chi_{B_{n-1}}(\omega)}{n^{\theta'}}. \tag{2.53}$$

If we show that, for almost all ω,

$$\sum \frac{\chi_{B_{n-1}}(\omega)}{n^{\theta'}} = \infty, \tag{2.54}$$

then (2.53) and Lemma 2.14 imply that $\sum \chi_{F_n}(\omega) = \infty$ for almost all ω which as noted above gives us the theorem because F_n occurring infinitely often for almost all ω implies that E_{t_n} occurs infinitely often for almost all ω. To prove (2.54) it will suffice, since $\theta'<1$, to demonstrate that, for almost all ω, $\Sigma\,[1-\chi_{B_{n-1}}(\omega)]<\infty$. From the Borel–Cantelli lemma, this last statement will follow if we show that

$$\sum P_0\{\tilde{B}_n\} < \infty. \tag{2.55}$$

It is a result of Lévy that, for symmetric stable processes of index α, $P_0\{\sup_{0\leqq s\leqq t} |x(s)| \geqq a\} = O(t)$ as $t \to 0$. But,

$$\begin{aligned} P_0\{\tilde{B}_n\} &= P_0\{\tau_{(-a\rho_{n+1}, a\rho_{n+1})} \leqq t_n\} \\ &= P_0\Big\{\sup_{0\leqq s\leqq t_n} |x(s)| \geqq a\rho_{n+1}\Big\} = P_0\Big\{\sup_{0\leqq s\leqq t_n/\rho_{n+1}^{\alpha}} |x(s)| \geqq a\Big\} \end{aligned} \tag{2.56}$$

and, since for all n, $(n+1)^k - n^k \geqq n^{k-1}$, we see that

$$\begin{aligned} \frac{t_n}{\rho_{n+1}^{\alpha}} &= \left(\frac{t_n}{t_{n+1}}\right)(\log\log t_{n+1}) \\ &= \exp\{[n^k-(n+1)^k]\}\log\log t_{n+1} \\ &\leqq \exp\{-n^{k-1}\}k\log(n+1), \end{aligned}$$

which goes to zero as $n\to\infty$ since $k>1$. Hence,

$$P_0\{\tilde{B}_n\} = O\left(\frac{t_n}{\rho_{n+1}^{\alpha}}\right) \leqq O(\exp\{-n^{k-1}\}k\log(n+1)),$$

so that we have (2.55) and the proof of the theorem is complete.

Using the following lemma we will be able to prove Theorem 2.9.

LEMMA 2.16. *Let $\mathscr{D}$ be the set of probability measures β on R_1 for which $I(\beta)<1$ and for each of which there exists some $K<\infty$ such that $\beta\{|x|\leqq K\}=1$. Let $C=\{\beta\in M : I(\beta)\leqq 1\}$. Then,*

$$C\subset\bar{\mathscr{D}}\,. \tag{2.57}$$

Proof: Let $\beta\in C$, and for $\lambda>1$, define $\beta_\lambda(A)=\beta((1/\lambda)A)$. As noted earlier, the I-function for a symmetric stable process of index α has the property that $I(\beta_\lambda)=(1/\lambda^\alpha)I(\beta)$. Thus, since $I(\beta)\leqq 1$, we see that $I(\beta_\lambda)<1$. Moreover, as $\lambda\to 1$, $\beta_\lambda\to\beta$ so to prove (2.57) it suffices to show that $\{\beta\in M : I(\beta)<1\}\subset\bar{\mathscr{D}}$. Let ϕ be a smooth probability density with $I(\phi)<\infty$. Let $\beta\in M$ with $\beta(R_1)=a<1$ and $I(\beta)<1$. For $0<\lambda\leqq 1$, define γ_λ by $d\gamma_\lambda=d\beta+(1-a)\lambda\phi(\lambda x)\,dx$. Now γ_λ is a probability measure and, as $\lambda\to 0$, $\gamma_\lambda\to\beta$. Also, by convexity, $I(\gamma_\lambda)\leqq I(\beta)+(1-a)\lambda^\alpha I(\phi)$, so that $I(\gamma_\lambda)<1$ for small enough λ. Thus to prove (2.57) it suffices to show that

$$\{\beta\in\mathscr{M} : I(\beta)<1\}\subset\bar{\mathscr{D}}\,, \tag{2.58}$$

where $\mathscr{M}\subset M$ is the set of probability measures on R_1. To prove (2.58) we take a $\beta\in\mathscr{M}$ with $I(\beta)<1$ and we must produce a sequence $\{\beta_n\}\in\mathscr{M}$, each β_n having compact support, such that $I(\beta_n)<1$ and $\beta_n\to\beta$. That such a sequence exists follows from the proof of Lemma 3.8 in [5].

From the definition of $\mathscr{D}$ in Lemma 2.16 we see that $\mathscr{D}\subset C$ and since C is closed in M (because I is lower semicontinuous) we have $\bar{\mathscr{D}}\subset C$. Thus the preceding lemma implies $C=\bar{\mathscr{D}}$. We can now give the

Proof of Theorem 2.9: First of all, it follows from Theorem 2.15 that if β is a probability measure supported on $(-a,a)$ such that $I(\beta)<1$ and G is a weak neighborhood of β, then, for almost all ω,

$$\omega\in\{\omega : \hat{L}_t(\omega,\cdot)\in G\}$$

for a sequence of times increasing to infinity. This means that for almost all ω the set of limit points of $\hat{L}_t(\omega,\cdot)$ as $t\to\infty$ contains $\mathscr{D}$, i.e., for almost all ω,

$$\bigcap_T \overline{\bigcup_{t\geqq T}\{\hat{L}_t(\omega,\cdot)\}}\supset\mathscr{D}\,. \tag{2.59}$$

Since the set on the left of (2.59) is closed, and since we just showed that

$\bar{\mathscr{D}} = C$, we conclude that, for almost all ω,

$$\bigcap_{T} \overline{\bigcup_{t \geqq T} \{\hat{L}_t(\omega, \cdot)\}} \supset C \tag{2.60}$$

which is Theorem 2.9.

As noted in (2.36), we now know that, for almost all ω,

$$C \subset \bigcap_{T} \overline{\bigcup_{t \geqq T} \{\hat{L}_t(\omega, \cdot)\}} \subset \bigcap_{T} \overline{\bigcup_{t \geqq T} C_t^{\omega}} \subset C,$$

and in particular,

$$\bigcap_{T} \overline{\bigcup_{t \geqq T} \hat{L}_t(\omega, \cdot)} = C. \tag{2.61}$$

If Φ is a functional on M which is continuous in the vague topology, then (2.61) implies that the set of limit points of $\Phi(\hat{L}_t(\omega, \cdot))$ as $t \to \infty$ is exactly the set $\{\Phi(\beta)\}$ with $\beta \in C$. Thus we have proved

THEOREM 2.17. *Let $\{x(s),\ 0 \leqq s < \infty\}$ be a symmetric stable process of index α, $0 < \alpha \leqq 2$, and let*

$$\hat{L}_t(\omega, A) = \frac{1}{t} \int_0^t \chi_A\left(\left(\frac{\log \log t}{t}\right)^{1/\alpha} x(s)\right) ds .$$

Let Φ be a functional on M, the space of measures β on R_1 such that $\beta(R_1) \leqq 1$, which is continuous in the vague topology. Then, for almost all ω,

$$\overline{\lim_{t \to \infty}} \Phi(\hat{L}_t(\omega, \cdot)) = \sup_{\beta \in C} \Phi(\beta) , \tag{2.62}$$

where $C = \{\beta \in M : I(\beta) \leqq 1\}$ and $I(\beta)$ is defined by (2.10).

From the preceding remarks and Theorem 2.17, the following corollaries are self-evident.

COROLLARY 2.18. *If Φ is a functional on M which is lower semicontinuous in the vague topology, then, for almost all ω,*

$$\overline{\lim_{t \to \infty}} \Phi(\hat{L}_t(\omega, \cdot)) \geqq \sup_{\beta \in C} \Phi(\beta) , \tag{2.63}$$

and if Φ is a functional on M which is upper semicontinuous in the vague topology, then, for almost all ω,

$$\overline{\lim_{t\to\infty}}\, \Phi(\hat{L}_t(\omega,\cdot)) \leqq \sup_{\beta\in C} \Phi(\beta)\,. \tag{2.64}$$

COROLLARY 2.19. *For $\beta \in M$, let $\beta_x(A)=\beta(A-x)$, and $\Phi_x(\beta)=\Phi(\beta_x)$. Then, if Φ is a functional on M which is continuous in the vague topology, we have for almost all ω*

$$\overline{\lim_{t\to\infty}}\, \sup_x \Phi_x(\hat{L}_t(\omega,\cdot)) = \sup_{\beta\in C} \Phi(\beta)\,. \tag{2.65}$$

To see (2.65) we note from (2.36) that

$$\bigcap_T \overline{\bigcup_{t\geqq T} C_t^\omega} = C\,, \tag{2.66}$$

and (2.66) and the definition of C_t^ω gives us (2.65).

3. The Strong Topology

Again we consider a symmetric stable process $\{x(s),\ 0 \leqq s < \infty\}$ of index α, but now we assume $1<\alpha\leqq 2$. *The reason for restricting* $\alpha > 1$ is that in this case $L_t(\omega,\cdot)$ has a continuous density with respect to Lebesgue measure (cf. Boylan [10], Kesten and Getoor [11]) which we denote by $l_t(\omega, y)$. In this section we shall prove the laws of the iterated logarithm for $\hat{l}_t(\omega, y)$ which is a suitably scaled version of $l_t(\omega, y)$. For $l_t(\omega, y)$ the topology will be uniform convergence on compact subsets of the real line rather than the vague convergence which we employed for $L_t(\omega,\cdot)$. In the functional form of the theorem, analogous to Theorem 2.17 of the last section, this allows for a richer class of functionals.

We start with

LEMMA 3.1. *Let $\phi(x)$ vanish outside of $|x|\leqq h$ and satisfy $\int_{-\infty}^{\infty} \phi(x)\,dx=0$, $\int_{-\infty}^{\infty} |\phi(x)|\,dx \leqq c<\infty$. For a given $1<\alpha\leqq 2$, pick $0<\theta<\alpha-1$ and define $\gamma=2-(\theta+2)/\alpha$. Then, for any positive integer n and $0<u\leqq 1$,*

$$\sup_x E_x\left\{\left(\int_0^u \phi(x(s))\,ds\right)^{2n}\right\} \leqq \frac{(2n)!\,B^n h^{n\theta}}{\Gamma(1+n\gamma)}\,, \tag{3.1}$$

where B is a constant independent of n.

Proof: The transition probability density for a symmetric stable process of index α is given by $(1/2\pi)\int_{-\infty}^{\infty} e^{-i\xi(x-y)}e^{-t|\xi|^\alpha}\,d\xi$ and so is of the form $t^{-1/\alpha}\Phi_\alpha(t^{-1/\alpha}(x-y))$, where

$$\Phi(x)=\Phi_\alpha(x)=\frac{1}{2\pi}\int_{-\infty}^{\infty} e^{-i\xi x}e^{-|\xi|^\alpha}\,d\xi .$$

Thus, we can write

$$E_x\left\{\left(\int_0^u \phi(x(s))\,ds\right)^{2n}\right\}$$

$$\tag{3.2} =(2n)!\int\cdots\int_{0\leqq s_1\leqq\cdots\leqq s_{2n}\leqq u}\int_{-\infty}^{\infty}\cdots\int_{-\infty}^{\infty}\prod_{j=0}^{2n-1}\frac{\phi(x_{j+1})}{(s_{j+1}-s_j)^{1/\alpha}}$$
$$\times\Phi\left(\frac{x_{j+1}-x_j}{(s_{j+1}-s_j)^{1/\alpha}}\right)dx_1\cdots dx_{2n}\,ds_1\cdots ds_{2n},$$

where $s_0=0$ and $x_0=x$. Since $\int_{-\infty}^{\infty}\phi(x_j)\,dx_j=0$ for all $1\leqq j\leqq 2n$, we can express (3.2) in the form

$$E_x\left\{\left(\int_0^u \phi(x(s))\,ds\right)^{2n}\right\}$$

$$\tag{3.3} =(2n)!\int\cdots\int_{0\leqq s_1\leqq s_2\leqq\cdots\leqq s_{2n}\leqq u}\int_{-\infty}^{\infty}\cdots\int_{-\infty}^{\infty}\prod_{j=0}^{2n-1}\frac{\phi(x_{j+1})}{(s_{j+1}-s_j)^{1/\alpha}}$$
$$\times\prod_{k=0}^{n-1}\left\{\Phi\left(\frac{x_{2k+1}-x_{2k}}{(s_{2k+1}-s_{2k})^{1/\alpha}}\right)\Phi\left(\frac{x_{2k+2}-x_{2k+1}}{(s_{2k+2}-s_{2k+1})^{1/\alpha}}\right)\right.$$
$$\left.-\Phi^2(0)\right\}dx_1\cdots dx_{2n}\,ds_1\cdots ds_{2n}.$$

Now $\Phi(x)$ is a bounded function on R_1, and for every $0\leqq\theta\leqq 2$ there exists a constant C_θ such that $|\Phi(x)-\Phi(0)|/|x|^\theta\leqq C_\theta$ for $0\leqq x<\infty$. Thus, if we denote the bound on $\Phi(x)$ by K, we have, for the θ picked in this lemma,

$$
\begin{aligned}
&\left|\Phi\left(\frac{x_{2k+1}-x_{2k}}{(s_{2k+1}-s_{2k})^{1/\alpha}}\right)\Phi\left(\frac{x_{2k+2}-x_{2k+1}}{(s_{2k+2}-s_{2k+1})^{1/\alpha}}\right)-\Phi^2(0)\right| \\
&\qquad \leqq K\left\{\left|\Phi\left(\frac{x_{2k+1}-x_{2k}}{(s_{2k-1}-s_{2k})^{1/\alpha}}\right)-\Phi(0)\right|\right. \\
&\qquad\qquad \left.+\left|\Phi\left(\frac{x_{2k+2}-x_{2k+1}}{(s_{2k+2}-s_{2k+1})^{1/\alpha}}\right)-\Phi(0)\right|\right\} \\
&\qquad \leqq KC_\theta\left[\frac{|x_{2k+1}-x_{2k}|^\theta}{|s_{2k+1}-s_{2k}|^{\theta/\alpha}}+\frac{|x_{2k+2}-x_{2k+1}|^\theta}{|s_{2k+2}-s_{2k+1}|^{\theta/\alpha}}\right] \\
&\qquad \leqq KC_\theta(2h)^\theta\left[\frac{1}{|s_{2k+1}-s_{2k}|^{\theta/\alpha}}+\frac{1}{|s_{2k+2}-s_{2k+1}|^{\theta/\alpha}}\right].
\end{aligned}
\tag{3.4}
$$

By hypothesis, $\int_{-\infty}^{\infty} |\phi(x)|\, dx \leqq x < \infty$. Using this and (3.4) in (3.3) we obtain

$$
\begin{aligned}
&\sup_x E_x\left\{\left(\int_0^u \phi(x(s))\, ds\right)^{2n}\right\} \\
&\qquad \leqq (2n)!A_\theta^n h^{n\theta} \int_{0\leqq s_1\leqq s_2\leqq\cdots\leqq s_{2n}\leqq 1}\cdots\int \prod_{j=0}^{2n-1}\frac{1}{(s_{j+1}-s_j)^{1/\alpha}} \\
&\qquad\qquad \times\prod_{k=0}^{n-1}\left(\frac{1}{|s_{2k+1}-s_{2k}|^{\theta/\alpha}}+\frac{1}{|s_{2k+2}-s_{2k+1}|^{\theta/\alpha}}\right) ds_1\, ds_2\cdots ds_{2n},
\end{aligned}
\tag{3.5}
$$

where A_θ is a constant depending on c, θ, and K but independent of n and where we used the fact that $0<u\leqq 1$. The right side of inequality (3.5) can be rewritten as

$$
\begin{aligned}
&(2n)!A_\theta^n h^{n\theta} \overset{(2n)}{\int_{0\leqq s_1\leqq s_2\leqq\cdots\leqq s_{2n}\leqq 1}\cdots\int} \prod_{j=0}^{n-1}\left[\frac{1}{(s_{2j+1}-s_{2j})^{\theta/\alpha}}+\frac{1}{(s_{2j+2}-s_{2j+1})^{\theta/\alpha}}\right] \\
&\qquad \times\frac{1}{(s_{2j+1}-s_{2j})^{1/\alpha}}\frac{1}{(s_{2j+2}-s_{2j+1})^{1/\alpha}}\, ds_1\, ds_2\cdots ds_{2n}.
\end{aligned}
\tag{3.6}
$$

If, in the integral

$$\int_x^y \left[\frac{1}{(z-x)^{\theta/\alpha}}+\frac{1}{(y-z)^{\theta/\alpha}}\right]\frac{1}{(z-x)^{1/\alpha}}\cdot\frac{1}{(y-z)^{1/\alpha}}\,dz,$$

we let $z=x+\xi(y-x)$ so that $0\leqq\xi\leqq 1$, we obtain

$$\frac{1}{(y-x)^{(\theta/\alpha)+(2/\alpha)-1}}\int_0^1\left[\frac{1}{\xi^{\theta/\alpha}}+\frac{1}{(1-\xi)^{\theta/\alpha}}\right]\frac{1}{\xi^{1/\alpha}}\cdot\frac{1}{(1-\xi)^{1/\alpha}}\,d\xi. \tag{3.7}$$

Since $\theta+1<\alpha$,

$$B_{\theta,\alpha}=\int_0^1\left[\frac{1}{\xi^{\theta/\alpha}}+\frac{1}{(1-\xi)^{\theta/\alpha}}\right]\frac{1}{\xi^{1/\alpha}}\cdot\frac{1}{(1-\xi)^{1/\alpha}}\,d\xi$$

is a finite constant and using this and (3.7) in (3.6) we obtain

$$\sup_x E_x\left\{\left(\int_0^u \phi(x(s))\,ds\right)^{2n}\right\}$$

$$\leqq(2n)!\,A_\theta^n h^{n\theta}B_{\theta,\alpha}^n\underset{0\leqq s_2\leqq s_4\leqq\cdots\leqq s_{2n}\leqq 1}{\overset{(n)}{\int\cdots\int}}\prod_{j=0}^{n-1}\frac{1}{(s_{2j+2}-s_{2j})^{\theta/\alpha+2/\alpha-1}}\,ds_2\,ds_4\cdots ds_{2n}$$

$$=(2n)!\,A_\theta^n h^{n\theta}B_{\theta,\alpha}^n\frac{[\Gamma(\gamma)]^n}{\Gamma(1+n\gamma)},$$

where $\gamma=1-(\theta/\alpha+2\alpha-1)=2-(\theta+2)/\alpha$. This completes the proof of the lemma.

LEMMA 3.2. *With the function ϕ and the constants θ and γ determined as in Lemma 3.1, we can find a constant $k_0>0$ such that, for all x and $0<u\leqq 1$,*

$$E_x\left\{\exp\left\{k_0\left|\frac{1}{h^{\theta/2}}\int_0^u\phi(x(s))\,ds\right|^\rho\right\}\right\}\leqq M<\infty,$$

where $\rho=2/(2-\gamma)$.

Proof: From the definition of γ we see that $0<\gamma<1$ and therefore

$1<\rho<2$. Consider, for $k>0$,

$$E_x\left\{\exp\left\{k\left|\int_0^u \phi(x(s)\,ds\right|^\rho\right\}\right\}=E_x\left\{\sum_{n=0}^\infty \frac{k^n\left|\int_0^u \phi(x(s))\,ds\right|^{n\rho}}{n!}\right\}$$

$$\leqq \sum_{n=0}^\infty \frac{k^n}{n!}\left[E_x\left\{\left(\int_0^u \phi(x(s))\,ds\right)^{2n}\right\}\right]^{\rho/2}$$

$$\leqq \sum_{n=0}^\infty \frac{k^n}{n!}\left[\frac{(2n)!\,B^n h^{n\theta}}{\Gamma(1+n\gamma)}\right]^{\rho/2}.$$

Let $k_0>0$ and set $k=k_0h^{-\theta\rho/2}$. Then this last inequality becomes

$$E_x\left\{\exp\left\{k_0\left|\frac{1}{h^{\theta/2}}\int_0^u \phi(x(s))\,ds\right|^\rho\right\}\right\}\leqq \sum_{n=0}^\infty \frac{k_0^n B^{n\rho/2}[2n!]^{\rho/2}}{n![\Gamma(1+n\gamma)]^{\rho/2}}. \tag{3.8}$$

But $[2n!]^{\rho/2}/n!\,[\Gamma(1+n\gamma)]^{\rho/2}\sim n^{n\rho}/(n^{n+1/2}n^{n\gamma\rho/2})$, and since $n+n\gamma\rho/2=n\rho$ we see that by choosing $k_0>0$ small enough we can make the series on the right of (3.8) converge, proving the lemma.

LEMMA 3.3. *Let θ and ρ be determined as in the preceding lemma. There exists a constant k_1 such that for almost all ω*

$$\iint_{|y_1-y_2|\leqq 1}\left(\exp\left\{k_1\left(\frac{|l_1(\omega,y_1)-l_1(\omega,y_2)|}{|y_1-y_2|^{\theta/2}}\right)^\rho\right\}-1\right)dy_1\,dy_2=M(\omega)<\infty, \tag{3.9}$$

and moreover, $E\{M(\omega)\}\leqq K<\infty$.

Proof: By definition, $l_1(\omega,y)=\int_0^1 \delta(x(s)-y)\,ds$; thus if in Lemma 3.2 we choose $\phi(x)=\delta_{y_1}(x)-\delta_{y_2}(x)$, $h=\frac{1}{2}|y_1-y_2|$, and $u=1$, we have, for all x,

$$E_x\left\{\exp\left\{k_1\left(\frac{|l_1(\omega,y_1)-l_1(\omega,y_2)|}{|y_1-y_2|^{\theta/2}}\right)^\rho\right\}\right\}\leqq K_1<\infty, \tag{3.10}$$

where $k_1 = k_0 2^{\theta\rho/2}$. We shall show now that

(3.11)
$$\iint_{|y_1-y_2|\leqq 1} E_0\left\{\left(\exp\left\{k_1\left(\frac{|l_1(\omega, y_1)-l_1(\omega, y_2)|}{|y_1-y_2|^{\theta/2}}\right)^{\rho}\right\}-1\right)\right\} dy_1\, dy_2 < \infty.$$

Since

$$\{y_1, y_2 : |y_1-y_2|\leqq 1\} \subset \bigcup_{n=-\infty}^{\infty} \{y_1, y_2 : |y_1-n|\leqq 1, |y_2-n|\leqq 1\},$$

to prove (3.11) it will suffice to show that

(3.12)
$$\sum_{n=-\infty}^{\infty} \iint_{\substack{|y_1-n|\leqq 1\\ |y_2-n|\leqq 1}} E_0\left\{\exp\left\{k_1\left(\frac{|l_1(\omega, y_1)-l_1(\omega, y_2)|}{|y_1-y_2|^{\theta/2}}\right)^{\rho}\right\}-1\right\} dy_1\, dy_2 < \infty.$$

Fix n, y_1 and y_2 in such a way that $|y_1-n|\leqq 1$ and $|y_2-n|\leqq 1$. Let $\tau_n = \inf\{t : x(t)\in[n-1, n+1]\}$. Then, using the strong Markov property once again, we have

$$E_0\left\{\exp\left\{k_1\left(\frac{|l_1(\omega, y_1)-l_1(\omega, y_2)|}{|y_1-y_2|^{\theta/2}}\right)^{\rho}\right\}-1\right\}$$

(3.13)
$$= E_0\left\{\chi_{\tau_n<1}(\omega)\left(\exp\left\{k_1\left(\frac{|l_1(\omega, y_1)-l_1(\omega, y_2)|}{|y_1-y_2|^{\theta/2}}\right)^{\rho}\right\}-1\right)\right\}$$

$$= E_Q\left\{\chi_{\tau_n<1}(\omega) E_{x(\tau_n)}\left\{\exp\left\{k_1\left(\frac{|l'_{1-\tau_n}(\omega, y_1)-l'_{1-\tau_n}(\omega, y_2)|}{|y_1-y_2|^{\theta/2}}\right)^{\rho}\right\}-1\right\}\right\},$$

where $l'_t(\omega, y) = \int_0^t \delta(x(s)-y)\, ds$. We use Lemma 3.2 again with $\phi(x) = \delta_{y_1}(x)-\delta_{y_2}(x)$, $h=\frac{1}{2}|y_1-y_2|$, $u = 1-\tau_n$ to conclude that

$$E_{x(\tau_n)}\left\{\exp\left\{k_1\left(\frac{l'_{1-\tau_n}(\omega, y_1)-l'_{1-\tau_n}(\omega, y_2)|}{|y_1-y_2|^{\theta/2}}\right)^{\rho}\right\}-1\right\} \leqq K_2 < \infty.$$

Hence, from (3.13) we see that

$$E_0\left\{\exp\left\{k_1\left(\frac{|l_1(\omega, y_1)-l_1(\omega, y_2)|}{|y_1-y_2|^{\theta/2}}\right)^p\right\}-1\right\} \tag{3.14}$$
$$\leqq K_2E_0\{\chi_{\tau_n<1}(\omega)\}=K_2P_0\{\tau_n<1\}.$$

Thus,

$$\iint\limits_{\substack{|y_1-n|\leqq 1\\ |y_2-n|\leqq 1}} E_0\left\{\exp\left\{k_1\left(\frac{|l_1(\omega, y_1)-l_1(\omega, y_2)|}{|y_1-y_2|^{\theta/2}}\right)^p\right\}-1\right\} dy_1\, dy_2 \leqq 4K_2P_0\{\tau_n<1\},$$

and to show that (3.12) (and hence (3.11)) holds it suffices to prove that $\sum_{n=-\infty}^{\infty} P_0\{\tau_n<1\}<\infty$. But this last follows from (2.3) of Lemma 2.1 so that we have established (3.11). Applying Fubini's theorem to (3.11), we conclude that, for almost all ω,

$$\iint\limits_{|y_1-y_2|\leqq 1} \left(\exp\left\{k_1\left(\frac{|l_1(\omega, y_1)-l_1(\omega, y_2)|}{|y_1-y_2|^{\theta/2}}\right)^p\right\}-1\right) dy_1\, dy_2 \leqq M(\omega)<\infty$$

and $E\{M(\omega)\}\leqq K<\infty$, which is the lemma.

In the following Theorem 3.5 we shall make essential use of a result of Garsia, Rodemich, and Rumsey [8] which we state as

LEMMA 3.4. *Let $f(x)$ be defined and continuous on $[a, b]$. Let $\psi(u)$ be a non-negative even function on $(-\infty, \infty)$ such that $\psi(u)$ is nondecreasing for $u\geqq 0$ and $\psi(\infty)=\infty$. Let $p(u)$ be a non-negative even continuous function on $[-1, 1]$ such that $p(0)=0$ and $p(u)$ is nondecreasing for $u\geqq 0$. If*

$$\int_a^b\int_a^b \psi\left(\frac{f(x)-f(y)}{p(x-y)}\right) dx\, dy \leqq B<\infty, \tag{3.15}$$

then, for all $s, t\in[a, b]$,

$$|f(s)-f(t)|\leqq 8\int_0^{|s-t|} \psi^{-1}\left(\frac{4B}{u^2}\right) dp(u). \tag{3.16}$$

We are now ready to prove

THEOREM 3.5. *For every* $a > 0$,

$$\lim_{\delta\to 0}\overline{\lim_{t\to\infty}}\frac{1}{t}\log P\left\{\sup_{|y_1-y_2|\le\delta}|l_t(\omega, y_1)-l_t(\omega, y_2)|\geqq a\right\}=-\infty. \tag{3.17}$$

Proof: First of all, using the scaling properties of symmetric stable processes of index α, we observe that

$$l_t(\omega, y)=\frac{1}{t}\int_0^t \delta(x(s)-y)\,ds=\int_0^1 \delta(x(ts)-y)\,ds$$

is distributed in the same way as

$$\begin{aligned}\int_0^1 \delta(t^{1/\alpha}x(s)-y)\,ds &= \int_0^1 \delta\left(t^{1/\alpha}\left(x(s)-\frac{y}{t^{1/\alpha}}\right)\right)ds\\ &=\frac{1}{t^{1/\alpha}}\int_0^1 \delta\left(x(s)-\frac{y}{t^{1/\alpha}}\right)ds\\ &=\frac{1}{t^{1/\alpha}}\,l_1\left(\omega, \frac{y}{t^{1/\alpha}}\right).\end{aligned}$$

Therefore, to show (3.17) is equivalent to showing

$$\lim_{\delta\to 0}\overline{\lim_{t\to\infty}}\frac{1}{t}\log P\left\{\sup_{|y_1-y_2|\le\delta/t^{1/\alpha}}|l_1(\omega, y_1)-l_1(\omega, y_2)|\geqq a t^{1/\alpha}\right\}=-\infty. \tag{3.18}$$

To prove (3.18) we use Lemma 3.4. Let $\psi(x)=\exp\{k_1|x|^\rho\}-1$, $p(y)=|y|^{\theta/2}$, and $f(y)=l_1(\omega, y)$, where k_1, ρ, and θ are as in Lemma 3.3. We note that $\psi(x)$ and $p(y)$ satisfy the hypotheses of Lemma 3.4. Moreover if we pick y_1 and y_2 such that $|y_1-y_2|\leqq\delta$, where δ is small enough, then hypothesis (3.15) of Lemma 3.4 for this ψ, p, and f follows from Lemma 3.3. We conclude from (3.16) that, for all y_1 and y_2 with $|y_1-y_2|\leqq\delta$,

$$|l_1(\omega, y_1)-l_1(\omega, y_2)|\leqq 8\int_0^\delta \psi^{-1}\left(\frac{4M(\omega)}{u^2}\right)dp(u), \tag{3.19}$$

where

$$M(\omega)=\iint\limits_{|y_1-y_2|\le 1}\left(\exp\left\{k_1\left(\frac{|l_1(\omega, y_1)-l_1(\omega, y_2)|}{|y_1-y_2|^{\theta/2}}\right)^\rho\right\}-1\right)dy_1\,dy_2.$$

Thus

$$\sup_{|y_1-y_2|\leqq\delta/t^{1/\alpha}} |l_1(\omega, y_1)-l_1(\omega, y_2)| \leqq 8\int_0^{\delta/t^{1/\alpha}} \psi^{-1}\left(\frac{4M(\omega)}{u^2}\right) dp(u)\,. \tag{3.20}$$

From the definition of the function $\psi(x)$ we see that $\psi^{-1}(4M(\omega)/u^2)=[1/k_1 \log(1+4M(\omega)/u^2)]^{1/\rho}$, and from the definition of $p(u)$, we see that (3.20) becomes

$$\begin{aligned}&\sup_{|y_1-y_2|\leqq\delta/t^{1/\alpha}} |l_1(\omega, y_1)-l_1(\omega, y_2)|\\ &\qquad\leqq \frac{4\theta}{k_1^{1/\rho}}\int_0^{\delta/t^{1/\alpha}} \left[\log\left(1+\frac{4M(\omega)}{u^2}\right)\right]^{1/\rho} u^{\theta/2-1}\,du\,.\end{aligned} \tag{3.21}$$

Therefore,

$$\begin{aligned}&P\left\{\sup_{|y_1-y_2|\leqq\delta/t^{1/\alpha}} |l_1(\omega, y_1)-l_1(\omega, y_2)| \geqq at^{1/\alpha}\right\}\\ &\qquad\leqq P\left\{\frac{4\theta}{k_1^{1/\rho}}\int_0^{\delta/t^{1/\alpha}} \left[\log\left(1+\frac{4M(\omega)}{u^2}\right)\right]^{1/\rho} u^{\theta/2-1}\,du \geqq at^{1/\alpha}\right\}.\end{aligned} \tag{3.22}$$

Now, in (3.18), a is arbitrary and we are letting $t\to\infty$ and $\delta\to 0$ so that in proving (3.18) it will suffice, in view of (3.22), to prove

$$\lim_{\delta\to 0}\overline{\lim_{t\to\infty}}\frac{1}{t}\log P\left\{\int_0^{\delta/t^{1/\alpha}} [\log 4M(\omega)]^{1/\rho}u^{\theta/2-1}\,du \geqq at^{1/\alpha}\right\} = -\infty\,. \tag{3.23}$$

Since $\theta>0$, we have, absorbing inessential constants into a, and noting that $\rho(1/\alpha+\theta/2\alpha)=\rho\cdot(2-\gamma)/2=1$,

$$\begin{aligned}&P\left\{\int_0^{\delta/t^{1/\alpha}} [\log 4M(\omega)]^{1/\rho}u^{\theta/2-1}\,du \geqq at^{1/\alpha}\right\}\\ &\qquad= P\left\{[\log 4M(\omega)]^{1/\rho}\left(\frac{\delta}{t^{1/\alpha}}\right)^{\theta/2} \geqq at^{1/\alpha}\right\}\\ &\qquad= P\left\{M(\omega)\geqq \tfrac{1}{4}\exp\left\{\frac{a^\rho t^{\rho(1/\alpha+\theta/2\alpha)}}{\delta^{\theta\rho/2}}\right\}\right\}\\ &\qquad= P\{M(\omega)\geqq \tfrac{1}{4}\exp\{a^\rho t/\delta^{\theta\rho/2}\}\,.\end{aligned} \tag{3.24}$$

But in Lemma 3.3 we showed that $E\{M(\omega)\} \leqq K < \infty$, and hence

$$P\{M(\omega) \geqq \tfrac{1}{4} \exp\{a^{\rho} t/\delta^{\theta\rho/2}\} \leqq \tfrac{1}{4} K \exp\{-a^{\rho} t/\delta^{\theta\rho/2}\}. \tag{3.25}$$

Since $\theta\rho > 0$, (3.25) and (3.24) imply (3.23) which we noted implies (3.18) and the theorem.

Again letting $\rho_t = (t/\log\log t))^{1/\alpha}$, we now define $\hat{l}_t(\omega, y) = \rho_t l_t(\omega, \rho_t y)$. We next prove

LEMMA 3.6. *Let $t_1 < t_2 < t_3 < \cdots$ be any sequence of times approaching infinity. Then, for all t satisfying $t_n \leqq t \leqq t_{n+1}$ and any $\delta > 0$,*

$$\begin{aligned} \sup_{|y_1 - y_2| \leqq \delta} |\hat{l}_t(\omega, y_1) - \hat{l}_t(\omega, y_2)| &\leqq \sup_{|y_1 - y_2| \leqq \rho_{t_{n+1}} \delta/\rho_{t_n}} |\hat{l}_{t_n}(\omega, y_1) - \hat{l}_{t_n}(\omega, y_2)| \\ &\quad + \frac{\rho t_n}{t_n} (t_{n+1} - t_n) \sup_y l_{t_n, t_{n+1}}(\omega, y), \end{aligned} \tag{3.26}$$

where $l_{s,t}(\omega, y) = (1/(t-s)) \int_s^t \delta(x(\sigma) - y)\, d\sigma$.

Proof: Clearly,

$$\begin{aligned} \hat{l}_t(\omega, y) &= \frac{\rho_t}{t} \cdot t l_t(\omega, \rho_t y) \\ &= \frac{\rho_t}{t} [t_n l_{t_n}(\omega, \rho_t y) + (t - t_n) l_{t_n, t}(\omega, \rho_t y)], \end{aligned}$$

so that

$$\begin{aligned} |\hat{l}_t(\omega, y_1) - \hat{l}_t(\omega, y_2)| &\leqq \frac{\rho_t}{t} t_n |l_{t_n}(\omega, \rho_t y_1) - l_{t_n}(\omega, \rho_t y_2)| + \frac{\rho_t (t - t_n)}{t} \sup_y l_{t_n, t}(\omega, y) \\ &= \frac{\rho_t t_n}{t \rho_{t_n}} \left| \rho_{t_n} l_{t_n}\left(\omega, \frac{\rho_t}{\rho_{t_n}} \rho_{t_n} y_1\right) - \rho_{t_n} l_{t_n}\left(\omega, \frac{\rho_t}{\rho_{t_n}} \rho_{t_n} y_2\right) \right| \\ &\quad + \frac{\rho_t (t - t_n)}{t} \sup_y l_{t_n, t}(\omega, y) \\ &= \frac{\rho_t t_n}{t \rho_{t_n}} \left| \hat{l}_{t_n}\left(\omega, \frac{\rho_t}{\rho_{t_n}} y_1\right) - \hat{l}_{t_n}\left(\omega, \frac{\rho_t}{\rho_{t_n}} y_2\right) \right| + \frac{\rho_t (t - t_n)}{t} \sup_y l_{t_n, t}(\omega, y). \end{aligned} \tag{3.27}$$

Thus,

$$\sup_{|y_1-y_2|\leqq\delta}|\hat{l}_t(\omega,y_1)-\hat{l}_t(\omega,y_2)|\leqq\frac{\rho_t t_n}{t\rho_{t_n}}\sup_{|y_1-y_2|<(\rho_t\delta/\rho_{t_n})}|\hat{l}_{t_n}(\omega,y_1)-\hat{l}_{t_n}(\omega,y_2)|$$
$$+\frac{\rho_t(t-t_n)}{t}\sup_y l_{t_n,t}(\omega,y)\,.$$

For sufficiently large t, ρ_t/t is monotone decreasing in t and ρ_t is monotone increasing in t. Since $t\geqq t_n$, this implies $\rho_t t_n/t\rho_{t_n}<1$, and since also $t\leqq t_{n+1}$, we obtain from (3.27)

$$\sup_{t_n\leqq t\leqq t_{n+1}}\sup_{|y_1-y_2|\leqq\delta}|\hat{l}_t(\omega,y_1)-\hat{l}_t(\omega,y_2)|\leqq\sup_{|y_1-y_2|\leqq(\rho_{t_{n+1}}\delta/\rho_{t_n})}|\hat{l}_{t_n}(\omega,y_1)-\hat{l}_{t_n}(\omega,y_2)|$$
$$+\frac{\rho_{t_n}(t_{n+1}-t_n)}{t_n}\sup_y l_{t_n,t_{n+1}}(\omega,y)$$

which is the lemma.

LEMMA 3.7. *For every* $a>0$,

$$\overline{\lim_{t\to\infty}}\frac{1}{t}\log P_0\left\{\sup_y l_t(\omega,y)\geqq a\right\}<0\,.$$

Proof: Let $f(y)$ be any probability density, i.e., $f\geqq 0$ and $\int_{-\infty}^{\infty}f(y)\,dy=1$, and assume that, for some $\delta>0$ and some $l<\infty$,

$$\sup_{|y_1-y_2|\leqq\delta}|f(y_1)-f(y_2)|\leqq l\,. \tag{3.28}$$

Let $M=\sup_y f(y)$, then $(M-l)2\delta\leqq 1$. This is because, on some interval $(x-\delta,x+\delta)$, (3.28) implies that f must be at least $M-l$ and hence $\int_{x-\delta}^{x+\delta}f(y)\,dy\geqq(M-l)2\delta$ so that indeed $(M-l)2\delta\leqq 1$ or $M\leqq l+1/2\delta$. Taking $l=\frac{1}{2}a$ and $\delta=1/a$, this remark clearly yields

$$P_0\left\{\sup_y l_t(\omega,y)\geqq a\right\}\leqq P_0\left\{\sup_{|y_1-y_2|\leqq 1/a}|l_t(\omega,y_1)-l_t(\omega,y_2)|\geqq\tfrac{1}{2}a\right\}. \tag{3.29}$$

Now $l_t(\omega, y)$ has the same distribution as $(1/t^{1/\alpha})l_1(\omega, y/t^{1/\alpha})$ and therefore

$$P_0\left\{\sup_{|y_1-y_2|\leqq 1/a}|l_t(\omega, y_1)-l_t(\omega, y_2)|\geqq \tfrac{1}{2}a\right\}$$

(3.30)

$$= P_0\left\{\sup_{|y_1-y_2|\leqq 1/at^{1/\alpha}}|l_1(\omega, y_1)-l_1(\omega, y_2)|\geqq \tfrac{1}{2}at^{1/\alpha}\right\}.$$

From (3.22)–(3.25) we see that

$$\overline{\lim_{t\to\infty}}\frac{1}{t}\log P_0\left\{\sup_{|y_1-y_2|\leqq 1/at^{1/\alpha}}|l_1(\omega, y_1)-l_1(\omega, y_2)|\geqq \tfrac{1}{2}at^{1/\alpha}\right\}\leqq\frac{-(\frac{1}{2}a)^\rho}{(1/a)^{\theta\rho}}<0. \tag{3.31}$$

Thus the lemma follows from (3.29)–(3.31),

THEOREM 3.8. *For each $a>0$, there exists $\delta>0$ such that*

$$P_0\left\{\overline{\lim_{t\to\infty}}\sup_{|y_1-y_2|\leqq\delta}|\hat{l}_t(\omega, y_1)-\hat{l}_t(\omega, y_2)|\geqq a\right\}=0. \tag{3.32}$$

Proof: Let $t_n=e^{\sqrt{n}}$ and note that $\rho_{t_{n+1}}/\rho_{t_n}\to 1$ as $n\to\infty$. Due to Lemma 3.6 and the Borel–Cantelli lemma it will clearly suffice to prove the following two statements in order to prove (3.32):

For any $a>0$, there exists $\delta>0$ such that

$$\sum_n P_0\left\{\sup_{|y_1-y_2|\leqq\delta}|\hat{l}_{t_n}(\omega, y_1)-\hat{l}_{t_n}(\omega, y_2)|\geqq a\right\}<\infty, \tag{3.33}$$

and, for any $a>0$,

$$\sum_n P_0\left\{\sup_y l_{t_n,t_{n+1}}(\omega, y)\geqq\frac{at_n}{\rho_{t_n}(t_{n+1}-t_n)}\right\}<\infty. \tag{3.34}$$

To show (3.33), we first observe that $\hat{l}_t(\omega,\cdot)$ has the same distribution as $l_{\log\log t}(\omega,\cdot)$ and, therefore, we must estimate

$$P_0\left\{\sup_{|y_1-y_2|\leqq\delta}|l_{\log\log t_n}(\omega, y_1)-l_{\log\log t_n}(\omega, y_2)|\geqq a\right\}.$$

For the given $a>0$, we can find, by Theorem 3.5, a $\delta>0$ such that

$$P_0\left\{\sup_{|y_1-y_2|\leqq\delta}|l_{\log\log t_n}(\omega, y_1)-l_{\log\log t_n}(\omega, y_2)|\geqq a\right\} \tag{3.35}$$
$$\leqq \exp\{-3\log\log t_n\}=\frac{1}{n^{3/2}},$$

and hence (3.33) follows.

To show (3.34), we first observe that, since the process has independent increments,

$$P_0\left\{\sup_y l_{t_n,t_{n+1}}(\omega, y)\geqq\frac{at_n}{\rho_{t_n}(t_{n+1}-t_n)}\right\} \tag{3.36}$$
$$=P_0\left\{\sup_y l_{t_{n+1}-t_n}(\omega, y)\geqq\frac{at_n}{\rho_{t_n}(t_{n+1}-t_n)}\right\}.$$

Also, since for any constant $\lambda>0$, $l_{\lambda t}(\omega, y)$ has the same distribution as $(1/\lambda^{1/\alpha})l_t(\omega,(1/\lambda^{1/\alpha})y)$ we see that $\sup_y l_{\lambda t}(\omega, y)$ has the same distribution as $(1/\lambda^{1/\alpha})\sup_y l_t(\omega, y)$ so that

$$P_0\left\{\sup_y l_{\lambda t}(\omega, y)\geqq a\right\}=P_0\left\{\sup_y l_t(\omega, y)\geqq a\lambda^{1/\alpha}\right\},$$

or equivalently

$$P_0\left\{\sup_y l_t(\omega, y)\geqq a\right\}=P_0\left\{\sup_y l_{t/\lambda}(\omega, y)\geqq a\lambda^{1/\alpha}\right\}. \tag{3.37}$$

In particular, let $\lambda=[\rho_{t_n}(t_{n+1}-t_n)/t_n]^\alpha$, we know from (3.37) that

$$P_0\left\{\sup_y l_{t_{n+1}-t_n}(\omega, y)\geqq\frac{at_n}{\rho_{t_n}(t_{n+1}-t_n)}\right\} \tag{3.38}$$
$$=P_0\left\{\sup_y l_{k_n}(\omega, y)\geqq a\right\},$$

where

$$
\begin{aligned}
k_n &= \frac{(t_{n+1}-t_n)t_n^\alpha}{[\rho_{t_n}(t_{n+1}-t_n)]^\alpha} = \frac{t_n^\alpha}{\rho_{t_n}^\alpha}\frac{1}{(t_{n+1}-t_n)^{\alpha-1}} \\
&= \frac{e^{\alpha\sqrt{n}}}{e^{\sqrt{n}}}\frac{\frac{1}{2}\log n}{(e^{\sqrt{n+1}}-e^{\sqrt{n}})^{\alpha-1}} \\
&= \frac{e^{\alpha\sqrt{n}}}{e^{\sqrt{n}}(e^{\sqrt{n}})^{\alpha-1}}\frac{\frac{1}{2}\log n}{(e^{\sqrt{n+1}-\sqrt{n}}-1)^{\alpha-1}} \\
&\geqq (\tfrac{1}{2}\log n)n^{(\alpha-1)/2}\cdot \text{const.} \geqq n^{(\alpha-1)/2}.
\end{aligned}
$$

But now Lemma 3.7 together with (3.36)–(3.38) yield (3.34). This completes Theorem 3.8.

Let $\mathscr{A}$ be the set of uniformly continuous subprobability densities on R_1, i.e.,

$$\mathscr{A} = \left\{ f : f \geqq 0, \int_{-\infty}^{\infty} f(y)\,dy \leqq 1 \text{ and } f \text{ is uniformly continuous on } (-\infty, \infty) \right\}.$$

We impose on $\mathscr{A}$ the topology of uniform convergence on compact subsets of R_1 and we note that $\hat{l}_t(\omega, \cdot) \in \mathscr{A}$.

We now define $B_{t,\omega}$ to be the set of all translates of $\hat{l}_t(\omega, \cdot)$, i.e.,

$$B_{t,\omega} = \{\hat{l}_t(\omega, \cdot + x), x \in R_1\},$$

and state

THEOREM 3.9. *For almost all ω, $B_{t,\omega}$ is conditionally compact in $\mathscr{A}$ as $t \to \infty$ and*

$$\bigcap_T \overline{\bigcup_{t \geqq T} B_{t,\omega}} \subset \{f \in \mathscr{A} : I(f) \leqq 1\}.$$

Moreover, for almost all ω,

$$\bigcap_T \overline{\bigcup_{t \geqq T} \{\hat{l}_t(\omega, \cdot)\}} \supset \{f \in \mathscr{A} : I(f) \leqq 1\}.$$

Proof: Pick a sequence $\{a_n\}$ such that $\lim_{n\to\infty} a_n = 0$. For each a_n there exists a δ_n and a set Ω_n with $P_0\{\Omega_n\} = 1$ such that, for $\omega \in \Omega_n$,

$$\overline{\lim_{t\to\infty}} \sup_{|y_1-y_2|\leqq\delta_n} |\hat{l}_t(\omega, y_1) - \hat{l}_t(\omega, y_2)| \leqq a_n.$$

This, of course, is precisely what we proved in Theorem 3.8. Let $\Omega_\infty = \bigcap_n \Omega_n$ and pick $\omega \in \Omega_\infty$. Let $\{t_k\}$ be a sequence such that $\lim_{k\to\infty} t_k = \infty$ and, for each k, pick $f_k \in B_{t_k,\omega}$. We note that

$$\lim_{\delta\to 0} \overline{\lim_{k\to\infty}} \sup_{|y_1-y_2|\leq\delta} |f_k(y_1)-f_k(y_2)| = 0 .$$

Therefore, by Ascoli's theorem, the f_k are totally bounded in $\mathscr{A}$ and hence $B_{t,\omega}$ is conditionally compact in $\mathscr{A}$ as $t\to\infty$. Thus, the set of limit points in $\mathscr{A}$ (in the topology imposed above on $\mathscr{A}$) is the same as the set of limit points in the vague topology which we imposed on the corresponding measures in the preceding section. Hence, using Theorems 2.8 and 2.9 we complete the proof of this theorem.

We can now state the analogue of Theorem 2.17 and its corollaries for the strong topology in $\mathscr{A}$.

THEOREM 3.10. *Let $\{x(s),\ 0\leqq s<\infty\}$ be a symmetric stable process of index α, $1<\alpha\leqq 2$, and let*

$$\hat{l}_t(\omega, y) = \left(\frac{t}{\log\log t}\right)^{1/\alpha} \frac{1}{t}\int_0^t \delta\left(x(s)-\left(\frac{t}{\log\log t}\right)^{1/\alpha} y\right) ds .$$

Let Φ be a functional on $\mathscr{A}$, the space of substochastic uniformly continuous densities on R_1, which is continuous in the topology of uniform convergence on compact subsets of R_1. Then, for almost all ω,

$$\overline{\lim_{t\to\infty}} \Phi(\hat{l}_t(\omega,\cdot)) = \sup_{f\in\{f\in\mathscr{A}: I(f)\leqq 1\}} \Phi(f) \tag{3.39}$$

and $I(f)$ is defined by (2.10).

We have, as in the preceding section, the following

COROLLARY 3.11. *If Φ is a functional on $\mathscr{A}$ which is lower semicontinuous, then, for almost all ω,*

$$\overline{\lim_{t\to\infty}} \Phi(\hat{l}_t(\omega,\cdot)) \geqq \sup_{f\in\{f\in\mathscr{A}: I(f)\leqq 1\}} \Phi(f) , \tag{3.40}$$

and if Φ *is a functional on* $\mathscr{A}$ *which is upper semicontinuous, then for almost all* ω,

$$\overline{\lim_{t\to\infty}}\, \Phi(\hat{l}_t(\omega,\cdot)) \leqq \sup_{f\in\{f\in\mathscr{A}:I(f)\leqq 1\}} \Phi(f)\,. \tag{3.41}$$

We also have, as before,

Corollary 3.12. *For* $f\in\mathscr{A}$, *let* $f_x(y)=f(y-x)$ *and* $\Phi_x(f)=\Phi(f_x)$. *Then, if* Φ *is a functional on* $\mathscr{A}$ *which is continuous in the topology of* $\mathscr{A}$, *we have, for almost all* ω,

$$\overline{\lim_{t\to\infty}}\, \sup_x \Phi_x(\hat{l}_t(\omega,\cdot)) = \sup_{f\in\{f\in\mathscr{A}:I(f)\leqq 1\}} \Phi(f)\,. \tag{3.42}$$

4. Examples

In this section we present a few examples to illustrate the theorems of the last two sections. Since we treated symmetric stable processes of index $0<\alpha\leqq 2$ all at once, the special case of Brownian motion, when $\alpha=2$, has been normalized to have variance 2 at time 1 instead of the customary variance 1. This should be noted in comparing the results with known results in the Brownian motion case. As in the previous sections, M denotes the class of measures λ on R such that $\lambda(R)\leqq 1$ and $\mathscr{M}\subset M$ denotes the class of probability measures on R.

Example 1. Let $V(x)$ be a continuous function on R such that $V(x)\to 0$ as $|x|\to\infty$. For each $\beta\in M$ consider the functional

$$\Phi(\beta)=\int V(x)\beta(dx)\,.$$

Let $\{x(s),\ 0\leqq s<\infty\}$ be a symmetric stable process of index α, $0<\alpha\leqq 2$, and let $I_\alpha(\beta)$ be the I-function corresponding to this process. From Theorem 2.17 we obtain in this special case

$$\overline{\lim_{t\to\infty}}\, \frac{1}{t}\int_0^t V\!\left(x(s)\left(\frac{\log\log t}{t}\right)^{1/\alpha}\right) ds = \sup_{\beta\in C_\alpha}\int V(x)\beta(dx) \text{ a.e.}, \tag{4.1}$$

where $C_\alpha=\{\beta : I_\alpha(\beta)\leqq 1\}$.

EXAMPLE 2. Let $V(x)$ be a continuous function on R with $V(x)\to\infty$ as $|x|\to\infty$. Consider the functional

$$\Phi(\beta)=\int V(x)\beta(dx)$$

for $\beta\in\mathcal{M}$ provided $\int V(x)\beta(dx)$ exists and is finite; otherwise, define $\Phi(\beta)=+\infty$. If $\beta\in M-\mathcal{M}$, we define $\Phi(\beta)=+\infty$ as well. It then follows from Corollary 2.18 that

$$\varliminf_{t\to\infty}\frac{1}{t}\int_0^t V\left(x(s)\left(\frac{\log\log t}{t}\right)^{1/\alpha}\right)ds=\inf_{\beta\in C_\alpha\cap\mathcal{M}}\int V(x)\beta(dx)\text{ a.e.}\tag{4.2}$$

To see this, we first note that Corollary 2.18 implies that, for almost all ω,

$$\varliminf_{t\to\infty}\frac{1}{t}\int_0^t V\left(x(s)\left(\frac{\log\log t}{t}\right)^{1/\alpha}\right)ds\geqq\inf_{\beta\in C_\alpha\cap\mathcal{M}}\int V(x)\beta(dx).\tag{4.3}$$

To show the reverse inequality, let $\beta_n\to\beta$, where all the β_n and β have a fixed compact support. Then $\Phi(\beta_n)\to\Phi(\beta)$ as $n\to\infty$, and from Theorem 2.15 it follows that for each β having compact support and such that $I_\alpha(\beta)<1$ we have, for almost all ω,

$$\varliminf_{t\to\infty}\frac{1}{t}\int_0^t V\left(x(s)\left(\frac{\log\log t}{t}\right)^{1/\alpha}\right)ds\leqq\int V(x)\beta(dx),$$

so that, for almost all ω,

$$\varliminf_{t\to\infty}\frac{1}{t}\int_0^t V\left(x(s)\left(\frac{\log\log t}{t}\right)^{1/\alpha}\right)ds\leqq\inf_{\substack{\beta\text{ such that }I_\alpha(\beta)<1,\\ \beta\text{ has compact support}}}\int V(x)\beta(dx).\tag{4.4}$$

It is routine to show that the right-hand side of inequality (4.4) is the same as the right-hand side of (4.2).

A special case of Example 2 is when $V(x)=|x|^\theta$ for some $\theta>0$. Then (4.2) becomes

$$\varliminf_{t\to\infty}\frac{(\log\log t)^{\theta/\alpha}}{t^{1+\theta/\alpha}}\int_0^t|x(s)|^\theta\,ds=A_{\alpha,\theta}\text{ a.e.,}\tag{4.5}$$

where

$$A_{\alpha,\theta}=\inf_{\beta\in C_\alpha\cap\mathcal{M}}\int|x|^\theta\beta(dx).$$

In particular when $\theta=\alpha=2$, the constant $A_{2,2}$ can be computed explicitly ($A_{2,2}=\frac{1}{2}$), and so, for Brownian motion paths $x(\cdot)$,

$$\varliminf_{t\to\infty} \frac{\log\log t}{t^2}\int_0^t |x(s)|^2\,ds = \tfrac{1}{2} \text{ a.e.} \tag{4.6}$$

One should compare (4.6) with Strassen's limit version

$$\varlimsup_{t\to\infty} \frac{1}{t^2\log\log t}\int_0^t |x(s)|^2\,ds = \frac{16}{\pi^2} \text{ a.e.} \tag{4.7}$$

There is no analogue of (4.7) for the stable case.

EXAMPLE 3. Consider, for $l>0$, the functional

$$\Phi_l(\beta)=\beta\{x : |x|\leqq l\}.$$

Although $\Phi_l(\beta)$ is not continuous, it is upper semicontinuous and

$$\Phi_l'(\beta)=\beta\{x : |x|< l\}$$

is lower semicontinuous. Now, if $I_\alpha(\beta)<\infty$ for any $0<\alpha\leqq 2$, then β is necessarily absolutely continuous with respect to Lebesgue measure. Therefore,

$$\sup_{\beta\in C_\alpha}\Phi_l(\beta)=\sup_{\beta\in C_\alpha}\Phi_l'(\beta)=k_{\alpha,l}.$$

We can conclude from Corollary 2.18 that for a stable process of order α

$$\varlimsup_{t\to\infty}\frac{1}{t}\operatorname{meas}\left\{s : 0\leqq s\leqq t, |x(s)|\leqq l\left(\frac{t}{\log\log t}\right)^{1/\alpha}\right\}=k_{\alpha,l} \text{ a.e.} \tag{4.8}$$

A computation of $k_{2,l}$ yields the following:

If $l<\frac{1}{2}\pi$, then

$$k_{2,l}=\frac{1+l^2(z_l^2-1)^2}{1+l^2z_l^2(z_l^2-1)^2},$$

where z_l is the root of the equation in z,

$$\frac{1}{z\tan lz}+l=lz^2.$$

If $l\geqq\frac{1}{2}\pi$, then $k_{2,l}=1$.

For $0<\alpha<2$, $k_{\alpha,l}$ is hard to compute but one can still conclude the existence of a smallest l_α such that, for $l \geqq l_\alpha$, $k_{\alpha,l}=1$.

EXAMPLE 4. If $l>l_\alpha$, we conclude from Example 3 that for the symmetric stable process of order α

$$\overline{\lim_{t\to\infty}} \frac{1}{t} \operatorname{meas}\left\{s : 0 \leqq s \leqq t, |x(s)| \leqq l\left(\frac{t}{\log\log t}\right)^{1/\alpha}\right\} = 1 \text{ a.e.}, \tag{4.9}$$

and that l_α is the smallest value of l for which statement (4.9) is true. Using Theorem 2.15 we can see that

$$\underline{\lim_{t\to\infty}} \left(\frac{\log\log t}{t}\right)^{1/\alpha} \sup_{0\leqq s\leqq t} |x(s)| = l_\alpha \text{ a.e.} \tag{4.10}$$

As we saw in Example 3, $l_2 = \frac{1}{2}\pi$ and (4.10) becomes (1.2), the so-called "other law of the iterated logarithm" (see [2] and the remark at the beginning of this section on Brownian motion normalization).

EXAMPLE 5. We now take $1<\alpha\leqq 2$ and consider the local time, $l_t(\omega, y)$, and the theorems of Section 3. In particular if we consider the functional $\Phi(f)=f(0)$, we have from Theorem 3.10

$$\overline{\lim_{t\to\infty}} \left(\frac{t}{\log\log t}\right)^{1/\alpha} \frac{1}{t} l_t(\omega, 0) = d_\alpha \text{ a.e.}, \tag{4.11}$$

where $d_\alpha = \sup_{\substack{f\in\mathcal{A}\\ I_\alpha(f)\leqq 1}} f(0)$. A computation of d_α can be carried out and yields

$$d_\alpha = \frac{\Gamma(1/\alpha)\Gamma(1-1/\alpha)}{\pi(\alpha-1)^{1-1/\alpha}},$$

for $1<\alpha\leqq 2$. We see that when $\alpha=2$, $d_\alpha=1$ and (4.11) becomes (1.4). Using Corollary 3.12 one can prove that

$$\overline{\lim_{t\to\infty}} \left(\frac{t}{\log\log t}\right)^{1/\alpha} \frac{1}{t} \sup_y l_t(\omega, y) = d_\alpha \text{ a.e.} \tag{4.12}$$

Bibliography

[1] Chung, K. L., *On the maximum partial sums of sequences of independent random variables*, T.A.M.S. 64, 1948, pp. 205–233.

[2] Jain, N. C., and Pruitt, W. E., *The other law of the iterated logarithm*, Annals of Prob. 3, 1975, pp. 1046–1049.

[3] Strassen, V., *An invariance principle for the law of the iterated logarithm*, Z. Wahrscheinlich. 3, 1964, pp. 227–246.

[4] Kesten, H., *An iterated logarithm law for local time*, Duke Math. J. 32 (3), 1965, pp. 447–456.

[5] Donsker, M. D., and Varadhan, S. R. S., *Asymptotic evaluation of certain Wiener integrals for large time*, in *Functional Integration and Its Applications*, Proceedings of the International Conference, London; Clarendon Press, Oxford, 1975, pp. 15–33.

[6] Donsker, M. D., and Varadhan, S. R. S., *Asymptotic evaluation of certain Markov process expectations for large time, I*, Comm. Pure Appl. Math. 28, 1975, pp. 1–47.

[7] Donsker, M. D., and Varadhan, S. R. S., *Asymptotic evaluation of certain Markov process expectations for large time, II*, Comm. Pure Appl. Math. 28, 1975, pp. 279–301.

[8] Donsker, M. D., and Varadhan, S. R. S., *Asymptotic evaluation of certain Markov process expectations for large time, III*, Comm. Pure Appl. Math. 29, 1976, pp. 389–461.

[9] Garsia, A. M., Rodemich, E., and Rumsey, H., *A real variable lemma and the continuity of paths of some Gaussian processes*, Indiana Univ. Math. J. 20 (6), 1970, pp. 565–578.

[10] Boylan, E., *Local times for a class of Markov processes*, Illinois J. Math. 8 (1) 1964, pp. 19–39.

[11] Getoor, R. K., and Kesten, H., *Continuity of local times for Markov processes*, Comp. Math. 24 (3), 1972, pp. 277–303.

Received March, 1977.

Symposia Mathematica, Vol. XXI, Academic Press, 1977, 313-318

SOME PROBLEMS OF LARGE DEVIATIONS (*) (**)

M. D. DONSKER and S. R. S. VARADHAN

Let $\xi_1, \xi_2, \ldots, \xi_n, \ldots$ be a sequence of independent identically distributed random variables with a common distribution function $F(x)$ on the line. Let $M(\theta)$ be the moment generating function

$$M(\theta) = \int \exp[\theta x]\, dF(x)$$

which we assume to be finite for all θ. The problem of estimating asymptotically the probability of large deviations for the sample mean

$$\eta_n = \frac{\xi_1 + \ldots + \xi_n}{n}$$

has been studied in great detail. Among other results Cramér proved in [1] that for each $a > m = \int x\, dF(x)$

$$\lim_{n\to\infty} \frac{1}{n} \log P[\eta_n \geqslant a] = -h^+(a)$$

exists and

$$h^+(a) = a\theta_1 - \log M(\theta_1)$$

where θ_1 is the root of $M'(\theta_1)/M(\theta_1) = a$. A similar formula holds for

$$\lim_{n\to\infty} \frac{1}{n} \log P[\eta_n \leqslant a] = -h^- \qquad \text{for } a < m\,.$$

(*) I risultati conseguiti in questo lavoro sono stati esposti da S. R. S. Varadhan nella conferenza tenuta il 23 giugno 1975.

(**) This work was supported by the National Science Foundation, Grant No. NFS-MPS74-01921-A01.

A convenient way of restating these results in a form that generalizes is to start with the function

$$h(a) = \sup_{\theta} [\theta a - \log M(\theta)] \tag{1}$$

and then Cramér's result becomes

THEOREM 1: For every closed set $C \subset R$

$$\limsup_{n\to\infty} \frac{1}{n} \log P[\eta_n \in C] \leqslant - \inf_{a\in C} h(a)$$

and for every open set $G \subset R$

$$\liminf_{n\to\infty} \frac{1}{n} \log P[\eta_n \in G] \geqslant - \inf_{a\in G} h(a) .$$

If $A \subset R$ is such that

$$\inf_{a\in A^0} h(a) = \inf_{a\in A} h(a) = \inf_{a\in \bar{A}} h(a)$$

where A^0 and $\bar{A}$ are respectively the interior and the closure of A then

$$\lim_{n\to\infty} \frac{1}{n} \log P[\eta_n \in A] = - \inf_{a\in A} h(a) .$$

Stated in the above manner the theorem can be generalized. Let V be a vector space and $\xi_1, \ldots, \xi_n, \ldots$ a sequence of independent identically distributed random variables with values in V, having a common distribution $\mu(dx)$. For each θ in V^*, the dual of V we can of course consider

$$M(\theta) = \int \exp [\langle \theta, x\rangle] \mu(dx)$$

and then define

$$h(a) = \sup_{\theta\in V^*} [\langle \theta, a\rangle - \log M(\theta)] . \tag{2}$$

The analog of Theorem 1 if V is finite dimensional is easy to prove and is carried out in [7]. If V is infinite dimensional additional assumptions are needed. We shall now assume that V is a separable Banach space and that $\mu(dx)$ on V is such that

$$\int \exp[\sigma\|x\|]\mu(dx) < \infty \qquad \text{for all } \sigma > 0 .$$

Then the exact analog of Theorem 1 is valid.

THEOREM 2: Let $C \subset V$ be closed and $G \subset V$ be open. Then

$$\limsup_{n\to\infty} \frac{1}{n} \log P[\eta_n \in C] \leqslant - \inf_{a\in C} h(a) ,$$

$$\liminf_{n\to\infty} \frac{1}{n} \log P[\eta_n \in G] \geqslant - \inf_{a\in G} h(a) .$$

We shall give an interesting application of Theorem 2 and then sketch a proof of Theorem 2 based on some recent results of the authors [2].

Suppose the common distribution $\mu(dx)$ is a Gaussian distribution on V with mean zero and a covariance $S(\theta, \varphi) = E[\langle\theta, x\rangle\langle\varphi, x\rangle]$ then the distribution of η_n is also Gaussian. In fact its distribution is the same as that of $\xi_1/\sqrt{n}$, with mean zero and covariance $(1/n)S(\theta, \varphi)$. Therefore

$$P[\eta_n \in A] = \mu[\sqrt{n}A] .$$

Furthermore, from a Theorem of Skorohod [6] for any Gaussian μ, $\int \exp[\sigma\|x\|]\mu(dx) < \infty$ for some $\sigma > 0$. From the fact that the distribution of ξ_1 is the same as that of $\xi_1 + \ldots + \xi_n/\sqrt{n}$ one can read off that $\int \exp[\sigma\|x\|]\mu(dx) < \infty$ for all $\sigma > 0$. This in turn makes Theorem 2 applicable and leads to

THEOREM 3: If μ is a Gaussian measure on V with mean zero, then for any $A \subset V$ with

$$\inf_{a\in A^0} h(a) = \inf_{a\in A} h(a) = \inf_{a\in \bar{A}} h(a) ,$$

$$\lim_{n\to\infty} \frac{1}{n^2} \log \mu(nA) = - \inf_{a\in A} h(a) .$$

Since the moment generating function $M(\theta)$ is just $\exp[\frac{1}{2}S(\theta, \theta)]$ for a Gaussian measure one obtains

$$h(a) = \sup_{\theta} [\langle a, \theta\rangle - \tfrac{1}{2}S(\theta, \theta)] .$$

This makes $h(a)$ the dual quadratic form to S. If V is a Hilbert space $h(a)$ will be the quadratic form $\frac{1}{2}Q(a, a)$ where the operators representing Q and S are inverses of each other. Such a result for the case of the Wiener measure is contained in Schilder [5]. For a Gaussian process on $C[0, 1]$ with Hölder continuous covariance function it was proved by Pincus [4]. Here we see that it is valid for any Gaussian process on V.

If we specialize further and set $A = \{x: \|x\| \leqslant 1\}$ then A is closed and $A^0 = \{x: \|x\| > 1\}$. It is easy to check that

$$\inf_{a \in A} h(a) = \inf_{a \in A^0} h(a) .$$

Hence we have

THEOREM 4.

$$\lim_{n\to\infty} \frac{1}{n^2} \log \mu[\|x\| \geqslant n] = -\inf_{a\in A} h(a) .$$

The computation of $\inf_{a\in A} h(a)$ is interesting. If we let $C = \sup_{\|\theta\| \leqslant 1} S(\theta, \theta)$ then one can verify that

$$\inf_{a\in A} h(a) = \frac{1}{2C} .$$

This result was proved by Shepp and Marcus [3].

We now turn to the proof of Theorem 2. Suppose V for the moment is only looked at as a complete separable metric space. Then given a finite set $\xi_1, \ldots, \xi_n$ of random variables with values in V we can form the sample distribution

$$\lambda_n(\omega, A) = \frac{1}{n}[\chi_A(\xi_1) + \ldots + \chi_A(\xi_n)] .$$

In our case where $\xi_1, \xi_2, \ldots, \xi_n, \ldots$ are independent and have a common distribution μ the ergodic theorem asserts that $\lambda_n(\omega, \cdot)$ is close to $\mu(\cdot)$ weakly with probability close to 1. We therefore introduce the measure Q_n on the space $\mathcal{M}$ of all probability distributions on V. The space $\mathcal{M}$, with weak convergence of probability measures as the convergence notion is a complete separable metric space. We note that as $n \to \infty$, $Q_n \Rightarrow \delta_\mu$ the measure degenerate at the measure μ. We are now interested in its rate.

The role of the $h(\cdot)$ function is now played by the following function $I(\lambda)$. For any λ such that $\lambda \ll \mu$ and $f(x) = d\lambda/d\mu$ with

$$\int |f(x) \log f(x)| \mu(dx) < \infty$$

we define

$$I(\lambda) = \int f(x) \log f(x)\, \mu(dx) .$$

In all other cases we define

$$I(\lambda) = \infty .$$

With this definition $I(\lambda)$ is a nonnegative lower semicontinuous functional on $\mathcal{M}$. It has the additional property that for each $l < \infty$

$$\{\lambda : I(\lambda) \leqslant l\}$$

is a compact subset of $\mathcal{M}$. The following theorem is proved in [2].

THEOREM 5: For $C \subset \mathcal{M}$ closed and $G \subset \mathcal{M}$ open we have

$$\limsup_{n\to\infty} \frac{1}{n} \log Q_n(C) \leqslant - \inf_{\lambda\in C} I(\lambda) ,$$

$$\liminf_{n\to\infty} \frac{1}{n} \log Q_n(G) \geqslant - \inf_{\lambda\in G} I(\lambda) .$$

We shall now indicate how Theorem 2 follows from Theorem 5. Assume for the moment that μ is concentrated in some sphere of finite radius B. Then all the sample distributions will be concentrated on B and therefore we need consider only those measures λ on $\mathcal{M}$ with $\lambda(B) = 1$. Let us consider the map T from such measures λ on $\mathcal{M}$ into V defined by

$$T\lambda = \int x\, d\lambda .$$

Clearly if λ is the sample distribution then $T\lambda$ is the sample mean Since every distribution involved has support in B, T is clearly a continuous map from $\mathcal{M}$ into V. We also have

$$\text{Prob}\,\{\eta_n \in A\} = Q_n\{T^{-1}A\} .$$

Since T is continuous we have that $T^{-1}A$ is closed or open according to A is closed or open in V. By Theorem 5 we conclude that for C closed in V

$$\limsup_{n\to\infty} \frac{1}{n} \log Q_n(T^{-1} C) \leqslant - \inf_{\lambda\in T^{-1}C} I(\lambda) .$$

If G is open in V

$$\liminf_{n\to\infty}\frac{1}{n}\log Q_n(T^{-1}G)\geqslant-\inf_{\lambda\in T^{-1}G} I(\lambda)\,.$$

We can calculate

$$\inf_{\lambda\in T^{-1}A} I(\lambda)=\inf_{a\in A}\inf_{\lambda:\,T\lambda=a} I(\lambda)$$

and verify that

$$h(a)=\inf_{\lambda:\,T\lambda=a} I(\lambda)$$

so that

$$\inf_{\lambda\in T^{-1}A} I(\lambda)=\inf_{a\in A} h(a)\,.$$

The case when μ does not have bounded support is technically more involved. But standard truncation of μ to a large sphere and passing to the limit works. The details are carried out in [2].

We have shown that starting from Theorem 5, Theorems 2, 3 and 4 can be obtained after some elementary computation.

Testo pervenuto il 19 giugno 1975.
Bozze licenziate il 1° dicembre 1976.

REFERENCES

[1] H. Cramér, *On a new limit theorem in the theory of probability*; Colloquium on the Theory of Probability, 1937, Hermann and Co., Paris.

[2] M. D. Donsker and S. R. S. Varadhan, *Asymptotic evaluation of certain Markov process expectations for large time - III*, Comm. Pure App. Math. Vol. XXIX, 4 (1976), pp. 389-461.

[3] M. B. Marcus and L. A. Shepp, *Sample behaviour of certain Gaussian processes*, Proceedings of the Sixth Berkeley Symposium, 1970, Vol. II, pp. 423-441.

[4] M. Pincus, *Gaussian processes and Hammerstein integral equations*, Trans. Amer. Math. Soc., 134 (1968), pp. 193-216.

[5] M. Schilder, *Some asymptotic formulae for Wiener integrals*, Trans. Amer. Math. Soc., 125 (1966), pp. 63-85.

[6] A. V. Skorohod, *A note on Gaussian measures in Banach space*, Theory of Probability and its Applications (English translation), Vol. 15 (1970), p. 508.

[7] S. R. S. Varadhan, *Asymptotic probabilities and differential equations*, Comm. Pure Appl. Math., Vol. 19 (1966), pp. 261-286.

Comm. Pure Appl. Math. 32 (1979) 721-747

On the Number of Distinct Sites Visited by a Random Walk*

M. D. DONSKER AND S. R. S. VARADHAN

1. Introduction

Consider a spatially homogeneous random walk on Z^d with one-step transition probability distribution π, i.e., to each point $j \in Z^d$ is assigned a non-negative probability π_j, $\sum_{j\in Z^d} \pi_j = 1$. If we let $X_1, X_2, X_3, \ldots$ be independent random variables taking values in Z^d and having common distribution π, then $Y_n = X_1 + X_2 + \cdots + X_n$ is the position of the random walk after n steps. Let $D_n = \{j \in Z^d : j = Y_r \text{ for some } r,\ 1 \leqq r \leqq n\}$, and let $D_n^{\#}$ be the cardinality of D_n, i.e., $D_n^{\#}$ is the number of distinct sites in Z^d visited by the random walk in the first n steps.

For $\nu > 0$, consider $E[\exp\{-\nu D_n^{\#}\}]$, where $E\{\ \}$ denotes expectation over the random walks starting from the origin. In this paper we study the asymptotic behavior of $E[\exp\{-\nu D_n^{\#}\}]$ as $n \to \infty$. This behavior depends, of course, on what we assume about π; therefore we state now precisely what these assumptions are. It is convenient to make these hypotheses on the characteristic function of the distribution π. For $t \in R_d$, let $\hat{\pi}(t) = \sum_{j\in Z^d} \pi_j e^{i\langle j,t\rangle}$ which is then periodic of period 2π in each of its d variables, $t = (t_1, t_2, \cdots, t_d)$. About $\hat{\pi}(t)$ we assume the following:

(i) $\hat{\pi}(t) = 1$ if and only if $t = 2\pi(n_1, n_2, \cdots, n_d)$, where $n_1, n_2, \cdots, n_d$ are integers.

(ii) $\hat{\pi}(t) = 1 - \Omega(t) + o(|t|^\alpha)$ as $|t| \to 0$, where $0 < \alpha \leqq 2$ and $e^{-\Omega(t)}$ is the characteristic function of a symmetric stable law of index α in R_d which is non-degenerate, i.e.,

$$\Omega(t) = \begin{cases} \sum a_{ij} t_i t_j \text{ for some} & \\ \text{positive definite matrix } \{a_{ij}\}, & \alpha = 2, \\ \displaystyle\int_{S^{d-1}} \int_0^\infty [1 - \cos\langle t, ry\rangle] \frac{M(dy)\, dr}{r^{\alpha+1}}, & 0 < \alpha < 2. \end{cases}$$

* The research for this paper was done at the Courant Institute of Mathematical Sciences and supported by the National Science Foundation under NSF Grant No. MCS-77-02687.

In the expression for $\Omega(t)$, M is a symmetric measure on the unit ball S^{d-1} in R_d $(M(dy)=M(d(-y)))$, and the assumption of non-degeneracy means that the support of M spans R_d.

Let $\{\pi_j^n\}$ be the n-step transition probabilities. It follows from hypotheses (i) and (ii) above that for any $j \in Z^d$ there exists an n_0 such that $\pi_j^{n_0} > 0$. Moreover, from hypothesis (ii) it follows that as $n \to \infty$

$$\left[\hat{\pi}\left(\frac{t}{c_n}\right)\right]^n \to e^{-k\Omega(t)},$$

provided $n/c_n^\alpha \to k$ as $n \to \infty$. In other words, hypotheses (i) and (ii) imply that the random walk is irreducible and that the distribution π belongs to the domain of normal attraction of a non-degenerate symmetric stable law of index α, $0<\alpha\leqq 2$.

Let $\{x(s),\ 0\leqq s<\infty\}$ be a symmetric stable process in R_d of index $\alpha, 0<\alpha\leqq 2$, corresponding to $\Omega(t)$ and let L_Ω be its infinitesimal generator. Our main result is that under the hypotheses above we have

THEOREM 1. *For* $\nu>0$,

$$\lim_{n\to\infty} \frac{1}{n^{d/(d+\alpha)}} \log E[\exp\{-\nu D_n^\#\}] = -k(\nu, L), \tag{1.1}$$

where

$$k(\nu, L) = k(\nu, L, d, \alpha) = \nu^{\alpha/(d+\alpha)} \left(\frac{d+\alpha}{\alpha}\right)\left(\frac{\alpha\lambda_L}{d}\right)^{d/(d+\alpha)},$$

and

$$\lambda_L = \inf_{G:|G|=1} \lambda(G),$$

the infimum being over all open sets G *in* R_d *of unit volume and* $\lambda(G)$ *the smallest eigenvalue of* $-L_\Omega$ *with Dirichlet boundary conditions for* G.

In [1] (Lemma 3.10), we showed that the constant λ_L appearing in Theorem 1 is always strictly positive. We note also that $k(\nu, L)$ in Theorem 1 is the same constant that appears in the asymptotics for the symmetric stable sausage (Theorem 3 in [1]) which is obviously the continuous analogue of the problem treated here. Consider the special case of Theorem 1 where $\alpha=2$ and the infinitesimal generator $L_\Omega=\frac{1}{2}\Delta$. Then $\inf_{G:|G|=1}\lambda(G)$ is attained for the sphere of unit volume (Lemma 3.13) in [1]) so that in this special case

$$k(\nu, L) = \nu^{2/(d+2)}\left(\frac{d+2}{2}\right)\left(\frac{2\gamma_d}{d}\right)^{d/(d+2)}, \tag{1.2}$$

where γ_d is the lowest eigenvalue of $-\frac{1}{2}\Delta$ for the sphere of unit volume in d dimensions with zero boundary values. The prototype random walk in Z^d, where transition is possible only to adjacent lattice points with equal probability, satisfies hypotheses (i) and (ii) for $\alpha = 2$ and Theorem 1 applies with $k(\nu, L)$ given in this special case by (1.2).

It is clear that in the asymptotic evaluation of $E[\exp\{-D_n^{\#}\}]$ as $n \to \infty$ the random walks which contribute most to the integral are those for which $D_n^{\#}$ are small so that what is required here are estimates on the probabilities of large deviations for $D_n^{\#}$.

2. Notation, Local Limit Theorem, and Upper Estimates

Certain estimates required for proving Theorem 1 are best made when the random walk is on a compact space; thus we find it convenient to map the walk on Z^d into an appropriately selected lattice in the d-dimensional torus T_d. In this mapping the periodicity of the walk must be taken into account.

Let $p = \text{g.c.d. } \{n: \pi_0^n \neq 0\}$ be the period, and let $\mathcal{T}_0 = \{t \in R_d : |\hat{\pi}(t)| = 1\}$ so that $\mathcal{T}_0 \supset \{2\pi(n_1, n_2, \cdots, n_d)\}$ where $n_1, n_2, \cdots, n_d$ are all integers. If $H \subset Z^d$ is the annihilator of $\mathcal{T}_0$, i.e.,

$$H = \{j \in Z^d : e^{i\langle j, t\rangle} = 1 \text{ for all } t \in \mathcal{T}_0\},$$

then H is a subgroup of Z^d and Z^d/H is cyclic with p elements. Moreover, if j is such that $\pi_j > 0$, then $H, H+j, \cdots, H+(p-1)j$ are the cyclically moving subclasses of the random walk.

Let $\lambda > 0$ be given, and let $\{a_n\}$ be a sequence of integers such that $p \mid a_n$ for every n, and

$$\frac{a_n}{n^{1/(d+\alpha)}} \to \lambda \quad \text{as} \quad n \to \infty. \tag{2.1}$$

Consider $R_d/Z^d = T_d$, the d-dimensional torus. We can think of T_d as the d-dimensional cube, $|x_i| \leqq \frac{1}{2}$, $i = 1, 2, \cdots, d$, with opposite sides identified. For each n, let

$$\mathscr{L}_d^{(n)} = \left\{ x \in T_d : x = \frac{j}{a_n} \text{ for some } j \in z^d \right\},$$

and, for $x \in \mathscr{L}_d^{(n)}$, let

$$\tilde{\pi}_n(x) = \sum_{j : j/a_n = x (\mathrm{mod}\, 1)} \pi_j .$$

Thus, $\tilde{\pi}_n(x)$ serves as the one-step transition probabilities for a random walk on the subgroup $\mathscr{L}_d^{(n)} \subset T_d$. Moreover, since $p \mid a_n$, the random walk on $\mathscr{L}_d^{(n)}$ is also periodic of period p, the cyclically moving subclasses being the images of H, $H+j, \cdots, H+(p-1)j$ under the canonical mapping of Z^d into $\mathscr{L}_d^{(n)}$. We shall denote these cyclically moving subclasses by $\tilde{H}_{d,r}^{(n)}$ for $r \geqq 1$ where $\tilde{H}_{d,r+p}^{(n)} = \tilde{H}_{d,r}^{(n)}$.

For any $\rho > 0$, let

$$\Phi(\rho, x) = \frac{1}{(2\pi)^d} \int_{R_d} e^{-\rho\Omega(t)} \cos \langle t, x \rangle \, dt,$$

and let $\tilde{\Phi}(\rho, x) = \sum_{j \in Z^d} \Phi(\rho, x+j)$. For $x \in \mathscr{L}_d^{(n)}$, denote by $\tilde{\pi}_n^l(x)$ the l-step transition probability for the random walk on $\mathscr{L}_d^{(n)}$, i.e.,

$$\tilde{\pi}_n^l(x) = \sum_{\substack{j \in Z^d \\ j/a_n = x (\mathrm{mod}\, 1)}} \pi_j^l .$$

From our assumptions, the following local limit theorem follows. The proof is omitted since it conforms to the usual pattern for local limit theorems.

THEOREM 2.

$$\lim_{\substack{l \to \infty \\ n \to \infty \\ ln^{\alpha/(\alpha+d)} \to p}} \sup_{x \in \tilde{H}_{d,l}^{(n)}} \left| a_n^d \tilde{\pi}_n^l(x) - p \tilde{\Phi}\left(\frac{\rho}{\lambda^\alpha}, x\right) \right| = 0 . \tag{2.2}$$

Let $\tilde{X}_1, \tilde{X}_2, \cdots$ be independent random variables taking values in $\mathscr{L}_d^{(n)}$ with common distribution $\tilde{\pi}_n(x)$, and let $\tilde{Y}_l = \tilde{X}_1 + \tilde{X}_2 + \cdots + \tilde{X}_l$ (addition on T_d). Denote by $\tilde{D}_n^{\#}(l)$ the number of distinct sites in $\mathscr{L}_d^{(n)}$ visited by

$\tilde{Y}_1, \tilde{Y}_2, \cdots, \tilde{Y}_l$. We recall that

$$D_n = \{j \in Z^d : j = Y_r \text{ for some } 1 \leqq r \leqq n\},$$

so that if we let

$$\bar{D}_n = \{x \in T_d : j/a_n = x \ (\text{mod } 1) \text{ for some } j \in D_n\},$$

we observe that $\bar{D}_n \subset \mathcal{L}_d^{(n)}$ and $\bar{D}_n^{\#} \leqq D_n^{\#}$. What is important is to note that the distribution of $\bar{D}_n^{\#}$ is the same as the distribution of $\tilde{D}_n^{\#}(n)$. We explain now why this is important for getting upper estimates on the expectation appearing in Theorem 1.

For A, a Borel set in T_d, we consider

$$\tilde{L}_l(\omega, A) = \frac{1}{l} \sum_{j=1}^{l} \chi_A(\tilde{Y}_j), \tag{2.3}$$

i.e., the proportion of visits to A made by the random walk in $\mathcal{L}_d^{(n)}$ in the first l steps. We note that, for fixed ω, l, and n, $\tilde{L}_l(\omega, \cdot)$ is a probability measure on T_d. Let $\mathcal{M}(T_d)$ be the space of all probability measures on T_d endowed with the weak topology. We shall subsequently consider the distribution of $\tilde{L}_l(\omega, \cdot)$ in $\mathcal{M}(T_d)$, i.e., let $P_l^{(n)}$ be a probability measure on $\mathcal{M}(T_d)$ defined by

$$P_l^{(n)}(C) = P\{\tilde{L}_l(\omega, \cdot) \in C\},$$

where C is a set of measures in $\mathcal{M}(T_d)$ and P is the probability measure on the random walks in $\mathcal{L}_d^{(n)}$ starting from the origin.

For $x \in T_d$, define

$$\tilde{L}_l^{(n)}(\omega, x) = \int_{T_d} \chi_n(x - y) \tilde{L}_l(\omega, dy), \tag{2.4}$$

where

$$\chi_n(x) = \begin{cases} a_n^d & \text{if } |x_i| \leqq \dfrac{1}{2a_n} \text{ for all } 1 \leqq i \leqq d, \\ 0 & \text{otherwise}. \end{cases} \tag{2.5}$$

Let $|x : \tilde{L}_l^{(n)}(\omega, x) > 0|$ denote the Lebesgue measure of $\{x: \tilde{L}_l^{(n)}(\omega, x) > 0\}$.

We observe that

$$|x : \tilde{L}_l^{(n)}(\omega, x) > 0| = \frac{1}{a_n^d} \tilde{D}_n^{\#}(l). \tag{2.6}$$

For fixed ω, l, and n, $\tilde{L}_l^{(n)}(\omega, \cdot)$ is a probability density on T_d. Let $L_1(T_d)$ be the space of all probability densities on T_d endowed with the L_1 topology. Let $Q_l^{(n)}$ be the probability measure on $L_1(T_d)$ defined as the distribution of $\tilde{L}_l^{(n)}(w, \cdot)$ on $L_1(T_d)$. In other words, if B is a set in $L_1(T_d)$ we define

$$Q_l^{(n)}(B) = P[\tilde{L}_l^{(n)}(\omega, \cdot) \in B], \tag{2.7}$$

where P is again the probability measure on the random walks in $L_d^{(n)}$ starting from the origin. Clearly,

$$Q_l^{(n)}[f \in L_1(T_d) : f \in B] = P_l^{(n)}[\mu \in \mathcal{M}(T_d) : \chi_n * \mu \in B].$$

From the comments and definitions above, we now see that

$$E\{\exp\{-\nu D_n^{\#}\}\} \leqq E\{\exp\{-\nu \tilde{D}_n^{\#}\}\} = E^{Q_n^{(n)}}\{\exp\{-\nu a_n^d |x : f(x) > 0|\}\}. \tag{2.8}$$

Corresponding to the $\Omega(t)$ which appears in the hypotheses on π, there is a symmetric stable process $\{x(s),\ 0 \leqq s \leqq \infty\}$ in R_d the infinitesimal generator of which acting on smooth functions on R_d is given by

$$L_\Omega u = \begin{cases} \sum a_{ij} \dfrac{\partial^2 u}{\partial x_i \partial x_j}, & \alpha = 2, \\ \displaystyle\int_{S^{d-1}} \int_0^\infty \left[\frac{u(x+ry) + u(x-ry)}{2} - u(x) \right] \frac{M(dy)\, dr}{r^{1+\alpha}}, & 0 < \alpha < 2. \end{cases} \tag{2.9}$$

The process $\{x(s), 0 \leqq s < \infty\}$ can be projected onto the torus T_d. Denote it by $\{\tilde{x}(s),\ 0 \leqq s < \infty\}$; then this is again a process with independent increments the generator $\tilde{L}_\Omega$ of which acts on smooth functions on T_d. The relation between L_Ω and $\tilde{L}_\Omega$ is very simple: Take a function u on T_d, view it as a periodic function on R_d, apply L_Ω to this periodic extension thereby getting a periodic function which is now $\tilde{L}_\Omega$ applied to u.

Let $\mathcal{U}$ be the space of positive functions $u \in C^\infty(T_d)$ each of which is strictly bounded away from zero. For $\mu \in \mathcal{M}(T_d)$, we define

$$I_{\tilde{L}}(\mu) = -\inf_{u \in \mathcal{U}} \int_{T_d} \left(\frac{\tilde{L}_\Omega u}{u} \right)(x) \mu(dx), \tag{2.10}$$

and if μ is given by a density f it will be convenient to denote $I_{\tilde{L}}(\mu)$ by $I_{\tilde{L}}(f)$.

For the λ picked just before (2.1) we define

$$I_\lambda(f)=\frac{1}{\lambda^{\alpha+d}}I_L(f).$$

We shall prove below (Theorem 6) that if $C\subset L_1(T_d)$ is closed, then

$$\overline{\lim_{n\to\infty}}\frac{1}{a_n^d}\log Q_n^{(n)}(C)\leqq-\inf_{f\in C}I_\lambda(f). \tag{2.11}$$

Since $|x : f(x)>0|$ is a lower semicontinuous function on $L_1(T_d)$, it follows from (2.11) (see corollary to Theorem 6 below) that for $\nu>0$

$$\begin{aligned}&\overline{\lim_{n\to\infty}}\frac{1}{a_n^d}\log E^{Q_n^{(n)}}\{\exp\{-\nu a_n^d|x : f(x)>0|\}\}\\&\qquad\leqq-\inf_{g\in L_1(T_d)}\{\nu|x : g(x)>0|+I_\lambda(g)\}.\end{aligned} \tag{2.12}$$

Recalling (2.1) and using (2.8) we then conclude that

$$\overline{\lim_{n\to\infty}}\frac{1}{n^{d/(d+\alpha)}}\log E[\exp\{-\nu D_n^{\#}\}]\leqq-\lambda^d\inf_{g\in L_1(T_d)}\{\nu|x : g(x)>0|+I_\lambda(g)\}. \tag{2.13}$$

But (2.13) holds for all $\lambda>0$ so that

$$\begin{aligned}&\overline{\lim_{n\to\infty}}\frac{1}{n^{d/(d+\alpha)}}\log E[\exp\{-\nu D_n^{\#}\}]\\&\qquad\leqq-\sup_{\lambda>0}\lambda^d\left[\inf_{g\in L_1(T_d)}\{\nu|x : g(x)>0|+I_\lambda(g)\}\right],\end{aligned} \tag{2.14}$$

and at the end of Section 4 we shall show that

$$\sup_{\lambda>0}\left[\lambda^d\inf_{g\in L_1(T_d)}\{v|x : g(x)>0)+I_\lambda(g)\}\right]=k(\nu,L), \tag{2.15}$$

where $k(\nu, L)$ is defined in the statement of Theorem 1. From (2.14) and (2.15) we have

$$\overline{\lim_{n\to\infty}}\frac{1}{n^{d/(d+\alpha)}}\log E[\exp\{-\nu D_n^{\#}\}]\leqq-k(\nu,L). \tag{2.16}$$

In Section 5 we shall show

$$\varliminf_{n\to\infty} \frac{1}{n^{d/(d+\alpha)}} \log E[\exp\{-\nu D_n^{\#}\}] \geqq -k(\nu, L) \tag{2.17}$$

and thus will have proved Theorem 1.

3. Preliminary Results

We first prove

THEOREM 3. *If C is closed in $\mathcal{M}(T_d)$, then*

$$\overline{\lim_{n\to\infty}} \frac{1}{a_n^d} \log P_n^{(n)}(C) \leqq -\inf_{\mu\in C} I_\lambda(\mu) . \tag{3.1}$$

Proof: Let $u \in \mathcal{U}$ and for $x \in T_d$ define the operator

$$(\tilde{\pi}_n u)(x) = \sum_{y\in\mathcal{L}_d^{(n)}} \tilde{\pi}_n(y)\, u(x+y) .$$

Let

$$V_n(x) = \log\left(\frac{\tilde{\pi}_n u}{u}\right)(x) = \log\left(1 + \frac{\tilde{\pi}_n u - u}{u}\right). \tag{3.2}$$

From elementary properties of Markov processes we deduce (c.f. (2.2) in [2]) that, for each $u \in \mathcal{U}$ and all n,

$$E^{P_n^{(n)}}\left\{\exp\left\{-n\int_{T_d} V_n(x)\mu(dx)\right\}\right\} \leqq \frac{\sup_{x\in T_d} u(x)}{\inf_{x\in T_d} u(x)}. \tag{3.3}$$

With $L_\Omega u$ as defined in (2.9), we now assert that, for all $u \in \mathcal{U}$,

$$\tilde{\pi}_n u - u = \frac{1}{a_n^\alpha} \tilde{L}_\Omega u + o\left(\frac{1}{a_n^\alpha}\right) \tag{3.4}$$

uniformly for $x \in T_d$ as $n \to \infty$. To see (3.4), it suffices to verify it for functions u of the form $e^{i\langle t,x\rangle}$ on T_d, where t is a multiple of 2π since linear combinations of such functions are dense in $\mathcal{U}$. It is easy to check that, whether $\alpha = 2$ or $0 < \alpha < 2$, for these special functions u, $\tilde{L}_\Omega u = -e^{i\langle t,x\rangle}\Omega(t)$

and $\tilde{\pi}_n u = e^{i\langle t,x\rangle}\hat{\pi}(t/a_n)$. Hence

$$\tilde{\pi}_n u - u = e^{i\langle t,x\rangle}\left(\hat{\pi}\left(\frac{t}{a_n}\right) - 1\right),$$

and verifying (3.4) for these special functions u is equivalent to showing

$$\hat{\pi}\left(\frac{t}{a_n}\right) - 1 = -\frac{1}{a_n^\alpha}\,\Omega(t) + o\left(\frac{1}{a_n^\alpha}\right). \tag{3.5}$$

But $\Omega(t)$ has the property that $\Omega(t/a_n) = (1/a_n^\alpha)\Omega(t)$, and therefore (3.5) follows from hypothesis (ii) on $\hat{\pi}(t)$ in the introduction.

From (3.4), for any $u \in \mathcal{U}$,

$$\frac{\tilde{\pi}_n u - u}{u} = \frac{1}{a_n^\alpha}\,\frac{\tilde{L}_\Omega u}{u} + o\left(\frac{1}{a_n^\alpha}\right)$$

and hence, as $n \to \infty$,

$$\begin{aligned} V_n(x) = \log\left(1 + \frac{\tilde{\pi}_n u - u}{u}\right) &= \frac{\tilde{\pi}_n u - u}{u} + O\left(\frac{1}{a_n^{2\alpha}}\right) \\ &= \frac{1}{a_n^\alpha}\,\frac{\tilde{L}_\Omega u}{u} + o\left(\frac{1}{a_n^\alpha}\right) \end{aligned}$$

uniformly for $x \in T_d$. Thus, for each $u \in \mathcal{U}$, there is a sequence of constants $\delta_n \to 0$ such that

$$V_n(x) \leqq \frac{1}{a_n^\alpha}\left[\frac{\tilde{L}_\Omega u}{u} + \delta_n\right]. \tag{3.6}$$

From (3.3) and (3.6), for each $u \in \mathcal{U}$,

$$E^{P_n^{(n)}}\left\{\exp\left\{-\frac{n}{a_n^\alpha}\int_{T_d}\left(\frac{\tilde{L}_\Omega u}{u}\right)(x)\mu(dx)\right\}\right\} \leqq \frac{\sup_{x\in T_d} u(x)}{\inf_{x\in T_d} u(x)}\exp\left\{\frac{n\delta_n}{a_n^\alpha}\right\}, \tag{3.7}$$

This implies that, for each $u \in \mathcal{U}$,

$$P_n^{(n)}(C) \leqq \frac{\sup_{x\in T_d} u(x)}{\inf_{x\in T_d} u(x)}\exp\left\{\frac{n\delta_n}{a_n^\alpha} + \frac{n}{a_n^\alpha}\sup_{\mu\in C}\int_{T_d}\left(\frac{\tilde{L}_\Omega u}{u}\right)(x)\mu(dx)\right\}$$

and, therefore, letting $b_n = n/a_n^\alpha$, we have

$$\overline{\lim_{n\to\infty}} \frac{1}{b_n} \log P_n^{(n)}(C) \leqq \sup_{\mu \in C} \int_{T_d} \left(\frac{\tilde{L}_\Omega u}{u}\right)(x)\mu(dx), \tag{3.8}$$

where we note from (2.1) that

$$b_n = \frac{n}{a_n^\alpha} \sim \frac{n}{\lambda^\alpha n^{\alpha/(d+\alpha)}} = \frac{n^{d/(d+\alpha)}}{\lambda^\alpha}. \tag{3.9}$$

Now (3.8) holds for all $u \in \mathcal{U}$ and hence

$$\overline{\lim_{n\to\infty}} \frac{1}{b_n} \log P_n^{(n)}(C) \leqq \inf_{u \in \mathcal{U}} \sup_{\mu \in C} \int_{T_d} \left(\frac{\tilde{L}_\Omega u}{u}\right)(x)\mu(dx). \tag{3.10}$$

Since C is closed in $\mathcal{M}(T_d)$ and $\mathcal{M}(T_d)$ is compact in the weak topology, we see that C is compact, and it follows (c.f. [4] just before (2.13)) that

$$\overline{\lim_{n\to\infty}} \frac{1}{b_n} \log P_n^{(n)}(C) \leqq \sup_{\mu \in C} \inf_{u \in \mathcal{U}} \int_{T_d} \left(\frac{\tilde{L}_\Omega u}{u}\right)(x)\mu(dx). \tag{3.11}$$

Thus, from (3.11) and the definition of $I_{\tilde{L}}(\mu)$ we get

$$\begin{aligned}\overline{\lim_{n\to\infty}} \frac{1}{b_n} \log P_n^{(n)}(C) &\leqq \sup_{\mu \in C} (-I_{\tilde{L}}(\mu)) \\ &= -\inf_{\mu \in C} I_{\tilde{L}}(\mu).\end{aligned} \tag{3.12}$$

From (3.9) and (2.1) we see that $b_n \sim a_n^d/\lambda^{\alpha+d}$ which allows us to rewrite (3.12) as

$$\begin{aligned}\overline{\lim_{n\to\infty}} \frac{1}{a_n^d} \log P_n^{(n)}(C) &\leqq -\inf_{\mu \in C} \frac{1}{\lambda^{\alpha+d}} I_{\tilde{L}}(\mu) \\ &= -\inf_{\mu \in C} I_\lambda(\mu),\end{aligned}$$

which completes the proof.

THEOREM 4. *Let C be closed in $\mathcal{M}(T_d)$. Then*

$$\overline{\lim_{n\to\infty}} \frac{1}{a_n^d} \log Q_n^{(n)}(C) \leqq -\inf_{\mu \in C} I_\lambda(\mu).$$

Proof: Let C be weakly closed in $\mathcal{M}(T_d)$. With χ_n defined by (2.5), we observe that $\chi_n * \mu$ converges weakly to μ as $n \to \infty$ uniformly for $\mu \in \mathcal{M}(T_d)$. Thus, for any $\varepsilon > 0$ there exists n_0 such that, for $n \geqq n_0$, $\chi_n * \mu \in C$ implies $\mu \in C^\varepsilon$, where C^ε is the set of measures in $\mathcal{M}(T_d)$ within distance ε (Prokhorov distance) of C. Therefore, for $n \geqq n_0$, $Q_n^{(n)}(C) \leqq P_n^{(n)}(C^\varepsilon)$, and letting $\overline{C^\varepsilon}$ denote the weak closure of C^ε we have from Theorem 3,

$$\overline{\lim_{n\to\infty}} \frac{1}{a_n^d} \log Q_n^{(n)}(C) \leqq \overline{\lim_{n\to\infty}} \frac{1}{a_n^d} \log P_n^{(n)}(\bar{C}^\varepsilon) \leqq -\inf_{\mu \in \overline{C^\varepsilon}} I_\lambda(\mu). \tag{3.13}$$

This last holds for every $\varepsilon > 0$, and since $I_\lambda(\mu)$ is lower semicontinuous and $\mathcal{M}(T_d)$ is compact in the weak topology, we obtain the theorem by letting $\varepsilon \to 0$ in (3.13).

Let $\varepsilon > 0$ be given and let $\psi_\varepsilon(x)$ be a C^∞ function on T_d with $\psi_\varepsilon(x) = \psi_\varepsilon(-x)$, $\psi_\varepsilon(x) \geqq 0$, $\int_{T_d} \psi_\varepsilon(x)\, dx = 1$, and $\psi_\varepsilon(x) \to \delta(x)$ as $\varepsilon \to 0$. Let $K_\varepsilon : L_1(T_d) \to L_1(T_d)$ be defined by

$$(K_\varepsilon f)(x) = \int_{T_d} \psi_\varepsilon(x-y) f(y)\, dy. \tag{3.14}$$

We now prove

THEOREM 5. *For every $\delta > 0$,*

$$\overline{\lim_{\varepsilon\to 0}}\ \overline{\lim_{n\to\infty}} \frac{1}{a_n^d} \log Q_n^{(n)}\left[f : \int_{T_d} |(K_\varepsilon f)(x) - f(x)|\, dx \geqq \delta\right] = -\infty. \tag{3.15}$$

Proof: From the definition of $Q_n^{(n)}$ measure on $L_1(T_d)$ and $P_n^{(n)}$ measure on $\mathcal{M}(T_d)$, we have

$$\begin{aligned} &Q_n^{(n)}\left[f : \int_{T_d} |(K_\varepsilon f)(x) - f(x)|\, dx \geqq \delta\right] \\ &= P_n^{(n)}\left[\mu : \int_{T_d} |(K_\varepsilon \chi_n * \mu)(x) - (\chi_n * \mu)(x)|\, dx \geqq \delta\right] \\ &= P_n^{(n)}\left[\mu : \sup_{\substack{g \in C(T_d) \\ \|g\| \leq 1}} \int_{T_d} [(K_\varepsilon \chi_n * \mu)(x) - (\chi_n * \mu)(x)] g(x)\, dx \geqq \delta\right]. \end{aligned} \tag{3.16}$$

Writing $\chi_n * g = \chi_n g$ and using the symmetry of χ_n and ψ_ε, we see that the last expression in (3.16) is the same as

$$(3.17)\qquad P_n^{(n)}\left[\mu : \sup_{\substack{g\in C(T_d)\\ \|g\|\leq 1}} \int_{T_d} [(K_\varepsilon - I)\chi_n g]\mu(dx) \geqq \delta\right].$$

Moreover, if we let $M_{n,\varepsilon} \subset C(T_d)$ be the set of all functions $V(x)$ of the form $V = (K_\varepsilon - I)\chi_n g$, where $g \in C(T_d)$ with $\|g\| \leqq 1$, then (3.17) is the same as (see definition of $P_l^{(n)}$ measure just after (2.3))

$$(3.18)\qquad P\left[\sup_{V\in M_{n,\varepsilon}} [V(\tilde{Y}_1) + V(\tilde{Y}_2) + \cdots + V(\tilde{Y}_n)] \geqq n\delta\right],$$

where, as before, $\tilde{Y}_k = \tilde{X}_1 + \tilde{X}_2 + \cdots + \tilde{X}_k$ and $\tilde{X}_1, \tilde{X}_2, \cdots$ are independent random variables with values in $\mathscr{L}_d^{(n)}$ having common distribution $\tilde{\pi}_n(x)$. Thus

$$(3.19)\qquad \begin{aligned} &Q_n^{(n)}\left[f : \int_{T_d} |(K_\varepsilon f)(x) - f(x)|\, dx \geqq \delta\right] \\ &\qquad = P\left[\sup_{V\in M_{n,\varepsilon}} [V(\tilde{Y}_1) + V(\tilde{Y}_2) + \cdots + V(\tilde{Y}_n)] \geqq n\delta\right]. \end{aligned}$$

In Lemma 4.1 below we show that, given n, ε, δ, there exist functions V_1, $V_2, \cdots, V_J$ all in $M_{n,\varepsilon}$ such that, for any function $V \in M_{n,\varepsilon}$,

$$\inf_{1\leq i\leq J} \sup_{x\in\mathscr{L}_d^{(n)}} |V(x) - V_i(x)| \leqq \tfrac{1}{2}\delta,$$

and, moreover, $J = J(n, \varepsilon, \delta) \leqq ([8/\delta] + 1)^{a_n^d}$. In other words, in Lemma 4.1 we show that M_n is totally bounded, and we estimate the number of spheres of radius $\frac{1}{2}\delta$ it takes to cover $M_{n,\varepsilon}$. Thus

$$(3.20)\qquad \begin{aligned} &P\left[\sup_{V\in M_{n,\varepsilon}} [V(\tilde{Y}_1) + V(\tilde{Y}_2) + \cdots + V(\tilde{Y}_n)] \geqq n\delta\right] \\ &\quad \leqq P\left[\sup_{1\leq i\leq J} [V_i(\tilde{Y}_1) + V_i(\tilde{Y}_2) + \cdots + V_i(\tilde{Y}_n)] \geqq \tfrac{1}{2}n\delta\right] \\ &\quad \leqq \left(\left[\frac{8}{\delta}\right] + 1\right)^{a_n^d} \sup_{1\leq i\leq J} P[[V_i(\tilde{Y}_1) + V_i(\tilde{Y}_2) + \cdots + V_i(\tilde{Y}_n)] \geqq \tfrac{1}{2}n\delta]. \end{aligned}$$

In Lemma 4.4 below we shall show that there exists an n_0 such that $n \geqq n_0$ implies, for any $\theta > 0$, and any V of the type being considered here,

$$E[\exp\{\theta[V(\tilde{Y}_1)+V(\tilde{Y}_2)+\cdots+V(\tilde{Y}_n)]\}] < c\exp\{4\theta(n^{\alpha/(\alpha+d)}+p)\}\exp\{n\lambda_n(\theta V)\}, \tag{3.21}$$

where c is a constant and

$$\lambda_n(\theta V) = \sup_{\mu\in\mathcal{M}(T_d)}\left[\theta\int_{T_d} V(x)\mu(dx) - \tilde{I}_n(\mu)\right]$$

with

$$\tilde{I}_n(\mu) = -\inf_{u\in\mathcal{U}_1}\int_{T_d}\log\left(\frac{\tilde{\pi}_n u}{u}\right)(x)\mu(dx),$$

$\mathcal{U}_1$ being the set of functions $u \in C(T_d)$ such that $u > 0$ on T_d. Thus, from (3.19), (3.20), and (3.21) we obtain, for $n \geqq n_0$ and any $\theta > 0$,

$$Q_n^{(n)}\left[f : \int_{T_d} |(K_\epsilon f)(x) - f(x)|\,dx \geqq \delta\right] \leqq \left(\left[\frac{8}{\delta}\right]+1\right)^{a_n^d} c\exp\{4\theta(n^{\alpha/(\alpha+d)}+p\}\exp\{-\tfrac{1}{2}n\delta\theta\}\sup_{1\leqq i\leqq J}\exp\{n\lambda_n(\theta V_i)\}. \tag{3.22}$$

In (3.22) choose $\theta = \tau/n^{\alpha/(d+\alpha)}$ where $\tau > 0$; then (3.22) becomes

$$Q_n^{(n)}\left[f : \int_{T_d} |(K_\epsilon f)(x) - f(x)|\,dx \geqq \delta\right] \leqq \left(\left[\frac{8}{\delta}\right]+1\right)^{a_n^d} c\exp\left\{4\tau + \frac{4p\tau}{n^{\alpha/(\alpha+d)}}\right\}\exp\{-\tfrac{1}{2}n^{d/(\alpha+d)}\tau\delta\} \times \sup_{1\leqq i\leqq J}\exp\left\{n\lambda_n\left(\frac{\tau}{n^{\alpha/(\alpha+d)}}V_i\right)\right\}. \tag{3.23}$$

Recalling that $|V| \leqq 2$ for all V under consideration, $\tilde{I}_n(\mu) > 0$, and

$$\lambda_n\left(\frac{\tau}{n^{\alpha/(\alpha+d)}}V\right) = \sup_{\mu\in\mathcal{M}(T_d)}\left[\frac{\tau}{n^{\alpha/(\alpha+d)}}\int_{T_d} V(x)\mu(dx) - \tilde{I}_n(\mu)\right], \tag{3.24}$$

we see that

$$\frac{-2\tau}{n^{\alpha/(\alpha+d)}} \leqq \lambda_n\left(\frac{\tau}{n^{\alpha/(\alpha+d)}} V\right) \leqq \frac{2\tau}{n^{\alpha/(\alpha+d)}}.$$

Hence, in (3.24) nothing is changed if, in taking the supremum over all μ, we take it instead over all μ such that $\tilde{I}_n(\mu) \leqq \frac{\tau}{n^{\alpha/(\alpha+d)}}$, i.e.,

$$\lambda_n\left(\frac{\tau}{n^{\alpha/(\alpha+d)}} V\right) \leqq \sup_{0<\sigma\leqq\tau} \sup_{\mu:\tilde{I}_n(\mu)=\sigma/n^{\alpha/(\alpha+d)}} \left[\frac{\tau}{n^{\alpha/(\alpha+d)}} \int_{T_d} V(x)\mu(dx) - \frac{\sigma}{n^{\alpha/(\alpha+d)}}\right]. \tag{3.25}$$

But, it follows from Lemma 4.5 below that, for any of the V under consideration, i.e., $V=(K-I)\chi_n g$ where $g \in C(T_d)$ with $\|g\| \leqq 1$, and any $t>0$,

$$\sup_{\mu:\tilde{I}_n(\mu)=\sigma/n^{\alpha/(\alpha+d)}} \frac{\tau}{n^{\alpha/(\alpha+d)}} \int_{T_d} V(x)\mu(dx) \leqq \frac{\tau}{n^{\alpha/(\alpha+d)}}[2h(t\sigma)+2\Delta_t(n)+k_t(\varepsilon)], \tag{3.26}$$

where $h(\sigma)\to 0$ as $\sigma\to 0$, $\Delta_t(n)\to 0$ as $n\to\infty$ for all $t>0$, and $k_t(\varepsilon)\to 0$ as $\varepsilon\to 0$ for all $t>0$. The functions $h(\sigma)$, $\Delta_t(n)$ and $k_t(\varepsilon)$ are all independent of V.

Thus, from (3.25) and (3.26), we get

$$\begin{aligned} \lambda_n\left(\frac{\tau}{n^{\alpha/(\alpha+d)}} V\right) &\leqq \sup_{0<\sigma\leqq\tau} \left[\frac{\tau}{n^{\alpha/(\alpha+d)}}\{2h(t\sigma)+2\Delta_t(n)+k_t(\theta)\} - \frac{\sigma}{n^{\alpha/(\alpha+d)}}\right] \\ &= \frac{1}{n^{\alpha/(\alpha+d)}} \sup_{0<\sigma\leqq t} [\tau\{2h(t\sigma)+2\Delta_t(n)+k_t(\varepsilon)\}-\sigma]. \end{aligned} \tag{3.27}$$

From (2.1), $n^{d/(d+\alpha)}/a_n^d \to 1/\lambda^d$ as $n\to\infty$, and hence from (3.23) and (3.27) we obtain, for all $t>0$ and all $\tau>0$,

$$\begin{aligned} &\overline{\lim_{n\to\infty}} \frac{1}{a_n^d} \log Q_n^{(n)}\left[f : \int_{T_d} |(K_\varepsilon f)(x)-f(x)|\, dx \geqq \sigma\right] \\ &\leqq \log\left(\left[\frac{8}{\delta}\right]+1\right) - \frac{\delta\tau}{2\lambda^d} + \frac{1}{\lambda^d} \overline{\lim_{n\to\infty}} \sup_{0<\sigma\leqq\tau} [\tau\{2h(t\sigma)+2\Delta_t(n)+k_t(\varepsilon)\}-\sigma] \\ &= \log\left(\left[\frac{8}{\delta}\right]+1\right) - \frac{\delta\tau}{2\lambda^d} + \frac{1}{\lambda^d} \sup_{0<\sigma\leqq\tau} [2\tau h(t\sigma)+\tau k_t(\varepsilon)-\sigma]. \end{aligned} \tag{3.28}$$

From (3.28) we get, for all $t>0$, $\tau>0$,

$$\overline{\lim_{\varepsilon\to 0}}\ \overline{\lim_{n\to\infty}} \frac{1}{a_n^d} \log Q_n^{(n)}\left[f : \int_{T_d} |(K_\varepsilon f)(x)-f(x)|\,dx \geqq \delta\right] \leqq \frac{1}{\lambda^d}\left[\sup_{0<\sigma\leqq\tau} \{2\tau h(t\sigma)-\sigma\} - \tfrac{1}{2}\delta\tau\right] + \log\left(\left[\frac{8}{\delta}\right]+1\right). \tag{3.29}$$

In (3.29) let $\sigma'=\sigma/\tau$ and the right side of inequality (3.29) becomes

$$\begin{aligned} &\frac{\tau}{\lambda^d}\left[\sup_{0<\sigma'\leqq 1} \{2h(t\tau\sigma')-\sigma'\} - \tfrac{1}{2}\delta\right] + \log\left(\left[\frac{8}{\delta}\right]+1\right) \\ &\leqq \frac{\tau}{\lambda^d}\left[\sup_{0<\sigma'\leqq 1} 2h(t\tau\sigma') - \tfrac{1}{2}\delta\right] + \log\left(\left[\frac{8}{\delta}\right]+1\right) \\ &\leqq \frac{\tau}{\lambda^d}[2h(t\tau)-\tfrac{1}{2}\delta] + \log\left(\left[\frac{8}{\delta}\right]+1\right), \end{aligned} \tag{3.30}$$

where the last inequality follows from the fact that without loss of generality we could pick the function h to go to 0 monotonically.

From (3.29) and (3.30), letting $t\to 0$, we get

$$\overline{\lim_{\varepsilon\to 0}}\ \overline{\lim_{n\to\infty}} \frac{1}{a_n^d} \log Q_n^{(n)}\left[f : \int_{T_d} |(K_\varepsilon f)(x)-f(x)|\,dx \geqq \sigma\right] \leqq -\frac{\delta\tau}{2\lambda^d} + \log\left(\left[\frac{8}{\delta}\right]+1\right), \tag{3.31}$$

and finally, if we let $\tau\to\infty$ in (3.31), we obtain (3.15).

4. Lemmas for Theorem 5 and Proof of Theorem 6

We now prove five lemmas used in the proof of Theorem 5 in the preceding section.

LEMMA 4.1. *Let $g\in C(T_d)$ with $\|g\|\leqq 1$ and let $M_{n,\varepsilon}\subset C(T_d)$ be the set of all functions $V(x)$ of the form $V=(K_\varepsilon - I)\chi_n g$. For any $\delta>0$ there exist functions $V_1, V_2, \cdots, V_J$ in $M_{n,\varepsilon}$ such that, for any $V\in M_{n,\varepsilon}$,*

$$\inf_{1\leqq i\leqq J}\ \sup_{x\in\mathcal{L}_d^{(n)}} |V(x)-V_i(x)| \leqq \tfrac{1}{2}\delta, \tag{4.1}$$

and, moreover, $J=J(n,\varepsilon,\delta)\leqq([8/\delta]+1)^{a_n^d}$.

Proof: We note that $|V|\leqq 2$ for any $V\in M_{n,\varepsilon}$ and also that $\mathscr{L}_d^{(n)}$ has a_n^d points. Subdivide the interval $[-2,2]$ into subintervals of length $\frac{1}{2}\delta$ so that there are at most $[8/\delta]+1$ subintervals. Now $[-2,2]^{a_n^d}$ consists of cubes of sides $\frac{1}{2}\delta$ and there are at most $([8/\delta]+1)^{a_n^d}$ such cubes. But, any $V\in M_{n,\varepsilon}$, as a function on $\mathscr{L}_d^{(n)}$, can be thought of as an element of $[-2,2]^{a_n^d}$. Thus, $M_{n,\varepsilon}\subset[-2,2]^{a_n^d}$. Think of the subcubes constructed above as closed and, for a particular cube C, pick one of its elements if $M_{n,\varepsilon}\cap C\neq\emptyset$. In this way we pick $V_1, V_2,\cdots, V_J$, where J is the number of cubes having non-empty intersection with $M_{n,\varepsilon}$. Thus $J=J(n,\varepsilon,\sigma)\leqq([8/\delta]+1)^{a_n^d}$. Moreover, if $V\in M_{n,\varepsilon}$, then $V\in[-2,2]^{a_n^d}$ and must belong to some cube, i.e., the intersection of $M_{n,\varepsilon}$ with that cube is non-empty and hence we must have picked some element V_i from that cube which means (4.1) holds.

LEMMA 4.2. *Let $l=[n^{\alpha/(d+\alpha)}]$. There exists an integer n_0 and there exists a constant $c<\infty$ such that, for $n\geqq n_0$,*

$$\tilde{\pi}_n^l(x)\leqq c\tilde{\pi}_n^{l+q}(y) \tag{4.2}$$

for all $x\in\mathscr{L}_d^{(n)}$, $y\in\mathscr{L}_d^{(n)}$, $y-x\in\tilde{H}_{d,q}^{(n)}$, where $0\leqq q\leqq p-1$.

Proof: Suppose first that $x\notin\tilde{H}_{d,l}^{(n)}$; then $y\notin\tilde{H}_{d,l+q}^{(n)}$ so that $\tilde{\pi}_n^l(x)=0$ and $\tilde{\pi}_n^{l+q}(y)=0$ which means (4.2) holds. Next, suppose $x\in\tilde{H}_{d,l}^{(n)}$ which implies $y\in\tilde{H}_{d,l+q}^{(n)}$. From Theorem 2 (with $\rho=1$ and noting that $\tilde{\Phi}$ is strictly positive for $x\in T_d$) we can choose n_0 such that $n\geqq n_0$ implies

$$\sup_{x\in\tilde{H}_{d,l}^{(n)}}\left|a_n^d\tilde{\pi}_n^l(x)-p\tilde{\Phi}\left(\frac{1}{x^\alpha},x\right)\right|\leqq\tfrac{1}{8}\inf_{x\in T_d}p\tilde{\Phi}\left(\frac{1}{\lambda^\alpha},x\right),$$

and

$$\sup_{y\in\tilde{H}_{d,l+q}^{(n)}}\left|a_n^d\tilde{\pi}_n^{l+q}(y)-p\tilde{\Phi}\left(\frac{1}{\lambda^\alpha},y\right)\right|\leqq\tfrac{1}{8}\inf_{x\in T_d}p\tilde{\Phi}\left(\frac{1}{\lambda^\alpha},x\right).$$

Hence, for all $x\in\tilde{H}_{d,l}^{(n)}$, we have

$$a_n^d\tilde{\pi}_n^l(x)\leqq\tfrac{9}{8}\sup_{x\in T_d}p\tilde{\Phi}\left(\frac{1}{\lambda^\alpha},x\right),$$

and, for all $y\in\tilde{H}_{d,l+q}^{(n)}$, we have

$$a_n^d\tilde{\pi}_n^{l+q}(y)\geqq\tfrac{7}{8}\inf_{x\in T_d}p\tilde{\Phi}\left(\frac{1}{\lambda^\alpha},x\right).$$

Letting

$$c=\frac{9}{7}\frac{\sup_{x\in T_d}\bar{\Phi}(1/\lambda^\alpha,x)}{\inf_{x\in T_d}\bar{\Phi}(1/\lambda^\alpha,x)},$$

we have the lemma.

LEMMA 4.3. *Let x and y be any two points in $\mathscr{L}_d^{(n)}$. There exists an integer n_0 such that $n\geqq n_0$ implies, for any $\theta>0$,*

$$\begin{aligned}E_y[\exp\{\theta[V(\tilde{Y}_1)+\cdots+V(\tilde{Y}_n)]\}]\\ \leqq C_{n,\theta}E_x[\exp\{\theta[V(\tilde{Y}_1)+\cdots+V(\tilde{Y}_n)]\}],\end{aligned}\tag{4.3}$$

where $C_{n,\theta}=c\exp\{4\theta n^{\alpha/(\alpha+d)}+p\}$. The constant c is the constant of Lemma 4.2 and, indeed, n_0 is the same as picked in Lemma 4.2. The notation $E_x\{\ \}$ denotes expectation where the starting point is x.

Proof: Choose l and n_0 as in Lemma 4.2. Since x and y are in $\mathscr{L}_d^{(n)}$, $y-x\in\tilde{H}_{d,q}^{(n)}$ for some q, $0\leqq q\leqq p-1$. Using the fact that $|V|\leqq 2$, we have

$$\begin{aligned}E_y[\exp\{\theta[V(\tilde{Y}_1)+\cdots+V(\tilde{Y}_n)]\}]&\leqq e^{2\theta l}E_y[\exp\{\theta[V(\tilde{Y}_{l+1})+\cdots+V(\tilde{Y}_n)]\}]\\ &=e^{2\theta l}\sum_{z\in\mathscr{L}_d^{(n)}}\tilde{\pi}_n^l(z-y)E_z[\exp\{\theta[V(\tilde{Y}_1)+\cdots+V(\tilde{Y}_{n-l})]\}].\end{aligned}\tag{4.4}$$

Since $(z-x)-(z-y)=y-x\in\tilde{H}_{d,q}^{(n)}$, we have from Lemma 4.2 and (4.4),

$$\begin{aligned}&E_y[\exp\{\theta[V(\tilde{Y}_1)+\cdots+V(\tilde{Y}_n)]\}]\\ &\quad\leqq ce^{2\theta l}\sum_{z\in\mathscr{L}_d^{(n)}}\tilde{\pi}_n^{l+q}(z-x)E_z[\exp\{\theta[V(\tilde{Y}_1)+\cdots+V(\tilde{Y}_{n-l})]\}]\\ &\quad=ce^{2\theta l}E_x[\exp\{\theta[V(\tilde{Y}_{l+q+1})+\cdots+V(\tilde{Y}_{n+q})]\}]\\ &\quad\leqq ce^{4\theta(l+q)}E_x[\exp\{\theta[V(\tilde{Y}_1)+\cdots+V(\tilde{Y}_n)]\}]\\ &\quad\leqq ce^{4\theta(l+p)}E_x[\exp\{\theta[V(\tilde{Y}_1)+\cdots+V(\tilde{Y}_n)]\}],\end{aligned}$$

which is the lemma.

LEMMA 4.4. *There exists an integer n_0 such that $n\geqq n_0$ implies, for any $\theta>0$,*

$$\begin{aligned}E[\exp\{\theta[V(\tilde{Y}_1)+\cdots+V(\tilde{Y}_n)]\}]=E_0[\exp\{\theta[V(\tilde{Y}_1)+\cdots+V(\tilde{Y}_n)]\}]\\ \leqq C_{n,\theta}\exp\{n\lambda_n(\theta V)\},\end{aligned}\tag{4.5}$$

Where $C_{n,\theta}$ *is the constant in Lemma* 4.3, *and*

$$\lambda_n(\theta V) = \sup_{\mu \in \mathcal{M}(T_d)} \left[\theta \int_{T_d} V(x)\mu(dx) - \tilde{I}_n(\mu)\right] \tag{4.6}$$

with (*c.f.* [2])

$$\tilde{I}_n(\mu) = -\inf_{u \in \mathcal{U}_1} \int_{T_d} \log\left(\frac{\tilde{\pi}_n u}{u}\right)(x)\mu(dx)\,, \tag{4.7}$$

and $\mathcal{U}_1$ *is the set of functions* $u \in C(T_d)$ *such that* $u > 0$ *on* T_d.

Proof: From Lemma 4.3, there exists an integer n_0 such that $n \geqq n_0$ implies, for any $\theta > 0$,

$$\begin{aligned} &E[\exp\{\theta[V(\tilde{Y}_1) + \cdots + V(\tilde{Y}_n)]\}] \\ &\qquad \leqq C_{n,\theta} \inf_{x \in T_d} E_x[\exp\{\theta[V(\tilde{Y}_1) + \cdots + V(\tilde{Y}_n)]\}]\,. \end{aligned} \tag{4.8}$$

It follows from the proof of Theorem 1 in [2] that

$$\inf_{x \in T_d} E_x[\exp\{\theta[V(\tilde{Y}_1) + \cdots + V(\tilde{Y}_n)]\}] \leqq \exp\{n\lambda_n(\theta V)\}\,,$$

and hence we have (4.5).

LEMMA 4.5. *Let* μ *be a probability measure on* $\mathcal{L}_d^{(n)}$ *such that* $\tilde{I}_n(\mu) \leqq \sigma/n^{\alpha/(\alpha+d)}$, where $\sigma > 0$. *Let* $V = (K_\varepsilon - I)\chi_n g$, *where* $\|g\| \leqq B$. *Then, for any* $t > 0$,

$$\left|\int_{T_d} V(x)\mu(dx)\right| \leqq B[2h(t\sigma) + 2\Delta_t(n) + k_t(\varepsilon)]\,, \tag{4.9}$$

where $h(\sigma) \to 0$ *as* $\sigma \to 0$, $\Delta_t(n) \to 0$ *as* $n \to \infty$ *for all* $t > 0$, *and* $k_t(\varepsilon) \to 0$ *as* $\varepsilon \to 0$ *for all* $t > 0$.

Proof: For any $t > 0$, let $n_t = [tn^{\alpha/(d+\alpha)}]$ and define

$$\tilde{\pi}_{n,t}(x) = \frac{\tilde{\pi}_n^{n_t}(x) + \tilde{\pi}_n^{n_t - 1}(x) + \cdots + \tilde{\pi}_n^{n_t-(p-1)}(x)}{p}\,.$$

Let

$$\tilde{I}_{n,t}(\mu) = -\inf_{u\in\mathcal{U}_1} \int_{T_d} \log\left(\frac{\tilde{\pi}_{n,t}u}{u}\right)(x)\mu(dx)\,, \tag{4.10}$$

and let

$$\tilde{\phi}_{n,t}(x) = \sum_{y\in\mathcal{L}_d^{(n)}} \chi_n(x-y)\tilde{\pi}_{n,t}(y)\,.$$

Since $\|g\| \leqq B$, we note that $|V(x)| \leqq 2B$ on T_d. Let $\mu\tilde{\pi}_{n,t}$ be defined by: for $x \in \mathcal{L}_d^{(n)}$

$$(\mu\tilde{\pi}_{n,t})(x) = \int_{T_d} \tilde{\pi}_{n,t}(x-y)\mu(dy)\,, \tag{4.11}$$

and write

$$\int_{T_d} V(x)\mu(dx) = \int_{T_d} V(x)(\mu - \mu\tilde{\pi}_{n,t})(dx) + \int_{T_d} V(x)\mu\tilde{\pi}_{n,t}(dx)\,. \tag{4.12}$$

The I-function has the property that if

$$I_1(\mu) = -\inf_{u\in\mathcal{U}_1} \int_{T_d} \log\left(\frac{\pi_1 u}{u}\right)\mu(dx)\,,$$

$$I_2(\mu) = -\inf_{u\in\mathcal{U}_1} \int_{T_d} \log\left(\frac{\pi_2 u}{u}\right)\mu(dx)\,,$$

and if $\pi_3 = a\pi_1 + (1-a)\pi_2$, where $0 < a < 1$, then

$$I_3(\mu) = -\inf_{u\in\mathcal{U}_1} \int_{T_d} \log\left(\frac{\pi_3 u}{u}\right)\mu(dx) \leqq aI_1(\mu) + (1-a)I_2(\mu)\,. \tag{4.13}$$

Also, if $\pi_1^{(n)}$ indicates the n-step transition probabilities, then

$$I_1^{(n)}(\mu) = -\inf_{u\in\mathcal{U}_1} \int_{T_d} \log\left(\frac{\pi_1^{(n)} u}{u}\right)\mu(dx) \leqq nI_1(\mu)\,. \tag{4.14}$$

Using (4.13) and (4.14) we see from (4.10) that

$$\tilde{I}_{n,t}(\mu) \leqq \sigma t\,. \tag{4.15}$$

By Lemma 4.1 in [2] it follows from (4.15) and (4.11) that (in the variation norm)

$$\|\mu\tilde{\pi}_{n,t}-\mu\|\leqq h(\sigma t)\,, \tag{4.16}$$

where $h(\sigma)\to 0$ as $\sigma\to 0$. Thus, from (4.16) and (4.12), we get

$$\begin{aligned}\left|\int_{T_d} V(x)\mu(dx)\right| &\leqq 2Bh(\sigma t)+B\int_{T_d}\left|\int_{T_d}[\tilde{\phi}_{n,t}(x-y)-\tilde{\phi}_{n,t}(x)]\psi_\varepsilon(y)\,dy\right|dx\\ &\leqq 2Bh(\sigma t)+B\sup_{|y|\leqq\varepsilon}\int_{T_d}|\tilde{\phi}_{n,t}(x-y)-\tilde{\phi}_{n,t}(x)|\,dx\\ &\leqq 2Bh(\sigma t)+B\sup_{|y|\leqq\varepsilon}\Bigg[\int_{T_d}\left|\tilde{\phi}_{n,t}(x-y)-\tilde{\Phi}\Big(\frac{t}{\lambda^\alpha},x-y\Big)\right|dx\\ &\qquad+\int_{T_d}\left|\tilde{\phi}_{n,t}(x)-\tilde{\Phi}\Big(\frac{t}{\lambda^\alpha},x\Big)\right|dx\\ &\qquad+\int_{T_d}\left|\tilde{\Phi}\Big(\frac{t}{\lambda^\alpha},x-y\Big)-\tilde{\Phi}\Big(\frac{t}{\lambda^\alpha},x\Big)\right|dx\Bigg]\\ &\leqq 2Bh(\sigma t)+2B\Delta_t(n)+Bk_t(\varepsilon)\,,\end{aligned} \tag{4.17}$$

where

$$\Delta_t(n)=\int_{T_d}\left|\tilde{\phi}_{n,t}(x)-\tilde{\Phi}\Big(\frac{t}{\lambda^\alpha},x\Big)\right|dx\,,$$

and

$$k_t(\varepsilon)=\sup_{|y|\leqq\varepsilon}\int_{T_d}\left|\tilde{\Phi}\Big(\frac{t}{\lambda^\alpha},x-y\Big)-\tilde{\Phi}\Big(\frac{t}{\lambda^\alpha},x\Big)\right|dx\,.$$

Since $\tilde{\Phi}$ is a density function on T_d, we see that, for each $t>0$, $k_t(\varepsilon)\to 0$ as $\varepsilon\to 0$.

To show that, for each $t>0$, $\Delta_t(n)\to 0$ as $n\to\infty$, it suffices to show, for each $t>0$,

$$\operatorname*{ess\,sup}_{x\in T_d}\left|\tilde{\phi}_{n,t}(x)-\tilde{\Phi}\Big(\frac{t}{\lambda^\alpha},x\Big)\right|\to 0\quad\text{as}\quad n\to\infty\,, \tag{4.18}$$

i.e.,

$$\operatorname*{ess\,sup}_{x\in T_d}\left|\sum_{y\in\mathcal{L}_d^{(n)}}\chi_n(x-y)\tilde{\pi}_{n,t}(y)-\tilde{\Phi}\Big(\frac{t}{\lambda^\alpha},x\Big)\right|\to 0\quad\text{as}\quad n\to\infty\,.$$

Let $A_n = \{x \in T_d : |x_i| < 1/2a_n,\ 1 \leqq i \leqq d\}$. On A_n, the function χ_n is equal to a_n^d; thus, if $x \in A_n + z$ for some $z \in \mathscr{L}_d^{(n)}$, then $\tilde{\phi}_{n,t}(x) = a_n^d \tilde{\pi}_{n,t}(z)$. Observe also that meas$(T_d - \bigcup_{z \in \mathscr{L}_d^{(n)}} \{A_n + z\}) = 0$. Since $z \in \mathscr{L}_d^{(n)}$, we have $z \in \tilde{H}_{d,q}^{(n)}$ for some q, $0 \leqq q \leqq p-1$. Consequently,

$$\tilde{\phi}_{n,t}(x) = \frac{a_n^d}{p}\, \tilde{\pi}_n^{n_t - r}(z)$$

for some r, $0 \leqq r \leqq p-1$, and $z \in \tilde{H}_{d,n_t-r}^{(n)}$. Now, from Theorem 2,

$$\begin{aligned} \tilde{\phi}_{n,t}(x) &= \bar{\Phi}\left(\frac{t}{\lambda^\alpha}, z\right) + o(1) \quad \text{as} \quad n \to \infty \\ &= \left(\frac{t}{\lambda^\alpha}, x\right) + o(1) \quad \text{as} \quad n \to \infty, \end{aligned}$$

which gives us (4.18) and the lemma.

THEOREM 6. *Let $C \subset L_1(T_d)$ be closed. Then*

$$\overline{\lim_{n \to \infty}} \frac{1}{a_n^d} \log Q_n^{(n)}(C) \leqq -\inf_{f \in C} I_\lambda(f)\,. \tag{4.19}$$

Proof: Consider the mapping from $\mathcal{M}(T_d)$ with its weak topology to $L_1(T_d)$ with its strong topology given by $\mu \to K_\varepsilon \mu$, where

$$(K_\varepsilon \mu)(x) = \int_{T_d} \psi_\varepsilon(x-y)\mu(dy)$$

with ψ_ε defined just before (3.14). This map is continuous and hence $\{\mu \in \mathcal{M}(T_d) : K_\varepsilon \mu \in C\}$ is closed in $\mathcal{M}(T_d)$. Thus, from Theorem 4 we get

$$\overline{\lim_{n \to \infty}} \frac{1}{a_n^d} \log Q_n^{(n)}[f : K_\varepsilon f \in C] \leqq - \inf_{f : K_\varepsilon f \in C} I_\lambda(f)\,. \tag{4.20}$$

For any $\delta > 0$, let $C_\delta = \{g \in L_1(T_d) : \|g - f\| < \delta$ for some $f \in C\}$ and let $\bar{C}_\delta$ be the L_1 closure of C_δ. From (4.20), applied to this closed set $\bar{C}_\delta$ instead of C, we have

$$\overline{\lim_{n \to \infty}} \frac{1}{a_n^d} \log Q_n^{(n)}[f : K_\varepsilon f \in \bar{C}_\delta] \leqq - \inf_{f : K_\varepsilon f \in \bar{C}_\delta} I_\lambda(f)\,. \tag{4.21}$$

Consequently, for any set C closed in $L_1(T_d)$, we have, for all $\delta > 0$ and all $\varepsilon > 0$,

$$\begin{aligned} &\overline{\lim_{n\to\infty}} \frac{1}{a_n^d} \log Q_n^{(n)}(C) \\ &\leqq \overline{\lim_{n\to\infty}} \frac{1}{a_n^d} \log \{Q_n^{(n)}[f : K_\varepsilon f \in \bar{C}_\delta] + Q_n^{(n)}[f : \|K_\varepsilon f - f\| \geqq \delta]\} \\ &\leqq \max\left\{ - \inf_{f:K_\varepsilon f \in \bar{C}_\delta} I_\lambda(f), \overline{\lim_{n\to\infty}} \frac{1}{a_n^d} \log Q_n^{(n)}[f : \|K_\varepsilon f - f\| \geqq \delta]\right\}. \end{aligned} \tag{4.22}$$

To prove the theorem it suffices then to show that

$$\begin{aligned} &\underline{\lim_{\delta\to 0}}\, \underline{\lim_{\varepsilon\to 0}} \max\left\{ - \inf_{f:K_\varepsilon f \in \bar{C}_\delta} I_\lambda(f), \overline{\lim_{n\to\infty}} \frac{1}{a_n^d} \log Q_n^{(n)}[f : \|K_\varepsilon f - f\| \geqq \delta]\right\} \\ &\qquad\qquad \leqq - \inf_{f\in C} I_\lambda(f). \end{aligned} \tag{4.23}$$

In view of Theorem 5, this means showing

$$\overline{\lim_{\delta\to 0}}\, \overline{\lim_{\varepsilon\to 0}} \inf_{f:K_\varepsilon f \in \bar{C}_\delta} I_\lambda(f) \geqq \inf_{f\in C} I_\lambda(f). \tag{4.24}$$

In fact, we prove even more, namely that

$$\underline{\lim_{\substack{\varepsilon\to 0\\ \delta\to 0}}} \inf_{f:K_\varepsilon f \in \bar{C}_\delta} I_\lambda(f) \geqq \inf_{f\in C} I_\lambda(f). \tag{4.25}$$

To see that (4.25) is true, let $\{f_n\}$ be a sequence in $L_1(T_d)$ such that $I_\lambda(f_n) \to l$ and $K_\varepsilon f_n \in \bar{C}_{\delta_n}$, where $\varepsilon_n \to 0$ and $\delta_n \to 0$; we must show that there exists $f \in C$ such that $I_\lambda(f) \leqq l$. If $l = \infty$, there is noting to prove; thus assume $l < \infty$. But, the set of f in $L_1(T_d)$ such that $I_\lambda(f) \leqq l$ is compact in $L_1(T_d)$, and the result follows from this compactness and the fact that $I_\lambda(f)$ is lower semicontinuous. This completes the proof of Theorem 6.

From Theorem 6, using the arguments of Section 3 in [3], (c.f. remarks just after (2.11) above) we obtain

COROLLARY TO THEOREM 6. *For any* $\nu > 0$,

$$\begin{aligned} &\overline{\lim_{n\to\infty}} \frac{1}{a_n^d} \log E^{Q_n^{(n)}}[\exp\{-\nu a_n^d |x : f(x) > 0|\}] \\ &\qquad\qquad \leqq - \inf_{g\in L_1(T_d)} \{\nu |x : g(x) > 0| + I_\lambda(g)\}. \end{aligned} \tag{4.26}$$

As pointed out in (2.14), this implies

$$\overline{\lim_{n\to\infty}} \frac{1}{n^{d/(d+\alpha)}} \log E[\exp\{-\nu D_n^{\#}\}] \leqq -\sup_{\lambda>0}\left\{\lambda^d \inf_{g\in L_1(T_d)} [\nu|x : g(x)>0| + I_\lambda(g)]\right\}. \tag{4.27}$$

We shall now show that

$$\sup_{\lambda>0}\left\{\lambda^d \inf_{g\in L_1(T_d)} [\nu|x : g(x)>0| + I_\lambda(g)]\right\} \geqq k(\nu, L), \tag{4.28}$$

as defined in the statement of Theorem 1. Thus, we will have shown that

$$\overline{\lim_{n\to\infty}} \frac{1}{n^{d/(d+\alpha)}} \log E[\exp\{-\nu D_n^{\#}\}] \leqq -k(\nu, L). \tag{4.29}$$

To see that (4.28), holds we note first that, from the definition of $I_\lambda(g)$,

$$\sup_{\lambda>0}\left\{\lambda^d \inf_{g\in L_1(T_d)} [\nu|x : g(x)>0| + I_\lambda(g)]\right\} = \sup_{\lambda>0} \inf_{g\in L_1(T_d)} \left\{\nu\lambda^d|x : g(x)>0)| + \frac{1}{\lambda^\alpha} I_{\tilde{L}}(g)\right\}. \tag{4.30}$$

Let T_d^λ be the torus in $R_d : |x_1| \leqq \frac{1}{2}\lambda$, $i = 1, 2, \cdots, d$, with opposite sides identified. Let

$$h(x) = \frac{1}{\lambda^d} g\left(\frac{x}{\lambda}\right).$$

We can then rewrite the expression on the right of (4.30) as

$$\sup_{\lambda>0} \inf_{h\in L_1(T_d)} \{\nu|x : h(x)>0| + I_{\tilde{L}}^\lambda(g)\}, \tag{4.31}$$

where $I_{\tilde{L}}^\lambda(h)$ is the I-function for the symmetric stable process on the torus T_d^λ with generator $\tilde{L}$. It now follows from Lemmas 3.5, 3.6, and 3.9 of [1] that

$$\sup_{\lambda>0} \inf_{h\in I_1(T_d^\lambda)} \{\nu|h(x)>0| + I_{\tilde{L}}^\lambda\} \geqq k(\nu, L). \tag{4.32}$$

Hence from (4.30), (4.31), and (4.32) we obtain (4.28).

To prove Theorem 1 it remains to show that

$$\varliminf_{n\to\infty} \frac{1}{n^{d/(d+\alpha)}} \log E[\exp\{-\nu D_n^\#\}] \geqq -k(\nu, L) \tag{4.33}$$

which is the object of the next section.

5. The Lower Bound

Let G be a bounded open set in R^d containing the origin such that the Lebesgue measure of its boundary is zero. Consider the original random walk on Z^d and recall the definition

$$D_n = \{j \in Z^d : j = Y_r \text{ for some } r,\ 1 \leqq r \leqq n\}.$$

Let us pick an open set U with compact closure $\bar{U}$ such that

$$G \subset \bar{G} \subset U \subset \bar{U}$$

and $|\bar{U}| \leqq |G| + \eta$, where $\eta > 0$ is a given arbitrarily small positive number. Let $b_n = n^{1/(\alpha+d)}$ and define the set

$$b_n U = \{y \in R^d : y = b_n x \text{ for some } x \in U\};$$

we have, clearly, by letting $(b_n U)^\#$ denote the number of lattice points in the set $b_n U$,

$$\begin{aligned} E[\exp\{-\nu D_n^\#\}] &\geqq E[\exp\{-\nu D_n^\#\};\ D_n \subset b_n U] \\ &\geqq \exp\{-\nu (b_n U)^\#\} P[D_n \subset b_n U]. \end{aligned} \tag{5.1}$$

In Lemma 5.1 below we shall show that

$$\varliminf_{n\to\infty} \frac{1}{n^{d/(\alpha+d)}} \log P[D_n \subset b_n U] \geqq -\lambda(G). \tag{5.2}$$

where $\lambda(G)$ is the smallest eigenvalue of $-L_\Omega$ with Dirichlet boundary conditions for G. Using (5.2) we obtain from (5.1)

$$\begin{aligned} \varliminf_{n\to\infty} \frac{1}{n^{d/(d+\alpha)}} \log E[\exp\{-\nu D_n^\#\}] &\geqq -\nu \varlimsup_{n\to\infty} \frac{(b_n U)^\#}{b_n^d} - \lambda(G) \\ &\geqq -\nu |U| - \lambda(G) \\ &\geqq -\nu |G| - \lambda(G) - \nu\eta ; \end{aligned}$$

since $\eta>0$ was arbitrary we conclude that

$$\varliminf_{n\to\infty} \frac{1}{n^{d/(\alpha+d)}} \log E[\exp\{-\nu D_n^{\#}\}] \geqq -\nu|G| - \lambda(G)\,. \tag{5.3}$$

Denoting by $\mathscr{G}$ the class of all bounded open sets that are continuity sets for Lebesgue measures we conclude from (5.3) that

$$\varliminf_{n\to\infty} \frac{1}{n^{d/(\alpha+d)}} \log E[\exp\{-\nu D_n^{\#}\}] \geqq -\inf_{G\in\mathscr{G}} [\nu|G| + \lambda(G)]\,. \tag{5.4}$$

It follows from the proof of Lemma 3.9 in [1] that

$$\inf_{G\in\mathscr{G}} [\nu|G| + \lambda(G)] = k(\nu, L_\Omega)\,, \tag{5.5}$$

as defined in the statement of Theorem 1. Thus (5.4) and (5.5) yield (4.33) and the proof of Theorem 1 is complete except for (5.2) which we state and prove as a lemma.

LEMMA 5.1.

$$\varliminf_{n\to\infty} \frac{1}{n^{d/(\alpha+d)}} \log P[D_n \subset b_n U] \geqq -\lambda(G)\,.$$

Proof: For $j \in Z^d$ consider

$$P_j\left[\frac{Y_k}{b_n} \in U, \text{ for } 1 \leqq k \leqq l_n, \frac{Y_{l_n}}{b_n} \in G\right],$$

where $P_j[\ ]$ denotes probabilities over the random walk starting from j. Let

$$q(U, G, l_n, n) = \inf_{j\in b_n G} P_j\left[\frac{Y_k}{b_n} \in U, 1 \leqq k \leqq l_n, \frac{Y_{l_n}}{b_n} \in G\right]. \tag{5.6}$$

Here l_n is a sequence of integers tending to ∞ such that $l_n/n^{\alpha/(\alpha+d)} \to \lambda > 0$, N_n is a sequence of integers such that $N_n/n^{d/(\alpha+d)} \to 1/\lambda$ with $l_n N_n \geqq n$ for every n, and $\lambda > 0$ is an arbitrary positive number. It follows from the Markov property that

$$\begin{aligned} P\{D_n \subset b_n U\} &= P\left[\frac{Y_k}{b_n} \in U \text{ for } 1 \leqq k \leqq n\right] \\ &\geqq [q(U, G, l_n, n)]^{N_n}\,. \end{aligned} \tag{5.7}$$

From (5.7) we get

$$(5.8)\qquad \varliminf_{n\to\infty} \frac{1}{n^{d/(\alpha+d)}} \log P[D_n \subset b_n U] \geqq \frac{1}{\lambda} \varliminf_{n\to\infty} \log q(U, G, l_n, n).$$

If $\{x(s),\ 0 \leqq s < \infty\}$ is the symmetric stable process in R_d of index α corresponding to $\Omega(t)$, then from (5.6) and the invariance principle we see that

$$(5.9)\qquad \varliminf_{n\to\infty} q(U, G, l_n, n) \geqq \inf_{x\in G} P_x[x(s) \in U,\ 0 \leqq s \leqq \lambda,\ x(\lambda) \in G].$$

Thus from (5.8) and (5.9), we have

$$(5.10)\qquad \begin{aligned} &\varliminf_{n\to\infty} \frac{1}{n^{d/(\alpha+d)}} \log P[D_n \subset b_n U] \\ &\qquad \geqq \frac{1}{\lambda} \log\left[\inf_{X\in G} P_x[x(s) \in U,\, 0 \leqq s \leqq \lambda,\, x(\lambda) \in G]\right]. \end{aligned}$$

The left-hand side being independent of λ,

$$(5.11)\qquad \begin{aligned} &\varliminf_{n\to\infty} \frac{1}{n^{d/(\alpha+d)}} \log P\{D_n \subset b_n U\} \\ &\qquad \geqq \varlimsup_{\lambda\to\infty} \frac{1}{\lambda} \log\left[\inf_{x\in \bar G} P_x[x(s) \in U,\, 0 \leqq s < \lambda,\, x(\lambda) \in G]\right]. \end{aligned}$$

Now

$$\begin{aligned} &\inf_{x\in \bar G} P_x[x(s) \in U,\, 0 \leqq s \leqq \lambda,\, x(\lambda) \in G] \\ &\qquad \geqq \inf_{x\in \bar G} P_x[x(s) \in V,\, 0 \leqq s \leqq \lambda - 1] \inf_{x\in \bar V} P_x[x(s) \in U,\, 0 \leqq s \leqq 1,\, x(1) \in G], \end{aligned}$$

where V is any open set such that

$$\bar G \subset V \subset \bar V \subset U.$$

Since the second term is positive for any stable process and does not depend on λ, we conclude that

$$\varliminf_{\lambda\to\infty} \frac{1}{\lambda} \log \left[\inf_{x\in \bar{G}} P_x[x(s)\in U, 0\leqq s\leqq \lambda, x(\lambda)\in G] \right]$$

$$\geqq \varliminf_{\lambda\to\infty} \frac{1}{\lambda} \log \left[\inf_{x\in \bar{G}} P_x[x(s)\in V, 0\leqq s\leqq \lambda - 1] \right]$$

$$= \varliminf_{\lambda\to\infty} \frac{1}{\lambda} \log \left[\inf_{x\in \bar{G}} P_x[x(s)\in V, 0\leqq s\leqq \lambda] \right].$$

It now follows from Theorem (8.1) of [5] that

$$\varliminf_{n\to\infty} \frac{1}{n^{d/(d+\alpha)}} \log P\{D_n \subset b_n U\} \geqq -\lambda(G)$$

and the proof of Lemma 5.1 is complete.

Bibliography

[1] Donsker, M. D., and Varadhan, S. R. S., *Asymptotics for the Wiener sausage,* Comm. Pure Appl. Math. 28, 1975 pp. 525–565.

[2] Donsker, M. D., and Varadhan, S. R. S., *Asymptotic evaluation of certain Markov process expectations for large time, I,* Comm. Pure Appl. Math. 28, 1975, pp. 1–47.

[3] Varadhan, S. R. S., *Asymptotic probabilities and differential equations,* Comm. Pure Appl. Math. 19, 1966, pp. 261–286.

[4] Donsker, M. D., and Varadhan, S. R. S., *Asymptotic evaluation of certain Weiener integrals for large time,* Functional Integration and Its Applications, Proceedings of the International Conference, London; Clarendon Press, Oxford, 1975, pp. 15–33.

[5] Donsker, M. D., and Varadhan, S. R. S., *Asymptotic evaluation of certain Markov process expectations for large time,* III, Comm. Pure Appl. Math. 29. 1976, pp. 389–461.

Received November, 1978.

Comm. Pure Appl. Math. 33 (1980) 365-393

A Law of the Iterated Logarithm for Total Occupation Times of Transient Brownian Motion*

M. D. DONSKER AND S. R. S. VARADHAN

Dedicated to Fritz John on the Occasion of his 70th Birthday, June 14, 1980

1. Introduction

Let $\{\beta(s), 0 \leqq s < \infty\}$ be Brownian motion in R_d starting from the origin with $d \geqq 3$, and let $T_d(\lambda, \omega)$ be the total time that a particular path $\omega = \beta(\cdot)$ occupies the sphere with center at the origin of radius λ. In [1] Ciesielski and Taylor showed that, for almost all Brownian paths,

$$\overline{\lim_{\lambda \downarrow 0}} \frac{T_d(\lambda, \omega)}{\lambda^2 \log\log(1/\lambda)} = \frac{2}{p_d^2}, \tag{1.1}$$

where p_d is the first positive zero of $J_\nu(x)$ with $\nu = \frac{1}{2}d - 2$.

In this paper, motivated by (1.1), the authors prove a Strassen type law of the iterated logarithm (see Theorems 1.1 and 1.2 below) for total Brownian occupation times in three or more dimensions. These theorems involve the I-function introduced by the authors in their recent work and which we now describe in the present context.

Let $\{\beta(s), 0 \leqq s < \infty\}$ be Brownian motion in R_d starting from $x \in R_d$ with $d \geqq 3$. For E, a Borel set in R_d, define

$$\sigma(E, \omega) = \int_0^\infty \chi_E(\beta(s))\, ds,$$

i.e., the total time the set E is occupied by the particular path $\omega = \beta(\cdot)$. For each ω, $\sigma(\cdot, \omega)$ is a σ-finite measure on R_d. If we let M denote the space of all σ-finite measures on R_d, we can, for each positive integer n and each $x \in R_d$, define a probability measures Q_n^x on M as follows: if $A \subset M$, then

$$Q_n^x(A) = P_x\{\omega : (1/n)\sigma(\cdot, \omega) \in A\}, \tag{1.2}$$

where $P_x\{\ \}$ is the Brownian measure on the paths $\omega = \beta(\cdot)$ in R_d starting from x.

---

* The research for this paper was done at the Courant Institute of Mathematical Sciences and supported by the National Science Foundation under NSF Grant No. MCS-77-02687. Reproduction in whole or in part is permitted for any purpose of the United States Government.

Let U be the space of functions $u(x) \in C^2$ on R_d such that

$$0 < \inf_{x \in R_d} u(x) \leqq \sup_{x \in R_d} u(x) < \infty$$

and Δu has compact support. For each σ-finite measure $\sigma \in M$, we define

$$I(\sigma) = -\inf_{u \in U} \int_{R_d} \left(\frac{\frac{1}{2}\Delta u}{u}\right)(x)\sigma(dx). \tag{1.3}$$

In Section 3 of this paper we show that the asymptotic behavior of the measure Q_n^x as $n \to \infty$ is described by the I-function of (1.3). Roughly speaking, as $n \to \infty$,

$$Q_n^x(A) \sim \exp\{-n \inf_{\sigma \in A} I(\sigma)\}, \tag{1.4}$$

so that the importance of the I-function is that it gives us the constant in the exponential decay of the Q_n^x measure as $n \to \infty$. The precise description of this asymptotic behavior is given in Section 3.

In Section 2 of the paper we show that the σ-finite measures $\sigma \in M$ for which $I(\sigma) < \infty$ have the property that $\sigma\{x : |x| \leqq R\} = o(R^2)$ as $R \to \infty$. Moreover (cf. Theorem 2.1 below), in this case, σ is absolutely continuous with respect to Lebesgue measure in R_d, and the I-function defined by (1.3) takes on a simple form; namely, if, in this case, we set $\sigma(dx) = \phi^2(x)\,dx$, we have

$$I(\sigma) = \tfrac{1}{2} \int_{R_d} |\nabla \phi|^2\,dx. \tag{1.5}$$

On the space M we impose the vague topology, i.e., if $\sigma_0 \in M$, r is a positive integer, and $\phi_1, \phi_2, \cdots, \phi_r$ are functions in $C_0^\infty(R_d)$, then for $\varepsilon > 0$ a neighborhood of σ_0 is defined by

$$\begin{aligned} N(\sigma_0) &= N(\sigma_0, r, \phi_1, \cdots, \phi_r, \varepsilon) \\ &= \left\{\sigma \in M : \left|\int_{R_d} \phi_i(x)\sigma(dx) - \int_{R_d} \phi_i(x)\sigma_0(dx)\right| < \varepsilon, 1 \leqq i \leqq r\right\}. \end{aligned}$$

For $\lambda \geqq 3$, let

$$Z_\lambda(A, \omega) = \frac{\sigma(\lambda A, \omega)}{\lambda^2 \log\log \lambda},$$

so that, for each $\omega = \beta(\cdot)$, $\{Z_\lambda(\cdot, \omega)\}$ is a family (in the parameter λ) of σ-finite measures on R_d. We show in Section 4 that, for almost all ω, $\{Z_\lambda(\cdot, \omega)\}$ is conditionally compact, and if we let C_ω be the cluster set for the family $\{Z_\lambda(\cdot, \omega)\}$ (as $\lambda \to \infty$), then the main result of Section 4 is to identify C_ω. Indeed, if we let

$$\mathscr{B} = \{\sigma \in M : I(\sigma) \leqq 1\},$$

then Lemma 4.3 and Theorem 4.4 show that, for almost all ω, $C_\omega = \mathscr{B}$.

This immediately gives us

THEOREM 1.1. *Let $\{\beta(s), 0 \leqq s < \infty\}$ be Brownian motion in R_d, $d \geqq 3$, starting from the origin, and for any Borel set $A \subset R_d$, let*

$$\sigma(A, \omega) = \int_0^\infty \chi_A(\beta(s))\, ds.$$

Define, for $\lambda \geqq 3$,

$$Z_\lambda(A, \omega) = \frac{\sigma(\lambda A, \omega)}{\lambda^2 \log\log \lambda}.$$

Let $\Phi(\sigma)$ be continuous (in the vague topology) at the points of $\mathscr{B} = \{\sigma \in M : I(\sigma) \leqq 1\}$. Then, for almost all Brownian paths ω,

$$\varlimsup_{\lambda \to \infty} \Phi(Z_\lambda(\cdot, \omega)) = \sup_{\sigma : I(\sigma) \leqq 1} \Phi(\sigma). \tag{1.6}$$

If we let

$$Z_\lambda^*(A, \omega) = \frac{\sigma(\lambda A, \omega)}{\lambda^2 \log\log(1/\lambda)},$$

then by a proof directly analogous to the proof of Theorem 4.4 we can show that as $\lambda \to 0$ the cluster set of the family $\{Z_\lambda^*(\cdot, \omega)\}$ is also the same set $\mathscr{B}$ as above. This yields the following theorem.

THEOREM 1.2. *Let $\{\beta(s), 0 \leqq s < \infty\}$ be Brownian motion in R_d, $d \geqq 3$, starting from the origin, and for any Borel set $A \subset R_d$, let*

$$\sigma(A, \omega) = \int_0^\infty \chi_A(\beta(s))\, ds.$$

Define

$$Z_\lambda^*(A,\omega)=\frac{\sigma(\lambda A,\omega)}{\lambda^2\log\log(1/\lambda)}.$$

Let $\Phi(\sigma)$ be continuous (in the vague topology) at the points of $\mathcal{B}=\{\sigma\in M: I(\sigma)\leqq 1\}$. Then, for almost all Brownian paths ω,

(1.7) $$\overline{\lim_{\lambda\to 0^+}}\,\Phi(Z_\lambda^*(\cdot,\omega))=\sup_{\sigma:I(\sigma)\leqq 1}\Phi(\sigma).$$

If, in (1.7), we let $\Phi(\sigma)=\sigma(s_d(0,1))$, where $s_d(0,1)$ is the unit sphere in R_d with center at 0, then the left side of (1.7) becomes the left side of (1.1). As pointed out above, if $I(\sigma)<\infty$, then σ is absolutely continuous with respect to Lebesgue measure and so this $\Phi(\sigma)$ is continuous at the points of $\mathcal{B}$. It is an easy calculation, since $I(\sigma)$ on $\mathcal{B}$ is given by (1.5), to see that the variational problem on the right of (1.7) yields the right side of (1.1).

2. σ-Finite Measures and Corresponding *I*-Functions

In (1.3) we defined, for each σ-finite measure σ on R_d, the I-function, $I(\sigma)$. We observe easily from this definition that Lebesgue measure on R_d, $d\geqq 3$, does not have a finite I-function. In this section we determine the properties of a σ-finite measure on R_d, $d\geqq 3$, for which the corresponding I-function is finite, and moreover in this case we find a simple expression for the I-function. These results are needed in later sections of this paper. The following theorem gives exact conditions for the finiteness of the I-function.

THEOREM 2.1. *Let $\sigma\in M$ be such that $I(\sigma)<\infty$. Then,*

(i) $\sigma[x:|x|\leqq R]=o(R^2)$ *as* $R\to\infty$,

(ii) *there exists $\phi(x)$ on R_d such that $\sigma(dx)=\phi^2(x)\,dx$,*

and

(iii) $\frac{1}{2}\int_{R_d}|\nabla\phi|^2\,dx=I(\sigma)$.

Conversely, if $\phi\in H^1(\text{loc})$ on R_d, $\int_{R_d}|\nabla\phi|^2\,dx<\infty$, and $\int_{|x|\leqq R}\phi^2(x)\,dx=o(R^2)$ as $R\to\infty$, then defining $\sigma(dx)=\phi^2(x)\,dx$ we have $I(\sigma)=\frac{1}{2}\int_{R_d}|\nabla\phi|^2\,dx$.

Theorem 2.1 follows from a succession of lemmas. Statement (i) of the theorem is the Corollary to Lemma 2.3. Statement (ii) of the theorem follows from Lemma 2.9, and statement (iii) follows from (i) and Lemmas 2.7 and 2.9. The converse part of the theorem follows from Lemmas 2.7 and 2.9.

LEMMA 2.2. *Let $\sigma \in M$ and let $G(x, y)$ be Green's function for $-\frac{1}{2}\Delta$ on R_d. Choose a smooth function $V \geqq 0$ such that*

$$u(x) = \int_{R_d} G(x, y)\, V(y)\, dy < \infty .$$

Then,

$$\int_{R_d} \frac{V(x)}{u(x)} \sigma(dx) \leqq I(\sigma) . \tag{2.1}$$

Proof: Let $\psi(y)$ be a smooth function on R_d satisfying $0 \leqq \psi \leqq 1$, and

$$\psi(y) = \begin{cases} 1 & \text{for } |y| \leqq 1 , \\ 0 & \text{for } |y| \geqq 2 . \end{cases}$$

Let $R > 0$, $\varepsilon > 0$, and define $\psi_R(y) = \psi(y/R)$ and

$$u_{R,\varepsilon}(x) = \varepsilon + \int_{R_d} G(x, y)\, V(y)\, \psi_R(y)\, dy .$$

We note that $u_{R,\varepsilon} \in U$, and from the definition of $I(\sigma)$, we have

$$\int_{R_d} \frac{V(x)\, \psi_R(x)}{u_{R,\varepsilon}(x)} \sigma(dx) \leqq I(\sigma) . \tag{2.2}$$

Letting $R \to \infty$ and $\varepsilon \to 0$, we obtain (2.1) from (2.2) and Fatou's lemma.

LEMMA 2.3. *If $\sigma \in M$ is such that $I(\sigma) = l < \infty$, then*

$$\int_{R_d} \frac{\sigma(dx)}{r^2} \leqq \frac{8l}{(d-2)^2} , \tag{2.3}$$

where r is the radius vector in R_d.

Proof: Let $\varepsilon > 0$, $\alpha = \frac{1}{4}(2-d)$, and define $u_\varepsilon(x) = (\varepsilon^2 + r^2)^\alpha$. Since

$$\Delta f(r) = f''(r) + \frac{d-1}{r} f'(r) ,$$

we have

$$
\begin{aligned}
-\tfrac{1}{2}\Delta u_\varepsilon(x) &= -\tfrac{1}{2}\Big[2\alpha(\varepsilon^2+r^2)^{\alpha-1}+4\alpha r^2(\alpha-1)(\varepsilon^2+r^2)^{\alpha-2} \\
&\qquad +2\Big(\frac{d-1}{r}\Big)\alpha r(\varepsilon^2+r^2)^{\alpha-1}\Big] \\
&=(\varepsilon^2+r^2)^{\alpha-2}\Big[-\alpha(\varepsilon^2+r^2)-2\alpha r^2(\alpha-1)-\Big(\frac{d-1}{r}\Big)\alpha r(\varepsilon^2+r^2)\Big] \\
&=(\varepsilon^2+r^2)^{\alpha-2}[\tfrac{1}{4}d(d-2)\varepsilon^2+\tfrac{1}{8}(d-2)^2r^2].
\end{aligned}
$$

If we set $V_\varepsilon(x)=-\tfrac{1}{2}\Delta u_\varepsilon(x)$, then $V_\varepsilon(x)\geqq 0$ is a smooth function. Since $\alpha<0$, $u_\varepsilon(x)\to 0$ as $|x|\to\infty$, and therefore

$$u_\varepsilon(x)=\int_{R_d} G(x,y)\,V_\varepsilon(y)\,dy.$$

Hence, from Lemma 2.2,

$$\int_{R_d}\frac{[\tfrac{1}{4}d(d-2)\varepsilon^2+\tfrac{1}{8}(d-2)^2r^2]}{(\varepsilon^2+r^2)^2}\,\sigma(dx)\leqq I(\sigma). \tag{2.4}$$

Letting $\varepsilon\to 0$ we obtain (2.3) from (2.4) and Fatou's lemma.

COROLLARY. *If $\sigma\in M$ is such that $I(\sigma)=l<\infty$, then*

$$\sigma[x:|x|\leqq R]=o(R^2)\quad \textit{as}\quad R\to\infty. \tag{2.5}$$

Proof: Let $\psi(R)=\sigma[x:|x|\leqq R]$. From Lemma 2.3 we see that

$$\int_{R_d}\frac{\sigma(dx)}{r^2}=\int_0^\infty\frac{d\psi(R)}{R^2}\leqq\frac{8l}{(d-2)^2}.$$

Hence, for any $R\geqq 1$,

$$\int_1^R\frac{d\psi(y)}{y^2}\leqq\frac{8l}{(d-2)^2},$$

and integrating by parts, we obtain

$$\frac{\psi(R)}{R^2}-\psi(1)+2\int_1^R\frac{\psi(y)}{y^3}\,dy\leqq\frac{8l}{(d-2)^2}.$$

Since $\psi(R)/R^2>0$, we conclude from this last inequality that $\int_1^\infty (\psi(y)/y^3)\,dy<\infty$. Let $\varepsilon>0$ be given and choose R so large that $\int_R^\infty (\psi(y)/y^3)\,dy<\varepsilon$. Since $\psi(y)$ is monotone increasing, we have

$$\frac{\psi(R)}{2R^2}=\psi(R)\int_R^\infty \frac{dy}{y^3}\leqq\int_R^\infty \frac{\psi(y)}{y^3}\,dy<\varepsilon,$$

which implies, since $\varepsilon>0$ is arbitrary, that $\psi(R)=o(R^2)$ as $R\to\infty$.

LEMMA 2.4. *Let $\sigma\in M$ be such that $I(\sigma)=l<\infty$, and assume $\sigma(dx)=f(x)\,dx$, where $f>0$, $f\in C^\infty$, and $f\to 0$ as $|x|\to\infty$. Then, if $\phi(x)=(f(x))^{1/2}$, we have*

$$\tfrac{1}{2}\int_{R_d} |\nabla\phi|^2\,dx\leqq l. \tag{2.6}$$

Proof: For all $u\in U$,

$$\tfrac{1}{2}\int_{R_d} \left(\frac{\Delta u}{u}\right)(x)f(x)\,dx\geqq -l. \tag{2.7}$$

If $h\in C_0^\infty$ and λ is real, then $u=e^{\lambda h}\in U$ so that from (2.7) we have

$$\tfrac{1}{2}\left[\lambda\int_{R_d} (\Delta h)f\,dx+\lambda^2\int_{R_d} |\nabla h|^2 f\,dx\right]\geqq -l. \tag{2.8}$$

Since (2.8) holds for all real λ, we obtain from the obvious discriminant inequality and integration by parts:

$$\left|\int_{R_d} \langle\nabla h,\nabla f\rangle\,dx\right|^2\leqq 8l\int_{R_d} |\nabla h|^2 f\,dx,$$

or

$$\left|\int_{R_d} \langle\nabla h,\nabla\phi\rangle\phi\,dx\right|^2\leqq 2l\int_{R_d} |\nabla h|^2\phi^2\,dx. \tag{2.9}$$

Now, for $n=1,2,3,\cdots$, let

$$\psi_n(y)=\begin{cases}0 & \text{for } 0<y\leqq 1/2n,\\ 1/y & \text{for } \quad y\geqq 1/n,\end{cases}$$

where $0 \leqq \psi_n(y) \leqq 1/y$ for all y and ψ_n is smooth. Let $\Psi_n(y) = \int_0^y \psi_n(s)\,ds$ and define $h_n(x) = \Psi_n(\phi(x))$; then we observe that $h_n \in C_0^\infty$ and that it must satisfy (2.9), i.e.,

$$\left|\int_{R_d} \psi_n(\phi(x))\,|\nabla\phi|^2\phi\,dx\right|^2 \leqq 2l \int_{R_d} \psi_n^2(\phi(x))\,|\nabla\phi|^2\phi^2\,dx$$
$$\leqq 2l \int_{R_d} \psi_n(\phi(x))\,|\nabla\phi|^2\phi\,dx,$$

where the last inequality follows from the fact that $\psi_n(y) \leqq 1/y$. Hence,

$$\int_{R_d} \psi_n(\phi(x))\,|\nabla\phi|^2\phi\,dx \leqq 2l,$$

and letting $n \to \infty$ we obtain (2.6) from this last inequality and Fatou's lemma.

LEMMA 2.5. *If $\sigma \in M$ is such that $I(\sigma) < \infty$, then*

$$\lim_{|x|\to\infty} \sigma[B(x, R)] = 0 \tag{2.10}$$

for all $0 < R < \infty$, where $B(x, R)$ is the ball in R_d of radius R with center at x.

Proof: Assume the contrary, and let $a_n = 4^n + 2(R+1)$ for $n = 1, 2, 3, \cdots$. Then, there exists a sequence $\{x_n\}$ such that $|x_m - x_n| \geqq a_{n\vee m}$ and, for all n, $\sigma[B(x_n, R)] \geqq \delta > 0$. Set

$$f(x) = \begin{cases} 1 & \text{for} \quad x \in B(0, R), \\ 0 & \text{for} \quad x \notin B(0, R+1), \end{cases}$$

where $f \geqq 0$ for all x and is in C^∞ on R_d. Let $G(x, y)$ be Green's function for $-\frac{1}{2}\Delta$ on R_d, and define

$$u(x) = \int_{R_d} G(x, y) f(y)\,dy.$$

Since f has compact support, there exist constants C_1 and C_2 such that

$$\sup_{x\in R_d} u(x) \leqq C_1, \tag{2.11}$$
$$u(x) \leqq \frac{C_2}{|x|^{d-2}} \quad \text{for all} \quad x \in R_d.$$

Consider $\sum_{i=1}^{n} u(x-x_i)$. This sum is bounded for al x and it approaches 0 as $|x|\to\infty$. It is harmonic outside of $\bigcup_{j=1}^{n} B(x_j, R+1)$ so that, by the maximum principle,

$$\sup_{x\in R_d} \sum_{j=1}^{n} u(x-x_j) = \sup_{x\in \bigcup_{j=1}^{n} B(x_j, R+1)} \sum_{j=1}^{n} u(x-x_j)\,. \tag{2.12}$$

Since $a_n = 4^n + 2(R+1)$ and $|x_m - x_n| \geqq a_{n\vee m}$ we have from (2.11) that

$$\begin{aligned}\sup_{x\in \bigcup_{j=1}^{n} B(x_j, R+1)} \sum_{j=1}^{n} u(x-x_j) &\leqq C_1 + C_2 \sup_i \sum_{j\neq i} \frac{1}{(|x_i - x_j| - R - 1)^{d-2}} \\ &\leqq C_1 + C_2 \sup_i \sum_{j\neq 1} \frac{1}{(4^{j\vee i})^{d-2}} \\ &\leqq C_1 + C_2 \sum_{j=1}^{n} \frac{1}{4^{j(d-2)}} \leqq C_1 + C_2 \sum_{j=1}^{\infty} \frac{1}{4^{j(d-2)}}\,.\end{aligned} \tag{2.13}$$

Therefore, in view of (2.12) and (2.13), there exists a constant K independent of n and x such that $\sum_{j=1}^{n} u(x-x_j) \leqq K$.

Using Lemma 2.2 we conclude that, for all n,

$$\begin{aligned} I(\sigma) &\geqq \int_{R_d} \frac{\sum_{j=1}^{n} f(x-x_j)}{\sum_{j=1}^{n} u(x-x_j)}\, \sigma(dx) \\ &\geqq \frac{1}{K} \sum_{j=1}^{n} \sigma[B(x_j, R)] \geqq \frac{n\delta}{K}\,,\end{aligned}$$

which implies $I(\sigma)=\infty$ contradicting our hypothesis.

LEMMA 2.6. *If $f=\phi^2$ and $\frac{1}{2}\int_{R_d} |\nabla\phi|^2\, dx = l$, where ϕ has compact support, and if we define $\sigma(dx) = f(x)\, dx$, then*

$$I(\sigma) \leqq l\,. \tag{2.14}$$

Proof: For any $u\in U$ consider $\frac{1}{2}\int_{R_d} (\Delta u/u)(x) f(x)\, dx$ which becomes, on writing $u=e^h$, $\frac{1}{2}\int_{R_d} (\Delta h + |\nabla h|^2)\phi^2\, dx$. Since ϕ has compact support, we see that

$$\begin{aligned}\tfrac{1}{2}\int_{R_d}(\Delta h+|\nabla h|^2)\phi^2\,dx &= \int_{R_d}\langle\nabla h,\nabla\phi\rangle\phi\,dx+\tfrac{1}{2}\int_{R_d}|\nabla h|^2\,\phi^2\,dx\\ &\geqq -\int_{R_d}|\nabla h|\,|\nabla\phi|\,\phi\,dx+\tfrac{1}{2}\int_{R_d}|\nabla h|^2\,\phi^2\,dx\\ &\geqq -\left(\int_{R_d}|\nabla h|^2\,\phi^2\,dx\right)^{1/2}\left(\int_{R_d}|\nabla\phi|^2\,dx\right)^{1/2}\\ &\quad+\tfrac{1}{2}\int_{R_d}|\nabla h|^2\,\phi^2\,dx\,.\end{aligned}$$

Thus, if we let $Q=\int_{R_d}|\nabla h|^2\,\phi^2\,dx$, we have for any $u\in U$, with h defined by $u=e^h$,

$$\tag{2.15}\begin{aligned}\tfrac{1}{2}\int_{R_d}\left(\frac{\Delta u}{u}\right)(x)\,f(x)\,dx &\geqq -(2lQ)^{1/2}+\tfrac{1}{2}Q\\ &= -l+[(\tfrac{1}{2}Q)^{1/2}-l^{1/2}]^2\geqq -l\,.\end{aligned}$$

Since (2.15) holds for all $u\in U$, we conclude that $I(\sigma)\leqq l$.

LEMMA 2.7. *If* $f=\phi^2$,

$$\tfrac{1}{2}\int_{R_d}|\nabla\phi|^2\,dx=l\quad\textit{and}\quad\int_{|x|\leqq R}\phi^2\,dx=o(R^2)$$

as $R\to\infty$, *letting* $\sigma(dx)=f(x)dx$, *we have* $I(\sigma)\leqq l$. *Moreover, there exists a sequence* $\{\phi_n\}$ *with compact support such that*

$$\tfrac{1}{2}\int_{R_d}|\nabla\phi_n|^2\,dx\to l\quad\textit{as}\quad n\to\infty\,,$$

and

$$\int_B|\phi_n-\phi|^2\,dx\to 0\quad\textit{as}\quad n\to\infty\,,$$

for all balls in R_d.

Proof: Let

$$\psi(x)=\begin{cases}1 & \text{for}\quad |x|\leqq 1\,,\\ 0 & \text{for}\quad |x|\geqq 2\,,\end{cases}$$

with $0 \leqq \psi(x) \leqq 1$ for all x and smooth. Define $\phi_n(x) = \phi(x)\,\psi(x/n)$. Then, $\nabla\phi_n = \psi(x/n)\,\nabla\phi + (1/n)\phi(\nabla\psi)(x/n)$ and, for a constant C,

$$\begin{aligned}\left(\int_{R_d} |\nabla\phi_n - \nabla\phi|^2\,dx\right)^{1/2} &\leqq \left(\int_{R_d} |\nabla\phi|^2 \left|1-\psi\left(\frac{x}{n}\right)\right|^2 dx\right)^{1/2} \\ &\quad + \left(\frac{1}{n^2}\int_{R_d} \phi^2 \left|\nabla\psi\left(\frac{x}{n}\right)\right|^2 dx\right)^{1/2} \\ &\leqq \left(\int_{|x|\geqq n} |\nabla\phi|^2\,dx\right)^{1/2} + \left(\frac{C^2}{n^2}\int_{|x|\leqq 2n} \phi^2(x)\,dx\right)^{1/2}.\end{aligned} \tag{2.16}$$

The hypotheses on ϕ imply that the right side of (2.16) approaches 0 as $n \to \infty$, and we conclude that $\frac{1}{2}\int_{R_d} |\nabla\phi_n|^2\,dx \to l$ as $n \to \infty$. From the definition of ϕ_n we see that, for all balls B,

$$\lim_{n\to\infty} \int_B |\phi_n - \phi|^2\,dx = 0\,.$$

Finally, if we define $\sigma_n(dx) = \phi_n^2(x)\,dx$, then Lemma 2.6 implies that

$$I(\sigma_n) \leqq \tfrac{1}{2}\int_{R_d} |\nabla\phi_n|^2\,dx\,.$$

Thus, if we let $n \to \infty$ and use the lower semicontinuity of the I-function we see that $I(\sigma) \leqq l$.

Using Lemma 2.7 and Lemma 2.4 we can prove a sharper form of Lemma 2.4, i.e., the function f appearing in Lemma 2.4 is now assumed only to be non-negative whereas in Lemma 2.4 it must be strictly positive.

LEMMA 2.8. *Let $\sigma \in M$ be such that $I(\sigma) = l < \infty$. Assume $\sigma(dx) = f(x)\,dx$, where $f(x) \geqq 0$, $f \in C^\infty$, and $f \to 0$ as $|x| \to \infty$. Then, letting $\phi = f^{1/2}$, we have*

$$\tfrac{1}{2}\int_{R_d} |\nabla\phi|^2\,dx \leqq l. \tag{2.17}$$

Proof: Let $f_0(x) = \exp\{-\frac{1}{2}|x|^2\}$ and let $\sigma_0(dx) = f_0(x)\,dx$. From Lemma 2.7 we conclude that $I(\sigma_0) = k < \infty$. For $\varepsilon > 0$ we obtain from the convexity property of the I-function that

$$I(\sigma + \varepsilon\sigma_0) \leqq I(\sigma) + \varepsilon I(\sigma_0) = l + \varepsilon k\,.$$

Hence, if we define $f_\varepsilon = f + \varepsilon f_0$, then f_ε satisfies the hypotheses of Lemma 2.4 so that $\phi_\varepsilon(x) = (f(x) + \varepsilon f_0(x))^{1/2}$ satisfies

$$\frac{1}{2}\int_{R_d} |\nabla \phi_\varepsilon|^2 \, dx \leqq l + \varepsilon k .$$

If we let $\varepsilon \to 0$ in this last inequality, we obtain (2.17) from Fatou's lemma.

LEMMA 2.9. *Let $\sigma \in M$ be such that $I(\sigma) = l < \infty$. Then σ is absolutely continuous with respect to Lebesgue measure on R_d, $\sigma(dx) = f(x)\,dx$, and if we set $\phi(x) = (f(x))^{1/2}$,*

$$\frac{1}{2}\int_{R_d} |\nabla \phi|^2 \, dx \leqq l .$$

Proof: Let $g(x) \geqq 0$, $g \in C^\infty$, $\operatorname{supp} g \subset B(0, 1)$, and $\int_{R_d} g(x)\,dx = 1$. For $\varepsilon > 0$, let $g_\varepsilon(x) = \varepsilon^{-d} g(x/\varepsilon)$ and define $\sigma_\varepsilon = \sigma * g_\varepsilon$. Then $\sigma_\varepsilon(dx) = f_\varepsilon(x)\,dx$, where

$$f_\varepsilon(x) = \int_{R_d} g_\varepsilon(x - y)\, \sigma(dy) .$$

From the fact that the I-function is translation invariant and convex we have $I(\sigma_\varepsilon) \leqq I(\sigma) = l$. We note that $f_\varepsilon \in C^\infty$, $f_\varepsilon \geqq 0$, and $f_\varepsilon \to 0$ as $|x| \to \infty$. Therefore, if $\phi_\varepsilon = (f_\varepsilon)^{1/2}$ we have, by Lemma 2.8,

$$\frac{1}{2}\int_{R_d} |\nabla \phi_\varepsilon|^2 \, dx \leqq l$$

for all $\varepsilon > 0$. Hence, letting $\varepsilon \to 0$ we conclude that σ is absolutely continuous with respect to Lebesgue measure, i.e., for some f, $\sigma(dx) = f(x)\,dx$ and, moreover, if $\phi = f^{1/2}$, then

$$\frac{1}{2}\int_{R_d} |\nabla \phi|^2 \, dx \leqq l .$$

3. Asymptotics

Let $\{\beta(s), 0 \leqq s < \infty\}$ be Brownian motion in R_d starting from $x \in R_d$ with $d \geqq 3$. For E, a Borel set in R_d, define

$$\sigma(E, \omega) = \int_0^\infty \chi_E(\beta(s))\, ds , \tag{3.1}$$

i.e., the total time the set E is occupied by the particular path $\omega=\beta(\cdot)$. Clearly, for each ω, $\sigma(\cdot,\omega)$ is a σ-finite measure on R_d, i.e., an element of M. For each $n=1,2,3,\cdots$ and each $x\in R_d$ we define a probability measure Q_n^x on M as follows: If A is a set of σ-finite measures on R_d, i.e., $A\subset M$, then

$$Q_n^x(A)=P_x\{\omega:(1/n)\sigma(\cdot,\omega)\in A\}, \tag{3.2}$$

where $P_x\{\ \}$ denotes Brownian motion measure on the space of paths $\omega=\beta(\cdot)$ starting from x.

In this section we shall examine the asymptotic behavior of the measure Q_n^x as $n\to\infty$. This will involve the I-function defined by (1.3) for all σ-finite measures in M. Here all topological notions in M refer to the vague topology on M described in the introduction.

THEOREM 3.1. *For each $K\subset M$ that is compact,*

$$\overline{\lim_{n\to\infty}}\frac{1}{n}\log Q_n^x(K)\leqq-\inf_{\sigma\in K}I(\sigma), \tag{3.3}$$

uniformly in x.

Proof: Assume $u\in U$ and consider, for any $t>0$,

$$u(\beta(t))\exp\left\{-\int_0^t\tfrac{1}{2}\left(\frac{\Delta u}{u}\right)(\beta(s))\,ds\right\}. \tag{3.4}$$

From the Feynman–Kac formula we see that

$$E_x\left\{\exp\left\{-\int_0^t\tfrac{1}{2}\left(\frac{\Delta u}{u}\right)(\beta(s))\,ds\right\}u(\beta(t))\right\}=u(x), \tag{3.5}$$

where, as usual, $E_x\{\ \}$ denotes integration over the Brownian motion paths $\beta(\cdot)$ starting at x. Equation (3.5) is just the statement that (3.4) is a martingale. We conclude that for any $u\in U$,

$$E_x\left\{\exp\left\{-\int_0^t\tfrac{1}{2}\left(\frac{\Delta u}{u}\right)(\beta(s))\,ds\right\}\right\}\leqq\frac{\sup_{x\in R_d}u(x)}{\inf_{x\in R_d}u(x)}. \tag{3.6}$$

Since $d\geqq 3$, Brownian motion is transient and, since $u\in U$, Δu has compact support. Therefore, if we let $t\to\infty$ in (3.6) we obtain from Fatou's lemma

$$E_x\left\{\exp\left\{-\int_0^\infty \tfrac{1}{2}\left(\frac{\Delta u}{u}\right)(\beta(s))\,ds\right\}\right\} \leqq \frac{\sup_{x\in R_d} u(x)}{\inf_{x\in R_d} u(x)}. \tag{3.7}$$

From the definition of Q_n^x measure on M we can rewrite (3.7) as

$$E^{Q_n^x}\left\{\exp\left\{-n\int_{R_d} \tfrac{1}{2}\left(\frac{\Delta u}{u}\right)(x)\sigma(dx)\right\}\right\} \leqq \frac{\sup_{x\in R_d} u(x)}{\inf_{x\in R_d} u(x)}, \tag{3.8}$$

where $E^{Q_n^x}\{\ \}$ denotes integration over the space M with respect to Q_n^x measure so that in (3.8) the variable of integration is σ.

From (3.8) we obtain by the usual estimate that, for the set K,

$$Q_n^x(K) \leqq \frac{\sup_{x\in R_d} u(x)}{\inf_{x\in R_d} u(x)} \exp\left\{n \sup_{\sigma\in K}\int_{R_d} \tfrac{1}{2}\left(\frac{\Delta u}{u}\right)(x)\sigma(dx)\right\}. \tag{3.9}$$

Since (3.9) holds for all $u\in U$ and all $x\in R_d$, we see that

$$\begin{aligned} \overline{\lim_{n\to\infty}} \sup_{x\in R_d} \frac{1}{n}\log Q_n^x(K) &\leqq \sup_{\sigma\in K}\int_{R_d} \tfrac{1}{2}\left(\frac{\Delta u}{u}\right)(x)\,\sigma(dx) \\ &\leqq \inf_{u\in U}\sup_{\sigma\in K}\int_{R_d} \tfrac{1}{2}\left(\frac{\Delta u}{u}\right)(x)\,\sigma(dx). \end{aligned} \tag{3.10}$$

Since K is compact, we can use the argument of [2] (after (2.10)) to conclude that (3.10) implies

$$\begin{aligned} \overline{\lim_{n\to\infty}} \sup_{x\in R_d} \frac{1}{n}\log Q_n^x(K) &\leqq -\inf_{\sigma\in K}\left[-\inf_{u\in U}\int_{R_d} \tfrac{1}{2}\left(\frac{\Delta u}{u}\right)(x)\,\sigma(dx)\right] \\ &= -\inf_{\sigma\in K} I(\sigma). \end{aligned} \tag{3.11}$$

In the following we denote the complement of a set A by A^c.

THEOREM 3.2. *For each $l<\infty$, there exists a compact set $K\subset M$ such that*

$$\overline{\lim_{n\to\infty}} \frac{1}{n}\log \sup_{x\in R_d} Q_n^x(K^c) \leqq -l. \tag{3.12}$$

Proof: Let r denote the radius vector in R_d and define $u(x) = (1+r^2)^{(2-d)/4}+1$. As in the proof of Lemma 2.3,

$$-\tfrac{1}{2}\Delta u(x) = (1+r^2)^{(2-d)/4-2}\left[\tfrac{1}{4}d(d-2)+\tfrac{1}{8}(d-2)^2 r^2\right],$$

so that if we let $V(x)=\frac{1}{2}(\Delta u/u)(x)$, then

$$\frac{\frac{1}{4}d(d-2)+\frac{1}{8}(d-2)^2r^2}{(1+r^2)^2+(1+r^2)^{2+(d-2)/4}}>0\,.$$

As in Theorem 3.1, for this particular u and with $V(x)$ as defined, we see that

$$u(\beta(t))\exp\left\{\int_0^t V(\beta(s))\,ds\right\}$$

is a martingale, and we conclude, as in the previous theorem, that

$$E^{Q_n^x}\left\{\exp\left\{n\int_{R_d} V(x)\sigma(dx)\right\}\right\}\leqq\frac{\sup_{x\in R_d}u(x)}{\inf_{x\in R_d}u(x)}\,. \tag{3.11)'$$

From (3.11)′ we see that

$$Q_n^x\left\{\sigma:\int_{R_d} V(x)\,\sigma(dx)>l\right\}\leqq\frac{\sup_{x\in R_d}u(x)}{\inf_{x\in R_d}u(x)}\,e^{-nl}\,. \tag{3.12)'$$

If we define

$$K_l=\left\{\sigma:\int_{R_d} V(x)\sigma(dx)\leqq l\right\},$$

then we note that K_l is compact and from (3.12)′ that

$$\overline{\lim_{n\to\infty}}\frac{1}{n}\log\sup_{x\in R_d} Q_n^x(K_l^c)\leqq -l\,.$$

THEOREM 3.3. *For each $C\subset M$ that is closed,*

$$\overline{\lim_{n\to\infty}}\frac{1}{n}\log\sup_{x\in R_d} Q_n^x(C)\leqq -\inf_{\sigma\in C} I(\sigma)\,. \tag{3.13}$$

Proof: Let $l>0$ be given and choose K_l as in Theorem 3.2. Since $C\subset(C\cap K_l)\cup K_l^c$, $Q_n^x(C)\leqq Q_n^x(C\cap K_l)+Q_n^x(K_l^c)$. Thus, from Theorems 3.1 and 3.2 we see that

$$\begin{aligned}\overline{\lim_{n\to\infty}}\frac{1}{n}\log\sup_{x\in R_d} Q_n^x(C)&\leqq\max\left[-\inf_{\sigma\in C\cap K_l} I(\sigma),-l\right]\\&\leqq\max\left[-\inf_{\sigma\in C} I(\sigma),-l\right].\end{aligned} \tag{3.14}$$

Since (3.14) holds for all $l>0$, we obtain (3.13) if we let $l\to\infty$.

Next, we seek asymptotic lower estimates for the Q_n^x measure of open sets in M. We begin with the following theorem.

THEOREM 3.4. *Let N be a neighborhood of $0\in M$ (the zero measure in M). Then,*

$$\lim_{n\to\infty} Q_n^x(N)=1\,, \tag{3.15}$$

uniformly in x.

Proof: Set $C=N^c$; from Theorem 3.3 we have

$$\overline{\lim_{n\to\infty}}\frac{1}{n}\log\sup_{x\in R_d} Q_n^x(N^c)\leqq -\inf_{\sigma\in N^c} I(\sigma)\,. \tag{3.16}$$

From the properties of the I-function and Lemma 2.3 we see that $\inf_{\sigma\in N^c} I(\sigma)>0$. Therefore, (3.16) implies that $Q_n^x(N)\to 1$ uniformly in x as $n\to\infty$.

THEOREM 3.5. *Let N be a neighborhood of some $\sigma_0\in M$ with $I(\sigma_0)=l<\infty$. Then,*

$$\underline{\lim_{n\to\infty}}\frac{1}{n}\log Q_n^x(N)\geqq -l\,, \tag{3.17}$$

uniformly for x in compact sets of R_d.

Proof: Let $\varepsilon>0$ be given. In the neighborhood N of σ_0 there exists a measure σ_1 such that $I(\sigma_1)\leqq l+\varepsilon$ and $\sigma_1(R_d)=T<\infty$. This follows from Theorem 2.1 and Lemma 2.7 wherein we constructed an approximating sequence with compact support. Thus in proving Theorem 3.5 we can without loss of generality assume that σ_0 in the statement of the theorem is actually a finite measure, i.e., $\sigma_0(R_d)=T<\infty$. We can then write $\sigma_0=T\mu_0$, where μ_0 is a probability measure on R_d, and note from the definition of the I-function that if $I(\sigma_0)=l$, then $I(\mu_0)=l/T$.

Since N is a neighborhood of $\sigma_0=T\mu_0$ we can find a neighborhood G_1 of 0 in M and a neighborhood G_2 of μ_0 such that if $\sigma\in G_1$ and $\mu\in G_2$, then $\sigma+T\mu\in N$. Also, for any integer n we can write

$$\begin{aligned}\sigma(A,\omega)&=\int_0^\infty \chi_A(\beta(s))\,ds=\int_0^{nT}\chi_A(\beta(s))\,ds+\int_{nT}^\infty \chi_A(\beta(s))\,ds\\&=\sigma_{[0,nT]}(A,\omega)+\sigma_{[nT,\infty]}(A,\omega)\,.\end{aligned} \tag{3.18}$$

Hence, with G_1 and G_2 as defined above, we observe that

$$(3.19)\quad P_x\left\{\frac{1}{n}\sigma(\cdot,\omega)\in N\right\}\geqq P_x\left\{\frac{1}{nT}\sigma_{[0,nT]}(\cdot,\omega)\in G_2, \frac{1}{n}\sigma_{[nT,\infty]}(\cdot,\omega)\in G_1\right\}.$$

But, from the Markov property, we have

$$(3.20)\quad \begin{aligned} &P_x\left\{\frac{1}{nT}\sigma_{[0,nT]}(\cdot,\omega)\in G_2, \frac{1}{n}\sigma_{[nT,\infty]}(\cdot,\omega)\in G_1\right\} \\ &\geqq P_x\left\{\frac{1}{nT}\sigma_{[0,nT]}(\cdot,\omega)\in G_2\right\}\inf_{x\in R_d} P_x\left\{\frac{1}{n}\sigma(\cdot,\omega)\in G_1\right\}. \end{aligned}$$

Thus,

$$(3.21)\quad \begin{aligned} Q_n^x(N) &= P_x\left\{\frac{1}{n}\sigma(\cdot,\omega)\in N\right\} \\ &\geqq P_x\left\{\frac{1}{nT}\sigma_{[0,nT]}(\cdot,\omega)\in G_2\right\}\inf_{x\in R_d} P_x\left\{\frac{1}{n}\sigma(\cdot,\omega)\in G_1\right\}. \end{aligned}$$

But,

$$P_x\left\{\frac{1}{n}\sigma(\cdot,\omega)\in G_1\right\}=Q_n^x(G_1)$$

and G_1 is a neighborhood of 0 in M so, by Theorem 3.4, $\lim_{n\to\infty} Q_n^x(G_1)=1$ uniformly in x. Using this in (3.21) we obtain

$$(3.22)\quad \varliminf_{n\to\infty}\frac{1}{n}\log Q_n^x(N)\geqq \varliminf_{n\to\infty}\frac{1}{n}\log P_x\left\{\frac{1}{nT}\sigma_{[0,nT]}(\cdot,\omega)\in G_2\right\}.$$

Now in [2] (Lemma 1) we showed that

$$(3.23)\quad \varliminf_{n\to\infty}\frac{1}{n}\log P_x\left\{\frac{1}{nT}\sigma_{[0,nT]}(\cdot,\omega)\in G_2\right\}\geqq -TI(\mu_0)=-l,$$

uniformly for x in compact sets. Thus, (3.22) and (3.23) yield the theorem.

We have as an immediate corollary

COROLLARY. *If G is an open set in M, then*

$$\varliminf_{n\to\infty} \frac{1}{n} \log Q_n^x(G) \geqq -\inf_{\sigma\in G} I(\sigma)\,, \tag{3.24}$$

uniformly for x in compact sets.

4. Strong Limit Theorems

It is an elementary scaling property of Brownian motion that the probability distribution of $(1/\lambda)\beta(s)$ is the same as the probability distribution of $\beta(s/\lambda^2)$. Using the notation $\stackrel{d}{=}$ to mean "has the same distribution as" we observe that

$$\begin{aligned}\sigma(\lambda A, \omega) &= \int_0^\infty \chi_{\lambda A}(\beta(s))\, ds = \int_0^\infty \chi_A\Big(\frac{1}{\lambda}\beta(s)\Big)\, ds \\ &\stackrel{d}{=} \int_0^\infty \chi_A\Big(\beta\Big(\frac{s}{\lambda^2}\Big)\Big)\, ds \\ &\stackrel{d}{=} \lambda^2 \sigma(A, \omega)\,.\end{aligned} \tag{4.1}$$

For $\lambda \geqq 3$, let

$$Z_\lambda(A, \omega) = \frac{\sigma(\lambda A, \omega)}{\lambda^2 \log\log \lambda}\,,$$

so that for each ω, i.e., each Brownian motion path $\beta(\cdot)$, $Z_\lambda(\cdot, \omega)$ is a family, in the parameter λ, of σ-finite measures on R_d. In this section we show, first of all, that, for almost all ω, $\{Z_\lambda(\cdot, \omega)\}$ is conditionally compact. Let C_ω, for each ω, denote the cluster set for the family $\{Z_\lambda(\cdot, \omega)\}$ of σ-finite measures.

As we saw in Section 2, if a σ-finite measure σ has a finite I-function, $I(\sigma)$, then σ has a density with respect to Lebesgue measure on R_d. Denoting this density by f and writing $I(f)$ instead of $I(\sigma)$ we consider the set $\mathscr{B} = \{f : I(f) \leqq 1\}$, and the main result of this section will be that, for almost all ω, $C_\omega = \mathscr{B}$. As indicated in the introduction, the strong limit theorem we seek then follows.

We start with the following lemma:

LEMMA 4.1. *For almost all* ω,

$$\sup_{\lambda \geqq 3} Z_\lambda(K, \omega) < \infty \tag{4.2}$$

for each compact set $K \subset R_d$.

Proof: It suffices to show the result for balls in R_d; so, as before, let $B = B(0, R)$ denote the ball in R_d with center at 0 of radius R. Let $\rho > 1$ and, for positive integers n, define

$$X_n = X_n(\omega) = \sup_{\rho^n \leqq \lambda \leqq \rho^{n+1}} Z_\lambda(B, \omega),$$

so that $\{X_n\}$ is a sequence of random variables (on the Brownian motion paths in R_d), and we want to show that this sequence is bounded almost surely.

We observe that

$$\begin{aligned} X_n = \sup_{\rho^n \leqq \lambda \leqq \rho^{n+1}} \frac{\sigma(\lambda B(0, R), \omega)}{\lambda^2 \log\log \lambda} &\leqq \frac{\sigma(B(0, \rho^{n+1} R), \omega)}{\rho^{2n} \log\log \rho^n} \\ &= \frac{\sigma(B(0, \rho^{n+1} R), \omega)}{\rho^{2n}(\log n + \log\log \rho)}. \end{aligned} \tag{4.3}$$

Let

$$I_0 = \inf_{f : \int_{B(0,R)} f dx \geqq 1} I(f).$$

From Lemma 2.3 we see that $I_0 > 0$. Pick α so that $\alpha I_0/\rho^2 > 1$, and from (4.3) and (4.1) we have

$$\begin{aligned} P\{X_n \geqq \alpha\} &\leqq P\left\{\frac{\sigma(B(0, \rho^{n+1} R), \omega)}{\rho^{2n+2}} \geqq \frac{\alpha}{\rho^2}(\log n + \log\log \rho)\right\} \\ &= P\left\{\sigma(B(0, R), \omega) \geqq \frac{\alpha}{\rho^2}(\log n + \log\log \rho)\right\}. \end{aligned} \tag{4.4}$$

Using the same kind of estimate as used in the proof of Theorem 3.1 we see that

$$(4.5)\qquad \begin{aligned} &P\left\{\sigma(B(0, R), \omega) \geqq \frac{\alpha}{\rho^2}(\log n + \log\log\rho)\right\} \\ &\qquad \leqq A \exp\left\{-\frac{\alpha}{\rho^2}(\log n + \log\log\rho)\right\} I_0 , \end{aligned}$$

where A is a constant.

Thus, (4.4) and (4.5) imply, since $\alpha I_0/\rho^2 > 1$, that $\sum P\{X_n \geqq \alpha\} < \infty$. It follows now from the Borel–Cantelli lemma that the sequence $\{X_n\}$ is bounded almost surely.

COROLLARY. *For almost all ω, the family of σ-finite measures $\{Z_\lambda(\cdot, \omega)\}$, $\lambda \geqq 3$, is conditionally compact.*

With C_ω and $\mathscr{B}$ as defined at the beginning of this section, we next show the inclusion $C_\omega \subset \mathscr{B}$ for almost all ω. To do this we use a lemma the proof of which is immediate from the definition of the vague topology.

LEMMA 4.2. *Given $\sigma \in M$ and a neighborhood N_1 of σ, there exists another neighborhood N of σ and an $\varepsilon > 0$ such that if $\sigma_1 \in N$, $|\alpha - 1| < \varepsilon$, and $|\beta - 1| < \varepsilon$, then*

$$(4.6)\qquad \sigma_2(\cdot) \equiv \beta\sigma_1(\alpha\,\cdot) \in N_1 .$$

LEMMA 4.3. *For almost all ω,*

$$(4.7)\qquad C_\omega \subset \mathscr{B}.$$

Proof: Let $\sigma^* \notin \mathscr{B}$. We must show that there exists a neighborhood N of σ^* such that

$$(4.8)\qquad \lim_{\lambda_0 \to \infty} P\{Z_\lambda(\cdot, \omega) \in N \quad \text{for some} \quad \lambda \geqq \lambda_0\} = 0 .$$

Since $\sigma^* \notin \mathscr{B}$, $I(\sigma^*) > 1$, i.e., $I(\sigma^*) = 1 + \Delta$, where $\Delta > 0$. Since the I-function is lower-semicontinuous, there exists a neighborhood N_1 of σ^* such that $\inf_{\sigma' \in \bar{N}_1} I(\sigma') \geqq 1 + \frac{1}{2}\Delta$, where $\bar{N}_1$ denotes the vague closure of N_1. Let $\rho = 1 + \varepsilon_0$, where $\varepsilon_0 > 0$, and let n be a positive integer. Pick $\lambda \in [\rho^n, \rho^{n+1}]$ and define

$$\sigma_1(\cdot, \omega) = Z_\lambda(\cdot, \omega) = \frac{\sigma(\lambda \cdot, \omega)}{\lambda^2 \log \log \lambda},$$

$$\sigma_2(\cdot, \omega) = Z_{\rho^{n+1}}(\cdot, \omega) = \frac{\sigma(\rho^{n+1} \cdot, \omega)}{\rho^{2(n+1)} \log \log \rho^{n+1}}.$$

Using (4.1) we see that

$$\sigma_2(\cdot, \omega) \overset{d}{=} \beta \sigma_1(\alpha \cdot, \omega), \tag{4.9}$$

where

$$\beta = \frac{\lambda^2 \log \log \lambda}{\rho^{2(n+1)} \log \log \rho^{n+1}} \qquad \text{and} \qquad \alpha = \frac{\rho^{n+1}}{\lambda}.$$

Since $\lambda \in [\rho^n, \rho^{n+1}]$ we see that $1 \leqq \alpha \leqq \rho$, and

$$\frac{1}{\rho^2} \frac{\log n + \log \log \rho}{\log (n+1) + \log \log \rho} \leqq \beta \leqq 1.$$

Hence, by choosing ε_0 close to 0 and by taking n large enough we can, according to (4.9) and Lemma 4.2, find a neighborhood N of σ^* such that if $\sigma_1 = Z_\lambda(\cdot, \omega) \in N$, then $\sigma_2 = Z_{\rho^{n+1}}(\cdot, \omega) \in N_1$. Thus, to show (4.8) it suffices to show that

$$\lim_{l \to \infty} P\{Z_{\rho^n}(\cdot, \omega) \in N_1 \quad \text{for some} \quad n \geqq l\} = 0. \tag{4.10}$$

We recall that N_1 is a neighborhood of σ^* such that $\inf_{\sigma' \in \bar{N}_1} I(\sigma') \geqq 1 + \frac{1}{2}\Delta$. Thus, using (4.1) and again the exponential estimates of Theorem 3.1 (or Theorem 3.4), we have, for sufficiently large n,

$$\begin{aligned} P\{Z_{\rho^n}(\cdot, \omega) \in N_1\} &= P\left\{\frac{\sigma(\rho^n \cdot, \omega)}{\rho^{2n} \log \log \rho^n} \in N_1\right\} \\ &= P\left\{\frac{\sigma(\cdot, \omega)}{\log \log \rho^n} \in N_1\right\} \\ &\leqq P\left\{\frac{\sigma(\cdot, \omega)}{\log \log \rho^n} \in \bar{N}_1\right\} \\ &\leqq A \exp\{-(\log \log \rho^n)(1 + \tfrac{1}{4}\Delta)\} \\ &\leqq \frac{A}{n^{(1+\Delta/4)}}, \end{aligned} \tag{4.11}$$

where A is a constant.

This last inequality implies that $\sum P\{Z_{\rho^n}(\cdot, \omega) \in N_1\}$ converges so that by the Borel–Cantelli lemma we have (4.10).

Now we wish to show that $\mathscr{B} \subset C_\omega$.

THEOREM 4.4. *If f is such that $I(f) \leqq 1$, then, for almost all ω, $f \in C_\omega$.*

Theorem 4.4 follows from a succession of lemmas all of which involve certain sequences and random times which we now define once and for all.

First of all we note that, since the cluster set C_ω is closed, the theorem follows if we show that for every f such that $I(f) < 1$, for almost all ω, $f \in C_\omega$. Since $I(f) < 1$, we can pick $\alpha > 1$ and $\delta > 0$ so that

$$\alpha(I(f) + \delta) < 1 . \tag{4.12}$$

For this α and each positive integer n, let

$$\lambda_n = \exp\{n^\alpha\}, \qquad a_n = n^2 \exp\{n^\alpha\}, \qquad c_n = 2n^2 \exp\{n^\alpha\}, \tag{4.13}$$

so that

$$c_{n-1} < \lambda_n < a_n < c_n < \lambda_{n+1} . \tag{4.14}$$

Consider Brownian motion in R_d starting from the origin and let τ_{a_n} be the first exit time from the ball $B(0, a_n)$, and τ_{c_n} the first exit time from the ball $B(0, c_n)$. For any set $A \subset R_d$ and any positive integer n, define

$$\begin{aligned} \alpha_n'(A, \omega) &= \int_0^{\tau_{a_n}} \chi_A(\beta(s))\, ds , \\ \beta_n'(A, \omega) &= \int_{\tau_{c_{n-1}}}^{\tau_{a_n}} \chi_A(\beta(s))\, ds . \end{aligned} \tag{4.15}$$

Thus, $\alpha_n'(A, \omega)$ is the amount of time a particular Brownian motion path $\omega = \beta(\cdot)$ starting from 0 spends in the set A up to its first exit time from the ball $B(0, a_n)$, whereas $\beta_n'(A, \omega)$ is the amount of time the path spends in A between the first exit time from $B(0, c_{n-1})$ and the first exit time from the larger ball $B(0, a_n)$. Let

$$\begin{aligned} \alpha_n(A, \omega) &= \frac{\alpha_n'(\lambda_n A, \omega)}{\lambda_n^2 \log\log \lambda_n} , \\ \beta_n(A, \omega) &= \frac{\beta_n'(\lambda_n A, \omega)}{\lambda_n^2 \log\log \lambda_n} , \end{aligned} \tag{4.16}$$

and recall that

$$Z_{\lambda_n}(A, \omega) = \frac{\sigma(\lambda_n A, \omega)}{\lambda_n^2 \log\log \lambda_n}.$$

Let N be a neighborhood of f. To prove Theorem 4.4 it clearly suffices to show that

$$P\{Z_{\lambda_n}(\cdot, \omega) \in N \text{ for infinitely many } n\} = 1\,, \tag{4.17}$$

and the following lemmas serve that purpose.

LEMMA 4.5. *Let N be a neighborhood of f, then*

$$P\{Z_{\lambda_n}(\cdot, \omega) \notin N,\ \alpha_n(\cdot, \omega) \in N \textit{ for infinitely many } n\} = 0\,. \tag{4.18}$$

Proof: From the Borel–Cantelli lemma it suffices to show that

$$\sum P\{Z_{\lambda_n}(\cdot, \omega) \notin N\},\ \alpha_n(\cdot, \omega) \in N\} < \infty\,.$$

Let us specify the vague neighborhood N of f, i.e., for some positive integer r, $\varepsilon > 0$, and $\phi_i \in C_0^\infty(R_d)$, $i = 1, 2, \cdots, r$,

$$\begin{aligned} N = N(f) &= N(f;\, r, \varepsilon, \phi_1, \cdots, \phi_r) \\ &= \left\{\sigma \in M : \left| \int_{R_d} \phi_i(x)\, \sigma(dx) - \int_{R_d} \phi_i(x)\, f(x)\, dx \right| < \varepsilon,\ i = 1, 2, \cdots, r \right\}. \end{aligned}$$

We note, first of all, that

$$\begin{aligned} &P\{Z_{\lambda_n}(\cdot, \omega) \notin N,\ \alpha_n(\cdot, \omega) \in N\} \\ &\quad \leqq P\left\{ \int_{R_d} \phi_i(x)\, Z_{\lambda_n}(dx, \omega) \neq \int_{R_d} \phi_i(x)\, \alpha_n(dx, \omega) \text{ for some } 1 \leqq i \leqq r \right\}. \end{aligned} \tag{4.19}$$

Also, the event

$$\left\{ \int_{R_d} \phi_i(x)\, Z_{\lambda_n}(dx, \omega) \neq \int_{R_d} \phi_i(x) \alpha_n(dx, \omega) \text{ for some } 1 \leqq i \leqq r \right\}$$

$$\begin{aligned} &\left\{ \int_{R_d} \phi_i\left(\frac{x}{\lambda_n}\right) \sigma(dx, \omega) \neq \int_{R_d} \phi_i\left(\frac{x}{\lambda_n}\right) \alpha_n'(dx, \omega) \quad \text{for some} \quad 1 \leqq i \leqq r \right\} \\ &= \left\{ \int_0^\infty \phi_i\left(\frac{\beta(s)}{\lambda_n}\right) ds \neq \int_0^{T_{a_n}} \phi_i\left(\frac{\beta(s)}{\lambda_n}\right) ds \quad \text{for some} \quad 1 \leqq i \leqq r \right\} \\ &= \left\{ \int_{T_{a_n}}^\infty \phi_i\left(\frac{\beta(s)}{\lambda_n}\right) ds \neq 0 \quad \text{for some} \quad 1 \leqq i \leqq r \right\}. \end{aligned} \tag{4.20}$$

Let

$$R = \inf\{l : \phi_j(x) = 0 \text{ for } |x| > l, j = 1, 2, \cdots, r\},$$

and let $\theta_{\lambda_n R}$ be the last exit time for the sphere of radius $\lambda_n R$. Since, for n large enough, $\lambda_n R < a_n$ and since the support of $\phi_i(x) \subset B(0, R)$ for all $1 \leqq i \leqq r$, we see that the last event in (4.20) can occur only if $\theta_{\lambda_n R} > \tau_{a_n}$. Thus, (4.19), (4.20), and this last remark imply that, for n large enough,

$$P\{Z_{\lambda_n}(\cdot, \omega) \notin N, \alpha_n(\cdot, \omega) \in N\} \leqq P\{\theta_{\lambda_n R} > \tau_{a_n}\}. \tag{4.21}$$

Let $\tilde{\tau}_{\lambda_n R}$ denote the first hitting time of the ball $B(0, \lambda_n R)$. From the strong Markov property we obtain

$$P\{\theta_{\lambda_n R} > \tau_{a_n}\} = E\{P_{\beta(\tau_{a_n})}[\tilde{\tau}_{\lambda_n R} < \infty]\}. \tag{4.22}$$

It is an elementary property of d-dimensional Brownian motion that if we are given two spheres in R_d centered at the origin of radii R_1 and R_2 with $R_1 < R_2$, then the probability of ever hitting the inner sphere of radius R_1 if the path starts on the outer sphere of radius R_2 is $(R_1/R_2)^{d-2}$. Therefore,

$$E\{P_{\beta(\tau_{a_n})}[\tilde{\tau}_{\lambda_n R} < \infty]\} = \left(\frac{\lambda_n R}{a_n}\right)^{d-2}, \tag{4.23}$$

and from (4.21), (4.22) and (4.23) we see that for n large enough

$$P\{Z_{\lambda_n}(\cdot, \omega) \notin N, \alpha_n(\cdot, \omega) \in N\} \leqq \left(\frac{\lambda_n R}{a_n}\right)^{d-2}. \tag{4.24}$$

Since $d \geqq 3$ and $\lambda_n/a_n = 1/n^2$, we obtain (4.18) and the lemma.

In the preceding lemma we specified the neighborhood $N = N(f)$. Now define $N_1 \subset N$ to be the neighborhood

$$N_1 = \left\{\sigma \in M : \left|\int_{R_d} \phi_i(x)\sigma(dx) - \int_{R_d} \phi_i(x) f(x)\, dx\right| < \tfrac{2}{3}\varepsilon, i = 1, 2, \cdots, r\right\}.$$

LEMMA 4.6. *We have*

$$P\{\beta_n(\cdot, \omega) \in N_1, \alpha_n(\cdot, \omega) \notin N \text{ for infinitely many } n\} = 0. \tag{4.25}$$

Proof: If, for some n and some ω, $\beta_n(\cdot,\omega)\in N_1$, but $\alpha_n(\cdot,\omega)\notin N$, then, for at least one $1\leqq i\leqq r$,

$$\left|\int_{R_d}\phi_i(x)\,\beta_n(dx,\omega)-\int_{R_d}\phi_i(x)\,\alpha_n(dx,\omega)\right|\geqq\tfrac{1}{3}\varepsilon,$$

which means

$$\left|\int_{\tau_{c_{n-1}}}^{\tau_{a_n}}\phi_i\left(\frac{\beta(s)}{\lambda_n}\right)ds-\int_0^{\tau_{a_n}}\phi_i\left(\frac{\beta(s)}{\lambda_n}\right)ds\right|\geqq\tfrac{1}{3}\varepsilon\lambda_n^2\log\log\lambda_n,$$

i.e.,

$$\left|\int_0^{\tau_{c_{n-1}}}\phi_i\left(\frac{\beta(s)}{n}\right)ds\right|\geqq\tfrac{1}{3}\varepsilon\lambda_n^2\log\log\lambda_n. \tag{4.26}$$

Let

$$\Lambda=\sup_{\substack{1\leqq i\leqq r\\ x\in R_d}}|\phi_i(x)|;$$

then, (4.26) implies that

$$\Lambda\tau_{c_{n-1}}\geqq\tfrac{1}{3}\varepsilon\lambda_n^2\log\log\lambda_n.$$

Hence, for each positive integer n,

$$P\{\beta_n(\cdot,\omega)\in N_1,\,\alpha_n(\cdot,\omega)\notin N\}\leqq P\left\{\tau_{c_{n-1}}\geqq\frac{\varepsilon}{3\Lambda}\lambda_n^2\log\log\lambda_n\right\}. \tag{4.27}$$

For Brownian motion starting at the origin it is well known that if τ_R is the first exit time from $B(0,R)$, then there exist constants $k_1>0$ and $k_2>0$ such that $P\{\tau_R>l\}\leqq k_2\exp\{-k_1 l/R^2\}$. Hence, for n large enough,

$$\begin{aligned}P\left\{\tau_{c_{n-1}}\geqq\frac{\varepsilon}{3\Lambda}\lambda_n^2\log\log\lambda_n\right\}&\leqq k_2\exp\left\{\frac{-\varepsilon k_1\lambda_n^2\log\log\lambda_n}{3\Lambda c_{n-1}^2}\right\}\\ &\leqq k_2\exp\left\{\frac{-\varepsilon k_1\lambda_n^2}{3\Lambda c_{n-1}^2}\right\}\\ &\leqq A\frac{c_{n-1}^2}{\lambda_n^2},\end{aligned} \tag{4.28}$$

where A is a constant. But from (4.13) we see that $\sum(c_{n-1}^2/\lambda_n^2)<\infty$, and therefore from the Borel–Cantelli lemma, (4.27), and (4.28) we obtain (4.25).

Let $N_2 \subset N_1 \subset N$ be the neighborhood of f specified by

$$N_2=\left\{\sigma\in M:\left|\int_{R_d}\phi_i(x)\,\sigma(dx)-\int_{R_d}\phi_i(x)\,f(x)\,dx\right|<\tfrac{1}{3}\varepsilon,\ i=1,2,\cdots,r\right\}.$$

LEMMA 4.7.

$$\sum P\{Z_{\lambda_n}(\cdot,\omega)\in N_2\}=\infty. \tag{4.29}$$

Proof: First of all, from (4.1),

$$\begin{aligned}P\{Z_{\lambda_n}(\cdot,\omega)\in N_2\}&=P\left\{\frac{\sigma(\lambda_n\cdot,\omega)}{\lambda_n^2\log\log\lambda_n}\in N_2\right\}\\&=P\left\{\frac{\sigma(\cdot,\omega)}{\log\log\lambda_n}\in N_2\right\}.\end{aligned}$$

From Theorem 3.5 we see that, for n large enough and for the $\delta>0$ picked in (4.12),

$$P\left\{\frac{\sigma(\cdot,\omega)}{\log\log\lambda_n}\in N_2\right\}\geqq\exp\{-(\log\log\lambda_n)(I(f)+\delta)\}. \tag{4.30}$$

Since $\lambda_n=\exp\{n^\alpha\}$, we conclude that

$$P\{Z_{\lambda_n}(\cdot,\omega)\in N_2\}\geqq\exp\{-(\alpha\log n)(I(f)+\delta)\}=\frac{1}{n^{\alpha[I(f)+\delta]}}. \tag{4.31}$$

Since α and δ were picked in (4.12) so that $\alpha[I(f)+\delta]<1$, we see that (4.29) holds.

LEMMA 4.8.

$$\sum P\{\alpha_n(\cdot,\omega)\in N_2\}=\infty. \tag{4.32}$$

Proof: Exactly as in Lemma 4.5 we see that

$$\sum P\{Z_{\lambda_n}(\cdot,\omega)\notin N_2,\ \alpha_n(\cdot,\omega)\in N_2\}<\infty,$$

and since

$$P\{\alpha_n(\cdot,\omega)\in N_2\}\geqq P\{Z_{\lambda_n}(\cdot,\omega)\in N_2\}-P\{Z_{\lambda_n}(\cdot,\omega)\notin N_2,\ \alpha_n(\cdot,\omega)\in N_2\}, \tag{4.33}$$

(4.32) follows from (4.33) and Lemma 4.7.

LEMMA 4.9.

$$\sum P\{\beta_n(\cdot, \omega) \in N_1\} = \infty . \tag{4.34}$$

Proof: The same argument that was used in Lemma 4.6 shows that

$$\sum P\{\beta_n(\cdot, \omega) \notin N_1, \alpha_n(\cdot, \omega) \in N_2\} < \infty .$$

Since,

$$P\{\beta_n(\cdot, \omega) \in N_1\} \geqq P\{\alpha_n(\cdot, \omega) \in N_2\} - P\{\beta_n(\cdot, \omega) \notin N_1, \alpha_n(\cdot, \omega) \in N_2\} , \tag{4.35}$$

we see that Lemma 4.8 and (4.35) imply (4.34).

LEMMA 4.10. *For $n = 1, 2, \cdots$, let $E_n = \{\omega : \beta_n(\cdot, \omega) \in N_1\}$. Then, for $n > m$ there exists a constant c, independent of m and n, such that*

$$P\{E_n \cap E_m\} \leqq cP\{E_m\}P\{E_n\} . \tag{4.36}$$

Proof: The event E_n occurs only between the times $\tau_{c_{n-1}}$ and τ_{a_n} whereas the event E_m occurs only between the times $\tau_{c_{m-1}}$ and τ_{a_m}. Since $n > m$, if we denote by $\gamma_{\tau_{a_m}}$ the σ-field corresponding to the stopping time τ_{a_m}, then we have, from the strong Markov property,

$$\begin{aligned} P\{E_n \cap E_m\} &= E\{P[E_n \cap E_m \mid \gamma_{\tau_{a_m}}]\} \\ &= E\{\chi_{E_m}(\omega) \cdot P_{\beta(\tau_{a_m})}[E_n]\} \\ &= E\{\chi_{E_m}(\omega) E_{\beta(\tau_{a_m})}\{P_{\beta(\tau_{c_{n-1}})}[E_n]\}. \end{aligned} \tag{4.37}$$

But,

$$E_{\beta(\tau_{a_m})}\{P_{\beta(\tau_{c_{n-1}})}[E_n]\} = \int_{S(0, c_{n-1})} \Pi^{c_{n-1}}(\beta(\tau_{a_m}), y) P_y[E_n]\, dy , \tag{4.38}$$

where $\Pi^{c_{n-1}}(x, y)$ is the Poisson kernel for the sphere of radius c_{n-1}, $S(0, c_{n-1})$ is the surface of that sphere, and dy is normalized so that the total surface measure is 1. If now we let

$$\Gamma_{m,n} = \sup_{\substack{|x| = a_m \\ |y| = c_{n-1}}} \Pi^{c_{n-1}}(x, y) ,$$

then (4.37) and (4.38) imply

$$P\{E_n \cap E_m\} \leqq \Gamma_{m,n} E\left\{\chi_{E_m}(\omega) \int_{S(0,c_{n-1})} P_y[E_n]\, dy\right\} \tag{4.39}$$
$$= \Gamma_{m,n} P\{E_m\} P\{E_n\}\,.$$

Since $a_{n-1}/c_{n-1} = \frac{1}{2}$, we see that $a_m/c_{n-1} < \frac{1}{2}$ for all $n > m$, and therefore from the Poisson kernel, $\Pi^{c_{n-1}}(x, y)$, we observe that there exists an absolute constant c such that $\Gamma_{m,n} \leqq c$ which yields the lemma.

LEMMA 4.11.

$$P\{\beta_n(\cdot, \omega) \in N_1 \textit{ for infinitely many } n\} = 1\,. \tag{4.40}$$

Proof: Lemmas 4.9 and 4.10 establish the nonobvious hypotheses of a refined form of the Borel–Cantelli lemma stated below and (4.40) is the conclusion of that lemma.

The following lemma is contained in [3] (p. 425) as well as in [1] (Theorem A).

LEMMA 4.12. *Let E_n, $n = 1, 2, \cdots$, be a sequence of events such that $P\{E_n\} \to 0$ as $n \to \infty$, while $\sum_{n=1}^{\infty} P\{E_n\}$ diverges and $P\{\limsup E_n\} = 0$ on 1. If there exists an absolute constant c such that, for $n \neq r$,*

$$P\{E_n \cap E_r\} \leqq cP\{E_n\}\, P\{E_r\}\,,$$

then infinitely many of the events E_n occur with probability 1.

Proof of Theorem 4.4: Combining Lemmas 4.11 and 4.6 we conclude that

$$P\{\alpha_n(\cdot, \omega) \in N \text{ for infinitely many } n\} = 1\,, \tag{4.41}$$

and (4.41) together with Lemma 4.5 imply

$$P\{Z_{\lambda_n}(\cdot, \omega) \in N \text{ for infinitely many } n\} = 1$$

which, as noted in (4.10), establishes the theorem.

Bibliography

[1] Ciesielski, Z., and Taylor, S. J., *First passage times and sojourn times for Brownian motion in space and the exact Hausdorff measure of the sample path*, Trans. Am. Math. Soc. 103, 1962, pp. 434–450.

[2] Donsker, M. D., and Varadhan, S. R. S., *Asymptotic evaluation of certain Wiener integrals for large time, in Proceedings of the International Conference on Functional Integration and its Applications*, London: Clarendon Press, Oxford, 1975, pp. 15–33.

[3] Erdös, P., *On the law of the iterated logarithm*, Ann. Math. 43, 1942, pp. 419–486.

Received November, 1979.

Proceedings of the International Congress of Mathematicians (Helsinki, 1978), Acad. Sci. Fennica, 1980, 755-762

Some Problems of Large Deviations

S. R. S. Varadhan

1. Introduction. The problem of large deviations arises naturally in probability theory in different contexts. We shall first look at some typical examples.

If $X_1, X_2, \ldots, X_n, \ldots$ is a sequence of independent and identically distributed random variables with mean zero, then for any $a>0$

$$\lim_{n\to\infty} P\left[\frac{X_1+\ldots+X_n}{n} \geqslant a\right] = 0 \tag{1.1}$$

by the law of large numbers. Under the assumption that for every real θ

$$M(\theta) = E[e^{\theta X_i}] < \infty$$

Cramér [3] showed that the probability in (1.1) has the specific behavior:

$$\lim_{n\to\infty} \left\{P\left[\frac{X_1+\ldots+X_n}{n} \geqslant a\right]\right\}^{1/n} = \varrho_a \tag{1.2}$$

exists and is given by

$$\varrho_a = \inf_{\theta\geqq 0} M(\theta) e^{-\theta a}.$$

For the second example we consider the process $x_\varepsilon(t)=\varepsilon\beta(t)$ defined in the time interval $0\leqslant t\leqslant 1$, where $\beta(\cdot)$ is the standard one-dimensional Brownian motion. We denote by P_ε the measure corresponding to $x_\varepsilon(t)$, which lives on the space $C_0[0, 1]$ of continuous functions vanishing at $t=0$. As $\varepsilon\to 0$, $x_\varepsilon(t)$ tends to the zero function uniformly and therefore

$$\lim_{\varepsilon\to 0} P_\varepsilon(A) = 0 \tag{1.3}$$

for sets A that are disjoint from some neighborhood of the zero function. One can again show that for a large class of sets $A\subset C_0[0, 1]$

(1.4) $$\lim_{\varepsilon\to 0} 2\varepsilon^2 \log P_\varepsilon(A) = -h(A)$$

exists where

(1.5) $$h(A) = \inf_{f\in A} \int_0^1 [f'(t)]^2\,dt$$

and the infimum is taken over absolutely continuous functions in f in A with a square integrable derivative. Such a result in a slightly different form can be found in **[14]**.

For the third example we will consider a Markov chain on a finite state space X and transition probabilities π_{ij}. For simplicity we shall assume that $\pi_{ij}>0$ for all i and j. Let $\vec{q}=\{q_i\}$ be the unique invariant probability vector. If $X_0, X_1, \ldots, X_n,\ldots$ is a realization of the Markov chain, we denote by $f_i^{(n)}$ the frequency of occurrence of the state i during the first n steps of the Markov chain and by $p_i^{(n)}$ the proportion of visits to the state i during the same time. The vector $\vec{p}^{(n)}$ with components $\{p_i^{(n)}\}$ is a random probability distribution on X and is the empirical distribution based on the first n steps. By the ergodic theorem, for large values of n, the vector $\vec{p}^{(n)}$ is close to the invariant probability vector $\vec{q}$ with a very high probability. The theorem on large deviations states in this context that for suitable sets A in space of probability distributions on X

(1.6) $$\lim_{n\to\infty} \frac{1}{n} \log P\{\vec{p}^{(n)}\in A\} = -h(A)$$

exists, where

(1.7) $$h(A) = \inf_{p\in A} I(\vec{p})$$

and the function $I(\vec{p})$ is defined for probability distributions $\vec{p}$ on X by

(1.8) $$I(\vec{p}) = -\inf_{u>0} \sum_{i\in X} p_i \log\left(\frac{\sum_{j\in X} \pi_{ij} u_j}{u_i}\right)$$

where the infimum is taken over vectors $u=\{u_i\}$ which have strictly positive components. The function $I(\cdot)$, while it is not explicit, is a nonnegative convex function of $\vec{p}$ which vanishes only when $\vec{p}$ equals the invariant probability $\vec{q}$.

The first example can be generalized by allowing the random variables to take values in more general linear spaces. The second can be generalized to cover processes $x_\varepsilon(t)$ governed by stochastic differential equations of the form

$$dx_\varepsilon(t) = \varepsilon\sigma\big(x_\varepsilon(t)\big)\,d\beta(t) + b\big(x_\varepsilon(t)\big)\,dt,$$
$$x_\varepsilon(0) = x.$$

Actually one can even combine the two classes in a single large class of problems of large deviations for Markov processes. See for instance **[15]**–**[20]**, **[1]**, **[2]**, and **[13]**.

As I am hoping that Professor Ventcel will cover this ground in some detail I shall proceed to a discussion of the ideas connected with the third example.

2. Large deviation for occupation times. Let the state space X of the Markov chain be a complete separable metric space instead of a finite set. Let the transition probabilities be given by $\pi(x, dy)$. We shall denote by π, the corresponding operator on the space $B(X)$ of bounded measurable functions on X, defined by

$$(\pi f)(x) = \int f(y)\pi(x, dy). \tag{2.1}$$

We denote by $\mathcal{M}$ the space of all probability distributions on X and we will view $\mathcal{M}$ as a complete separable metric space with weak convergence as the underlying convergence notion. We make the following assumptions on $\pi(x, dy)$.

(H1) *(Feller Property).* If f is bounded and continuous then so is πf.

(H2) *(Strong transitivity).* There is a reference measure $\alpha(dy)$ on X such that $\alpha(dy)$ and $\pi(x, dy)$ are mutually absolutely continuous for every $x \in X$.

(The above condition can be relaxed somewhat.)

For $\mu \in \mathcal{M}$ we define $I(\mu)$ by

$$I(\mu) = -\inf_{u \in U} \int_X \log\left(\frac{\pi u}{u}\right)(x)\mu(dx) \tag{2.2}$$

where U consists of functions in the space $C(X)$ of bounded continuous functions which have a positive lower bound. One can verify that $I(\cdot)$ is a convex, nonnegative, lower semicontinuous functional on $\mathcal{M}$ which vanishes only at invariant probability distributions. The chain of course may not possess an invariant probability distribution in which case $I(\cdot)$ is never zero.

We shall denote by ω, a realization $X_0, X_1, \ldots, X_n, \ldots$ of the chain and by $L_n(\omega, \cdot)$ the empirical distribution based on the first n steps

$$L_n(\omega, E) = \frac{1}{n} \sum_{j=1}^{n} \chi_E(X_j), \quad E \subset X. \tag{2.3}$$

Fixing an arbitrary starting point $x \in X$ for the chain we have the measure P_x corresponding to the process on the space Ω of all possible realizations. One can view $L_n(\cdot, \cdot)$ as a map of Ω into $\mathcal{M}$ and this induces a measure $Q_{n,x}$ on $\mathcal{M}$ which is the distribution of the empirical distribution. If A is a subset of $\mathcal{M}$ then we are interested in the asymptotic behavior of $Q_{n,x}(A)$ as $n \to \infty$. We have the following results:

If $C \subset \mathcal{M}$ is compact, then

$$\limsup_{n\to\infty} \frac{1}{n} \log Q_{n,x}[C] \leqslant -\inf_{\mu \in C} I(\mu). \tag{2.4}$$

If $G \subset \mathcal{M}$ is open, then

$$\limsup_{n\to\infty} \frac{1}{n} \log Q_{n,x}[G] \geqslant -\inf_{\mu \in G} I(\mu). \tag{2.5}$$

The reason that (2.4) can be established only for compact sets is that we have no assumption of positive recurrence. Perhaps the measures $Q_{n,x}$ are dissipative and (2.4) really measures the rate of dissipation. It is even possible that $\inf_\mu I(\mu)>0$ and in that case (2.4) cannot hold when C is taken to be the whole space $\mathcal{M}$. However when X is not compact, we can impose a strong positive recurrence condition that will enable us to prove (2.4) for closed sets.

There are continuous time analogs to these results for Markov processes $x(t)$ with transition probabilities $p(t, x, dy)$ on a state space X. We have the corresponding operators

$$(T_t f)(x) = \int f(y)\, p(t, x, dy) \tag{2.6}$$

we make the following assumptions regarding the semigroup $\{T_t\}$.

(H1*) T_t maps the space of bounded continuous functions into itself.

(H2*) There is a reference measure $\alpha(dy)$ such that $\alpha(dy)$ and $p(t, x, dy)$ are mutually absolutely continuous for every $t>0$ and $x\in X$.

(H3*) The strongly continuous center of the semigroup T_t is sufficiently large.

We denote by L the infinitesimal generator acting on the domain $\mathcal{D}\subset C(X)$. We denote by $\mathcal{D}^+$ those functions in $\mathcal{D}$ with a positive lower bound. We define the I-function as the analog of (2.2) by

$$I(\mu) = -\inf_{u\in\mathcal{D}^+} \int \left(\frac{Lu}{u}\right)(x)\, \mu(dx). \tag{2.7}$$

We look at the occupation distribution

$$L_t(\omega, E) = \frac{1}{t}\int_0^t \chi_E(x(s))\, ds \tag{2.8}$$

as the analog of (2.3) and the distribution $Q_{t,x}$ of $L_t(\cdot,\cdot)$ as the analog of $Q_{n,x}$. We then have the exact analogs of (2.4) and (2.5).

For C compact in $\mathcal{M}$,

$$\limsup_{t\to\infty} \frac{1}{t} \log Q_{t,x}(C) \leqslant -\inf_{\mu\in C} I(\mu). \tag{2.9}$$

For G open in $\mathcal{M}$,

$$\liminf_{t\to\infty} \frac{1}{t} \log Q_{t,x}(G) \geqslant -\inf_{\mu\in G} I(\mu). \tag{2.10}$$

Again if X is not compact one can impose a strong recurrence condition and obtain (2.9) for closed sets instead of just compact sets.

If the transition probabilities $p(t, x, dy)$ are such that they have a symmetric density with respect to the reference measure $\alpha(dy)$ then the semigroup T_t is a family of self adjoint contractions in the Hilbert space of functions that are square integrable with respect to the measure $\alpha(dy)$. Then the infinitesimal generator L can be thought of as a nonpositive self adjoint operator. Therefore the operator $\sqrt{-L}$ is well defined. In such a context one can show that $I(\mu)$ is finite if and only if μ is absolutely continuous with respect to α and the square integrable function $(d\mu/d\alpha)^{1/2}$ is in the domain of $\sqrt{-L}$. In such a case

$$I(\mu) = \left\| \sqrt{-L}\left(\frac{d\mu}{d\alpha}\right)^{1/2} \right\|^2 \tag{2.11}$$

where $\| \ \|$ is the mean square norm with respect to α. For the Brownian motion in R^d this becomes

$$I(f) = I(\mu) = \frac{1}{8}\int \frac{|\nabla f|^2}{f}\,dx = \frac{1}{2}\int |\nabla \sqrt{f}|^2\,dx \tag{2.12}$$

where $\mu(dx)=f(x)\,dx$.

The details of these results can be found in [4], [5], [21].

3. Connections with the principal eigenvalue. Let us consider the case when X is compact. If $x(t)$ is the Markov process corresponding to the generator L and $I(\cdot)$ is the corresponding I-functional defined by (2.7) then one can show [see [15]] that

$$\begin{aligned}\lambda(V) &= \lim_{t\to\infty} \frac{1}{t} \log E_x \left[\exp\left\{\int_0^t V(x(s))\,dx\right\}\right] \\ &= \sup_{\mu\in\mathcal{M}} \left[\int V(x)\,\mu(dx) - I(\mu)\right]\end{aligned} \tag{3.1}$$

for all $V\in C(X)$. The quantity $\lambda(V)$ can also be identified as the principal eigenvalue (i.e. the point in the spectrum with the largest real part) of the operator $L+V$. From the maximum principle for L one can conclude that $\lambda(V)$ is a convex functional of V and $I(\cdot)$ is its conjugate convex functional. This is the explanation of why $I(\cdot)$ is readily computable in the self-adjoint case. The variational formula (3.1) in that case reduces to the classical Rayleigh–Ritz variational formula for the principal eigenvalue of self adjoint operators. See in this context [6] and [12].

4. Applications. If $\beta(s)$ is the d-dimensional Brownian motion then the Wiener sausage $C_\varepsilon^t(\omega)$ for the trajectory $\omega=\beta(\cdot)$ up to time t is the set

$$C_\varepsilon^t(\omega) = \left\{y\colon \inf_{0\leqq s\leqq t} |y-\beta(s)| < \varepsilon\right\}.$$

A problem that comes from statistical mechanics [11] is the behavior of

$$E[\exp\{-\nu|C_\varepsilon^t(\omega)|\}]$$

where $|A|$ stands for the Lebesgue measure of the set A. Using the techniques developed in the preceding section (see [7], [8]) one can show that

$$\lim_{t\to\infty} \frac{1}{t^{d/(d+2)}} \log E[\exp\{-\nu|C_\varepsilon^t(\omega)|\}] = -k(\nu, d) \tag{4.1}$$

exists and compute $k(v, d)$ as

(4.2) $$k(v, d) = \inf_{\substack{f \in L_1(R^d) \\ f \geqq 0}} [v|x: f(x) > 0| + I(f)]$$

where $I(f)$ is given by (2.12).

One can consider a random walk on the lattice Z^d in R^d of points with integral coordinates. If the distribution of the single step has mean zero and covariance identity then under mild irreducibility assumptions the number $D(n)$ of distinct sites visited by the random walk in the first n steps can be shown to satisfy

(4.3) $$\lim_{n \to \infty} \frac{1}{n^{d/(d+2)}} \log E[\exp\{-vD(n)\}] = -k(v, d)$$

where $k(v, d)$ is again given by (4.2). There are analogs when the Brownian motion is replaced by a symmetric stable process (see [9] for details).

Another application is to derive laws of the iterated logarithm for local times. Let $\omega = \beta(\cdot)$ be one dimensional Brownian motion. Let

$$l(t, y) = \int_0^t \delta(\beta(s) - y)\, ds$$

be the local time of the Brownian path as a function of t and y. We define

$$L(t, y) = \frac{1}{\sqrt{t \log\log t}}\, l\left(t, y\sqrt{\frac{t}{\log\log t}}\right).$$

We can view $L(t, y)$ as a random probability density on the line. One can show that the set of limit points of $L(t, \cdot)$ as $t \to \infty$ coincides as functions of y (in the topology of uniform convergence on compact sets) almost surely with the set of subprobability densities $f(\cdot)$ satisfying

$$\int f(x)\, dx \leqslant 1 \quad \text{and} \quad \frac{1}{8} \int \frac{[f'(x)]^2}{f(x)}\, dx \leqslant 1.$$

These results and similar ones for certain stable processes can be found in [10].

5. General remarks. In the case of continuous time processes the I-functional plays the role of the Dirichlet integral to which it reduces in the self adjoint case. For instance $G \subset X$ is an open set with compact closure and one is interested in the exponential decay rate of $P_x[x(s) \in G, 0 \leqslant s \leqslant t]$ as $t \to \infty$. One can under mild conditions compute it as

$$\lambda(G) = \inf_{\substack{\mu \in \mathcal{M} \\ \mu(G)=1}} I(\mu).$$

There is a close connection, at least in spirit, to the theory developed by Lanford in [13] in the context of statistical mechanics.

If we start with a Markov process and reverse it in time using an invariant probability distribution then the new reversed Markov process has the same I-functional as the old one. It is an interesting question to examine how much information about the process can be recovered by a knowledge of its I-functional.

In case the process is transient one can consider the total occupation time i.e.

$$L(\omega, E) = \int_0^\infty \chi_E(x(s))\, ds$$

which will be a σ-finite measure on the state space. The tail behavior of its distribution on the space of all σ-finite measures on the state space should again be related to the I-functional. The details of the connection are being currently worked out.

References

1. R. Azencott and G. Ruget, *Mélanges d'équation différentielles et grand écart à la loi des grands nombres,* Z. Wahrscheinlichkeitstheorie und Verw. Gebiete **38** (1977), 1—54.

2. R. R. Bahadur and S. L. Zabell, *Large deviations of the sample mean in general vector spaces* (preprint).

3. H. Carmér, *On a new limit theorem in the theory of probability:* Colloquium on the Theory of Probability, Hermann, Paris, 1937.

4. M. D. Donsker and S. R. S. Varadhan, *Asymptotic evaluation of certain Markov process expectations for large time,* I, Comm. Pure Appl. Math. **28** (1975), 1—47.

5. *Asymptotic evaluation of certain Markov process expectations for large time,* III, Comm. Pure Appl. Math. **29** (1977), 389—461.

6. *On a variational formula for the principal eigenvalue for operators with maximum principle,* Proc. Nat. Acad. Sci. U.S.A. **72** (1975), 780—783.

7. *Asymptotic evaluation of certain Markov process expectations for large time,* II, Comm. Pure Appl. Math. **28** (1975), 279—301.

8. *Asymptotics for the Wiener sausage,* Comm. Pure Appl. Math. **28** (1975), 525—565.

9. *On the number of distinct sites visited by a random walk* (preprint).

10. *On laws of the iterated logarithms for local times,* Comm. Pure Appl. Math. **30** (1977), 707—753.

11. M. Kac, *Probabilistic methods in some problems of scattering theory,* Rocky Mountain J. Math. **4** (1974), 511—538.

12. T. G. Kurtz, *A variational formula for the growth rate of a positive operator semigroup* (preprint).

13. O. E. Lanford, *Entropy and equilibrium states in classical statistical mechanics,* Statistical Mechanics and Mathematical Problems, Lecture notes in Physics, vol. 20, Springer-Verlag, Berlin and New York, 1971, pp. 1—113.

14. M. Schilder, *Some asymptotic formulae for Wiener integrals,* Trans. Amer. Math. Soc. **125** (1966), 63—85.

15. S. R. S. Varadhan, *Asymptotic probabilities and differential equations,* Comm. Pure Appl. Math. **19** (1966), 261—286.

16. *On the behavior of the fundamental solution of the heat equation with variable coefficients,* Comm. Pure Appl. Math. **20** (1967), 431—455.

17. ——— *Diffusion processes in a small time interval,* Comm. Pure Appl. Math. **20** (1967), 659—685.

18. A. D. Ventcel, *Rough limit theorems on large deviations for Markov stochastic processes,* I, Theor. Probability Appl. **21** (1976), 227—242.

19. ——— *Rough limit theorems on large deviations for Markov stochastic processes,* II, Theor. Probability Appl. **21** (1976), 499—512.

20. A. D. Ventcel and M. I. Freidlin, *On small random perturbations of dynamical systems,* Russian Math. Surveys **25** (1970), 1—55.

21. J. Gärtner, *On large deviations from the Invariant Measure,* Theor. Probability Appl. **22** (1977), 24—39.

COURANT INSTITUTE OF MATHEMATICAL SCIENCES
NEW YORK UNIVERSITY
NEW YORK, NEW YORK 10012, U.S.A.

PHYSICS REPORTS (Review Section of Physics Letters) 77, No. 3 (1981) 235–237. North-Holland Publishing Company

The Polaron Problem and Large Deviations*

M.D. DONSKER and S.R.S. VARADHAN

Courant Institute of Mathematical Sciences, 251 Mercer Street, New York, N.Y. 10012, U.S.A.

Let $\{x_t(s), 0 \le s \le t\}$ be three dimensional Brownian motion tied down at the ends of the time interval, i.e., $x_t(0) = x_t(t) = 0$. Let $\alpha > 0$ and consider the following function space integral:

$$A(t, \alpha) = E\left\{\exp\left[\alpha \int_0^t \int_0^t \frac{e^{-|\sigma - s|}}{\|x_t(\sigma) - x_t(s)\|} \, ds \, d\sigma\right]\right\}. \tag{1}$$

The "Polaron Problem" which arises in statistical mechanics [1] is to evaluate $G(\alpha) = \lim_{t\to\infty}(1/t) \log A(t, \alpha)$. Now, $G(\alpha)$ is complicated to evaluate and even to estimate, but a conjecture by Pekar is that

$$\lim_{\alpha\to\infty} \frac{G(\alpha)}{\alpha^2} = c$$

where

$$c = \sup_{\substack{\varphi \in L^2(R^3) \\ \|\varphi\|_2 = 1}} \left[2 \int \int \frac{\varphi^2(x)\, \varphi^2(y)}{\|x - y\|} \, dx \, dy - \frac{1}{2} \int |\nabla \varphi|^2 \, dx\right].$$

Using methods developed by us in [2, 3], we succeeded in finding a "fairly explicit" expression for $G(\alpha)$ and then used this expression to prove rigorously the conjecture of Pekar. The details of that argument will be found in [4]. In this note we briefly describe some of the ideas lying behind our work on asymptotics.

In determining the asymptotic behavior of (1) as $t \to \infty$, it is clear that the major contribution is coming from those Brownian motion paths which tend to stay near themselves as time goes on. Since "typical" Brownian motion paths do not behave that way, determination of the asymptotics of (1) requires a large deviation result. In the Wiener sausage [5], and other applications [6–10], a common feature is that the asymptotic behavior results from the effect of atypical paths so that what is required is precise estimates of large deviations. Let us examine in a simplified context what we mean by a large deviation result.

Consider, for example, a complete separable metric space X and a sequence $\{Q_n\}$ of probability measures on it. As $n \to \infty$, Q_n behaves in one of two different ways:

*The research for this paper was done at the Courant Institute of Mathematical Sciences and supported by the National Science Foundation under NSF Grant No. MCS-77-02687.

Case 1. $Q_n \Rightarrow \delta_{x_0}$ for some $x_0 \in X$.
Case 2. Q_n is dissipative so that $Q_n(K) \to 0$ as $n \to \infty$ for every compact set $K \subset X$.
In most cases of interest in case 1, the space X is itself compact, whereas in case 2, X must necessarily be non-compact. Now, by a large deviation property for Q_n, we mean in case 1:

There exists a function $I(x)$: $X \to [0, \infty]$ such that
(i) $I(x)$ is lower semicontinuous.
(ii) $[x : I(x) \leq l]$ is compact in X for each $l < \infty$, and

$$\limsup_{n\to\infty} \frac{1}{n} \log Q_n(C) \leq - \inf_{x \in C} I(x) \qquad \text{for C closed},$$

$$\liminf_{n\to\infty} \frac{1}{n} \log Q_n(G) \geq - \inf_{x \in G} I(x) \qquad \text{for G open}.$$

In case 2 we demand only that there exists a function $I(x)$: $X \to [0, \infty]$ such that $I(x)$ is lower semicontinuous and

$$\limsup_{n\to\infty} \frac{1}{n} \log Q_n(K) \leq - \inf_{x \in K} I(x) \qquad \text{for K compact}$$

$$\liminf_{n\to\infty} \frac{1}{n} \log Q_n(G) \geq - \inf_{x \in G} I(x) \qquad \text{for G open}.$$

In case 1, if we have an $I(x)$ with the stated properties, then one can prove [11] that for any F: $X \to R$ that is bounded and continuous,

$$\lim_{n\to\infty} \frac{1}{n} \log \int \exp[n\,F(x)]\, dQ_n = \sup_x [F(x) - I(x)]. \tag{2}$$

In certain situations in case 2, (2) also holds but only for a more restricted class of functions F (see e.g. [12]).

Thus, in a particular context the problem is to determine the I-function. Let us illustrate in the simplest possible example. The reader is referred to our papers cited above for other contexts. Let $\xi_1, \xi_2, \ldots$ be a sequence of independent random variables with values in $[0, 1]$ having a common distribution λ on $[0, 1]$. Let $a = E(\xi) = \int_0^1 x\, \lambda(dx)$, and let $\eta_n = (\xi_1 + \xi_2 + \cdots + \xi_n)/n$, so that the strong law of large numbers states that $\eta_n \to a$ with probability one. If we let Q_n be the probability distribution of η_n, then $Q_n \Rightarrow \delta_a$ on $[0, 1]$. Now, a case 1 large deviation result was found in this classical case by Cramér [13] with $I_1(x) = \sup_\theta [\theta x - \log \int e^{\theta x} \lambda(dx)]$. In this situation (2) becomes

$$\lim_{n\to\infty} \frac{1}{n} \log E\{\exp[nF(\eta_n)]\} = \sup_x [F(x) - I_1(x)]. \tag{3}$$

We can look at this classical problem at a more complex level. Given $\xi_1, \xi_2, \ldots$, consider for each n the sample distribution, i.e., the random measure $\mu_{n,\omega} = (\delta_{\xi_1} + \delta_{\xi_2} + \cdots + \delta_{\xi_n})/n$ as a random variable in the space $\mathcal{M}$ of probability distributions on $[0, 1]$. Then, $\mu_{n,\omega}$ will have its probability distribution which we will denote by Q_n. Q_n is now a probability measure on $\mathcal{M}$. By the ergodic theorem (Glivenko–Cantelli theorem in this context), we know $\mu_{n,\omega} \to \lambda$ almost surely, and therefore $Q_n \Rightarrow \delta_\lambda$ on $\mathcal{M}$. We have again a case 1 situation, and now the I-function must be found as a function on $\mathcal{M}$, the space of probability measures on $[0, 1]$. In [2-III] the authors determined that $I_2(\mu) = \int \log(\mathrm{d}\mu/\mathrm{d}\lambda)\,\mathrm{d}\mu$ is the appropriate I-function. Hence (2) becomes

$$\lim_{n\to\infty} \frac{1}{n} \log E\{\exp[nF(\mu_{n,\omega})]\} = \sup_{\mu\in\mathcal{M}} [F(\mu) - I_2(\mu)] . \tag{4}$$

Similar large deviation results can be found for Markov chains instead of independent random variables as in the illustrations above. Whether we are in case 1 or in case 2 in the case of Markov chains depends on the recurrence properties of the chain. Instead of discrete time we can consider Markov processes with continuous time and then instead of n getting large it is t which becomes infinite. In all these cases the I-function can be found and asymptotic evaluation of the corresponding expectations obtained.

References

[1] R.P. Feynman and A.R. Hibbs, Quantum Mechanics and Path Integrals (1965).

[2] M.D. Donsker and S.R.S. Varadhan, Asymptotic evaluation of certain Markov process expectations for large time, I, Comm. Pure Appl. Math. 28 (1975) 1–47; II, Comm. Pure Appl. Math. 28 (1975) 279–301; III, Comm. Pure Appl. Math. 29 (1976) 389–461.

[3] M.D. Donsker and S.R.S. Varadhan, Asymptotic evaluation of certain Markov process expectations for large time, IV, to appear in Comm. Pure Appl. Math.

[4] M.D. Donsker and S.R.S. Varadhan, Asymptotics for the Polaron problem, to appear.

[5] M.D. Donsker and S.R.S. Varadhan, Asymptotics for the Wiener sausage, Comm. Pure Appl. Math. 28 (1975) 525–565.

[6] M.D. Donsker and S.R.S. Varadhan, On the number of distinct sites visited by a random walk, Comm. Pure Appl. Math. 32 (1979) 721–747.

[7] M.D. Donsker and S.R.S. Varadhan, On a variational formula for the principal eigenvalue for operators with maximum principle, Proc. Nat. Acad. Sci. U.S.A., Vol. 72, No. 3 (March 1975) pp. 780–783.

[8] M.D. Donsker and S.R.S. Varadhan, On the principal eigenvalue of second-order elliptic differential operators, Comm. Pure Appl. Math. 29 (1976) 595–621.

[9] M.D. Donsker and S.R.S. Varadhan, On laws of the iterated logarithm for local times, Comm. Pure Appl. Math. 30 (1977) 707–753.

[10] M.D. Donsker and S.R.S. Varadhan, A law of the iterated logarithm for total occupation times of transient Brownian motion, Comm. Pure Appl. Math. 33 (1980) 365–393.

[11] S.R.S. Varadhan, Asymptotic probabilities and differential equations, Comm. Pure Appl. Math. 19 (1966) 261, 286.

[12] M.D. Donsker and S.R.S. Varadhan, Asymptotic evaluation of certain Wiener integrals for large time, Functional Integration and its Applications, in: Proc. Intern. Conf. 1974, ed. A.M. Arthurs (Clarendon Press, Oxford, 1975).

[13] H. Cramer, On a new limit theorem in the theory of probability, Colloquium on the Theory of Probability, 1937 (Hermann and Co., Paris).

Comm. Pure Appl. Math. 36 (1983) 183-212

Asymptotic Evaluation of Certain Markov Process Expectations for Large Time. IV

M. D. DONSKER AND S. R. S. VARADHAN

1. Introduction

In this paper we extend our asymptotic results [1], [2] to a more general setting. Let Ω be a space of functions $\omega(\cdot)$ on $-\infty<t<\infty$ with values in a Polish space X. We assume Ω consists of functions with discontinuities only of the first kind, normalized to be right continuous, and with convergence induced by the Skorohod topology on bounded intervals. In this case, Ω is itself a Polish space. Denote by Ω_t^+ the corresponding space of functions on $[t,\infty)$ with values in X. We denote by $\mathcal{F}_t^s$ the σ-field in Ω generated by $\omega(\sigma)$ for $s\leqq\sigma\leqq t$. We denote the translation map on Ω by θ_t, i.e., $(\theta_t\omega)(s)=\omega(s+t)$.

Let $P_{0,x}$ be a Markov process on Ω_0^+ starting from $x\in X$ satisfying the hypothesis that the mapping $x\to P_{0,x}$ is weakly continuous (which implies the Feller property for the process $P_{0,x}$).

Let $\omega\in\Omega$ and, for each $t>0$, define ω_t by

$$\omega_t(s)=\omega(s), \qquad 0\leqq s<t,$$

$$\omega_t(s+t)=\omega_t(s) \quad \text{for all} \quad s\in(-\infty,\infty).$$

Now, $(\theta_s\omega_t)(\tau)=\omega_t(s+\tau)$, $0\leqq s\leqq t$, and for any set $A\subset\Omega$ we define

$$R_{t,\omega}(A)=\frac{1}{t}\int_0^t \chi_A(\theta_s\omega_t)\,ds. \tag{1.1}$$

If we let $\mathcal{M}_S(\Omega)$ denote the space of stationary processes on Ω, we see that, for each $\omega\in\Omega$ and each $t>0$, $R_{t,\omega}(\cdot)\in\mathcal{M}_S(\Omega)$. This is because, for any $\sigma>0$,

$$R_{t,\omega}(\theta_\sigma A)=\frac{1}{t}\int_0^t \chi_{\theta_\sigma A}(\theta_s\omega_t)\,ds=\frac{1}{t}\int_0^t \chi_A(\theta_{s-\sigma}\omega_t)\,ds$$

$$=\frac{1}{t}\int_{-\sigma}^{t-\sigma} \chi_A(\theta_s\omega_t)\,ds=\frac{1}{t}\int_0^t \chi_A(\theta_s\omega_t)\,ds=R_{t,\omega}(A).$$

Thus, for fixed $t>0$, $R_{t,\omega}$ is a mapping of $\Omega\to\mathcal{M}_S(W)$ and as such is $\mathcal{F}_t^0$ measurable. For each $t>0$ and each $x\in X$, we use this mapping to induce a probability measure $\Gamma_{t,x}$ on $\mathcal{M}_S(\Omega)$ by defining $\Gamma_{t,x}=P_{0,x}R_{t,\omega}^{-1}$, i.e., if $B\subset\mathcal{M}_S(\Omega)$, then

$$\Gamma_{t,x}(B)=P_{0,x}\{\omega\in\Omega\colon R_{t,\omega}\in B\}. \tag{1.2}$$

If $P_{0,x}$ is ergodic with invariant measure $\nu(dx)$ on X, and if $\bar{Q} \in \mathcal{M}_S(\Omega)$ is the stationary Markov process with ν as its marginal distribution, then, by the ergodic theorem, as $t \to \infty$,

$$\Gamma_{t,x} \Rightarrow \delta_{\bar{Q}} \quad \text{for all} \quad x. \tag{1.3}$$

Here, $\delta_{\bar{Q}}$ is the Dirac measure on $\mathcal{M}_S(\Omega)$ concentrated at $\bar{Q}$. What this paper is concerned with are the probabilities of large deviations for the measure $\Gamma_{t,x}$. If $G \subset \mathcal{M}_S(\Omega)$ is open and contains $\bar{Q}$, then, as just noted, $\lim_{t\to\infty} \Gamma_{t,x}(G) = 1$. On the other hand, if A is a set whose closure does not contain $\bar{Q}$, $\Gamma_{t,x}(A) \to 0$ as $t \to \infty$ and the question we are concerned with here is the rate at which it goes to 0.

In Sections 2 and 3 of this paper we define, for any $Q \in \mathcal{M}_S(\Omega)$, the entropy, $H(Q)$, of the stationary process Q with respect to the Markov process $P_{0,x}$. We then show that this entropy function governs the rate at which $\Gamma_{t,x}(G) \to 0$ as described above. In particular, we show, under suitable hypotheses to be detailed below, that for open sets $G \subset \mathcal{M}_S(\Omega)$,

$$\varliminf_{t\to\infty} \frac{1}{t} \log \Gamma_{t,x}(G) \geqq - \inf_{Q \in G} H(Q). \tag{1.4}$$

We also show under suitable hypotheses to be detailed below that, for closed sets $C \subset \mathcal{M}_S(\Omega)$,

$$\varlimsup_{t\to\infty} \frac{1}{t} \log \Gamma_{t,x}(C) \leqq - \inf_{Q \in C} H(Q). \tag{1.5}$$

Now, $H(Q) \geqq 0$ and it turns out that $H(\bar{Q}) = 0$ so that (1.4) and (1.5) are consistent with (1.3).

If $F(Q)$ is a continuous real-valued function on $\mathcal{M}_S(\Omega)$, then (1.4) and (1.5) imply (see [3])

$$\lim_{t\to\infty} \frac{1}{t} \log E^{\Gamma_{t,x}}\{e^{tF(Q)}\} = \sup_{Q \in \mathcal{M}_S(\Omega)} [F(Q) - H(Q)]. \tag{1.6}$$

An equivalent form of (1.6), which is responsible for the title of this paper, is

$$\lim_{t\to\infty} \frac{1}{t} \log E^{P_{0,x}}\{\exp\{tF(R_{t,\omega})\}\} = \sup_{Q \in \mathcal{M}_S(\Omega)} [F(Q) - H(Q)]. \tag{1.7}$$

Before giving examples of (1.7) we want to make two remarks which will relate (1.7) to our earlier results.

First of all, our earlier asymptotic results involved functions of the occupation distribution: for $\omega \in \Omega$, $t > 0$, and $A \subset X$,

$$L_{t,\omega}(A) = \frac{1}{t} \int_0^t \chi_A(\omega(s))\, ds,$$

i.e., $L_{t,\omega}(A)$ is the proportion of time up to t that the particular path $\omega(\cdot)$ occupies the set $A \subset X$. For fixed $t > 0$, and $\omega \in \Omega$, $L_{t,\omega}(\cdot)$ is then a probability measure on X. We denote the space of probability measures on X by $\mathcal{M}(X)$. Now, the relation between $L_{t,\omega} \in \mathcal{M}(X)$ and $R_{t,\omega} \in \mathcal{M}_S(\Omega)$ is made clear by observing that the marginal distribution of $R_{t,\omega}$ is precisely $L_{t,\omega}$.

Furthermore, in our earlier papers the asymptotic rate involving functionals of $L_{t,\omega}$ was governed by the I-function (for the Markov process $P_{0,x}$) which was defined on $\mathcal{M}(X)$ (see, e.g., [2]). To see the relation between that I-function and the entropy function $H(Q)$ of this paper, we prove in Section 6 that

$$\inf_{\substack{Q \in \mathcal{M}_S(\Omega) \\ q(Q)=\mu}} H(Q) = I(\mu), \tag{1.8}$$

where, in (1.8), the notation $q(Q)=\mu$ means the marginal of the stationary measure Q is μ. We refer to relation (1.8) as the contraction principle.

We now mention a few examples to illustrate (1.7) and the remarks just made.

EXAMPLE 1. Let $V(\cdot)$ be a continuous real-valued function on X and define

$$F(Q) = E^Q\{V(\omega(0))\}.$$

Then,

$$F(R_{t,\omega}) = \int_\Omega V(\omega'(0)) R_{t,\omega}(d\omega') = \frac{1}{t}\int_0^t V(\omega(s))\, ds,$$

and (1.7) becomes, in this case,

$$\lim_{t\to\infty} \frac{1}{t} \log E^{P_{0,x}}\left\{\exp\left\{\int_0^t V(\omega(s))\, ds\right\}\right\} = \sup_{Q \in \mathcal{M}_S(\Omega)} [E^Q\{V(\omega(0))\} - H(Q)]. \tag{1.9}$$

Using (1.8), this becomes

$$\lim_{t\to\infty} \frac{1}{t} \log E^{P_{0,x}}\left\{\exp\left\{\int_0^t V(\omega)(s))\, ds\right\}\right\} = \sup_{\mu \in \mathcal{M}(X)} \left[\int V(x)\mu(dx) - I(\mu)\right], \tag{1.10}$$

where, of course, $I(\mu)$ in (1.10) is the I-function associated with the Markov process $P_{0,x}$. Formula (1.10) agrees with earlier results of the authors.

EXAMPLE 2. Let $F(Q) = E^Q\{V(\omega(0), \ \omega(1))\}$, where $V: X \times X \to R$ is bounded and continuous. Now,

$$F(R_{t,\omega}) = \frac{1}{t}\int_0^{t-1} V(\omega(s), \omega(s+1))\, ds + O\left(\frac{1}{t}\right) \quad \text{as} \quad t \to \infty,$$

and from this and (1.7) it follows that

$$\begin{aligned} \log_{t\to\infty} \frac{1}{t} \log E^{P_{0,x}}\left\{\exp\left(\int_0^t V(\omega(s), \omega(s+1))\, ds\right)\right\} \\ = \sup_{Q \in \mathcal{M}_S(\Omega)} [E^Q\{V(\omega(0), \omega(1))\} - H(Q)]. \end{aligned} \tag{1.11}$$

EXAMPLE 3. Let P_t and $E^{P_t}\{\ \}$ denote, respectively, the probability measure and expectation with respect to three-dimensional Brownian motion $\omega(\cdot)$ tied down at both ends, i.e., $\omega(0) = \omega(t) = 0$. For $\alpha > 0$, let

$$G(\alpha, t) = E^{P_t}\left\{\exp\left\{\alpha \int_0^t \int_0^t \frac{e^{-|\sigma - s|}}{|\omega(\sigma) - \omega(s)|} d\sigma\, ds\right\}\right\}.$$

A long standing problem in statistical mechanics (cf. [4]), the "polaron problem" has been to show

$$\lim_{t\to\infty} \frac{1}{t} \log G(\alpha, t) = g(\alpha) \text{ exists,} \tag{1.12}$$

and, moreover, according to a conjecture of Pekar [5],

$$\lim_{\alpha\to\infty} \frac{g(\alpha)}{\alpha^2} = g_0 \text{ exists,} \tag{1.13}$$

with

$$g_0 = \sup_{\substack{\phi \in L_2(R_3) \\ \|\phi\| = 1}} \left[2 \int\int \frac{\phi^2(x)\phi^2(y)}{|x - y|} dx\, dy - \tfrac{1}{2} \int |\nabla\phi|^2 dx\right]. \tag{1.14}$$

Using the results of the present paper, the authors succeeded in proving (1.12) obtaining as an expression for $g(\alpha)$ a variational formula over all stationary processes as in (1.7). This infinite-dimensional variational formula turns out to be just explicit enough to allow us to also prove Pekar's conjecture (1.13) obtaining indeed the expression (1.14) for g_0. The details of this example will be found in a forthcoming paper [6].

Now, we wish to make precise the hypotheses imposed in order to prove (1.4) and (1.5). For the upper estimate (Theorem 4.6 below) we first comment that if the space X happens to be compact, then no further hypotheses on the Markov process $P_{0,x}$ need to be imposed in order to obtain (1.5), but if X is not compact we need to impose the following hypothesis on $P_{0,x}$.

There exists a sequence $\{u_n(x)\}$ of functions in $\mathcal{D}(L)$ (the domain of the infinitesimal generator L of the Markov process $P_{0,x}$) with the five properties:

(i) $u_n(x) \geqq c > 0$ for all x and n.

(ii) There exists, for every compact set $K \subset X$, a constant C_K such that $\sup_{x\in K} \sup_n u_n(x) \leqq C_K$.

(iii) $-(Lu_n/u_n)(x) \equiv V_n(x) \geqq -C$ for all n and all x.

(iv) There exists a function $V(x)$ such that, for all $x \in X$, $\lim_{n\to\infty} V_n(x) = V(x)$.

(v) $\{x: V(x) \leqq l\}$ is compact in X for every $l < \infty$.

In Section 7 of this paper we show that the Ornstein–Uhlenbeck process in d-dimensions, for example, satisfies these 5 properties. In the polaron problem we use the upper estimate (1.5) for precisely this case of the Ornstein–Uhlenbeck process.

For the lower estimate (1.4) (see Theorem 5.5 below), we impose two restrictions on $P_{0,x}$. Let $p(t, x, dy)$ be the transition probability associated with $P_{0,x}$. We shall assume that there exists a density function for $p(1, x, dy)$ with respect to a reference measure α on X, i.e., for all $x \in X$ there exists a function $p(1, x, y) > 0$ for almost all y (α measure) such that for all $x \in X$:

I: $p(1, x, dy) = p(1, x, y)\alpha(dy)$,

and

II: $p(1, x, \cdot)$ as a mapping from $X \to L_1(\alpha)$ is continuous.

In the polaron problem, for example, we need the lower estimate (1.4) for the case of ordinary three-dimensional Brownian motion and this satisfies I and II.

As for the asymptotic formula (1.6) or equivalently (1.7) we see that this holds for any Markov process $P_{0,x}$ satisfying both (i)–(v) and I–II above.

2. Some Properties of Entropy

Let (X, Σ) be a measurable space and let λ and μ be the probability measures on (X, Σ). Let $\mathscr{B}(\Sigma)$ be the space of bounded measurable functions on (X, Σ). We define the entropy of μ with respect to λ by

$$h(\lambda:\mu) = \inf\left\{c: \int \Phi(x)\mu(dx) \leqq c + \log \int e^{\Phi(x)}\lambda(dx) \quad \text{for all} \quad \Phi \in \mathscr{B}(\Sigma)\right\}. \tag{2.1}$$

If X is a Polish space and Σ is the Borel σ-field, then replacing $\mathscr{B}(\Sigma)$ by $C(X)$ in (2.1) gives the same infimum (see [2-III]).

From the definition (2.1) we see that, for fixed λ, $h(\lambda; \mu)$ is a non-negative, convex function of μ and $0 \leqq h(\lambda; \mu) \leqq \infty$. Moreover, if X is a Polish space, then $h(\lambda; \mu)$ is lower semicontinuous in μ in the weak topology.

The following basic theorem was proved in our earlier work (see [2-III]).

THEOREM 2.1. *For any (X, Σ) and any two probability measures λ, μ on (X, Σ), $h(\lambda; \mu)$ is finite if and only if*

(a) $\mu \ll \lambda$,

and

(b) *$d\mu/d\lambda = f(x)$ is such that $f(x) \log f(x) \in L_1(\lambda)$.*

Under these circumstances,

$$h(\lambda; \mu) = \int \log f(x)\mu(dx)$$

$$= \int f(x) \log f(x)\lambda(dx).$$

Remark. Suppose $\mu \ll \lambda$ and $d\mu/d\lambda = f(x)$. Since $f(x)\log f(x)$ is bounded for small $f(x)$, there is no difficulty when $\log f(x)$ is negative, i.e., $f(x)(\log(f(x))^-$ is always integrable with respect to λ and thus $f(x)\log f(x) \in L_1(\lambda)$ if and only if $f(x)(\log f(x))^+ \in L_1(\lambda)$.

We shall also use the following lemma (proved in [2-III]) which we state here for convenience.

LEMMA 2.2. *Let $h(\lambda:\mu) \leqq c < \infty$ and let $A \subset X$ be any set with $\lambda(A) = \varepsilon > 0$. Then, $\mu(A) \leqq \phi(c, \varepsilon)$, where $\phi(c, \varepsilon)$ is a universal function tending to 0 as $\varepsilon \to 0$ for every $c < \infty$.*

If $\mathscr{F} \subset \Sigma$ is a sub σ-field in the measurable space (X, Σ), and if we take $\Phi \in \mathscr{B}(\mathscr{F})$ instead of $\Phi \in \mathscr{B}(\Sigma)$ in definition (2.1), then we shall denote the corresponding entropy by $h_{\mathscr{F}}(\lambda\,;\mu)$. This is, of course, the same as restricting λ and μ to $(X, \mathscr{F})$.

LEMMA 2.3. *Let (X, Σ) be a Polish space and let $\mathscr{F}_1 \subset \mathscr{F}_2 \subset \Sigma$ be sub σ-fields. Denote by λ_ω, μ_ω the regular conditional probability distributions (r.c.p.d.) of λ and μ given $\mathscr{F}_1$. Then,*

$$h_{\mathscr{F}_2}(\lambda\,;\mu) = h_{\mathscr{F}_1}(\lambda\,;\mu) + E^{\mu}\{h_{\mathscr{F}_2}(\lambda_\omega\,;\mu_\omega)\}. \tag{2.2}$$

Proof: If $h_{\mathscr{F}_1}(\lambda\,;\mu) = \infty$, then since $h_{\mathscr{F}_2}(\lambda\,;\mu) \geqq h_{\mathscr{F}_1}(\lambda\,;\mu)$ we have $h_{\mathscr{F}_2}(\lambda\,;\mu) = \infty$ also and there is nothing to prove. Thus, we assume $h_{\mathscr{F}_1}(\lambda\,;\mu) < \infty$ and thus, from Theorem 2.1, $\mu \ll \lambda$ on $\mathscr{F}_1$. This means λ_ω is defined almost everywhere with respect to μ measure and therefore the expectation on the right of (2.2) is well defined. Let $\Phi \in \mathscr{B}(\mathscr{F}_2)$ so that from the definition of $h_{\mathscr{F}_2}(\lambda_\omega\,;\mu_\omega)$ we have

$$\begin{aligned} E^{\mu}\{\Phi\} &= E^{\mu}\{E^{\mu_\omega}\{\Phi\}\}, \\ &\leqq E^{\mu}\{h_{\mathscr{F}_2}(\lambda_\omega\,;\mu_\omega) + \log E^{\lambda_\omega}\{e^{\Phi}\}\}. \end{aligned} \tag{2.3}$$

If we let $\psi(\omega) = \log E^{\lambda_\omega}\{e^{\Phi}\}$, then $\psi(\omega) \in \mathscr{B}(\mathscr{F}_1)$ and so from the definition of $h_{\mathscr{F}_1}(\lambda\,;\mu)$ and (2.3) we get

$$\begin{aligned} E^{\mu}\{\Phi\} &\leqq E^{\mu}\{h_{\mathscr{F}_2}(\lambda_\omega\,;\mu_\omega)\} + E^{\mu}\{\psi(\omega)\} \\ &\leqq E^{\mu}\{h_{\mathscr{F}_2}(\lambda_\omega\,;\mu_\omega)\} + h_{\mathscr{F}_1}(\lambda\,;\mu) + \log E^{\lambda}\{e^{\psi(\omega)}\} \\ &= E^{\mu}\{h_{\mathscr{F}_2}(\lambda_\omega\,;\mu_\omega)\} + h_{\mathscr{F}_1}(\lambda\,;\mu) + \log E^{\lambda}\{e^{\Phi}\}. \end{aligned} \tag{2.4}$$

Using again the definition of $h_{\mathscr{F}_2}(\lambda\,;\mu)$, we conclude from (2.4) that

$$h_{\mathscr{F}_2}(\lambda\,;\mu) \leqq E^{\mu}\{h_{\mathscr{F}_2}(\lambda_\omega\,;\mu_\omega)\} + h_{\mathscr{F}_1}(\lambda\,;\mu). \tag{2.5}$$

Now we must show this inequality in the reverse direction. The inequality desired is trivial if $h_{\mathscr{F}_2}(\lambda\,;\mu) = \infty$; thus assume $h_{\mathscr{F}_2}(\lambda\,;\mu) < \infty$ which means by Theorem

2.1 that $\mu \ll \lambda$ on $\mathcal{F}_2$. Let $g(\omega) = d\mu/d\lambda|_{\mathcal{F}_2}$. As for the regular conditional probability distributions, we see that $\mu_\omega \ll \lambda_\omega$ on $\mathcal{F}_2$ for almost all ω (λ measure). Moreover,

$$g(\omega) = \frac{d\mu}{d\lambda}\bigg|_{\mathcal{F}_1}(\omega)\,\frac{d\mu_\omega}{d\lambda_\omega}\bigg|_{\mathcal{F}_2}(\omega),$$

and hence, from Theorem 2.1,

$$\begin{aligned} E^\mu\{\log g(\omega)\} &= E^\mu\left\{\log\frac{d\mu}{d\lambda}\bigg|_{\mathcal{F}_1}\right\} + E^\mu\left\{\log\frac{d\mu_\omega}{d\lambda_\omega}\bigg|_{\mathcal{F}_2}\right\}, \\ &= E^\mu\left\{\log\frac{d\mu}{d\lambda}\bigg|_{\mathcal{F}_1}\right\} + E^\mu\left\{E^{\mu_\omega}\left\{\log\frac{d\mu_\omega}{d\lambda_\omega}\bigg|_{\mathcal{F}_2}\right\}\right\} \\ &= E^\mu\left\{\log\frac{d\mu}{d\lambda}\bigg|_{\mathcal{F}_1}\right\} + E^\mu\{h_{\mathcal{F}_2}(\lambda_\omega;\mu_\omega)\} \\ &= h_{\mathcal{F}_1}(\lambda;\mu) + E^\mu\{h_{\mathcal{F}_2}(\lambda_\omega;\mu_\omega)\}. \end{aligned} \tag{2.6}$$

Again using Theorem 2.1 we get from (2.6)

$$h_{\mathcal{F}_2}(\lambda;\mu) = h_{\mathcal{F}_1}(\lambda,\mu) + E^\mu\{h_{\mathcal{F}_2}(\lambda_\omega;\mu_\omega)\}.$$

3. Entropy of Processes

Let Ω be a space of functions $\omega(\cdot)$ on $-\infty < t < \infty$ with values in a Polish space X. We shall assume that either Ω consists of continuous functions with uniform convergence on bounded intervals or it consists of functions with discontinuities only of the first kind and with convergence induced by the Skorohod topology on bounded intervals. In either case, Ω itself is a Polish space. For each $\omega \in \Omega$, we denote by $\omega(t)$ the value of the function $\omega(\cdot)$ at time t. We denote by Ω_t^+ the corresponding space of functions on $[t, \infty)$ with values in X and we denote by $\mathcal{F}_t^s$ the σ-fields in Ω generated by $\omega(\sigma)$ for $s \leqq \sigma \leqq t$. The space Ω_t^+ can, of course, be naturally identified with the space Ω and the σ-field $\mathcal{F}_\infty^t$. We shall use the notation θ_t for the translation map on Ω, i.e., $(\theta_t\omega)(s) = \omega(t+s)$.

Let Q be a stationary process on Ω and let $\{P_{t,x}\}$ be a homogeneous Markov family of measures on Ω_t^+. We define $P_{t,\omega(t)}$ to be $P_{t,x}$ with the starting point $x = \omega(t)$. We denote by $\{Q_{t,\omega}\}$ the regular conditional probability distributions of Q given $\mathcal{F}_t^{-\infty}$. We now define the entropy of the stationary process Q with respect to the Markov process $P_{0,x}$ at time $t > 0$ by

$$H(t, Q) = E^Q\{h_{\mathcal{F}_t^0}(P_{0,\omega(0)}; Q_{0,\omega})\}. \tag{3.1}$$

Here, $h_{\mathcal{F}_t^0}(P_{0,\omega(0)}; Q_{0,\omega})$ is the entropy introduced in the previous section. It is easy to check that $H(t, Q)$ is well defined for all $t > 0$ with $0 \leqq H(t, Q) \leqq \infty$.

THEOREM 3.1. *Either $H(t, Q) \equiv \infty$ for all $t>0$ or there exists a constant $H(Q)<\infty$ such that*

$$H(t, Q) = tH(Q) \quad \text{for all} \quad t>0. \tag{3.2}$$

Proof: Pick $0<t_1<t_2$ and consider $h_{\mathscr{F}^0_{t_2}}(P_{0,\omega(0)}; Q_{0,\omega})$. From Lemma 2.3,

$$\begin{aligned} h_{\mathscr{F}^0_{t_2}}(P_{0,\omega(0)}; Q_{0,\omega}) &= h_{\mathscr{F}^0_{t_1}}(P_{0,\omega(0)}; Q_{0,\omega}) \\ &\quad + E^{Q_{0,\omega}}\{h_{\mathscr{F}^0_{t_2}}((P_{0,\omega(0)})_{t,\bar{\omega}}; (Q_{0,\omega})_{t,\bar{\omega}})\}, \end{aligned} \tag{3.3}$$

where $(P_{0,\omega(0)})_{t_1,\bar{\omega}}$ means the r.c.p.d. of $P_{0,\omega(0)}$ with respect to the σ-field $\mathscr{F}^0_{t_1}$ and where $\bar{\omega}$ is the variable, i.e., $E^{Q_{0,\omega}}\{\ \}$ indicates integration with respect to $\bar{\omega}$ under $Q_{0,\omega}$ measure. Of course, $(Q_{0,\omega})_{t_1,\bar{\omega}}$ is the analogous r.c.p.d.

It will be convenient to express (3.3) in an alternative notation. Let $\bar{\omega}$ be a trajectory in Ω, $t \in (-\infty, \infty)$, and P be a measure on $\mathscr{F}^s_\infty$ with $s \leq t$. Suppose $P\{\omega: \omega(t) = \bar{\omega}(t)\} = 1$. Then, we define $\delta_{\bar{\omega}} \otimes_t P$ to be a new measure on Ω such that

$$\begin{aligned} &(\delta_{\bar{\omega}} \otimes_t P)\{\omega(t_1) \in A_1, \omega(t_2) \in A_2, \cdots, \omega(t_n) \in A_n\} \\ &\quad = \chi_{A_1}(\bar{\omega}(t_1)) \cdots \chi_{A_k}(\bar{\omega}(t_k)) P\{\omega(t_{k+1}) \in A_{k+1}, \cdots, \omega(t_n) \in A_n\}, \end{aligned}$$

where $t_1<t_2<\cdots<t_k \leq t \leq t_{k+1}<t_{k+2}<\cdots<t_n$. Now, $Q_{0,\omega}\{\bar{\omega}: \bar{\omega}(0) = \omega(0)\} = 1$, and $(P_{0,\omega(0)})_{t_1,\bar{\omega}} = \delta_{\bar{\omega}} \otimes_{t_1} P_{t_1,\bar{\omega}}(t_1)$ on $\mathscr{F}^0_\infty$. Moreover, $(Q_{0,\omega})_{t_1,\bar{\omega}} = Q_{t_1,\bar{\omega}} = \delta_{\bar{\omega}} \otimes_{t_1} Q_{t_1,\bar{\omega}}$ on $\mathscr{F}^0_\infty$. Thus, we can rewrite (3.3) as

$$\begin{aligned} h_{\mathscr{F}^0_{t_2}}(Q_{0,\omega(0)}; Q_{0,\omega}) &= h_{\mathscr{F}^0_{t_1}}(P_{0,\omega(0)}; Q_{0,\omega}) \\ &\quad + E^{Q_{0,\omega}}\{h_{\mathscr{F}^0_{t_2}}(\delta_{\bar{\omega}} \otimes_{t_1} P_{t_1,\bar{\omega}(t_1)}; \delta_{\bar{\omega}} \otimes_{t_1} Q_{t_1,\bar{\omega}})\}. \end{aligned} \tag{3.4}$$

We now take expectations on both sides of (3.4) with respect to Q measure (ω is the variable of integration), and noting that

$$E^Q\{E^{Q_{0,\omega}}\{h_{\mathscr{F}^{t_1}_{t_2}}(P_{t_1,\bar{\omega}(t_1)}; Q_{t_1,\bar{\omega}})\} = E^Q\{h_{\mathscr{F}^{t_1}_{t_2}}(P_{t_1,\omega(t_1)}; Q_{t_1,\omega})\},$$

we obtain

$$\begin{aligned} E^Q\{h_{\mathscr{F}^0_{t_2}}(P_{0,\omega(0)}; Q_{0,\omega})\} &= E^Q\{h_{\mathscr{F}^0_{t_1}}(P_{0,\omega(0)}; Q_{0,\omega})\} \\ &\quad + E^Q\{h_{\mathscr{F}^{t_1}_{t_2}}(P_{t_1,\omega(t_1)}; Q_{t_1,\omega})\}. \end{aligned} \tag{3.5}$$

By definition (3.1), the left side of (3.5) is $H(t_2, Q)$ and the first term on the right of (3.5) is $H(t_1, Q)$. By the stationarity of the Q process and the homogeneity of the Markov process we see that the second term on the right of (3.5) is $H(t_2-t_1, Q)$. Thus, $H(t, Q)$ is additive in t and from its definition it is clearly monotone in t. Hence, $H(t, Q)$ is linear in t unless it is identically infinite.

We shall call $H(Q)$ the entropy of the stationary process Q with respect to the Markov process $P_{0,x}$.

THEOREM 3.2. *Let $P_\omega = \delta_\omega \otimes_0 P_{0,\omega(0)}$. For all $t>0$,*

$$H(t, Q) = \sup_{\Phi \in \mathscr{B}(\mathscr{F}_t^{-\infty})} [E^Q\{\Phi\} - E^Q\{\log E^{P_\omega}\{e^\Phi\}\}]. \tag{3.6}$$

Proof: Let $\Phi \in \mathscr{B}(\mathscr{F}_t^{-\infty})$. Using (2.1), (3.1), and the fact that $Q_{0,\omega} = \delta_\omega \otimes_0 Q_{0,\omega}$, we obtain

$$\begin{aligned} E^Q\{\Phi\} &= E^Q\{E^{Q_{0,\omega}}\{\Phi\}\} \\ &\leqq E^Q\{h_{\mathscr{F}_t^{-\infty}}(P_\omega; Q_{0,\omega})\} + E^Q\{\log E^{P_\omega}\{e^\Phi\}\} \\ &= E^Q\{h_{\mathscr{F}_t^0}(P_{0,\omega(0)}; Q_{0,\omega})\} + E^Q\{\log E^{P_\omega}\{e^\Phi\}\} \\ &= H(t, Q) + E^Q\{\log E^{P_\omega}\{e^\Phi\}\}. \end{aligned}$$

Thus,

$$\sup_{\Phi \in \mathscr{B}(\mathscr{F}_t^{-\infty})} [E^Q\{\Phi\} - E^Q\{\log E^{P_\omega}\{e^\Phi\}\}] \leqq H(t, Q). \tag{3.7}$$

To show the inequality the other way, let

$$\sup_{\Phi \in \mathscr{B}(\mathscr{F}_t^{-\infty})} [E^Q\{\Phi\} - E^Q\{\log E^{P_\omega}\{e^\Phi\}\}] = l.$$

Then, by Jensen's inequality,

$$\sup_{\Phi \in \mathscr{B}(\mathscr{F}_t^{-\infty})} [E^Q\{\Phi\} - \log E^Q\{E^{P_\omega}\{e^\Phi\}\}] \leqq l. \tag{3.8}$$

Let $\bar{Q}$ measure be defined by $\int P_\omega Q(d\omega)$. Then, from (3.8) we have, for every $\Phi \in \mathscr{B}(\mathscr{F}_t^{-\infty})$,

$$E^Q\{\Phi\} - \log E^{\bar{Q}}\{e^\Phi\} \leqq l,$$

which implies from (2.1) that

$$h_{\mathscr{F}_t^{-\infty}}(\bar{Q}; Q) \leqq l.$$

Using Lemma 2.3, we then get

$$h_{\mathscr{F}_t^{-\infty}}(\bar{Q}; Q) = h_{\mathscr{F}_0^{-\infty}}(\bar{Q}; Q) + E^Q\{h_{\mathscr{F}_t^{-\infty}}(\bar{Q}_\omega; Q_{0,\omega})\} \leqq l. \tag{3.9}$$

But, by the definition of P_ω, $\bar{Q} = Q$ on $\mathscr{F}_0^{-\infty}$, and therefore $h_{\mathscr{F}_0^{-\infty}}(\bar{Q}; Q) = 0$. Moreover, $\bar{Q}_\omega = P_\omega = \delta_\omega \otimes_0 P_{0,\omega(0)}$, and $Q_{0,\omega} = \delta_\omega \otimes_0 Q_{0,\omega}$ so that

$$E^Q\{h_{\mathscr{F}_t^0}(P_{0,\omega(0)}; Q_{0,\omega})\} = E^Q\{h_{\mathscr{F}_t^{-\infty}}(\bar{Q}_\omega; Q_{0,\omega})\}.$$

Hence, from (3.9) we obtain

$$H(t, Q) = E^Q\{h_{\mathscr{F}_t^0}(P_{0,\omega(0)}; Q_{0,\omega})\} \leqq l$$

which completes the proof.

THEOREM 3.3. *Under the hypothesis that the mapping $x \to P_{t,x}$ is weakly continuous, $H(Q)$ is lower semicontinuous and convex in Q.*

Proof: From Theorems 3.1 and 3.2,

$$H(Q) = H(1, Q) = \sup_{\Phi \in \mathscr{B}(\mathscr{F}_1^{-\infty})} [E^Q\{\Phi\} - E^Q\{\log E^{P_\omega}\{e^\Phi\}\}]. \tag{3.10}$$

The supremum on the right of (3.10) is not changed if we restrict $\Phi \in \mathcal{B}(\mathcal{F}_1^{-\infty}) \cap C(\Omega)$. Thus, to show the lower semicontinuity of $H(Q)$, it suffices to show that, for each $\Phi \in \mathcal{B}(\mathcal{F}_1^{-\infty}) \cap C(\Omega)$, $E^Q\{\Phi\} - E^Q\{\log E^{P_\omega}\{e^\Phi\}\}$ is continuous as a functional of Q. Clearly, $E^Q\{\Phi\}$ is a continuous functional of Q. Thus let $\psi(\omega) = E^{P_\omega}\{e^\Phi\}$ and it suffices to show that $\psi(\omega)$ is continuous almost everywhere with respect to Q measure for an arbitrary stationary Q. To show this it suffices to show that if $\omega_n \to \omega$ in Ω, then $P_{\omega_n} \Rightarrow P_\omega$. Assume, then that $\omega_n \to \omega$ and that ω has no jump at the origin, i.e., $\omega(0^+) = \omega(0^-)$. Then, $\omega_n(0) \to \omega(0)$ and hence $P_{\omega_n} \Rightarrow P_\omega$. It remains to observe that, for all stationary processes Q, $Q\{\omega: \omega$ has a jump at $0\} = 0$.

The convexity of $H(Q)$ follows from (3.10) when we note that $H(Q)$ is the supremum of a collection of linear functionals of Q.

Next (Theorem 3.5 below) we prove the surprising fact that $H(Q)$ is linear in Q, but first we need a lemma.

LEMMA 3.4. *Let $\mathcal{M}_\Omega$ be the space of probability measures on Ω. There exists a conditional probability $R_\omega: \Omega \to \mathcal{M}_\Omega$ such that $R_\omega = \delta_\omega \otimes_0 R_\omega$ and, for all stationary measures Q on Ω, $Q_{0,\omega} = R_\omega$ almost everywhere with respect to Q measure.*

Proof: As before, let $\mathcal{M}_S(\Omega)$ be the space of stationary measures on Ω and let $\mathcal{M}_E(\Omega) \subset \mathcal{M}_S(\Omega)$ be the subset of $\mathcal{M}_S(\Omega)$ consisting of the ergodic measures. $\mathcal{M}_E(\Omega)$ is the set of extremals of $\mathcal{M}_S(\Omega)$. It follows from an argument of Oxtoby [7] that there exists a subset $\Omega_0 \subset \Omega$ which is $\mathcal{F}_0^{-\infty}$ measurable and a $\mathcal{F}_0^{-\infty}$ measurable map $\pi_\omega: \Omega_0 \to \mathcal{M}_E(\Omega)$ with the following properties: $Q(\Omega_0) = 1$ for all $Q \in \mathcal{M}_S(\Omega)$ and $Q\{\omega: \pi_\omega = Q\} = 1$ for all $Q \in \mathcal{M}_E(\Omega)$.

Now, denote by $R(Q, \omega)$ the r.c.p.d. of Q given $\mathcal{F}_0^{-\infty}$. It is easily seen that this r.c.p.d. can always be selected so that it is jointly measurable in Q and ω. We define the desired R_ω by $R_\omega = R(\pi_\omega, \omega)$. Then, for all $Q \in \mathcal{M}_E(\Omega)$,

$$\int R_\omega Q(d\omega) = \int R(Q, \omega) Q(d\omega) = Q.$$

But, $R_\omega = R(\pi_\omega, \omega)$ is independent of Q and so we have a linear relation true for extremals and hence true for all $Q \in \mathcal{M}_S(\Omega)$.

THEOREM 3.5. *$H(Q)$ is linear in Q.*

Proof: From Theorem 3.1 and the preceding lemma,

$$\begin{aligned} H(Q) = H(1, Q) &= E^Q\{h_{\mathcal{F}_1^0}(P_{0,\omega(0)}; Q_{0,\omega})\} \\ &= E^Q\{h_{\mathcal{F}_1^0}(P_{0,\omega(0)}; R_\omega)\}, \end{aligned} \tag{3.11}$$

where R_ω, as defined in the preceding lemma, is independent of Q. Thus, the integrand on the right of (3.11) is independent of Q and so $H(Q)$ is linear in Q.

In Theorem 3.2 we proved that, for $P_\omega = \delta_\omega \otimes_0 P_{0,\omega(0)}$ and for all $t > 0$,

$$H(t, Q) = \sup_{\Phi \in \mathscr{B}(\mathscr{F}_t^{-\infty})} [E^Q\{\Phi\} - E^Q\{\log E^{P_\omega}\{e^\Phi\}\}].$$

Now define

$$\begin{aligned} \bar{H}(t, Q) &= \sup_{\Phi \in \mathscr{B}(\mathscr{F}_t^0)} [E^Q\{\Phi\} - E^Q\{\log E^{P_{0,\omega(0)}}\{e^\Phi\}\}] \\ &= \sup_{\Phi \in \mathscr{B}(\mathscr{F}_t^0) \cap C(\Omega)} [E^Q\{\Phi\} - E^Q\{\log E^{P_{0,\omega(0)}}\{e^\Phi\}\}]. \end{aligned} \tag{3.12}$$

Clearly, $\bar{H}(t, Q) \leqq H(t, Q) = tH(Q)$.

THEOREM 3.6.

$$\lim_{T \to \infty} \frac{\bar{H}(t, Q)}{t} = H(Q). \tag{3.13}$$

Proof: Define $\bar{P}$ measure on $\mathscr{F}_\infty^0$ by $\bar{P} = \int P_{0,\omega(0)} Q(d\omega)$. By Jensen's inequality and (2.1),

$$\begin{aligned} \bar{H}(t, Q) &\geqq \sup_{\Phi \in \mathscr{B}(\mathscr{F}_t^0)} [E^Q\{\Phi\} - \log E^Q\{E^{P_{0,\omega(0)}}\{e^\Phi\}\}] \\ &= \sup_{\Phi \in \mathscr{B}(\mathscr{F}_t^0)} [E^Q\{\Phi\} - \log E^{\bar{P}}\{e^\Phi\}] \\ &= h_{\mathscr{F}_t^0}(\bar{P}; Q). \end{aligned}$$

Hence, to show (3.13), it suffices to show

$$\underline{\lim}_{t \to \infty} \frac{1}{t} h_{\mathscr{F}_t^0}(\bar{P}; Q) \geqq H(Q). \tag{3.14}$$

From Lemma 2.3,

$$h_{\mathscr{F}_{t+1}^0}(\bar{P}; Q) - h_{\mathscr{F}_t^0}(\bar{P}; Q) = E^Q\{h_{\mathscr{F}_{t+1}^0}(\bar{P}_{t,\omega}^0; Q_{t,\omega}^0)\}, \tag{3.15}$$

where $\bar{P}_{t,\omega}^0$ and $Q_{t,\omega}^0$ are the r.c.p.d. of $\bar{P}$ and Q, respectively, given $\mathscr{F}_t^0$. But,

$$\begin{aligned} E^Q\{h_{\mathscr{F}_{t+1}^0}(\bar{P}_{t,\omega}^0; Q_{t,\omega}^0)\} &= E^Q\{h_{\mathscr{F}_{t+1}^0}(\delta_\omega \otimes_t \bar{P}_{t,\omega}^0; \delta_\omega \otimes_t Q_{t,\omega}^0)\} \\ &= E^Q\{h_{\mathscr{F}_{t+1}^t}(\bar{P}_{t,\omega}^0; Q_{t,\omega}^0)\} \\ &= E^Q\{h_{\mathscr{F}_{t+1}^t}(P_{t,\omega(t)}; Q_{t,\omega}^0)\} \\ &= E^Q\{h_{\mathscr{F}_1^0}(P_{0,\omega(0)}; Q_{0,\omega}^{-t})\}, \end{aligned} \tag{3.16}$$

using the homogeneity of the Markov family and the stationarity of the Q process. By the martingale convergence theorem, $Q_{0,\omega}^{-t} \Rightarrow Q_{0,\omega}$ as $t \to \infty$ almost everywhere with respect to Q measure. As noted earlier, $h(\lambda; \mu)$ is lower-semicontinuous in μ, and hence

$$\underline{\lim}_{t \to \infty} h_{\mathscr{F}_1^0}(P_{0,\omega(0)}; Q_{0,\omega}^{-t}) \geqq h_{\mathscr{F}_1^0}(P_{0,\omega(0)}; Q_{0,\omega}). \tag{3.17}$$

Thus, by Fatou's lemma, (3.15), (3.16), and (3.17),

$$\varliminf_{t\to\infty} [h_{\mathcal{F}^0_{t+1}}(\bar{P}; Q) - h_{\mathcal{F}^0_1}(\bar{P}; Q)] \geqq E^Q\{h_{\mathcal{F}^0_1}(P_{0,\omega(0)}; Q_{0,\omega})\} \tag{3.18}$$
$$= H(1, Q) = H(Q).$$

Using the usual Caesaro argument, we obtain (3.14) and (3.18). This completes the proof of Theorem 3.6.

Let $C_S(\Omega)$ be the space of functions Φ on Ω which are bounded and measurable on Ω and such that, for all $Q \in \mathcal{M}_S(\Omega)$,

$$Q\{\text{discontinuities of } \Phi\} = 0.$$

For $\Phi \in C_S(\Omega)$, the linear functional of Q, $\int \Phi Q(d\omega)$, is then continuous in Q.

LEMMA 3.7. *Let* $Y_1 = \{\Phi: \Phi \in \mathcal{B}(\mathcal{F}^0_t), E^{P_{0,x}}\{e^{\Phi}\} \leqq 1$, *for all* $x\}$ *and let* $Y_2 = \{\Phi: \Phi \in \mathcal{B}(\mathcal{F}^0_t) \cap C_S(\Omega), E^{P_{0,x}}\{e^{\Phi}\} \leqq 1$, *for all* $x\}$. *Then, with* $\bar{H}(t, Q)$ *defined by* (3.12),

$$\bar{H}(t, Q) = \sup_{\Phi \in Y_1} E^Q\{\Phi\} = \sup_{\Phi \in Y_2} E^Q\{\Phi\}. \tag{3.19}$$

Proof: From (3.12),

$$\bar{H}(t, Q) = \sup_{\Phi \in \mathcal{B}(\mathcal{F}^0_t)} [E^Q\{\Phi\} - E^Q\{\log E^{P_{0,\omega(0)}}\{e^{\Phi}\}\}]$$
$$\geqq \sup_{\Phi \in Y_1} E^Q\{\Phi\} \geqq \sup_{\Phi \in Y_2} E^Q\{\Phi\}.$$

Conversely, suppose $\sup_{\Phi \in Y_2} E^Q\{\Phi\} = l$. Let $\psi \in \mathcal{B}(\mathcal{F}^0_t) \cap C(\Omega)$ and define $\bar{\psi} = \log E^{P_{0,x}}\{e^{\psi}\}$. Clearly, $\bar{\psi} \in C(R_1)$ and therefore $\bar{\psi}(\omega(0))$ is continuous as a function of ω on the set where ω is continuous at 0. This is a set of ω of Q measure 1 for every $Q \in \mathcal{M}_S(\Omega)$, i.e., $\bar{\psi}(\omega(0)) \in C_S(\Omega)$. Moreover,

$$E^{P_{0,x}}\{e^{\psi - \bar{\psi}(\omega(0))}\} = e^{-\bar{\psi}(x)} E^{P_{0,x}}\{e^{\psi}\}$$
$$= e^{-\bar{\psi}(x)} e^{\bar{\psi}(x)} = 1.$$

Thus, if we let $\Phi = \psi(\omega) - \bar{\psi}(\omega(0))$, then $\Phi \in \mathcal{B}(\mathcal{F}^0_t) \cap C_S(\Omega)$ and, as we just noted, $E^{P_{0,x}}\{e^{\Phi}\} = 1$. Therefore, Φ, as defined, is in Y_2 and so $E^Q\{\Phi\} \leqq l$. Hence,

$$E^Q\{\psi\} = E^Q\{\Phi\} + E^Q\{\bar{\psi}(\omega(0))\}$$
$$\leqq l + E^Q\{\bar{\psi}(\omega(0))\}. \tag{3.20}$$

But,

$$\bar{H}(t, Q) = \sup_{\psi \in \mathcal{B}(\mathcal{F}^0_t) \cap C(\Omega)} [E^Q\{\psi\} - E^Q\{\log E^{P_{0,\omega(0)}}\{e^{\psi}\}\}]$$
$$= \sup_{\psi \in \mathcal{B}(\mathcal{F}^0_t) \cap C(\Omega)} [E^Q\{\psi\} - E^Q\{\bar{\psi}(\omega(0))\}].$$

Consequently, (3.20) implies $\bar{H}(t, Q) \leqq l$ and the proof is complete.

4. Upper Bounds

In this section we obtain upper bounds on $\Gamma_{t,x}(A)$ as $t\to\infty$ for appropriate sets A. The main results are Lemma 4.4 and Theorem 4.6. We need some preparatory lemmas.

LEMMA 4.1. *Let $\phi:\Omega\to R_1$ be F_T^0 measurable and such that $E^{P_{0,x}}\{e^{\phi(\omega)}\}\leqq 1$ for all x. Then,*

$$E^{P_{0,x}}\left\{\exp\left\{\frac{1}{T}\int_0^t \phi(\theta_s\omega)\,ds\right\}\right\}\leqq 1 \quad \text{for all} \quad t. \tag{4.1}$$

Proof: Define

$$\psi_s(\omega)=\sum_{\substack{k:k\geqq 0\\ s+kT\leqq t}}\phi(\theta_{s+kT}\omega)$$

so that we can rewrite the left side of (4.1) as $E^{P_{0,x}}\{\exp\{(1/T)\int_0^T \psi_s(\omega)\,ds\}\}$. From Jensen's inequality,

$$E^{P_{0,x}}\left\{\exp\left\{\frac{1}{T}\int_0^T \psi_s(\omega)\,ds\right\}\right\}\leqq E^{P_{0,x}}\left\{\frac{1}{T}\int_0^T \exp\{\psi_s(\omega)\}\,ds\right\}. \tag{4.2}$$

Since $\phi(\theta_{s+kT}\omega)$ is $\mathscr{F}_{s+(k+1)T}^{s+kT}$ measurable, we have from the hypothesis on ϕ and successive conditioning that

$$E^{P_{0,x}}\{\exp\{\psi_s(\omega)\}\}=E^{P_{0,x}}\left\{\exp\left\{\sum_k \phi(\theta_{s+kT}\omega)\right\}\right\}\leqq 1. \tag{4.3}$$

Inequalities (4.3) and (4.2) give us (4.1).

COROLLARY. *Let $\Phi\in\mathscr{B}(\mathscr{F}_T^0)$ and be such that $E^{P_{0,x}}\{e^{\Phi(\omega)}\}\leqq 1$ for all x. Then, for all t,*

$$E^{\Gamma_{t,x}}\left\{\exp\left\{\frac{t}{T}\int_\Omega \Phi(\omega)Q(d\omega)\right\}\right\}\leqq \exp\left\{2\sup_{\omega\in\Omega}|\phi(\omega)|\right\}. \tag{4.4}$$

Proof: By definition of $\Gamma_{t,x}$ measure,

$$E^{\Gamma_{t,x}}\left\{\exp\left\{\frac{t}{T}\int_\Omega \Phi(\omega)Q(d\omega)\right\}\right\}=E^{P_{0,x}}\left\{\exp\left\{\frac{t}{T}\int_\Omega \Phi(\omega')R_{t,\omega}(d\omega')\right\}\right\},$$

and since

$$\left|\int_0^t \Phi(\theta_s,\omega)\,ds-t\int_\Omega \Phi(\omega')R_{t,\omega}(d\omega')\right|\leqq 2T\sup_{\omega\in\Omega}|\Phi(\omega)|, \tag{4.5}$$

we have from Lemma 4.1, for all t,

$$E^{\Gamma_{t,x}}\left\{\exp\left\{\frac{t}{T}\int_\Omega \Phi(\omega)Q(d\omega)\right\}\right\}\leqq \exp\left\{2\sup_{\omega\in\Omega}|\Phi(\omega)|\right\}$$

In what follows we shall use the notation

$$J(A)=\overline{\lim_{t\to\infty}}\frac{1}{t}\log\sup_{x\in X}\Gamma_{t,x}(A)$$

for $A\subset\mathcal{M}_S(\Omega)$.

LEMMA 4.2. *Let E_T be the set of $\Phi\in\mathcal{B}(\mathcal{F}_T^0)\cap C_S(\Omega)$ such that, for all x, $E^{P_{0,x}}\{e^{\Phi(\omega)}\}\leqq 1$. Then, for any set $A\subset\mathcal{M}_S(\Omega)$,*

$$J(A)\leqq-\sup_{\substack{l:A_1,A_2,\cdots,A_l\\ A\subset\bigcup_{i=1}^l A_i}}\ \inf_{1\leqq j\leqq l}\ \sup_{T>0}\ \sup_{\Phi\in E_T}\ \inf_{Q\in A_j}\frac{1}{T}\int_\Omega\Phi(\omega)Q(d\omega). \tag{4.6}$$

Proof: From (4.4) we see that, for any set $A\subset\mathcal{M}_S(\Omega)$ and any $\Phi\in E_T$,

$$\Gamma_{t,x}(A)\leqq\exp\left\{2\sup_{\omega\in\Omega}|\Phi(\omega)|\right\}\cdot\exp\left\{-\frac{t}{T}\inf_{Q\in A}\int_\Omega\Phi(\omega)Q(d\omega)\right\}$$

Thus,

$$J(A)\leqq-\inf_{Q\in A}\frac{1}{T}\int_\Omega\Phi(\omega)Q(d\omega). \tag{4.7}$$

Since (4.7) holds for all $\Phi\in E_T$ and all $T>0$,

$$J(A)\leqq-\sup_{T>0}\ \sup_{\Phi\in E_T}\ \inf_{Q\in A}\frac{1}{T}\int_\Omega\Phi(\omega)Q(d\omega).$$

From the definition of $J(A)$ we see that $J(A\cup B)\leqq\max(J(A),J(B))$. Thus, if $A_1,A_2,\cdots,A_l$ are sets in $\mathcal{M}_S(\Omega)$ with $A\subset\bigcup_{j=1}^l A_j$, then $J(A)\leqq\max_{1\leqq j\leqq l}\{J(A_j)\}$. Finally, taking the supremum over all such finite coverings of A we obtain (4.6).

LEMMA 4.3. *Let A be compact in $\mathcal{M}_S(\Omega)$. For any $\varepsilon>0$, there exists an l and open sets $G_1,G_2,\cdots,G_l$ in $\mathcal{M}_S(\Omega)$ such that $A\subset\bigcup_{j=1}^l G_j$ and*

$$\inf_{1\leqq j\leqq l}\ \sup_{T>0}\ \sup_{\Phi\in E_T}\ \inf_{Q\in G_j}\frac{1}{T}\int_\Omega\Phi(\omega)Q(d\omega)\geqq\inf_{Q\in A}H(Q)-\varepsilon. \tag{4.8}$$

In particular, for any compact set A in $\mathcal{M}_S(\Omega)$,

$$\sup_{\substack{l:A_1,A_2,\cdots,A_l\\ A\subset\bigcup_{j=1}^l A_j}}\ \inf_{1\leqq j\leqq l}\ \sup_{T>0}\ \sup_{\Phi\in E_T}\ \inf_{\phi\in A_j}\frac{1}{T}\int_\Omega\Phi(\omega)Q(d\omega)\geqq\inf_{Q\in A}H(Q). \tag{4.9}$$

Proof: Let $\inf_{Q\in A}H(Q)=\eta$. From Theorem 3.6 and Lemma 3.7 it follows that, given $Q\in A$ and $\varepsilon>0$, there exists T_Q and $\Phi_Q\in E_{T_Q}$ such that

$$\frac{1}{T_Q}\int_\Omega\Phi_Q(\omega)Q(d\omega)\geqq\eta-\tfrac{1}{2}\varepsilon. \tag{4.10}$$

Since $\Phi_Q \in C_S(\Omega)$, the integral on the left of (4.10) is a continuous functional of Q which implies that there exists a neighborhood G_Q of Q in $\mathcal{M}_S(\Omega)$ such that

$$\frac{1}{T_Q}\int_\Omega \Phi_Q(\omega)Q(d\omega) \geqq \eta - \varepsilon \tag{4.11}$$

for all Q in G_Q. The neighborhoods $\{G_Q\}$ form an open covering of the compact set A and therefore there exists $G_{Q_1}, G_{Q_2}, \cdots, G_{Q_l}$ such that $A \subset \bigcup_{j=1}^{l} G_{Q_j}$. This and (4.11) yield (4.8) and hence (4.9).

LEMMA 4.4. *Let A be closed in $\mathcal{M}_S(\Omega)$ and such that the family of one-dimensional marginals of Q as Q varies over A forms a tight family of measures on X. Then*

$$J(A) \leqq -\inf_{Q\in A} H(Q). \tag{4.12}$$

Proof: Let $\mathcal{M}(X)$ be the space of probability measures on X. Let $A_M \subset \mathcal{M}(X)$ be the family of one-dimensional marginals of Q as Q varies over A. Since, by hypothesis, A_M is tight, given $\varepsilon_n \to 0$ there exist compact sets $K_n \subset X$ such that $\mu(K_n) \geqq 1 - \varepsilon_n$ for all n and all $\mu \in A_M$. We assumed that $x \to P_{0,x}$ is weakly continuous, and therefore, given $\eta_n \to 0$ there exists a sequence of compact sets $C_n \subset D_X[0, 1]$ such that $P_{0,x}(C_n) \geqq 1 - \eta_n$ for all $x \in K_n$. Thus, if we let $\nu > 0$, denote the complement of C_n by $\tilde{C}_n$, and take $x \in K_n$,

$$\begin{aligned} E^{P_{0,x}}\{\exp\{\nu\chi_{K_n}(\omega(0))\chi_{\tilde{C}_n}(\omega)\}\} &= E^{P_{0,x}}\{\exp\{\nu\chi_{\tilde{C}_n}(\omega)\}\} \\ &= e^{\nu}P_{0,x}\{\tilde{C}_n\} + P_{0,x}\{C_n\} \\ &= 1 + (e^{\nu} - 1)P_{0,x}\{\tilde{C}_n\} \\ &\leqq 1 + (e^{\nu} - 1)\eta_n. \end{aligned}$$

On the other hand, if $x \notin K_n$, $E^{P_{0,x}}\{\exp\{\nu\chi_{K_n}(\omega(0))\chi_{\tilde{C}_n}(\omega)\}\} = 1$. Hence, for all x and all $\nu > 0$,

$$E^{P_{0,x}}\{\exp\{\nu\chi_{K_n}(\omega(0))\chi_{\tilde{C}_n}(\omega)\}\} \leqq 1 + \eta_n(e^{\nu} - 1). \tag{4.13}$$

Let $\phi(\omega) = \nu\chi_{K_n}(\omega(0))\chi_{\tilde{C}_n}(\omega) - \log[1 + \eta_n(e^{\nu} - 1)]$ so that from (4.13) we observe that $E^{P_{0,x}}\{e^{\phi(\omega)}\} \leqq 1$, and thus, using Lemma 4.1 (with $T = 1$), we obtain

$$E^{P_{0,x}}\left\{\exp\left\{\nu\int_0^t \chi_{K_n}(\omega(0))\chi_{\tilde{C}_n}(\theta_s\omega)\,ds\right\}\right\} \leqq \exp\{t\log[1 + \eta_n(e^{\nu} - 1)]\}. \tag{4.14}$$

From (4.5),

$$\begin{aligned} \int_0^t \chi_{K_n}(\omega(s))\chi_{\tilde{C}_n}(\theta_s\omega)\,ds &= \int_0^t \chi_{\tilde{C}_n}(\theta_s\omega)\,ds - \int_0^t \chi_{\tilde{K}_n}(\omega(s))\chi_{\tilde{C}_n}(\theta_s\omega)\,ds \\ &\geqq \int_0^t \chi_{\tilde{C}_n}(\theta_s\omega)\,ds - \int_0^t \chi_{\tilde{K}_n}(\omega(s))\,ds \\ &\geqq tR_{t,\omega}(\tilde{C}_n) - 1 - tR_{t,\omega}\{\omega' : \omega'(0) \in \tilde{K}_n\}, \end{aligned}$$

which implies, in view of (4.14), and the definition of $\Gamma_{t,x}$ measure,

$$E^{\Gamma_{t,x}}\left\{\exp\left\{\nu t\left[\dot{Q}(\tilde{C}_n)-\frac{1}{t}-Q\{\omega:\omega(0)\in\tilde{K}_n\}\right]\right\}\right\} \leqq \exp\{t\log[1+\eta_n(e^{\nu}-1)]\}. \tag{4.15}$$

Since $Q\in A$, $Q\{\omega:\omega(0)\in\tilde{K}_n\}\leqq\varepsilon_n$ and, therefore, from (4.15), for each n,

$$\Gamma_{t,x}\left\{A\cap\left\{Q: Q(\tilde{C}_n)\geqq\frac{1}{t}+2\varepsilon_n\right\}\right\} \leqq \exp\{t\log[1+\eta_n(e^{\nu}-1)]-\varepsilon_n\nu t\}. \tag{4.16}$$

Inequality (4.16) holds for any $\nu>0$, sequence $\varepsilon_n\to0$, and sequence $\eta_n\to0$. In particular, let us choose, for $\lambda>0$, $\nu=\lambda n^2$, $\varepsilon_n=1/n$, and $\eta_n=\exp\{-\lambda n^2\}$. Then,

$$t\log[1+\eta_n(e^{\nu}-1)]=t\log[1+\exp\{-\lambda n^2\}(\exp\{\lambda n^2\}-1)],$$

and $\varepsilon_n\nu t=(1/n)\cdot\lambda n^2 t=\lambda nt$. Thus, from (4.16), for each n,

$$\Gamma_{t,x}\left\{A\cap\left\{Q: Q(\tilde{C}_n)\geqq\frac{1}{t}+\frac{2}{n}\right\}\right\} \leqq e^{t\log 2-\lambda nt}.$$

This implies

$$\Gamma_{t,x}\left\{A\cap\bigcup_{n=1}^{\infty}\left\{Q: Q(\tilde{C}_n)\geqq\frac{1}{t}+\frac{2}{n}\right\}\right\}\leqq e^{t\log 2}\sum_{n=1}^{\infty}e^{-\lambda nt} = e^{t\log 2}\left(\frac{e^{-\lambda t}}{1-e^{-\lambda t}}\right).$$

If we let $A_t=\{Q: Q(\tilde{C}_n)\leqq 1/t+2/n \text{ for all } n\}$, then we have just shown

$$\Gamma_{t,x}\{A\cap\tilde{A}_t\}\leqq e^{t\log 2}\left(\frac{e^{-\lambda t}}{1-e^{-\lambda t}}\right). \tag{4.17}$$

We should note that the set A_t depends on λ since the sets C_n depend on the choice of η_n, and η_n was selected in terms of λ. From (4.17) we get

$$\overline{\lim_{t\to\infty}}\frac{1}{t}\log\sup_{x\in X}\Gamma_{t,x}(A\cap\tilde{A}_t)\leqq\log 2-\lambda. \tag{4.18}$$

Let $A_\infty=\bigcap_{t>0}A_t=\{Q\in\mathcal{M}_S(\Omega): Q(\tilde{C}_n)\leqq 2/n, \text{ for all } n\}$. The restriction of Q to $\mathcal{F}_1^0$ constitutes a tight family of measures on $D_X[0,1]$. But, for stationary processes, tightness when restricted to some interval implies overall tightness and hence A_∞ is a compact set in $\mathcal{M}_S(\Omega)$. Since A is closed, $A\cap A_\infty$ is compact in $\mathcal{M}_S(\Omega)$. Thus, from Lemmas 4.2 and 4.3, $J(A\cap A_\infty)\leqq-\inf_{Q\in A\cap A_\infty}H(Q)\leqq$

$-\inf_{Q\in A} H(Q)$ which implies that for any $\varepsilon > 0$ there exists an open set $G_\varepsilon \supset A \cap A_\infty$ such that

$$J(G_\varepsilon) \leqq -\inf_{Q\in A} H(Q) + \varepsilon. \tag{4.19}$$

In Lemma 4.5 to follow, we show that, given any open set $G \subset \mathcal{M}_S(\Omega)$ such that $G \supset A \cap A_\infty$, there exists a t_0 such that $t \geqq t_0$ implies $A \cap A_t \subset G$. Using this, (4.18), and (4.19), we have, for any $\varepsilon > 0$, $\lambda > 0$, and $t \geqq t_0$,

$$\begin{aligned} \Gamma_{t,x}(A) &\leqq \Gamma_{t,x}(A \cap A_t) + \Gamma_{t,x}(A \cap \tilde{A}_t) \\ &\leqq \Gamma_{t,x}(G_\varepsilon) + \Gamma_{t,x}(A \cap \tilde{A}_t) \end{aligned}$$

which means

$$\begin{aligned} J(A) &\leqq \max\,(J(G_\varepsilon), \log 2 - \lambda) \\ &\leqq \max\left(-\inf_{Q\in A} H(Q) + \varepsilon, \log 2 - \lambda\right). \end{aligned} \tag{4.20}$$

Since the left side of (4.20) is independent of ε and λ we now let $\varepsilon \to 0$ and $\lambda \to \infty$ obtaining the desired (4.12).

LEMMA 4.5. *Let $A \subset \mathcal{M}_S(\Omega)$ be closed, $A_t = \{Q: Q(\tilde{C}_n) < 2/n + 1/t \text{ for all } n\}$, and $A_\infty = \bigcap_{t>0} A_t$. Let G be a neighborhood in $\mathcal{M}_S(\Omega)$ such that $G \supset A \cap A_\infty$. Then, there exists a t_0 such that $t \geqq t_0$ implies $A \cap A_t \subset G$.*

Proof: Let $\{t_k\} \to \infty$ and let $Q_k \in A \cap A_{t_k}$. Assume $Q_k \in \tilde{G}$ for every k. We want to show that this leads to a contradiction. Since $Q_k \in A_{t_k}$, we have $Q_k(\tilde{C}_n) < 2/n + 1/t_k$ for all n. Thus $\overline{\lim}_{k\to\infty} Q_k(\tilde{C}_n) \leqq 2/n$ for each n. This last implies that the family of stationary measures $\{Q_k\}$ is tight on [0, 1]. As noted earlier, this means the family $\{Q_k\}$ is tight in $\mathcal{M}_S(\Omega)$. Thus, if $\bar{Q}$ is any limit point, we must have $\bar{Q} \in A \cap A_\infty$ since $A \cap A_\infty$ is closed, and also $\bar{Q} \in \tilde{G}$ since $\tilde{G}$ is closed. Hence, we have contradicted the assumption, $A \cap A_\infty \subset G$.

We should note that Lemma 4.4 gives us the desired upper estimate for closed sets A in $\mathcal{M}_S(\Omega)$, i.e., $J(A) \leqq -\inf_{Q\in A} H(Q)$ but only under the additional hypothesis that the family A_M of one-dimensional marginals of Q as Q varies over A forms a tight family of measures in X. Now, of course, if X is itself compact, then this last hypothesis on A_M is automatically satisfied. However, if X is not compact, then we can replace the hypothesis of tightness on A_M by imposing additional hypotheses on the Markov process $P_{0,x}$. We still obtain the upper estimate cited above for closed sets but we must give up a bit on the uniformity in x. The estimates obtained above involving $J(A)$ were (by definition of $J(A)$) uniform for all $x \in X$. Now, under the additional hypotheses to be imposed on $P_{0,x}$ (when X is not compact) we will get the same upper bounds but they will hold uniformly only for x in compact subsets of X.

THEOREM 4.6. *Assume the Markov process $P_{0,x}$ satisfies hypotheses* (i)–(v). *Let A be a closed set in $\mathcal{M}_S(\Omega)$. Then, for any compact set K in X,*

$$\overline{\lim_{t\to\infty}} \frac{1}{t} \log \sup_{x\in K} \Gamma_{t,x}(A) \leqq - \inf_{Q\in A} H(Q). \tag{4.21}$$

Proof: By hypothesis there exists a sequence $\{u_n(x)\}$ of functions in the domain of L, the infinitesimal generator of the Markov process $P_{0,x}$, with properties (i)–(v) (see the introduction). Consider, then, the equation for $u(x, t)$:

$$u_t = Lu + V_n(x)u,$$
$$u(x, 0) = u_n(x).$$

By the Feynman-Kac formula the solution is given by

$$u(x, t) = E^{P_{0,x}}\left\{\exp\left\{\int_0^t V_n(\omega(s))\,ds\right\}u_n(\omega(t))\right\},$$

but, from hypothesis (iii), $-(Lu_n/u_n)(x) = V_n(x)$ so that the solution $u(x, t) \equiv u_n(x)$, and thus

$$E^{P_{s,x}}\left\{\exp\left\{\int_s^t V_n(\omega(s))\,ds\right\}u_n(\omega(t))\right\} = u_n(x). \tag{4.22}$$

From hypothesis (ii), for any compact set K there exists a constant C_K such that $\sup_{x\in K} \sup_n u_n(x) \leqq C_K$. Also, from hypothesis (i), there exists a constant c such that $u_n(x) \geqq c > 0$ for all x and n. Thus, from (4.22) we conclude that, for any compact set K,

$$\sup_{x\in K} \sup_n E^{P_{0,x}}\left\{\exp\left\{\int_0^t V_n(\omega(s))\,ds\right\}\right\} \leqq \frac{\sup_{x\in K} \sup_n u_n(x)}{\inf_{x\in X} u_n(x)} \leqq \frac{C_K}{c}.$$

Using hypothesis (iv) we get from Fatou's lemma that, for all t,

$$\sup_{x\in K} E^{P_{0,x}}\left\{\exp\left\{\int_0^t V(\omega(s))\,ds\right\}\right\} \leqq \frac{C_K}{c} = M. \tag{4.23}$$

By the definition of $\Gamma_{t,x}$ measure we have

$$E^{\Gamma_{t,x}}\{e^{tF(Q)}\} = E^{P_{0,x}}\{\exp\{tF(R_{t,\omega})\}\},$$

hence, in particular, if

$$F(R_{t,\omega}) = \int_\Omega V(\omega'(0))R_{t,\omega}(d\omega') = \frac{1}{t}\int_0^t V(\omega(s))\,ds,$$

then $F(Q) = \int_X V(y)\mu(dy)$, where μ is the one-dimensional marginal of Q. Thus,

(4.23) becomes

$$\sup_{x\in K} E^{\Gamma_{t,x}}\left\{\exp\left\{t\int_X V(y)\mu(dy)\right\}\right\}\leqq M. \tag{4.24}$$

Let $K_l=\{x\in X\colon V(x)\leqq l\}$. Using hypothesis (iii) and the constant C which appears there, we get from (4.24)

$$\sup_{x\in K} E^{\Gamma_{t,x}}\{\exp\{t[l\mu(\tilde{K}_l)-C]\}\}\leqq M.$$

Thus, for all $x\in K$,

$$\Gamma_{t,x}\{Q\colon \mu(\tilde{K}_l)\geqq\delta_l\}\leqq M\exp\{t(C-l\delta_l)\}. \tag{4.25}$$

In particular, take $\lambda>0$ and choose $l_n=Cn+\lambda n^2$, $\delta_{l_n}=1/n$. From (4.25) we get, for all $x\in K$,

$$\Gamma_{t,x}\left\{Q\colon \mu(\tilde{K}_{l_n})\geqq\frac{1}{n}\right\}\leqq Me^{-\lambda nt},$$

which implies that, for all $x\in K$,

$$\Gamma_{t,x}\left\{\bigcup_{n=1}^{\infty}\left\{Q\colon \mu(\tilde{K}_{l_n})\geqq\frac{1}{n}\right\}\right\}\leqq M\frac{e^{-\lambda t}}{1-e^{-\lambda t}}. \tag{4.26}$$

If we let

$$A_\lambda=\bigcap_{n=1}^{\infty}\left\{Q\colon \mu(\tilde{K}_{l_n})<\frac{1}{n}\right\},$$

then (4.26) says that

$$\Gamma_{t,x}(\tilde{A}_\lambda)\leqq M\frac{e^{-\lambda t}}{1-e^{-\lambda t}}. \tag{4.27}$$

Now, let A be a closed set in $\mathcal{M}_S(\Omega)$. From the definition of A_λ we see that the marginals of Q as Q varies over A_λ form a tight family. Since $A\cap A_\lambda$ is closed, we have from Lemma 4.4 that, for each $\lambda>0$,

$$J(A\cap A_\lambda)\leqq-\inf_{Q\in A\cap A_\lambda}H(Q)\leqq-\inf_{Q\in A}H(Q). \tag{4.28}$$

Since $A\subset(A\cap A_\lambda)\cup\tilde{A}_\lambda$, we conclude from (4.27) and (4.28) that, for every $\lambda>0$,

$$\overline{\lim_{t\to\infty}}\frac{1}{t}\log\sup_{x\in K}\Gamma_{t,x}(A)\leqq\max\left(-\inf_{Q\in A}H(Q),-\lambda\right).$$

If we let $\lambda\to\infty$ in this last inequality, we get (4.21).

5. Lower Bounds

We need additional hypotheses on the Markov process $P_{0,x}$ in order to obtain the desired lower bounds. Let $p(t, x, dy)$ be the transition probability associated with $P_{0,x}$. We shall assume that there exists a density function for $p(1, x, dy)$ with respect to a reference measure α on X, i.e., there exists, for all $x \in X$, a function $p(1, x, y) > 0$ for almost all y (α measure) such that for all $x \in X$:

I: $p(1, x, dy) = p(1, x, y)\alpha(dy)$,

and

II: $p(1, x, \cdot)$ as a mapping from $X \to L_1(\alpha)$ is continuous.

We begin with three lemmas. The main result of this section is Theorem 5.5.

LEMMA 5.1. *Let $Q \in \mathcal{M}_S(\Omega)$ be such that $H(Q) < \infty$ and let μ be the marginal distribution of Q. Then, $\mu \ll \alpha$.*

Proof: Let $A \subset X$ be such that $\alpha(A) = 0$. We want to show that if $Q \in \mathcal{M}_S(\Omega)$ such that $H(Q) < \infty$, then $\mu(A) = 0$, where μ is the marginal of Q. From hypothesis I, for all $x \in X$,

$$P_{0,x}(\omega : \omega(1) \in A) = \int_A p(1, x, y)\alpha(dy) = 0. \tag{5.1}$$

Since $H(Q) < \infty$ and $H(1, Q) = H(Q)$, we know from (3.1) and Theorem 2.1 that $Q_{0,\omega}$ is absolutely continuous with respect to $P_{0,\omega(0)}$ for almost all ω (Q measure). Thus, from (5.1) we conclude that for almost all ω (Q measure) $Q_{0,\omega}\{\omega' : \omega'(1) \in A\} = 0$. Hence, $E^Q\{Q_{0,\omega}\{\omega' : \omega'(1) \in A\}\} = 0$, i.e., $\mu(A) = 0$.

LEMMA 5.2. *The reference measure α on X is absolutely continuous with respect to $p(1, x, \cdot)$ uniformly for x in compact subsets of X. In particular, if $H(Q) < \infty$, and μ is the marginal of Q, then $\mu \ll p(1, x, \cdot)$ uniformly for x in compact sets.*

Proof: The second statement of the lemma follows from the first and Lemma 5.1; thus we want to prove $\alpha \ll p(1, x, \cdot)$ uniformly for x in compact sets. Let $A \subset X$ such that $\alpha(A) = \delta > 0$. By hypothesis *I*, $p(1, x, A) \geqq \varepsilon(x, \delta)$. By hypothesis II, $p(1, x, \cdot)$ as a mapping from $X \to L_1(\alpha)$ is continuous and therefore $p(1, x', A) \geqq \varepsilon(x, \delta)/2$ for all x' in a neighborhood of x. The result now follows by the usual compactness argument.

LEMMA 5.3. *Let $Q \in \mathcal{M}_S(\Omega)$ be such that $H(Q) < \infty$ and let μ be the marginal of Q. Let $0 \leqq \phi_n(y) \leqq 1$ for $n = 1, 2, \cdots$, where $y \in X$. Assume*

$$\int_X \phi_n(y)\mu(dy) \geqq c_1 > 0 \quad \text{for} \quad n = 1, 2, \cdots.$$

Then, for any compact set $K \subset X$, there exists $c_2 > 0$ such that, for all $x \in K$, and all $n = 1, 2, 3, \cdots$,

$$\int_X \phi_n(y) p(1, x, dy) \geqq c_2.$$

Proof: Assume the contrary, i.e., for some sequence $\{x_n\} \in K$,

$$\lim_{n\to\infty} \int_X \phi_n(y) p(1, x_n, dy) = 0.$$

This implies $p(1, x_n, \{y \in X : \phi_n(y) \geqq \varepsilon\}) \to 0$ as $n \to \infty$ for every $\varepsilon > 0$. From Lemma 5.2 we conclude that $\mu\{y \in X: \phi_n(y) \geqq \varepsilon\} \to 0$ as $n \to \infty$ for every $\varepsilon > 0$. Since $\phi_n(y) \leqq 1$, the bounded convergence theorem then implies $\int_X \phi_n(y)\mu(dy) \to 0$ as $n \to \infty$ which is a contradiction.

THEOREM 5.4. *Let $Q \in \mathcal{M}_S(\Omega)$ be ergodic and such that $H(Q) < \infty$. Let K_1 be a compact set in X such that $\alpha(K_1) > 0$, and let K_2 be any compact set in X. For any neighborhood N of Q in $\mathcal{M}_S(\Omega)$, we have uniformly, for $x \in K_2$,*

$$\varliminf_{t\to\infty} \frac{1}{t} \log P_{0,x}\{\omega : R_{t,\omega} \in N, \omega(t) \in K_1\} \geqq -H(Q). \tag{5.2}$$

Proof: As before, let $Q_{0,\omega}$ be the r.c.p.d. of Q given $\mathcal{F}_0^{-\infty}$. Let

$$\psi(\omega, t) = \log \left| \frac{dQ_{0,\omega}}{dP_{0,\omega(0)}} \right|_{\mathcal{F}_t^0},$$

so that we have

$$\psi(\omega, t+s) = \psi(\omega, t) + \psi(\theta_t\omega, s)$$

and also, from (3.1) and Theorem 2.1,

$$E^Q\{\psi(\omega, t)\} = tH(Q).$$

Hence, by the ergodic theorem, for almost all ω (Q measure)

$$\lim_{t\to\infty} \frac{\psi(\omega, t)}{t} = H(Q). \tag{5.3}$$

Stated in another way, (5.3) says that

$$\lim_{t\to\infty} \frac{\psi(\omega', t)}{t} = H(Q) \tag{5.4}$$

almost everywhere ω'($Q_{0,\omega}$ measure) for almost all ω (Q measure). Again, from the ergodic theorem, for almost all ω (Q measure), and any neighborhood N of Q,

$$\lim_{t\to\infty} Q_{0,\omega}\{\omega' : R_{t,\omega'} \in N\} = 1. \tag{5.5}$$

With μ denoting the marginal of Q, let K_3 be a compact set in X such that $\mu(K_3) \geqq \frac{1}{2}$. Let $D_t = \{\omega: R_{t,\omega} \in N, \omega(t) \in K_3\}$ and consider $P_{0,\omega(0)}\{D_t\}$. Although $Q_{0,\omega}$ is absolutely continuous with respect to $P_{0,\omega(0)}$ the reverse is not necessarily true; however, for $\varepsilon > 0$,

$$P_{0,\omega(0)}\{D_t\} \geqq \int_{D_t} \frac{dP_{0,\omega(0)}}{dQ_{0,\omega}}\bigg|_{\mathscr{F}_t^0} dQ_{0,\omega}$$

$$= \int_{D_t} e^{-\psi(\omega',t)} Q_{0,\omega}(d\omega')$$

$$\geqq e^{-t(H(Q)+\varepsilon)} Q_{0,\omega}\left\{D_t\left\{\omega': \frac{1}{t}\psi(\omega', t) \leqq H(Q)+\varepsilon\right\}\right\}.$$

If we let $F_t = D_t \cap \{\omega': (1/t)\psi(\omega', t) \leqq H(Q)+\varepsilon\}$, and $\phi(t, x) = [P_{0,x}\{D_t\} \exp\{t(H(Q)+\varepsilon)\}] \wedge 1$, then we have just shown that

$$\phi(t, \omega(0)) \geqq Q_{0,\omega}\{F_t\}. \tag{5.6}$$

Taking expectations on both sides of (5.6) with respect to Q measure we get

$$\int_X \phi(t, x)\mu(dx) \geqq E^Q\{Q_{0,\omega}\{F_t\}\} = Q\{F_t\}. \tag{5.7}$$

Moreover, it follows from (5.4) and (5.5) that

$$\lim_{t\to\infty} Q\{F_t\} = \mu(K_3)$$

and so, from (5.7) and the choice of K_3, we obtain

$$\underline{\lim}_{t\to\infty} \int_X \phi(t, x)\mu(dx) \geqq \tfrac{1}{2}. \tag{5.8}$$

Given the neighborhood N of Q in $\mathscr{M}_S(\Omega)$, we can pick another neighborhood N_1 such that $\bar{N}_1 \subset N$ and we pick $T < \infty$ and $\varepsilon > 0$ such that, for any $Q' \in N_1$ and any Q'' satisfying $\|Q' - Q''\|_{\mathscr{F}_T^0} < \varepsilon$, it follows that $Q'' \in N$. Here the norm is the variation norm. Now $\|R_{t,\omega} - R_{t-2,\theta_1\omega}\|_{\mathscr{F}_t^0} \leqq (T+2)/t$ so that by taking t large enough this norm can be made less than ε. Therefore, for the neighborhood N_1 just picked and for t large enough,

$$\begin{aligned}
&P_{0,x}\{R_{t,\omega} \in N, \omega(t) \in K_1\} \\
&\quad \geqq P_{0,x}\{R_{t-2,\theta_1\omega} \in N_1, \omega(t) \in K_1\} \\
&\quad = \int_X \int_X p(1, x, dy) P_{0,y}\{R_{t-2,\omega} \in N_1, \omega(t-2) \in dz\} p(1, z, K_1) \\
&\quad \geqq \left[\int_X p(1, x, dy) P_{0,y}\{R_{t-2,\omega} \in N_1, \omega(t-2) \in K_3\right] \cdot W,
\end{aligned} \tag{5.9}$$

where $W=\inf_{z\in K_3} p(1, z, K_1)$. Using the notation introduced earlier in this proof we can rewrite (5.9) as

$$e^{t(H(Q)+\varepsilon)}P_{0,x}\{R_{t,\omega}\in N, \omega(t)\in K_1\}\geqq W\int_X \phi(t-2, y)p(1, x, dy). \tag{5.10}$$

From (5.8), (5.10), and Lemma 5.3 we have uniformly, for $x\in K_2$,

$$\varliminf_{t\to\infty} \frac{1}{t}\log P_{0,x}\{R_{t,\omega}\in N, \omega(t)\in K_1\}\geqq -H(Q). \tag{5.11}$$

In obtaining (5.11) we also used the fact that since $\alpha(K_1)>0$ and K_3 is compact we have $W>0$.

THEOREM 5.5. *Let $Q\in\mathcal{M}_S(\Omega)$ be such that $H(Q)<\infty$ and let N be any neighborhood of Q; then, uniformly for x in compact subsets of X,*

$$\varliminf_{t\to\infty} \frac{1}{t}\log \Gamma_{t,x}(N)\geqq -H(Q). \tag{5.12}$$

Thus, if G is open in $\mathcal{M}_S(\Omega)$, we have, uniformly for x in compact subsets of X,

$$\varliminf_{t\to\infty} \frac{1}{t}\log \Gamma_{t,x}(G)\geqq -\inf_{Q\in G} H(Q). \tag{5.13}$$

Proof: We can assume without loss of generality that $Q=\sum_{j=1}^{l}\pi_j Q_j$ where each Q_j is an ergodic member of $\mathcal{M}_S(\Omega)$ and the π_j are non-negative constants whose sum is 1. Since $H(Q)$ is linear by Theorem 3.5 we have $H(Q)=\sum_{j=1}^{l}\pi_j H(Q_j)$. Divide the interval $[0, t]$ into subintervals of length $\pi_j t$ and let t_j, $j=1, 2, \cdots, l$, be the right-hand endpoints of these subintervals. Let $K\subset X$ be compact. For each of the ergodic stationary processes Q_j, $j=1, 2, \cdots l$, let N_j be a neighborhood of Q_j such that if $Q_j'\in N_j$ for $j=1, 2, \cdots, l$ and $\|Q''-\sum_{j=1}^{l}\pi_j Q_j'\|_{\mathcal{F}_T^0}<\varepsilon$ for some choice of T and ε, then $Q''\in N$. Then,

$$\begin{aligned}\Gamma_{t,x}(N)&=P_{0,x}\{\omega: R_{t,\omega}\in N\}\\ &\geqq P_{0,x}\{\omega: R_{t_j-t_{j-1}}, \theta_{t_{j-1}}\omega\in N_j, j=1, 2, \cdots, l\}\\ &\geqq P_{0,x}\{\omega: R_{t_j-t_{j-1}}, \theta_{t_{j-1}}\omega\in N_j, \omega(t_j)\in K, j=1, 2, \cdots, l\}.\end{aligned}$$

Hence, if K_0 is the compact set in X for which we want to show uniformity in the estimate, we get from the preceding inequality

$$\begin{aligned}\inf_{x\in K_0}\Gamma_{t,x}(N)\geqq& \inf_{x\in K_0} P_{0,x}\{\omega: R_{t_1,\omega}\in N_1, \omega(t_1)\in K\}\\ &\cdot\prod_{j=1}^{l-1}\inf_{x\in K} P_{0,x}\{\omega: R_{t_j-t_{j-1}}, \omega\in N_j, \omega(t_j-t_{j-1})\in K\}.\end{aligned} \tag{5.14}$$

Now, if we apply Theorem 5.4 to (5.14) using the fact that the Q_i are ergodic and the previously noted linearity of $H(Q)$ we obtain (5.12).

6. The Contraction Principle

In this section we show the connection between the entropy function $H(Q)$ and the I-function introduced by the authors in their earlier papers. The relationship is clear from the following theorem. For the definition of the I-function see [2-III].

THEOREM 6.1. *For $Q \in \mathcal{M}_S(\Omega)$ let $q(Q)$ denote its one-dimensional marginal. Then,*

$$\inf_{Q: q(Q)=\mu} H(Q) = I(\mu). \tag{6.1}$$

Proof: Let $Q \in \mathcal{M}_S(\Omega)$ such that $H(Q)=l$ and $q(Q)=\mu$. Let $\Phi \in \mathcal{B}(\mathcal{F}_t^0)$ and be such that

$$E^{P_{0,x}}\{e^{\Phi}\} \leqq 1 \quad \text{for all} \quad x. \tag{6.2}$$

We earlier defined (in (3.12))

$$\bar{H}(t, Q) = \sup_{\Phi \in \mathcal{B}(\mathcal{F}_t^0)} [E^Q\{\Phi\} - E^Q\{\log E^{P_{0,\omega(0)}}\{e^{\Phi}\}\}],$$

and noted that $\bar{H}(t, Q) \leqq tH(Q)$. Thus, for the Φ selected,

$$E^Q\{\Phi\} \leqq lt. \tag{6.3}$$

In particular, let $\mathcal{U}$ be the space of functions $u \in \mathcal{B}(X)$ for each of which there exist constants c and C such that, for all x, $0 < c \leqq u \leqq C < \infty$ and, for $h > 0$ and $u \in \mathcal{U}$, take

$$\Phi(\omega) = \log \frac{u(\omega(h))}{(T_h u)(\omega(0))},$$

where T_h is the semigroup associated with the Markov process $P_{0,x}$. Now, this particular Φ is an element of $\mathcal{B}(\mathcal{F}_h^0)$ and direct calculation shows that Φ satisfies condition (6.2). Thus, from (6.3) we conclude

$$E^Q\left\{\log \frac{u(\omega(h))}{(T_h u)(\omega(0))}\right\} \leqq lh,$$

i.e.,

$$\begin{aligned} &E^Q\{\log u(\omega(h))\} - E^Q\{\log (T_h u)(\omega(0))\} \\ &= \int_X \log u(x)\mu(dx) - \int_X \log (T_h u)(x)\mu(dx) \\ &= \int_X \log \frac{u(x)}{(T_h u)(x)} \mu(dx) \leqq lh. \end{aligned} \tag{6.4}$$

Since (6.4) holds for every $u \in \mathcal{U}$ we get

$$\sup_{u \in \mathcal{U}} \int_X \log \frac{u(x)}{(T_h u)(x)} \mu(dx) \leqq lh. \tag{6.5}$$

In the notation of [2-III], (6.5) states that $(1/h)I_h(\mu) \leqq l$. But in [2-III] we proved $\lim_{h \to 0} (1/h)I_h(\mu) = I(\mu)$, and therefore from (6.5) we conclude that $I(\mu) \leqq l$. Since this was true for any Q such that $q(Q) = \mu$ we have

$$I(\mu) \leqq \inf_{Q:q(Q)=\mu} H(Q). \tag{6.6}$$

We want now to show the inequality in the other direction. Let μ be a probability measure on X and suppose $I(\mu) = l < \infty$. With $I_h(\mu)$ defined as above in (6.5), we know from [2-III] that, for all $h > 0$, $I_h(\mu) \leqq hI(\mu) = hl$. In [2-III] we proved that there exists a bivariate distribution $\rho(dx, dy)$ on $X \times X$ such that both marginals of ρ are equal to μ, and if $\rho_x(dy)$ denotes the r.c.p.d. of the second component given the first, then for all $h > 0$

$$\int_X h(p(h, x, \cdot); \rho_x(\cdot))\mu(dx) = I_h(\mu) \leqq hl. \tag{6.7}$$

In (6.7), $h(\alpha; \beta)$ is, of course, the entropy of β with respect to α, as introduced in Section 2, and $p(t, x, \cdot)$ is the transition function for the Markov process $P_{0,x}$.

Let n be a positive integer, let $h = 1/n$ and consider the grid $\{jh\}$, $j = 0, \pm 1, \pm 2, \cdots$. Define a Markov chain on this grid with stationary marginal μ and transition probability $\rho_x(dy)$. Given a sample of this Markov chain, let us interpolate between the grid points by random trajectories whose distribution is $P^h_{x,y}$, where $P^h_{x,\omega(h)}$ is the r.c.p.d. of $P_{0,x}$ given $\mathcal{F}^h_h$ and where, in the interpolation, x and y are the end points, i.e., values of the Markov chain at adjacent grid points. Let $Q^{(h)}$ be the measure induced by this interpolation procedure from the Markov chain on the grid. We note that $P^h_{x,y}$ is defined for each x for almost all y with respect to $p(h, x, \cdot)$ measure, but $\rho_x(\cdot)$ is absolutely continuous with respect to $p(h, x, \cdot)$ for almost all x (μ measure) and hence $P^h_{x,y}$ is defined almost everywhere with respect to $\rho(dx, dy)$ measure. Thus the interpolation procedure is well defined.

Let P_μ be the Markov process on $\mathcal{F}^0_\infty$ with initial distribution μ, i.e., $P_\mu = \int P_{0,x}\mu(dx)$. Consider $h_{\mathcal{F}^0_1}(P_\mu; Q^{(h)})$. Let $\mathcal{F}^{(h)}$ be the σ-field generated by $\omega(jh)$, $0 \leqq j \leqq 1/h = n$. Then, by Lemma 2.3,

$$h_{\mathcal{F}^0_1}(P_\mu; Q^{(h)}) = h_{\mathcal{F}^{(h)}}(P_\mu; Q^{(h)}) + E^{Q^{(h)}}\{h_{\mathcal{F}^0_1}(P_{\mu,\omega}; Q^{(h)}_\omega)\}, \tag{6.8}$$

where $P_{\mu,\omega}$ and $Q^{(h)}_\omega$ are, respectively, the r.c.p.d. of P_μ and $Q^{(h)}$ given $\mathcal{F}^{(h)}$. Notice that, by construction, $P_{\mu,\omega} = Q^{(h)}_\omega$ on $\mathcal{F}^0_1$ for almost all ω with respect to $Q^{(h)}$ measure, and therefore $E^{Q^{(h)}}\{h_{\mathcal{F}^0_1}(P_{\mu,\omega}; Q^{(h)}_\omega)\} = 0$. Hence, from (6.8) we get

$$h_{\mathcal{F}^0_1}(P_\mu; Q^{(h)}) = h_{\mathcal{F}^{(h)}}(P_\mu; Q^{(h)}). \tag{6.9}$$

In Lemma 6.2 to follow we show, using (6.7), that $h_{\mathcal{F}^{(h)}}(P_\mu; Q^{(h)}) \leqq nhl = l$. Hence, from (6.9), we conclude that $h_{\mathcal{F}_1^0}(P_\mu; Q^{(h)}) \leqq l$. This last inequality and Lemma 2.2 imply that the family of measures $\{Q^{(h)}\}$ is tight. Thus, let Q be any weak limit of $Q^{(h)}$ as $h \to 0$ ($n \to \infty$) and we see that Q is stationary with marginal μ and, moreover, $H(Q) \leqq l$. Hence, $\inf_{Q: q(Q)=\mu} H(Q) \leqq l = I(\mu)$ and the proof is complete except for the lemma.

LEMMA 6.2. *Let $\pi(x, dy)$ be any Markov chain transition probability with state space X and let μ be a probability measure on X. Let $\rho(dx, dy)$ be a probability measure on $X \times X$ with both marginals equal to μ, conditional measure $\rho_x(dy)$, and let*

$$\tau = \int_X h(\pi(x, \cdot); \rho_x(\cdot))\mu(dx).$$

Let P_μ be the π-chain with initial distribution μ and let R_μ be the stationary ρ-chain with initial distribution μ. Let $\mathcal{F}_n$ be the σ-field generated by $x_0, x_1, \cdots, x_n$, the first n steps of the chain. Then

$$h_{\mathcal{F}_n}(P_\mu; R_\mu) = n\tau. \tag{6.10}$$

Proof: If $\tau = \infty$ the result is obvious, so assume $\tau < \infty$. This means $\rho_x(dy) \ll \pi(x, dy)$ for almost all x (μ measure). Let

$$\theta(x, y) = \frac{d\rho_x(\cdot)}{d\pi(x, \cdot)},$$

and we see that

$$\left.\frac{dR_\mu}{dP_\mu}\right|_{\mathcal{F}_n} = \theta(x_0, x_1)\theta(x_1, x_2) \cdots \theta(x_{n-1}, x_n).$$

Now,

$$h_{\mathcal{F}_n}(P_\mu; R_\mu) = E^{R_\mu}\left\{\log \left.\frac{dR_\mu}{dP_\mu}\right|_{\mathcal{F}_n}\right\}$$

$$= \sum_{i=1}^{n} E^{R_\mu}\{\log \theta(x_{i-1}, x_i)\}.$$

But R_μ is stationary, and hence

$$\sum_{i=1}^{n} E^{R_\mu}\{\log \theta(x_{i-1}, x_i)\} = nE^{R_\mu}\{\log \theta(x_0, x_1)\}$$

$$= n\tau.$$

We finish this section with two theorems which, although not used directly in this paper, are used in applying this theory to the polaron problem in [6]. We include them here since the contraction principle is used in the proof.

THEOREM 6.3. *Let $\mathcal{A}$ be a family of stationary processes satisfying*
(i) *the family $\mathcal{A}_M$ of one-dimensional marginals is tight,*
(ii) *$H(Q) \leqq l < \infty$ for all $Q \in \mathcal{A}$.*
Then, $\mathcal{A}$ is tight in $\mathcal{M}_S(\Omega)$.

Proof: Let $Q \in \mathcal{A} \subset \mathcal{M}_S(\Omega)$ with marginal μ and let $\Phi \in \mathcal{B}(\mathcal{F}_T^0)$. Then,

$$E^Q\{\Phi\} \leqq TH(Q) + \int \log E^{P_{0,x}}\{e^{\Phi}\}\mu(dx). \tag{6.11}$$

Let $K \subset X$ be compact and let $\varepsilon > 0$ be given. Since we have assumed everywhere that $x \to P_{0,x}$ is weakly continuous, there exists a set $C \subset \Omega$ which is compact in $D_X[0, 1]$ such that $P_{0,x}(\tilde{C}) < \varepsilon$ for all $x \in K$. As a special case of (6.11) choose, for $\lambda > 0$,

$$\Phi(\omega) = \lambda \int_0^{T-1} \chi_K(\omega(s))\chi_{\tilde{C}}(\theta_s\omega)\, ds.$$

Now, this Φ is an element of $\mathcal{B}(\mathcal{F}_T^0)$, and from (6.11)

$$E^Q\{\Phi\} \leqq Tl + \int \log E^{P_{0,x}}\left\{\exp\left\{\lambda \int_0^{T-1} \chi_K(\omega(s))\chi_{\tilde{C}}(\theta_s\omega)\, ds\right\}\right\}\mu(dx) \tag{6.12}$$

since $H(Q) \leqq l$. On the other hand, for this Φ,

$$\begin{aligned} E^Q\{\Phi\} &= \lambda \int_0^{T-1} E^Q\{\chi_K(\omega(s))\chi_{\tilde{C}}(\theta_s\omega)\}\, ds \\ &\geqq \lambda(T-1)[Q(\tilde{C}) - \mu(\tilde{K})]. \end{aligned} \tag{6.13}$$

From (6.12) and (6.13) we get

$$\begin{aligned} Q(\tilde{C}) &\leqq \mu(\tilde{K}) + \frac{Tl}{\lambda(T-1)} \\ &+ \frac{1}{\lambda(T-1)} \int \log E^{P_{0,x}}\left\{\exp\left\{\lambda \int_0^{T-1} \chi_K(\omega(s))\chi_{\tilde{C}}(\theta_s\omega)\, ds\right\}\right\}\mu(dx). \end{aligned} \tag{6.14}$$

In particular, take $T = 2$ and using the same estimates as in (4.13) we get

$$Q(\tilde{C}) \leqq \mu(\tilde{K}) + \frac{2l}{\lambda} + \frac{1}{\lambda}\log[1 + \varepsilon(e^{\lambda} - 1)]. \tag{6.15}$$

Since the one-dimensional marginals $\mathcal{A}_M$ form a tight family, one can clearly choose K, ε, λ to make the right side of (6.15) as small as one pleases for all $Q \in \mathcal{A}$ simultaneously, showing $\mathcal{A}$ to be tight in $\mathcal{M}_S(\Omega)$.

THEOREM 6.4. *Assume the Markov process $P_{0,x}$ satisfies hypotheses* (i)–(v) *(in the introduction). Let $\mathcal{A}$ be a family in $\mathcal{M}_S(\Omega)$ such that $H(Q) \leqq l < \infty$ for all $Q \in \mathcal{A}$. Then, $\mathcal{A}$ is tight in $\mathcal{M}_S(\Omega)$.*

Proof: If $\mu \in \mathscr{A}_M$, the family of one-dimensional marginals in $\mathscr{A}$, then, by the contraction principle,

$$I(\mu) = \inf_{Q:q(Q)=\mu} H(Q).$$

But since $\mu \in \mathscr{A}_M$, there is a Q with $H(Q) \leqq l$ and marginal μ so that $I(\mu) \leqq l$. This is true for all $\mu \in \mathscr{A}_M$. We proved in [2-III] that if, under hypotheses (i)–(v), for all $\mu \in \mathscr{A}_M$, $I(\mu) \leqq l$, then $\mathscr{A}_M$ is tight. The result now follows from Theorem 6.3.

7. Special Properties of the Ornstein-Uhlenbeck Process

In the case when X is not compact, to get appropriate upper bounds required imposing hypotheses (i)–(v) on the Markov process $P_{0,x}$, as we saw in Section 4. We show now that the O–U process in d-dimensions satisfies (i)–(v). Also in this section we prove (Theorem 7.1) an entropy identity for the O–U process which we use in the polaron problem in [6].

For the O–U process in d-dimensions, the infinitesimal generator is

$$L = \tfrac{1}{2}\Delta - kx \cdot \nabla, \tag{7.1}$$

where $k > 0$ is an arbitrary positive constant.

To show (i)–(v) are satisfied, define $\theta(x) = x$ for $|x| \leqq 1$, bounded on R_1, θ' and θ'' uniformly bounded on R_1, $\theta' \geqq 0$, and θ chosen to be odd. Define

$$u_n(x) = \prod_{i=1}^{d} \cosh(n\theta(x_i/n)).$$

Clearly, $u_n(x) \geqq 1$ for all x and all n so that hypothesis (i) is satisfied. As $n \to \infty$, $u_n(x) \to \prod_{i=1}^{d} \cosh x_i$ uniformly on compact sets which means hypothesis (ii) is satisfied.

Using (7.1), we find, in $d = 1$,

$$\begin{aligned} -\frac{Lu_n(x)}{u_n} &= V_n(x) \\ &= -\left\{\frac{1}{2}\left[\theta'\left(\frac{x}{n}\right)\right]^2 + \frac{1}{2n}\theta''\left(\frac{x}{n}\right)\tanh\left(n\theta\left(\frac{x}{n}\right)\right)\right. \\ &\qquad \left. - kx\theta'\left(\frac{x}{n}\right)\tanh\left(n\theta\left(\frac{x}{n}\right)\right)\right\} \end{aligned}$$

and hence, showing that there is a constant c such that $V_n(x) \geqq -c$ for all n and all x is equivalent to showing that the expression inside the curly braces is bounded. But this follows from the fact that θ' and θ'' are bounded and $x \tanh(n\theta(x/n)) \geqq 0$ for all x. A similar argument holds in d-dimensions. So hypothesis (iii) holds.

If we let $f(x) = \cosh(n\theta(x/n))$ and $\psi_n(x) = (\frac{1}{2}f'' - kxf')/f$, then in d-dimensions, $V_n(x) = -\sum_{i=1}^d \psi_n(x_i)$. As $n \to \infty$,

$$V_n(x) \to -\sum_{i=1}^{d} (\tfrac{1}{2} - kx_i \tanh x_i).$$

Thus, if we define

$$V(x) = \sum_{i=1}^{d} (kx_i \tanh x_i - \tfrac{1}{2}),$$

we see that $V_n(x) \to V(x)$ as $n \to \infty$ for all x. Hence, hypothesis (iv) is satisfied. Hypothesis (v) is satisfied because there is a constant c_1 such that, for all x,

$$V(x) \geqq \sum_{i=1}^{d} k|x_i| - c_1.$$

THEOREM 7.1. *Let $H_k(Q)$ be the entropy of a d-dimensional stationary process Q with respect to the d-dimensional O–U process with constant $k > 0$. Let $H_0(Q)$ be the entropy of Q with respect to d-dimensional Brownian motion. Then,*

$$H_k(Q) = H_0(Q) + \tfrac{1}{2}k^2 \int \|x\|^2 \mu(dx) - \tfrac{1}{2}dk, \tag{7.2}$$

where μ is the marginal of Q.

Proof: By definition,

$$H_k(Q) = E^Q\{h_{\mathcal{F}_1^0}(P^k_{0,\omega(0)}; Q_{0,\omega})\},$$

where $P^k_{0,x}$ is the O–U process with parameter k. Brownian motion is $P^0_{0,x}$ and we now use the fact that $P^k_{0,x}$ and $P^0_{0,x}$ are mutually absolutely continuous. Indeed, for any pair $\alpha \approx \beta$,

$$\begin{aligned} h_\Sigma(\alpha; \lambda) &= E^\lambda\left\{\log \frac{d\lambda}{d\alpha}\Big|_\Sigma\right\} \\ &= E^\lambda\left\{\log \frac{d\lambda}{d\beta}\Big|_\Sigma + \log \frac{d\beta}{d\alpha}\Big|_\Sigma\right\} \\ &= h_\Sigma(\beta, \lambda) + E^\lambda\left\{\log \frac{d\beta}{d\alpha}\Big|_\Sigma\right\}. \end{aligned}$$

Apply this with $\lambda = Q_{0,\omega}$, $\alpha = P^k_{0,\omega(0)}$, $\beta = P^0_{0,\omega(0)}$ and we get ($\Sigma = \mathcal{F}_1^0$)

$$H_k(Q) = H_0(Q) + E^Q\left\{E^{Q_{0,\omega}}\left\{\log \frac{dP^0_{0,\omega(0)}}{dP^k_{0,\omega(0)}}\Big|_{\mathcal{F}_1^0}\right\}\right\}. \tag{7.3}$$

From the Cameron-Martin formula,

$$\log \frac{dP^0_{0,\omega(0)}}{dP^k_{0,\omega(0)}}\Big|_{\mathcal{F}_1^0} = \tfrac{1}{2}k\|\omega(1)\|^2 + \tfrac{1}{2}k^2 \int_0^1 \|\omega(s)\|^2\, ds - \tfrac{1}{2}kd - \tfrac{1}{2}k\|\omega(0)\|^2$$

and since Q is stationary we have

$$E^Q\left\{E^{Q_{0,\omega}}\left\{\log \frac{dP^0_{0,\omega(0)}}{dP^k_{0,\omega(0)}}\bigg|_{\mathcal{F}^0_1}\right\}\right\} = E^Q\left\{\log \frac{dP^0_{0,\omega(0)}}{dP^k_{0,\omega(0)}}\bigg|_{\mathcal{F}^0_1}\right\}$$

$$= E^Q\left\{\tfrac{1}{2}k^2 \int_0^1 \|\omega(s)\|^2\, ds\right\} - \tfrac{1}{2}kd$$

$$= \tfrac{1}{2}k^2 \int \|x\|^2 \mu(dx) - \tfrac{1}{2}kd$$

which gives us (7.2).

Acknowledgment. This work was supported by the National Science Foundation, Grant No. MCS-8002568.

Bibliography

[1] Donsker, M. D., and Varadhan, S. R. S., *Asymptotic Evaluation of Certain Wiener Integrals for Large Time*, Functional Integration and its Applications, Proceedings of the International Conference held at the Cumberland Lodge, Windsor Great Park, London, in April 1974. A. M. Arthurs, Ed., Clarendon, Oxford, 1975.

[2] Donsker, M. D., and Varadhan, S. R. S., *Asymptotic evaluation of certain Markov process expectations for large time, I*, Comm. Pure Appl. Math. 28, 1975, pp. 279–301; III, Comm. Pure Appl. Math. 29, 1976, pp. 389–461.

[3] Varadhan, S. R. S., *Asymptotic probabilities and differential equations*, Comm. Pure Appl. Math. 19, 1966, pp. 261–286.

[4] Feynman, R. P., *Statistical Mechanics, a Set of Lectures*, Benjamin, New York, 1972.

[5] Pekar, G. S., *Research on Electron Theory of Crystals*, Translation Series No. AEC-tr. 5575, Physics AEC, Division of Technical Information, Washington, DC, 1963.

[6] Donsker, M. D., and Varadhan, S. R. S., *Asymptotics for the polaron*, to appear.

[7] Oxtoby, J. C., *Ergodic sets*, Bull. Amer. Math. Soc. 58, 1952, pp. 116–136.

Received August, 1982.

Comm. Pure Appl. Math. 36 (1983) 505-528

Asymptotics for the Polaron

M. D. DONSKER

AND

S. R. S. VARADHAN

1. Introduction

Let P_t and $E^{P_t}\{\ \}$ denote, respectively, the probability measure and expectation with respect to three-dimensional Brownian motion $x(\cdot)$ tied down at both ends, i.e., with $x(0)=x(t)=0$. For $\alpha>0$, let

$$G(\alpha, t)=E^{P_t}\left\{\exp\left\{\alpha\int_0^t\int_0^t \frac{e^{-|\sigma-s|}}{|x(\sigma)-x(s)|}\,d\sigma\,ds\right\}\right\}.$$

A long standing problem in statistical mechanics (cf. [1]), the "polaron problem," has been to show that

$$\lim_{t\to\infty}\frac{1}{t}\log G(\alpha,t)=g(\alpha)\quad\text{exists,}\tag{1.1}$$

and, moreover, according to a conjecture of Pekar [2], that

$$\lim_{\alpha\to\infty}\frac{g(\alpha)}{\alpha^2}=g_0\quad\text{exists,}\tag{1.2}$$

with

$$g_0=\sup_{\substack{\phi\in L_2(R^3)\\ \|\phi\|=1}}\left[2\iint\frac{\phi^2(x)\phi^2(y)}{|x-y|}\,dx\,dy-\tfrac{1}{2}\int|\nabla\phi|^2\,dx\right].\tag{1.3}$$

In this paper we prove (1.1) obtaining an expression for $g(\alpha)$ which is explicit enough to allow us to prove also the conjecture (1.2) of Pekar, g_0 being indeed given by (1.3). We make use of large deviation results obtained in our earlier papers [3] and, in particular, [4]. That the polaron problem depends on sharp large deviation theorems is natural, since in determining the asymptotic behavior of $G(\alpha,t)$ for large t it is clear that the three-dimensional Brownian motion paths which contribute the most are those which make $|x(\sigma)-x(s)|$ small. However σ and s must not be so far apart that the contribution is killed by $e^{-|\sigma-s|}$. Thus, the influential paths are those which tend to stay awhile near where they have just been. Since this is not the way "typical" Brownian motion paths behave, we are dealing with probabilities of large deviations.

Our expression for $g(\alpha)$ is an infinite-dimensional variational formula. The fact that, as the coupling constant α becomes infinite, this variational formula behaves as indicated in (1.2) and (1.3) finally involving only a finite-dimensional variational formula is perhaps of significance in other problems of statistical mechanics, e.g., justification of mean field approximations.

The authors are grateful to Elliot Lieb who suspected that the polaron problem might yield to our methods and brought the problem to our attention. Recently in [5] Adamouski, Gerlach and Lesche provide a convincing argument of Pekar's conjecture using some of the earlier results of the present authors. For discussion of the existence and uniqueness of a function ϕ yielding the supremum in (1.3) see Lieb [6].

In Section 2 we give an outline of the proof of (1.1) and indicate the nature of the variational expression for $g(\alpha)$. Also in Section 2 we show formally how this last expression can be handled as $\alpha \to \infty$ yielding (1.2) and (1.3). In Sections 3 and 4 we present the proofs of the heuristic arguments described in Section 2.

2. Outline of the Results

To prove (1.1) we perform successive reductions of the problem. First of all we show that the asymptotic behavior of $G(\alpha, t)$ as $t \to \infty$ is the same if we replace tied-down Brownian motion P_t by standard Brownian motion P. This is justified because for large t the terminal condition makes no essential difference. The proof of these remarks is contained in Theorem 3.3 below.

Next we approximate the integrand in $G(\alpha, t)$ as follows:

$$\begin{aligned}\alpha \int_0^t \int_0^t \frac{e^{-|\sigma-s|}}{|x(\sigma)-x(s)|}\, d\sigma\, ds &= 2\alpha \int_0^t \int_s^t \frac{e^{-(\sigma-s)}}{|x(\sigma)-x(s)|}\, d\sigma\, ds \\ &\cong 2\alpha \int_0^t ds \int_s^\infty \frac{e^{-(\sigma-s)}}{|x(\sigma)-x(s)|}\, d\sigma \\ &= 2\alpha \int_0^t ds \int_0^\infty \frac{e^{-\tau}}{|x(s+\tau)-x(s)|}\, d\tau \\ &= \int_0^t F(\theta_s \omega)\, ds,\end{aligned}$$

where

$$F(\omega) = 2\alpha \int_0^\infty \frac{e^{-\tau}}{|x(\tau)-x(0)|}\, d\tau, \qquad \omega = x(\cdot),$$

and $\theta_s \omega = \omega(\cdot + s)$ is the translation operator. To prove (1.1) then becomes the problem of finding

$$\lim_{t\to\infty} \frac{1}{t} \log E^P\left\{\exp\left\{\int_0^t F(\theta_s\, \omega)\, ds\right\}\right\}, \tag{2.1}$$

where $E^P\{\ \}$ denotes expectation with respect to standard Brownian motion in R^3.

Let Ω be the space of trajectories on $-\infty<t<\infty$. Let Q be any stationary stochastic process with values in R^3 and let $\mathcal{F}_0^{-\infty}$ be the σ-field generated by the process $x(t)$ for $t\leqq 0$. Let $\mathcal{F}_T^0$ be the σ-field generated by the process $x(t)$ for $0\leqq t\leqq T$. We can think of $\mathcal{F}_0^{-\infty}$ and $\mathcal{F}_T^0$ as sub σ-fields of $\mathcal{F}$, the entire σ-field on which Q is a translation invariant measure.

Let Q_ω^T be the regular conditional probability of Q on $\mathcal{F}_T^0$ given $\mathcal{F}_0^{-\infty}$. Denote by $P_{\omega(0)}^T$ the measure on $\mathcal{F}_T^0$ corresponding to a Markov process P starting at time 0 from $\omega(0)$. Let

$$h_T(\omega)=\int \log R_\omega^T(\omega')Q_\omega^T(d\omega')$$

be the entropy where $R_\omega^T(\cdot)$ is the Radon-Nikodym derivative of Q_ω^T with respect to $P_{\omega(0)}^T$ on $\mathcal{F}_T^0$. We then define

$$h(T)=\int h_T(\omega)Q(d\omega).$$

It turns out that either $h(T)=\infty$ for all $T>0$ or $h(T)=HT$ for some constant H which depends on Q. We therefore write $h(T)=T\cdot H(Q)$ for some $0\leqq H(Q)\leqq\infty$ and call $H(Q)$ the entropy of the stationary process Q with respect to the Markov process P. All the details about this entropy are contained in [4] and what is important about $H(Q)$ is that it governs the probabilities of large deviations which is the main topic of [4]. Indeed, we show in [4] that, under suitable hypotheses,

$$\lim_{t\to\infty}\frac{1}{t}\log E^P\left\{\exp\left\{\int_0^t F(\theta_s\omega)\,ds\right\}\right\}=\sup_Q\,[E^Q\{F(\omega)\}-H(Q)], \tag{2.2}$$

where the supremum is taken over all stationary processes Q, $E^P\{\ \}$ is the expectation with respect to the Markov process P, and F is a bounded continuous function on Ω. What we mean by "suitable hypotheses" above is that the Markov process P must satisfy certain restrictions in order to obtain

$$\overline{\lim_{t\to\infty}}\frac{1}{t}\log E^P\left\{\exp\left\{\int_0^t F(\theta_s\omega)\,ds\right\}\right\}\leqq\sup_Q\,[E^Q\{F(\omega)\}-H(Q)] \tag{2.3}$$

and must satisfy certain other restrictions in order to obtain

$$\underline{\lim_{t\to\infty}}\frac{1}{t}\log E^P\left\{\exp\left\{\int_0^t F(\theta_s\omega)\,ds\right\}\right\}\geqq\sup_Q\,[E^Q\{F(\omega)\}-H(Q)]. \tag{2.4}$$

In [4] we show that (2.3) holds if the Markov process P satisfies some strong positive recurrence conditions which in that paper we list as hypotheses (i)–(v).

We show there that conditions (i)–(v) are satisfied by the Ornstein–Uhlenbeck process in particular and so in the present paper we use (2.3) for the Ornstein–Uhlenbeck process and then use that result to get the corresponding upper estimate on (2.1).

For (2.4) we found in [4] two hypotheses needed for the Markov process P which we called I–II. Ordinary Brownian motion satisfies these and therefore we use (2.4) directly to get a lower estimate on (2.1).

Thus, with

$$F(\omega)=2\alpha\int_0^\infty \frac{e^{-\tau}}{|x(\tau)-x(0)|}\,d\tau,$$

and P denoting Brownian motion, we establish in Section 3 below that (1.1) holds by showing

$$\begin{aligned} g(\alpha) &= \lim_{t\to\infty}\frac{1}{t}\log E^P\left\{\exp\left\{\int_0^t F(\theta_s\omega)\,ds\right\}\right\} \\ &= \sup_Q\{E^Q\{F(\omega)\}-H(Q)\}, \end{aligned} \tag{2.5}$$

where the supremum is taken over all stationary processes and $H(Q)$ is the entropy of the stationary process Q with respect to ordinary Brownian motion.

In Section 4 of this paper we show that the variational formula for $g(\alpha)$ on the right of (2.5) can be used to verify Pekar's conjecture (1.2) and (1.3). First of all, using Brownian scaling we show in Section 4 that

$$g_0=\lim_{\alpha\to\infty}\frac{g(\alpha)}{\alpha^2}=\lim_{\alpha\to\infty}\sup_Q\left[E^Q\left\{2\int_0^\infty \frac{e^{-t}}{|x(\alpha^2 t)-x(0)|}\,dt\right\}-H(Q)\right]. \tag{2.6}$$

In [4] we prove that $H(Q)$ is linear in Q and therefore the supremum involved in (2.6) is over linear functionals of Q and is therefore attained at an extremal which is an ergodic process. Using then the ergodic theorem, we show that roughly speaking

$$\begin{aligned} \lim_{\alpha\to\infty} E^Q\left\{2\int_0^\infty \frac{e^{-t}}{|x(\alpha^2 t)-x(0)|}\,dt\right\} &= \lim_{\alpha\to\infty} 2\left[\int_0^\infty e^{-t}\,dt\iint\frac{1}{|x-y|}\lambda_Q^{\alpha^2 t}(dx,dy)\right] \\ &= 2\int_0^\infty e^{-t}\,dt\iint\frac{1}{|x-y|}\mu_Q(dx)\mu_Q(dy). \end{aligned}$$

Here $\lambda_Q^t(dx,dy)$ is the joint distribution of $x(0)$, $x(t)$ under Q and μ_Q is the marginal distribution of $x(\cdot)$ under Q. We prove in Section 4 that

$$g_0=\lim_{\alpha\to\infty}\frac{g(\alpha)}{\alpha^2}=\sup_Q\left[2\iint\frac{\mu_Q(dx)\mu_Q(dy)}{|x-y|}-H(Q)\right]. \tag{2.7}$$

In [4] we proved the contraction principle for the entropy, namely

$$\inf_{Q:\mu_Q=\mu} H(Q)=I(\mu),$$

where $I(\mu)$ is the I function introduced by the authors in [3]. Here, of course, the I function is for Brownian motion and we have shown in the earlier papers that in that case $I(\mu)<\infty$ if and only if $\mu(dx)=\phi^2(x)\,dx$ for some $\phi\in L^2(R^3)$ with $\|\phi\|_2=1$ and $\int|\nabla\phi|^2\,dx<\infty$. In such a case, $I(\mu)=\frac{1}{2}\int|\nabla\phi|^2\,dx$. Thus, from (2.7),

$$\begin{aligned} g_0&=\sup_{\mu}\left[2\iint\frac{\mu(dx)\mu(dy)}{|x-y|}-I(\mu)\right]\\ &=\sup_{\substack{\phi\in L_2(R^3)\\ \|\phi\|=1}}\left[2\iint\frac{\phi^2(x)\phi^2(y)}{|x-y|}\,dx\,dy-\frac{1}{2}\int|\nabla\phi|^2\,dx\right].\end{aligned}$$

3. Proof of (1.1)

As before, we let P_t be the measure corresponding to tied-down $(x(0)=x(t)=0)$ Brownian motion in R^3 and P be the measure corresponding to free $(x(0)=0)$ Brownian motion in R^3. Let

$$z(\alpha,\lambda,t_1,t_2)=\alpha\int_{t_1}^{t_2}\int_{t_1}^{t_2}\frac{\lambda\,e^{-\lambda|\sigma-s|}}{|x(\sigma)-x(s)|}\,d\sigma\,ds,$$

where $\alpha\geqq 0$, $\lambda>0$, and $0\leqq t_1<t_2$. The object of this section is to prove

THEOREM 3.1.

$$\lim_{t\to\infty}\frac{1}{t}\log E^{P_t}\{\exp\{z(\alpha,\lambda,0,t)\}\}=\Phi(\alpha,\lambda), \tag{3.1}$$

where

$$\Phi(\alpha,\lambda)=\sup_{Q}\left[E^{Q}\left\{2\alpha\int_0^\infty\frac{\lambda\,e^{-\lambda\sigma}}{|x(\sigma)-x(0)|}\,d\sigma\right\}-H(Q)\right], \tag{3.2}$$

the supremum being taken over all stationary processes Q in R^3, and where $H(Q)$ is the entropy of the stationary process Q relative to the Brownian motion P (see Section 2 above for definitions).

It should be noted that statement (1.1) in the introduction is just (3.1) with $\lambda=1$, i.e., $g(\alpha)=\Phi(\alpha,1)$. To prove Theorem 3.1 it will suffice, as will be shown in Theorem 3.3, to prove

THEOREM 3.2

$$\lim_{t\to\infty}\frac{1}{t}\log E^{P}\{\exp\{z(\alpha,\lambda,0,t)\}\}=\Phi(\alpha,\lambda), \tag{3.3}$$

where $\Phi(\alpha,\lambda)$ is given by (3.2).

We first show

THEOREM 3.3. *Theorem* 3.2 *implies Theorem* 3.1.

The proof of this last theorem requires a few preliminaries.

LEMMA 3.4. *If ρ_1 and ρ_2 are covariance functions for two Gaussian processes in R^3 with mean* 0, *and if $\rho_1 \leqq \rho_2$ in the sense of quadratic forms, then*

$$E^{\rho_1}\{\exp\{z(\alpha,\lambda,t_1,t_2)\}\} \geqq E^{\rho_2}\{\exp\{z(\alpha,\lambda,t_1,t_2)\}\}. \tag{3.4}$$

Proof: Since $\alpha \geqq 0$, to show (3.4) it suffices to check that for each $n \geqq 1$,

$$E^{\rho_1}\left\{\left(\int_0^t\int_0^t \frac{e^{-\lambda|\sigma-s|}}{|x(\sigma)-x(s)|}\,d\sigma\,ds\right)^n\right\} \geqq E^{\rho_2}\left\{\left(\int_0^t\int_0^t \frac{e^{-\lambda|\sigma-s|}}{|x(\sigma)-x(s)|}\,d\sigma\,ds\right)^n\right\}.$$

But for this it suffices to show that

$$\begin{aligned} &E^{\rho_1}\left\{\frac{1}{|x(\sigma_1)-x(s_1)|}\cdot\frac{1}{|x(\sigma_2)-x(s_2)|}\cdots\frac{1}{|x(\sigma_n)-x(s_n)|}\right\} \\ &\quad \geqq E^{\rho_2}\left\{\frac{1}{|x(\sigma_1)-x(s_1)|}\cdot\frac{1}{|x(\sigma_2)-x(s_2)|}\cdots\frac{1}{|x(\sigma_n)-x(s_n)|}\right\}, \end{aligned} \tag{3.5}$$

for arbitrary choices of $\sigma_1, \sigma_2, \cdots, \sigma_n, s_1, s_2, \cdots, s_n$ in $[0, t]$.

For each $\delta > 0$, $1/(\delta^2+|x|^2)^{1/2}$ is positive definite and since the product of positive definite functions is also positive definite there exists a positive measure μ on R^{6n} such that

$$\begin{aligned} \phi(\xi) &= \frac{1}{(\delta^2+|x(\sigma_1)-x(s_1)|^2)^{1/2}}\cdots\frac{1}{(\delta^2+|x(\sigma_n)-x(s_n)|^2)^{1/2}} \\ &= \int e^{i\langle\xi,\eta\rangle}\mu(d\eta), \end{aligned}$$

where $\xi = (x(\sigma_1), x(s_1), \cdots, x(\sigma_n), x(s_n)) \in R^{6n}$. Thus

$$\begin{aligned} E^{\rho_1}\{\phi(\xi)\} &= \int e^{-\langle\rho_1\eta,\eta\rangle/2}\mu(d\eta) \\ &\geqq \int e^{-\langle\rho_2\eta,\eta\rangle/2}\mu(d\eta) \\ &= E^{\rho_2}\{\phi(\xi)\}, \end{aligned}$$

and (3.5) follows by letting $\delta \to 0$.

We shall later use the following obvious

Corollary 3.5. *Let*

$$z_\delta(\alpha,\lambda,t_1,t_2) = \alpha\int_{t_1}^{t_2}\int_{t_1}^{t_2}\frac{\lambda e^{-\lambda|\sigma-s|}}{(\delta^2+|x(\sigma)-x(s)|^2)^{1/2}}\,d\sigma\,ds.$$

If ρ_1 and ρ_2 are as in the lemma, then

$$E^{\rho_1}\{\exp\{z_\delta(\alpha,\lambda,t_1,t_2)\}\} \geqq E^{\rho_2}\{\exp\{z_\delta(\alpha,\lambda,t_1,t_2)\}\}.$$

COROLLARY 3.6. *With P_t and P as defined at the beginning of this section,*

$$E^{P_t}\{\exp\{z(\alpha,\lambda,t_1,t_2)\}\} \geqq E^{P}\{\exp\{z(\alpha,\lambda,t_1,t_2)\}\}. \tag{3.6}$$

Proof: In view of Lemma 3.4 it suffices to show that the covariance function ρ_t of the P_t process is less than or equal to the covariance function ρ of the P process in the sense of quadratic forms. But, $\rho(s,u)_{ij} = \delta_{ij}\min(s,u)$ and $\rho_t(s,u)_{ij} = \delta_{ij}[\min(s,u) - su/t]$. Thus, for any $f^i(s)$, $i = 1, 2, 3$,

$$\begin{aligned}\sum_{i,j=1}^{3}\int_0^\infty\int_0^\infty [\rho(s,u)_{ij} - \rho_t(s,u)_{ij}] f^i(s) f^j(u)\,ds\,du \\ = \sum_{i=1}^{3}\frac{1}{t}\left(\int_0^\infty s f^i(s)\,ds\right)^2 \geqq 0.\end{aligned}$$

We are now ready for

Proof of Theorem 3.3: From Corollary 3.6 and (3.3) we have

$$\underline{\lim}_{t\to\infty}\frac{1}{t}\log E^{P_t}\{\exp\{z(\alpha,\lambda,0,t)\}\} \geqq \Phi(\alpha,\lambda);$$

thus we want to show that (3.3) implies the corresponding upper estimate,

$$\overline{\lim}_{t\to\infty}\frac{1}{t}\log E^{P_t}\{\exp\{z(\alpha,\lambda,0,t)\}\} \leqq \Phi(\alpha,\lambda). \tag{3.7}$$

To do this, we note first of all that

$$\begin{aligned} z(\alpha,\lambda,0,t) &= z(\alpha,\lambda,0,t-1) + z(\alpha,\lambda,t-1,t) \\ &\quad + 2\alpha\int_0^{t-1}\int_{t-1}^{t}\frac{\lambda e^{-\lambda|\sigma-s|}}{|x(\sigma)-x(s)|}\,d\sigma\,ds. \end{aligned} \tag{3.8}$$

Since $1/|x|$ is positive definite, there exists a measure μ on R^3 such that $1/|x| = \int e^{i(\xi,x)}\mu(d\xi)$, and so

$$\begin{aligned}&\int_0^{t-1}\int_{t-1}^{t}\frac{\lambda e^{-\lambda|\sigma-s|}}{|x(\sigma)-x(s)|}\,d\sigma\,ds \\ &= \int_0^{t-1}\lambda\,d\sigma\int_{t-1}^{t}\frac{e^{-\lambda(s-\sigma)}}{|x(\sigma)-x(s)|}\,ds \\ &= \int\mu(d\xi)\int_0^{t-1}\lambda e^{\lambda\sigma}\,d\sigma\int_{t-1}^{t}e^{-\lambda s}e^{i(\xi,x(\sigma)-x(s))}\,ds\end{aligned}$$

$$
\begin{aligned}
&= \int \mu(d\xi) \int_0^{t-1} \lambda\, e^{\lambda(\sigma-t+1)} e^{i\langle \xi, x(\sigma)\rangle} d\sigma \int_{t-1}^{t} e^{\lambda(t-1-s)} e^{-i\langle \xi, x(s)\rangle} ds \\
(3.9)\qquad &\leqq \left[\int \mu(d\xi) \left| \int_0^{t-1} \lambda\, e^{\lambda(\sigma-t+1)} e^{i\langle \xi, x(\sigma)\rangle} d\sigma\right|^2\right]^{1/2} \\
&\quad \times \left[\int \mu(d\xi) \left| \int_{t-1}^{t} e^{\lambda(t-1-s)} e^{-i\langle \xi, x(s)\rangle} ds\right|^2\right]^{1/2} \\
&= \left[\int \mu(d\xi) \int_0^{t-1}\int_0^{t-1} \lambda^2 e^{\lambda[\sigma+\tau-2(t-1)]} e^{i\langle \xi, x(\sigma)-x(\tau)\rangle} d\sigma\, d\tau\right]^{1/2} \\
&\quad \times \left[\int \mu(d\xi) \int_{t-1}^{t}\int_{t-1}^{t} e^{\lambda[2(t-1)-s-u]} e^{-i\langle \xi, x(s)-x(u)\rangle} ds\, du\right]^{1/2} \\
&= \left[\int_0^{t-1}\int_0^{t-1} \frac{\lambda\, e^{[\sigma+\tau-2(t-1)]}}{|x(\sigma)-x(\tau)|} d\sigma\, d\tau\right]^{1/2} \\
&\quad \times \left[\int_{t-1}^{t}\int_{t-1}^{t} \frac{\lambda\, e^{\lambda[2(t-1)-s-u]}}{|x(s)-x(u)|} ds\, du\right]^{1/2}.
\end{aligned}
$$

Now, $\sigma \leqq t-1$ and $\tau \leqq t-1$ imply $\sigma+\tau-2(t-1) \leqq -|\sigma-\tau|$. Similarly, $t-1 \leqq s$ and $t-1 \leqq u$ imply $2(t-1)-s-u \leqq -|s-u|$. Thus, from (3.9) we conclude that

$$
\begin{aligned}
\int_0^{t-1}\int_{t-1}^{t} \frac{\lambda\, e^{-\lambda|\sigma-s|}}{|x(\sigma)-x(s)|} d\sigma\, ds \leqq & \left[\int_0^{t-1}\int_0^{t-1} \frac{\lambda\, e^{-\lambda|\sigma-\tau|}}{|x(\sigma)-x(\tau)|} d\sigma\, d\tau\right]^{1/2} \\
& \times \left[\int_{t-1}^{t}\int_{t-1}^{t} \frac{\lambda\, e^{-\lambda|s-u|}}{|x(s)-x(u)|} ds\, du\right]^{1/2}.
\end{aligned}
$$

Using this in (3.8), we have

$$
\begin{aligned}
z(\alpha, \lambda, 0, t) \leqq{}& z(\alpha, \lambda, 0, t-1) + z(\lambda, \lambda, t-1, t) \\
&+ 2[z(\alpha, \lambda, 0, t-1) z(\alpha, \lambda, t-1, t)]^{1/2},
\end{aligned}
$$

and since $2\sqrt{xy} = 2\sqrt{\theta x \cdot y/\theta} \leqq \theta x + (1/\theta) y$ for any $\theta > 0$, we get

$$
z(\alpha, \lambda, 0, t) \leqq (1+\theta) z(\alpha, \lambda, 0, t-1) + (1+1/\theta) z(\alpha, \lambda, t-1, t), \tag{3.10}
$$

for any $\theta > 0$.

Using (3.10) and the Hölder inequality, we get for $p>1$, $q>1$, such that $1/p+1/q=1$,

$$
\begin{aligned}
&E^{P_t}\left\{\exp\left\{\alpha \int_0^t\int_0^t \frac{\lambda\, e^{-\lambda|\sigma-s|}}{|x(\sigma)-x(s)|} d\sigma\, ds\right\}\right\} \\
&\quad \leqq E^{P_t}\left\{\exp\left\{\alpha(1+\theta) \int_0^{t-1}\int_0^{t-1} \frac{\lambda\, e^{-\lambda|\sigma-\tau|}}{|x(\sigma)-x(\tau)|} d\sigma\, d\tau\right\}\right.
\end{aligned}
$$

$$\times \exp\left\{\alpha\left(1+\frac{1}{\theta}\right)\int_{t-1}^{t}\int_{t-1}^{t}\frac{\lambda e^{-\lambda|s-u|}}{|x(s)-x(u)|}\,ds\,du\right\}\Bigg\}$$

$$\leqq\left[E^{P_t}\left\{\exp\left\{p\alpha(1+\theta)\int_0^{t-1}\int_0^{t-1}\frac{\lambda e^{-\lambda|\sigma-\tau|}}{|x(\sigma)-x(\tau)|}\,d\sigma\,d\tau\right\}\right\}\right]^{1/p}$$

$$\times\left[E^{P_t}\left\{\exp\left\{q\alpha\left(1+\frac{1}{\theta}\right)\int_{t-1}^{t}\int_{t-1}^{t}\frac{\lambda e^{-\lambda|s-u|}}{|x(s)-x(u)|}\,ds\,du\right\}\right\}\right]^{1/q}.$$

For tied-down Brownian motion, the measure P_t is reversible in time, i.e., the probability distribution of

$$\int_{t-1}^{t}\int_{t-1}^{t}\frac{\lambda e^{-\lambda|s-u|}}{|x(s)-x(u)|}\,ds\,du$$

is the same as the probability distribution of

$$\int_0^1\int_0^1\frac{\lambda e^{-\lambda|s-u|}}{|x(s)-x(u)|}\,ds\,du$$

so that the last inequality can be rewritten as

$$E^{P_t}\left\{\exp\left\{\alpha\int_0^t\int_0^t\frac{\lambda e^{-\lambda|\sigma-s|}}{|x(\sigma)-x(s)|}\,d\sigma\,ds\right\}\right\}$$

$$\leqq\left[E^{P_t}\left\{\exp\left\{p\alpha(1+\theta)\int_0^{t-1}\int_0^{t-1}\frac{\lambda e^{-\lambda|\sigma-\tau|}}{|x(\sigma)-x(\tau)|}\,d\sigma\,d\tau\right\}\right\}\right]^{1/p} \tag{3.11}$$

$$\times\left[E^{P_t}\left\{\exp\left\{q\alpha\left(1+\frac{1}{\theta}\right)\int_0^1\int_0^1\frac{\lambda e^{-\lambda|s-u|}}{|x(s)-x(u)|}\,ds\,du\right\}\right\}\right]^{1/q}.$$

For any $t'<t$, the measure P_t is absolutely continuous with respect to the measure P on the σ-field up to time t' which we shall denote by $\mathcal{F}_{t'}$. Moreover,

$$\left.\frac{dP_t}{dP}\right|_{\mathcal{F}_{t'}}=\left(\frac{t}{t-t'}\right)^{3/2}\exp\left\{-\frac{\|x(t')\|^2}{2(t-t')}\right\}\leqq\left(\frac{t}{t-t'}\right)^{3/2}.$$

Thus, for any functional G measurable with respect to $\mathcal{F}_{t'}$, $E^{P_t}\{G[x]\}\leqq(t/(t-t'))^{3/2}E^{P}\{G[x]\}$. Using this in (3.11), we obtain

$$E^{P_t}\left\{\exp\left\{\alpha\int_0^t\int_0^t\frac{\lambda e^{-\lambda|\sigma-s|}}{|x(\sigma)-x(s)|}\,d\sigma\,ds\right\}\right\}$$

$$\leqq t^{3/2p}\left[E^{P}\left\{\exp\left\{p\alpha(1+\theta)\int_0^{t-1}\int_0^{t-1}\frac{\lambda e^{-\lambda|\sigma-\tau|}}{|x(\sigma)-x(\tau)|}\,d\sigma\,d\tau\right\}\right\}\right]^{1/p} \tag{3.12}$$

$$\times\left(\frac{t}{t-1}\right)^{3/2q}\left[E^{P}\left\{\exp\left\{q\alpha\left(1+\frac{1}{\theta}\right)\int_0^1\int_0^1\frac{\lambda e^{-\lambda|s-u|}}{|x(s)-x(u)|}\,ds\,du\right\}\right\}\right]^{1/q}.$$

Hence, from (3.3) and (3.12) we get, for each $\theta > 0$ and $p > 1$,

$$\overline{\lim_{t\to\infty}} \frac{1}{t} \log E^{P_t}\left\{\exp\left\{\alpha \int_0^t \int_0^t \frac{\lambda\, e^{-\lambda|\sigma-s|}}{|x(\sigma)-x(s)|}\, d\sigma\, ds\right\}\right\} \leq \frac{1}{p}\Phi(p\alpha(1+\theta), \lambda). \tag{3.13}$$

Now, Φ is a finite convex function of α and therefore continuous, which means that (3.13) implies (3.7) by letting $p \to 1$ and $\theta \to 0$. This completes the proof of Theorem 3.3.

Before proving Theorem 3.2 we need some lemmas.

LEMMA 3.7. *For any $\delta > 0$, let*

$$V_\delta(x) = \frac{1}{|x|} - \frac{1}{(\delta^2 + |x|^2)^{1/2}}.$$

Let P_x denote Brownian motion measure in R^3 for paths starting at x. Then,

$$\lim_{\delta\to 0} \sup_x E^{P_x}\left\{\int_0^1 V_\delta(x(\sigma))\, d\sigma\right\} = 0. \tag{3.14}$$

Proof: Since

$$V_\delta(x) \leq \frac{1}{|x|} - \frac{1}{\delta + |x|} = \frac{\delta}{|x|(\delta + |x|)},$$

it suffices to show that

$$\lim_{\delta\to 0} \sup_x E^{P_x}\left\{\int_0^1 \frac{\delta}{|x(\sigma)|(\delta + |x(\sigma)|)}\, d\sigma\right\} = 0.$$

Using Brownian scaling, we have, for any $0 < \varepsilon < 1$,

$$\begin{aligned}
\int_0^1 E^{P_x}\left\{\frac{\delta}{|x(\sigma)|(\delta + |x(\sigma)|)}\right\} d\sigma &= \int_0^1 \frac{d\sigma}{\sqrt{\sigma}} E^{P_{x/\sqrt{\sigma}}}\left\{\frac{\delta}{|x(1)|(\delta + \sqrt{\sigma}|x(1)|)}\right\} \\
&\leq \left(\int_0^\varepsilon \frac{d\sigma}{\sqrt{\sigma}}\right) \sup_x E^{P_x}\left\{\frac{1}{|x(1)|}\right\} \\
&\quad + \left(\int_\varepsilon^1 \frac{d\sigma}{\sqrt{\sigma}}\right) \sup_x E^{P_x}\left\{\frac{\delta}{\sqrt{\varepsilon}|x(1)|^2}\right\} \\
&= I_1 + I_2.
\end{aligned}$$

But

$$\begin{aligned}
E^{P_x}\left\{\frac{1}{|x(1)|}\right\} &= \frac{1}{(2\pi)^{3/2}} \int \frac{\exp\{-\frac{1}{2}|y-x|^2\}}{|y|}\, dy \\
&\leq c_1\left(\int_{|y|\leq 1} \frac{dy}{|y|} + \int_{|y|\geq 1} \frac{\exp\{-\frac{1}{2}|y-x|^2\}}{2}\, dy\right) \\
&\leq c_1\left(\int_{|y|\leq 1} \frac{dy}{|y|} + \int \exp\{-\tfrac{1}{2}|y|^2\}/dy\right).
\end{aligned}$$

Thus, $I_1 \leqq c_2\sqrt{\varepsilon}$. Similarly $I_2 \leqq c_3\delta/\sqrt{\varepsilon}$, and hence $\overline{\lim}_{\delta\to 0}(I_1+I_2) \leqq c_2\sqrt{\varepsilon}$ which implies the lemma.

LEMMA 3.8. *With $V_\delta(x)$ as defined in the preceding lemma, for any $\alpha > 0$ and $\lambda > 0$,*

$$\lim_{\delta\to 0}\overline{\lim_{t\to\infty}}\frac{1}{t}\log E^P\left\{\exp\left\{\alpha\int_0^t\int_0^t V_\delta(x(\sigma)-x(s))\,e^{-\lambda|\sigma-s|}\lambda\,d\sigma\,ds\right\}\right\}=0. \tag{3.15}$$

Proof: For simplicity of exposition we set $\alpha = \frac{1}{2}$ and $\lambda = 1$. Let $G(\omega) = \int_0^\infty e^{-u}V_\delta(x(u)-x(0))\,du$, and let

$$\begin{aligned} F(\omega) &= \tfrac{1}{2}\int_0^t\int_0^t V_\delta(x(\sigma)-x(s))\,e^{-|\sigma-s|}\,d\sigma\,ds \\ &= \int_0^t ds\int_s^t e^{-(\sigma-s)}V_\delta(x(\sigma)-x(s))\,d\sigma. \end{aligned}$$

Using the notation introduced in Section 2,

$$G(\theta_s\omega) = \int_0^\infty e^{-u}V_\delta(x(u+s)-x(s))\,du = \int_s^\infty e^{-(\sigma-s)}V_\delta(x(\sigma)-x(s))\,d\sigma,$$

and so $F(\omega) \leqq \int_0^t G(\theta_s\omega)\,ds$. Let $G_n(\omega) = \int_n^{n+1} V_\delta(x(u)-x(0))\,du$ and note that $G(\omega) \leqq \sum_{n=0}^\infty e^{-n}G_n(\omega)$. Using these inequalities and Hölder's inequality, we get

$$\begin{aligned} E^P\{\exp\{F(\omega)\}\} &\leqq E^P\left\{\exp\left\{\sum_{n=0}^\infty e^{-n}\int_0^t G_n(\theta_s\omega)\,ds\right\}\right\} \\ &\leqq \prod_{n=0}^\infty\left(E^P\left\{\exp\left\{2^{n+1}e^{-n}\int_0^t G_n(\theta_s\omega)\,ds\right\}\right\}\right)^{1/2^{n+1}}. \end{aligned}$$

Thus, letting $\beta = \sup_n (n+1)2^{n+1}e^{-n} < \infty$, we have

$$\begin{aligned} \log E^P\{\exp\{F(\omega)\}\} &\leqq \sum_{n=0}^\infty\frac{1}{2^{n+1}}\log E^P\left\{\exp\left\{2^{n+1}e^{-n}\int_0^t G_n(\theta_s\omega)\,ds\right\}\right\} \\ &\leqq \sup_n \log E^P\left\{\exp\left\{\frac{\beta}{n+1}\int_0^t G_n(\theta_s\omega)\,ds\right\}\right\}. \end{aligned} \tag{3.16}$$

It follows from Lemma 4.1 of [4] that

$$\log E^P\left\{\exp\left\{\frac{\beta}{n+1}\int_0^t G_n(\theta_s\omega)\,ds\right\}\right\} \leqq t\log\sup_x E^{P_x}\{\exp\{\beta G_n(\omega)\}\}, \tag{3.17}$$

where P_x denotes ordinary Brownian motion in R^3 starting from x. By Lemma 3.7, we can take $\delta > 0$ so small that $\beta\sup_x E^{P_x}\{\int_0^1 V_\delta(x(\sigma))\,d\sigma\} < 1$. For such

$\delta > 0$, we have

$$\sup_x E^{P_x}\{\exp\{\beta G_n(\omega)\}\} = \sup_x E^{P_x}\left\{\exp\left\{\beta \int_n^{n+1} V_\delta(x(\sigma) - x(0))\, d\sigma\right\}\right\}$$

$$\leqq \sup_x E^{P_x}\left\{\exp\left\{\beta \int_0^1 V_\delta(x(\sigma))\, d\sigma\right\}\right\} \tag{3.18}$$

$$\leqq \frac{1}{1 - \beta \sup_x E^{P_x}\{\int_0^1 V_\delta(x(\sigma))\, d\sigma\}}.$$

From (3.16), (3.17), and (3.18) we obtain, for sufficiently small $\delta > 0$,

$$\overline{\lim_{t\to\infty}} \frac{1}{t} \log E^P\{\exp\{F(\omega)\}\} \leqq \log\left(\frac{1}{1 - \beta \sup_x E^{P_x}\{\int_0^1 V_\delta(x(\sigma))\, d\sigma\}}\right).$$

Now, (3.15) follows by letting $\delta \to 0$ and applying Lemma 3.7.

LEMMA 3.9. *Let R_1 and R_2 be two Gaussian processes with mean 0 and covariance functions ρ_1 and ρ_2, respectively. If $\mathscr{F}_T^0$ denotes the σ-field up to time T and if $dR_2/dR_1|_{\mathscr{F}_T^0}$ is bounded for each T, then $\rho_2 \leqq \rho_1$ in the sense of quadratic forms.*

Proof: For any function $f(s)$ we have

$$E^{R_2}\left\{\exp\left\{\int_0^T f(s)x(s)\, ds\right\}\right\} = E^{R_1}\left\{\exp\left\{\int_0^T f(s)x(s)\, ds\right\} \frac{dR_2}{dR_1}\bigg|_{\mathscr{F}_T^0}\right\}$$

$$\leqq c_T E^{R_1}\left\{\exp\left\{\int_0^T f(s)x(s)\, ds\right\}\right\}$$

for some constant c_T. This implies that

$$\exp\{\tfrac{1}{2}\langle \rho_2 f, f\rangle_T\} \leqq c_T \exp\{\tfrac{1}{2}\langle \rho_1 f, f\rangle_T\}.$$

Since f is arbitrary, for any a,

$$\exp\{\tfrac{1}{2}a^2\langle \rho_2 f, f\rangle_T\} \leqq c_T \exp\{\tfrac{1}{2}a^2\langle \rho_1 f, f\rangle_T\},$$

and hence

$$\exp\{\tfrac{1}{2}\langle \rho_2 f, f\rangle_T\} \leqq c_T^{1/a^2} \exp\{\tfrac{1}{2}\langle \rho_1 f, f\rangle_T\}.$$

Letting $a \to \infty$, we obtain $\langle \rho_2 f, f\rangle_T \leqq \langle \rho_1 f, f\rangle_T$ for any T.

LEMMA 3.10. *Let P^ε denote the probability measure corresponding to the Ornstein–Uhlenbeck* (O-U) *process in R^3 starting from the origin with generator $\frac{1}{2}\Delta - \varepsilon x \cdot \nabla$. Then,*

$$E^P\{\exp\{z_\delta(\alpha, \lambda, 0, t)\}\} \leqq E^{P^\varepsilon}\{\exp\{z_\delta(\alpha, \lambda, 0, t)\}\}, \tag{3.19}$$

where P, as before, is ordinary Brownian motion in R^3 starting from the origin and $z_\delta(\alpha, \lambda, 0, t)$ is defined in Corollary 3.5.

Proof: By the Cameron–Martin–Girsanov formula, we have, for any $T<\infty$,

$$\left.\frac{dP^\varepsilon}{dP}\right|_{\mathcal{F}_T^0} = \exp\left\{-\varepsilon\int_0^T x(s)\,dx(s) - \tfrac{1}{2}\varepsilon^2\int_0^T |x(s)|^2\,ds\right\}$$

$$= \exp\left\{-\tfrac{1}{2}\varepsilon|x(T)|^2 + \tfrac{3}{2}\varepsilon T - \tfrac{1}{2}\varepsilon^2\int_0^T |x(s)|^2\,ds\right\}$$

$$\leqq \exp\{\tfrac{3}{2}\varepsilon T\}.$$

From Lemma 3.9 we conclude that $\rho_{P_\varepsilon} \leqq \rho_P$ in the sense of quadratic forms and (3.19) then follows from Corollary 3.5.

With entropy as defined in Section 2 we now state a lemma the proof of which is given in [4] (Theorem 7.1 of that paper).

LEMMA 3.11. *Let $H_\varepsilon(Q)$ be the entropy of the stationary process Q (in R^3) relative to the Ornstein–Uhlenbeck process P^ε, and let $H(Q)$ be the entropy of Q relative to the Brownian motion P. Then,*

$$H_\varepsilon(Q) = H(Q) + \tfrac{1}{2}\varepsilon^2\int |\omega(0)|^2 Q(d\omega) - \tfrac{3}{2}\varepsilon. \tag{3.20}$$

LEMMA 3.12. *For each $\varepsilon>0$,*

$$\begin{aligned}&\overline{\lim_{t\to\infty}}\frac{1}{t}\log E^P\{\exp\{z_\delta(\alpha,\lambda,0,t)\}\}\\ &\leqq \sup_Q\left[E^Q\left\{2\alpha\int_0^\infty \frac{\lambda e^{-\lambda\sigma}}{(\delta^2+|x(\sigma)-x(0)|^2)^{1/2}}\,d\sigma\right\} - H_\varepsilon(Q)\right],\end{aligned} \tag{3.21}$$

where the supremum is taken over all stationary processes Q in R^3 and where $H_\varepsilon(Q)$ is the entropy of Q relative to the O–U *process P^ε defined in Lemma* 3.10. *Moreover, for each $\alpha>0$, $\lambda>0$, and $\delta>0$,*

$$\overline{\lim_{t\to\infty}}\frac{1}{t}\log E^P\{\exp\{z_\delta(\alpha,\lambda,0,t)\}\} \leqq \Phi(\alpha,\lambda), \tag{3.22}$$

where $\Phi(\alpha,\lambda)$ is given by (3.2).

Proof: In order to show (3.21), it suffices, in view of Lemma 3.10, to show that

$$\begin{aligned}&\overline{\lim_{t\to\infty}}\frac{1}{t}\log E^{P^\varepsilon}\{\exp\{z_\delta(\alpha,\lambda,0,t)\}\}\\ &\leqq \sup_Q\left[E^Q\left\{2\alpha\int_0^\infty \frac{\lambda e^{-\lambda\sigma}}{(\delta^2+|x(\sigma)-x(0)|^2)^{1/2}}\,d\sigma\right\} - H_\varepsilon(Q)\right].\end{aligned} \tag{3.23}$$

Let

$$F_\delta(\omega)=2\alpha\int_0^\infty \frac{\lambda\, e^{-\lambda\sigma}}{(\delta^2+|\omega(\sigma)-\omega(0)|^2)^{1/2}}\,d\sigma.$$

Since stationary processes have no fixed discontinuities, $\int F_\delta(\omega)Q(d\omega)$ is a bounded continuous function from $\mathcal{M}_S(\Omega)$ to R^1. Using the notation of [4], we observe that

$$\begin{aligned}\int F_\delta(\omega')R_{t,\omega}(d\omega') &= \frac{1}{t}\int_0^t F_\delta(\theta_s\omega_t)\,ds\\ &= \frac{2\alpha}{t}\int_0^t ds\int_0^\infty \frac{\lambda e^{-\lambda\sigma}}{(\delta^2+|\omega_t(\sigma+s)-\omega_t(s)|^2)^{1/2}}\,d\sigma\\ &= \frac{2\alpha}{t}\int_0^t ds\Bigg[\int_0^{t-s}\frac{\lambda\, e^{-\lambda\sigma}}{(\delta^2+|x(\sigma+s)-x(s)|^2)^{1/2}}\,d\sigma\\ &\quad+\int_{t-s}^\infty \frac{\lambda\, e^{-\lambda\sigma}}{(\delta^2+|\omega_t(\sigma+s)-\omega_t(s)|^2)^{1/2}}\,d\sigma\Bigg].\end{aligned}$$

Therefore,

$$\left|t\int F_\delta(\omega')R_{t,\omega}(d\omega')-z_\delta(\alpha,\lambda,0,t)\right|\leqq \frac{2\alpha}{\delta}\int_0^t ds\int_{t-s}^\infty \lambda\, e^{-\lambda\sigma}\,d\sigma\leqq\frac{2\alpha}{\delta\lambda},\tag{3.24}$$

and hence,

$$E^{P^\varepsilon}\{\exp\{z_\delta(\alpha,\lambda,0,t)\}\}\leqq E^{P^\varepsilon}\left\{\exp\left\{t\int F_\delta(\omega')R_{t,\omega}(d\omega')\right\}\right\}e^{2\alpha/\delta\lambda}.\tag{3.25}$$

The Ornstein–Uhlenbeck process satisfies the hypotheses (i)–(v) for Theorem 4.6 of [4] (see e.g. Section 7 of [4]) and thus it follows from (3.25) that

$$\overline{\lim_{t\to\infty}}\frac{1}{t}\log E^{P^\varepsilon}\{\exp\{z_\delta(\alpha,\lambda,0,t)\}\}\leqq\sup_Q\,[E^Q\{F_\delta(\omega)\}-H_\varepsilon(Q)],$$

which is (3.23). Hence, we have (3.21) for every $\varepsilon>0$ and so, using Lemma 3.11, we get

$$\begin{aligned}&\overline{\lim_{t\to\infty}}\frac{1}{t}\log E^P\{\exp\{z_\delta(\alpha,\lambda,0,t)\}\}\\ &\leqq\inf_{\varepsilon>0}\sup_Q\left[E^Q\left\{2\alpha\int_0^\infty\frac{\lambda\, e^{-\lambda\sigma}}{|x(\sigma)-x(0)|}\,d\sigma\right\}-H(Q)+\tfrac{3}{2}\varepsilon\right].\\ &=\Phi(\alpha,\lambda).\end{aligned}$$

We are now ready to prove Theorem 3.2.

Proof of Theorem 3.2: First of all, let $U_\delta=1/(\delta^2+|x|^2)^{1/2}$ and, as before,

$V_\delta = 1/|x| - 1/(\delta^2 + |x|^2)^{1/2}$, so that $1/|x| = U_\delta + V_\delta$. Using Hölder's inequality and noting that

$$z_\delta(\alpha, \lambda, 0, t) = \alpha \int_0^t \int_0^t U_\delta(x(\sigma) - x(s)\lambda\, e^{-\lambda|\sigma - s|}\, d\sigma\, ds,$$

we have

$$\begin{aligned} &E^P\{\exp\{z(\alpha, \lambda, 0, t)\}\} \\ &\quad\leqq \left(E^P\left\{\exp\left\{p\alpha \int_0^t \int_0^t U_\delta(x(\sigma) - x(s))\, e^{-\lambda|\sigma - s|}\lambda\, d\sigma\, ds\right\}\right\}\right)^{1/p} \\ &\qquad\times \left(E^P\left\{\exp\left\{q\alpha \int_0^t \int_0^t V_\delta(x(\sigma) - x(s))\, e^{-\lambda|\sigma - s|}\lambda\, d\sigma\, ds\right\}\right\}\right)^{1/q}, \end{aligned}$$

and

$$\begin{aligned} &\overline{\lim_{t\to\infty}} \frac{1}{t} \log E^P\{\exp\{z(\alpha. \lambda, 0, t)\}\} \\ (3.26)\qquad &\leqq \frac{1}{p} \overline{\lim_{t\to\infty}} \frac{1}{t} \log E^P\{\exp\{z_\delta(p\alpha, \lambda, 0, t)\}\} \\ &\quad + \frac{1}{q} \overline{\lim_{t\to\infty}} \frac{1}{t} \log E^P\left\{\exp\left\{q\alpha \int_0^t \int_0^t V_\delta(x(\sigma) - x(s))\, e^{-\lambda|\sigma - s|}\lambda\, d\sigma\, ds\right\}\right\}. \end{aligned}$$

From (3.26) and (3.22) we then get

$$\begin{aligned} \overline{\lim_{t\to\infty}} \frac{1}{t} \log E^P\{\exp\{z(\alpha, \lambda, 0, t)\}\} &\leqq \frac{1}{p} \Phi(p\alpha, \lambda) \\ &\quad + \frac{1}{q} \overline{\lim_{t\to\infty}} \frac{1}{t} \log E^P\left\{\exp\left\{q\alpha \int_0^t \int_0^t V_\delta(x(\sigma) - x(s))\, e^{-\lambda|\sigma - s|}\lambda\, d\sigma\, ds\right\}\right\}, \end{aligned}$$

which, recalling Lemma 3.8 and letting $\delta \to 0$, yields

$$(3.27)\qquad \overline{\lim_{t\to\infty}} \frac{1}{t} \log E^P\{\exp\{z(\alpha, \lambda, 0, t)\}\} \leqq \frac{1}{p} \Phi(p\alpha, \lambda).$$

As noted earlier, $\Phi(\alpha, \lambda)$ is continuous in α and therefore if we let $p \to 1$ we obtain from (3.27) the desired upper estimate.

For the corresponding lower estimate we use again (3.24) to get

$$\begin{aligned} &E^P\{\exp\{z(\alpha, \lambda, 0, t)\}\} \geqq E^P\{\exp\{z_\delta(\alpha, \lambda, 0, t)\}\} \\ (3.28)\qquad &\geqq E^P\left\{\exp\left\{t \int F_\delta(\omega')R_{t,\omega}(d\omega')\right\}\right\} e^{-2\alpha/\delta\lambda}. \end{aligned}$$

Now Brownian motion satisfies hypotheses I–II of Theorem 5.5 in [4] and thus we obtain from (3.28)

$$\begin{aligned}\underline{\lim}_{t\to\infty}\frac{1}{t}\log E^{P}\{\exp\{z(\alpha,\lambda,0,t)\}\}&\geqq\underline{\lim}_{t\to\infty}\frac{1}{t}\log E^{P}\Big\{\exp\Big\{t\int F_{\delta}(\omega')R_{t,\omega}(d\omega')\Big\}\Big\}\\ &\geqq\sup_{Q}\,[E^{Q}\{F_{\delta}(\omega)\}-H(Q)].\end{aligned}\tag{3.29}$$

Letting $\delta\to 0$ in (3.29) we get the desired lower estimate and hence complete the proof of Theorem 3.2.

4. Proof of Pekar's Conjecture

In Section 3 we proved that if

$$G(\alpha,t)=E^{P_{t}}\Big\{\exp\Big\{\alpha\int_0^t\int_0^t\frac{e^{-|\sigma-s|}}{|x(\sigma)-x(s)|}\,d\sigma\,ds\Big\}\Big\},$$

then

$$\lim_{t\to\infty}\frac{1}{t}\log G(\alpha,t)=g(\alpha)\quad\text{exists,}$$

and in fact,

$$g(\alpha)=\Phi(\alpha,1)=\sup_{Q}\Big[E^{Q}\Big\{2\alpha\int_0^\infty\frac{e^{-\sigma}}{|x(\sigma)-x(0)|}\,d\sigma\Big\}-H(Q)\Big].$$

In this section we want to prove Pekar's conjecture that

$$\lim_{\alpha\to\infty}\frac{g(\alpha)}{\alpha^2}=g_0\quad\text{exists}$$

and

$$g_0=\sup_{\substack{\phi\in L_2(R^3)\\ \|\phi\|=1}}\Big[2\iint\frac{\phi^2(x)\phi^2(y)}{|x-y|}\,dx\,dy-\tfrac{1}{2}\int|\nabla\phi|^2\Big].$$

First of all, if we let

$$K(\alpha,\lambda,t)=E^{P}\{\exp\{z(\alpha,\lambda,0,t\}\},$$

then using Brownian scaling we observe that, for any $\tau>0$, $K(\alpha,\lambda,t)=K(\alpha\sqrt{\tau},\lambda\tau,t/\tau)$. Thus, from Theorem 3.2 we conclude that, for any $\tau>0$, $\Phi(\alpha,\lambda)=(1/\tau)\Phi(\alpha\sqrt{\tau},\lambda\tau)$. In particular, this means $\Phi(\alpha,\lambda)=\alpha^2\Phi(1,\lambda/\alpha^2)$, and so $g(\alpha)/\alpha^2=1/\alpha^2\Phi(\alpha,1)=\Phi(1,1/\alpha^2)$. Hence, to prove Pekar's conjecture, it suffices to show

$$\lim_{\lambda\to 0}\Phi(1,\lambda)=g_0=\sup_{\substack{\phi\in L_2(R^3)\\ \|\phi\|=1}}\Big[2\iint\frac{\phi^2(x)\phi^2(y)}{|x-y|}\,dx\,dy-\tfrac{1}{2}\int|\nabla\phi|^2\Big],\tag{4.1}$$

where

$$\Phi(1,\lambda)=\sup_Q \left[E^Q\left\{2\int_0^\infty \frac{\lambda e^{-\lambda\sigma}}{|x(\sigma)-x(0)|}\,d\sigma\right\}-H(Q)\right].$$

We first prove a theorem (Theorem 4.1 below) in an abstract setting which we shall use in the proof of (4.1).

Let X be a Polish space and let T_t, $t\geqq 0$, be a one-parameter family of homeomorphisms of X. Let $\mathcal{M}$ be the space of probability measures on X and $\mathcal{M}_I\subset\mathcal{M}$ be the measures on X invariant under T_t.

For each $\lambda>0$, define a map $\alpha_{\lambda,x}: X\to\mathcal{M}$ by

$$\alpha_{\lambda,x}(A)=\lambda\int_0^\infty e^{-\lambda t}\chi_A(T_t x)\,dt,$$

where $A\subset X$. For $P\in\mathcal{M}_I$, define $\tilde{P}_\lambda=P\alpha_{\lambda,x}^{-1}$ so that $\tilde{P}_\lambda$ is a probability measure on $\mathcal{M}$, i.e., if $G\subset\mathcal{M}$, then

$$\tilde{P}_\lambda(G)=P\{x\in X:\alpha_{\lambda,x}\in G\}.$$

We will have several occasions to use an elementary identity implied by these definitions. Let $A\subset X$, $P\in\mathcal{M}_I$, and $\lambda>0$, then

$$\begin{aligned}E^P\{\alpha_{\lambda,x}(A)\}&=\lambda\int_0^\infty e^{-\lambda t}E^P\{\chi_A(T_t x)\}\,dt\\ &=\lambda\int_0^\infty e^{-\lambda t}E^P\{\chi_A(x)\}\,dt=\lambda\int_0^\infty e^{-\lambda t}P(A)\,dt=P(A).\end{aligned}\tag{4.2}$$

In terms of the $\tilde{P}_\lambda$ measure on $\mathcal{M}$ we can express (4.2) as

$$E^{\tilde{P}_\lambda}\{Q(A)\}=P(A).\tag{4.3}$$

In what follows let $\phi(x,y)\in C(X\times X)$ and assume further that $\phi(x,y)$ is symmetric.

Theorem 4.1. *Let $P_n\in M_I$ for $n=1,2,\cdots$ and assume $P_n\Rightarrow P$. Let $\lambda_n\to 0$ as $n\to\infty$ and define $\tilde{P}_{n,\lambda_n}=P_n\alpha_{\lambda_n,x}^{-1}$,* i.e., *if $G\subset\mathcal{M}$, $\tilde{P}_{n,\lambda_n}(G)=P_n\{x\in X:\alpha_{\lambda_n,x}\in G\}$. Then:*

(i) *$\{\tilde{P}_{n,\lambda_n}\}$ is a tight family of measures on $\mathcal{M}$.*

(ii) *If we choose a subsequence n_j, $j=1,2,\cdots$, such that $\tilde{P}_{n_j,\lambda_{n_j}}\Rightarrow R$, then*

$$\text{(a)}\quad R(\mathcal{M}_I)=1,$$

and, moreover,

$$\text{(b)}\quad \int_{\mathcal{M}_I} Q R(dQ)=P.$$

(iii) $\lim_{j\to\infty}E^{P_{n_j}}\{\lambda_{n_j}\int_0^\infty\phi(x,T_t x)\exp\{-\lambda_{n_j}t\}\,dt\}$

$=\int_{\mathcal{M}_I}\iint\phi(x,y)Q(dx)Q(dy)R(dQ).$

Proof: since $\{P_n\}$ is a tight family of measures on X, there exists for every $l>0$ a compact set $K_l \subset X$ such that $P_n(K_l) \geqq 1-1/2^l$ for all $n=1, 2, \cdots$. Thus, using (4.3), we see that, for every $l>0$,

$$E^{\tilde{P}_{n,\lambda_n}}\{Q(K_l)\} \geqq 1-\frac{1}{2^l}, \qquad n=1, 2, 3, \cdots, \tag{4.4}$$

and from this we obtain in the usual way

$$\tilde{P}_{n,\lambda_n}\{Q\colon Q(K_l) \geqq 1-1/2^{l/2}\} \geqq 1-\frac{1}{2^{l/2}}, \tag{4.5}$$

for $n=1, 2, \cdots$. From (4.5) we have, for any $l_0>0$,

$$\tilde{P}_{n,\lambda_n}\{Q\colon Q(K_l) \geqq 1-1/2^{l/2} \text{ for all } l \geqq l_0\} \geqq 1-\sum_{l \geqq l_0} \frac{1}{2^{l/2}}. \tag{4.6}$$

But, for any $l_0>0$, the set of Q such that $Q(K_l) \geqq 1-1/2^{l/2}$ for all $l \geqq l_0$ is a compact subset of $\mathcal{M}$. We have therefore shown that $\{\tilde{P}_{n,\lambda_n}\}$ is a tight family of measures on $\mathcal{M}$.

For $s>0$ and $f \in c(X)$ we note that

$$\begin{aligned}
&\int_X f(y)\alpha_{\lambda,x}(dy) - \int_X f(T_s y)\alpha_{\lambda,x}(dy) \\
&\qquad = \lambda \int_0^\infty e^{-\lambda t} f(T_t x)\, dt - \lambda \int_0^\infty e^{-\lambda t} f(T_{t+s} x)\, dt \\
&\qquad = \lambda \int_0^\infty e^{-\lambda t} f(T_t x)\, dt - \lambda \int_s^\infty e^{-\lambda(t-s)} f(T_t x)\, dt \\
&\qquad = \lambda \int_0^s e^{-\lambda t} f(T_t x)\, dt - \lambda \int_s^\infty f(T_t x)[e^{-\lambda(t-s)} - e^{-\lambda t}]\, dt,
\end{aligned}$$

and therefore,

$$\left| \int_X f(y)\alpha_{\lambda,x}(dy) - \int_X f(T_s y)\alpha_{\lambda,x}(dy) \right| \leqq \lambda s \|f\| + \|f\|(1-e^{-\lambda s}).$$

Thus, for $s>0$, $f \in c(X)$, and $n=1, 2, \cdots$,

$$\int_X \left| \int_X f(y)\alpha_{\lambda_n,x}(dy) - \int_X f(T_s y)\alpha_{\lambda_n,x}(dy) \right| P_n(dx) \leqq \lambda_n s \|f\| + \|f\|(1-\exp\{-\lambda_n s\}),$$

or equivalently,

$$\begin{aligned}
&\int_{\mathcal{M}} \left| \int_X f(x)Q(dx) - \int_X f(T_s x)Q(dx) \right| \tilde{P}_{n,\lambda_n}(dQ) \\
&\qquad \leqq \lambda_n s \|f\| + \|f\|(1-\exp\{-\lambda_n s\}).
\end{aligned} \tag{4.7}$$

Now we choose a subsequence n_j, $j = 1, 2, \cdots$, such that $\tilde{P}_{n_j,\lambda_{n_j}} \Rightarrow R$ as $j \to \infty$. Since both $\int_X f(x)Q(dx)$ and $\int_X f(T_s x)Q(dx)$ are bounded continuous functions of Q on $\mathcal{M}$ and since $\lambda_n \to 0$, we have from (4.7), replacing n by n_j and letting $j \to \infty$,

$$\int_{\mathcal{M}} \left| \int_X f(x)Q(dx) - \int_X f(T_s x)Q(dx) \right| R(dQ) = 0. \tag{4.8}$$

Thus, for each $s > 0$ and each $f \in c(X)$, (4.8) implies

$$R\left\{Q: \int_X f(x)Q(dx) = \int_X f(T_s x)Q(dx)\right\} = 1.$$

By the usual measure-theoretic arguments we can then conclude that $R(\mathcal{M}_I) = 1$.

From (4.3) we see that

$$\int_{\mathcal{M}} Q\tilde{P}_{n_j,\lambda_{n_j}}(dQ) = P_{n_j}$$

and thus if we let $j \to \infty$ we get $\int_{\mathcal{M}} QR(dQ) = P$. We have just proved that $R(\mathcal{M}_I) = 1$ and hence we conclude that in fact $\int_{\mathcal{M}_I} QR(dQ) = P$.

Finally, to prove statement (iii) of the theorem, let $P \in \mathcal{M}_I$ and using the symmetry of ϕ observe that

$$\begin{aligned} E^P\left\{\lambda \int_0^\infty \phi(x, T_t x)\, e^{-\lambda t}\, dt\right\} &= E^P\left\{\lambda \int_0^\infty \phi(T_t x, x)\, e^{-\lambda t}\, dt\right\} \\ &= E^P\left\{\lambda \int_0^\infty \phi(x, T_{-t} x)\, e^{-\lambda t}\, dt\right\} \\ &= E^P\left\{\lambda \int_{-\infty}^0 \phi(x, T_t x)\, e^{\lambda t}\, dt\right\}, \end{aligned}$$

i.e.,

$$E^P\left\{\lambda \int_0^\infty \phi(x, T_t x) e^{-\lambda t}\, dt\right\} = E^P\left\{\tfrac{1}{2}\lambda \int_{-\infty}^\infty \phi(x, T_t x)\, e^{-\lambda |t|}\, dt\right\}. \tag{4.9}$$

Let ξ_1 and ξ_2 be independent random variables each having probability density $\lambda\, e^{-\lambda t}$ for $t \geqq 0$. It is then trivial to prove that $\xi_1 - \xi_2$ has probability density $\frac{1}{2}\lambda\, e^{-\lambda|t|}$, $-\infty < t < \infty$. Using this remark and the fact that $P \in \mathcal{M}_I$, we get

$$\begin{aligned} E^P\left\{\tfrac{1}{2}\lambda \int_{-\infty}^\infty \phi(x, T_t x)\, e^{-\lambda|t|}\, dt\right\} &= E^P\left\{\lambda^2 \int_0^\infty \int_0^\infty \phi(x, T_{\sigma - t} x)\, e^{-\lambda\sigma - \lambda t}\, d\sigma\, dt\right\} \\ &= E^P\left\{\lambda^2 \int_0^\infty \int_0^\infty \phi(T_t x, T_\sigma x)\, e^{-\lambda\sigma - \lambda t}\, d\sigma\, dt\right\}. \end{aligned} \tag{4.10}$$

From (4.9), (4.10), and the assumption that n_j is a sequence such that $\tilde{P}_{n_j,\lambda_{n_j}} \Rightarrow R$ as $j \to \infty$ we obtain

$$\begin{aligned}
&\lim_{j\to\infty} E^{P_{n_j}}\left\{\lambda_{n_j}\int_0^\infty \phi(x, T_t x)\exp\{-\lambda_{n_j}t\}\,dt\right\}\\
&= \lim_{j\to\infty} E^{P_{n_j}}\left\{\lambda_{n_j}^2\int_0^\infty\int_0^\infty \phi(T_t x, T_\sigma x)\exp\{-\lambda_{n_j}\sigma - \lambda_{n_j}t\}\,d\sigma\,dt\right\}\\
&= \lim_{j\to\infty} E^{P_{n_j}}\left\{\int_X\int_X \phi(y, z)\alpha_{\lambda_{n_j},x}(dy)\alpha_{\lambda_{n_j},x}(dz)\right\}\\
&= \lim_{j\to\infty} E^{\tilde{P}_{n_j,\lambda_{n_j}}}\left\{\int_X\int_X \phi(x, y)Q(dx)Q(dy)\right\}\\
&= E^R\left\{\int_X\int_X \phi(x, y)Q(dx)Q(dy)\right\}\\
&= \int_{\mathcal{M}_I}\int_X\int_X \phi(x, y)Q(dx)Q(dy)R(dQ).
\end{aligned}$$

This completes the proof of Theorem 4.1.

LEMMA 4.2. *Let $U_\delta(x) = 1/(\delta^2 + |x|^2)^{1/2}$ and let $H_\varepsilon(Q)$ be the entropy of the stationary process Q with respect to the O–U process (defined in Lemma 3.10). Then, for each $\varepsilon > 0$,*

$$\begin{aligned}
&\overline{\lim_{\lambda\to 0}}\sup_Q\left[2E^Q\left\{\lambda\int_0^\infty e^{-\lambda t}U_\delta(x(t)-x(0))\,dt\right\} - H_\varepsilon(Q)\right]\\
&\leqq \sup_{\substack{\phi\in L_2(R^3)\\ \|\phi\|=1}}\left[2\iint \phi^2(x)\phi^2(y)U_\delta(x-y)\,dx\,dy - \tfrac{1}{2}\int |\nabla\phi|^2\,dx + \varepsilon\right].
\end{aligned} \tag{4.11}$$

Proof: Let

$$N(\lambda) = \sup_Q\left[2E^Q\left\{\lambda\int_0^\infty e^{-\lambda t}U_\delta(x(t)-x(0))\,dt\right\} - H_\varepsilon(Q)\right],$$

and, for each $\lambda > 0$, choose Q_λ such that

$$2E^{Q_\lambda}\left\{\lambda\int_0^\infty e^{-\lambda t}U_\delta(x(t)-x(0))\,dt\right\} - H_\varepsilon(Q_\lambda) \geqq N(\lambda) - \lambda. \tag{4.12}$$

From (4.12) we see that $\sup_{\lambda\leq 1} H_\varepsilon(Q_\lambda) < \infty$. Now the O–U process satisfies hypotheses (i)–(v) (see Section 2 of this paper); moreover, in [4] (Theorem 6.4) we showed that a family of stationary measures having uniformly bounded entropies with respect to a Markov process satisfying (i)–(v) is a tight family of measures. Thus, $\{Q_\lambda\}_{\lambda\leqq 1}$ is tight. Let $\overline{\lim}_{\lambda\to 0} N(\lambda)=N$. Choosing subsequences, using the tightness just established, and denoting Q_{λ_n} by Q_n, we conclude that, for some sequence $\lambda_n \to 0$,

$$\lim_{n\to\infty}\left[2E^{Q_n}\left\{\lambda_n\int_0^\infty \exp\{-\lambda_n t\}U_\delta(x(t)-x(0))\,dt\right\}-H_\varepsilon(Q_n)\right]=N \tag{4.13}$$

and $Q_n \Rightarrow Q'$.

Now we want to use Theorem 4.1. To do so, let the X of that theorem be the space of trajectories $x(\cdot)$ and let $\phi(x(\cdot), y(\cdot))=U_\delta(x(0)-y(0))$ which is then a symmetric, bounded, and continuous function on $X\times X$. Let $\tilde{P}_{n,\lambda_n}$ be defined as in Theorem 4.1 (with Q_n playing the role of P_n); we then conclude that $\{\tilde{P}_{n,\lambda_n}\}$ is a tight family of measures on $\mathcal{M}$ so that, choosing another subsequence if necessary,

$$\tilde{P}_{n,\lambda_n}\Rightarrow R, \tag{4.14}$$

where R is some probability measure on $\mathcal{M}$. From (iii) of Theorem 4.1 and (4.13) we then conclude that

$$N=2\int_{\mathcal{M}_I}\iint U_\delta(x(0)-y(0))Q(dx(\cdot))Q(dy(\cdot))R(dQ)-\lim_{n\to\infty} H_\varepsilon(Q_n). \tag{4.15}$$

But $H_\varepsilon(Q)$ is lower semicontinuous (see [4], Theorem 3.3) and $Q_n\Rightarrow Q'$ so that from (4.15) we get

$$N\leqq 2\int_{\mathcal{M}_I}\iint U_\delta(x(0)-y(0))Q(dx(\cdot))Q(dy(\cdot))R(dQ)-H_\varepsilon(Q'). \tag{4.16}$$

From (iib) of Theorem 4.1 we know that $Q'=\int_{\mathcal{M}_I} QR(dQ)$ and therefore we get

$$\begin{aligned} N\leqq 2\int_{\mathcal{M}_I}\iint & U_\delta(x(0)-y(0))Q(dx(\cdot))Q(dy(\cdot))R(dQ)\\ & -H_\varepsilon\left(\int_{\mathcal{M}_I} QR(dQ)\right).\end{aligned} \tag{4.17}$$

In [4] (Theorem 3.5) we proved that $H(Q)$ is linear in Q and therefore, from (4.17),

$$N\leqq\int_{\mathcal{M}_I}\left[2\iint U_\delta(x(0)-y(0))Q(dx(\cdot))Q(dy(\cdot))-H_\varepsilon(Q)\right]R(dQ). \tag{4.18}$$

By part (iia) of Theorem 4.1 we know that $R(\mathcal{M}_I)=1$ and so (4.18) implies

$$N\leqq\sup_Q\left[2\iint U_\delta(x(0)-y(0))Q(dx(\cdot))Q(dy(\cdot))-H_\varepsilon(Q)\right]. \tag{4.19}$$

Let μ be the marginal distribution of the stationary measure Q. In [4] (Section 6) we established the contraction principle which states

$$\inf_{Q:q(Q)=\mu} H(Q)=I(\mu),$$

where $q(Q)$ denotes the marginal distribution of Q and $I(\mu)$ is the I-function introduced by the authors in [3]. Thus, (4.19) implies

$$N \leqq \sup_{\mu}\left[2\iint U_\delta(x-y)\mu(dx)\mu(dy)-I_\varepsilon(\mu)\right], \tag{4.20}$$

where $I_\varepsilon(\mu)$ is the I-function corresponding to the O–U process with parameter ε. If we let $I(\mu)$ be the I-function corresponding to standard Brownian motion, then $I_\varepsilon(\mu)\geqq I(\mu)-\varepsilon$. Moreover, in our earlier papers, we showed that $I(\mu)$ is finite if and only if μ has a density ϕ^2 in which case $I(\mu)=I(\phi)=\frac{1}{2}\int|\nabla\phi|^2\,dx$. Thus, from (4.20), for all $\varepsilon>0$ and $\delta>0$,

$$N \leqq \sup_{\substack{\phi\in L_2(R^3)\\ \|\phi\|=1}}\left[2\iint \frac{\phi^2(x)\,\phi^2(y)}{(|x-y|^2+\delta^2)^{1/2}}\,dx\,dy-\frac{1}{2}\int|\nabla\phi|^2\,dx+\varepsilon\right], \tag{4.21}$$

which is (4.11).

LEMMA 4.3. *We have*

$$\overline{\lim_{\lambda\to 0}}\,\Phi(1,\lambda)\leqq \sup_{\substack{\phi\in L_2(R^3)\\ \|\phi\|=1}}\left[2\iint\frac{\phi^2(x)\phi^2(y)}{|x-y|}\,dx\,dy-\frac{1}{2}\int|\nabla\phi|^2\,dx\right]. \tag{4.22}$$

Proof: By Theorem 3.2,

$$\Phi(1,\lambda)=\lim_{t\to\infty}\frac{1}{t}\log E^P\{\exp\{z(1,\lambda,0,t\}\}. \tag{4.23}$$

But from inequality (3.26) we have, for any $0<p<1$ and q such that $1/p+1/q=1$, and any $\delta>0$,

$$\begin{aligned}
&\overline{\lim_{t\to\infty}}\,\frac{1}{t}\log E^P\{\exp\{z(1,\lambda,0,t)\}\}\\
&\leqq \frac{1}{p}\,\overline{\lim_{t\to\infty}}\,\frac{1}{t}\log E^P\{\exp\{z_\delta(p,\lambda,0,t)\}\} \qquad (4.24)\\
&\quad+\frac{1}{q}\,\overline{\lim_{t\to\infty}}\,\frac{1}{t}\log E^P\left\{\exp\left\{q\int_0^t\int_0^t V_\delta(x(\sigma)-x(s))\lambda\,e^{-\lambda|\sigma-s|}\,d\sigma\,ds\right\}\right\}.
\end{aligned}$$

From (3.21) we have, for each $\varepsilon > 0$, and each $\delta > 0$,

$$\overline{\lim_{t\to\infty}} \frac{1}{t} \log E^P\{\exp\{z_\delta(p, \lambda, 0, t)\}\}$$
(4.25)
$$\leqq \sup_Q \left[E^Q\left\{2p \int_0^\infty \frac{\lambda\, e^{-\lambda\sigma}}{(\delta^2 + |x(\sigma) - x(0)|^2)^{1/2}}\, d\sigma\right\} - H_\varepsilon(Q)\right].$$

Thus, letting $\delta \to 0$ and $p \to 1$, we see from (4.23), (4.24), (4.25), and Lemma 3.8 that

$$\Phi(1, \lambda) \leqq \sup_Q \left[E^Q\left\{2 \int_0^\infty \frac{\lambda\, e^{-\lambda\sigma}}{|x(\sigma) - x(0)|}\, d\sigma\right\} - H_\varepsilon(Q)\right]. \tag{4.26}$$

If we apply Lemma 4.2 (with $\delta = 0$) to (4.26) and then let $\varepsilon \to 0$ we obtain (4.22).

LEMMA 4.4.

$$\underline{\lim_{\lambda\to 0}}\, \Phi(1, \lambda) \geqq \sup_{\substack{\phi \in L_2(R^3)\\ \|\phi\| = 1}} \left\{2 \iint \frac{\phi^2(x)\phi^2(y)}{|x - y|}\, dx\, dy - \tfrac{1}{2} \int |\nabla\phi|^2\, dx\right\}. \tag{4.27}$$

Proof: Let $\phi \in L_2(R^3)$ with $\|\phi\| = 1$ and consider the stationary Markov process Q_ϕ whose generator is $\frac{1}{2}\Delta + (\nabla\phi/\phi) \cdot \nabla$. The invariant measure for this process is ϕ^2 and the process is ergodic (in fact, mixing). Therefore,

$$\Phi(1, \lambda) = \sup_Q \left[E^Q\left\{2 \int_0^\infty \frac{\lambda\, e^{-\lambda\sigma}}{|x(\sigma) - x(0)|}\, d\sigma\right\} - H(Q)\right]$$
(4.28)
$$\geqq E^{Q_\phi}\left\{2 \int_0^\infty \frac{\lambda\, e^{-\lambda\sigma}}{|x(\sigma) - x(0)|}\, d\sigma\right\} - H(Q_\phi).$$

The ergodic theorem yields

$$\lim_{\lambda\to 0} E^{Q_\phi}\left\{2 \int_0^\infty \frac{\lambda\, e^{-\lambda\sigma}}{|x(\sigma) - x(p)|}\, d\sigma\right\} = \lim_{\lambda\to 0} W^{Q_\phi}\left\{2 \int_0^\infty \frac{e^{-u}}{|x(u/\lambda) - x(0)|}\, du\right\}.$$
(4.29)
$$= 2 \iint \frac{\phi^2(x)\phi^2(y)}{|x - y|}\, dx\, dy.$$

Moreover, for the process Q_ϕ, $H(Q_\phi) = \frac{1}{2}\int |\nabla\phi|^2\, dx$ (see [4], Section 6). Hence,

$$\underline{\lim_{\lambda\to 0}}\, \Phi(1, \lambda) \geqq 2 \iint \frac{\phi^2(x)\phi^2(y)}{|x - y|}\, dx\, dy - \tfrac{1}{2} \int |\nabla\phi|^2\, dx, \tag{4.30}$$

and the lemma follows since (4.30) holds for any $\phi \in L^2(R^3)$ with $\|\phi\| = 1$.

Combining (4.22) and (4.27) we have proved (4.1) which as pointed out at the beginning of this section implies Pekar's conjecture.

Acknowledgment The work in this paper was supported in part by the National Science Foundation under Grant No. NSF MCS-80-02568.

Bibliography

[1] Feynman, R. P., *Statistical Mechanics*, W. A. Benjamin, Reading, MA, 1972.

[2] Pekar, S. I., *Theory of Polarons*, Zh. Eksperim. i Teor. Fiz. 19, 1949.

[3] Donsker, M. D., and Varadhan, S. R. S., *Asymptotic evaluation of certain Markov process expectations for large time*, I, II, III, Comm. Pure Appl. Math. 28, 1975, pp. 1–47; 28, 1975, pp. 279–301; 29, 1976, pp. 389–461.

[4] Donsker, M. D., and Varadhan, S. R. S., *Asymptotic evaluation of certain Markov process expectations for large time*, IV, Comm. Pure Appl. Math. 36, 1983, pp. 183–212.

[5] Adamowski, J., Gerlach, B., and Leschke, H., *Strong coupling limit of polaron energy, revisited*, Physical Letter, Sept. 1980, pp. 249–251.

[6] Lieb, E. H. *Existence and uniqueness of the minimizing solution of Choquard's nonlinear equation*, Studies in Applied Mathematics 57, 1977, pp. 93–105.

Received November, 1982.

Commun. Math. Phys. 97, 187–210 (1985)

Communications in Mathematical Physics

Large Deviations for Stationary Gaussian Processes*

M. D. Donsker and S. R. S. Varadhan

Courant Institute of Mathematical Sciences, New York University, New York, NY 10012, USA

Abstract. In their previous work on large deviations the authors always assumed the base process to be Markovian whereas here they consider the base process to be stationary Gaussian. Similar large deviation results are obtained under natural hypotheses on the spectral density function of the base process. A rather explicit formula for the entropy involved is also obtained.

1. Introduction

Let $\{X_k\}$, $-\infty<k<\infty$, be a stationary Gaussian process with $E\{X_k\}=0$ and $E\{X_0X_j\}=\varrho_j=\frac{1}{2\pi}\int_0^{2\pi} e^{ij\theta}f(\theta)d\theta$. We assume that the spectral density function $f(\theta)$ is continuous on $[0,2\pi]$, $f(0)=f(2\pi)$, and

$$\int_0^{2\pi} \log f(\theta)d\theta > -\infty . \tag{1.1}$$

Let $\Omega=\prod_{j=-\infty}^{\infty} \mathbb{R}_j$ where, for each j, $\mathbb{R}_j$ is the real line, i.e., Ω is the space of doubly infinite sequences of real numbers. We specify a point $\omega\in\Omega$ by $\omega=\{x_k\}$, $-\infty<k<\infty$, and let $\omega(j)=x_j$ for $-\infty<j<\infty$. The process $\{X_k\}$ induces a probability measure P on Ω. We will denote integration over Ω with respect to P measure by $E^P\{\}$.

For each positive integer n and each $\omega\in\Omega$, let $\omega^{(n)}$ be the point in Ω obtained by the periodic extension in both directions of the elements $x_1, x_2, \ldots, x_n$ of ω, i.e., if $\omega=\{x_k\}$, $-\infty<k<\infty$, then $\omega^{(n)}$ is the point

$$\ldots, x_1, \ldots, x_{n-1}, x_n, x_1, x_2, \ldots, x_n, x_1, x_2, \ldots, x_n, \ldots .$$

---

* The research in this paper was supported by the National Science Foundation, Grant No. MCS-80-02568

Let T be the shift operator acting on the points of Ω. For each $\omega \in \Omega$ and each positive integer n, we define a probability measure on Ω, call it $\pi_n(\omega)$, as follows:

$$\pi_n(\omega) = \frac{1}{n}(\delta_{\omega^{(n)}} + \delta_{T\omega^{(n)}} + \ldots + \delta_{T^{n-1}\omega^{(n)}}). \tag{1.2}$$

In other words, given $\omega \in \Omega$ and $n \geqq 1$, $\pi_n(\omega)$ is the probability measure on Ω which assigns mass $1/n$ to $\omega^{(n)}$ and to each of its $(n-1)$ translates. We note that $\pi_n(\omega)$ is a translation invariant measure on Ω, i.e., a stationary measure on Ω. Let $\underline{M}_S$ be the space of all stationary measures on Ω and impose on $\underline{M}_S$ the topology of weak convergence. We can use the mapping $\pi_n : \Omega \to \underline{M}_S$ and the measure P on Ω to construct a probability measure on $\underline{M}_S$ by defining, for each n, $Q_n = \pi_n P^{-1}$, i.e., if A is a set of stationary measures in $\underline{M}_S$, then

$$Q_n(A) = P\{\omega \in \Omega : \pi_n(\omega) \in A\}. \tag{1.3}$$

The assumptions we made on the Gaussian process $\{X_k\}$ imply that it is an ergodic process, and it then follows from the ergodic theorem that, for almost all ω (P-measure), the measure $\pi_n(\omega)$ converges weakly to the measure P as $n \to \infty$, i.e.,

$$P\{\omega \in \Omega : \pi_n(\omega) \Rightarrow P\} = 1. \tag{1.4}$$

Thus, with reference to the Q_n measure on $\underline{M}_S$ given by (1.3), we expect that if the set A contains the measure P, then $Q_n(A) \to 1$ as $n \to \infty$, whereas if P is not in the closure of A, we expect $Q_n(A) \to 0$ as $n \to \infty$. In this paper we show that in this latter case $Q_n(A)$ approaches zero exponentially fast as $n \to \infty$ and we determine the constant in the exponential rate. To be more specific, let R denote a general element of $\underline{M}_S$, i.e., a stationary measure on Ω, and recall that f is the spectral density of our basic stationary Gaussian process $\{X_k\}$. We define a functional $H_f(R)$ which is, in fact, the entropy of the stationary process R with respect to the stationary Gaussian process $\{X_k\}$, and $H_f(R)$: $\underline{M}_S \to [0, \infty]$ is such that:

If $C \subset M_S$ is closed,

$$\overline{\lim_{n \to \infty}} \frac{1}{n} \log Q_n(C) \leqq - \inf_{R \in C} H_f(R), \tag{1.5}$$

and if $G \subset \underline{M}_S$ is open,

$$\underline{\lim_{n \to \infty}} \frac{1}{n} \log Q_n(G) \geqq - \inf_{R \in G} H_f(R). \tag{1.6}$$

Consistent with remarks made above it will indeed be true that $H_f(P) = 0$. We give an explicit formula for $H_f(R)$ in (1.9) below.

Let $\Phi : \underline{M}_S \to \mathbb{R}$ be bounded and continuous, and let $E^{Q_n}\{\,\}$ denote integration over $\underline{M}_S$ with respect to Q_n measure. From (1.5) and (1.6) it follows easily that

$$\lim_{n \to \infty} \frac{1}{n} \log E^{Q_n}\{e^{n\Phi(R)}\} = \sup_{R \in \underline{M}_S} [\Phi(R) - H_f(R)]. \tag{1.7}$$

Since $Q_n = \pi_n P^{-1}$, equivalent to (1.7) is

$$\lim_{n \to \infty} \frac{1}{n} \log E^P\{e^{n\Phi(\pi_n(\omega))}\} = \sup_{R \in M_S} [\Phi(R) - H_f(R)]. \tag{1.8}$$

The main results of this paper then are (1.5), (1.6), their implication (1.8), and the formula (1.9) for $H_f(R)$ which we discuss now.

Let $R \in \underline{M}_S$ and let, for $A \subset \mathbb{R}$, $R(A|\omega) = R(X_0 \in A | X_{-1}, X_{-2}, \ldots)$ be the regular conditional probability distribution of X_0 given the entire past. Denote by $r(y|\omega)$ the corresponding density. If $G(\theta)$ is the spectral measure of the stationary process R, then the formula for $H_f(R)$ is

$$H_f(R) = E^R\left\{\int_{-\infty}^{\infty} r(y|\omega)\log r(y|\omega)dy\right\} + \frac{1}{2}\log 2\pi + \frac{1}{4\pi}\int_0^{2\pi}\frac{dG(\theta)}{f(\theta)} + \frac{1}{4\pi}\int_0^{2\pi}\log f(\theta)d\theta\,, \tag{1.9}$$

where it is understood that $H_f(R)$ is defined to be $+\infty$ if for any reason we cannot define any of the ingredients in (1.9).

The authors have developed a theory of large deviations for Markov processes (see [1, 2], for theoretical results and e.g. [3–5] for some applications thereof). In theory, our methods should apply to rather general stationary processes, but it is difficult to see in any great generality what the natural hypotheses to impose on a stationary process are in order to obtain the analogues of (1.5), (1.6), and (1.8), much less to obtain the analogue of (1.9) in anything like explicit form. For recent work in this direction see Orey [6]. In this paper we consider the special case of a Gaussian stationary process because as seen above the hypotheses needed are natural and we obtain an explicit formula for the corresponding entropy.

In Sect. 2 of this paper we prove (1.5) and (1.6) but in doing so we must use properties of $H_f(R)$ and other entropies which occur in the proofs. So as not to interrupt the argument we prove these properties in Sect. 3.

2. Proofs of the Main Theorems

As in the introduction, let $\{X_k\}$ be a stationary Gaussian process and $E\{X_0X_j\} = \varrho_j = \frac{1}{2\pi}\int_0^{2\pi} e^{ij\theta}f(\theta)d\theta$, where we assume that the spectral density function $f(\theta)$ is continuous on $[0, 2\pi]$, $f(0) = f(2\pi)$, and $\int_0^{2\pi}\log f(\theta)d\theta > -\infty$.

Let $\{\xi_k\}$, $-\infty < k < \infty$, be a sequence of independent Gaussian random variables each with mean 0 and variance 1. If we let $\{a_n\}$ be the Fourier coefficients of $\sqrt{f(\theta)}$, i.e., $\sqrt{f(\theta)} = \sum_{n=-\infty}^{\infty} a_n e^{in\theta}$, then we can write

$$X_k = \sum_{n=-\infty}^{\infty} a_{n-k}\xi_k = \sum_{n=-\infty}^{\infty} a_n\xi_{n+k}\,. \tag{2.1}$$

Let N be a positive integer and define

$$h_N(\theta) = \sum_{|j|<N} a_j\left(1 - \frac{|j|}{N}\right)e^{ij\theta}\,, \tag{2.2}$$

where the a_j's are defined in the preceding paragraph. Since $f(\theta)$ is continuous and $f(0)=f(2\pi)$, we have by Fejer's theorem,

$$\lim_{N\to\infty} \sup_{0\leq\theta\leq 2\pi} |h_N(\theta)-\sqrt{f(\theta)}|=0. \tag{2.3}$$

For each N, define the process

$$X_k^N = \sum_{|j|<N} a_j\left(1-\frac{|j|}{N}\right)\xi_{j+k}. \tag{2.4}$$

If we let $f_N(\theta)=h_N^2(\theta)$, then $f_N(\theta)$ is the spectral density of the process $\{X_k^N\}$.

In the introduction we noted that the process $\{X_k\}$ induces a probability measure P on Ω. Similarly the process $\{X_k^N\}$ just defined induces a probability measure P^N on Ω. Indeed, the family $\{\xi_k\}$ of independent, Gaussian, mean 0, variance 1, random variables itself induces a probability measure on Ω which we will denote by μ. Each of these three stationary measures on Ω together with the mapping $\pi_n:\Omega\to\underline{M}_S$ of the introduction generates a probability measure on $\underline{M}_S$. We have already introduced $Q_n=P\pi_n^{-1}$. Analogously, we define $\nu_n=\mu\pi_n^{-1}$ and for each N, $Q_n^N=P^N\Pi_n^{-1}$.

For convenience let $b_k=a_k\left(1-\frac{|k|}{N}\right)$ and define the mapping $F_N:\Omega\to\Omega$ by

$$(F_N(\omega))(j)=\sum_{|k|<N} b_k x_{j+k}, \tag{2.5}$$

so that F_N is a moving average map of Ω into itself. The mapping $F_N:\Omega\to\Omega$ induces a corresponding map $\tilde{F}_N:\underline{M}_S\to\underline{M}_S$.

We also have the mapping $F:\Omega\to\Omega$ given by

$$(F(\omega))(j)=\sum_{k=-\infty}^{\infty} a_k x_{j+k}, \tag{2.6}$$

which exists for almost all ω (μ measure on Ω), since $\sum_{n=-\infty}^{\infty} a_n^2<\infty$. We note that for sets $A\subset\underline{M}_S$:

$$\begin{aligned} Q_n(A)&=P\{\omega\in\Omega:\pi_n(\omega)\in A\}=\mu\{\omega\in\Omega:\pi_n\cdot F(\omega)\in A\},\\ Q_n^N(A)&=P^N\{\omega\in\Omega:\pi_n(\omega)\in A\}=\mu\{\omega\in\Omega:\pi_n\cdot F_N(\omega)\in A\},\\ \nu_n(A)&=\mu\{\omega\in\Omega:\pi_n(\omega)\in A\}. \end{aligned} \tag{2.7}$$

It will be important to introduce also the probability measure $\tilde{Q}_n^N$ on $\underline{M}_S$ given by $\tilde{Q}_n^N=\nu_n\tilde{F}_n^{-1}$ so that

$$\tilde{Q}_n^N(A)=\nu_n\{R\in\underline{M}_S:\tilde{F}_N(R)\in A\}=\mu\{\omega\in\Omega:\tilde{F}_N\cdot\pi_n(\omega)\in A\}. \tag{2.8}$$

To deduce the asymptotic behavior of Q_n as $n\to\infty$, i.e., to prove (1.5) and (1.6), we first determine the asymptotic behavior of Q_n^N as $n\to\infty$ and then let $N\to\infty$. To obtain the asymptotics of Q_n^N we compare it with $\tilde{Q}_n^N$ the asymptotic behavior of which as $n\to\infty$ follows from our earlier work and which we state below as Theorem 2.2. Looking at (2.8) and (2.7) we see that to compare $\tilde{Q}_n^N$ and Q_n^N means comparing $\pi_n \cdot F_N(\omega)$ with $\tilde{F}_N \cdot \pi_n(\omega)$. Now,

$$\tilde{F}_N \cdot \pi_n(\omega) = \frac{1}{n}[\delta_{F_N(\omega^{(n)})} + \delta_{F_N(T\omega^{(n)})} + \ldots + \delta_{F_N(T^{n-1}\omega^{(n)})}], \tag{2.9}$$

and

$$\pi_n \cdot F_N(\omega) = \frac{1}{n}[\delta_{(F_N(\omega))^{(n)}} + \delta_{T(F_N(\omega))^{(n)}} + \ldots + \delta_{T^{n-1}(F_N(\omega))^{(n)}}]. \tag{2.10}$$

Thus, to compare the two measures $\tilde{F}_N \cdot \pi_n$ and $\pi_n \cdot F_N$, we need to determine when $T^{j-1}(F_N(\omega))^{(n)} = F_N(T^{j-1}\omega^{(n)})$ with $1 \leqq j \leqq n$. It is clear that the difference of the two measures $\pi_n \cdot F_N$ and $\tilde{F}_N \cdot \pi_n$ can be estimated in the variation norm over the σ-field generated by the first L components. We obtain

$$\|\pi_n \cdot F_N - \tilde{F}_N \cdot \pi_n\|_{F_L^1} \leqq \frac{2N+L}{n}. \tag{2.11}$$

In the space $\underline{M}_S$ we have the topology of weak convergence. This topology can be induced by the following metric. Let R_1 and R_2 be in $\underline{M}_S$ and let L be a positive integer. Denote by R_1^L and R_2^L the respective L-dimensional measures one obtains by considering the joint distribution of L successive random variables in the processes R_1 and R_2. We then define the distance between R_1 and R_2 in $\underline{M}_S$ by

$$d(R_1, R_2) = \sum_{L=1}^{\infty} \frac{1}{2^L} d_L(R_1^L, R_2^L), \tag{2.12}$$

where d_L is the Prohorov distance for L-dimensional measures. This metric in $\underline{M}_S$ yields the topology of weak convergence in $\underline{M}_S$. We now can prove the following lemma.

Lemma 2.1. *For any $\delta > 0$ and large enough n,*

$$\mu\{\omega \in \Omega : d(\pi_n \cdot F_N, \tilde{F}_N \cdot \pi_n) \geqq \varepsilon\} = 0. \tag{2.13}$$

Proof. From (2.12), (2.11), and the fact that the variation norm dominates the Prohorov norm in L-dimensions, we have for $n > N$,

$$d(\pi_n \cdot F_N, \tilde{F}_n \cdot \pi_n) \leqq \sum_{L=1}^{\infty} \frac{1}{2^L} \|\pi_n \cdot F_N - \tilde{F}_N \cdot \pi_n\|_{\underline{F}_L^1} \leqq \frac{1}{n} \sum_{L=1}^{\infty} \frac{2N+L}{2^L}.$$

Since the infinite series on the right converges, we can make the distance on the left as small as we please by taking n large enough which implies (2.13).

As noted in (2.8),

$$\tilde{Q}_n^N(A) = \nu_n\{R \in \underline{M}_S : \tilde{F}_N(R) \in A\} = \mu\{\omega \in \Omega : \tilde{F}_N \cdot \pi_n(\omega) \in A\}.$$

Since μ measure on Ω comes from the sequence $\{\xi_k\}$ of independent Gaussian variables with mean 0 and variance 1, the results of [2-IV] give us the asymptotics of ν_n measure on $\underline{M}_S$ and hence of $\tilde{Q}_n^N$ measure on $\underline{M}_S$.

In the introduction we defined for $R \in \underline{M}_S$ the density $r(y|\omega)$ of the regular condition probability distribution of X_0 given the entire past. For $R \in \underline{M}_S$ define

$$I(R) = E^R\left\{\int_{-\infty}^{\infty} r(y|\omega)\log\left(\frac{r(y|\omega)}{\frac{1}{\sqrt{2\pi}}e^{-y^2/2}}\right)dy\right\}, \tag{2.14}$$

and define for $R \in \underline{M}_S$

$$H_{f_N}(R) = \inf_{R' \in \underline{M}_S : R'\tilde{F}_N^{-1} = R} I(R'). \tag{2.15}$$

Then in Sect. 3, formula (3.21), we show

$$H_{f_N}(R) = E^R\left\{\int_{-\infty}^{\infty} r(y|\omega)\log r(y|\omega)dy\right\} + \frac{1}{2}\log 2\pi$$
$$+ \frac{1}{4\pi}\int_0^{2\pi}\frac{dG(\theta)}{f_N(\theta)} + \frac{1}{4\pi}\int_0^{2\pi}\log f_N(\theta)d\theta,$$

where $G(\theta)$ is the spectral measure of R and $f_N(\theta)$ (see just after (2.4)) is the spectral density of the $\{X_R^N\}$ process.

Thus, from [2-IV] and the contraction principle discussed in that paper, we have

Theorem 2.2. *If $C \subset \underline{M}_S$ is closed,*

$$\overline{\lim_{n\to\infty}} \frac{1}{n}\log \tilde{Q}_n^N(C) \leqq -\inf_{R \in C} H_{f_N}(R), \tag{2.16}$$

and if $G \subset \underline{M}_S$ is open,

$$\underline{\lim_{n\to\infty}} \frac{1}{n}\log \tilde{Q}_n^N(G) \geqq -\inf_{R \in G} H_{f_N}(R). \tag{2.17}$$

Using this and Lemma 2.1 we can prove the following theorem.

Theorem 2.3. *If $C \subset \underline{M}_S$ is closed,*

$$\overline{\lim_{n\to\infty}} \frac{1}{n}\log Q_n^N(C) \leqq -\inf_{R \in C} H_{f_N}(R), \tag{2.18}$$

and if $G \subset \underline{M}_S$ is open,

$$\underline{\lim_{n\to\infty}} \frac{1}{n}\log Q_n^N(G) \geqq -\inf_{R \in G} H_{f_N}(R). \tag{2.19}$$

Proof. Let $\delta > 0$ and $C^\delta = \bigcup_{R \in C} S(R,\delta)$ where $S(R,\delta)$ is the ball in $\underline{M}_S$ of radius δ centered at R. From (2.7) and (2.8) we see that

$$\begin{aligned} Q_n^N(C) &= \mu\{\omega \in \Omega : \pi_n \cdot F_N(\omega) \in C\} \\ &\leqq \mu\{\omega \in \Omega : \tilde{F}_N \cdot \pi_n(\omega) \in C^\delta\} + \mu\{\omega \in \Omega : d(\pi_n \cdot F_N, \tilde{F}_n \cdot \pi_n) \geqq \delta\} \\ &= \tilde{Q}_n^N(C^\delta) + \mu\{\omega \in \Omega : d(\pi_n \cdot F_N, \tilde{F}_N \cdot \pi_n) \geqq \delta\}. \end{aligned}$$

Let $\bar{C}^\delta$ be the closure of C^δ. Then, from Lemma 2.1 and Theorem 2.2,

$$\overline{\lim_{n\to\infty}} \frac{1}{n} \log Q_n^N(C) \leqq \overline{\lim_{n\to\infty}} \frac{1}{n} \log \tilde{Q}_n^N(\bar{C}^\delta) \leqq - \inf_{R\in\bar{C}^\delta} H_{f_N}(R). \tag{2.20}$$

Since C is closed, $H_{f_N}(R)$ is lower semicontinuous, and $\{R: H_{f_N}(R) \leqq K\}$ is tight, (cf. Lemma 3.10) we obtain (2.18) from (2.20) on letting $\delta \to 0$.

For the lower bound, let G be open in $\underline{M}_S$ and let $R \in G$. Pick $\delta > 0$ and an open set V such that $R \in V \subset V^\delta \subset G$. Then, again noting (2.7) and (2.8),

$$\begin{aligned} Q_n^N(G) &\geqq Q_n^N(V^\delta) = \mu\{\omega\in\Omega: \pi_n \cdot F_N(\omega) \in V^\delta\} \\ &\geqq \mu\{\omega\in\Omega: \tilde{F}_N \cdot \pi_n(\omega) \in V\} - \mu\{\omega\in\Omega: d(\pi_n\cdot F_N, \tilde{F}_N\cdot\pi_n) \geqq \delta\} \\ &= \tilde{Q}_n^N(V) - \mu\{\omega\in\Omega: d(\pi_n\cdot F_N, \tilde{F}_N\cdot\pi_n)\geqq\delta\}. \end{aligned}$$

Using again Lemma 2.1 and Theorem 2.2,

$$\underline{\lim_{n\to\infty}} \frac{1}{n} \log Q_n^N(G) \geqq \underline{\lim_{n\to\infty}} \frac{1}{n} \log \tilde{Q}_n^N(V) \geqq - \inf_{R'\in V} H_{f_N}(R') \geqq -H_{f_N}(R). \tag{2.21}$$

Since inequality (2.21) holds for any $R \in G$ we obtain (2.19).

Before proving the next theorem we need some preliminary lemmas.

Lemma 2.4. *Let $\{X_k\}$ and $\{X_k^N\}$ be defined respectively by (2.1) and (2.4). There exists a function $\varepsilon(N) = o(1)$ as $N\to\infty$ such that if $0 < \lambda < \frac{1}{2\varepsilon(N)}$, then for all n,*

$$\frac{1}{n} \log E^\mu \left\{ \exp\left\{ \lambda \sum_{k=1}^n (X_k - X_k^N)^2 \right\}\right\} \leqq -\frac{1}{2}\log(1 - 2\lambda\varepsilon(N)). \tag{2.22}$$

Proof. First of all, let $\{Y_k\}$ be any stationary Gaussian process with mean 0, covariance $\gamma_j = \frac{1}{2\pi}\int_0^{2\pi} e^{ij\theta} g(\theta)d\theta$, and strictly positive definite covariance matrix

$$A_n = \begin{pmatrix} \gamma_0 & \gamma_1 & \cdots & \gamma_{n-1} \\ \gamma_1 & \gamma_0 & \cdots & \gamma_{n-2} \\ \cdots & \cdots & & \cdots \\ \gamma_{n-1} & \cdots & & \gamma_0 \end{pmatrix}.$$

Let $c = \sup_{0\leqq\theta\leqq 2\pi} |g(\theta)|$. We show now that if $\lambda < \frac{1}{2c}$, then

$$E\left\{\exp\left\{\lambda\sum_{k=1}^n Y_k^2\right\}\right\} \leqq (1-2\lambda c)^{-n/2}. \tag{2.23}$$

To see this, first observe that if c_n is the maximum eigenvalue of A_n,

$$\begin{aligned} E\left\{\exp\left\{\lambda\sum_{k=1}^n Y_k^2\right\}\right\} &= \frac{1}{\sqrt{|A_n|}}\frac{1}{(2\pi)^{n/2}} \int_{-\infty}^{\infty}\!\!\ldots\!\int \exp\left\{\lambda\sum_{k=1}^n y_k^2 - \frac{1}{2} Y' A_n^{-1} Y\right\} dy_1 \ldots dy_n \\ &= \frac{1}{\sqrt{|A_n|}}\frac{1}{\sqrt{|A_n^{-1} - 2\lambda I|}} = \frac{1}{\sqrt{|I - 2\lambda A_n|}} \leqq (1-2\lambda c_n)^{-n/2}, \end{aligned}$$

providing $\lambda < 1/2c_n$. To estimate c_n, consider

$$\sum_{j,k=1}^{n} \gamma_{k-j}\eta_k\bar{\eta}_j = \frac{1}{2\pi}\int_0^{2\pi}\left|\sum_{k=1}^{n}\eta_k e^{ik\theta}\right|^2 g(\theta)d\theta \leqq \sup_{0\leqq\theta\leqq 2\pi}|g(\theta)|\sum_{j=1}^{n}|\eta_j|^2,$$

which implies $c_n \leqq \sup_{0\leqq\theta\leqq 2\pi}|g(\theta)| = c$. Thus, $(1-2\lambda c_n)^{-n/2} \leqq (1-2\lambda c)^{-n/2}$ and $\lambda < 1/2c$ implies $\lambda < 1/2c_n$, so we have (2.23). Now, in particular let

$$y_k = X_k^N - X_k = \sum_{|j|<N} a_j\left(1-\frac{|j|}{N}\right)\xi_{j+k} - \sum_{j=-\infty}^{\infty} a_j\xi_{j+k}.$$

The spectral density $g(\theta)$ for the process $\{Y_k\}$ is now given by

$$g(\theta) = \left|\sum_{j=-\infty}^{\infty} a_j e^{ij\theta} - \sum_{|j|\leqq N} a_j\left(1-\frac{|j|}{N}\right)e^{ij\theta}\right|^2 = |\sqrt{f(\theta)} - h_N(\theta)|^2.$$

Let $\varepsilon(N) = \sup_{0\leqq\theta\leqq 2\pi}|g(\theta)| = \sup_{0\leqq\theta\leqq 2\pi}|\sqrt{f(\theta)} - h_N(\theta)|^2$. From (2.3) we see $\varepsilon(N) = o(1)$ as $N\to\infty$, and from (2.23) if $0<\lambda<1/2\varepsilon(N)$, then we have for all n,

$$\frac{1}{n}\log E\left\{\exp\left\{\lambda\sum_{k=1}^{n}(x_k - x_k^N)^2\right\}\right\} \leqq -\frac{1}{2}\log(1-2\lambda\varepsilon(N)),$$

which is the lemma.

Lemma 2.5. *Let $\omega\in\Omega$ be given by $\omega=\{x_k\}$, $-\infty<k<\infty$ and let $\omega'\in\Omega$ be given by $\omega'=\{x_k+\varepsilon_k\}$, $-\infty<k<\infty$. Then, there is a constant c such that*

$$d(\pi_n(\omega), \pi_n(\omega')) \leqq c\sqrt{\frac{\varepsilon_1^2+\varepsilon_2^2+\ldots+\varepsilon_n^2}{n}}. \tag{2.24}$$

Proof. By definition [cf. (2.12)],

$$d(\pi_n(\omega), \pi_n(\omega')) = \sum_{L=1}^{\infty}\frac{1}{2^L}d_L([\pi_n(\omega)]_L^1, [\pi_n(\omega')]_L^1). \tag{2.25}$$

Let α and β be two L-dimensional prob distributions and $\hat{\alpha}$ and $\hat{\beta}$ their characteristic functions. Since uniform convergence of characteristic functions on compact sets in L-dimensions is equivalent to weak convergence in L-dimensions which is equivalent to convergence in the Prohorov metric d_L in L-dimensions, we can write

$$d_L(\alpha,\beta) = \sum_{k=1}^{\infty}\frac{1}{2^k}\sup_{|\eta|\leqq k}|\hat{\alpha}(\eta) - \hat{\beta}(\eta)|. \tag{2.26}$$

In particular, if $\alpha = \delta_{(x_1,x_2,\ldots,x_L)}$ and $\beta = \delta_{(x_1+\varepsilon_1,\ldots,x_L+\varepsilon_L)}$, we have from (2.26),

$$\begin{aligned} d_L(\alpha,\beta) &= \sum_{k=1}^{\infty}\frac{1}{2^k}\sup_{|\eta|\leqq k}|e^{i\langle x,\eta\rangle} - e^{i\langle x+\varepsilon,\eta\rangle}| \\ &= \sum_{k=1}^{\infty}\frac{1}{2^k}\sup_{|\eta|\leqq k}|e^{i\langle\varepsilon,\eta\rangle} - 1| \\ &\leqq \|\varepsilon\|\sum_{k=1}^{\infty}\frac{k}{2^k} = 2\|\varepsilon\|. \end{aligned}$$

Thus, from the definition of π_n,

$$d_L([\pi_n(\omega)]_L^1, [\pi_n(\omega')]_L^1)$$

$$\leqq \frac{2}{n}\left(\sqrt{\varepsilon_1^2+\varepsilon_2^2+\ldots+\varepsilon_L^2}+\sqrt{\varepsilon_2^2+\ldots+\varepsilon_{1+L}^2}+\ldots+\sqrt{\varepsilon_n^2+\ldots+\varepsilon_{n+L-1}^2}\right),$$

where $\varepsilon_{n+j}=\varepsilon_j$. Using Minkowski's inequality on this last, we get

$$d_L([\pi_n(\omega)]_L^1, [\pi_n(\omega')]_L^1)$$

$$\leqq 2\sqrt{\frac{1}{n}[(\varepsilon_1^2+\ldots+\varepsilon_L^2)+(\varepsilon_2^2+\ldots+\varepsilon_{2+L}^2)+\ldots+(\varepsilon_n^2+\ldots+\varepsilon_{n+L}^2)]}$$

$$=2\sqrt{\frac{L}{n}(\varepsilon_1^2+\varepsilon_2^2+\ldots+\varepsilon_n^2)}\,.$$

Finally, from (2.25) and this last inequality,

$$d(\pi_n(\omega), \pi_n(\omega'))\leqq \sum_{L=1}^{\infty}\frac{1}{2^L}\cdot 2\sqrt{\frac{L(\varepsilon_1^2+\ldots+\varepsilon_n^2)}{n}}=c\sqrt{\frac{\varepsilon_1^2+\ldots+\varepsilon_n^2}{n}}\,.$$

Lemma 2.6. *Let F_N and F be defined by (2.5) and (2.6) respectively. For any $\delta>0$,*

$$\mu\{\omega\in\Omega: d(\pi_n\cdot F_N, \pi_n\cdot F)\geqq\delta\}\leqq\mu\left\{\omega\in\Omega: \sum_{k=1}^{n}\frac{(X_k^N-X_k)^2}{n}\geqq\frac{\delta^2}{c^2}\right\}, \quad (2.27)$$

where c is the constant in Lemma 2.5.

Proof. From (2.1) and (2.4), $X_k=\sum_{n=-\infty}^{\infty}a_n\xi_{n+k}$ and $x_k^N=\sum_{|j|<N}a_j\left(1-\frac{|j|}{N}\right)\xi_{j+k}$. If we let $\varepsilon_k=X_k-X_k^N$, then $X_k=X_k^N+\varepsilon_k$. Thus, from Lemma 2.5,

$$d(\pi_n\cdot F_N, \pi_n\cdot F)\leqq c\sqrt{\frac{1}{n}\sum_{k=1}^{n}(X_k-x_k^N)^2}\,,$$

and (2.27) follows.

Lemma 2.7. *For every $\delta>0$,*

$$\lim_{N\to\infty}\overline{\lim_{n\to\infty}}\frac{1}{n}\log\mu\{\omega\varepsilon\,\Omega: d(\pi_n\cdot F_N, \pi_n\cdot F)\geqq\delta\}=-\infty\,. \quad (2.28)$$

Proof. For fixed N, and $\lambda<1/2\varepsilon(N)$, where $\varepsilon(N)$ is defined above in Lemma 2.4, we have from (2.27) and (2.24) for any $\delta>0$,

$$\mu\{\omega\in\Omega: d(\pi_n\cdot F_N, \pi_n\cdot F)\geqq\delta\}\leqq\mu\left\{\omega\in\Omega:\frac{1}{n}\sum_{k=1}^{n}(X_k^N-X_k)^2\geqq\frac{\delta^2}{c^2}\right\}$$

$$\leqq E^{\mu}\left\{\exp n\lambda\left(\frac{1}{n}\sum_{k=1}^{n}(X_k^N-X_k)^2\right)\right\}e^{-n\delta^2\lambda/c^2}\leqq e^{-n\delta^2\lambda/c^2}e^{-\frac{n}{2}\log(1-2\lambda\varepsilon(N))}\,.$$

Therefore, for any $\delta>0$ and $\lambda<1/2\varepsilon(N)$ with N fixed,

$$\overline{\lim_{n\to\infty}}\frac{1}{n}\log\mu\{\omega\in\Omega: d(\pi_n\cdot F_N, \pi_n\cdot F)\geqq\delta\}\leqq-\frac{\delta^2\lambda}{c^2}-\frac{1}{2}\log(1-2\lambda\varepsilon(N)). \quad (2.29)$$

Since $\varepsilon(N)=o(1)$ as $N\to\infty$, we get from (2.29) that for all $\delta>0$ and all $\lambda>0$,

$$\overline{\lim_{N\to\infty}}\ \overline{\lim_{n\to\infty}}\ \frac{1}{n}\log\mu\{\omega\in\Omega: d(\pi_n\cdot F_N,\pi_n\cdot F)\geqq\delta\}\leqq-\frac{\delta^2\lambda}{c^2}.$$

In this last if we let $\lambda\to\infty$ we obtain (2.28).

We are now ready to prove (1.5) and (1.6) of the introduction.

Theorem 2.8. *If $C\subset \underline{M}_S$ is closed, then*

$$\overline{\lim_{n\to\infty}}\ \frac{1}{n}\log Q_n(C)\leqq-\inf_{R\in C}H_f(R), \tag{2.30}$$

and if $G\subset\underline{M}_S$ is open, then

$$\underline{\lim}_{n\to\infty}\ \frac{1}{n}\log Q_n(G)\geqq-\inf_{R\in G}H_f(R), \tag{2.31}$$

where $H_f(R)$ is defined by (1.9).

Proof. Let C be closed in $\underline{M}_S$. Then,

$$Q_n(C)=P\{\omega\in\Omega:\pi_n(\omega)\in C\}=\mu\{\omega\in\Omega:\pi_n\cdot F(\omega)\in C\}.$$

For any $\delta>0$, we have

$$\begin{aligned}Q_n(C)&\leqq\mu\{\omega\in\Omega:\pi_n\cdot F_n\in C^\delta\}+\mu\{\omega\in\Omega:d(\pi_n\cdot F_N,\pi_n\cdot F)\geqq\delta\}\\&=P^N\{\omega\in\Omega:\pi_n(\omega)\in C^\delta\}+\mu\{\omega\in\Omega:d(\pi_n\cdot F_N,\pi_n\cdot F)\geqq\delta\}\\&=Q_n^N(C^\delta)+\mu\{\omega\in\Omega:d(\pi_n\cdot F_N,\pi_n\cdot F)\geqq\delta\}\\&\leqq Q_n^N(\bar{C}^\delta)+\mu\{\omega\in\Omega:d(\pi_n\cdot F_N,\pi_n\cdot F)\geqq\delta\}.\end{aligned}$$

From Theorem 2.3, this last inequality implies

$$\overline{\lim_{n\to\infty}}\ \frac{1}{n}\log Q_n(C)\leqq\max\Big(-\inf_{R\in\bar{C}^\delta}H_{f_N}(R),A_N(\delta)\Big),$$

where $A_N(\delta)=\overline{\lim_{n\to\infty}}\ \frac{1}{n}\log\mu\{\omega\in\Omega:d(\pi_n\cdot F_N,\pi_n\cdot F)\geqq\delta\}$. From Lemma 2.7 we then obtain,

$$\overline{\lim_{n\to\infty}}\ \frac{1}{n}\log Q_n(C)\leqq-\underline{\lim}_{N\to\infty}\inf_{R\in\bar{C}^\delta}H_{f_N}(R). \tag{2.32}$$

From Lemma 3.10 below it follows that

$$\underline{\lim}_{N\to\infty}\inf_{R\in\bar{C}^\delta}H_{f_N}(R)=\inf_{R\in\bar{C}^\delta}H_f(R), \tag{2.33}$$

where $H_f(R)$ is given by (1.9). Using (2.33) in (2.32) we get

$$\overline{\lim_{n\to\infty}}\ \frac{1}{n}\log Q_n(C)\leqq-\inf_{R\in\bar{C}^\delta}H_f(R). \tag{2.34}$$

Since C is closed and $H_f(R)$ is lower semicontinuous (2.34) implies (2.30) on letting $\delta\to 0$.

For the lower bound, let G be open in $\underline{M}_S$ and let R be any element of G. Pick $\delta > 0$ and an open set V such that $R \in V \subset V^\delta \subset G$. Then

$$\begin{aligned} Q_n(G) &\geqq Q_n(V^\delta) = \mu\{\omega \in \Omega : \pi_n \cdot F(\omega) \in V^\delta\} \\ &\geqq \mu\{\omega \in \Omega : \pi_n \cdot F_N \in V\} - \mu\{\omega \in \Omega : d(\pi_n \cdot F_N, \pi_n \cdot F) \geqq \delta\} \\ &= Q_n^N(V) - \mu\{\omega \in \Omega : d(\pi_n \cdot F_N, \pi_n \cdot F) \geqq \delta\}. \end{aligned} \tag{2.35}$$

Without loss of generality we can assume that $H_f(R) < \infty$ for the $R \in G$ we picked.

As before, let

$$A_N(\delta) = \overline{\lim_{n\to\infty}} \frac{1}{n} \log \mu\{\omega \in \Omega : d(\pi_n \cdot F_N, \pi_n \cdot F) \geqq \delta\}.$$

From Lemma 2.7 we can choose N so large that $A_N(\delta) \leqq -(H_f(R)+3)$, i.e., we can choose N so large that

$$\mu\{\omega \in \Omega : d(\pi_n \cdot F_N, \pi_n \cdot F) \geqq \delta\} \leqq e^{-n(H_f(R)+3)+o(n)}. \tag{2.36}$$

Also, since V is open we obtain from Theorem 2.3,

$$\underline{\lim_{n\to\infty}} \frac{1}{n} \log Q_n^N(V) \geqq -\inf_{R' \in V} H_{f_N}(R'). \tag{2.37}$$

In Lemma 3.9 below we prove that for an $R \in \underline{M}_S$ such that $H_f(R) < \infty$, there exists a sequence $\{R_N\}$ in $\underline{M}_S$ which converges weakly to R and for which $\lim_{N\to\infty} H_{f_N}(R_N) = H_f(R)$. If we let $0 < \eta < 1$ be given, then from (2.37) and Lemma 3, since V is a neighborhood of R, we can take N so large that

$$\underline{\lim_{n\to\infty}} \frac{1}{n} \log Q_n^N(V) \geqq -\inf_{R' \in V} H_{f_N}(R') \geqq -H_{f_N}(R_N) \geqq -H_f(R) - \eta.$$

In other words, for N large enough,

$$Q_n^N(V) \geqq e^{-n(H_f(R)+\eta)+o(n)}. \tag{2.38}$$

Using (2.38) and (2.36) in (2.35) we see that for N large enough

$$Q_n(G) \geqq e^{-n(H_f(R)+\eta)+o(n)} - e^{-n(H_f(R)+3)+o(n)}.$$

Since $\eta < 1$ the first expression on the right is the controlling asymptotic term and we get,

$$\underline{\lim_{n\to\infty}} \frac{1}{n} \log Q_n(G) \geqq -H_f(R) - \eta. \tag{2.39}$$

Since η is arbitrary and R is an arbitrary element of G, we conclude finally that

$$\underline{\lim_{n\to\infty}} \frac{1}{n} \log Q_n(G) \geqq -\inf_{R \in G} H_f(R),$$

which completes the proof of the theorem.

3. Properties of Entropy

In this section we derive the formula for $H_{f_N}(R)$ occurring in Theorem 2.2. We also prove those properties of $H_{f_N}(R)$ and $H_f(R)$ which we called upon in obtaining the upper and lower bounds in Theorem 2.8.

We will be somewhat more general at first and start with $I(R)$ as defined by (2.16). Note, that for $R \in \underline{M}_S$, and with $\phi(y) = \frac{1}{\sqrt{2\pi}} e^{-y^2/2}$,

$$\begin{aligned} I(R) &= E^R\left\{\int_{-\infty}^{\infty} r(y|\omega)\log\left(\frac{r(y|\omega)}{\phi(y)}\right)dy\right\} \\ &= E^R\left\{\int_{-\infty}^{\infty} r(y|\omega)\log r(y|\omega)dy\right\} + \frac{1}{2}\log 2\pi + \frac{1}{2}E^R\{\omega^2(0)\}. \end{aligned} \tag{3.1}$$

Let $\{a_k\}$, $-\infty < k < \infty$ be any sequence of reals such that $\sum_{j=1}^{\infty} |a_j| < \infty$ and $|h(\theta)| = \left|\sum_{j=-\infty}^{\infty} a_j e^{ij\theta}\right| \neq 0$ for all $\theta \in [0, 2\pi]$. Consider the mapping $\tau: \Omega \to \Omega$ based on such a sequence $\{a_k\}$ given by $(\tau\omega)(k) = \sum_{j=-\infty}^{\infty} a_j \omega(j+k)$. We will call such mappings τ "admissible."

For τ such an admissible mapping, define for $R \in \underline{M}_S$

$$H^\tau(R) = \inf_{R': R'\tau^{-1} = R} I(R'). \tag{3.2}$$

We prove below that if R has spectral measure $G(\theta)$, then

$$H^\tau(R) = I(R) + \frac{1}{4\pi}\int_0^{2\pi} \frac{dG(\theta)}{|h(\theta)|^2} - \frac{1}{4\pi}\int_0^{2\pi} dG(\theta) + \frac{1}{2\pi}\int_0^{2\pi} \log|h(\theta)d\theta. \tag{3.3}$$

Before proving (3.3) we need to examine how $I(R\tau^{-1})$ is related to $I(R)$ for an admissible map τ. We start with some preliminary lemmas. The first is easy and we omit the proof. The second is a classic result found in standard references on matrices.

Lemma 3.1. *Let $R \in \underline{M}_S$ and let $R_n(y_1, y_2, \ldots, y_n)$ be the n-dimensional joint density for the R process, i.e., $R_n(y_1, y_2, \ldots, y_n)dy_1 \ldots dy_n = R\{X_1 \in dy_1, \ldots, X_n \in dy_n\}$. Define, with $\phi(y) = \frac{1}{\sqrt{2\pi}} e^{-y^2/2}$,*

$$I_n(R) = \int_{-\infty}^{\infty}\cdots\int R_n(y_1, y_2, \ldots, y_n)\log\left(\frac{R_n(y_1, y_2, \ldots, y_n)}{\prod_{j=1}^{n}\phi(y_j)}\right)dy_1 \ldots dy_n.$$

Then, $I_n(R)-I_{n-1}(R)$ is monotone increasing in n, $\lim_{n\to\infty}(I_n(R)-I_{n-1}(R))=I(R)$ so that also

$$\lim_{n\to\infty}\frac{I_n(R)}{n}=I(R).$$

Lemma 3.2. *Let D be the determinant of a circulant matrix of order n, i.e.,*

$$D=\begin{vmatrix} d_1 & d_2 & \dots & d_n & \\ d_n & d_1 & d_2 \dots & d_{n-1} & \\ \vdots & & & & \\ d_2 & d_3 & \dots & d_n & d_1 \end{vmatrix}.$$

Then

$$D=\prod_{\xi\in Z_n}(d_1+d_2\xi+\dots+d_n\xi^{n-1}), \tag{3.4}$$

where Z_n is the set of nth roots of unity.

Lemma 3.3. *Let $f(x)$ be a probability density in d-dimensions and let $g(x)$ be the d-dimensional Gaussian density, $g(x)=\frac{1}{(2\pi)^{d/2}}e^{-(1/2)\|x\|^2}$. Let A be a nonsingular matrix. Then,*

$$\int_{R^d}\left(\frac{f(A^{-1}x)}{|A|}\right)\log\left(\frac{f(A^{-1}x)/|A|}{g(x)}\right)dx=\int_{R^d}f(y)\log\frac{f(y)}{g(y)}dy$$
$$+\frac{1}{2}\int_{R^d}(\|Ay\|^2-\|y\|^2)f(y)dy-\log|A|. \tag{3.5}$$

Proof. We observe that if $\alpha(dx)=f(x)dx$, then $(\alpha A^{-1})(dx)=\frac{f(A^{-1}x)}{|A|}dx$. Now, letting $x=Ay$,

$$\int_{R^d}\left(\frac{f(A^{-1}x)}{|A|}\right)\log\left(\frac{f(A^{-1}x)/|A|}{g(x)}\right)dx=\int_{R^d}f(y)\log\left(\frac{f(y)}{g(Ay)}\right)dx-\log|A|$$
$$=\int_{R^d}f(y)\log\left(\frac{f(y)}{g(y)}\right)dy+\int_{R^d}f(y)\log\left(\frac{g(y)}{g(Ay)}\right)dy-\log|A|$$
$$=\int_{R^d}f(y)\log\frac{f(y)}{g(y)}dy+\frac{1}{2}\int_{R^d}(\|Ay\|^2-\|y\|^2)f(y)dy-\log|A|.$$

As a preparatory lemma we consider admissible maps of a special kind – where only a finite number of the elements in $\{a_k\}$ are different from 0.

Lemma 3.4. *For any $R\in \underline{M}_S$, let $I(R)$ be defined by (3.1) and let $\tau:\Omega\to\Omega$ be defined by $(\tau\omega)(j)=\sum_{k=-\infty}^{\infty}a_k\omega(k+j)$, where, for some K, $a_k=0$ if $|k|>K$ and $|h(\theta)|=\left\|\sum_{k=-\infty}^{\infty}a_ke^{ik\theta}\right\|\neq 0$ for all $\theta\in[0,2\pi]$. Then,*

$$I(R\tau^{-1})\leq I(R)+\frac{1}{2}[E^{R\tau^{-1}}\{\omega^2(0)\}-E^R\{\omega^2(0)\}-\frac{1}{2\pi}\int_0^{2\pi}\log|h(\theta)|d\theta. \tag{3.6}$$

Proof. Let $n>K$ where K is such that $a_k=0$ if $|k|>K$. Consider the $(2n+1)\times(2n+1)$ circulant matrix

$$A_{n+1}=\begin{vmatrix} a_0 & a_1\ldots a_n & a_{-n}\ldots a_{-1} \\ a_{-1} & a_0\ldots a_{n-1} & \ldots a_{-2} \\ a_1 & a_2 \quad a_{-n} & a_0 \end{vmatrix}.$$

Let E_{2n+1} be $(2n+1)$ dimensional space and define a mapping $E_{2n+1}\to E_{2n+1}$ by $x\to A_{2n+1}x$. If $\omega\in\Omega$, let $\Gamma_{2n+1}\omega=(\omega(j))_{|j|\leqq n}$, i.e., if $\omega=\{x_k\}$, $-\infty<k<\infty$, then Γ_{2n+1} is the section:

$$x_{-n}, x_{-n+1}, \ldots, x_{-1}, x_0, x_1, \ldots, x_{n-1}, x_n.$$

It is easy to see that if $|i|\leqq n-K$, then

$$(\tau\omega)(i)=(A_{2n+1}\Gamma_{2n+1}\omega)(i). \tag{3.7}$$

Define the density $R'_{2n+1}(y_{-n}, \ldots, y_0, \ldots, y_n)$ by

$$R'_{2n+1}(y_{-n}, \ldots, y_0, \ldots, y_n)dy_{-n}\ldots dy_0\ldots dy_n=R\tau^{-1}\Gamma^{-1}_{2n+1},$$

and similarly define the density $\hat{R}'_{2n+1}$ by

$$\hat{R}'_{2n+1}(y_{-n}, \ldots, y_0, \ldots, y_n)dy_{-n}, \ldots, dy_0\ldots dy_n=R\Gamma^{-1}_{2n+1}A^{-1}_{2n+1}.$$

From the definition of $I_n(R)$ in Lemma 3.1,

$$I_{2k+1}(R\tau^{-1})=\int_{-\infty}^{\infty}\!\!\cdots\!\int R'_{2k+1}(y_{-k}, \ldots, y_0, \ldots, y_k) \cdot\log\left(\frac{R'_{2k+1}(y_{-k}, \ldots, y_0, \ldots, y_k)}{\left(\frac{1}{\sqrt{2\pi}}\right)^{2k+1}\exp\left\{-1\frac{1}{2}\sum_{j=-k}^{k} y_j^2\right\}}\right)dy_{-k}dy_0\ldots dy_k. \tag{3.8}$$

From (3.7) we see that $R\tau^{-1}\Gamma^{-1}_{2n+1}=R\Gamma^{-1}_{2l+1}A^{-1}_{2l+1}\Gamma^{-1}_{2n+1}$ provided $l\geqq n+K$. In particular, if we choose $n=k$, $l=k+K$, we have

$$R\tau^{-1}\Gamma^{-1}_{2k+1}=R\Gamma^{-1}_{2(k+K)+1}A^{-1}_{2(k+K)+1}\Gamma^{-1}_{2k+1}.$$

Therefore, from (3.8) we obtain

$$I_{2k+1}(R\tau^{-1})\leqq\int_{-\infty}^{\infty}\!\!\cdots\!\int \hat{R}'_{2(k+K)+1}(y_{-k-k}, \ldots, y_0, \ldots, y_{k+K}) \cdot\log\frac{\hat{R}'_{2(k+K)+1}(y_{-K+k}, \ldots, y_0, \ldots, y_{k+K})}{\left(\frac{1}{\sqrt{2\pi}}\right)^{2(k+K)+1}\exp\left\{-\frac{1}{2}\sum_{j=-k-K}^{k+K} y_j^2\right\}}\,dy_{-k-K}\ldots dy_{k+K}. \tag{3.9}$$

From Lemma 3.3 and (3.9) we get the estimate

$$I_{2k+1}(R\tau^{-1})\leqq I_{2(k+K)+1}(R)-\log|A_{2(k+K)+1}| +\frac{1}{2}E^R\{\|\mathrm{A}_{2(k+K)+1}\Gamma_{2(k+K)+1}\omega\|^2-\|\Gamma_{2(k+K)+1}\omega\|^2\}. \tag{3.10}$$

Since R is stationary and $a_k=0$ if $|k|>K$,

$$\lim_{n\to\infty}\frac{1}{2n+1}E^R\{\|A_{2n+1}\Gamma_{2n+1}\omega\|^2\}=E^R\left\{\left|\sum_{j=-\infty}^{\infty}a_j\omega(j)\right|^2\right\}=E^{R\tau^{-1}}\{\omega^2(0)\},$$

and

$$\lim_{n\to\infty}\frac{1}{2n+1}E^R\{\|\Gamma_{2n+1}\omega\|^2\}=E^R\{\omega^2(0)\}.$$

Let $c=\lim_{n\to\infty}\frac{1}{2n+1}\log|A_{2n+1}|$, divide both sides of (3.10) by $2k+1$, let $k\to\infty$, and use Lemma 3.1 to obtain

$$I(R\tau^{-1})\leqq I(R)-c+\frac{1}{2}[E^{R\tau^{-1}}\{\omega^2(0)\}-E^R\{\omega^2(0)\}]. \tag{3.11}$$

Thus, it remains to show that

$$c=\frac{1}{2\pi}\int_0^{2\pi}\log|h(\theta)|d\theta,$$

where $h(\theta)=\sum_{k=-\infty}^{\infty}a_ke^{ik\theta}$. Since A_{2n+1} is a circulant matrix, Lemma 3.2 applies. We note that if $\xi\in Z_{2n+1}$ then $\xi^{-j}=\xi^{2n+1-j}$. Since $a_k=0$ if $|k|>K$, we get

$$\begin{aligned}\lim_{n\to\infty}\frac{1}{2n+1}\log|A_{2n+1}|&=\lim_{n\to\infty}\frac{1}{2n+1}\log\prod_{\xi\in Z_{2n+1}}\left(\sum_{|j|\leqq n}a_j\xi^j\right)\\&=\lim_{n\to\infty}\frac{1}{2n+1}\sum_{\xi\in Z_{2n+1}}\log\sum_{j=-\infty}^{\infty}a_j\xi^j=\frac{1}{2\pi}\int_0^{2\pi}\log\left|\sum_{j=-\infty}^{\infty}a_je^{ij\theta}\right|d\theta,\end{aligned}$$

where, of course, in the last equality we used the properties of roots of unity and that the last expression is finite since $|h(\theta)|\neq 0$ for all $\theta\in[0,2\pi]$ by hypothesis.

In the preceding lemma we showed inequality (3.6) under the hypothesis that the admissible mapping τ was in fact based on coefficients $\{a_k\}$ only a finite number of which were different from zero. Using that result we can now show that for any admissible mapping equality actually holds in (3.6).

Lemma 3.5. *Let τ be any admissible mapping. Then, for any $R\in \underline{M}_S$,*

$$I(R\tau^{-1})=I(R)+\frac{1}{2}[E^{R\tau^{-1}}\{\omega^2(0)\}-E^R\{\omega^2(0)\}]-\frac{1}{2\pi}\int_0^{2\pi}\log\left|\sum_{k=-\infty}^{\infty}a_ke^{ik\theta}\right|d\theta. \tag{3.12}$$

Proof. First define the map τ_n by $(\tau_n\omega)(j)=\sum_{k=-n}^{n}a_k\omega(k+j)$ with the same a_k's as in the statement of the lemma. We apply Lemma 3.4 to τ_n so that for any $R\in\underline{M}_S$,

$$\begin{aligned}I(R\tau_n^{-1})\leqq I(R)&+\frac{1}{2}[E^{R\tau_n^{-1}}\{\omega^2(0)\}-E^R\{\omega^2(0)\}]\\&-\frac{1}{2\pi}\int_0^{2\pi}\log\left|\sum_{k=-n}^{n}a_ke^{ik\theta}\right|d\theta.\end{aligned} \tag{3.13}$$

Since $I(R)$ is lower semicontinuous on $\underline{M}_S$, $I(R\tau^{-1}) \leq \varliminf_{n\to\infty} I(R\tau_n^{-1})$. Moreover, $\lim_{n\to\infty} E^{R\tau_n^{-1}}\{\omega^2(0)\} = E^{R\tau^{-1}}\{\omega^2(0)\}$, and

$$\lim_{n\to\infty} \frac{1}{2\pi}\int_0^{2\pi} \log\left|\sum_{k=-n}^{n} a_k e^{ik\theta}\right| d\theta = \frac{1}{2\pi}\int_0^{2\pi} \log\left|\sum_{k=-\infty}^{\infty} a_k e^{ik\theta}\right| d\theta,$$

since $\left|\sum_{k=-\infty}^{\infty} a_k e^{ik\theta}\right| \neq 0$ for all $\theta \in [0, 2\pi]$. Thus, from (3.13) we conclude, for $R \in \underline{M}_S$,

$$I(R\tau^{-1}) \leq I(R) + \frac{1}{2}[E^{R\tau^{-1}}\{\omega^2(0)\} - E^R\{\omega^2(0)\}] - \frac{1}{2\pi}\int_0^{2\pi} \log\left|\sum_{k=-\infty}^{\infty} a_k e^{ik\theta}\right| d\theta \tag{3.14}$$

for any admissible mapping τ. Let $g(\theta) = \sum_{j=-\infty}^{\infty} a_j e^{ij\theta}$. By our hypothesis this is an absolutely convergent Fourier series and $g(\theta) \neq 0$ for any $\theta \in [0, 2\pi]$. Wiener's theorem says that the L^2-function $1/g(\theta)$ has a Fourier series $\frac{1}{g(\theta)} = \sum_{j=-\infty}^{\infty} b_j e^{ij\theta}$ and $\sum_{j=-\infty}^{\infty} |b_j| < \infty$. Of course, $\left|\sum_{j=-\infty}^{\infty} b_j e^{ij\theta}\right| \neq 0$ for all $\theta \in [0, 2\pi]$.

Consider then the admissible map σ given by $(\sigma\omega)(j) = \sum_{k=-\infty}^{\infty} b_k \omega(k+j)$ and note that $\sigma = \tau^{-1}$. Apply (3.14) to σ and we get

$$I(R\sigma^{-1}) \leq I(R) + \frac{1}{2}[E^{R\sigma^{-1}}\{\omega^2(0)\} - E^R\{\omega^2(0)\}] + \frac{1}{2\pi}\int_0^{2\pi} \log\left|\sum_{k=-\infty}^{\infty} a_k e^{ik\theta}\right| d\theta, \tag{3.15}$$

where the sign in the last term on the right is now positive since $\left|\sum_{k=-\infty}^{\infty} a_k e^{ik\theta}\right| = \frac{1}{\left|\sum_{k=-\infty}^{\infty} b_k e^{ik\theta}\right|}$. But (3.15) is true for all $R \in \underline{M}_S$ and, therefore, true in particular if we replace R in (3.15) by $R\tau^{-1}$ yielding, since $\sigma = \tau^{-1}$,

$$I(R\tau^{-1}\sigma^{-1}) = I(R) \leq I(R\tau^{-1}) + \frac{1}{2}[E^R\{\omega^2(0)\} - E^{R\tau^{-1}}\{\omega^2(0)\}] + \frac{1}{2\pi}\int_0^{2\pi} \log\left|\sum_{k=-\infty}^{\infty} a_k e^{ik\theta}\right| d\theta,$$

or

$$I(R\tau^{-1}) \geq I(R) + \frac{1}{2}[E^{R\tau^{-1}}\{\omega^2(0)\} - E^R\{\omega^2(0)\}] - \frac{1}{2\pi}\int_0^{2\pi} \log\left|\sum_{k=-\infty}^{\infty} a_k e^{ik\theta}\right| d\theta. \tag{3.16}$$

Inequalities (3.14) and (3.16) give us (3.12) for any admissible map τ which is the lemma.

From the preceding lemma we see that for any admissible mapping τ, $I(R)<\infty$ if and only if $I(R\tau^{-1})<\infty$. Since an admissible mapping τ is $1-1$ we observe that for any τ in this class, and with $\sigma=\tau^{-1}$,

$$H^{\tau}(R)=\inf_{R':R'\tau^{-1}=R} I(R')=I(R\sigma^{-1})$$

$$=I(R)+\frac{1}{2}[E^{R\sigma^{-1}}\{\omega^2(0)\}-E^{R}\{\omega^2(0)\}]+\frac{1}{2\pi}\int_0^{2\pi}\log|h(\theta)|d\theta\,, \quad (3.17)$$

where $h(\theta)=\sum_{j=-\infty}^{\infty} a_j e^{ij\theta}$, and τ is given by $(\tau\omega)(j)=\sum_{k=-\infty}^{\infty} a_k\omega(k+j)$. Note, that in (3.17) the sign before the last term on the right is + [cf. (3.12)] because $\sigma=\tau^{-1}$ and the sequence $\{a_k\}$ belongs to τ.

Moreover if $R\in \underline{M}_S$ has spectral measure $G(\theta)$, then $\frac{1}{2\pi}\int_0^{2\pi} dG(\theta)=E^R\{\omega^2(0)\}$ and also since $\sigma=\tau^{-1}$ we have $E^{R\sigma^{-1}}\{\omega^2(0)\}=\frac{1}{2\pi}\int_0^{2\pi}\frac{dG(\theta)}{|h(\theta)|^2}$. Hence, we can rewrite (3.17) in the form

$$H^{\tau}(R)=I(R)+\frac{1}{4\pi}\int_0^{2\pi}\frac{dG(\theta)}{|h(\theta)|^2}-\frac{1}{4\pi}\int_0^{2\pi} dG(\theta)+\frac{1}{2\pi}\int_0^{2\pi}\log|h(\theta)|d\theta\,, \quad (3.18)$$

which is formula (3.3) at the beginning of this section.

In formula (3.18) which holds for any admissible τ, replace $I(R)$ by the last line in (3.1), using again that $E^R\{w^2(0)\}=\frac{1}{2\pi}\int_0^{2\pi} dG(\theta)$, so

$$H^{\tau}(R)=E^R\left\{\int_{-\infty}^{\infty} r(y|\omega)\log r(y|\omega)dy\right\}+\frac{1}{2}\log 2\pi+\frac{1}{4\pi}\int_0^{2\pi}\frac{dG(\theta)}{|h(\theta)|^2}+\frac{1}{2\pi}\int_0^{2\pi}\log|h(\theta)|d\theta\,. \quad (3.19)$$

Now, in particular, consider the mapping $F_N:\Omega\to\Omega$ given by (2.5) or more precisely, its induced map $\tilde{F}_N$ in $\underline{M}_S$. This is an admissible map since

$$(F_N\omega)(j)=\sum_{|k|<N} a_k\left(1-\frac{|k|}{N}\right)\omega(k+j)\,,$$

and therefore at most a finite number of the coefficients defining this map are different from 0. Moreover, $h_N(\theta)=\sum_{|k|<N} a_k\left(1-\frac{|k|}{N}\right)e^{ik\theta}$, defined in (2.2), never vanishes since by Fejer's theorem,

$$h_N(\theta)=\int_0^{2\pi}\sqrt{f(\phi)}\,\psi_N(\theta-\phi)d\phi\,,$$

where

$$\psi_N(\theta) = \sum_{|k|<N} \left(1 - \frac{|k|}{N}\right) e^{ik\theta},$$

is strictly positive. If as before we let $f_N(\theta) = |h_N(\theta)|^2$, then $f_N(\theta)$ is the spectral density of the process $\{X_k^N\}$, and so with $\tau = F_N$ and $H_{f_N}(R) = H^\tau(R)$, we get from (3.19)

$$H_{f_N}(R) = E^R\left\{\int_{-\infty}^{\infty} r(y|\omega)\log r(y|\omega)dy\right\} + \frac{1}{2}\log 2\pi + \frac{1}{4\pi}\int_0^{2\pi} \frac{dG(\theta)}{f_N(\theta)} + \frac{1}{4\pi}\int_0^{2\pi} \log f_N(\theta)d\theta. \quad (3.20)$$

Formula (3.20) is the same as in Sect. 2 just after (2.17).

Finally, we must see in what sense $H_{f_N}(R)$ approximates $H_f(R)$ as defined by formula (1.9). We begin with the following well known lemma.

Lemma 3.6. *Let $\phi(x)$ be a probability density on $\mathbb{R}$ with two moments and let $\sigma^2(\phi)$ be the variance of ϕ. Then, for any $\alpha > 0$,*

$$\int_{-\infty}^{\infty} \phi(y)\log\phi(y)dy \geqq -\frac{1}{2}\log 2\pi\alpha - \frac{1}{2\alpha}\sigma^2(\phi). \quad (3.21)$$

Proof. Let $\phi_{\sigma,\mu} = \frac{1}{\sigma\sqrt{2\pi}}\exp\left\{-\frac{(y-\mu)^2}{2\sigma^2}\right\}$. We note

$$\int_{-\infty}^{\infty} \phi(y)\log\phi(y)dy = \int_{-\infty}^{\infty} \phi(y)\log\left(\frac{\phi(y)}{\phi_{\sigma,\mu}(y)}\right)dy + \int_{-\infty}^{\infty} \phi(y)\log\phi_{\sigma,\mu}(y)dy.$$

The first term on the right is nonnegative being the entropy of one probability density with respect to another. Now, $\log\phi_{\sigma,\mu}(y) = -\log(\sigma\sqrt{2\pi}) - \frac{1}{2\sigma^2}(y-\mu)^2$, and therefore

$$\int_{-\infty}^{\infty} \phi(y)\log\phi(y)dy \geqq -\frac{1}{2}\log(2\pi\sigma^2) - \frac{1}{2\sigma^2}\int_{-\infty}^{\infty} (y-\mu)^2\phi(y)dy.$$

If, in particular, we choose $\mu = \int_{-\infty}^{\infty} y\phi(y)dy$ and $\sigma^2 = \alpha$ we obtain (3.21).

Lemma 3.7. *Let τ be an admissible map and for $R \in \underline{M}_S$ let*

$$e(R) = \int_{-\infty}^{\infty} r(y|\omega)\log r(y|\omega)dy.$$

If $R' = R\tau^{-1}$, then

$$e(R') = e(R) - \frac{1}{2\pi}\int_0^{2\pi} \log|h(\theta)|d\theta, \quad (3.22)$$

where $h(\theta)=\sum_{j=-\infty}^{\infty} a_j e^{ij\theta}$ *and* $\{a_j\}$, $-\infty<j<\infty$ *are the coefficients defining the map* τ.

Proof. From (3.12) and (3.1)

$$I(R')=I(R)+\frac{1}{2}[E^{R'}\{\omega^2(0)\}-E^R\{\omega^2(0)\}]-\frac{1}{2\pi}\int_0^{2\pi}\log|h(\theta)|d\theta$$
$$=e(R)+\frac{1}{2}\log 2\pi+\frac{1}{2}E^{R'}\{\omega^2(0)\}-\frac{1}{2\pi}\int_0^{2\pi}\log|h(\theta)|d\theta.$$

Using (3.1) on the left of this last equation we see that

$$e(R')+\frac{1}{2}\log 2\pi+\frac{1}{2}E^{R'}\{\omega^2(0)\}$$
$$=e(R)+\frac{1}{2}\log 2\pi+\frac{1}{2}E^{R'}\{\omega^2(0)\}-\frac{1}{2\pi}\int_0^{2\pi}\log|h(\theta)|d\theta,$$

from which (3.22) follows.

Lemma 3.8. *With* $f_N(\theta)$ *and* $f(\theta)$ *as in Sect. 2,*

$$\lim_{N\to\infty}\int_0^{2\pi}\log f_N(\theta)d\theta=\int_0^{2\pi}\log f(\theta)d\theta. \tag{3.23}$$

Proof. It suffices to show, since $f_N(\theta)=h_N^2(\theta)$, that

$$\lim_{N\to\infty}\int_0^{2\pi}\log h_N(\theta)d\theta=\int_0^{2\pi}\log\sqrt{f(\theta)}d\theta.$$

Now, $h_N(\theta)=\sqrt{f}*\psi_N$, where ψ_N is Fejer's kernel. Therefore, by Jensen's inequality,

$$\log h_N(\theta)=\log(\sqrt{f}*\psi_N)\geqq\log\sqrt{f}*\psi_N,$$

so that

$$\int_0^{2\pi}\log h_N(\theta)d\theta\geqq\int_0^{2\pi}(\log\sqrt{f}*\psi_N)d\theta=\int_0^{2\pi}\log\sqrt{f(\theta)}d\theta,$$

and hence

$$\lim_{N\to\infty}\int_0^{2\pi}\log h_N(\theta)d\theta\geqq\int_0^{2\pi}\log\sqrt{f(\theta)}d\theta.$$

On the other hand, from (2.3) we have that $\{h_N(\theta)\}$ are uniformly bounded and $\log h_N(\theta)\to\log\sqrt{f}$, so that from Fatou's lemma

$$\overline{\lim_{N\to\infty}}\int_0^{2\pi}\log h_N(\theta)d\theta\leqq\int_0^{2\pi}\log\sqrt{f(\theta)}d\theta.$$

Lemma 3.9. *Let $R \in \underline{M}_S$ such that $H_f(R) < \infty$; then there exists a sequence $\{R_N\} \in \underline{M}_S$ such that $\{R_N\}$ converges weakly to R and*

$$\lim_{N\to\infty} H_{f_N}(R_N) = H_f(R). \tag{3.24}$$

Proof. From its definition (1.9),

$$H_f(R) = e(R) + \frac{1}{2}\log 2\pi + \frac{1}{4\pi}\int_0^{2\pi}\frac{dG(\theta)}{f(\theta)} + \frac{1}{4\pi}\int_0^{2\pi}\log f(\theta)d\theta, \tag{3.25}$$

where $G(\theta)$ is the spectral measure of R and $e(R)$ is as defined in Lemma 3.7. From (3.20), for any $R' \in \underline{M}_S$,

$$H_{f_N}(R') = e(R') + \frac{1}{2}\log 2\pi + \frac{1}{4\pi}\int_0^{2\pi}\frac{dG'(\theta)}{f_N(\theta)} + \frac{1}{4\pi}\int_0^{2\pi}\log f_N(\theta)d\theta, \tag{3.26}$$

where $G'(\theta)$ is the spectral measure of R'. The way we construct the desired sequence $\{R_N\}$ is to define $R_N = R\tau_N^{-1}$, where $\{\tau_N\}$ is a suitably chosen sequence of admissible mappings which first of all will approach the identity map as $N\to\infty$, so that R_N will converge weaky to R. Secondly, $\{\tau_N\}$ will have to be selected so that (3.24) holds. Choosing τ_N means choosing a suitable sequence of coefficients. Let $g_N(\theta) = \sum_{j=-\infty}^{\infty} c_k^{(N)} e^{ik\theta}$, where $\{c_k^{(N)}\}$ are the defining coefficients of the admissible map τ_N. Taking into account Lemma 3.7 and 3.8, we see that for τ_N to satisfy the necessary requirements, $g_N(\theta)$ must satisfy:

(1) $$\lim_{N\to\infty}\int_0^{2\pi}\log|g_N(\theta)|d\theta = 0,$$

(2) $$\lim_{N\to\infty}\int_0^{2\pi}\frac{|g_N(\theta)|^2}{f_N(\theta)}dG(\theta) = \int_0^{2\pi}\frac{dG(\theta)}{f(\theta)},$$

(3) $$\lim_{N\to\infty}\int_0^{2\pi}|g_N(\theta)-1|^2 dG(\theta) = 0.$$

Condition (1), because of (3.22), guarantees $e(R_N)\to e(R)$. Since the spectral measure of R_N, call it $dG_N(\theta) = |g_N(\theta)|^2 dG(\theta)$, we see that (2), (1) and Lemma 3.8 insure that $H_{f_N}(R_N)\to H_f(R)$. Condition (3) insures that τ_N approaches the identity because it is the same as requiring $E^R\{|\omega(0)-(\tau_N\omega)(0)|^2\}\to 0$ as $N\to\infty$.

Let $g_{N,\varepsilon}(\theta) = \dfrac{\sqrt{f_N(\theta)}}{\sqrt{f_N(\theta)+\varepsilon}}$. We will show

(1)′ $$\lim_{\varepsilon\to 0}\lim_{N\to\infty}\left|\int_0^{2\pi}\log|g_{N,\varepsilon}(\theta)|d\theta\right| = 0,$$

(2)′ $$\lim_{\varepsilon\to 0}\lim_{N\to\infty}\left|\int_0^{2\pi}\frac{|g_{N,\varepsilon}(\theta)|^2 dG(\theta)}{f_N(\theta)} - \int_0^{2\pi}\frac{dG(\theta)}{f(\theta)}\right| = 0,$$

and

(3)′ $$\lim_{\varepsilon\to 0}\lim_{N\to\infty}\int_0^{2\pi}|g_{N,\varepsilon}(\theta)-1|^2 dG(\theta) = 0.$$

This will mean there exists a sequence $\varepsilon_N \to 0$ as $N \to \infty$ such that if we define $g_N(\theta) = g_{N,\varepsilon_N}(\theta)$, then (1)′, (2)′, and (3)′ imply (1), (2), (3) and hence the lemma. Because of (2.3), $g_{N,\varepsilon}(\theta) \to \sqrt{f(\theta)}/(\sqrt{f(\theta)}+\varepsilon)$ uniformly for $\theta \in [0, 2\pi]$. Also $\frac{|g_{N,\varepsilon}(\theta)|^2}{f_N(\theta)} \to \frac{1}{(\sqrt{f(\theta)}+\varepsilon)^2}$ uniformly for $\theta \in [0, 2\pi]$. Thus,

$$\lim_{N\to\infty} \int_0^{2\pi} \log|g_{N,\varepsilon}(\theta) d\theta = \int_0^{2\pi} \log\left(\frac{\sqrt{f(\theta)}}{\sqrt{f(\theta)}+\varepsilon}\right) d\theta\,, \tag{3.27}$$

$$\lim_{N\to\infty} \int_0^{2\pi} \frac{|g_{N,\varepsilon}(\theta)|^2 dG(\theta)}{f_N(\theta)} = \int_0^{2\pi} \frac{dG(\theta)}{(\sqrt{f(\theta)}+\varepsilon)^2}\,, \tag{3.28}$$

and

$$\lim_{N\to\infty} \int_0^{2\pi} |g_{N,\varepsilon}(\theta) - 1|^2 dG(\theta) = \int_0^{2\pi} \left|\frac{\sqrt{f(\theta)}}{\sqrt{f(\theta)}+\varepsilon} - 1\right|^2 dG(\theta)\,. \tag{3.29}$$

In (3.27), since by hypothesis $\int_0^{2\pi} \log f(\theta) d\theta > -\infty$, we see $f(\theta) > 0$ almost everywhere on $[0, 2\pi]$. Thus

$$\lim_{\varepsilon\to 0} \int_0^{2\pi} \log\left(\frac{\sqrt{f(\theta)}}{(\sqrt{f(\theta)})+\varepsilon}\right) d\theta = 0\,,$$

which shows (1)′. By the monotone convergence theorem

$$\lim_{\varepsilon\to 0} \int_0^{2\pi} \frac{dG(\theta)}{((\sqrt{f(\theta)})+\varepsilon)^2} = \int_0^{2\pi} \frac{dG(\theta)}{f(\theta)}\,,$$

which shows (2)′. Finally by the bounded convergence theorem

$$\lim_{\varepsilon\to 0} \int_0^{2\pi} \left|\frac{\sqrt{f(\theta)}}{\sqrt{f(\theta)}+\varepsilon} - 1\right|^2 dG(\theta) = 0\,,$$

which shows (3)′ and completes the proof.

The preceding lemma was needed to complete the proof of the lower bound in Theorem 2.8. We now prove a lemma providing what was needed for the upper bound.

Lemma 3.10. *Let $H_{f_N}(R)$ be given by (3.20) and $H_f(R)$ be given by (1.9). Then for any $l > 0$,*

$$\bigcup_{N=1}^{\infty} \{R \in \underline{M}_S : H_{f_N}(R) \leqq l\} \tag{3.30}$$

is tight in $\underline{M}_S$.

If $\{R_N\}$ is any sequence in $\underline{M}_S$ converging weakly to $R \in \underline{M}_S$, then

$$\underline{\lim}_{N\to\infty} H_{f_N}(R_N) \geqq H_f(R)\,. \tag{3.31}$$

Proof. From (2.3) we have that $f_N(\theta)\to f(\theta)$ uniformly in $[0,2\pi]$. From Lemma 3.8 we have that $\lim_{N\to\infty}\int_0^{2\pi}\log f_N(\theta)d\theta=\int_0^{2\pi}\log f(\theta)d\theta$. Since

$$H_{f_N}(R)=E^R\left\{\int_{-\infty}^{\infty} r(y|\omega)\log r(y|\omega)dy\right\}$$
$$+\frac{1}{2}\log 2\pi+\frac{1}{4\pi}\int_0^{2\pi}\frac{dG(\theta)}{f_N(\theta)}+\frac{1}{4\pi}\int_0^{2\pi}\log f_N(\theta)d\theta\,,$$

and

$$H_f(R)=E^R\left\{\int_{-\infty}^{\infty} r(y|\omega)\log r(y|\omega)dy\right\}$$
$$+\frac{1}{2}\log 2\pi+\frac{1}{4\pi}\int_0^{2\pi}\frac{dG(\theta)}{f(\theta)}+\frac{1}{4\pi}\int_0^{2\pi}\log f(\theta)d\theta\,,$$

it suffices, in order to show (3.30) and (3.31), to show

$$\bigcup_{N=1}^{\infty}\{R\in \underline{M}_S:\Phi_N(R)\leqq l\} \tag{3.32}$$

is tight in $\underline{M}_S$, and, for any sequence $\{R_N\}$ converging to R,

$$\varliminf_{N\to\infty}\Phi_N(R_N)\geqq\Phi(R)\,, \tag{3.33}$$

where

$$\Phi_N(R)=E^R\left\{\int_{-\infty}^{\infty} r(y|\omega)\log r(y|\omega)dy\right\}+\frac{1}{4\pi}\int_0^{2\pi}\frac{dG(\theta)}{f_N(\theta)}\,,$$

and

$$\Phi(R)=E^R\left\{\int_{-\infty}^{\infty} r(y|\omega)\log r(y|\omega)dy\right\}+\frac{1}{4\pi}\int_0^{2\pi}\frac{dG(\theta)}{f(\theta)}\,.$$

We start with Lemma 3.6 where the ϕ of that lemma is $r(y|\omega)$, giving us for any $\alpha>0$,

$$E^R\left\{\int_{-\infty}^{\infty} r(y|\omega)\log r(y|\omega)dy\right\}\geqq-\frac{1}{2}\log 2\pi\alpha-\frac{1}{2\alpha}\cdot\frac{1}{2\pi}\int_0^{2\pi}dG(\theta)\,. \tag{3.34}$$

Since $f_N(\theta)\to f(\theta)$, there exists a c such that $f_N(\theta)\leqq c$ on $[0,2\pi]$. Thus, from (3.34)

$$E^R\left\{\int_{-\infty}^{\infty} r(y|\omega)\log r(y|\omega)dy\right\}\geqq-\frac{1}{2}\log 2\pi\alpha-\frac{1}{2\alpha}\cdot\frac{1}{2\pi}\int_0^{2\pi}\frac{f_N(\theta)dG(\theta)}{f_N(\theta)}$$
$$\geqq-\frac{1}{2}\log 2\pi\alpha-\frac{1}{2\alpha}\frac{c}{2\pi}\int_0^{2\pi}\frac{dG(\theta)}{f_N(\theta)}\,.$$

In this last if we let $\alpha=c/\varepsilon$ for $0<\varepsilon<1$, we get

$$E^R\left\{\int_{-\infty}^{\infty} r(y|\omega)\log r(y|\omega)dy\right\}\geqq-\frac{1}{2}\log\frac{2\pi c}{\varepsilon}-\frac{\varepsilon}{4\pi}\int_0^{2\pi}\frac{dG(\theta)}{f_N(\theta)}\,,$$

so that if $\Phi_N(R)\leqq l$, then

$$\frac{1}{4\pi}\int_0^{2\pi}\frac{dG(\theta)}{f_N(\theta)}\leqq\frac{l+\frac{1}{2}\log\frac{2\pi c}{\varepsilon}}{1-\varepsilon}.$$

Since $f_N(\theta)$ is bounded by c on $[0,2\pi]$, this last implies

$$\frac{1}{2\pi}\int_0^{2\pi}dG(\theta)\leqq c\left(\frac{l+\frac{1}{2}\log\frac{2\pi c}{\varepsilon}}{1-\varepsilon}\right).$$

This means

$$E^R\{\omega^2(0)\}\leqq c\left(\frac{l+\frac{1}{2}\log\frac{2\pi c}{\varepsilon}}{1-\varepsilon}\right),$$

and since the constant on the right depends only on l and not on N or R, we see that $\bigcup_{N=1}^{\infty}\{R\in \underline{M}_S:\Phi_N(R)\leqq l\}$ is tight, which is (3.22).

To prove (3.23) we want to show that if $\{R_N\}$ converges weakly to R and $\Phi_N(R_N)\leqq M$, then $\Phi(R)\leqq M$.

We just proved that if $\Phi_N(R)\leqq l$ then $E^R\{\omega^2(0)\}$ is bounded by a constant independent of N. Thus if $\Phi_N(R_N)\leqq M$, we conclude that the second moments of $\{R_N\}$ are uniformly bounded which in turn implies the total mass of the spectral measure of R_N is uniformly bounded, i.e., $\int_0^{2\pi}dG_N(\theta)$ is uniformly bounded. Hence, by Helly's theorem (choosing a subsequence if necessary) $\{G_N\}$ converges weakly to $G'=G+G_1$, where G is the spectral measure of R and G_1 is nonnegative.

By hypothesis

$$\Phi_N(R_N)=E^{R_N}\left\{\int_{-\infty}^{\infty}r_N(y|\omega)\log r_N(y|\omega)dy\right\}+\frac{1}{4\pi}\int_0^{2\pi}\frac{dG_N(\theta)}{f_N(\theta)}\leqq M. \tag{3.35}$$

In (3.1) if, instead of entropy of R with respect to the stationary process generated by independent Gaussian, mean 0 variance 1 variables, we let $I_\lambda(R)$ be entropy of R with respect to the same stationary Gaussian process but with variance λ for $\lambda>0$, then, of course,

$$I_\lambda(R)=E^R\left\{\int_{-\infty}^{\infty}r(y|\omega)\log r(y|\omega)dy\right\}+\frac{1}{2}\log 2\pi\lambda+\frac{1}{2\lambda}E^R\{\omega^2(0)\}.$$

From (3.35), we see that for any $\lambda>0$,

$$I_\lambda(R_N)+\frac{1}{4\pi}\int_0^{2\pi}\frac{dG_N(\theta)}{f_N(\theta)}\leqq M+\frac{1}{2}\log 2\pi\lambda+\frac{1}{2\lambda}E^{R_N}\{\omega^2(0)\}. \tag{3.36}$$

Now, $I_\lambda(R)$ is lower semicontinuous on $\underline{M}_S$, and so since $R_N\Rightarrow R$, $f_N(\theta)\to f(\theta)$, $G_N\Rightarrow G'=G+G_1$, we conclude that

$$I_\lambda(R)+\frac{1}{4\pi}\int_0^{2\pi}\frac{dG'(\theta)}{f(\theta)}\leqq M+\frac{1}{2}\log 2\pi\lambda+\frac{1}{4\pi\lambda}\int_0^{2\pi}dG'(\theta). \tag{3.37}$$

Therefore, using the expression for $I_\lambda(R)$,

$$E^R\left\{\int_{-\infty}^{\infty} r(y|\omega)\log r(y|\omega)\right\} + \frac{1}{2}\log 2\pi\lambda + \frac{1}{4\pi\lambda}\int_0^{2\pi} dG(\theta) + \frac{1}{4\pi}\int_0^{2\pi}\frac{dG'(\theta)}{f(\theta)}$$

$$\leqq M + \frac{1}{2}\log 2\pi\lambda + \frac{1}{4\pi\lambda}\int_0^{2\pi} dG'(\theta).$$

Since this last is true for any $\lambda > 0$, we let $\lambda \to \infty$, after cancelling $\frac{1}{2}\log 2\pi\lambda$ from both sides, to obtain

$$E^R\left\{\int_{-\infty}^{\infty} r(y|\omega)\log r(y|\omega)dy\right\} + \frac{1}{4\pi}\int_0^{2\pi}\frac{dG'(\theta)}{f(\theta)} \leqq M.$$

Since $G'(\theta) = G(\theta) + G_1(\theta)$ and $G_1(\theta)$ is nonnegative, we conclude that in fact

$$\Phi(R) = E^R\left\{\int_{-\infty}^{\infty} r(y|\omega)\log r(y|\omega)dy\right\} + \frac{1}{4\pi}\int_0^{2\pi}\frac{dG(\theta)}{f(\theta)} \leqq M,$$

which completes the proof.

References

1. Donsker, M.D., Varadhan, S.R.S.: Asymptotic evaluation of certain Wiener integrals for large time. Functional Integration and its Applications, Proceedings of the International Conference held at the Cumberland Lodge, Windsor Great Park, London, in April 1974, A. M. Arthurs, (ed.). Oxford: Clarendon 1975
2. Donsker, M.D., Varadhan, S.R.S.: Asymptotic evaluation of certain Markov process expectations for large time, I. Commun. Pure Appl. Math. **28**, 1–47 (1976); II. Commun. Pure Appl. Math. **28**, 279–301 (1975); III. Commun. Pure Appl. Math. **29**, 389–461 (1976); IV. Commun. Pure Appl. Math. **36**, 183–212 (1983)
3. Donsker, M.D., Varadhan, S.R.S.: Asymptotics for the Wiener Sausage. Commun. Pure Appl. Math. **28**, 525–565 (1975)
4. Donsker, M.D., Varadhan, S.R.S.: On laws of the iterated logarithm for local times. Commun. Pure Appl. Math. **30**, 707–753 (1977)
5. Donsker, M.D., Varadhan, S.R.S.: Asymptotics for the Polaron. Commun. Pure Appl. Math. **36**, 505–528 (1983)
6. Orey, S.: Large deviations and Shanon-McMillan theorems (preprint)

Communicated by A. Jaffe

Received July 6, 1984

EXPOSITIONES MATHEMATICAE

Expo. Math. 3 (1985), 251–272

Large deviations and applications

S.R.S. Varadhan

1. Introduction

Let $V(x)$ be a periodic continuous function of period 2π on the line. We consider the solution $u(t,x)$ of the initial value problem

$$\frac{\partial u}{\partial t}=\frac{1}{2}\frac{\partial^2 u}{\partial x^2}+V(x)\,u \qquad (1.1)$$
$$u(0,x)=1$$

The Feynman-Kac formula expresses the solution of (1.1) as a function space integral

$$u(t,x)=E_x\left\{\exp\left[\int_0^t V(x(s))\,ds\right]\right\} \qquad (1.2)$$

where E_x refers to the expectation with respect to the standard Brownian motion starting from the point $x \in R$ at time $t=0$. Since the function $V(x)$ as well as the solution $u(t,x)$ are both periodic in x of period 2π the problem could be considered on the circle Z rather than the real line R.

$$\frac{\partial u}{\partial t}=\frac{1}{2}\frac{\partial^2 u}{\partial \theta^2}+V(\theta)\,u \qquad \text{for } t>0,\ \theta \in Z \qquad (1.3)$$
$$u(0,\theta)=1$$

If we denote by $\theta(t)$ the Brownian motion on the circle then the solution $u(t,\theta)$ has again a representation by Feynman-Kac's formula has

$$u(t,\theta)=E_\theta\left[\exp\left\{\int_0^t V(\theta(s))\,ds\right\}\right] \qquad (1.4)$$

where E_θ is the expectation with respect to Brownian motion on Z starting from $\theta \in Z$ at time $t=0$.

The operator $\frac{1}{2}\frac{\partial}{\partial\theta^2}+V(\theta)$ is an operator on $L_2(Z)$ with a compact resolvent. Moreover the operator is self adjoint and bounded above. It therefore has a sequence $\{\lambda_j\}$ of eigenvalues tending to $-\infty$ and a sequence $\phi_j(\cdot)$ of eigenfunctions. The largest eigenvalue λ_1 is of multiplicity one and the corresponding eigenfunction is strictly positive. Denoting by $\langle\ ,\ \rangle$ the inner product in $L_2(Z)$ we can obtain an expansion for the solution $u(t,\theta)$ in terms of the eigenvalues and eigenfunctions.

$$u(t,\theta)=\sum_{j=1}^{\infty} e^{\lambda_j t}\,\phi_j(\theta)\,\langle\phi_j,1\rangle \tag{1.5}$$

From the formula it is not difficult to see that

$$\lim_{t\to\infty}\frac{1}{t}\log u(t,\theta)=\lambda_1 \qquad \text{for all } \theta\in Z. \tag{1.6}$$

For the largest eigenvalue λ_1 we also have the classical variational formula

$$\lambda_1=\sup_{\substack{\phi\in L_2(Z)\\ \|\phi\|_2=1}}\{\textstyle\int V(\theta)\,\phi^2(\theta)\,d\theta-\frac{1}{2}\int[\phi'(\theta)]^2\,d\theta\} \tag{1.7}$$

since the supremum is attained at a positive ϕ we can take $\phi(\theta)=\sqrt{f(\theta)}$ and vary $f(\cdot)$ over all L_1 functions, that are nonnegative and have total integral 1. The formula then becomes

$$\lambda_1=\sup_{\substack{f\in L_1(Z)\\ f\geq 0,\,\int f\,d\theta=1}}\left[\int V(\theta)\,f(\theta)\,d\theta-\frac{1}{8}\int\frac{[f'(\theta)]^2}{f(\theta)}\,d\theta\right] \tag{1.8}$$

We would like to derive (1.8) directly from (1.4) without passing through the differential equation (1.3) or the interpretation of the limit in (1.4) as an eigenvalue. The reasons for doing this are many. We would like to replace $\exp\left[\int_0^t V(\theta(s))\,ds\right]$ by a more general functional $F_t(\omega)$ depending on the trajectory $\theta(\cdot)$. We aim to calculate limits of the form

$$\lim_{t\to\infty}\frac{1}{t}\log E^P\{\exp[F_t(\omega)]\} \tag{1.9}$$

for suitable functionals $F_t(\omega)$. We would like to generalize formula (1.4) by taking expectations with respect to processes other than the Brownian motion. If one replaces the Brownian motion by a more general reversible Markov process we still have a connection with a self adjoint operator for the special functional $\exp\left[\int_0^t V(\theta(s))\,ds\right]$. If the process is Markov but not reversible, we lose the self adjointness of the operator and hence the analogue of the variational formula does not seem to be valid in any obvious sense. If the process P is a stationary process without any Markov property there is no differential operator in the background at all. A direct derivation of formula (1.8) from (1.4) would therefore give us a clue as to how to generalize the formula in situations where there are no differential equations floating around.

Let us fix the starting point $\theta=0$ for simplicity and denote by P the corresponding Brownian motion measure. Let us define

$$\lambda(t, A)=P\left\{\theta(\cdot): \frac{1}{t}\int_0^t V(\theta(s))\,ds \in A\right\}$$

Then

$$u(t,0)=\int_{-\infty}^{\infty} \exp[t\,y]\,\lambda(t,dy) \tag{1.10}$$

If we denote by

$$m=\frac{1}{2\pi}\int_0^{2\pi} V(y)\,dy$$

then by the ergodic theorem for the Brownian motion on the circle, which has normalized Lebesgue measure as the invariant measure

$$\lambda(t,A)\to 0 \quad \text{as } t\to\infty.$$

for any closed set A, not containing the number m.

Suppose there exists a function $h(y)$ such that decay rate of the probability $\lambda(t,dy)$ locally near the point y is given by $\exp[-t\,h(y)]$ for every $y\in R$. We will make this precise later. If we think of the integral (1.10) as a sum

$$u(t,0)=\sum_y \exp[t\,y]\,\lambda(t,dy)$$

then approximately

$$u(t,0)\sim\sum_y \exp[t\,y]\exp[-t\,h(y)] \tag{1.11}$$

It is now obvious that

$$\lim_{t\to\infty}\frac{1}{t}\log u(t,0)=\sup_y\,[y-h(y)]. \tag{1.12}$$

although we have demonstrated the possibility of a variational formula (1.12) for the rate of growth of $u(t,0)$ we have lost the formula (1.8) in the process. The potential V is hidden in the function $h(y)$.

To remedy this situation we go up one level. We introduce the local time or the occupation time

$$L_t(\omega, A)=\frac{1}{t}\int_0^t \chi_A(\theta(s))\,ds \tag{1.13}$$

$L_t(\omega, A)$ is the proportion of time the Brownian trajectory $\theta(\cdot)$ spends in the set A during the time period $[0, t]$. $L_t(\omega, \cdot)$ is a random probability measure on Z and since we have a measure P on the space of trajectories, for each $t>0$ we get a probability measure Q_t on the space M of all probability measures on Z. Q_t is nothing but the distribution of $L_t(\omega, \cdot)$ where ω is distributed according to P. We can now rewrite $u(t, 0)$ in the form

$$u(t,0)=E^{Q_t}\{\exp[t \int V(\theta)\,\mu(d\theta)]\} \tag{1.14}$$

Here μ is the variable of integration for Q_t. By the ergodic theorem the Q_t measure approaches the δ-measure at the point $\mu_0 \in M$ where μ_0 is the invariant measure i.e. the normalized Lebesgue measure. If we have a rate of decay $\exp[-t\,I(\mu)]$ for the probability $Q_t(A)$, locally around $\mu \in M$ then an argument similar to (1.11) leads to

$$\lim_{t\to\infty} \frac{1}{t} \log u(t,0)=\sup_{\mu\in M} [\int V(\theta)\,\mu(d\theta)-I(\mu)] \tag{1.15}$$

Suppose it were also true that

$$I(\mu)=\frac{1}{8}\int \frac{[f'(\theta)]^2}{f(\theta)}\,d\theta$$

if μ has a nice density $f(\cdot)$ and

$$I(\mu)=\infty \quad \text{otherwise.}$$

Then (1.5) reduces to (1.8). The same sort of reasoning will also lead to a formula for integrals $\alpha(t)$ of the form

$$\alpha(t)=E^P\{\exp[t\,F(L(t,\cdot))]\}=E^{Q_t}\{\exp[t\,F(\mu)]\}$$

which will look like

$$\lim_{t\to\infty} \frac{1}{t} \log \alpha(t)=\sup_{\mu\in M} [F(\mu)-I(\mu)]. \tag{1.16}$$

Clearly if we take $F(\mu)=\int V(\theta)\,\mu(d\theta)$, a linear functional, we are back to (1.15) or (1.8). But we could take more general F and we would get formulae that cannot be obtained by differential equations.

The variational formula (1.16) is very similar to the energy-entropy formulae in statistical mechanics. See for instance [13].

Let us consider an example that is slightly different. Let

$$a(t)=E^P\left[\exp\left[\int_0^t V(\theta(s), \theta(s+1))\,ds\right]\right]$$

where $V(\theta_1, \theta_2)$ is a function on $Z \times Z$. The function $\int_0^t V(\theta(s), \theta(s+1))\,ds$ cannot be expressed in terms of the local time $L_t(\omega, \cdot)$ alone. We need to build a theory that is at one level higher than the one that we described just now, in order to study the growth rate of $a(t)$.

Given a segment of a trajectory $\theta(s)$ for $0 \le s \le t$, we extend it periodically to obtain a trajectory $\theta_t(s)$ for $-\infty < s < \infty$. We look at the periodic trajectory $\theta_t(\cdot)$ as well as its translates $\theta_t(\cdot + s)$ for $0 \le s \le t$. We define a measure R_t^ω on the space of all trajectories by

$$R_t^\omega = \frac{1}{t} \int_0^t \delta_{\theta_t(\cdot + s)}\, ds$$

In other words R_t^ω is the orbital measure on the random periodic orbit of $\theta_t(\cdot)$ obtained by time translations. R_t^ω is a random invariant measure (with respect to time translations) on the space of trajectories. Moreover

$$\int V(\theta(0), \theta(1))\, dR_t^\omega = \frac{1}{t} \int_0^t V(\theta(s), \theta(s+1))\, ds$$

provided the addition $\theta + 1$ on the right hand side is interpreted modulo t. The integration on the left is over the space of trajectories $\theta(\cdot)$ continuous except for jumps introduced by the periodization of $\theta(\cdot)$. Since the effect of periodization is only felt near the end the effect is small as t becomes large. The growth rate of $a(t)$ is therefore the same as that of

$$\hat{a}(t) = E^P \{\exp[t \int V(\theta(0), \theta(1))\, dR_t^\omega]\}$$

If we denote by π_t the distribution of R_t^ω then

$$\hat{a}(t) = E^{\pi_t} \{\exp[t \int V(\theta(0), \theta(1))\, dR]\}$$

Here R is an arbitrary stationary process i.e. a translation invariant measure on the space of trajectories on $(-\infty, \infty)$ with values in Z which are continuous except for jumps and R is the variable of integration with respect to π_t. In other words π_t is a mesare on the space of stationary stochastic processes. If we denote by R_0 the Brownian motion viewed as a stationary stochastic process on the time interval $(-\infty, \infty)$ then the ergodic theorem asserts that π_t is close to the δ-measure at R_0 for large t. If we now postulate once again a local decay rate $H(R)$ for π_t probabilities away from R_0, we clearly could predict a variational formula

$$\lim_{t \to \infty} \frac{1}{t} \log a(t) = \sup_R [E^R \{V(\theta(0), \theta(1))\} - H(R)] \tag{1.17}$$

We have three functions $H(R)$, $I(\mu)$ and $h(y)$ giving local decay rates. There are clearly relations between them. Since the sets

$$\left\{\omega : \frac{1}{t} \int_0^t V(\theta(s))\, ds = y\right\}$$

$$\{\omega\colon \int V(\theta)\, L_t(\omega, d\theta) = y\}$$
$$\{\omega\colon \int V(\theta(0))\, dR_t^{\omega} = y\}$$

are all the same we can expect

$$I(\mu) = \inf_{R:\, q(R) = \mu} H(R) \tag{1.18}$$

when $q(R)$ is the common one dimensional marginal of the stationary process R.

$$h(y) = \inf_{\mu:\, \int V(\theta)\, \mu(d\theta) = y} I(\mu) \tag{1.19}$$

Formulae like (1.18) and (1.19) are referred to as the contraction principle.

To avoid technical difficulties we will build the theory first for discrete time stochastic processes and then for continuous time processes. We will deal mostly with Markov processes except for the special case of stationary Gaussian processes in Section 10. In Sections 12–14 we will consider some applications.

2. Mathematical Formulation

We consider a sequence P_n of probability measures on a Polish space X which converges weakly to the degenerate distribution or point mass at some point $x_0 \in X$. For closed sets A disjoint from x_0, $P_n(A) \to 0$ as $n \to \infty$ and we are interested in the rate at which $P_n(A) \to 0$ as $n \to \infty$. The examples we consider will be those for which this rate is exponential in n and our interest will be in the exponential constant which of course will depend on the set A.

Formally if we assume that $P_n(A)$ behaves like $\exp[-n\,\rho(A)]$ for large n, then from the additivity properties of $P_n(A)$ one can conclude that

$$\rho(A \cup B) = \operatorname{Min}[\rho(A), \rho(B)].$$

A natural guess for the set function $\rho(A)$ is then a function of the form

$$\rho(A) = \inf_{x \in A} I(x)$$

where $I(x)$ is a point function i.e. a map from $X \to [0, \infty]$. The function $I(x)$ controls the local rate of decay of the probability measures P_n at the point x.

For technical reasons it is convenient to formulate the connection between the sequence $\{P_n\}$ and the function $I(x)$ in the following terms.

Definition: We say that the sequence $\{P_n\}$ satisfies the large deviation principle with a rate function $I(x)$ if

$$\limsup_{n \to \infty} \frac{1}{n} \log P_n(A) \le -\inf_{x \in A} I(x) \quad \text{for } A \text{ closed;}$$

$$\liminf_{n \to \infty} \frac{1}{n} \log P_n(G) \ge -\inf_{x \in G} I(x) \quad \text{for } G \text{ open.}$$

The rate function $I(x)$ is supposed to be a function on X with the following properties

(i) $0 \le I(x) \le \infty$ for all $x \in X$
(ii) $I(x)$ is lower-semi-continuous.
(iii) $\{x: I(x) \le l\}$ is totally bounded in X for each $l < \infty$.

We will look at several examples of this type of a situation in the following sections.

3. Some Elementary Facts

Suppose $\{P_n\}$ is a sequence of probability measures satisfying a large deviation principle with a rate function $I(x)$. Then it is straightforward to conclude the following implications:

(a) Suppose $A \subset X$ is a set such that

$$\inf_{x \in A^0} I(x) = \inf_{x \in A} I(x) = \inf_{x \in \bar{A}} I(x)$$

where A^0, $\bar{A}$ are the interior and the closure of A. Then

$$\lim_{n \to \infty} \frac{1}{n} \log P_n(A) = -\inf_{x \in A} I(x). \tag{3.1}$$

(b) Suppose $F(x)$ is a bounded continuous function on X, then

$$\lim_{n \to \infty} \frac{1}{n} \log \int \exp[nF(x)]\, dP_n = \sup_{x \in X} [F(x) - I(x)]. \tag{3.2}$$

(c) More generally suppose $F_n(x)$ is a sequence of uniformly bounded continuous functions coverging uniformly on compact sets to a limiting function $F(x)$, then

$$\lim_{n \to \infty} \frac{1}{n} \log \int \exp[nF_n(x)]\, dP_n = \sup_{x \in X} [F(x) - I(x)]. \tag{3.3}$$

(d) Suppose Φ is a continuous map from $X \to Y$ where Y is another Polish space then the family $Q_n = P_n \Phi^{-1}$ satisfies a large deviation principle on Y with a rate function $\hat{I}(y)$ given by

$$\hat{I}(y) = \inf_{x:\, \Phi(x) = y} I(x) \tag{3.4}$$

More generally we can have Q_n given by $Q_n = P_n \Phi_n^{-1}$ where Φ_n tends to Φ as $n \to \infty$ uniformly on compact subsets of X. It is understood here that if $[x: \Phi(x) = y]$ is empty then $\hat{I}(y)$ is taken to be $+\infty$. We will refer to this as the Contraction Principle.

These implications (a) through (d) are direct consequences of the definition. They are proved in [15] or [16].

4. Cramér's Theorem

The simplest example of large deviation is to consider the sequence P_n of measures on R defined by

$$P_n(A) = \text{Prob}\left\{\frac{X_1 + \dots X_n}{n} \in A\right\}$$

where $X_1, \dots, X_n \dots$ are an independent identically distributed sequence of random variables with a common distribution α. Cramér in [2] showed that the large deviation principle holds for P_n with a rate function

$$I(x) = \sup_{\theta} [\theta x - \log M(\theta)] \tag{4.1}$$

where

$$M(\theta) = \int_R e^{\theta x} \alpha(dx) \tag{4.2}$$

If we take α to be the standard Gaussian distribution then $M(\theta) = \exp\left[\frac{\theta^2}{2}\right]$ and $I(x) = x^2/2$. P_n in this case is easily seen to be given by

$$dP_n = \sqrt{\frac{n}{2\pi}} \exp\left[-\frac{nx^2}{2}\right] dx$$

from which one can easily deduce the large deviation principle. If we take $F(x) = \theta x$ and try to evaluate

$$\begin{aligned}\lim_{n\to\infty} \frac{1}{n} \log \int e^{n\theta x} dP_n &= \lim_{n\to\infty} \frac{1}{n} \log E \exp[\theta(X_1 + \dots X_n)] \\ &= \lim_{n\to\infty} \frac{1}{n} \log(E \exp[\theta x_1])^n \\ &= \log E \exp[\theta X_1] \\ &= \log M(\theta)\end{aligned}$$

by formula (3.2) we obtain

$$\log M(\theta) = \sup_{\theta} [\theta x - I(x)] \tag{4.3}$$

Formulae (4.1) and (4.3) express the duality between the pair of conjugate convex functions $I(x)$ and $\log M(\theta)$.

The extensions of Cramér's theorem to the case of vector valued random variables have been considered. See [8] for the case of a Banach space or [1] for the case of topological vector spaces.

5. Empirical Distribution Functions

If $X_1, X_2, \dots X_n, \dots$ is a sequence of independent identically distributed random variables with a common distribution α and we look at the empirical distribution of the first n observation we obtain the random measure $\frac{1}{n}(\delta_{X_1}+\ldots\delta_{X_n})$ on R. If we dente by $\boldsymbol{M}$ the space of probability measures on R, then $\frac{1}{n}(\delta_{X_1}+\ldots\delta_{X_n})$ is a random element of $\boldsymbol{M}$ and we can denote by Q_n the probability distribution on $\boldsymbol{M}$ which is the distribution of the empirical distribution. As $n\to\infty$, the Glivenko-Cantelli theorem asserts that $Q_n\Rightarrow\delta_\alpha$ as $n\to\infty$. The principle of large deviation is valid in this situation with a rate function $I_\alpha(\mu)$ defined for $\mu\in\boldsymbol{M}$ as follows:

If $\mu\ll\alpha$ with $f(x)=d\mu/d\alpha$ satisfying $\int|f(x)\log f(x)|\,\alpha(dx)<\infty$ then

$$I_\alpha(\mu)=\int f(x)\log f(x)\,\alpha(dx) \tag{5.1}$$

otherwise $I_\alpha(\mu)=\infty$.

The formula (5.1) is clearly the same as relative entropy or the Kullback-Liebler information number. If we look at the map from $\boldsymbol{M}\to R$ given by $\mu\to\int x\,d\mu$, then the contraction principle of Section 3, gives us that

$$\inf_{\mu:\,\int x\,d\mu=a} I_\alpha(\mu)=\sup_\theta[\theta a-\log\int e^{\theta x}\,\alpha(dx)]. \tag{5.2}$$

To see this, we note that the mean of the empirical distribution $\frac{1}{n}(\delta_{X_1}+\ldots+\delta_{X_n})$ is nothing but $\frac{1}{n}(X_1+\ldots+X_n)$. The left side of (5.2) is computed by the contraction principle and the right side by Cramér's theorem.

In some sense the large deviation principle for the empirical distribution is a special case of Cramér's theorem when the random variable takes values in the topological vector space of all signed measures on R. θ then belongs to the dual space of bounded continuous functions

$$\begin{aligned} M(\theta)&=E\exp\langle\theta,\delta_X\rangle\\ &=E\exp[\theta(X)]\\ &=\int\exp[\theta(x)]\,\alpha(dx)\\ I(\mu)&=\sup[\int\theta(x)\,\mu(dx)-\log M(\theta)]. \end{aligned} \tag{5.2}$$

It is not very hard to compute (5.2) and verify that it yields again (5.1). For some of the properties of $I_\alpha(\mu)$ and its relation to Cramér's theorem see [4] and [8].

An interesting special case is when α is a probability distribution giving mass $\alpha_1,\ldots,\alpha_k$ to k points $x_1,\ldots,x_k$, then the empirical distribution is the normalized frequency count. What we have then is essentially the large deviation principle for the multinomial probability distribution with cell probabilities $\alpha_1,\ldots\alpha_k$. If we denote by $\mu_1,\ldots,\mu_k$ the probabilities of another distribution on these same points then

$$\text{Prob}\left[\frac{1}{n}(f_1, \ldots, f_k) \text{ is nearly equal to } (\mu_1, \ldots, \mu_k)\right]$$

can be computed by the multinomial distribution and using Stirling's formula one can see that the log of the above probability is nearly

$$-n\sum_{i=1}^{k} \mu_i \log \frac{\mu_i}{\alpha_i}.$$

6. Empirical Processes

The empirical distribution of the first n members of the sequence $X_1, X_2, \ldots$ does not convey any information about the order in which $X_1, \ldots, X_n$ occur. In order to take care of this we define the empirical process as follows:

Consider for any n-tuple $X_1, \ldots, X_n$ the two sided periodic extension of $X_1, \ldots, X_n$ given by

$$\xi_n = (\ldots, X_1, \ldots, X_n, X_1, \ldots, X_n, \ldots).$$

Although ξ_n is ambiguous because of the location of the starting point X_1, the periodic orbit $\xi_n, T\xi_n, \ldots, T^{n-1}\xi_n$ defined by ξ_n is well defined. Here T denotes the shift on the space of sequences.

Given the orbit one can of course define a measure $\frac{1}{n}(\delta_{\xi_n} + \ldots + \delta_{T^{n-1}\xi_n})$ on the orbit. Since the orbit is periodic we now have a stationary measure or a stationary stochastic process on the space of sequences. The stationary process is random because the n-tuple $X_1 \ldots X_n$ is random. We have now a random stationary process which we will call the empirical process. If R_n is the empirical process then the one dimensional marginal of R is $\frac{1}{n}(\delta_{X_n} + \ldots + \delta_{X_n})$ or the empirical distribution. The two dimensional marginal is easily seen to be $\frac{1}{n}(\delta_{X_1, X_2} + \ldots + \delta_{X_{n-1}, X_n} + \delta_{X_n, X_1})$. Except for the last term it is the distribution of consecutive pairs and for large n the effect of the last term can be ignored. We now have a probability distribution π_n on the space of all stationary processes Ω_S and as $n \to \infty$, $\pi_n \Rightarrow \delta_{P_0}$ where P_0 is the infinite product measure with factor α. Clearly $P_0 \in \Omega_S$ and the ergodic theorems assure us that $\pi_n \Rightarrow \delta_{P_0}$ as $n \to \infty$. Note that weak convergence in Ω_S is essentially convergence of finite dimensional distributions.

One can naturally ask if there is a principle of large deviations for π_n on Ω_S. The answer is yes and the rate function is given by

$$H_{P_0}(Q) = E^Q\{I_\alpha(Q_\omega)\}. \tag{6.1}$$

Here Q_ω is the regular conditional probability of X_0 given the past $X_{-1}, X_{-2}, \ldots$ under Q and is a distribution on R for each ω representing the past information.

For each ω, $I_\alpha(Q_\omega)$ is calculated by formula (5.1). This is a nonnegative function of ω and $H_{P_0}(Q)$ is the integral with respect to Q.

When we replace R by a finite set with k elements and take α to be the uniform distribution on these k elements then $\log k - H_{P_0}(Q)$ is nothing but Shannon's entropy of the stationary process Q and the principle of large deviation is essentially the Shannon-Breimann-McMillan theorem of information theory.

We have a map q from Ω_S to $\boldsymbol{M}$ which assigns to any stationary process Q its common one dimensional marginal distribution μ. The map q applied to the empirical process gives the empiricial distribution. The contraction principle should yield

$$I_\alpha(\mu) = \inf_{Q:\, q(Q)=\mu} H_{P_0}(Q). \tag{6.2}$$

One can verify this formula by noting that infimum on the right is attained when Q is the product measure with factor μ. In such a case $Q_\omega = \mu$ for all ω and one can compute from (6.1) that for this special Q, $H_{P_0}(Q) = I_\alpha(\mu)$.

As an application of these ideas let us try to evaluate

$$\lim_{n\to\infty} \frac{1}{n} \log E \exp\{F(X_1, \dots, X_k) + \dots + F(X_n, \dots, X_{n+k-1})\}$$

where $X_1, \dots, X_n, \dots$ is a sequence of independent identically distributed random variables with a common distribution α. We assume that F is a bounded continuous function of $X_1, \dots, X_k$. If we define a functional $\Phi(Q)$ on Ω_S by

$$\Phi(Q) = E^Q F(X_1, \dots, X_k)$$

then $\log E\{\exp\{F(X_1, \dots, X_k) + \dots + F(X_n, \dots, X_{n+k-1})\}$ differs from $\log \int \exp[n\Phi(Q)]\, \pi_n(dQ)$ by only the effect of periodization and this is a bounded effect. When we divide by n, this disappears as $n \to \infty$. If we apply formula (3.2) we can conclude that

$$\lim_{n\to\infty} \frac{1}{n} \log E\{\exp[F(X_1, \dots, X_k) + \dots + F(X_n, \dots, X_{n+k-1})]\}$$
$$= \sup_Q [E^Q\{F(X_1, \dots, X_k)\} - H_{P_0}(Q)].$$

7. The Dependent Case

It is not unreasonable to want to replace the underlying stochastic process of independent random variables $X_1, X_2, \dots$ with common distribution α, with an ergodic stationary process. In other words we change the product measure P_0 to a stationary ergodic measure P. It also seems reasonable at first glance that since the ergodic theorem assures us that $\pi_n \Rightarrow \delta_P$ there might be a principle of large deviation lurking around. The rate function should be given by

$$H_P(Q) = E^Q[I_{P_\omega}(Q_\omega)] \tag{7.1}$$

where Q_ω, P_ω are respectively the regular conditional probability distributions of X_0 given the past $X_{-1}, \ldots, X_{-k}, \ldots$ under Q and P respectively. One can convince oneself that (7.1) is the logical extension of (6.1) when P is not a product measure.

However there is a proverbial fly in the ointment. Q_ω as a conditional probability is defined only almost everywhere with respect to Q just as P_ω is only defined almost everywhere with respect to P. Since Q and P, being stationary processes want to be orthogonal most of the time formula (7.1) does not make sense. Since the expectation is with repect to Q the problem is to define P_ω almost everywhere with respect to Q.

Clearly one cannot do this under great generality but we have to make special assumptions on the nature of P so that P_ω may be defined almost everywhere with respect to Q for any $Q \in \Omega_S$.

If P is a stationary Markov process with transition probability $\pi(x, dy)$ then

$$P_\omega(A) = \pi(X_{-1}, A)$$

and this is clearly well defined everywhere. It seems therefore likely that if we restrict P to Markov processes we may get somewhere. Another possibility is to take P to be a Gaussian process in which case P_ω is a Gaussian measure with mean

$$m(\omega) = \sum_{j=1}^{\infty} a_{-j} X_{-j}$$

and variance θ_0 equal to the prediction error. One can then see for which processes Q the expression $m(\omega)$ makes reasonable sense. These ideas will be explored in the next two sections.

8. The Markov Case

Let $\pi(x, dy)$ be the transition probabilities of a Markov chain on X. For any initial point $X_0 = x$, we can construct the probability measure P_x on the space of sequences $X_0 = X_1, X_2, \ldots$.

Irrespective of whether there exists a stationary Markov process with transition probability $\pi(x, dy)$, we can construct the measure π_n^x on Ω_S as in Section 7. The starting point x enters because the probability measure on n-tuples is induced by P_x. If there is no finite invariant measure, we expect π_n^x to dissipate as $n \to \infty$.

If the Markov chain is positive recurrent and ergodic then $\pi_n \Rightarrow \delta_P$ where P is the unique stationary Markov process with $\pi(x, dy)$ as transition probability.

We make the following assumptions on $\pi(x, dy)$.

(i) For each bounded continuous function $f(y)$ on X, $(\pi f)(x)=\int f(y)\,\pi(x,dy)$ is a bounded continuous function of x.

(ii) There exists a function $u_0(x)$ on X such that $u_0(x)\geq 1$, $(\pi u_0)(x)=\int u(y)\,\pi(x,dy)$ is bounded on compact subsets of X and for each $l<\infty$

$$\left\{x: \frac{u_0(x)}{(\pi u_0)(x)}\leq l\right\}$$

is a totally bounded subset of X.

(iii) There exists a reference measure α on X such that $\pi(x,dy)$ has a density $\pi(x,y)$ with respect to α. $\pi(x,y)$ is positive almost everywhere with respect to α and the mapping $x\to\pi(x,y)$ is continuous as a map from X into $L_1(\alpha)$.

Assumption (ii) guarantees the existence of a finite invariant measure and assumption (iii) guarantees the uniqueness of the invariant measure and the ergodicity of the stationary process. Assumption (iii) can be weakened somewhat.

Under these conditions one can prove the principle of large deviations with a rate function given by (7.1). The details for the discrete Markov chain case have not appeared in print. But the details for the continuous time analogue have appeared in [5] and the discrete time situation, if anything, is easier.

9. Continuous Time Markov Processes

We start with a continuous time Markov process on a Polish state space X. The process is specified by transition probabilities $p(t,x,dy)$. These transition probabilities generate a semigroup T_t on the space of bounded measurable functions on X. We will assume that the process is Feller, i.e. that the semigroup maps the space $C(X)$ of bounded continuous functions into itself. We will also assume that the process can be realized in the Skorohod space $D_X[0,\infty)$ of right continuous functions with left limits on $[0,\infty)$ with values in X. For each starting point $x\in X$ we have the measure P_X on $D_X[0,\infty)$ which is again a Polish space and we assume that the map $x\to P_x$ is weakly continuous. Given a realization of the process $x(\cdot)$ in $[0,T]$ i.e. a function $x(s)$ for $0\leq s\leq T$ we can extend the function periodically both sides and construct a periodic family $x(s+t)$, $0\leq t\leq T$ of functions on $(-\infty,\infty)$. Denoting this function by $x_t(s)$ we can construct on the space $D_X(-\infty,\infty)$ of trajectories a random measure $\frac{1}{T}\int_0^T \delta_{x_t}\,dt$. This is a translation invariant random measure and therefore induces a measure π_T^x on the space of all stationary stochastic processes. Here x is the starting point and comes in because the measure on the set of trajectories $D_X[0,T]$ is P_x which is constructed from the transition probabilities and the starting point x.

One can ask if there is a large deviation principle here governed by a rate function similar to (7.1). If Q is any stationary process and Q_ω is the regular conditional probability of Q given the past i.e. given $x(s)$ for $-\infty < s \leq 0$ then we denote by Q_ω^T the restriction of Q_ω to the σ-field generated by $x(s)$ for $0 \leq s \leq T$. We also have the measure $P_{x(0,\omega)}^T$ which is the Markov measure corresponding to our transition probabilities starting from $x = x(0, \omega)$ restricted to the same σ-field: If we define

$$H_P^T(Q) = E^Q[I_{P_{x(0,\omega)}^T}(Q_\omega^T)] \tag{9.1}$$

then it turns out that

$$H_P^T(Q) = T H_P(Q) \quad \text{for} \quad T > 0 \tag{9.2}$$

for some $H_P(Q)$ which may be infinite. The large deviation principle holds for π_T^x with this rate function under suitable conditions: the conditions are

(i) The map $x \to P_x$ is weakly continuous.
(ii) There exists a sequence of functions $u_n(x)$ in the domain of the infinitesimal generator L of the semigroup T_t of the Markov process such that

(a) $u_n(x) \geq 1$

(b) For every compact set $W \subset X$

$$\sup_{x \in W} \sup_n u_n(x) < \infty.$$

(c) For each $x \in X$

$$\lim_{n \to \infty} \left(\frac{L u_n}{u_n}\right)(x) = V(x) \qquad \text{exists.}$$

(d) For some $A < \infty$

$$\sup_n \sup_x \left(\frac{L u_n}{u_n}\right)(x) \leq A.$$

(e) $\{x: V(x) \geq l\}$ is a compact set in X for each $l > -\infty$.

(iii) The transition probability $p(1, x, dy)$ is equivalent to a fixed reference measure α with density $p(1, x, y)$ and the map $x \to p(1, x, y)$ is a continuous map of X into $L_1(\alpha)$.

These are analogues of conditions (i) to (iii) in the Markov chain case introduced in Section 8. The proofs and details appear in [5]. Condition (ii) should be interpreted as a coerciveness condition. It requires the generator L to be such that $(L - V)u = 0$ has a solution $u \geq 1$ for some V satisfying (e).

10. The Gaussian Case

Let $\{X_n\}$ be a Gaussian process with mean zero and covariance $EX_iX_j = \rho_{j-i}$. We assume that the covariance arises from a spectral density

$$\rho_n = \frac{1}{2\pi} \int_0^{2\pi} e^{in\theta} f(\theta)\, d\theta.$$

We assume that the process is purely nondeterministic i.e. $\int_0^{2\pi} \log f(\theta)\, d\theta > -\infty$. We further make the regularity assumption that the spectral density is a continuous periodic function of θ of period 2π. Under these conditions we can show that a principle of large deviation holds and formula (7.1) can be simplified to yield the following explicit expression for the rate functional:

$$H_P(Q) = E^Q \int_{-\infty}^{\infty} q(x, \omega) \log q(x, \omega)\, dx$$
$$+ \frac{1}{2} \log 2\pi + \frac{1}{4\pi} \int_0^{2\pi} \log f(\theta)\, d\theta + \frac{1}{4\pi} \int \frac{dG(\theta)}{f(\theta)}.$$

Here $q(x, \omega)$ is the conditional probability density of X_0, under Q given the past history $X_{-1}, X_{-2}, \ldots$. We take $H_P(Q) = \infty$, unless the stationary process Q has two moments and if Q has two moments then $E^Q \int_{-\infty}^{\infty} q(x, \omega) \log q(x, \omega)\, dx$ is well defined as a real number which could be $+\infty$ but never $-\infty$. Here dG is the spectral measure of the process Q i.e.

$$E^Q X_0 X_n = \frac{1}{2\pi} \int e^{in\theta}\, dG(\theta).$$

The details of this situation will appear in [10].

11. The Case of Brownian Motion on R^d

If we take for our underlying Markov process the standard Brownian motion on R^d it satisfies the conditions (i) and (iii) of Section 9 but not condition (ii). In fact it does not have a finite invariant measure. In terms of the rate functions $I(\mu)$ and $H(P)$ this means that $I(\mu)$ and $H(P)$ are strictly positive and never zero. There is always local decay of probability. The upper estimate for closed sets require putting together local estimates for a global estimate and this cannot be done. However the global lower estimate is really a local estimate and can be derived without assumption (ii). The upper estimate can still be derived for compact sets and this can sometimes be used in special problems. In later sections we will finesse the issue by modifying the Brownian motion process either by considering it on a large torus or by introducing a small restoring drift. Condition (ii) will then be satisfied and we have to remove the effect of the modification by a special argument.

12. Application: Laws of the Iterated Logarithm

If $x(t)$, $0 \leq t < \infty$ is the standard one dimensional Brownian motion it is well known that it has a local time as a continuous function. In other words the random measure

$$L_t(\omega, A) = \frac{1}{t} \int_0^t \chi_A(x(s))\, ds$$

has for almost all ω a continuous density $l_t(\omega, y)$ with respect to the Lebesgue measure.

$$l_t(\omega, y) = \frac{1}{t} \int_0^t \delta(x(s) - y)\, ds$$

and

$$L_t(\omega, A) = \int_A l_t(\omega, y)\, dy.$$

Brownian scaling property implies that the distribution of $\sqrt{t}\, l_t(\omega, \sqrt{t}\, y)$ is independent of t and is the same as that of $l_1(\omega, y)$. However for $t \to \infty$, $l_t(\omega, y)$ tends to be more and more dispersed. It is likely to have a zero limit in the topology of uniform convergence on compact sets but not in the topology of $L_1(R)$.

Let us define

$$\hat{l}_t(\omega, y) = \sqrt{\frac{t}{\log\log t}}\, l_t\left(\omega, \sqrt{\frac{t}{\log\log t}}\, y\right).$$

Statistically this has the same distribution as $l_{\log\log t}(\omega, y)$ and therefore will dissipate slowly. By our local decay rates

$$\text{Prob}\,\{l_{\log\log t}(\omega, \cdot) \sim f\} \simeq \exp[-\log\log t\, I(f)].$$

If we sample at times $t_n = \rho^n$ for some $\rho > 1$

$$\text{Prob}\,\{l_{\log\log t_n}(\omega, \cdot) \sim f\} \simeq \exp[-[\log n + \log\log \rho]\, I(f)] \sim \frac{1}{n^{I(f)}}.$$

Then by an application of the Borel Cantelli lemma and by standard arguments used in proving laws of the iterated logarithm one can conclude that the family $\{\hat{l}_t(\omega, \cdot)\}$ is conditionally compact as a family in the space of continuous functions with the topology of uniform convergence on compact sets and its limits points are precisely the set C of subprobability densities $f(x)$ with $\frac{1}{8} \int \frac{(f')^2}{f}\, dx \leq 1$.

In particular if Φ is a reasonable function on the space of nonnegative functions with integral at most one then

$$\limsup_{t \to \infty} \Phi(\hat{l}_t(\omega, \cdot)) = \sup_{f \in C} \phi(f) \qquad \text{a.e.}$$

we can try various functionals for $\phi(f)$. For example if we take $\Phi(f)=f(0)$ we get

$$\limsup_{t\to\infty} \frac{1}{\sqrt{t \log\log t}} \int_0^t \delta(x(s))\,ds = \sup_C f(0) = \sqrt{2} \quad \text{a.e.}$$

As another example if we take

$$\phi(f) = \inf\left\{a\colon \int_{-a}^{a} f(x)\,dx = 1\right\}$$

by an additional argument we can show that

$$\liminf_{t\to\infty} (\sup_{0\le s\le t} |x(s)|) \left(\frac{\sqrt{\log\log t}}{t}\right) = \inf_{f\in C} \phi(f) = \frac{\pi}{2\sqrt{2}} \quad \text{a.e.}$$

for these laws of the iterated logarithms and related references see [7]. For further generalizations regarding laws of iterated logarithm see [12] and [14].

13. Application: Wiener Sausage

The following problem comes up in the study of the behavior of the density of states for the Laplace operator in R^d with a random distribution of holes. The operator has Dirichlet boundary conditions at these holes. The centers of these holes have a Poisson point process for their distribution with some uniform intensity and the holes have some fixed size $\varepsilon > 0$.

Let $x(s)$, $0 \le s \le t$ be the d-dimensional Brownian motion conditioned to start at time 0 from the origin and end at time t again at the origin. Let C_t be the random set which is the range of this random Brownian trajectory. Let C_t^ε be the sausage around C_t of radius ε. That is to say

$$C_t^\varepsilon = \{y\colon |y - x| < \varepsilon \text{ for some } x \in C_t\}.$$

The set C_t^ε has some d-dimensional volume which we denote by $|C_t^\varepsilon|$. We let

$$a(t) = E_t\{\exp[-v|C_t^\varepsilon|]\} \tag{13.1}$$

where E_t refers to expectation with respect to the conditional Brownian motion in the interval $[0, t]$. The problem is to show that

$$\lim_{t\to\infty} \frac{1}{t^{\frac{d}{d+2}}} \log a(t) = -k_d(v) \tag{13.2}$$

exists in each dimension and is finite and nonzero.

As a first step it is straightforward to prove that the conditional Brownian motion can be replaced by the free Brownian motion starting at the origin for large times with negligible error so that it is sufficient to prove that

$$\lim_{t\to\infty} \frac{1}{t^{\frac{d}{d+2}}} \log b(t) = -k_d(v) \tag{13.3}$$

where

$$b(t) = E\{\exp[-v|C_t^\varepsilon|]\} \tag{13.4}$$

where E is expectation with respect to the ordinary Brownian motion starting from the origin.

By the Brownian scaling property one can show that

$$b(t) = E\{\exp[-v t^{d/(d+2)} |C_{t^{d/(d+2)}}^{\varepsilon t^{-1/(d+2)}}|]\}. \tag{13.5}$$

It is therefore sufficient to show that

$$\lim_{t\to\infty} \frac{1}{t} \log c(t) = -k_d(v) \tag{13.6}$$

where

$$c(t) = E\{\exp[-vt|C_t^{\varepsilon t^{-1/d}}|]\}. \tag{13.7}$$

We have just replaced $t^{d/(d+2)}$ by the new parameter t.

Formally we have a family of functionals on probability measures μ on R^d defined by

$$F_t(\mu) = |C_\mu^{\varepsilon t^{-1/\alpha}}| \tag{13.8}$$

where C_μ is the support of the measure μ. The range of a continuous trajectory is the support of its occupation distribution.

$$\lim_{t\to\infty} F_t(\mu) = F(\mu) = |C_\mu|. \tag{13.9}$$

The expression $C(t)$ is given by

$$c(t) = E\{\exp[-vt F_t(L_t(\omega, \cdot))]\}.$$

It is not unreasonable to expect that

$$\lim_{t\to\infty} \frac{1}{t} \log c(t) = \sup_f [-vF(f) - I(f)] = -\inf_f [F(f) + I(f)].$$

Since $\inf_{\text{supp } f \subset A} I(f) = \lambda(A)$ where $\lambda(A)$ is the lowest eigenvalue of $-\frac{1}{2}\Delta$ on A with Dirichlet boundary conditions, we can reduce the variational expression is (13.10) to

$$k_d(v) = \inf_A \{v|A| + \lambda(A)\}. \tag{13.11}$$

The infimum is seen to be attained over spheres by means of an isoperimetric inequality and this reduces to just a variation over the radius of the sphere.

$$k_d(v) = \inf_{v>0} \left\{ v r^d v_1 + \frac{\lambda_1}{r^2} \right\} \tag{13.12}$$

where v_1 is he volume of the unit sphere in d-dimensions and λ_1 is the lowest eigenvalue of $-\frac{1}{2}\Delta$ in the unit sphere with zero boundary conditions.

The analysis however is filled with difficulties. First of all the lower bound is fairly easy. If S_r is the sphere of radius r, then by insisting that the Brownian motion not leave the sphere for time t we can make $|C_t^{\varepsilon t^{-1/\alpha}}|$ be at most $v_1(r+\varepsilon t^{-1/d})^d$. The probability of this is approximately $\exp\left[-t\frac{\lambda_1}{r^2}\right]$ for large t. This gives the lower bound since r can be chosen arbitrarily.

The upper bound is technically hard because although the contribution when the range is fixed to be a given set has the right behavior the totality of such sets is huge.

It is here that we need to use our theory or more precisely some refinements of it. The Brownian motion on infinite space lacks the coerciveness condition imposed for upper bounds. We look instead at the Brownian motion on the d-dimensional torus of size l. Fortunately the volume of any set projected to the torus is at most the volume of the set so that any upper bound for any torus is automatically an upper bound for $c(t)$. We denote by

$$c_l(t) = E_l\{\exp[-vt|C_t^{\varepsilon t^{-1/d}}|]\} \tag{13.12}$$

where E_l is Brownian motion on the torus of size l in d dimensions. If we show

$$\lim_{t\to\infty} \frac{1}{t} \log c_l(t) = -\inf_A [v|A| + \lambda_l(A)] = k_{d,l}(v)$$

then it is not hard to show that

$$\lim_{l\to\infty} k_{d,l}(v) = k_d(v)$$

by standard methods.

Although our results apply to Brownian motion on the d torus, the basic topology we use is weak convergence of probability measures. The functional $F(\mu) = |\text{supp}\,\mu|$ is very unstable in this topology. It is however stable, at least lower semi continuous in the L_1 topology on probability densities. The sausage of radius $\varepsilon t^{-1/\alpha}$ offers a certain mollification. Consider a mollifier $\phi_t(x) = t\phi(t^{1/d}x)$ on the torus, where ϕ is supported in a sphere of radius ε around the origin. Then

$$|C_\mu^{\varepsilon t^{-1/\alpha}}| = |C_{\mu * \phi_t}|.$$

We now need to show that if we define a random density f_t by $f_t = L_t * \phi_t$ then the large deviation principle holds for the distributions of f_t in the L_1-topology. L_t here stands for the occupation time of a Brownian trajectory for duration t. This has been done in [3]. Once we have the basic large deviation principle in the L_1 topology the lower semi continuity of $F(f)$ as a functional of f in the L_1 topology is enough to give the upper bound. The details appear in [6].

14. Application: The Polaron Problem

In a statistical mechanical problem, see [11] for details, the following question comes up. Let $x(s)$, $0 \leq s \leq t$ be the 3-dimensional Brownian motion conditioned to start at time 0 and end at time t at the origin. We want to look at the integral

$$a(t)=E_t\left[\exp\left[\alpha \int_0^t \int_0^t \frac{e^{-(\sigma-s)}}{|x(\sigma)-x(s)|}\, d\sigma\, ds\right]\right]. \tag{14.1}$$

The problem is to show that

$$\lim_{t\to\infty} \frac{1}{t} \log a(t) = g(\alpha) \tag{4.2}$$

exists and that

$$\lim_{\alpha\to\infty} \frac{g(\alpha)}{\alpha^2} = g_0 \tag{14.3}$$

exists and is given by

$$g_0 = \sup_{\substack{\phi\in L_2(R^2)\\ \|\phi\|_2=1}} \left[2\iint \frac{\phi^2(x)\,\phi^2(y)}{|x-y|}\, dx\, dy - \tfrac{1}{2}\int |\nabla\phi|^2\, dx\right]. \tag{14.4}$$

First we replace the conditioned Brownian motion by free Brownian motion.

$$b(t)=E\left\{\exp\left[\alpha \int_0^t \int_0^t \frac{e^{-|\sigma-s|}}{|x(\sigma)-x(s)|}\, d\sigma\, ds\right]\right\} \tag{4.5}$$

where E is the expectation with respect to free Brownian motion. The difference is shown to be negligible for large t.

$$\int_0^t \int_0^t \frac{e^{-(\sigma-s)}}{|x(\sigma)-x(s)|}\, d\sigma\, ds = 2\int_0^t ds \int_0^t \frac{e^{-(\sigma-s)}}{|x(\sigma)-x(s)|}\, d\sigma \simeq 2\int_0^t F(\theta_s \sigma)\, ds \tag{14.6}$$

where

$$F(\omega)=\int_0^\infty \frac{e^{-\sigma}}{|x(\sigma)-x(0)|}\, d\sigma$$

and θ_s is the translation of the ω trajectory by s in time.

In view of the exponential factor the second integral can be taken up to ∞ for large t. We therefore look at

$$c(t)=E\left\{\exp\left[2\alpha\int_0^t F(\theta_s\omega)\,ds\right]\right\}. \tag{14.7}$$

Our general theory of Section 9 will tell us that

$$\lim_{t\to\infty}\frac{1}{t}\log c(t)=g(\alpha)=\sup_R\left\{E^R\left[2\alpha\int_0^\infty\frac{e^{-\sigma}}{|x(\sigma)-x(0)|}\right]-H(R)\right\} \tag{14.8}$$

where the supremum is over all stationary processes. By using Brownian scaling

$$\frac{g(\alpha)}{\alpha^2}=\sup_R\left[2E^R\left\{\int_0^\infty\frac{e^{-\sigma}\,d\sigma}{|x(\alpha^2\sigma)-x(0)|}\right\}-H(R)\right]. \tag{14.9}$$

By the ergodic theorem, if R is ergodic

$$\lim_{\alpha\to\infty}E^R\left\{\int_0^\infty\frac{e^{-\sigma}}{|x(\alpha^2\sigma)-x(0)|}\,d\sigma\right\}=\iint\frac{r(dx)\,r(dy)}{|x-y|}$$

where $r(dx)$ is the one dimensional marginal of R. Essentially $x(\alpha^2\sigma)$ and $x(\sigma)$ are nearly independent for large α. It is quite reasonable therefore to expect that

$$\lim_{\alpha\to\infty}\frac{g(\alpha)}{\alpha^2}=g_0=\sup_R\left[2\iint\frac{r(dx)\,r(dy)}{|x-y|}-H(R)\right].$$

By the contraction principle if we fix $r(dx)$ at a density $f(x)\,dx$ then

$$\inf_{R:\,r(dx)=f(x)\,dx}H(R)=I(f)=\frac{1}{8}\int\frac{|\nabla f|^2}{f}\,dx.$$

Therefore taking $f(x)=\phi^2(x)$ for some L_2 function $\phi(x)$ we conclude that

$$g_0=\sup_{\substack{\phi\in L_2(R^3)\\ \|\phi\|_2=1}}\left[2\iint\frac{\phi^2(x)\phi^2(y)}{|x-y|}\,dx\,dy-\frac{1}{2}\int\frac{|\nabla\phi|^2}{\phi}\,dx\right].$$

The derivation of (14.8) still has some difficulties. As usual the infinite volume creates problems with the upper bound. It turns out that replacing Brownian motion by the Ornstein-Uhlenbeck process with generator $\frac{1}{2}\Delta-\varepsilon\langle x\cdot\nabla\rangle$ one can finesse the problem. The details can be found in [9].

References

1. Bahadur, R.R. and Zabell, S., Large Deviation of the Sample Mean in General Vector Spaces, Annals of Prob. 7 (1979), 587–621.

2. Cramér, H., On a new limit theorem in the theory of probability: Colloquium on the Theory of Probability, Hermann, Paris (1937).
3. Donsker, M.D. and Varadhan, S.R.S., Asymptotic evaluation of certain Markov process expectations for large time, II, Comm. Pure Appl. Math. 28 (1975) 279–301.
4. Donsker, M.D., and Varadhan, S.R.S., Asymptotic evaluation of certain Markov process expectations for large time, III, Comm. Pure Appl. Math. 29 (1977), 389–461.
5. Donsker, M.D. and Varadhan, S.R.S., Asymptotic Evaluation of Certain Markov process expectations for large time, IV, Comm. Pure Appl. Math. 36 (1983), 183–212.
6. Donsker, M.D. and Varadhan, S.R.S., Asymptotics for the Wiener sausage, Comm. Pure Appl. Math. 28 (1975), 525–565.
7. Donsker, M.D. and Varadhan, S.R.S., On the laws of the iterated logarithms for local times, Comm. Pure Appl. Math. 30 (1977), 707–753.
8. Donsker, M.D. and Varadhan, S.R.S., some Problems of Large Deviations, Symposia Mathematica, Vol. XXI (1977) 313–318.
9. Donsker, M.D. and Varadhan, S.R.S., Asymptotics for the Polaron, Comm. Pure Appl. Math. 36 (1983), 505–528.
10. Donsker, M.D. and Varadhan, S.R.S., Large Deviations for Gaussian Processes (preprint).
11. Feynman, R.P., Statistical Mechanics, W.A. Benjamin, Reading, MA (1972)
12. Jain, N.C., A. Donsker-Varadhan type of invariance principle, Z. Wahrsch. Verw. Gebiete 59 (1982), 117–138.
13. Lanford, Oscar, Entropy and equilibrium states in classical statistical mechanics, Statistical Mechanics and Mathematical Problems, Lecture notes in Physics, Vol. 20, Springer-Verlag, Berlin and New York (1971), 1–113.
14. Mueller, Carl, Strassen's Law for Local Time, to appear.
15. Varadhan, S.R.S., Asymptotic probabilities and differential equations, Comm. Pure Appl. Math. 19 (1966), 261–286.
16. Varadhan, S.R.S., Large Deviations and Applications, CBMS-NSF Regional Conference Series, Number 46, SIAM (1984).

Received 29. 5. 84

Courant Institute
of Mathematical Sciences
New York University
251 Mercer St.
N.Y. 10012

Journal of Statistical Physics, Vol. 46, Nos. 5/6, 1987

Large Deviations for Noninteracting Infinite-Particle Systems

M. D. Donsker[1] and S. R. S. Varadhan[1]

Received November 25, 1986

A large deviation property is established for noninteracting infinite particle systems. Previous large deviation results obtained by the authors involved a single *I*-function because the cases treated always involved a unique invariant measure for the process. In the context of this paper there is an infinite family of invariant measures and a corresponding infinite family of *I*-functions governing the large deviations.

KEY WORDS: Large deviations; infinite particle systems; invariant measures; asymptotics for expectations.

1. INTRODUCTION

Let X be a countable set and let $X_0, X_1, X_2,...$ be a Markov chain with state space X and transition probabilities $\{\pi_{xy}\}$. Probabilities and expectations for this chain starting from a point $x \in X$ will be denoted by $P_x^\pi\{\ \}$ and $E_x^\pi\{\ \}$, respectively. We make five assumptions about this chain: it is irreducible, transient, the matrix $\{\pi_{xy}\}$ is doubly stochastic, and if for any finite set $F \subset X$ we define $\tau_F = \inf_{j \geqslant 1}\{X_j \in F\}$, then we assume $\lim_{x \to \infty} P_x^\pi\{\tau_F < \infty\} = 0$, i.e., for any $\varepsilon > 0$ there exists another finite set F_1 such that $F \subset F_1$ and $x \notin F_1$ implies $P_x^\pi\{\tau_F < \infty\} < \varepsilon$. Finally, we assume that the Green's function $G(x, y)$ for the chain has the property $\sup_{x \in X} G(x, x) < \infty$.

For each $x \in X$, let $n_0(x)$ be a nonnegative integer giving the number of particles at x initially, so that $\{n_0(x), x \in X\}$ is the initial configuration of particles. At time 1 all of these particles move independently according to the transition probabilities $\{\pi_{xy}\}$, giving us a new configuration

---

[1] Courant Institute of Mathematical Sciences, New York University, New York, New York.

$\{n_1(x), x \in X\}$. There is no interaction here; all the $n_0(x)$ particles at x leave independently of one another and for any other point $y \in X$, the $n_0(y)$ particles at y leave independently not only of each other, but of all the particles at x. At time 2 the process is repeated, giving us a new configuration $\{n_2(x), x \in X\}$ and so on. If we let $N = \{0, 1, 2,...\}$, then $Z = N^X$ is the space of all configurations, and if at time j we have a particular configuration $n(\cdot) \in Z$, we can ask for the transition probability $\hat{\pi}(n(\cdot), n'(\cdot))$ that at time $j+1$ we have configuration $n'(\cdot) \in Z$.

From our assumptions, the transition probability $\hat{\pi}(n(\cdot), n'(\cdot))$ is given through its moment generating function by

$$\int_Z \left\{\exp\left[-\sum_{x \in X} \lambda(x)\, n'(x)\right]\right\} \hat{\pi}(n(\cdot), dn'(\cdot)) = \prod_{x \in X} \left(\sum_{y \in X} \{\exp[-\lambda(y)]\}\, \pi_{xy}\right)^{n(x)} \tag{1.1}$$

where $\lambda: X \to \mathbb{R}$ vanishes outside some finite subset of X.

Important for us will be the family $\underline{A}$ of σ-finite invariant measures $\alpha(\cdot)$ for π_{xy}, i.e., for all $x \in X$, $0 \leqslant \alpha(x) < \infty$, $\sum_{x \in X} \alpha(x)\, \pi_{xy} = \alpha(y)$, and we assume $\alpha(\cdot) \not\equiv 0$. Since the chain is irreducible, if $\alpha(\cdot)$ vanishes at any point, it vanishes everywhere, so actually $\alpha(x) > 0$ for all $x \in X$.

For each $\alpha \in \underline{A}$ we define a measure P_α on Z as follows: for each $x \in X$ the number of particles at x, i.e., $n(x)$, is Poisson-distributed with mean $\alpha(x)$ and $\{n(x), x \in X\}$ are mutually independent.

It is well known and, using (1.1), easy to show that for each $\alpha \in \underline{A}$, P_α is an invariant measure for $\hat{\pi}(\cdot, \cdot)$. Also, using the Kolmogorov zero-one law, we can show that for each $\alpha \in \underline{A}$, P_α is an ergodic measure on Z.

For each $x \in X$ and any positive integer N, let

$$L_N(x) = \frac{n_0(x) + n_1(x) + \cdots + n_{N-1}(x)}{N}$$

i.e., the average number of particles at site x during the first N steps. Let Σ be the space of sequences of particle configurations on X, i.e., $\{n_j(\cdot), j = 0, 1, 2,...\}$. Any initial configuration $n(\cdot)$ of particles and $\hat{\pi}(\cdot, \cdot)$ generate a probability measure on Σ which we will denote by $\underline{P}_n$ [we will often in the sequel use $n(\cdot)$ instead of $n_0(\cdot)$ as a generic notation for the initial configuration]. Let $\underline{M}$ be the space of σ-finite measures σ on X. Then, for each N, L_N maps Σ into $\underline{M}$, and we use this mapping to define a probability measure $Q_{n,N}$ on $\underline{M}$ by $Q_{n,N} = \underline{P}_n \cdot L_N^{-1}$, i.e., if $A \subset \underline{M}$, then $Q_{n,N}(A) = \underline{P}_n\{L_N \in A\}$.

Since, for each $\alpha \in \underline{A}$, P_α is an invariant measure for $\hat{\pi}(\cdot, \cdot)$ and each P_α

is ergodic, it follows from the ergodic theorem that as $N \to \infty$, $Q_{n,N} \Rightarrow \delta_\alpha$ where, of course, δ_α is the Dirac measure on $\underline{M}$ concentrated at $\alpha(\cdot)$. What is essential is that the statement $Q_{n,N} \Rightarrow \delta_\alpha$ holds *for almost all* $n(\cdot) \in Z$ *with respect to* P_α*-measure*. This makes sense since if $\alpha_1 \not\equiv \alpha_2$, then P_{α_1} and P_{α_2} are mutually singular measures on Z.

In this paper we prove that a large-deviation property holds for the $Q_{n,N}$ measure, i.e., we find a functional $I_\alpha(\sigma)$ that for each $\alpha \in \underline{A}$ maps $\underline{M}$ into $[0, \infty]$ such that for each closed set $C \subset \underline{M}$,

$$\overline{\lim_{N \to \infty}} \frac{1}{N} \log Q_{n,N}(C) \leqslant - \inf_{\sigma \in C} I_\alpha(\sigma) \tag{1.2}$$

and for each open set $G \subset \underline{M}$,

$$\underline{\lim_{N \to \infty}} \frac{1}{N} \log Q_{n,N}(G) \geqslant - \inf_{\sigma \in G} I_\alpha(\sigma) \tag{1.3}$$

where both of the statements (1.2) and (1.3) hold for almost all $n(\cdot)$ with respect to P_α-measure on Z.

This last "almost all" statement emphasizes what is new in this paper relative to our previous work (see Refs. 1–3 for the theory, and Refs. 4–6 for some applications), namely that we have here more than one invariant measure. In our earlier papers we always imposed sufficient hypotheses on the underlying process so that there was a unique invariant measure and hence a single I-function governing large deviations. Here the underlying process has a whole family of invariant measures $\{P_\alpha, \alpha \in \underline{A}\}$, and we have an I-function therefore corresponding to each $\alpha \in \underline{A}$.

We obtain an explicit expression for $I_\alpha(\sigma)$, which we discuss now. Let $\underline{V}$ be the space of functions $V: X \to \mathbb{R}$ that vanish outside some finite subset of X and have the property that for some $x \in X$,

$$u_V(x) = u(x) = E_x^\pi \left\{ \exp \left[\sum_{j=0}^{\infty} V(X_j) \right] \right\} < \infty \tag{1.4}$$

It is an elementary consequence of the irreducibility hypothesis (see first few lines of Lemma 2.1 below) that $u(x) < \infty$ for all $x \in X$ if it is finite for any x.

Now, for $\alpha \in \underline{A}$ and $\sigma \in \underline{M}$, define

$$I_\alpha(\sigma) = \sup_{V \in \underline{V}} \left\{ \sum_{x \in X} \sigma(x) V(x) - \sum_{x \in X} \alpha(x)(1 - e^{-V(x)}) u(x) \right\} \tag{1.5}$$

and this is the I-function that enters into (1.2) and (1.3). For a given $\alpha \in \underline{A}$, $I_\alpha(\sigma)$ turns out to be finite if and only if σ behaves "asymptotically" like α. This is made explicit in Theorem 3.3 below. If $\alpha \in \underline{A}$ and $\beta \in \underline{A}$ and $\sigma \in \underline{M}$ exists such that both $I_\alpha(\sigma) < \infty$ and $I_\beta(\sigma) < \infty$, then $\alpha \equiv \beta$ (Lemma 3.4). In

Section 3 other properties of this I-function are established. In Section 2 we prove the upper bound (1.2) and in Section 4 we prove the lower bound (1.3).

Once such a large-deviation property for $Q_{n,N}$ is established, i.e., once one has an $I_\alpha(\sigma)$ for which (1.2) and (1.3) hold, then one easily obtains[7] an asymptotic result for certain expectations with respect to the $\underline{P}_n$ measure. To make this explicit, let Φ: $\underline{M} \to \mathbb{R}$ be bounded and continuous; then (1.2) and (1.3) imply

$$\lim_{N \to \infty} \frac{1}{N} \log E^{\underline{P}_n}\{e^{N\Phi(L_N(\cdot))}\} = \sup_{\sigma \in \underline{M}} [\Phi(\sigma) - I_\alpha(\sigma)] \tag{1.6}$$

for almost all $n(\cdot) \in Z$ (P_α-measure).

To obtain large-deviation properties where more than one invariant measure is involved, the present context of infinite-particle systems is natural. Here, in the simplest situation of noninteracting particles, we obtain an explicit I-function (1.5) in terms of which (1.2) and (1.3) are proved. In the case of infinite-particle systems with interaction the problem is to find the appropriate I-function in terms of which the analogues of (1.2) and (1.3) can be established.

Some interesting large-deviation results for infinite-particle systems with interaction have been found by Cox and Griffeath.[9] Lee[8] obtained a large-deviation property with explicit I-function for an infinite-particle system with no interaction, but where the particles move according to independent Brownian motions with constant drift.

2. THE UPPER BOUND

Using the definitions and notation of the introduction, let $V \in \underline{V}$ and C be a closed subset of $\underline{M}$. Then,

$$\begin{aligned}
&E^{\underline{P}_n}\left\{\exp\left[\sum_{j=0}^{N-1} \sum_{x \in X} n_j(x)\, V(x)\right]\right\} \\
&\quad = E^{\underline{P}_n}\left\{\exp\left[N \sum_{x \in X} L_N(x)\, V(x)\right]\right\} \\
&\quad = E^{Q_{n,N}}\left\{\exp\left[N \sum_{x \in X} \sigma(x)\, V(x)\right]\right\} \\
&\quad \geqslant E^{Q_{n,N}}\left\{\exp\left[N \sum_{x \in X} \sigma(x)\, V(x)\right]; \sigma \in C\right\} \\
&\quad \geqslant \left\{\exp\left[N \inf_{\sigma \in C} \sum_{x \in X} \sigma(x)\, V(x)\right]\right\} Q_{n,N}(C)
\end{aligned} \tag{2.1}$$

In Theorem 2.10 below we show

$$\lim_{N\to\infty} \frac{1}{N} \log E^{P_n}\left\{\exp\left[\sum_{j=0}^{N-1} \sum_{x\in X} n_j(x)\, V(x)\right]\right\}$$
$$= \sum_{x\in X} \alpha(x)(1-e^{-V(x)})\, u(x) \tag{2.2}$$

for almost all $n(\cdot)\in Z$ (P_α-measure). We recall from the introduction that

$$u(x) = E_x^\pi\left\{\exp\left[\sum_{j=0}^{\infty} V(X_j)\right]\right\}$$

and that if $\alpha_1 \not\equiv \alpha_2$, then P_{α_1} and P_{α_2} are mutually singular measures on Z.

Thus, from (2.1) and (2.2) we conclude

$$\overline{\lim_{N\to\infty}} \frac{1}{N} \log Q_{n,N}(C)$$
$$\leqslant -\inf_{\sigma\in C} \sum_{x\in X} \sigma(x)\, V(x) + \sum_{x\in X} \alpha(x)(1-e^{-V(x)})\, u(x)$$
$$= -\inf_{\sigma\in C}\left[\sum_{x\in X} \sigma(x)\, V(x) - \sum_{x\in X} \alpha(x)(1-e^{-V(x)})\, u(x)\right] \tag{2.3}$$

for almost all $n(\cdot)\in Z$ (P_α-measure). Since (2.3) holds for any $V\in \underline{V}$, we obtain

$$\overline{\lim_{N\to\infty}} \frac{1}{N} \log Q_{n,N}(C)$$
$$\leqslant -\sup_{V\in \underline{V}} \inf_{\sigma\in C}\left[\sum_{x\in X} \sigma(x)\, V(x) - \sum_{x\in X} \alpha(x)(1-e^{-V(x)})\, u(x)\right] \tag{2.4}$$

for almost all $n(\cdot)\in Z$ (P_α-measure). In Lemma 2.11 below we show that, because C is closed, (2.4) implies

$$\overline{\lim_{N\to\infty}} \frac{1}{N} \log Q_{n,N}(C)$$
$$\leqslant -\inf_{\sigma\in C} \sup_{V\in \underline{V}}\left[\sum_{x\in X} \sigma(x)\, V(x) - \sum_{x\in X} \alpha(x)(1-e^{-V(x)})\, u(x)\right]$$
$$= -\inf_{\sigma\in C} I_\alpha(\sigma) \tag{2.5}$$

for almost all $n(\cdot)\in Z$ (P_α-measure).

Now (2.5) is just the upper bound (1.2) of the introduction, and hence what needs to be proved in this section is (2.2) and Lemma 2.11. We prove (2.2) as Theorem 2.10 after a succession of preliminary lemmas.

Let $\underline{V}_1$ be the space of functions $V: X \to \mathbb{R}$ such that V vanishes outside some finite subset of X and

$$u(x) = E_x^\pi\left\{\exp\left[\sum_{j=0}^{\infty} V(X_j)\right]\right\} < \infty$$

for *all* $x \in X$.

For F a finite subset of X, let $\tau = \tau_F = \inf_{j \geq 1}\{X_j \in F\}$, i.e., the first entry time of the Markov chain into F, and let $\pi_{xy}^F = P_x^\pi\{X_\tau = y, \tau < \infty\}$. Let M_F be the matrix $\{e^{V(x)}\pi_{xy}^F, x, y \in F\}$, where V vanishes outside F, and let $\rho(M_F)$ be the spectral radius of M_F. Define $\underline{V}_2$ to be the space of functions $V: X \to \mathbb{R}$ such that V vanishes outside some finite set $F = F_V$ and $\rho(M_F) < 1$.

Lemma 2.1. $\underline{V} = \underline{V}_1 = \underline{V}_2$.

Proof. Let $V \in \underline{V}$ and $x \in X$ such that

$$u(x) = E_x^\pi\left\{\exp\left[\sum_{j=0}^{\infty} V(X_j)\right]\right\} < \infty$$

Since the chain is irreducible, for any $y \in X$ there is a positive integer k such that $\pi_{xy}^{(k)} > 0$. If we let

$$\|V\| = \sup_{z \in F = \operatorname{supp} V} |V(z)|$$

then

$$\begin{aligned}
\infty &> E_x^\pi\left\{\exp\left[\sum_{j=0}^{\infty} V(X_j)\right]\right\} \\
&\geq [\exp(-k\,\|V\|)]\, E_x^\pi\left\{\exp\left[\sum_{j=k}^{\infty} V(X_j)\right]\right\} \\
&\geq [\exp(-k\,\|V\|)]\, E_x^\pi\left\{\exp\left[\sum_{j=k}^{\infty} V(X_j)\right]; X_k = y\right\} \\
&= [\exp(-k\,\|V\|)]\, \pi_{xy}^{(k)} E_y^\pi\left\{\exp\left[\sum_{j=0}^{\infty} V(X_j)\right]\right\}
\end{aligned}$$

so that

$$E_y^\pi\left\{\exp\left[\sum_{j=0}^{\infty} V(X_j)\right]\right\} < \infty$$

which means $V \in \underline{V}_1$ and hence $\underline{V} = \underline{V}_1$.

To show $\underline{V}_1 \subset \underline{V}_2$, let $V \in \underline{V}_1$, so that

$$u(x) = E_x^\pi \left\{ \exp\left[\sum_{j=0}^{\infty} V(X_j) \right] \right\} < \infty$$

for all $x \in X$. Let $F \subset X$ be the finite set outside of which V vanishes, and with τ and π_{xy}^F as defined just before this lemma, let

$$\eta_x = P_x^\pi\{\tau = \infty\} = 1 - \sum_{y \in F} \pi_{xy}^F$$

We note that, since the chain is transient, $\eta_x > 0$ for at least one $x \in F$.

Now, for any $x \in F$, using the strong Markov property of the chain with respect to τ,

$$\begin{aligned}
u(x) &= E_x^\pi \left\{ \exp\left[\sum_{j=0}^{\infty} V(X_j) \right] \right\} \\
&= [\exp V(x)]\, E_x^\pi \left\{ \exp\left[\sum_{j=1}^{\infty} V(X_j) \right]; \tau < \infty \right\} \\
&\quad + [\exp V(x)]\, E_x^\pi \left\{ \exp\left[\sum_{j=1}^{\infty} V(X_j) \right]; \tau = \infty \right\} \\
&= [\exp V(x)]\, E_x^\pi \left\{ \exp\left[\sum_{j=1}^{\infty} V(X_j) \right]; \tau < \infty \right\} \\
&\quad + [\exp V(x)]\, P_x^\pi\{\tau = \infty\} \\
&= [\exp V(x)] \sum_{y \in F} \pi_{xy}^F E_y^\pi \left\{ \exp\left[\sum_{j=0}^{\infty} V(X_j) \right] \right\} + [\exp V(x)]\, \eta_x \\
&= [\exp V(x)] \sum_{y \in F} \pi_{xy}^F u(y) + [\exp V(x)]\, \eta_x
\end{aligned} \tag{2.6}$$

Let u be the column vector with entries $\{u(x),\ x \in F\}$, $e^V \eta$ be the column vector with entries $\{e^{V(x)} \eta_x,\ x \in F\}$; then, with M_F the matrix defined above, we can rewrite (2.6) as

$$(I - M_F)\, u = e^V \eta \tag{2.7}$$

Since (2.7) has the strictly positive solution

$$u(x) = E_x^\pi \left\{ \exp\left[\sum_{j=0}^{\infty} V(X_j) \right] \right\}, \qquad x \in F$$

we conclude that $\rho(M_F) < 1$, i.e., $V \in \underline{V}_2$.

Finally, to show $\underline{V}_2 \subset \underline{V}_1$, let $V \in \underline{V}_2$ with finite support F. Since we are now assuming $\rho(M_F) < 1$ and, as noted earlier, at least one entry in $e^V \eta$ is strictly positive, the matrix equation $(I - M_F)\, u = e^V \eta$ for u has a solution $\bar{u}(x)$, $x \in F$. Indeed,

$$\bar{u} = (I - M_F)^{-1}(e^V \eta) \tag{2.8}$$

and since the chain is irreducible, $\bar{u} = \{\bar{u}(x), x \in F\}$ has strictly positive entries. We extend the column vector $\bar{u}$ by defining for $y \notin F$,

$$\begin{aligned} \bar{u}(y) &= \sum_{z \in F} \pi^F_{yz} \bar{u}(z) + P^\pi_y\{\tau = \infty\} \\ &= \sum_{z \in F} \pi^F_{yz} \bar{u}(z) + 1 - \sum_{z \in F} \pi^F_{yz} \end{aligned} \tag{2.9}$$

so that now $\bar{u}$ is a column vector having entries for all the elements of X. Thus, for any $x \in X$,

$$\begin{aligned} \sum_{y \in X} \pi_{xy} \bar{u}(y) &= \sum_{y \in F} \pi_{xy} \bar{u}(y) + \sum_{y \notin F} \pi_{xy} \bar{u}(y) \\ &= \sum_{y \in F} \pi_{xy} \bar{u}(y) + \sum_{y \notin F} \pi_{xy} \sum_{z \in F} \pi^F_{yz} \bar{u}(z) + \sum_{y \notin F} \pi_{xy} P^\pi_y\{\tau = \infty\} \end{aligned} \tag{2.10}$$

But,

$$\begin{aligned} \sum_{z \in F} \pi^F_{xz} \bar{u}(z) &= E^\pi_x\{\bar{u}(X_\tau); \tau < \infty\} \\ &= E^\pi_x\{\bar{u}(X_\tau); \tau = 1\} + E^\pi_x\{\bar{u}(X_\tau); 2 \leqslant \tau < \infty\} \\ &= \sum_{y \in F} \pi_{xy} \bar{u}(y) + \sum_{y \notin F} \pi_{xy} \sum_{z \in F} \pi^F_{yz} \bar{u}(z) \end{aligned}$$

and using this in (2.10), we have, for any $x \in X$,

$$\begin{aligned} \sum_{y \in X} \pi_{xy} \bar{u}(y) &= \sum_{z \in F} \pi^F_{xz} \bar{u}(z) + \sum_{y \notin F} \pi_{xy} P^\pi_y\{\tau = \infty\} \\ &= \sum_{z \in F} \pi^F_{xz} \bar{u}(z) + P^\pi_x\{\tau = \infty\} \end{aligned} \tag{2.11}$$

First consider (2.11) for $x \notin F$. Using (2.9), we obtain

$$\sum_{y \in X} \pi_{xy} \bar{u}(y) = \bar{u}(x)$$

On the other hand, if $x \in F$, then (2.11) and (2.8) state that for $V \in \underline{V}_2$,

$$\begin{aligned} e^{V(x)} &\sum_{y \in X} \pi_{xy} \bar{u}(y) \\ &= e^{V(x)} \sum_{z \in F} \pi^F_{xy} \bar{u}(z) + e^{V(x)} P^\pi_x\{\tau = \infty\} \\ &= e^{V(x)} \sum_{z \in F} \pi^F_{xz} \bar{u}(z) + e^{V(x)} \eta_x \\ &= (M_F \bar{u})_x + (e^V \eta)_x = \bar{u}(x) \end{aligned}$$

Since V vanishes outside of F, we conclude that for all $x \in X$,

$$e^{V(x)} \sum_{y \in X} \pi_{xy} \bar{u}(y) = \bar{u}(x)$$

or, with the abbreviated notation $(\pi\bar{u})(x) = \sum_{y \in X} \pi_{xy} \bar{u}(y)$, we have for all $x \in X$

$$e^{V(x)} (\pi\bar{u})(x) = \bar{u}(x) \tag{2.12}$$

Identity (2.12) holds for a $V \in \underline{V}_2$ and $\bar{u}$ defined by (2.8) and (2.9). In this connection we should note that for $V \in \underline{V}$ the same identity holds for

$$\begin{aligned} u(x) &= E^\pi_x \left\{ \exp\left[\sum_{j=0}^{\infty} V(X_j) \right] \right\} \\ &= [\exp V(x)]\, E^\pi_x \left\{ \exp\left[\sum_{j=1}^{\infty} V(X_j) \right] \right\} \\ &= [\exp V(x)] \sum_{y \in X} \pi_{xy} E^\pi_y \left\{ \exp\left[\sum_{j=0}^{\infty} V(X_j) \right] \right\} \\ &= [\exp V(x)] (\pi u)(x) \end{aligned} \tag{2.13}$$

Since $\bar{u}(x)$ is strictly positive for $x \in F$, we have

$$0 < \inf_{x \in F} \bar{u}(x) \leqslant \sup_{x \in F} \bar{u}(x) < \infty$$

From this and (2.9) we see that for all $y \in X$

$$0 < \min[1, \inf_{z \in F} \bar{u}(z)] \leqslant \bar{u}(y) \leqslant \max[\sup_{z \in F} \bar{u}(z), 1] < \infty \tag{2.14}$$

Now, (2.12) implies that the sequence of random variables

$$Z_n = \bar{u}(X_n) \exp\left[\sum_{j=0}^{n-1} V(X_j) \right]$$

is a martingale because if $\underline{F}_{n-1}$ is the σ-field up to time $n-1$,

$$E_x^\pi\{Z_n|\underline{F}_{n-1}\} = \left\{\exp\left[\sum_{j=0}^{n-1} V(X_j)\right]\right\}(\pi\bar{u})(X_{n-1})$$
$$= \left\{\exp\left[\sum_{j=0}^{n-2} V(X_j)\right]\right\}\bar{u}(X_{n-1}) = Z_{n-1}$$

Thus, for all $x \in X$,

$$E_x^\pi\left\{\left[\exp\sum_{j=0}^{n-1} V(X_j)\right]\bar{u}(X_n)\right\} = \bar{u}(x) \tag{2.15}$$

and from this and (2.14) we conclude, for all $x \in X$,

$$E_x^\pi\left\{\exp\left[\sum_{j=0}^{n-1} V(X_j)\right]\right\} \leqslant \frac{\bar{u}(x)}{\inf_{y\in X}\bar{u}(y)} < \infty$$

Finally, from Fatou's lemma, this last inequality implies

$$u(x) = E_x^\pi\left\{\exp\left[\sum_{j=0}^{\infty} V(X_j)\right]\right\} < \infty$$

for all $x \in X$, i.e., $V \in \underline{V}_1$, and the lemma is proved.

Lemma 2.2. Let F be a finite subset of X and let $\phi(x) = P_x^\pi\{\inf_{j\geqslant 1}(X_j \in F) < \infty\}$. Then for all $x \in X$, $\lim_{n\to\infty} E_x^\pi\{\phi(X_n)\} = 0$.

Proof. This is obvious, since

$$E_x^\pi\{\phi(X_n)\} = P_x^\pi\{X_j \in F, j > n\}$$
$$= P_x^\pi\{\text{last visit of chain to } F \text{ occurs after time } n\}$$

and the last probability must go to 0 as $n \to \infty$ because the chain is transient.

Corollary. With $\bar{u}(x)$ defined for $x \in F$ by (2.8) and for $x \notin F$ by (2.9), we have for all $x \in X$,

$$\lim_{n\to\infty} E_x^\pi\{|\bar{u}(x_n) - 1|\} = 0$$

Proof. With $\phi(x)$ as defined in Lemma 2.2, we have for $x \notin F$,

$$\bar{u}(x) = \sum_{y\in F} \pi_{xy}^F \bar{u}(y) + \left(1 - \sum_{y\in F} \pi_{xy}^F\right)$$
$$= \sum_{y\in F} \pi_{xy}^F \bar{u}(y) + [1 - \phi(x)]$$

Thus,

$$E_x^\pi\{|\bar{u}(X_n)-1|\} \leqslant \sup_{y\in F} \bar{u}(y)\, E_x^\pi\{\phi(X_n)\} + E_x^\pi\{\phi(X_n)\}$$

and from the lemma, for all $x \in X$,

$$\lim_{n\to\infty} E_x^\pi\{|\bar{u}(X_n)-1|\} = 0$$

In this proof there was no loss in generality by taking $\bar{u}(x)$ as defined for $x \notin F$, since F is finite and the chain is transient.

Lemma 2.3. With $\bar{u}(x)$ defined by (2.8) for $x \in F$ and by (2.9) for $x \notin F$, we have $\bar{u}(x) = u(x)$ for all $x \in X$.

Proof. We showed in (2.15) that for *all* $x \in X$,

$$\bar{u}(x) = E_x^\pi \left\{ \left[\exp \sum_{j=0}^{N-1}{}' V(X_j) \right] \bar{u}(X_N) \right\}$$

and, because of the corollary to Lemma 2.2, the present lemma will follow if we can show that

$$\left\{ \left[\exp \sum_{j=0}^{N-1}{}' V(X_j) \right] \bar{u}(X_N) \right\}$$

are uniformly integrable. Now, by (2.14),

$$0 < \inf_{y\in X} \bar{u}(y) \leqslant \sup_{y\in X} \bar{u}(y) < \infty$$

so it suffices to show that $\{\exp \sum_{j=0}^{N-1}{}' V(X_j)\}$ are uniformly integrable. This follows if, for some $\lambda > 1$,

$$\sup_N E_x^\pi \left\{ \exp\left[\lambda \sum_{j=0}^{N-1}{}' V(X_j) \right] \right\} \leqslant \text{const}$$

But in Lemma 2.1 we showed that if $V \in \underline{V}$, then it is in $\underline{V}_2$, i.e., $\rho(M^F) < 1$, where M^F is the matrix $\{e^{V(x)}\pi_{xy}^F,\ x, y \in F\}$. Let $M_\lambda^F = \{e^{\lambda V(x)}\pi_{xy}^F,\ x, y \in F\}$. Since the matrix is finite, its spectral radius $\rho(M_\lambda^F)$ is a continuous function of λ. Since the spectral radius is strictly less than 1 when $\lambda = 1$, there then exists $\lambda > 1$ such that $\rho(M_\lambda^F) < 1$ also. But, again by Lemma 2.1, this means for that λ, $\lambda V \in \underline{V}$, which implies

$$\sup_N E_x^\pi \left\{ \exp\left[\lambda \sum_{j=0}^{N-1}{}' V(X_j) \right] \right\} \leqslant \text{const}$$

For any $\alpha \in \underline{A}$ we introduce the Markov chain time-reversed with respect to α, i.e., for $x, y \in X$, let $\pi_{xy}(\alpha) = \pi_{yx}\alpha(y)/\alpha(x)$. The $\pi_{xy}(\alpha)$ are transition probabilities, since

$$\sum_{y \in X} \pi_{xy}(\alpha) = \frac{1}{\alpha(x)} \sum_{y \in X} \pi_{yx}\alpha(y) = \frac{\alpha(x)}{\alpha(x)} = 1$$

Since $\pi_{xy}^{(k)}(\alpha) = \pi_{yx}^{(k)}\alpha(y)/\alpha(x)$, the time-reversed chain is also irreducible. Since the original chain is transient, it has a finite Green's function, i.e., for all $x, y \in X$, $G(x, y) = \sum_{k=0}^{\infty} \pi_{xy}^{(k)} < \infty$. But the Green's function for the reversed chain $G^{\alpha}(x, y)$ is given by

$$G^{\alpha}(x, y) = \sum_{k=0}^{\infty} \pi_{xy}^{(k)}(\alpha) = \sum_{k=0}^{\infty} \pi_{yx}^{(k)} \frac{\alpha(y)}{\alpha(x)} = \frac{\alpha(y)}{\alpha(x)} G(y, x) < \infty$$

which means the reversed chain is also transient. We denote probabilities and expectations with respect to the reversed chain by $P_x^{\pi(\alpha)}\{\ \}$ and $E_x^{\pi(\alpha)}\{\ \}$, respectively.

Let $\underline{V}^{\alpha}$ be the space of functions $V: X \to \mathbb{R}$ that vanish outside some finite set and such that for some $x \in X$,

$$E_x^{\pi(\alpha)} \left\{ \exp\left[\sum_{j=0}^{\infty} V(X_j) \right] \right\} < \infty$$

Now, in the proof of Lemma 2.1, we used only the hypotheses that the original chain was irreducible and transient. Hence, Lemma 2.1 applies also to the time-reversed chain and thus $\underline{V}^{\alpha} = \underline{V}_1^{\alpha} = \underline{V}_2^{\alpha}$.

Lemma 2.4. For any $\alpha \in \underline{A}$,

$$\underline{V} = \underline{V}^{\alpha}$$

Proof. Let $\alpha \in \underline{A}$, F be a finite subset of X, and $x, y \in F$. Then,

$$\pi_{xy}^{F}(\alpha) = \pi_{xy}(\alpha) + \sum_{z_1 \in F^c} \pi_{xz_1}(\alpha)\, \pi_{z_1 y}(\alpha) + \sum_{z_1, z_2 \in F^c} \pi_{xz_1}(\alpha)\, \pi_{z_1 z_2}(\alpha)\, \pi_{z_2 y}(\alpha) + \cdots$$

But

$$\pi_{z_1 z_2}(\alpha) = \pi_{z_2 z_1}\alpha(z_2)/\alpha(z_1)$$

so that

$$\pi^F_{xy}(\alpha) = \pi_{yx}\frac{\alpha(y)}{\alpha(x)} + \sum_{z_1 \in F^c} \pi_{yz_1}\frac{\alpha(y)}{\alpha(z_1)}\pi_{z_1x}\frac{\alpha(z_1)}{\alpha(x)} + \cdots$$
$$= \frac{\alpha(y)}{\alpha(x)}\pi^F_{yx}$$

Hence, for a V vanishing outside F, the matrix

$$M_F(\alpha) = \{e^{V(x)}\pi^F_{xy}(\alpha),\ x,\ y \in F\} = \left\{e^{V(x)}\pi^F_{yx}\frac{\alpha(y)}{\alpha(x)},\ x,\ y \in F\right\}$$

Since $M_F = \{e^{V(x)}\pi^F_{xy},\ x,\ y \in F\}$, we see that if we let D be the diagonal matrix

$$D = (e^{-V(y)}\alpha(y)\,\delta_{xy},\ x,\ y \in F)$$

then

$$M_F(\alpha) = D^{-1}M_F^TD$$

which implies $\rho(M_F(\alpha)) = \rho(M_F)$. From Lemma 2.1 and the remarks just preceding this lemma we conclude that $\underline{V} = \underline{V}^\alpha$.

Lemma 2.5. Let $V \in \underline{V}$ and define

$$u_N(x) = E^\pi_x\left\{\exp\left[\sum_{j=0}^{N-1} V(X_j)\right]\right\}$$

For $\alpha \in \underline{A}$,

$$\sup_N \sum_{x \in X} \alpha(x)\,|u_N(x) - u_{N-1}(x)| < \infty$$

Proof. It suffices to show

$$\sup_N \sum_{x \in X} \alpha(x)\,E^\pi_x\left\{\exp\left[\sum_{j=0}^{N-2} V(X_j)\right]|\exp[V(X_{N-1})] - 1|\right\} < \infty$$

and this will follow, since V has finite support, if we show that for every $y \in X$

$$\sup_N \sum_{x \in X} \alpha(x)\,E^\pi_x\left\{\left[\exp\sum_{j=0}^{N-2} V(X_j)\right]\delta_y(X_{N-1})\right\} < \infty \qquad (2.16)$$

From Lemma 2.4, $\underline{V} = \underline{V}^{\alpha}$, so we know

$$E_y^{\pi(\alpha)}\left\{\exp\left[\sum_{j=0}^{\infty} V(X_j)\right]\right\} < \infty$$

for every $y \in X$. Hence, for every $y \in X$,

$$\begin{aligned}\alpha(y) \sup_N E_y^{\pi(\alpha)}&\left\{\exp\left[\sum_{j=1}^{N-1} V(X_j)\right]\right\}\\ &= \alpha(y) \sup_N \sum_{x_1, x_2, \ldots, x_{N-1}} \left\{\exp\left[\sum_{j=1}^{N-1} V(x_j)\right]\right\}\\ &\quad \times \pi_{yx_1}(\alpha)\, \pi_{x_1x_2}(\alpha) \cdots \pi_{x_{n-2}x_{n-1}}(\alpha) < \infty\end{aligned}$$

Since $\pi_{xy}(\alpha) = \pi_{yx}\alpha(y)/\alpha(x)$, this last becomes

$$\begin{aligned}\sup_N \sum_{x_1, x_2, \ldots, x_{N-1}} &\alpha(x_{N-1}) \left\{\exp\left[\sum_{j=1}^{N-1} V(x_j)\right]\right\}\\ &\times \pi_{x_1 y} \pi_{x_2 x_1} \cdots \pi_{x_{N-1} x_{N-2}} < \infty \end{aligned} \tag{2.17}$$

Now, if we let $x_{N-1} = x$ and $x_{N-j} = x_{j-1}$, $j = 2, 3, \ldots, N-1$, (2.17) states that for $y \in X$,

$$\begin{aligned}\sup_N \sum_{x, x_1, x_2, \ldots, x_{N-2}} &\alpha(x)[\exp V(x)] \left\{\exp\left[\sum_{j=1}^{N-2} V(x_j)\right]\right\}\\ &\times \pi_{xx_1} \pi_{x_1x_2} \cdots \pi_{x_{N-2}y} < \infty\end{aligned}$$

i.e., for all $y \in X$,

$$\sup_N \sum_{x \in X} \alpha(x)\, E_x^{\pi}\left\{\left[\exp \sum_{j=0}^{N-2} V(X_j)\right] \delta_y(X_{N-1})\right\} < \infty$$

which is (2.16).

Lemma 2.6. $u_N(x) \to 1$ as $x \to \infty$ uniformly in N, i.e., for any $\delta > 0$ there exists a finite set F such that $x \notin F$ implies $|1 - u_N(x)| < \delta$ for all N

Proof. For any $V \in \underline{V}$,

$$E_x^{\pi}\left\{\left[\exp \sum_{j=0}^{N-1} V(X_j)\right] u(X_N)\right\} = u(x)$$

From Lemma 2.3 and (2.14) we then obtain

$$u_N(x) = E_x^\pi \left\{ \exp\left[\sum_{j=0}^{N-1} V(X_j) \right] \right\}$$

$$\leqslant \frac{\sup_{x \in X} u(x)}{\inf_{x \in X} u(x)} \leqslant \text{const} \quad \text{for all } N \text{ and all } x$$

Since, as noted in the proof of Lemma 2.3, $V \in \underline{V}$ implies $\lambda V \in \underline{V}$ for some $\lambda > 1$, we see that

$$\sup_{\substack{x \in X \\ N}} E_x^\pi \left\{ \exp\left[\lambda \sum_{j=0}^{N-1} V(X_j) \right] \right\} \leqslant \text{const}$$

for some $\lambda > 1$. This shows the uniform integrability of $\{\exp \sum_{j=0}^{N-1} V(X_j)\}$.

Now, one of our assumptions on the underlying chain is that $\lim_{x \to \infty} P_x^\pi\{\tau_F < \infty\} = 0$ for any finite subset $F \in X$. In particular, $\lim_{x \to \infty} P_x^\pi\{\tau_{\{y\}} < \infty\} = 0$, but if $y \neq x$, $P_x^\pi\{\tau_{\{y\}} < \infty\} = G(x, y)/G(y, y)$. Thus, we have $G(x, y) \to 0$ as $x \to \infty$ for every $y \in X$.

Moreover,

$$E_x^\pi \left\{ \left| \sum_{j=0}^{N-1} V(X_j) \right| \right\} \leqslant E_x^\pi \left\{ \sum_{j=0}^{\infty} |V(X_j)| \right\} = \sum_{y \in X} G(x, y) \, |V(y)|$$

and, since V has finite support, we see that

$$\sup_N E_x^\pi \left\{ \left| \sum_{j=0}^{N-1} V(X_j) \right| \right\} \to 0 \qquad \text{as} \quad x \to \infty$$

This, together with the uniform integrability shown above, implies the lemma.

Lemma 2.7. Let $V \in \underline{V}$ and $\alpha \in \underline{A}$. Then,

$$\lim_{N \to \infty} \left| \sum_{x \in X} \alpha(x)[\log u_N(x) - \log u_{N-1}(x)] - \sum_{x \in X} \alpha(x)[u_N(x) - u_{N-1}(x)] \right| = 0 \tag{2.18a}$$

and, for any real λ,

$$\lim_{N \to \infty} \left| \sum_{x \in X} \alpha(x)\{[u_N(x)]^\lambda - [u_{N-1}(x)]^\lambda\} - \lambda \sum_{x \in X} \alpha(x)[u_N(x) - u_{N-1}(x)] \right| = 0 \tag{2.18b}$$

Proof. Let $\varepsilon > 0$ be given. There exists a $\delta > 0$ such that $|1 - x| < \delta$ and $|1 - y| < \delta$ imply

$$|\log x - \log y - (x - y)| < \varepsilon\, |x - y|$$

For this δ, let F be the finite subset of X guaranteed by Lemma 2.6, so that $|1 - u_N(x)| < \delta$ for all N provided $x \notin F$. To show the first part of (2.18), it suffices to show

$$\begin{aligned}&\overline{\lim_{N\to\infty}} \Big| \sum_{x\in F} \alpha(x)[\log u_N(x) - \log u_{N-1}(x)] \\ &\quad - \sum_{x\in F} \alpha(x)[u_N(x) - u_{N-1}(x)] \Big| \\ &\quad + \overline{\lim_{N\to\infty}} \Big| \sum_{x\in F^c} \alpha(x)[\log u_N(x) - \log u_{N-1}(x)] \\ &\quad - \sum_{x\in F^c} \alpha(x)[u_N(x) - u_{N-1}(x)] \Big| = 0 \end{aligned} \tag{2.19}$$

Since $\lim_{N\to\infty} u_N(x) = u(x) > 0$ for all $x \in X$ and since summation over $x \in F$ is just a finite sum, the first term in (2.19) is zero. For the second term in (2.19) we get, from our choice of δ and F made above,

$$\begin{aligned}&\overline{\lim_{N\to\infty}} \Big| \sum_{x\in F^c} \alpha(x)[\log u_N(x) - \log u_{N-1}(x)] \\ &\quad - \sum_{x\in F^c} \alpha(x)[u_N(x) - u_{N-1}(x)] \Big| \\ &\quad \leqslant \overline{\lim_{N\to\infty}}\ \varepsilon \sum_{x\in F^c} \alpha(x)\, |u_N(x) - u_{N-1}(x)| \\ &\quad \leqslant \varepsilon \sup_N \sum_{x\in X} \alpha(x)\, |u_N(x) - u_{N-1}(x)| \end{aligned}$$

In Lemma 2.5 we showed that the multiple of ε in this last inequality is finite for any $V \in \underline{V}$ and any $\alpha \in \underline{A}$, so we have (2.19). The proof of the second statement in (2.18) is similar.

Lemma 2.8. For $V \in \underline{V}$, $\alpha \in \underline{A}$, and all $N = 1, 2, \ldots$,

$$\sum_{x\in X} \alpha(x)[(\pi u_N)(x) - u_N(x)] = 0 \tag{2.20}$$

Proof. If we can justify changing the order of summation, (2.20) follows immediately from the fact that α is an invariant measure, since

$$\sum_{x\in X} \alpha(x)[(\pi u_N)(x)-1] = \sum_{x\in X} \alpha(x)\left[\sum_{y\in X} \pi_{xy}u_N(y)-1\right]$$
$$= \sum_{x\in X} \alpha(x)[u_N(x)-1]$$

To justify the interchange, we show

$$\sum_{x\in X} \alpha(x)\,|(\pi u_N)(x)-1| < \infty$$

which is implied by

$$\sum_{x\in X} \alpha(x)\,|u_N(x)-1| < \infty \tag{2.21}$$

To show (2.21), we note that for any $x\in X$ and with $F=\text{supp}\,V$,

$$\begin{aligned}|u_N(x)-1| &= \left|E_x^\pi\left\{\exp\left[\sum_{j=0}^{N-1} V(X_j)\right]\right\}-1\right|\\ &\leqslant [\exp(N\,\|V\|)]\,P_x^\pi\{\text{at least on } X_j\in F,\ j=0,1,\dots,N-1\}\\ &= [\exp(N\,\|V\|)][\delta_x(F)+\pi^{(1)}(x,F)\\ &\quad+\pi^{(2)}(x,F)+\cdots+\pi^{(N-1)}(x,F)]\end{aligned}$$

where

$$\pi^{(k)}(x,F)=\sum_{y\in F}\pi_{xy}^{(k)}, \qquad k=1,2,\dots,N-1$$

Thus,

$$\begin{aligned}&\sum_{x\in X}\alpha(x)\,|u_N(x)-1|\\ &\quad\leqslant e^{N\|V\|}\sum_{x\in X}\alpha(x)[\delta_x(F)+\pi^{(1)}(x,F)+\cdots+\pi^{(N-1)}(x,F)]\end{aligned}$$

But $\sum_{x\in X}\alpha(x)\,\delta_x(F)=\alpha(F)$ and, for each $k=1,2,\dots,N-1$,

$$\sum_{x\in X}\alpha(x)\,\pi^{(k)}(x,F)=\sum_{x\in X}\alpha(x)\sum_{y\in F}\pi_{xy}^{(k)}=\sum_{y\in F}\alpha(y)=\alpha(F)$$

Since F is finite and α is a σ-finite measure,

$$\sum_{x \in X} \alpha(x)\, |u_N(x) - 1| \leqslant e^{N\|V\|} N\alpha(F) < \infty$$

Although we have just shown that for each N, $\sum_{x \in X} \alpha(x)\, |u_N(x) - 1| < \infty$, in fact $\sum_{x \in X} \alpha(x)\, |u(x) - 1|$ may be infinite.

Lemma 2.9. Let $V \in \underline{V}$ and $\alpha \in \underline{A}$. Then,

$$\lim_{N \to \infty} \sum_{x \in X} \alpha(x)[u_N(x) - u_{N-1}(x)] = \sum_{x \in X} \alpha(x)(1 - e^{-V(x)})\, u(x)$$

Proof. First of all,

$$\begin{aligned}
u_N(x) &= E_x^\pi \left\{ \exp\left[\sum_{j=0}^{N-1} V(X_j) \right] \right\} \\
&= [\exp V(x)]\, E_x^\pi \left\{ \exp\left[\sum_{j=1}^{N-1} V(X_j) \right] \right\} \\
&= [\exp V(x)] \sum_{y \in X} \pi_{xy} E_y^\pi \left\{ \exp\left[\sum_{j=0}^{N-2} V(X_j) \right] \right\} \\
&= [\exp V(x)] \sum_{y \in X} \pi_{xy} u_{N-1}(y) \\
&= [\exp V(x)](\pi u_{N-1})(x)
\end{aligned}$$

and therefore

$$\begin{aligned}
&\lim_{N \to \infty} \sum_{x \in X} \alpha(x)[u_N(x) - u_{N-1}(x)] \\
&\quad = \lim_{N \to \infty} \sum_{x \in X} \alpha(x)[e^{V(x)}(\pi u_{N-1})(x) - u_{N-1}(x)] \\
&\quad = \lim_{N \to \infty} \left\{ \sum_{x \in X} \alpha(x)(e^{V(x)} - 1)(\pi u_{N-1})(x) \right. \\
&\qquad \left. + \sum_{x \in X} \alpha(x)[(\pi u_{N-1})(x) - u_{N-1}(x)] \right\} \\
&\quad = \lim_{N \to \infty} \left\{ \sum_{x \in X} \alpha(x)(1 - e^{-V(x)})\, u_N(x) \right. \\
&\qquad \left. + \sum_{x \in X} \alpha(x)[(\pi u_{N-1})(x) - u_{N-1}(x)] \right\}
\end{aligned}$$

The first term in this last expression is only a finite sum, since V vanishes outside a finite set, and so in this first term we can bring the limit inside the summation. Moreover, the second term goes to 0 as $N \to \infty$, by Lemma 2.8. Hence, since $V \in \underline{V}$,

$$\lim_{N \to \infty} \sum_{x \in X} \alpha(x)[u_N(x) - u_{N-1}(x)] = \sum_{x \in X} \alpha(x)(1 - e^{-V(x)})\, u(x)$$

Theorem 2.10. For $V \in \underline{V}$ and $\alpha \in \underline{A}$,

$$\lim_{N \to \infty} \frac{1}{N} \log E^{P_n} \left\{ \exp \left[\sum_{j=0}^{N-1} \sum_{x \in X} n_j(x)\, V(x) \right] \right\}$$
$$= \sum_{x \in X} \alpha(x) \{1 - \exp[-V(x)]\}\, u(x) \tag{2.22}$$

for almost all $n(\cdot) \in Z$ (P_α-measure).

Remark. Although Theorem 2.10 is stated and proved in a form in which the null set depends on $V \in \underline{V}$, by an appropriate choice of a countable subset of $\underline{V}$ and by using standard arguments one can obtain the existence of a single null set that works simultaneously for all $V \in \underline{V}$.

Proof. From (1.1),

$$E^{P_n} \left\{ \exp \left[\sum_{j=0}^{N-1} \sum_{x \in X} n_j(x)\, V(x) \right] \right\}$$
$$= \prod_{x \in X} \left(E_x^{\pi} \left\{ \exp \left[\sum_{j=0}^{N-1} V(X_j) \right] \right\} \right)^{n(x)}$$

and therefore

$$\frac{1}{N} \log E^{P_n} \left\{ \exp \left[\sum_{j=0}^{N-1} \sum_{x \in X} n_j(x)\, V(x) \right] \right\}$$
$$= \frac{1}{N} \sum_{x \in X} n(x) \log E_x^{\pi} \left\{ \exp \left[\sum_{j=0}^{N-1} V(X_j) \right] \right\}$$
$$= \frac{1}{N} \sum_{x \in X} n(x) \log u_N(x)$$

To show that (2.22) holds for almost all $n(\cdot) \in Z$ (P_α-measure) means then to show

$$\lim_{N \to \infty} \frac{1}{N} \sum_{x \in X} n(x) \log u_N(x) = \sum_{x \in X} \alpha(x)(1 - e^{-V(x)})\, u(x)$$

for almost all $n(\cdot)\in Z$ (P_α-measure). For this it suffices to show that for all real λ

$$\lim_{N\to\infty}\frac{1}{N}\log E^{P_\alpha}\left\{\exp\left[\lambda\sum_{x\in X}n(x)\log u_N(x)\right]\right\}$$
$$=\lambda\sum_{x\in X}\alpha(x)\{1-\exp[-V(x)]\}\,u(x)$$

But

$$\frac{1}{N}\log E^{P_\alpha}\left\{\exp\left[\lambda\sum_{x\in X}n(x)\log u_N(x)\right]\right\}$$
$$=\frac{1}{N}\log\prod_{x\in X}\exp(\alpha(x)\{\exp[\lambda\log u_N(x)]-1\})$$
$$=\frac{1}{N}\sum_{x\in X}\alpha(x)\{[u_N(x)]^\lambda-1\}$$

and so we must show

$$\lim_{N\to\infty}\frac{1}{N}\sum_{x\in X}\alpha(x)\{[u_N(x)]^\lambda-1\}$$
$$=\lambda\sum_{x\in X}\alpha(x)(1-e^{-V(x)})\,u(x) \tag{2.23}$$

From the usual Caesaro argument, (2.23) is implied by

$$\lim_{N\to\infty}\sum_{x\in X}\alpha(x)\{[u_N(x)]^\lambda-[u_{N-1}(x)]^\lambda\}$$
$$=\lambda\sum_{x\in X}\alpha(x)(1-e^{-V(x)})\,u(x) \tag{2.24}$$

which follows from Lemmas 2.7 and 2.9.

Lemma 2.11. If $C\subset \underline{M}$ is closed and

$$\overline{\lim_{N\to\infty}}\frac{1}{N}\log Q_{n,N}(C)$$
$$\leqslant -\sup_{V\in\underline{V}}\inf_{\sigma\in C}\left[\sum_{x\in X}\sigma(x)\,V(x)-\sum_{x\in X}\alpha(x)(1-e^{-V(x)})\,u(x)\right]$$

then

$$\overline{\lim_{N\to\infty}} \frac{1}{N} \log Q_{n,N}(C)$$

$$\leqslant -\inf_{\sigma\in C} \sup_{V\in \underline{V}} \left[\sum_{x\in X} \sigma(x)\, V(x) - \sum_{x\in X} \alpha(x)(1-e^{-V(x)})\, u(x) \right]$$

$$= -\inf_{\sigma\in C} I_\alpha(\sigma)$$

Proof. First of all, one can obtain the desired result if C is compact, using methods developed in our earlier papers (cf. Ref. 1, for example).

We now describe how one can go from the result for compact sets to closed sets.

We begin by picking a $V>0$ such that

$$\sup_x \sum_y G(x, y)\, V(y) = \theta < 1$$

and

$$\sum_{x\in X} \alpha(x)\, V(x) \leqslant \text{const} < \infty$$

With such a V we then have by Portenko's lemma (see Lemma 3.1 below) that $u(x) = u_V(x) \leqslant 1/(1-\theta)$. This implies

$$\lim_{N\to\infty} \frac{1}{N} \log E^{P_\alpha} E^{\underline{P}_n} \left\{ \exp\left[\sum_{j=0}^{N-1} \sum_{x\in X} V(x)\, n_j(x) \right] \right\}$$

$$= \lim_{N\to\infty} \sum_{x\in X} \alpha(x)[u_N(x) - u_{N-1}(x)]$$

$$= \sum_{x\in X} \alpha(x)(1-e^{-V(x)})\, u(x)$$

$$\leqslant \frac{1}{1-\theta} \cdot \text{const} = K < \infty \tag{2.25}$$

Using a Chebycheff argument, we obtain from (2.25)

$$P_\alpha \left\{ n(\cdot){:}\; Q_{n,N} \left\{ \frac{1}{N} \sum_{j=0}^{N-1} \sum_{x\in X} V(x)\, n_j(x) \geqslant L \right\} \geqslant e^{-\beta N} \right\} \leqslant e^{KN+\beta N-LN} \tag{2.26}$$

Now, (2.26) implies

$$\overline{\lim_{L\to\infty}}\; \overline{\lim_{N\to\infty}} \frac{1}{N} \log Q_{n,N} \left\{ \frac{1}{N} \sum_{j=0}^{N-1} \sum_{x\in X} V(x)\, n_j(x) \geqslant L \right\} = -\infty \tag{2.27}$$

and (2.27), together with the fact that $\{\sigma{:}\, \sum_{x\in X} V(x)\, \sigma(x) \leqslant L\}$ is a compact set in the vague topology, implies the lemma.

3. PROPERTIES OF THE *I*-FUNCTION

In this section we prove various properties of $I_\alpha(\sigma)$ given by (1.5). Some of these results are necessary for establishing the lower bound (1.3) and will be used in the next section, where the lower bound is proved. Other properties of $I_\alpha(\sigma)$ proved here are of separate interest.

As noted earlier, the Green's function $G(x, y) < \infty$ for all $x, y \in X$, since the $\{\pi_{xy}\}$ chain is transient. If $y \neq x$, $P_x^\pi\{\tau_{\{y\}} < \infty\} = G(x, y)/G(y, y)$, and therefore our hypothesis on $P_x^\pi\{\tau_F < \infty\}$ implies

$$\begin{aligned} G(x, y) &\to 0 \quad \text{as} \quad y \to \infty \quad \text{for any } x \in X \\ G(x, y) &\to 0 \quad \text{as} \quad x \to \infty \quad \text{for any } y \in X \end{aligned} \tag{3.1}$$

We also recall that we are assuming

$$\sup_{x \in X} G(x, x) < \infty \tag{3.2}$$

Both (3.1) and (3.2) will be used in this section.

Lemma 3.1. Let $V: X \to R$ vanish outside some finite subset of X and be such that

$$\sup_{x \in X} \sum_{y \in X} G(x, y)\, |V(y)| \leqslant \theta < 1 \tag{3.3}$$

Then, $V \in \underline{V}$ and $|1 - u_V(x)| \leqslant \theta/(1-\theta)$ for each $x \in X$.

Proof. Let $r \geqslant 1$ and consider

$$\begin{aligned} &E_x^\pi\{[V(X_0) + V(X_1) + \cdots]^r\} \\ &\quad \leqslant E_x^\pi\{[|V(X_0)| + |V(X_1)| + \cdots]^r\} \\ &\quad = E_x^\pi\left\{\sum_{i_1, i_2, \ldots, i_r} |V(X_{i_1})|\, |V(X_{i_2})| \cdots |V(X_{i_r})|\right\} \\ &\quad \leqslant r!\, E_x^\pi\left\{\sum_{i_1 \leqslant i_2 \leqslant \cdots \leqslant i_r} |V(X_{i_1})|\, |V(X_{i_2})| \cdots |V(X_{i_r})|\right\} \\ &\quad \leqslant r! \sum_{i_1 \leqslant i_2 \leqslant \cdots \leqslant i_{r-1}} E_x^\pi\left\{|V(X_{i_1})|\, |V(X_{i_2})|\right. \\ &\qquad \times \cdots |V(X_{i_{r-1}})|\, \left. E_{X_{i_{r-1}}}^\pi\left(\sum_{j=0}^\infty |V(X_j)|\right)\right\} \\ &\quad \leqslant \theta r! \sum_{i_1 \leqslant i_2 \leqslant \cdots \leqslant i_{r-1}} E_x^\pi\{|V(X_{i_1})|\, |V(X_{i_2})| \cdots |V(X_{i_{r-1}})| \end{aligned}$$

the last inequality following from (3.3). If we keep repeating this process, we end up with

$$E_x^\pi\{[V(X_0)+V(X_1)+\cdots]^r\}\leqslant\theta^r r!$$

and thus

$$|1-u_V(x)|=\left|1-E_x^\pi\left\{\exp\left[\sum_{j=0}^{\infty}V(X_j)\right]\right\}\right|\leqslant\sum_{r=1}^{\infty}\frac{\theta^r r!}{r!}=\frac{\theta}{1-\theta}$$

which implies $u_V(x)<\infty$, so that $V\in \underline{V}$.

Lemma 3.2. Let $x_1, x_2,\ldots, x_N$ be a finite subset of X for which $G(x_i, x_i)\leqslant C$, $i=1, 2,\ldots, N$, $G(x_i, x_j)\leqslant\varepsilon$ for all $i\neq j$, $i, j=1, 2,\ldots, N$. Let $\gamma>0$ be so small that $\gamma[C+(N-1)\varepsilon]=\theta<1$. Let $\alpha\in\underline{A}$. For any $\sigma\in\underline{M}$ such that $I_\alpha(\sigma)<l$, we have

$$\begin{aligned}\frac{e^\gamma-1}{\gamma}&\left[\sum_{i=1}^{N}\alpha(x_i)\right]\left(1-\frac{\theta}{1-\theta}\right)-\frac{l}{\gamma}\\&\leqslant\sum_{i=1}^{N}\sigma(x_i)\\&\leqslant\frac{1-e^{-\gamma}}{\gamma}\left[\sum_{i=1}^{N}\alpha(x_i)\right]\left(1+\frac{\theta}{1-\theta}\right)+\frac{l}{\gamma}\end{aligned}\tag{3.4}$$

Proof. Let

$$V(x)=\begin{cases}\gamma & \text{if } x=x_i, \quad i=1, 2,\ldots, N\\ 0 & \text{otherwise}\end{cases}$$

Then, using the maximum principle,

$$\begin{aligned}\sup_{x\in X}\sum_{y\in X}&G(x, y)\,V(y)\\&=\sup_{1\leqslant i\leqslant N}\sum_{y\in X}G(x_i, y)\,V(y)\\&=\gamma\sup_{1\leqslant i\leqslant N}\sum_{j=1}^{N}G(x_i, x_j)\\&\leqslant\gamma[C+(N-1)\varepsilon]=\theta<1\end{aligned}$$

Hence, from Lemma 3.1, for the V just selected, $u_V(x)\leqslant 1+\theta/(1-\theta)$, and $V\in\underline{V}$. Since $\sigma\in\underline{M}$ is such that $I_\alpha(\sigma)\leqslant l$, we then have for this V,

$$\sum_{x\in X}\sigma(x)\,V(x)-\sum_{x\in X}\alpha(x)(1-e^{-V(x)})\,u_V(x)\leqslant l$$

i.e.,

$$\gamma \sum_{i=1}^{N} \sigma(x_i) - (1-e^{-\gamma})\left[\sum_{i=1}^{N} \alpha(x_i)\right]\left(1+\frac{\theta}{1-\theta}\right) \leqslant l$$

or

$$\sum_{i=1}^{N} \sigma(x_i) \leqslant \frac{1-e^{-\gamma}}{\gamma}\left[\sum_{i=1}^{N} \alpha(x_i)\right]\left(1+\frac{\theta}{1-\theta}\right)+\frac{l}{\gamma}$$

If we choose instead

$$V(x)=\begin{cases}-\gamma, & x=x_i, \quad i=1,2,\dots,N \\ 0, & \text{otherwise}\end{cases}$$

then again Lemma 3.1 implies this $V \in \underline{V}$ also and $u_V(x) \geqslant 1-\theta/(1-\theta)$, which, by the same argument as above, leads to the inequality on the left of (3.4).

Theorem 3.3. Let $\alpha \in \underline{A}$ and $\sigma \in \underline{M}$ such that $I_\alpha(\sigma)=l<\infty$. Then:

1. If, for some sequence $\{y_n\} \to \infty$, $\alpha(y_n) \to 0$, then $\sigma(y_n) \to 0$.
2. If, for some sequence $\{y_n\} \to \infty$, $\alpha(y_n) \to L>0$, then $\sigma(y_n) \to L$ also.
3. If, for some sequence $\{y_n\} \to \infty$, $\alpha(y_n) \to \infty$, then $\alpha(y_n)/\sigma(y_n) \to 1$.

These three statements combine into: If $\{y_n\} \to \infty$, then $[1+\alpha(y_n)]/[1+\sigma(y_n)] \to 1$.

Proof. We prove each of the three statements. First assume for some sequence $\{y_n\} \to \infty$, $\alpha(y_n) \to 0$. To prove statement 1, we assume the contrary, i.e., there exists $\delta>0$ such that $\sigma(y_n) \geqslant \delta$ for all n. From (3.2) let $\sup_{x \in X} G(x,x) < C$ and let $\varepsilon>0$ be given. Let γ_0 be so small that $\gamma<\gamma_0$ implies

$$\frac{1-e^{-\gamma}}{\gamma}\left(1+\frac{\gamma(C+1)}{1-\gamma(C+1)}\right)<2$$

Choose $\gamma>0$, but $\gamma<\min(1/2(C+1), \gamma_0)$. Choose N so large that $l/N\gamma<\varepsilon/2$. Because we are assuming here that $\alpha(y_n) \to 0$ and because of (3.1) we can choose N elements $x_1, x_2,\dots, x_N$ in X such that $G(x_i, x_j) \leqslant 1/N$ for $i \neq j$, $i, j=1,2,\dots,N$, and such that $\alpha(x_j) \leqslant \varepsilon/4$, $j=1,2,\dots,N$.

Now,

$$\gamma[C+(N-1)(1/N)] \leqslant \gamma[C+1]=\theta<1/2$$

and hence from Lemma 3.2,

$$N\delta \leqslant \sum_{i=1}^{N} \sigma(x_i) \leqslant \frac{1-e^{-\gamma}}{\gamma}\left(1+\frac{\gamma(C+1)}{1-\gamma(C+1)}\right) N\frac{\varepsilon}{4}+\frac{l}{\gamma} \leqslant \frac{N\varepsilon}{2}+\frac{l}{\gamma}$$

i.e.,

$$\delta \leqslant \frac{\varepsilon}{2}+\frac{l}{N\gamma} \leqslant \varepsilon$$

Since $\varepsilon > 0$ is arbitrary, we have a contradiction and statement 1 is proved.

To prove statement 2, suppose $\{y_n\} \to \infty$ such that $\alpha(y_n) \to L > 0$. Again we assume the contrary, i.e., there is a subsequence of $\{y_n\}$, which we will again label $\{y_n\}$, such that either $\sigma(y_n) \geqslant L' > L$ for all $n = 1, 2,\ldots$, or $\sigma(y_n) \leqslant L' < L$ for all $n = 1, 2,\ldots$. For the first case, let $\varepsilon > 0$ be given and choose $\gamma > 0$ so small that $\gamma[C+1] = \theta < 1/2$ and

$$\frac{1-e^{-\gamma}}{\gamma}\left[1+\frac{\gamma(C+1)}{1-\gamma(C+1)}\right] < 1+\varepsilon$$

Choose N so large that $l/N\gamma < \varepsilon$. For that N we can find points $x_1, x_2,\ldots, x_N$ such that $G(x_i, x_j) \leqslant 1/N$ for all $i \neq j$, $i, j = 1, 2,\ldots, N$, and such that $\alpha(x_j) \leqslant L+\varepsilon$. Thus, from Lemma 3.2,

$$NL' \leqslant \sum_{i=1}^{N} \sigma(x_i) \leqslant N(L+\varepsilon)(1+\varepsilon)+l/\gamma$$

or

$$L' \leqslant (L+\varepsilon)(1+\varepsilon)+\varepsilon$$

Since $\varepsilon > 0$ is arbitrary, this contradicts $L' > L$. A similar argument using the lower estimate in Lemma 3.2 takes care of the second case, $L' < L$, and statement 2 is proved.

Finally, to show statement 3, let $\{y_n\} \to \infty$ be a sequence such that $\alpha(y_n) \to \infty$. Again assuming the contrary, suppose there is a subsequence $\{y_n\}$ such that $\sigma(y_n) \geqslant \lambda\alpha(y_n)$ for some $\lambda > 1$, $n = 1, 2,\ldots$, or $\sigma(y_n) \leqslant \lambda\alpha(y_n)$ for some $\lambda < 1$, $n = 1, 2,\ldots$. Take the first case, and use the upper estimate of Lemma 3.2 with $N = 1$. Choose $\gamma > 0$ so small that $\gamma C = \theta < 1$ and

$$\frac{1-e^{-\gamma}}{\gamma}\left(1+\frac{\gamma C}{1-\gamma C}\right) < 1+\frac{\varepsilon}{2}$$

where $\varepsilon > 0$ is given. Choose $M > 0$ so large that $l/M\gamma < \varepsilon/2$. Now, there is an $x \in X$ for which $\alpha(x) > M$. Hence from Lemma 3.2 for this particular x,

$$\frac{\sigma(x)}{\alpha(x)} \leqslant \frac{1-e^{-\gamma}}{\gamma}\left(1+\frac{\gamma C}{1-\gamma C}\right)+\frac{l}{\alpha(x)\,\gamma}$$
$$\leqslant \left(1+\frac{\varepsilon}{2}\right)+\frac{l}{M\gamma} \leqslant 1+\varepsilon$$

and since $\varepsilon > 0$ is arbitrary, this contradicts $\sigma(x)/\alpha(x) \geqslant \lambda > 1$. Similarly, of course, for the other alternative, and the proof is complete.

Lemma 3.4. Let $\alpha \in \underline{A}$ and $\beta \in \underline{A}$. Assume for some $\sigma \in \underline{M}$ that both $I_\alpha(\sigma) < \infty$ and $I_\beta(\sigma) < \infty$. Then $\alpha \equiv \beta$.

Proof. First we want to prove the following: if $\alpha \in \underline{A}$ and $\beta \in \underline{A}$ and if $h(x) = \beta(x)/\alpha(x) \to 1$ as $x \to \infty$, then $\alpha \equiv \beta$.

As in Section 2, let $\pi_{xy}(\alpha) = \pi_{yx}\alpha(y)/\alpha(x)$ and recall that $\{\pi_{xy}(\alpha)\}$ are transition probabilities for an irreducible chain. Now,

$$\sum_{y \in X} \pi_{xy}(\alpha)\, h(y) = \sum_{y \in X} \pi_{yx} \frac{\alpha(y)}{\alpha(x)} \frac{\beta(y)}{\alpha(y)}$$
$$= \frac{1}{\alpha(x)} \sum_{y \in X} \pi_{yx}\beta(y) = \frac{\beta(x)}{\alpha(x)} = h(x)$$

so that h is a harmonic function of the chain time-reversed with respect to α. By hypothesis, $h(x) \to 1$ at ∞, but, for an irreducible Markov chain, if a bounded harmonic function approaches 1 at infinity, it is then identically 1.

Now, to prove this lemma, we have from Theorem 3.3 that $\{y_n\} \to \infty$ implies $[1+\sigma(y_n)]/[1+\alpha(y_n)] \to 1$ and also $[1+\sigma(y_n)]/[1+\beta(y_n)] \to 1$, i.e., $[1+\alpha(y_n)]/[1+\beta(y_n)] \to 1$. Since, by hypothesis, $\{\pi_{xy}\}$ is doubly stochastic, we see that 1 is an invariant measure for the π-chain, i.e., $1 \in \underline{A}$. Since the sum of two invariant measures is also an invariant measure, the statement at the beginning of this proof implies $1+\alpha(\cdot) \equiv 1+\beta(\cdot)$, i.e., $\alpha(\cdot) \equiv \beta(\cdot)$.

Lemma 3.5. Let $\alpha \in \underline{A}$ and $\sigma \in \underline{M}$ such that $I_\alpha(\sigma) < \infty$. Then, for all $y \in X$,

$$\lim_{k \to \infty} \sum_{x \in X} \sigma(x)\, \pi_{xy}^{(k)} = \alpha(y) \tag{3.5}$$

Proof. As noted in the preceding lemma, since $\{\pi_{xy}\}$ is doubly stochastic, 1 is an invariant measure, and therefore to show (3.5), it suffices to show

$$\lim_{k \to \infty} \sum_{x \in X} [\sigma(x)+1]\, \pi_{xy}^{(k)} = \alpha(y)+1 \tag{3.6}$$

for all $y \in X$. Let

$$h(x) = \frac{\sigma(x)+1}{\alpha(x)+1}, \qquad \tilde{\pi}_{xy} = \pi_{yx} \frac{\alpha(y)+1}{\alpha(x)+1}$$

Now

$$\tilde{\pi}^{(k)}_{xy} = \pi^{(k)}_{yx} \frac{\alpha(y)+1}{\alpha(x)+1}$$

and

$$\begin{aligned} \sum_{x \in X} [\sigma(x)+1]\, \pi^{(k)}_{xy} &= \sum_{x \in X} \frac{\alpha(y)+1}{\alpha(x)+1} \tilde{\pi}^{(k)}_{yx} [\sigma(x)+1] \\ &= [\alpha(y)+1] \sum_{x \in X} h(x)\, \tilde{\pi}^{(k)}_{yx} \\ &= [\alpha(y)+1]\, E^{\tilde{\pi}}_y \{h(X_k)\} \end{aligned}$$

As we have argued several times already, the $\tilde{\pi}$-chain is also transient, which implies, for every $y \in X$, $X_k \to \infty$ for almost all paths starting from y ($\tilde{\pi}$-measure). Moreover, from Theorem 3.3, $h(x_k) \to 1$ if $x_k \to \infty$, so by bounded convergence $\lim_{k \to \infty} E^{\tilde{\pi}}_y \{h(X_k)\} = 1$ and we have (3.6) for all $y \in X$ and the lemma.

We now prove a succession of lemmas leading up to Theorem 3.11, which will be used in Section 4 in order to prove the lower bound (1.3).

In these lemmas as well as in Theorem 3.11, F will denote a fixed, finite subset of X having elements $x_1, x_2, \ldots, x_k$. Let $\boldsymbol{\theta} = \{\theta_1, \theta_2, \ldots, \theta_k\}$ be a point in R^k and let $C = \{V\colon X \to \mathbb{R}\}$. Let $T\colon R^k \to C$ be the mapping given by

$$\begin{aligned} V(x_i) &= \theta_i, \qquad & i &= 1, 2, \ldots, k \\ V(x) &= 0, & x &\neq x_i \end{aligned}$$

Let $\underline{Q} \in R^k$ consist of those $\boldsymbol{\theta} \in R^k$ such that $T\boldsymbol{\theta} \in \underline{V}$. For $\alpha \in \underline{A}$ and $\boldsymbol{\theta} \in \underline{Q}$, define

$$\Phi(\boldsymbol{\theta}) = \Phi_F(\boldsymbol{\theta}) = \sum_{i=1}^{k} \alpha(x_i)(1 - e^{-\theta_i})\, u(\boldsymbol{\theta}; x_i) - \sum_{i=1}^{k} \alpha(x_i)\, \theta_i$$

where, of course,

$$u(\boldsymbol{\theta}; x_i) = u_V(x_i) = E^{\pi}_{x_i} \left\{ \exp \left[\sum_{j=0}^{\infty} V(X_j) \right] \right\}$$

with $V = T\boldsymbol{\theta}$.

In the first of these lemmas we use the notation of Section 2. Recall that $\pi_{xy}^F = P_x^\pi\{X_\tau = y; \tau < \infty\}$, where $\tau = \inf_{j\geq 1}\{X_j \in F\}$, and

$$\eta_x = 1 - \sum_{y\in F} \pi_{xy}^F = P_x^\pi\{X_j \notin F, j \geqslant 1\}$$

For the time-reversed chain with respect to an $\alpha \in \underline{A}$, i.e., the chain with transition probabilities $\pi_{xy}(\alpha) = \pi_{yx}\alpha(y)/\alpha(x)$, we use the notations

$$\pi_{xy}^F(\alpha) = P_x^{\pi(\alpha)}\{X_\tau = y; \tau < \infty\}$$

$$\eta_x(\alpha) = 1 - \sum_{y\in F} \pi_{xy}^F(\alpha) = P_x^{\pi(\alpha)}\{X_j \notin F, j \geqslant 1\}$$

as before.

Lemma 3.6. For $\alpha \in \underline{A}$ and $\boldsymbol{\theta} \in \underline{Q}$,

$$\Phi_F(\boldsymbol{\theta}) = \sum_{i=1}^k \alpha(x_i)\,\eta_{x_i}(\alpha)\,u(\boldsymbol{\theta}; x_i) - \sum_{i=1}^k \alpha(x_i)\,\eta_{x_i} - \sum_{i=1}^k \alpha(x_i)\,\theta_i \tag{3.7}$$

Proof. Since $\boldsymbol{\theta} \in \underline{Q}$, $V \in \underline{V}$ and therefore from (2.6) we have for any $x \in X$,

$$u_V(x) = e^{V(x)}\left[\sum_{y\in F} \pi_{xy}^F u_V(y) + \eta_x\right]$$

Thus,

$$\begin{aligned}
&\sum_{x\in X} \alpha(x)(1 - e^{-V(x)})\,u_V(x) \\
&\quad= \sum_{x\in F} \alpha(x)(1 - e^{-V(x)})\,u_V(x) \\
&\quad= \sum_{x\in F} \alpha(x)\,u_V(x) - \sum_{x\in F} \alpha(x)\left[\sum_{y\in F} \pi_{xy}^F u_V(y) + \eta_x\right] \\
&\quad= \sum_{y\in F}\left[\alpha(y) - \sum_{x\in F} \alpha(x)\,\pi_{xy}^F\right] u_V(y) - \sum_{x\in F} \alpha(x)\,\eta_x \qquad (3.8)
\end{aligned}$$

For x, $y \in F$,

$$\begin{aligned}
\pi_{xy}^F &= \pi_{xy} + \sum_{z_1\notin F} \pi_{xz_1}\pi_{z_1y} + \sum_{\substack{z_1\notin F\\ z_2\notin F}} \pi_{xz_1}\pi_{z_1z_2}\pi_{z_2y} + \cdots \\
&= \pi_{xy}^{(1)} + \pi_{xy}^{(2)} + \cdots + \pi_{xy}^{(n)} + \cdots
\end{aligned}$$

and therefore

$$\begin{aligned}
\alpha(y) - \sum_{x\in F} \alpha(x)\,\pi^F_{xy}
&= \sum_{x\in X} \alpha(x)\,\pi_{xy} - \sum_{x\in F} \alpha(x)\,\pi^F_{xy} \\
&= \sum_{x\notin F} \alpha(x)\,\pi_{xy} - \sum_{x\in F} \alpha(x)[\pi^F_{xy} - \pi_{xy}] \\
&= \sum_{x\notin F} \alpha(x)\,\pi_{xy} - \sum_{x\in F} \alpha(x) \sum_{n=2}^{\infty} \pi^{(n)}_{xy} \\
&= \sum_{x\notin F}\sum_{z_1\in X} \alpha(z_1)\,\pi_{z_1x}\pi_{xy} - \sum_{x\in F} \alpha(x) \sum_{n=2}^{\infty} \pi^{(n)}_{xy} \\
&= \sum_{x\notin F}\sum_{z_1\notin F} \alpha(z_1)\,\pi_{z_1x}\pi_{xy} \\
&\quad - \sum_{x\in F} \alpha(x)\left[\sum_{n=2}^{\infty} \pi^{(n)}_{xy} - \sum_{z_1\notin F} \pi_{xz_1}\pi_{z_1y}\right] \\
&= \sum_{x\notin F}\sum_{z_1\notin F} \alpha(z_1)\,\pi_{z_1x}\pi_{xy} - \sum_{x\in F} \alpha(x) \sum_{n=3}^{\infty} \pi^{(n)}_{xy}
\end{aligned}$$

If we repeat this process we arrive at the following expression, which holds for any $n = 1, 2, \ldots$:

$$\begin{aligned}
\alpha(y) - \sum_{x\in F} \alpha(x)\,\pi^F_{xy}
&= \sum_{z_1, z_2, \ldots, z_n \notin F} \alpha(z_1)\,\pi_{z_1z_2}\pi_{z_2z_3}\cdots\pi_{z_ny} \\
&\quad - \sum_{x\in F} \alpha(x) \sum_{k=n+1}^{\infty} \pi^{(k)}_{xy}
\end{aligned} \tag{3.9}$$

Letting $n \to \infty$ in (3.9), we see that (cf. proof of Lemma 2.3)

$$\begin{aligned}
\alpha(y) - \sum_{x\in F} \alpha(x)\,\pi^F_{xy}
&= \lim_{n\to\infty} \sum_{z_1, z_2, \ldots, z_n \notin F} \alpha(z_1)\,\pi_{z_1z_2}\pi_{z_2z_3}\cdots\pi_{z_ny} \\
&= \lim_{n\to\infty} \sum_{z_1, z_2, z_3, \ldots, z_n \notin F} \alpha(y)\,\pi_{yz_1}(\alpha)\,\pi_{z_1z_2}(\alpha)\cdots\pi_{z_{n-1}z_n}(\alpha) \\
&= \alpha(y)\,\eta_y(\alpha)
\end{aligned} \tag{3.10}$$

Using (3.10) in (3.8), we get

$$\sum_{x \in X} \alpha(x)(1-e^{-V(x)})\, u_V(x)$$

$$= \sum_{y \in F} \alpha(y)\, \eta_y(\alpha)\, u_V(y) - \sum_{x \in F} \alpha(x)\, \eta_x$$

Therefore,

$$\Phi_F(\boldsymbol{\theta}) = \sum_{i=1}^{k} \alpha(x_i)\, \eta_{x_i}(\alpha)\, u(\boldsymbol{\theta}; x_i) - \sum_{i=1}^{k} \alpha(x_i)\, \eta_{x_i} - \sum_{i=1}^{k} \alpha(x_i)\, \theta_i$$

which is the lemma.

Lemma 3.7. $\underline{O}$ is an open, convex set in R^k, $\Phi(\boldsymbol{\theta}) \geqslant 0$ for $\boldsymbol{\theta} \in \underline{O}$, and if $\tilde{\boldsymbol{\theta}} \in R^k$ such that $\tilde{\boldsymbol{\theta}} \in \partial \underline{O}$ and $\boldsymbol{\theta}^{(n)} \to \tilde{\boldsymbol{\theta}}$ as $n \to \infty$, then $\Phi(\boldsymbol{\theta}^{(n)}) \to \infty$.

Proof. In Section 2 we noted that if $V \in \underline{V}$, there exists $\lambda > 1$ such that $\lambda V \in \underline{V}$, which implies $\underline{O}$ is open. From Lemma 2.1, $\underline{V} = \underline{V}_2$, so $V \in \underline{V}$ implies $\rho(M_F) < 1$, where M_F is the matrix $\{e^{V(x)} \pi^F_{xy},\ x, y \in F\}$. Since here F is fixed, $\rho(M_F) = \rho(\boldsymbol{\theta})$ and $\boldsymbol{\theta} \in \underline{O}$ implies $\rho(\boldsymbol{\theta}) < 1$, i.e., the convex function $\log \rho(\boldsymbol{\theta}) < 0$ for $\boldsymbol{\theta} \in \underline{O}$, which implies $\underline{O}$ is convex. That $\Phi(\boldsymbol{\theta}) \geqslant 0$ for $\boldsymbol{\theta} \in \underline{O}$ follows by applying Jensen's inequality to the definition of $\Phi(\boldsymbol{\theta})$.

Although $\underline{O}$ is open and convex, it is not bounded, and indeed we later must examine the behavior of $\Phi(\boldsymbol{\theta})$ as $\boldsymbol{\theta}$ stays in $\underline{O}$ but approaches ∞. Here the sequence $\{\boldsymbol{\theta}^{(n)}\} \in \underline{O}$ converges to the finite point $\tilde{\boldsymbol{\theta}} \in \partial \underline{O}$. In the definition of $\Phi(\boldsymbol{\theta})$, the term

$$\sum_{i=1}^{k} \alpha(x_i)(1-e^{-\theta_i})\, u(\boldsymbol{\theta}; x_i)$$

is troublesome because, depending on the signs of the θ_i, some terms are positive and some negative, presenting possible cancellations. This problem is obviated by the decomposition in Lemma 3.6. Thus,

$$\Phi(\boldsymbol{\theta}^{(n)}) = \sum_{i=1}^{k} \alpha(x_i)\, \eta_{x_i}(\alpha)\, u(\boldsymbol{\theta}^{(n)}; x_i) - \sum_{i=1}^{k} \alpha(x_i)\, \eta_{x_i} - \sum_{i=1}^{k} \alpha(x_i)\, \theta_i^{(n)} \tag{3.11}$$

In (3.11) not all of the factors $\eta_{x_i}(\alpha)$, $i = 1, 2, \ldots, k$, can be zero, because the chain time-reversed with respect to α is transient. Since $\underline{O}$ is open, $\tilde{\boldsymbol{\theta}} \notin \underline{O}$, and hence $u(\tilde{\boldsymbol{\theta}}; x_i) = \infty$, $i = 1, 2, \ldots, k$. From Fatou's lemma, we see then that

$u(\boldsymbol{\theta}^{(n)}; x_i) \to \infty$ as $\boldsymbol{\theta}^{(n)} \to \tilde{\boldsymbol{\theta}}$ for $i = 1, 2, \ldots, k$. Thus, the first term in (3.11) becomes infinite as $\boldsymbol{\theta}^{(n)} \to \tilde{\boldsymbol{\theta}}$ and so $\Phi(\boldsymbol{\theta}^{(n)}) \to \infty$ as $\boldsymbol{\theta}^{(n)} \to \tilde{\boldsymbol{\theta}}$, completing the lemma.

Now we must examine the behavior of $\Phi(\boldsymbol{\theta})$ as $\boldsymbol{\theta} \to \infty$. There exist rays along which $\Phi(\boldsymbol{\theta})$ remains bounded and we have to take these into account in what follows. To this end, we define a point $\mathbf{a} \in R^k$ to be *special* if $\sup_{\lambda > 0} \Phi(\lambda \mathbf{a}) \leqslant \text{const}$. Let $S = \{\mathbf{a} \in R^k, \mathbf{a} \text{ special}\}$. Clearly, $S \subset \underline{O}$.

Lemma 3.8. If $\mathbf{a} \in S$, then $-\mathbf{a} \in S$ and $\Phi(\mathbf{a}) = 0$. S is a linear subspace of R^k and if $\mathbf{b} \in \underline{O}$ and $\mathbf{a} \in S$, then $\mathbf{b} + \mathbf{a} \in \underline{O}$ and $\Phi(\mathbf{b} + \mathbf{a}) = \Phi(\mathbf{b})$.

Proof. From the definition of $\Phi(\boldsymbol{\theta})$ we see that it is convex and $\Phi(\mathbf{0}) = 0$. Thus, if $\Phi(\lambda \mathbf{a}) \leqslant c$ for all $\lambda > 0$, then $\Phi(\lambda \mathbf{a}) \equiv 0$ for all $\lambda > 0$. Since Φ is analytic in $\boldsymbol{\theta}$, we have $\Phi(\lambda \mathbf{a}) \equiv 0$ for all real λ. To show that S is a linear subspace, it suffices to show that $\mathbf{a}_1 \in S$ and $\mathbf{a}_2 \in S$ imply $\frac{1}{2}(\mathbf{a}_1 + \mathbf{a}_2) \in S$. But

$$0 \leqslant \Phi\left(\frac{\mathbf{a}_1 + \mathbf{a}_2}{2}\right) \leqslant \frac{1}{2}[\Phi(\mathbf{a}_1) + \Phi(\mathbf{a}_2)] = 0$$

so not only is S a linear subspace, but $\Phi(\boldsymbol{\theta}) \equiv 0$ for $\boldsymbol{\theta} \in S$.

Let $\mathbf{b} \in \underline{O}$ and $\mathbf{a} \in S$. Since $\underline{O}$ is open, there exists $\varepsilon > 0$ small enough so that $[1/(1-\varepsilon)]\,\mathbf{b} \in \underline{O}$. By definition, $(1/\varepsilon)\,\mathbf{a} \in S \subset \underline{O}$ and since $\underline{O}$ is a convex set,

$$\mathbf{b} + \mathbf{a} = (1-\varepsilon)\frac{1}{1-\varepsilon}\mathbf{b} + \varepsilon\left(\frac{1}{\varepsilon}\right)\mathbf{a} \in \underline{O}$$

By the convexity of Φ and since $\Phi(\mathbf{a}) = 0$, we have

$$\Phi(\mathbf{a} + \mathbf{b}) \leqslant (1-\varepsilon)\,\Phi\left(\frac{\mathbf{b}}{1-\varepsilon}\right)$$

Convex functions being continuous, we obtain from this that

$$\Phi(\mathbf{a} + \mathbf{b}) \leqslant \Phi(\mathbf{b}) \qquad \text{for all} \quad \mathbf{a} \in S \text{ and } \mathbf{b} \in \underline{O} \tag{3.12}$$

In (3.12) we can replace $\mathbf{a}$ by $-\mathbf{a}$, giving $\Phi(\mathbf{b} - \mathbf{a}) \leqslant \Phi(\mathbf{b})$ for all $\mathbf{a} \in S$ and $\mathbf{b} \in \underline{O}$. But we just showed $\mathbf{a} + \mathbf{b} \in \underline{O}$ and therefore this last inequality gives $\Phi(\mathbf{b}) \leqslant \Phi(\mathbf{a} + \mathbf{b})$, which, together with (3.12), finishes the lemma.

We have noted that $S \subset \underline{O}$. Let $\underline{O}_0 = \{\boldsymbol{\theta}' \in \underline{O}, \boldsymbol{\theta}' \perp S\}$, so $\underline{O}_0$ is a closed subset of $\underline{O}$, and any $\boldsymbol{\theta} \in \underline{O}$ can be written $\boldsymbol{\theta} = \boldsymbol{\theta}' + \mathbf{a}$, where $\boldsymbol{\theta}' \in \underline{O}_0$ and $\mathbf{a} \in S$.

Lemma 3.9. Let $\{\boldsymbol{\theta}^{(n)}\} \in \underline{O}_0$ such that $\|\boldsymbol{\theta}^{(n)}\| \to \infty$. Then, $\Phi(\boldsymbol{\theta}^{(n)}) \to \infty$.

Proof. Assume the contrary; then, there exists a constant C such that $\Phi(\boldsymbol{\theta}^{(n)}) \leqslant C$ for all n. For $\lambda > 0$,

$$\begin{aligned}\Phi\left(\frac{\lambda\boldsymbol{\theta}^{(n)}}{\|\boldsymbol{\theta}^{(n)}\|}\right) &= \Phi\left(\frac{\lambda}{\|\boldsymbol{\theta}^{(n)}\|}\boldsymbol{\theta}^{(n)} + \left(1 - \frac{\lambda}{\|\boldsymbol{\theta}^{(n)}\|}\right)\cdot \mathbf{0}\right)\\ &\leqslant \frac{\lambda}{\|\boldsymbol{\theta}^{(n)}\|}\Phi(\boldsymbol{\theta}^{(n)})\\ &\leqslant \frac{C\lambda}{\|\boldsymbol{\theta}^{(n)}\|}\end{aligned}$$

which implies $\Phi(\lambda\boldsymbol{\theta}^{(n)}/\|\boldsymbol{\lambda}^{(n)}\|) \to 0$ as $n \to \infty$. Since Q_0 is closed and, for each n, $\|\boldsymbol{\theta}^{(n)}/\|\boldsymbol{\theta}^{(n)}\|\,\| = 1$, there exists a subsequence, which we will label $\boldsymbol{\theta}^{(n)}/\|\boldsymbol{\theta}^{(n)}\|$, converging to some element $\mathbf{a} \in Q_0$ such that $\|\mathbf{a}\| = 1$. Since Φ is continuous, we have for all $\lambda > 0$, $\Phi(\lambda\mathbf{a}) = 0$, which means $\mathbf{a}$ is special. But $\mathbf{a} \in Q_0$ means $\mathbf{a} \perp \mathbf{a}$, which is impossible, since $\mathbf{a}$ is a unit vector.

Lemma 3.10. Let $\alpha \in \underline{A}$ and $\sigma \in \underline{M}$ such that $I_\alpha(\sigma) < \infty$. If $\mathbf{a} \in S$, then

$$\sum_{i=1}^{k} \sigma(x_i)\, a_i = \sum_{i=1}^{k} \alpha(x_i)\, a_i$$

Proof.

$$\begin{aligned}I_\alpha(\sigma) &= \sup_{V \in \underline{V}}\left[\sum_{x \in X} \sigma(x)\, V(x) - \sum_{x \in X} \alpha(x)(1 - e^{-V(x)})\, u_V(x)\right]\\ &\geqslant \sup_{\boldsymbol{\theta} \in \underline{Q}}\left[\sum_{i=1}^{k} \sigma(x_i)\, \theta_i - \sum_{i=1}^{k} \alpha(x_i)(1 - e^{-\theta_i})\, u(\boldsymbol{\theta}; x_i)\right]\end{aligned}$$

since the first supremum is over V's vanishing outside any finite subset of X, whereas the second supremum is just over V's vanishing outside our given finite subset F.

Thus, using the fact that $\Phi(\mathbf{a}) = 0$ for $\mathbf{a} \in S$,

$$\begin{aligned}I_\alpha(\sigma) &\geqslant \sup_{\boldsymbol{\theta} \in \underline{Q}}\left[\sum_{i=1}^{k} \sigma(x_i)\, \theta_i - \sum_{i=1}^{k} \alpha(x_i)\, \theta_i - \Phi_F(\boldsymbol{\theta})\right]\\ &\geqslant \sup_{\mathbf{a} \in S}\left[\sum_{i=1}^{k} \sigma(x_i)\, a_i - \sum_{i=1}^{k} \alpha(x_i)\, a_i - \Phi_F(\mathbf{a})\right]\\ &= \sup_{\mathbf{a} \in S}\left[\sum_{i=1}^{k} \sigma(x_i)\, a_i - \sum_{i=1}^{k} \alpha(x_i)\, a_i\right]\end{aligned}$$

Since S is a linear subspace, this last supremum is either ∞ or 0. The former is ruled out by our hypothesis $I_\alpha(\sigma) < \infty$, and hence we have this lemma, because

$$\sup_{\mathbf{a} \in S} \sum_{i=1}^{k} [\sigma(x_i) - \alpha(x_i)]\, a_i = 0$$

implies

$$\sum_{i=1}^{k} [\sigma(x_i) - \alpha(x_i)]\, a_i = 0 \qquad \text{identically in } S$$

Theorem 3.11. Let $\alpha \in \underline{A}$ and $\sigma \in \underline{M}$ such that $I_\alpha(\sigma) < \infty$. Let $\sigma' = (1-\varepsilon)\,\sigma + \varepsilon\alpha$ for $\varepsilon > 0$ fixed, and let F be a finite subset of X. In the variational problem

$$\sup_{\boldsymbol{\theta} \in \underline{Q}} \left[\sum_{i=1}^{k} \sigma'(x_i)\, \theta_i - \sum_{i=1}^{k} \alpha(x_i)(1 - e^{-\theta_i})\, u(\boldsymbol{\theta}; x_i) \right] \tag{3.13}$$

the supremum is attained at a point $\boldsymbol{\theta}^* \in \underline{Q}$.

Proof. Since $\alpha \in \underline{A}$, $I_\alpha(\alpha) = 0$, and the convexity of the I-function implies

$$I_\alpha(\sigma') \geqslant \sup_{\boldsymbol{\theta} \in \underline{Q}} \left[\sum_{i=1}^{k} \sigma'(x_i)\, \theta_i - \sum_{i=1}^{k} \alpha(x_i)(1 - e^{-\theta_i})\, u(\boldsymbol{\theta}; x_i) \right]$$

we see that the supremum in (3.13) is finite. Also, since $\mathbf{0} \in \underline{Q}$, we see that the supremum in (3.13) is nonnegative.

With $\Phi_F(\boldsymbol{\theta})$ as defined earlier, we can write (3.13) as

$$\sup_{\boldsymbol{\theta} \in \underline{Q}} \left[\sum_{i=1}^{k} \sigma'(x_i)\, \theta_i - \sum_{i=1}^{k} \alpha(x_i)\, \theta_i - \Phi_F(\boldsymbol{\theta}) \right] \tag{3.14}$$

As noted before, for any $\boldsymbol{\theta} \in \underline{Q}$ we can write $\boldsymbol{\theta} = \mathbf{a} + \mathbf{b}$, where $\mathbf{a} \in S$ and $\mathbf{b} \in \underline{Q}_0 = \{\boldsymbol{\theta}' \in \underline{Q}\colon\ \boldsymbol{\theta}' \perp S\}$. From Lemma 3.8, $\Phi(\mathbf{a} + \mathbf{b}) = \Phi(\mathbf{b})$. Since $I_\alpha(\sigma') < \infty$, for this $\mathbf{a} \in S$ we have from Lemma 3.10 that $\sum_{i=1}^{k} \sigma'(x_i)\, a_i = \sum_{i=1}^{k} \alpha(x_i)\, a_i$. Thus, (3.14) becomes

$$\begin{aligned}
&\sup_{\mathbf{b} \in \underline{Q}_0} \left[\sum_{i=1}^{k} \sigma'(x_i)\, b_i - \sum_{i=1}^{k} \alpha(x_i)\, b_i - \Phi_F(\mathbf{b}) \right] \\
&= \sup_{\mathbf{b} \in \underline{Q}_0} \left[(1-\varepsilon) \sum_{i=1}^{k} \sigma(x_i)\, b_i \right. \\
&\qquad \left. - (1-\varepsilon) \sum_{i=1}^{k} \alpha(x_i)\, b_i - \sum_{i=1}^{k} \alpha(x_i)(1 - e^{-b_i})\, u(\mathbf{b}; x_i) + \sum_{i=1}^{k} \alpha(x_i)\, b_i \right] \\
&= \sup_{\mathbf{b} \in \underline{Q}_0} \left\{ (1-\varepsilon) \left[\sum_{i=1}^{k} \sigma(x_i)\, b_i - \sum_{i=1}^{k} \alpha(x_i)(1 - e^{-b_i})\, u(\mathbf{b}; x_i) \right] - \varepsilon \Phi_F(\mathbf{b}) \right\}
\end{aligned} \tag{3.15}$$

Let $\mathbf{b}^{(n)}$ be a maximizing sequence in $\underline{Q}_0$. There are three possibilities: (1) $\|\mathbf{b}^{(n)}\| \to \infty$, (2) $\mathbf{b}^{(n)} \to \mathbf{b} \in \partial \underline{Q}_0$, or (3) $\mathbf{b}^{(n)} \to \boldsymbol{\theta}^*$, a point in $\underline{Q}_0 \subset \underline{Q}$.

Assume possibility 1 first. By hypothesis,

$$\infty > I_\alpha(\sigma) \geqslant \sup_{\mathbf{b} \in \underline{Q}_0} \left[\sum_{i=1}^{k} \sigma(x_i)\, b_i - \sum_{i=1}^{k} \alpha(x_i)(1 - e^{-b_i})\, u(\mathbf{b}; x_i) \right]$$

so that in the last line of (3.15) the quantity in square brackets is bounded, whereas Lemma 3.9 gives us $\Phi_F(\mathbf{b}^{(n)}) \to \infty$. Hence, in the last line of (3.15) the quantity in curly braces goes to $-\infty$ as $\|\mathbf{b}^{(n)}\| \to \infty$, which contradicts the supremum in (3.13) being nonnegative.

Next, assume possibility 2 holds. From Lemma 3.7, $\Phi_F(\mathbf{b}^{(n)}) \to \infty$ if $\mathbf{b}^{(n)} \to \mathbf{b} \in \partial \underline{Q}_0 \subset \partial \underline{Q}$, and, just as above, the quantity in square brackets in the last line of (3.15) remains bounded. Thus, the quantity in curly braces in the last line of (3.15) again goes to $-\infty$ as $\mathbf{b}^{(n)} \to \mathbf{b} \in \partial \underline{Q}_0$, which gives the same contradiction, and the proof is complete.

4. THE LOWER BOUND

Let $\alpha \in \underline{A}$. To prove the lower bound (1.3), it suffices, since G is open, to show that for any $\sigma \in \underline{M}$ such that $I_\alpha(\sigma) < \infty$,

$$\varliminf_{N \to \infty} \frac{1}{N} \log Q_{n,N}(N_\sigma) \geqslant -I_\alpha(\sigma) \tag{4.1}$$

for almost all $n(\cdot) \in Z$ (P_α-measure), where N_σ is any $\underline{M}$-neighborhood of σ.

We can choose $\varepsilon > 0$ so small that $\sigma' = (1-\varepsilon)\,\sigma + \varepsilon\alpha$ is an element of N_σ and then we can choose an $\underline{M}$-neighborhood of σ', call it $N_{\sigma'}$, such that $N_{\sigma'} \subset N_\sigma$. Moreover, since the I-function is convex and $I_\alpha(\alpha) = 0$, we see that

$$I_\alpha(\sigma') \leqslant (1-\varepsilon)\, I_\alpha(\sigma) + \varepsilon I_\alpha(\alpha) \leqslant I_\alpha(\sigma)$$

Thus, to show (4.1), it suffices to show for σ' of the form above and any $\underline{M}$-neighborhood $N_{\sigma'}$ of σ' that

$$\varliminf_{N \to \infty} \frac{1}{N} \log Q_{n,N}(N_{\sigma'}) \geqslant -I_\alpha(\sigma') \tag{4.2}$$

for almost all $n(\cdot) \in Z$ (P_α-measure).

Sets in M of the form

$$B_{\sigma'} = \{\lambda(\cdot) \in \underline{M} \colon |\lambda(x) - \sigma'(x)| < \delta \text{ for all } x \in F, \text{ a finite subset of } X)$$

form a basis for the weak topology in $\underline{M}$, so that, to show (4.2), it suffices to show, for any set $B_{\sigma'}$ in the basis and $\sigma' = (1-\varepsilon)\,\sigma + \varepsilon\alpha$, that

$$\underline{\lim}_{N\to\infty} \frac{1}{N} \log Q_{n,N}(B_{\sigma'}) \geqslant -I_\alpha(\sigma') \tag{4.3}$$

for almost all $n(\cdot) \in Z$ (P_α-measure).

With this in mind, we let F be a finite subset of X having elements $x_1, x_2,..., x_k$. Let $\underline{Q} \subset R^k$ be the set of $\boldsymbol{\theta} \in R^k$ having the property that $V: X \to \mathbb{R}$ defined by

$$V(x) = \begin{cases} \theta_i & \text{if} \quad x = x_i, \quad i = 1, 2,..., k \\ 0 & \text{otherwise} \end{cases}$$

is in $\underline{V}$. As noted in Section 3, $\underline{Q}$ is an open set in R^k. For each $N = 1, 2,...,$ define a probability measure μ_N on R^k by: if $A \subset R^k$,

$$\mu_N(A) = \underline{P}_n\{(L_N(x_1), L_N(x_2),..., L_N(x_k)) \in A\}$$

Let

$$M_N(\boldsymbol{\theta}) = \int \exp(N\langle \boldsymbol{\theta}, \mathbf{y}\rangle)\, d\mu_N(\mathbf{y}) = E^{\underline{P}_n}\left\{\exp\left[N \sum_{i=1}^{k} \theta_i L_N(x_i)\right]\right\}$$

which exists if $\boldsymbol{\theta} \in \underline{Q}$. What is important is that, if $\alpha \in \underline{A}$ and $\boldsymbol{\theta} \in \underline{Q}$, we have from Lemma 2.10 that

$$\begin{aligned}
&\lim_{N\to\infty} \frac{1}{N} \log M_N(\boldsymbol{\theta}) \\
&\quad = \lim_{N\to\infty} \frac{1}{N} \log E^{P_n}\left\{\exp\left[N \sum_{i=1}^{k} \theta_i L_N(x_i)\right]\right\} \\
&\quad = \lim_{N\to\infty} \frac{1}{N} \log E^{P_n}\left\{\exp\left[\sum_{j=0}^{N-1} \sum_{i=1}^{k} \theta_i n_j(x_i)\right]\right\} \\
&\quad = \lim_{N\to\infty} \frac{1}{N} \log E^{P_n}\left\{\exp\left[\sum_{j=0}^{N-1} \sum_{x\in X} n_j(x)\, V(x)\right]\right\} \\
&\quad = \sum_{x\in X} \alpha(x)\{1 - \exp[-V(x)\}\, u_V(x) \\
&\quad = \sum_{i=1}^{k} \alpha(x_i)[1 - \exp(-\theta_i)]\, u(\boldsymbol{\theta}; x_i) = \psi(\boldsymbol{\theta})
\end{aligned}$$

for almost all $n(\cdot) \in Z$ (P_α-measure).

We note that $\psi(\boldsymbol{\theta})$ is continuously differentiable on $\underline{O}$. For $\alpha \in \underline{A}$, $\sigma \in \underline{M}$ such that $I_\alpha(\sigma) < \infty$, and with $\sigma' = (1-\varepsilon)\sigma + \varepsilon\alpha$, we proved (Theorem 3.11) that in the variational problem,

$$\sup_{\boldsymbol{\theta} \in \underline{O}} \left[\sum_{i=1}^{k} \sigma'(x_i)\,\theta_i - \psi(\boldsymbol{\theta}) \right]$$

the supremum is actually attained at a point $\boldsymbol{\theta}^* \in \underline{O}$. This, of course, implies $(\nabla\psi)(\boldsymbol{\theta}^*) = \boldsymbol{\sigma}'$, where $\boldsymbol{\sigma}' = (\sigma'(x_i),\ i = 1, 2,..., k)$.

All of these observations become important because of the following lemma, which is a standard, general result, the proof of which can be found, for example, in Ref. 10.

Lemma 4.1. Let $\{\mu_N\}$ be a family of probability measures on R^k and assume

$$M_N(\boldsymbol{\theta}) = \int \exp(N\langle \boldsymbol{\theta}, \mathbf{y}\rangle)\, d\mu_N(\mathbf{y})$$

exists for $\boldsymbol{\theta}$ belonging to an open set $O \subset R^k$. Assume also that

$$\lim_{N \to \infty} \frac{1}{N} \log M_N(\boldsymbol{\theta} = \psi(\boldsymbol{\theta})$$

exists for $\boldsymbol{\theta} \in O$ and that $\psi(\boldsymbol{\theta})$ is continuously differentiable in O. Finally, assume $\boldsymbol{\sigma}' \in R^k$ is such that $(\nabla\psi)(\boldsymbol{\theta}^*) = \boldsymbol{\sigma}'$ for some $\boldsymbol{\theta}^* \in O$. Then,

$$\varliminf_{N \to \infty} \frac{1}{N} \log \mu_N(N_{\boldsymbol{\sigma}'}) \geqslant -[\langle \boldsymbol{\sigma}', \boldsymbol{\theta}^* \rangle - \psi(\boldsymbol{\theta}^*)] \tag{4.4}$$

where $N_{\boldsymbol{\sigma}'}$ is an R^k neighborhood of $\boldsymbol{\sigma}'$.

Since, as we just noted, all the hypotheses of Lemma 4.1 are satisfied, we conclude that

$$\begin{aligned}
&\varliminf_{N \to \infty} \frac{1}{N} \log \mu_N(N_{\boldsymbol{\sigma}'}) \\
&\quad = \varliminf_{N \to \infty} \frac{1}{N} \log P_n\{(L_N(x_1), L_N(x_2),..., L_N(x_k)) \in N_{\boldsymbol{\sigma}'}\} \\
&\quad = \varliminf_{N \to \infty} \frac{1}{N} \log Q_{n,N}(N_{\boldsymbol{\sigma}'}) \\
&\quad \geqslant -[\langle \boldsymbol{\sigma}', \boldsymbol{\theta}^* \rangle - \psi(\boldsymbol{\theta}^*)] \\
&\quad = -\sup_{\boldsymbol{\theta} \in \underline{O}} \left[\sum_{i=1}^{k} \sigma'(x_i)\,\theta_i - \psi(\boldsymbol{\theta}) \right]
\end{aligned}$$

for almost all $n(\cdot) \in Z$ (P_α-measure). But

$$I_\alpha(\sigma') = \sup_{V \in \underline{V}} \left[\sum_{x \in X} \sigma'(x)\, V(x) - \sum_{x \in X} \alpha(x)(1 - e^{-V(x)})\, u_V(x) \right]$$

$$= \sup_{\theta \in \underline{Q}} \left[\sum_{i=1}^{k} \sigma'(x_i)\, \theta_i - \psi(\boldsymbol{\theta}) \right] \tag{4.5}$$

because the first supremum is over V's vanishing outside any finite subset of X, whereas the second supremum is just over V's vanishing outside a fixed, finite set F. From (4.4) and (4.5) we have

$$\varliminf_{N \to \infty} \frac{1}{N} \log Q_{n,N}(N_{\sigma'}) \geqslant -I_\alpha(\sigma') \tag{4.6}$$

for almost all $n(\cdot) \in Z$ (P_α-measure).

Now, since F was any finite subset of X, and since $N_{\sigma'}$ is any R^k-neighborhood of σ', we see that (4.6) implies (4.3) and hence the lower bound (1.3).

ACKNOWLEDGMENT

This work was supported by NSF grants NSF-MCS-83-01364 and NSF-DMS-86-00233.

REFERENCES

1. M. D. Donsker and S. R. S. Varadhan, Asymptotic evaluation of certain Wiener integrals for large time, in *Functional Integration and Its Applications* (Proceedings of the International Conference, London) (Clarendon Press, Oxford, 1975), pp. 15–33.
2. M. D. Donsker and S. R. S. Varadhan, Asymptotic evaluation of certain Wiener integrals for large time, I, *Commun. Pure Appl. Math.* **28**:1–47 (1975); II, *Commun. Pure Appl. Math.* **28**:279–301 (1975); III, *Commun. Pure Appl. Math.* **29**:389–461 (1976); IV, *Commun. Pure Appl. Math.* **36**:183–212 (1983).
3. M. D. Donsker and S. R. S. Varadhan, Large deviations for stationary Gaussian processes, *Commun. Math. Phys.* **97**:187–210 (1985).
4. M. D. Donsker and S. R. S. Varadhan, Asymptotics for the Wiener sausage, *Commun. Pure Appl. Math.* **28**:525–565 (1975).
5. M. D. Donsker and S. R. S. Varadhan, Asymptotics for the polaron, *Commun. Pure Appl. Math.* **36**:505–528 (1983).
6. M. D. Donsker and S. R. S. Varadhan, On laws of the iterated logarithm for local times, *Commun. Pure Appl. Math.* **30**:707–753 (1977).

7. S. R. S. Varadhan, Asymptotic probabilities and differential equations, *Commun. Pure Appl. Math.* **19**:261–286 (1966).
8. T. Y. Lee, Large deviation theory for empirical density of the noninteracting infinite particle systems, Thesis, Courant Institute, New York University (June 1986).
9. J. T. Cox and D. Griffeath, Large deviations for some infinite particle system occupation times, *Contemp. Math.* **41**:43–53 (1985).
10. R. S. Ellis, *Entropy, Large Deviations, and Statistical Mechanics* (Springer-Verlag, New York, 1985).

Comm. Pure Appl. Math. 44 (1991) 911-923

Some Familiar Examples for Which the Large Deviation Principle Does Not Hold

J. R. BAXTER AND N. C. JAIN
University of Minnesota

AND

S. R. S. VARADHAN
Courant Institute

This paper is dedicated to Natascha Brunswick.

Abstract

For a class of Markov processes (in continuous or discrete time) we show that if the full large deviation holds for normalized occupation time measures $L_t(\omega, \cdot)$ with some rate function J, then the lower semicontinuous regularization of J must agree with the rate function I introduced by M. D. Donsker and S. R. S. Varadhan. As a consequence we show that for processes such as Brownian motion the full large deviation principle for $L_t(\omega, \cdot)$ cannot hold with *any* rate function.

1. Introduction

Let S be a complete separable metric space. $\mathcal{M}(S)$ will denote the set of probability measures on the Borel subsets of S. The topology of weak convergence of probability measures in $\mathcal{M}(S)$ is metrizable by the Prokhorov metric, and with this metric $\mathcal{M}(S)$ is itself a complete separable metric space. By a ball in $\mathcal{M}(S)$ we will always mean a ball with respect to this metric. Let $\{\mu_t, t \in T\} \subset \mathcal{M}(S)$, where $T = [o, \infty)$ or $\{1, 2, \cdots\}$. This family is said to satisfy the *full large deviation principle* with rate function $J: S \to [o, \infty]$, if (1.1) and (1.2) hold:

(1.1) For any closed set $C \subset S$,

$$\limsup_{t \to \infty} \frac{1}{t} \log \mu_t(C) \leqq - \inf_{x \in C} J(x);$$

(1.2) for any open $G \subset S$,

$$\limsup_{t \to \infty} \frac{1}{t} \log \mu_t(G) \geqq - \inf_{x \in G} J(x).$$

We will say that $\{\mu_t, t \in T\}$ satisfies the *restricted large deviation principle* with rate function J if (1.1) is assumed to hold only for compact sets C. In what follows, X_t, $t \in T$, will denote a Markov process with state space E, a complete separable metric space, and with transition probability function $\pi(t, x, dy)$. If $T = [o, \infty)$, the normalized occupation time measure will be denoted by ($1_A(x) = 1$ if $x \in A$, o otherwise)

$$(1.3) \qquad L_t(\omega, A) = \frac{1}{t}\int_o^t 1_A(X_s(\omega))\, ds, \qquad t > o, A \in \mathcal{E},$$

where $\mathcal{E}$ denotes the σ-algebra of Borel subsets of E. The process will be assumed to be Feller, i.e., $\int f(y)\pi(t, x, dy)$ as a function of x belongs to $C_b(E)$, the space of bounded continuous functions on E, whenever $f \in C_b(E)$. Let L denote the infinitesimal generator (on the Banach space $C_b(E)$ with sup norm) of the process, and let $D(L)$ denote its domain. If $\mu \in \mathcal{M}(E)$, $I(\mu)$ will denote the functional introduced in [2]

$$(1.4) \qquad I(\mu) = -\inf_u \int \left(\frac{Lu}{u}\right)(x)\, d\mu(x),$$

where the inf is taken over $u \in D(L)$ such that $\inf u > o$.

If $T = \{1, 2, \cdots\}$, then the corresponding definitions are given by

$$(1.5) \qquad L_n(\omega, A) = \frac{1}{n}\sum_{j=o}^{n-1} 1_A(X_j(\omega)), \qquad n \geqq 1, A \in \mathcal{E}$$

and

$$(1.6) \qquad I(\mu) = -\inf_u \int \log\left(\frac{\pi u}{u}\right)(x)\, d\mu(x),$$

where the inf is taken over all $u \in C_b(E)$ such that $\inf u > o$.

For $x \in E$, as usual $P^x(\Lambda)$ will denote the probability of the event Λ when the process starts at x. For a given $x \in E$, A a Borel subset of $\mathcal{M}(E)$, let

$$(1.7) \qquad \mu_t(A) = P^x[L_t(\omega, \cdot) \in A], \qquad t \in T.$$

In many familiar situations such as when $X_t, t \geqq o$, is a symmetric stable process with $E = \mathbb{R}^d$, or $X_n, n = 1, 2, \cdots$ is a simple random walk, with $E = \mathbb{Z}^d$, the results of Donsker and Varadhan (see [3]) show that for every $x \in E$, $\{\mu_t, t \in T\}$ defined by (1.7) satisfies the *restricted* large deviation principle with rate function I (given by (1.4) or (1.6), as the case may be). Since π is Feller in these cases, I is always lower semicontinuous (l.s.c.) but does not have compact level sets. A natural question arises whether a rate function $J : \mathcal{M}(E) \to [o, \infty]$ can be found such that $\{\mu_t, t \in T\}$ satisfies the full large deviation principle ((1.1) and (1.2)) with rate function J. In this note we show that the answer to this question is negative. To accomplish this, we first prove a general uniqueness theorem (Theorem 2.1) for the rate function in the next section. This theorem says that if the process is Feller, J is a lower semicontinuous rate function such that, for some x, μ_t defined by (1.7) satisfies the large deviation principle, and if (1.2) also holds with I, then $J = I$. If we now want to show that for a certain Feller process the full large deviation principle cannot hold for any rate function, we first show that for some closed set $C \subset \mathcal{M}(E)$ the upper bound (1.1) is not satisfied with rate function I. By the uniqueness theorem this implies that the full large deviation principle cannot hold for $\{\mu_t\}$ with any lower semicontinuous rate function J. Remark 2.4 is then used to show that the full large deviation principle cannot hold with any rate function.

The examples mentioned above are either transient or null recurrent Markov processes. We also give an example of a positive recurrent process (in fact, a Doeblin process—see [4], p. 192, for the definition of Hypothesis (D)) for which the full large deviation principle cannot hold with any rate function. All these applications of the uniqueness theorem are given in Section 3.

2. Uniqueness of the Rate Function

The main results of the section are Theorems 2.1 and 2.4. Theorem 2.4 is the discrete time version of Theorem 2.1. Since the proofs are essentially the same in the two cases, we shall prove only Theorem 2.1. The theorems imply that if a full large deviation principle is to hold with a lower semicontinuous rate function J, and the lower bound holds with rate function I defined by (1.4) or (1.6), then $J = I$.

THEOREM 2.1. *Suppose $\pi(t, x, dy)$ is a Feller transition probability function. Suppose $J : \mathcal{M}(E) \to [0, \infty]$ is lower semicontinuous and I is given by* (1.4). *Assume that for some $x \in E$,*

for every closed ball $F \subset \mathcal{M}(E)$

(2.1)
$$\limsup_{t \to \infty} \frac{1}{t} \log P^x\{L_t(\omega, \cdot) \in F\} \leqq -\inf_{\mu \in F} J(\mu),$$

and

for every open ball $G \subset \mathcal{M}(E)$

(2.2)
$$\liminf_{t \to \infty} \frac{1}{t} \log P^x\{L_t(\omega, \cdot) \in G\} \geqq -\inf_{\mu \in G} (J \wedge I)(\mu),$$

then $J = I$.

Remark 2.2. In the above theorem we do not assume J to have compact level sets.

Proof: We know from [2], page 40, that for any measurable $C \subset \mathcal{M}(E)$, any $x \in E$

$$\limsup_{t \to \infty} \frac{1}{t} \log P^x\{L_t(\omega, \cdot) \in C\} \leqq \inf_u \sup_{\mu \in C} \int \frac{Lu}{u}\, d\mu , \tag{2.3}$$

where the inf is over $u \in D(L)$ such that $\inf u > 0$. Fix $\mu_o \in \mathcal{M}(E)$ and let $a < I(\mu_o)$. Then for some $u_o \in D(L)$ with $\inf u_o > o$, we have $\int (Lu_o/u_o)\, d\mu_o < -a$. Since Lu/u is a bounded continuous function for such u, there exists a neighborhood N of μ_o in $\mathcal{M}(E)$ such that

$$\sup_{\mu \in N} \int (Lu_o/u_o)\, d\mu \leqq -a ,$$

which implies

$$\inf_u \sup_{\mu \in N} \int (Lu/u)\, d\mu \leqq -a .$$

If $C = N$ in (2.3), we conclude the following: given $\mu \in \mathcal{M}(E)$, given $a < I(\mu)$, there exists a neighborhood N of μ such that

$$\limsup_{t \to \infty} \frac{1}{t} \log P^x\{L_t(w, \cdot) \in N\} \leqq -a . \tag{2.4}$$

Let $\mu \in \mathcal{M}(E)$ and let $a < I(\mu)$. Then (2.4) and (2.2) immediately imply that $J(\mu) \geqq a$, showing that $I \leqq J$. Since J is assumed to be lower semicontinuous, (2.1) implies that given $\mu \in \mathcal{M}(E)$ and $a < J(\mu)$, there exists a neighborhood N of μ for which (2.4) holds. Since (2.2) is symmetric in I and J, the above argument also shows that $J \leqq I$. This proves the theorem.

The discrete time analogue of Theorem 2.1 is the following theorem.

THEOREM 2.3. *Suppose $\pi(1, x, dy)$ is a Feller transition probability function. Suppose $J : \mathcal{M}(E) \to [o, \infty]$ is lower semicontinuous and I is given by* (1.6). *Assume that for some $x \in E$,*

for every closed ball $F \subset \mathcal{M}(E)$

$$\limsup_{n \to \infty} \frac{1}{n} \log P^x\{L_n(\omega, \cdot) \in F\} \leqq -\inf_{\mu \in F} J(\mu) , \tag{2.5}$$

and

$$\text{for every open ball} \quad G \subset \mathcal{M}(E)$$

(2.6)
$$\liminf_{n\to\infty} \frac{1}{n} \log P^x\{L_n(\omega, \cdot) \in G\} \geqq -\inf_{\mu\in G}(J \wedge I)(\mu),$$

then $J = I$; *here* $a \wedge b = \min(a, b)$.

Remark 2.4. In Theorems 2.1 and 2.4 if J is not assumed to be lower semicontinuous, then we can define $J_1(\mu) = \lim_{\varepsilon\to o}\inf_{\nu\in B_\varepsilon} J(\nu)$, where B_ε is a ball of radius ε with center μ. Then J_1 is lower semicontinuous, $J_1 \leqq J$ and satisfies the conditions of the theorems. Thus we then have $J_1 = I$.

3. Applications

We shall show that empiricals of symmetric stable processes cannot satisfy the full large deviation principle with any rate function. In discrete time we shall show the same to hold for simple random walk. When the space dimension $d = 1$, these are examples of "null recurrent" processes. We shall also give an example of a positive recurrent Markov chain for which the same conclusion holds. The basic strategy is to show that for μ_t, defined by (1.7), the upper bound in (1.1) does not hold for all closed sets with rate function I given by (1.4) (or (1.6)). Theorem 2.1 (or Theorem 2.3) will then show that the full large deviation principle cannot hold with any rate function.

We shall need the following technical result.

PROPOSITION 3.1. *Suppose* $\{\mu_n\}$ *is a sequence of probability measures on a complete separable metric space* S *such that for some* $\delta > o$, *all compact* $K \subset S$, $\limsup_n \mu_n(K) \leqq 1 - 3\delta$. *If* C *is a closed set such that* $\limsup_n \mu_n(C) \leqq \delta$, *then there exists a closed set* $C_1 \subset C^c$ *such that* $\limsup_n \mu_n(C_1) \geqq \delta$.

Before proving the proposition we observe that S is homeomorphic to a G_δ subset of the Hilbert cube. Therefore without any loss of generality S may be given an equivalent metric which makes it totally bounded, and if S^* = completion of S under this metric then S is a Borel subset of S^*. If ν is a measure on S and A is a Borel subset of S^*, then $\nu(A) := \nu(A \cap S)$ extends ν to S^*. Since S^* is compact, every subsequence of $\{\mu_n\}$ on S^* has a weakly convergent subsequence. We now prove a lemma.

LEMMA 3.2. *Let* S, S^*, *and* $\{\mu_n\}$ *be as above. If* $\nu_k = \mu_{n_k} \to \mu$ *weakly on* S^*, *then* $\mu(S) \leqq 1 - 3\delta$.

Proof: Let $K \subset S$ be a compact set. It is enough to show that $\mu(K) \leqq 1 - 3\delta$. K is also a compact subset of S^*. There exist open $U_j \subset S^*$, $j \geqq 1$, such that $U_j \searrow K$. Since $\nu_k \to \mu$ weakly, we have

$$\mu(K) \leqq \mu(U_j) \leqq \liminf_k \nu_k(U_j), \qquad j \geqq 1. \tag{3.1}$$

Therefore, given $\varepsilon > o$ there exist $k(j) \nearrow \infty$ and compact $K_j \subset U_j \cap S$ such that

$$\nu_{k(j)}(K_j) \geqq \mu(K) - \varepsilon, \qquad j \geqq 1. \tag{3.2}$$

Let $B = \bigcup_{j=1}^{\infty} K_j \cup K$. If U is an open subset of S^* which contains the compact set K, then $U \supset K_j$ for all j sufficiently large. Therefore B is a compact subset of S and by (3.2) we then have $\mu(K) \leqq \limsup_n \mu_n(B) + \varepsilon \leqq 1 - 3\delta + \varepsilon$ for all $\varepsilon > o$. Therefore $\mu(K) \leqq 1 - 3\delta$, and the lemma is proved.

Proof of Proposition 3.1: Let $\nu_k \to \mu$ weakly in S^* as above. By Lemma 3.2 we have

$$\mu(S) \leqq 1 - 3\delta. \tag{3.3}$$

Therefore there exists a compact $H \subset S^* - S$ such that $\mu(H) > 2\delta$. Let

$$h(x) = \frac{1}{\rho(x, H)}, \tag{3.4}$$

where ρ denotes the equivalent metric chosen above and $\rho(x, H) = \inf\{\rho(x, y) : y \in H\}$. For any $\alpha > o$, let $V_\alpha = \{x : h(x) < \alpha\} \cap S$. These are open subsets of S and $V_\alpha \nearrow S$ as $\alpha \nearrow \infty$. By weak convergence we have

$$\begin{aligned} \limsup_k \nu_k(V_\alpha) &\leqq \limsup_k \nu_k\{x \in S^* : h(x) \leqq \alpha\} \\ &\leqq \mu\{x \in S^* : h(x) \leqq \alpha\} \leqq \mu(H^c) < 1 - 2\delta. \end{aligned}$$

There exists $k(1)$ such that $\nu_{k(1)}(C \cup V_1) < 1 - \delta$, so we can find compact $K_1 \subset S - (C \cup V_1)$ such that $\nu_{k(1)}(K_1) > \delta$. There exists $m(1)$ such that $K_1 \subset V_{m(1)}$. Then we can find $k(2)$ such that $\nu_{k(2)}(C \cup V_{m(1)+1}) < 1 - 2\delta$, and then a compact $K_2 \subset S - (C \cup V_{m(1)+1})$, $K_2 \subset V_{m(2)}$, such that $\nu_{k(2)}(K_2) > \delta$. We proceed in this manner, and find $m(o) = 1$, $m(i+1) > m(i) + 1$, $k(i+1) > k(i)$, and compacts $K_{i+1} \subset S - (C \cup V_{m(i)+1})$, $K_i \subset V_{m(i)}$, and $\nu_{k(i)}(K_i) > \delta$ for $i \geqq 1$. Let $C_1 = \bigcup_{i=1}^{\infty} K_i$. Then $C_1 \subset S - C$, $\nu_{k(i)}(C_1) \geqq \delta$ for all i. It remains to check that C_1 is a closed subset of S. Let $\{x_n\} \subset C_1$, $x_n \to x \in S$ as $n \to \infty$. If infinitely many x_n's belong to some K_i, then $x \in K_i \subset C_1$, so we might as well assume that $x_n \in K_n$, $n \geqq 1$. Since h is a real-valued continuous function on S, we have $h(x_n) \to h(x)$.

On the other hand, $h(x_n) \in K_n \subset V_{m(n)} - V_{m(n-1)+1}$, i.e., $|h(x_n) - h(x_{n+1})| \geqq 1$, which is a contradiction and the assumption cannot be valid. This proves the proposition.

We now come to the question of a symmetric stable process. A process X_t, $t \geqq o$, with stationary, independent increments, taking values in $\mathbb{R}^d$, is symmetric stable if

$$E^o\{e^{i\langle X_t, u\rangle}\} = e^{-t|u|^\alpha}, \qquad u \in \mathbb{R}^d,$$

where $\langle \cdot, \cdot \rangle$ denotes the Euclidean inner product and $|\cdot|$ the corresponding norm in $\mathbb{R}^d$ and $o < \alpha \leqq 2$. The case $\alpha = 2$ corresponds to Brownian motion with variance parameter equal to 2.

THEOREM 3.3. *Let X_t, $t \geqq o$, be a symmetric stable process. Then there is no rate function $J : \mathcal{M}(\mathbb{R}^d) \to [o, \infty]$ such that the full large deviation principle holds for μ_t, defined by (1.7), with rate function J.*

Proof: Suppose the full large deviation principle holds with rate function J. If J_1 is defined in terms of J as in Remark 2.5, then J_1 is lower semicontinuous and satisfies the full large deviation principle. By the results in [3], the lower bound result holds for a symmetric stable process with rate function I. Since the transition function of the process satisfies the Feller property, it follows from Theorem 2.1 that $I = J_1$, i.e., I must satisfy the large deviation principle. The theorem will therefore be proved if we show that there exists a closed set $C \subset \mathcal{M}(\mathbb{R}^d)$ for which the upper bound cannot hold with rate function I. Suppose we can show that for any x

$$P^x\{\omega : I(L_1(\omega, \cdot)) = \infty\} = 1, \tag{3.5}$$

then by spatial homogeneity and the scaling property of the symmetric stable process we have $I(L_t(\omega, \cdot)) = \infty$, P^x-a.s. The set $C = \{\mu : I(\mu) \leqq 1\}$ is closed since I is lower semicontinuous, and we have $P^x\{L_t(\omega, \cdot) \in C\} = o$ for all $t > o$. We shall show that for any compact $K \subset \mathcal{M}(\mathbb{R}^d)$, any x

$$\lim_{t\to\infty} P^x\{L_t(\omega, \cdot) \in K\} = o. \tag{3.6}$$

Let $t_n \nearrow \infty$, and let $\mu_n(A) = P^x\{L_{t_n}(\omega, \cdot) \in A\}$, for $A \subset \mathcal{M}(\mathbb{R}^d)$. With $S = \mathcal{M}(\mathbb{R}^d)$, the sequence $\{\mu_n\}$ satisfies the conditions of Proposition 3.1 with $\delta = \frac{1}{3}$ and we conclude that there exists a closed set $C_1 \subset C^c$ such that $\limsup_n \mu_n(C_1) \geqq \frac{1}{3}$. It follows that

$$\limsup_{t\to\infty} \frac{1}{t} \log P^x\{L_t(\omega, \cdot) \in C_1\} \geqq o. \tag{3.7}$$

On the other hand, since $C_1 \subset C^c = \{\mu : I(\mu) > 1\}$, if the full large deviation principle were true with rate function I, we would have the lim sup in (3.7) less than or equal to -1, which is a contradiction. Therefore the full large deviation principle cannot hold with rate function I and the theorem is proved if (3.5) and (3.6) can be established.

The next lemma shows that (3.6) holds. The proof of (3.5) will be accomplished through Propositions 3.5, 3.7, and Theorem 3.8.

LEMMA 3.4. *Let $K \subset \mathcal{M}(\mathbb{R}^d)$ be compact. Then*

$$\lim_{t\to\infty} P^x\{L_t(\omega, \cdot) \in K\} = o, \qquad x \in \mathbb{R}^d.$$

Proof: Note that $L_t(\omega, F) \to o$ as $t \to \infty$, P^x-a.s., for any compact $F \subset \mathbb{R}^d$ and for any $x \in \mathbb{R}^d$ since the process is either transient or null-recurrent. Assume that there exist x, $K \subset \mathcal{M}(\mathbb{R}^d)$ compact, $\eta > o$, and $t_n \nearrow \infty$ such that

$$P^x\{L_{t_n}(\omega, \cdot) \in K\} \geqq \eta, \qquad n \geqq 1.$$

Then $P^x(\Lambda) \geqq \eta$, where $\Lambda = \{L_{t_n}(\omega, \cdot) \in K \text{ i.o.}\}$. Therefore, if $\omega \in \Lambda$, there exists a subsequence $\{t'_n\}$ such that $L_{t'_n}(\omega, \cdot) \in K$, i.e., $\{L_{t'_n}(\omega, \cdot), n \geqq 1\}$ is a tight family, which is a contradiction, so the lemma is proved.

We now proceed to prove (3.5). We first observe that if $\alpha \leqq 1$ or if $d \geqq 2$, then P^x-a.s., $L_t(\omega, \cdot)$ is singular with respect to Lebesgue measure in $\mathbb{R}^d$. Since the Lebesgue measure plays the role of a reference measure in [3], $I(L_t(\omega, \cdot)) < \infty \Rightarrow L_t(\omega, \cdot) \ll m$, where m denotes Lebesgue measure. Therefore without any loss of generality we may assume $d = 1$ and $1 < \alpha \leqq 2$. In this situation it is known (see [1]) that $(dL_t(\omega, \cdot)/dm)(x) = l_t(\omega, x)$ exists and has a jointly continuous version in (t, x).

The next proposition is essentially Theorem 1.3 in [5].

PROPOSITION 3.5. *Let Y_t, $t \geqq o$, be a Markov process, reversible with respect to invariant probability measure π. Assume that P_π is ergodic. Let $g \in L^2(\pi)$ be such that $\int g\, d\pi = o$. Assume that there exists C such that*

$$\left|\int fg\, d\pi\right| \leqq C\left(-\int fLf\, d\pi\right)^{1/2}, \qquad f \in D(L),$$

where $D(L)$ denotes the domain of the generator of the process. Then $t^{-1/2} \int_o^t g(Y_s)\, ds$ converges weakly to $N(o, \sigma^2)$, for almost every x with respect to π, where $\sigma^2 = 2(-L^{-1}g, g)$. In particular, if $Lg = -\lambda g$ for some $\lambda > o$, then $\sigma^2 = (2/\lambda)\|g\|^2$. Here $(\cdot, \cdot)$ and $\|\cdot\|$ denote the inner product and the norm, respectively, in $L^2(\pi)$.

Remark 3.6. In the context of Proposition 3.5, $Lf = o \Rightarrow f =$ constant, and since $(1, g) = o$, the expression for σ^2 is well defined.

We now return to the symmetric stable process X_t. Let n be a positive integer, and let $\theta_1, \theta_2, \cdots, \theta_n$ be real numbers linearly independent over the rationals. We define the process

$$Z_n(t) = (e^{i\theta_1 X_t}, \cdots, e^{i\theta_n X_t}) \tag{3.8}$$

which is a process with stationary independent increments with state space T^n, the n-dimensional torus. The normalized Haar measure, denoted by μ, is clearly the invariant measure for its transition probability function. If $k_1, \cdots, k_n$ are integers, let $\Psi_{k_1,\cdots,k_n} : T^n \to T^1$ denote the group character

$$\Psi_{k_1,\cdots,k_n}(e^{ix_1}, \cdots, e^{ix_n}) = e^{i(k_1x_1 + \cdots + k_nx_n)} . \tag{3.9}$$

If $L_{\theta_1,\cdots,\theta_n}$ denotes the infinitesimal generator of the Markov process $Z_n(t)$, $t \geqq o$, then

$$L_{\theta_1,\cdots,\theta_n}\Psi_{k_1,\cdots,k_n} = -|\theta_1 k_1 + \cdots + \theta_n k_n|^{\alpha}\Psi_{k_1,\cdots,k_n} . \tag{3.10}$$

Therefore $\Psi_{k_1,\cdots,k_n}$ are eigenfunctions for $L_{\theta_1,\cdots,\theta_n}$; $\{\Psi_{k_1,\cdots,k_n} : k_1, \cdots k_n \text{ integers}\}$ are a complete orthonormal system in $L^2(\mu)$. It follows from (3.10) that if $f(e^{ix_1}, \cdots, e^{ix_n}) = \cos x_j$, then

$$(L_{\theta_1,\cdots,\theta_n}f)(e^{ix_1}, \cdots, e^{ix_n}) = -|\theta_j|^{\alpha}\cos x_j , \qquad j = 1, \cdots, n . \tag{3.11}$$

Since $\theta_1, \cdots, \theta_n$ are chosen to be linearly independent over the rationals, (3.10) immediately implies that if $L_{\theta_1,\cdots,\theta_n}f = o$ then $f =$ constant. This implies that $Z_n(t)$ with initial distribution μ is a stationary ergodic process. Let $c_1, c_2, \cdots, c_n$ be real, and define

$$Y_n(t) = c_1\cos(\theta_1 X_t) + \cdots + c_n\cos(\theta_n X_t) . \tag{3.12}$$

We now prove

PROPOSITION 3.7. *Let $Y_n(t)$, $t \geqq o$, be defined by* (3.12). *Then $t^{-1/2} \times \int_o^t Y_n(s)\, ds$ converges weakly to $N(o, \sigma^2)$, as $t \to \infty$, for almost every $x(\mu)$, where $\sigma^2 = c_1^2|\theta_1|^{-\alpha} + \cdots + c_n^2|\theta_n|^{-\alpha}$. In particular, $t^{-1/2}\int_o^t (\cos(\theta_1 X_s), \cdots, \cos(\theta_n X_s))\, ds$ converges weakly to $N(o, \Sigma)$, as $t \to \infty$, where Σ is a diagonal matrix with the i-th diagonal element equal to $|\theta_i|^{-\alpha}$.*

Proof: We simply apply Proposition 3.5 to the process $Z_n(t)$ with $g(e^{ix_1}, \cdots, e^{ix_n}) = c_1\cos x_1 + \cdots + c_n\cos x_n$, and $\pi = \mu$. Then $g \in \text{Range}((-L)^{1/2})$, so the conditions of the proposition are satisfied. Since $2(-L^{-1}g, g) = c_1^2|\theta_1|^{-\alpha} + \cdots + c_n^2|\theta_n|^{-\alpha}$, the proposition is proved.

THEOREM 3.8. *For a symmetric stable process X_t, $t \geqq o$, we have $I(L_1(\omega, \cdot)) = \infty$, P^x-a.s., for all x.*

Proof: As we observed earlier, it suffices to consider $1 < \alpha \leqq 2$ and $d = 1$. Also, if we know the result for some x, P^x-a.s., then, I being invariant under translations of measures in the present context, we have the result for all x. Let

$$u(x) = \exp\left\{\frac{1}{\lambda^{\alpha/2}} \sum_{j=1}^{n} u_j \cos(\lambda\theta_j x)\right\}, \qquad x \in \mathbb{R}, \tag{3.13}$$

where $\lambda > o$, $u_1, u_2, \cdots, u_n$ are rational. If L denotes the generator of the stable process, then $u \in D(L)$. Let $v(x) = u(x/\lambda)$, then by the scaling property $(Lv)(\lambda x) = \lambda^{-\alpha}(Lu)(x)$. This together with the scaling property gives

$$\int_o^1 \frac{Lu}{u}(X_s)\,ds \overset{d}{=} \int_o^{\lambda^\alpha} \frac{Lv}{v}(X_s)\,ds, \tag{3.14}$$

where $Z_1 \overset{d}{=} Z_2$ means the random variables Z_1 and Z_2 have the same distribution. Let $h(x) = \sum_{j=1}^n u_j\cos(\theta_j x)$. If A is an upper bound for $|h|$, $|h'|$, and $|h''|$, then it is easy to verify that there exists $C > o$ such that $\|Lh^k\|_\infty \leqq CA^k k(k-1)$ for all positive integers k. Therefore

$$Lv = \lambda^{-\alpha/2}Lh + \tfrac{1}{2}\lambda^{-\alpha}Lh^2 + o(\lambda^{-\alpha})$$

where o does not depend on x. Therefore

$$\frac{Lv}{v} = \lambda^{-\alpha/2}Lh + \frac{1}{2}\lambda^{-\alpha}Lh^2 - \lambda^{-\alpha}hLh + o(\lambda^{-\alpha}), \tag{3.15}$$

and we conclude that

$$\begin{aligned} &\lim_{\lambda\to\infty} \int_o^1 \frac{Lu}{u}(X_s)\,ds \\ &\overset{d}{=} \lim_{\lambda\to\infty} \int_o^{\lambda^\alpha} \left\{\lambda^{-\alpha/2}Lh + \frac{1}{2}\lambda^{-\alpha}Lh^2 - \lambda^{-\alpha}hLh\right\}(X_s)\,ds, \end{aligned} \tag{3.16}$$

where these limits are to be understood in the sense of weak convergence. We will now identify the limit on the right side in (3.16). Since $(L\cos(\theta\cdot))(y) = -|\theta|^\alpha\cos(\theta y)$, by Proposition 3.7 we get

$$(3.17)\qquad \frac{1}{\lambda^{\alpha/2}}\int_o^{\lambda^\alpha} Lh(X_s)\,ds \to N(o,\sigma^2)$$

weakly as $\lambda \to \infty$, where $\sigma^2 = \sum_{j=1}^n u_j^2 |\theta_j|^\alpha$. Also

$$h^2 = \frac{1}{2}\sum_{j,k} u_j u_k[\cos(\theta_j+\theta_k)x + \cos(\theta_j-\theta_k)x]$$

and

$$hLh = -\frac{1}{2}\sum_{j,k} u_j u_k |\theta_k|^\alpha[\cos(\theta_j+\theta_k)x + \cos(\theta_j-\theta_k)x]\,,$$

hence by (3.17) only the constants in the diagonal terms of hLh contribute to the limit in (3.16) and we get

$$(3.18)\qquad \lim_{\lambda\to\infty}\int_o^1 \frac{-Lu}{u}(X_s)\,ds \overset{d}{=} Z - \frac{1}{2}\sum_{j=1}^n u_j^2|\theta_j|^\alpha\,,$$

where Z is a $N(o, \sum_{j=1}^n u_j^2|\theta_j|^\alpha)$ random variable.

To finish the argument, let $\theta = \{\theta_j\}$ be a sequence of real numbers such that for any n, $\theta_1, \theta_2, \cdots, \theta_n$ are linearly independent over rationals, and let $\{\xi_j\}$ be a sequence of independent $N(o, |\theta_j|^{-\alpha})$ random variables defined on some probability space. Consider the collection of random variables

$$\zeta_{u_1,u_2,\cdots,u_n;\theta_1,\theta_2,\cdots,\theta_n} := \sum_{j=1}^n u_j|\theta_j|^\alpha\xi_j - \frac{1}{2}\sum_{j=1}^n u_j^2|\theta_j|^\alpha\,,$$

where $u_1, u_2, \cdots, u_n$, $n \geqq 1$, are rational. Let

$$\eta = \sup_{\substack{\lambda>o\\ u_1,u_2,\cdots,u_n}} \int_o^1 \frac{-Lu}{u}(X_s)\,ds\,,$$

where u is given by (3.13) and the sup is taken over rational $u_1, \cdots, u_n$, and rational λ, then for any $a > o$, $n \geqq 1$,

$$P^o[\eta \leqq a] \leqq P^o[\sup_{\substack{u_1,u_2,\cdots,u_n\\ \text{rational}}} \zeta_{u_1,u_2,\cdots,u_n;\theta_1,\theta_2,\cdots,\theta_n} \leqq a]\,.$$

Since $\eta(\omega) \leqq I(L_1(\omega,\cdot))$, and $\sup \zeta = \frac{1}{2}\sum_{j=1}^n |\theta_j|^\alpha \xi_j^2$, we have

$$P^o[I(L_1(\omega,\cdot)) \leqq a] \leqq P^o\left[\frac{1}{2}\sum_{j=1}^n |\theta_j|^\alpha\xi_j^2 \leqq a\right],$$

the latter probability clearly tends to zero as $n \to \infty$. Therefore $P^o[I(L_1(\omega, \cdot)) \leqq a] = o$ for every $a > o$, and the theorem is proved.

We now consider the simple random walk example. Let $X_1, X_2, \cdots$ be a simple random walk on the integer lattice $\mathbb{Z}$, i.e., it is a Markov chain for which X_n is the n-th partial sum of a sequence of independent, identically distributed random variables $\eta_1, \eta_2, \cdots$, and $P[\eta_1 = 1] = P[\eta_1 = -1] = \frac{1}{2}$. Let $\pi(x, dy)$ denote the one-step transition probability function of the chain. If $\mu \in \mathcal{M}(\mathbb{Z})$, then according to (1.6) we have

$$I(\mu) = \sup_{u > o} \sum_{j=-\infty}^{\infty} \left(\log \frac{2u(j)}{u(j-1) + u(j+1)} \right) \mu(\{j\}) .$$

Let

$$C = \left\{ \mu : \mu(\{j\}) = \frac{1}{2m+1}, \quad j = o, 1, \cdots, 2m, \quad \text{for some } m \geqq 1 \right\} .$$

C is a closed subset of $\mathcal{M}(\mathbb{Z})$. If $\mu \in C$, and we take $u(2j+1) = 1$, $u(2j) = k \geqq 2$, for all j, then an elementary computation shows that $I(\mu) \geqq (2m+1)^{-1} \log k$, where μ is distributed over $o, 1, \cdots, 2m$. Therefore $I(\mu) = \infty$ whenever $\mu \in C$. For $j \geqq 1$, $n = 2r + 1$, $r \geqq 1$, we have

$$P^j\{L_n(\omega, \cdot) \in C\} \geqq P^j\{X_o = j, X_1 = j+1, \cdots, X_{n-1} = j + n - 1\} = 2^{-n} .$$

Therefore

$$-\log 2 \leqq \limsup_{n \to \infty} \frac{1}{n} \log P^j\{L_n(\omega, \cdot) \in C\} .$$

Since $\inf\{I(\mu) : \mu \in C\} = \infty$, the upper bound in the large deviation principle cannot hold for C with rate function I. By Theorem 2.4 and Remark 2.5 we conclude as in the proof of Theorem 3.3 that the full large deviation principle for the simple random walk cannot hold with any rate function.

The next example is one of a positive recurrent chain (Doeblin chain). Let the state space $E = \{o, 1, 2, \cdots\}$, and with $o < \rho < 1$ let the transition probability function be given by $\pi(x, x+1) = \rho$, $\pi(x, o) = 1 - \rho$ for all $x \in E$. Then for $\mu \in \mathcal{M}(E)$,

$$I(\mu) = \sup_{u > o} \sum_{j} \left(\log \frac{u(j)}{(1-\rho)u(o) + \rho u(j+1)} \right) \mu(\{j\}) .$$

Let

$$C = \left\{ \mu : \mu(\{j\}) = \frac{1}{m}, j = l + 1, \cdots, l + m, \quad \text{some } m \geqq 1, \quad \text{and } l \geqq 1 \right\}.$$

If we pick u such that $u(j) = 1$ for $j \neq l + 1$ and $u(l + 1) = k$, then $I(\mu) \geqq m^{-1}\log k$, where $\mu(\{j\}) = m^{-1}$ for $j = l + 1, \cdots, l + m$; so $I(\mu) = \infty$ whenever $\mu \in C$. For $j \geqq 1$,

$$P^j\{L_n(\omega, \cdot) \in C\} \geqq P^j\{X_o = j, X_1 = j + 1, \cdots, X_{n-1} = j + n - 1\}$$

$$\geqq \rho^{n-1}.$$

We then conclude, as in the simple random walk example, that the full large deviation principle for this chain cannot hold with any rate function.

Acknowledgment. This work was partially supported by National Science Foundation grants.

Bibliography

[1] Boylan, E. S., *Local times for a class of Markov processes*, Ill. Jour. Math. 8, 1969, pp. 19–39.

[2] Donsker, M. D., and Varadhan, S. R. S., *Asymptotic evaluation of certain Markov process expectations for large time I*, Comm. Pure Appl. Math. 28, 1975, pp. 1–47.

[3] Donsker, M. D., and Varadhan, S. R. S., *Asymptotic evaluation of certain Markov process expectations for large time III*, Comm. Pure Appl. Math. 29, 1976, pp. 389–461.

[4] Doob, J. L., *Stochastic Processes*, John Wiley & Sons, New York, 1953.

[5] Kipnis, C., and Varadhan, S. R. S., *Central limit theorem for additive functionals of reversible Markov processes and applications to simple exclusions*, Comm. Math. Phy. 104, 1986, pp. 1–19.

Received March 1990.

European Journal of Combinatorics 32 (2011) 1000–1017

Contents lists available at ScienceDirect

European Journal of Combinatorics

journal homepage: www.elsevier.com/locate/ejc

The large deviation principle for the Erdős-Rényi random graph

Sourav Chatterjee, S.R.S. Varadhan

Courant Institute of Mathematical Sciences, New York University, 251 Mercer Street, New York, NY 10012, United States

ARTICLE INFO

Article history:
Available online 6 May 2011

ABSTRACT

What does an Erdős-Rényi graph look like when a rare event happens? This paper answers this question when p is fixed and n tends to infinity by establishing a large deviation principle under an appropriate topology. The formulation and proof of the main result uses the recent development of the theory of graph limits by Lovász and coauthors and Szemerédi's regularity lemma from graph theory. As a basic application of the general principle, we work out large deviations for the number of triangles in $G(n, p)$. Surprisingly, even this simple example yields an interesting double phase transition.

1. Introduction

1.1. The Erdős–Rényi graph

Let $G(n, p)$ be the random graph on n vertices where each edge is added independently with probability p. This model has been the subject of extensive investigations since the pioneering work of [14], yielding a large body of literature (see [2,18] for partial surveys).

This paper studies the following basic aspect of Erdős–Rényi graphs: What does the graph look like if one knows that some rare event has happened? One way to comprehensively answer this question is to formulate a large deviation principle for the Erdős–Rényi graph, in the same way as Sanov's theorem [27] gives a large deviation principle for an i.i.d. sample.

The setting of Sanov's theorem conforms naturally to the abstract theory of large deviations (see Chapter 6 in [11]) because i.i.d. samples can be thought of as random probability measures, allowing them to be viewed as random elements of a single topological space irrespective of the sample size. The first hurdle in formulating such a program for random graphs is in constructing a single abstract space

E-mail address: sourav@cims.nyu.edu (S. Chatterjee).

in which all graphs can be embedded. Fortunately, this issue has been settled recently. In a sequence of papers [5–7,15,19–25] Laszlo Lovász with coauthors (listed here in order of frequency) Sós, Szegedy, Borgs, Chayes, Vesztergombi, Schrijver and Freedman have developed a beautiful, unifying limit theory (see also the related work of [12] which traces this back to work of [1,17].) This sheds light on topics such as graph homomorphisms, Szemerédi's regularity lemma, quasi-random graphs, graph testing and extremal graph theory, and has even found applications in statistics and related areas (see e.g. [10]). Their theory has been developed for dense graphs (number of edges comparable with the square of number of vertices) but parallel theories for sparse graphs are beginning to emerge [3].

1.2. Graph limits and graphons

The limit of a sequence of dense graphs can be defined as follows. We quote the definition verbatim from [22] (see also [6,7,12]). Let G_n be a sequence of simple graphs whose number of nodes tends to infinity. For every fixed simple graph H, let $|\operatorname{hom}(H, G)|$ denote the number of homomorphisms of H into G (i.e. edge-preserving maps $V(H) \to V(G)$, where $V(H)$ and $V(G)$ are the vertex sets). This number is normalized to get the homomorphism density

$$t(H, G) := \frac{|\operatorname{hom}(H, G)|}{|V(G)|^{|V(H)|}}. \tag{1}$$

This gives the probability that a random mapping $V(H) \to V(G)$ is a homomorphism.

Suppose that the graphs G_n become more and more similar in the sense that $t(H, G_n)$ tends to a limit $t(H)$ for every H. One way to define a limit of the sequence $\{G_n\}$ is to define an appropriate limit object from which the values $t(H)$ can be read off.

The main result of [22] (following the earlier equivalent work of [1,17]) is that indeed there is a natural "limit object" in the form of a function $f \in \mathcal{W}$, where $\mathcal{W}$ is the space of all measurable functions from $[0, 1]^2$ into $[0, 1]$ that satisfy $f(x, y) = f(y, x)$ for all x, y.

Conversely, every such function arises as the limit of an appropriate graph sequence. This limit object determines all the limits of subgraph densities: if H is a simple graph with $V(H) = [k] = \{1, \ldots, k\}$, then

$$t(H, f) = \int_{[0,1]^k} \prod_{(i,j)\in E(H)} f(x_i, x_j)\, dx_1 \cdots dx_k. \tag{2}$$

Here $E(H)$ denotes the edge set of H. A sequence of graphs $\{G_n\}_{n\geq 1}$ is said to converge to f if for every finite simple graph H,

$$\lim_{n\to\infty} t(H, G_n) = t(H, f). \tag{3}$$

Intuitively, the interval $[0, 1]$ represents a 'continuum' of vertices, and $f(x, y)$ denotes the probability of putting an edge between x and y. For example, for the Erdős–Rényi graph $G(n, p)$, if p is fixed and $n \to \infty$, then the limit graph is represented by the function that is identically equal to p on $[0, 1]^2$.

These limit objects, i.e. elements of $\mathcal{W}$, are called 'graphons' in [22,6,7]. A finite simple graph G on $\{1, \ldots, n\}$ can also be represented as a graphon f^G is a natural way, by defining

$$f^G(x, y) = \begin{cases} 1 & \text{if}(\lceil nx\rceil, \lceil ny\rceil)\text{is an edge in } G, \\ 0 & \text{otherwise.} \end{cases} \tag{4}$$

Note that this allows *all* simple graphs, irrespective of the number of vertices, to be represented as elements of a single abstract space, namely $\mathcal{W}$.

1.3. The cut metric

With the above representation, it turns out that the notion of convergence in terms of subgraph densities outlined above can be captured by an explicit metric on $\mathcal{W}$, the so-called 'cut distance' (originally defined for finite graphs by Frieze and Kannan [16]). We start with the space $\mathcal{W}$ of measurable functions $f(x, y)$ on $[0, 1]^2$ that satisfy $0 \le f(x, y) \le 1$ and $f(x, y) = f(y, x)$. We define the cut distance

$$d_\square(f, g) := \sup_{S,T \subseteq [0,1]} \left| \int_{S \times T} [f(x, y) - g(x, y)] \mathrm{d}x\mathrm{d}y \right|. \tag{5}$$

We introduce in $\mathcal{W}$ an equivalence relation. Let Σ be the space of measure preserving bijections $\sigma : [0, 1] \to [0, 1]$. Say that $f(x, y) \sim g(x, y)$ if $f(x, y) = g_\sigma(x, y) := g(\sigma x, \sigma y)$ for some $\sigma \in \Sigma$. Denote by $\widetilde{g}$ the closure in $(\mathcal{W}, d_\square)$ of the orbit $\{g_\sigma\}$. The quotient space is denoted by $\widetilde{\mathcal{W}}$ and τ denotes the natural map $g \to \widetilde{g}$. Since $d_\square$ is invariant under σ one can define on $\widetilde{\mathcal{W}}$, the natural distance $\delta_\square$ by

$$\delta_\square(\widetilde{f}, \widetilde{g}) := \inf_\sigma d_\square(f, g_\sigma) = \inf_\sigma d_\square(f_\sigma, g) = \inf_{\sigma_1, \sigma_2} d_\square(f_{\sigma_1}, g_{\sigma_2})$$

making $(\widetilde{\mathcal{W}}, \delta_\square)$ into a metric space. To any finite graph G, we associate f^G as in (4) and its orbit $\widetilde{G} = \tau f^G = \widetilde{f^G} \in \widetilde{\mathcal{W}}$. One of the key results of [6] is the following:

Theorem 1.1 (*Theorem 3.8 in [6]*). *A sequence of graphs $\{G_n\}_{n\ge1}$ converges to a limit $f \in \mathcal{W}$ in the sense defined in* (3) *if and only if $\delta_\square(\widetilde{G}_n, \widetilde{f}) \to 0$ as $n \to \infty$.*

Szemerédi's regularity lemma and the related deep results of Lovász and Szegedy will play a crucial role in this paper:

1.4. Szemerédi's lemma

Let $G = (V, E)$ be a simple graph, and let X, Y be subsets of V. Then we denote by $e_G(X, Y)$ the number of X-Y edges of G (edges whose endpoints belong to $X \cap Y$ are counted twice), and call

$$\rho_G(X, Y) := \frac{e_G(X, Y)}{|X||Y|}$$

the *density* of the pair (X, Y). Given some $\epsilon > 0$, we call a pair (A, B) of disjoint sets $A, B \subseteq V$ ϵ-regular if all $X \subseteq A$ and $Y \subseteq B$ with $|X| \ge \epsilon|A|$ and $|Y| \ge \epsilon|B|$ satisfy

$$|\rho_G(X, Y) - \rho_G(A, B)| \le \epsilon.$$

A partition $\{V_0, \ldots, V_K\}$ of V is called an ϵ-regular partition of G if it satisfies the following two conditions:

(i) $|V_0| \le \epsilon n$;
(ii) $|V_1| = |V_2| = \cdots = |V_K|$;
(iii) all but at most ϵK^2 of the pairs (V_i, V_j) with $1 \le i < j \le K$ are ϵ-regular.

Szemerédi's regularity lemma goes as follows.

Theorem 1.2 (*Szemerédi's lemma [28]*). *Given $\epsilon > 0$ and an integer $m \ge 1$ there exists an integer $M = M(\epsilon, m)$ such that every graph of order at least M admits an ϵ-regular partition $\{V_0, \ldots, V_K\}$ for some K in the range $m \le K \le M$.*

This result was proved by Szemerédi [28] in 1976 and has since found numerous applications in combinatorics, number theory and many other areas of discrete mathematics. The version presented above is from [13], Section 7.2.. Lovász and Szegedy proved the following related result.

Theorem 1.3 (*Theorem 5.1 in [23]*). *The metric space $(\widetilde{\mathcal{W}}, \delta_\square)$ is compact.*

2. The main result

2.1. The rate function

The main goal of this paper is to prove a large deviation principle for $G(n, p)$ when p is fixed and $n \to \infty$. The discussion in Section 1 gives a topological space (namely, $\widetilde{\mathcal{W}}$) that is suitable for this purpose. The next step is to define a rate function on this space. Let $I_p : [0, 1] \to \mathbb{R}$ be the function

$$\begin{aligned} I_p(u) &:= \frac{1}{2} u \log \frac{u}{p} + \frac{1}{2}(1-u) \log \frac{1-u}{1-p} \\ &= \frac{1}{2} \sup_{a,b\in\mathbb{R}} [au + b(1-u) - \log(pe^a + (1-p)e^b)] \\ &= \frac{1}{2} \sup_{a\in\mathbb{R}} [au - \log(pe^a + (1-p))]. \end{aligned} \tag{6}$$

The domain of the function I_p can be extended to $\mathcal{W}$ as

$$I_p(h) : = \int_0^1 \int_0^1 I_p(h(x, y))\, dx\, dy$$

$$= \frac{1}{2} \int_0^1 \int_0^1 \left[h(x, y) \log \frac{h(x, y)}{p} + (1 - h(x, y)) \log \frac{1 - h(x, y)}{1 - p} \right] dxdy \tag{7}$$

$$= \frac{1}{2} \sup_{a(\cdot,\cdot)} \left[\int a(x, y)h(x, y)\, dx\, dy - \int \log(pe^{a(x,y)} + (1 - p))\, dx\, dy \right]. \tag{8}$$

The following property of I_p is crucial.

Lemma 2.1. *The function I_p is well defined on $\widetilde{\mathcal{W}}$ and is lower semicontinuous under the cut metric $\delta_\square$ on $\widetilde{\mathcal{W}}$.*

Proof. The supremum in (8) can be taken over all bounded measurable functions a on $[0, 1]^2$. As the supremum of a family of affine linear functionals continuous in the weak topology, $I_p(h)$ is lower semi-continuous in the weak topology and therefore also in the topology of the metric $d_\square$. If $\sigma : [0, 1] \to [0, 1]$ is a measure preserving bijection then $I_p(h_\sigma) = I(h)$. By lower semi-continuity of $g \in \widetilde{h}$, $I_p(g) \le I_p(h)$. But $g \in \widetilde{h}$ implies $h \in \widetilde{g}$ so that $I_p(h) \le I_p(g)$. Hence, $I_p(g) = I_p(h)$ and $I_p(\cdot)$ is well defined and lower semi-continuous on $\widetilde{\mathcal{W}}$. □

2.2. The large deviation principle

The random graph $G(n, p)$ induces probability distributions $\mathbb{P}_{n,p}$ on the space $\mathcal{W}$ through the map $G \to f^G$ and $\widetilde{\mathbb{P}}_{n,p}$ on $\widetilde{\mathcal{W}}$ through the map $G \to f^G \to \widetilde{f}^G = \tau f^G$. The space $\mathcal{W}$ is compact in the weak topology and a large deviation principle for $\mathbb{P}_{n,p}$ on $\mathcal{W}$ in the weak topology with the lower-semicontinuous rate function $I_p(h)$ given by (7) is elementary but is not of much use since quantities like 'triangle counts' are not stable in the weak topology. We will state it for the record and find a use for it later.

Theorem 2.2. *The sequence $\mathbb{P}_{n,p}$ on $\mathcal{W}$ satisfies a large deviation principle in the weak topology. That is, for every weakly closed set $F \subset \mathcal{W}$*

$$\limsup_{n\to\infty} \frac{1}{n^2} \log \mathbb{P}_{n,p}(F) \le - \inf_{f\in F} I_p(f)$$

and for any open set U (again in the weak topology) in $\mathcal{W}$

$$\liminf_{n\to\infty} \frac{1}{n^2} \log \mathbb{P}_{n,p}(U) \ge - \inf_{f\in U} I_p(f).$$

Proof. The weak topology is defined through an arbitrary but finite number of linear functionals. Therefore the large deviation principle can be reduced to the large deviation behavior of a finite set of linear functionals $\{Z_\phi(f)\}$ given by

$$Z_\phi(f) = \langle \phi, f\rangle := \iint f(x,y)\phi(x,y)\mathrm{d}x\mathrm{d}y$$

under the measure $\mathbb{P}^{n,p}$. The limit

$$\lim_{n\to\infty} \frac{1}{n^2}\log \mathbb{E}^{\mathbb{P}_{n,p}}\left[\exp\left(n^2 \iint f(x,y)\phi(x,y)\mathrm{d}x\mathrm{d}y\right)\right]$$

is easily calculated to yield

$$\frac{1}{2}\iint \log(pe^{2\phi(x,y)} + (1-p))\mathrm{d}x\mathrm{d}y.$$

(Note that this is true only if ϕ is symmetric. However, since f is symmetric, it suffices to restrict attention to symmetric ϕ.) Therefore, an abstract Gärtner–Ellis Theorem (see e.g. Theorem 4.5.3 in [11]) gives the upper bound with rate function

$$I_p(f) = \sup_\phi \left[\langle\phi, f\rangle - \frac{1}{2}\iint \log(pe^{2\phi(x,y)} + (1-p))\mathrm{d}x\mathrm{d}y\right].$$

Note that this is the rate function I_p defined in (7) and (8). The supremum is attained at the function

$$\phi_f(x,y) := \frac{1}{2}\log\frac{f(x,y)}{p} - \frac{1}{2}\log\frac{1-f(x,y)}{1-p}.$$

Note that for any $g \neq f$,

$$\begin{aligned}&(\langle\phi_f, f\rangle - I_p(f)) - (\langle\phi_f, g\rangle - I_p(g))\\ &= \frac{1}{2}\iint \left(g(x,y)\log\frac{g(x,y)}{f(x,y)} + (1-g(x,y))\log\frac{1-g(x,y)}{1-f(x,y)}\right)\mathrm{d}x\,\mathrm{d}y > 0.\end{aligned}$$

This shows that every f is an exposed point of the lower semicontinuous rate function I_p, in the parlance of convex analysis. Therefore by the Gärtner–Ellis Theorem (see e.g. Theorem 4.5.20 in [11]) and the compactness of the weak topology, we get the lower bound. □

The large deviation principle for $\widetilde{\mathbb{P}}_{n,p}$ on $(\widetilde{\mathcal{W}}, \delta_\square)$ is much more useful and is the main result of this article.

Theorem 2.3. *For each fixed $p \in (0,1)$, the sequence $\widetilde{\mathbb{P}}_{n,p}$ obeys a large deviation principle in the space $\widetilde{\mathcal{W}}$ (equipped with the cut metric) with rate function I_p defined by (7). Explicitly, this means that for any closed set $\widetilde{F} \subseteq \widetilde{\mathcal{W}}$,*

$$\limsup_{n\to\infty} \frac{1}{n^2}\log\widetilde{\mathbb{P}}_{n,p}(\widetilde{F}) \le -\inf_{\widetilde{h}\in\widetilde{F}} I_p(\widetilde{h}). \tag{9}$$

and for any open set $\widetilde{U} \subseteq \widetilde{\mathcal{W}}$,

$$\liminf_{n\to\infty} \frac{1}{n^2}\log\widetilde{\mathbb{P}}_{n,p}(\widetilde{U}) \ge -\inf_{\widetilde{h}\in\widetilde{U}} I_p(\widetilde{h}). \tag{10}$$

For the upper bound, because $(\widetilde{\mathcal{W}}, \delta_\square)$ is compact, it is sufficient to prove that for any $\widetilde{h} \in \widetilde{\mathcal{W}}$,

$$\lim_{\eta\to 0}\limsup_{n\to\infty} \frac{1}{n^2}\log\widetilde{\mathbb{P}}_{n,p}(S_\square(\widetilde{h},\eta)) \le -I_p(\widetilde{h}).$$

For the lower bound we need to prove that if $\widetilde{h} \in \widetilde{\mathcal{W}}$ and $\eta > 0$ is arbitrary

$$\liminf_{n\to\infty} \frac{1}{n^2}\log\widetilde{\mathbb{P}}_{n,p}(S_\square(\widetilde{h},\eta)) \ge -I_p(\widetilde{h}),$$

where $S_\square(\widetilde{h},\eta) = \{\widetilde{g} : \delta_\square(\widetilde{g},\widetilde{h}) \le \eta\}$.

2.3. Proof of the upper bound in Theorem 2.3

Let $B(\widetilde{h}, \eta) \subset \mathcal{W}$ be defined as

$$B(\widetilde{h}, \eta) = \tau^{-1} S_{\square}(\widetilde{h}, \eta) \subset \mathcal{W}$$

i.e. the union of all the orbits from $S_{\square}(\widetilde{h}, \eta) \subset \widetilde{\mathcal{W}}$. We need to show that

$$\lim_{\eta \to 0} \limsup_{n \to \infty} \frac{1}{n^2} \log \mathbb{P}_{n,p}[B(\widetilde{h}, \eta)] \leq -I_p(\widetilde{h}). \tag{11}$$

Let the set of n vertices be partitioned into K subsets of size a with a remainder of size b, so that $n = Ka + b$. We assume that $b \leq \epsilon n$. We order the vertices so that $V_0 = \{1, 2, \ldots b\}$ and $V_i = \{b + (i-1)a + 1, \ldots, b + ia\}$ for $i = 1, 2, \ldots, K$. We map the vertices into subintervals of the unit interval, with the vertex r represented by the interval $[\frac{r-1}{n}, \frac{r}{n}]$. The sets V_i of vertices will then correspond to the intervals $E_0 = [0, \frac{b}{n}]$ for $i = 0$ and $E_i = [\frac{b+(i-1)a}{n}, \frac{b+ia}{n}]$ for $1 \leq i \leq K$. Let us denote by $\mathcal{V}_K$ the subset of $\mathcal{W}$ consisting of $g(x, y)$ that are of the form

$$g(x, y) = \sum_{i,j=1}^{K} p_{i,j} \mathbf{1}_{E_i}(x) \mathbf{1}_{E_j}(y)$$

where $\{p_{i,j}\}$, $1 \leq i, j \leq K$ is symmetric and satisfies $0 \leq p_{i,j} \leq 1$. For any pair m, M with $m < M$, we define $\mathcal{V}_{m,M} = \cup_{m \leq K \leq M} \mathcal{V}_K$. The following is a restatement of the Szemerédi regularity lemma.

Lemma 2.4. *Given any $\epsilon > 0$ and $m \geq 1$ such that $2m^{-1} < \epsilon$, there is $M = M(\epsilon, m)$ such that for any graph G, there exists a permutation π, i.e. a relabeling of the vertices of the graph, such that*

$$\inf_{g \in \mathcal{V}_{m,M}} d_{\square}(f^{\pi G}, g) \leq \epsilon.$$

Proof. Let ϵ' and m be given. According to Szemerédi's lemma, there is $M(\epsilon', m)$ such that, for some K in the range $m \leq K \leq M$, we can find a partition $V_0, \ldots V_K$ which is ϵ'-regular. After a permutation we can assume that the ordering of the vertices coincides with the ordering of the partitions. We define $p_{i,i} = p_{0,i} = p_{i,0} = 0$ and for $1 \leq i \neq j$, $p_{i,j} = \rho_G(V_i, V_j)$. This leads to

$$g(x, y) = \sum_{\substack{i,j=1 \\ i \neq j}}^{K} \rho_G(V_i, V_j) \mathbf{1}_{E_i}(x) \mathbf{1}_{E_j}(y)$$

to be compared with f^G when $V_0, \ldots, V_K$ is an ϵ' regular partition of G. Recall that

$$d_{\square}(f^G, g) = \sup_{S,T \subset [0,1]} \left| \iint_{S \times T} [f^G(x, y) - g(x, y)] \mathrm{d}x\, \mathrm{d}y \right|.$$

Since both f^G and g are constant on sets of the form $[\frac{i}{n}, \frac{i+1}{n}] \times [\frac{j}{n}, \frac{j+1}{n}]$ it is easy to see that S and T can be restricted to unions of intervals of the form $[\frac{i}{n}, \frac{i+1}{n}]$ i.e. subsets of $[0, 1]$ that represent subsets of vertices. These subsets will also be denoted by S and T. Now, given two such subsets S and T,

$$\iint_{S \times T} (f^G(x, y) - g(x, y)) \mathrm{d}x\, \mathrm{d}y = \sum_{0 \leq i,j \leq K} \iint_{(S \cap E_i) \times (T \cap E_j)} (f^G(x, y) - g(x, y)) \mathrm{d}x\, \mathrm{d}y.$$

Let A_{ij} denote the (i, j)th term in the above sum. Since f^G and g both take values in $[0, 1]$, therefore for each i and each j,

$$|A_{i0}| \leq |E_i||E_0|, \qquad |A_{0j}| \leq |E_0||E_j|.$$

For the same reason, if $|S \cap E_i| < \epsilon' |E_i|$ or $|T \cap E_j| < \epsilon' |E_j|$, then

$$|A_{ij}| \leq \epsilon' |E_i||E_j|.$$

If $1 \le i \ne j$, $|S \cap E_i| \ge \epsilon'|E_i|$, $|T \cap E_j| \ge \epsilon'|E_j|$ and the pair (V_i, V_j) is ϵ'-regular, then

$$\begin{aligned}|A_{ij}| &= \left|\frac{1}{n^2} e_G(S \cap V_i, T \cap V_j) - \rho_G(V_i, V_j)|S \cap E_i||T \cap E_j|\right| \\ &\le |\rho_G(S \cap V_i, S \cap V_j) - \rho_G(V_i, V_j)||S \cap E_i||T \cap E_j| \\ &\le \epsilon'|E_i||E_j|.\end{aligned}$$

Finally, if either $1 \le i = j$ or (V_i, V_j) is not ϵ'-regular (of which there are at most $K + 2\epsilon' K^2$ cases), then we have the trivial bound $|A_{ij}| \le \frac{a^2}{n^2} \le \frac{1}{K^2}$. A combination of the above estimates gives

$$\begin{aligned}\sum_{0 \le i,j \le K} |A_{ij}| &\le 2|E_0| + \epsilon' + (K + 2\epsilon' K^2)\frac{1}{K^2} \\ &\le (2\epsilon' + 2\epsilon' + \epsilon' + K^{-1}).\end{aligned}$$

Thus,

$$d_\square(f^G, g) \le 5\epsilon' + K^{-1} \le 5\epsilon' + m^{-1}$$

Since $m^{-1} < \epsilon/2$, we can choose ϵ' so that $5\epsilon' + m^{-1} < \epsilon$. □

Lemma 2.5. *Let ϵ, m and M be as in Lemma 2.4.*

$$\mathbb{P}_{n,p}(B(\widetilde{h}, \eta)) \le n!\, \mathbb{P}_{n,p}(B(\widetilde{h}, \eta) \cap B(\mathcal{V}_{m,M}, \epsilon))$$

where $B(\mathcal{V}_{m,M}, \epsilon) = \{g : \inf_{f \in \mathcal{V}_{m,M}} d_\square(g, f) \le \epsilon\}$.

Proof. The orbit under the permutation group has at most $n!$ elements and they all have the same probability under $\mathbb{P}_{n,p}$. Moreover by the Lemma 2.4 every orbit meets $B(\mathcal{V}_{m,M}, \epsilon)$, and $B(\widetilde{h}, \eta)$ is invariant under $\sigma \in \Sigma$ and therefore under π. Consequently

$$B(\widetilde{h}, \eta) \subset \bigcup_\pi \pi^{-1}(B(\widetilde{h}, \eta) \cap B(\mathcal{V}_{m,M}, \epsilon))$$

and the lemma follows.

Lemma 2.6. *There exists a function $\delta(\widetilde{h}, \epsilon)$, depending only on $\widetilde{h}$ and ϵ, with $\delta(\widetilde{h}, \epsilon) \to 0$ as $\epsilon \to 0$, such that for each arbitrary but fixed ϵ, m, M satisfying Lemma 2.4,*

$$\lim_{\eta \to 0} \limsup_{n \to \infty} \frac{1}{n^2} \log \mathbb{P}_{n,p}(B(\widetilde{h}, \eta) \cap B(\mathcal{V}_{m,M}, \epsilon)) \le -I_p(\widetilde{h}) + \delta(\widetilde{h}, \epsilon)$$

Proof. Since $\mathcal{V}_{m,M}$ is a finite union $\bigcup_{m \le K \le M} \mathcal{V}_K$ it is sufficient to prove that for each K

$$\lim_{\eta \to 0} \limsup_{n \to \infty} \frac{1}{n^2} \log \mathbb{P}_{n,p}(B(\widetilde{h}, \eta) \cap B(\mathcal{V}_K, \epsilon)) \le -I_p(\widetilde{h}) + \delta(\widetilde{h}, \epsilon)$$

and $\delta(\widetilde{h}, \epsilon) \to 0$ as $\epsilon \to 0$. For fixed K, $\mathcal{V}_K$ consists of a compact set of functions in $L_1([0, 1]^2)$ and can be covered by a finite number of spheres of radius ϵ in L_1 and therefore in $\mathcal{W}$. It is therefore sufficient to show that for fixed K and $g \in \mathcal{V}_K$

$$\lim_{\eta \to 0} \limsup_{n \to \infty} \frac{1}{n^2} \log \mathbb{P}_{n,p}(B(\widetilde{h}, \eta) \cap B(g, 2\epsilon)) \le -I_p(\widetilde{h}) + \delta(\widetilde{h}, \epsilon)$$

We can assume that $B(\widetilde{h}, \eta) \cap B(g, 2\epsilon) \ne \emptyset$. Therefore $g \in B(\widetilde{h}, \eta + 2\epsilon)$. Since $\eta \to 0$ we can assume $\eta < \epsilon$ so that $g \in B(\widetilde{h}, 3\epsilon)$. By lower semi-continuity of $I_p(\cdot)$, $I_p(f) \ge I_p(\widetilde{h}) - \delta(\widetilde{h}, \epsilon)$ on

$B(g,2\epsilon) \subset B(\widetilde{h},5\epsilon)$ and $\delta(\widetilde{h},\epsilon) \to 0$ as $\epsilon \to 0$. We note that $B(g,2\epsilon) \subset \mathcal{W}$ is weakly closed and therefore by the upper bound in Theorem 2.2,

$$\begin{aligned}\lim_{\eta\to 0}\limsup_{n\to\infty}\frac{1}{n^2}\log \mathbb{P}_{n,p}(B(\widetilde{h},\eta)\cap B(g,2\epsilon)) &\le \limsup_{n\to\infty}\frac{1}{n^2}\log \mathbb{P}_{n,p}(B(g,2\epsilon))\\ &\le -\inf_{f\in B(g,2\epsilon)} I_p(f) \le -\inf_{f\in B(\widetilde{h},5\epsilon)} I_p(f) \le -I_p(\widetilde{h}) + \delta(\widetilde{h},\epsilon)\end{aligned}$$

where $\delta(\widetilde{h},\epsilon) \to 0$ as $\epsilon \to 0$. □

Lemmas 2.5 and 2.6 Eq. (11), which proves the upper bound in Theorem 2.3.

2.4. *Proof of the lower bound in Theorem 2.3*

Let $h(x,y) \in \mathcal{W}$ be given. We define

$$p_{i,j}^{(n)} = n^2 \int\int_{[\frac{i-1}{n},\frac{i}{n}]\times[\frac{j-1}{n},\frac{j}{n}]} h(x,y)\mathrm{d}x\mathrm{d}y$$

and the corresponding function $h_n(x,y) \in \mathcal{W}$ by

$$h_n(x,y) = \sum_{i,j} p_{i,j}^{(n)} \mathbf{1}_{[\frac{i-1}{n},\frac{i}{n}]}(x)\mathbf{1}_{[\frac{j-1}{n},\frac{j}{n}]}(y).$$

Since $\|h_n - h\|_{L_1([0,1]^2)} \to 0$, it follows that $d_\square(h_n,h) \to 0$. It is therefore sufficient to prove that for any $\epsilon > 0$

$$\liminf_{n\to\infty}\frac{1}{n^2}\log \mathbb{P}(d_\square(f^{G(n,p)},h_n) \le \epsilon) \ge -I_p(h).$$

We define an inhomogeneous random graph where the edge connecting the vertices i and j is present with probability $p_{i,j}^{(n)}$. Different edges are independent. If $\xi_{i,j} = 1$ when the edge connecting i,j is present and 0 otherwise then $\xi_{i,j}$ are independent Binomial random variables with $\mathbb{P}(\xi_{i,j}=1) = p_{i,j}^{(n)}$. We denote by $\mathbb{P}_{n,h}$ the measure on $\mathcal{W}$ induced by

$$f_n(x,y) := \sum_{\substack{i,j\\ i\neq j}} \xi_{i,j}\mathbf{1}_{[\frac{i-1}{n},\frac{i}{n}]}(x)\mathbf{1}_{[\frac{j-1}{n},\frac{j}{n}]}(y)$$

If A and B are subsets of $\{1,\ldots,n\}$, it is straightforward to calculate

$$\begin{aligned}\psi_n(\lambda) &:= \frac{1}{n^2}\log \mathbb{E}^{\mathbb{P}_{n,h}}\left[\exp\left(\lambda\sum_{\substack{i\in A, j\in B\\ i\neq j}}(\xi_{i,j}-p_{i,j}^{(n)})\right)\right]\\ &= \frac{1}{n^2}\left[\sum_{\substack{i,j\in A\cap B\\ i>j}}\log \mathbb{E}^{\mathbb{P}_{n,h}}[\exp(2\lambda(\xi_{i,j}-p_{i,j}^{(n)}))]\right.\\ &\qquad\left. + \sum_{\substack{i\in A\cap B,\, j\in B\cap A^c\\ \text{or } i\in A\cap B^c,\, j\in A\cap B\\ \text{or } i\in A\cap B^c,\, j\in A^c\cap B}}\log \mathbb{E}^{\mathbb{P}_{n,h}}[\exp(\lambda(\xi_{i,j}-p_{i,j}^{(n)}))]\right].\end{aligned}$$

Each term in the sum is easily estimated by $\frac{\lambda^2}{2}$, providing an estimate of the form

$$\mathbb{P}_{n,h}\left(\left|\int\int_{\widetilde{A}\times\widetilde{B}}(f_n-h_n)\mathrm{d}x\mathrm{d}y\right| \ge \epsilon\right) \le \mathrm{e}^{-\frac{n^2\epsilon^2}{2}},$$

where $\widetilde{A} = \cup_{i\in A}[\frac{i-1}{n}, \frac{i}{n}]$ and $\widetilde{B}$ is defined similarly. Since the number of sets like $\widetilde{A} \times \widetilde{B}$ is only 2^{2n} it follows that

$$\mathbb{P}_{n,h}(d_\square(f_n, h_n) \geq \epsilon) \to 0$$

as $n \to \infty$. Now the lower bound is easily established by a simple tilting argument. Denoting by $B_{\epsilon,n}$ the set $\{f : d_\square(f, h_n) \leq \epsilon\}$

$$\mathbb{P}_{n,p}(B_{\epsilon,n}) = \int_{B_{\epsilon,n}} d\mathbb{P}_{n,p} = \int_{B_{\epsilon,n}} e^{-\log \frac{d\mathbb{P}_{n,h}}{d\mathbb{P}_{n,p}}} d\mathbb{P}_{n,h}$$

$$= \mathbb{P}_{n,h}(B_{\epsilon,n}) \frac{1}{\mathbb{P}_{n,h}(B_{\epsilon,n})} \int_{B_{\epsilon,n}} e^{-\log \frac{d\mathbb{P}_{n,h}}{d\mathbb{P}_{n,p}}} d\mathbb{P}_{n,h}.$$

By Jensen's inequality

$$\log \mathbb{P}_{n,p}(B_{\epsilon,n}) \geq \log \mathbb{P}_{n,h}(B_{\epsilon,n}) - \frac{1}{\mathbb{P}_{n,h}(B_{\epsilon,n})} \int_{B_{\epsilon,n}} \log \frac{d\mathbb{P}_{n,h}}{d\mathbb{P}_{n,p}} d\mathbb{P}_{n,h}.$$

Since $\mathbb{P}_{n,h}(B_{\epsilon,n}) \to 1$, it is easy to see that

$$\liminf_{n\to\infty} \frac{1}{n^2} \log \mathbb{P}_{n,p}(B_{\epsilon,n}) \geq -\lim_{n\to\infty} \frac{1}{n^2} \int \log \frac{d\mathbb{P}_{n,h}}{d\mathbb{P}_{n,p}} d\mathbb{P}_{n,h}$$

The entropy cost of tilting (i.e. the integral in the preceding display) is

$$\frac{1}{n^2} \sum_{i>j} \left(p_{i,j}^{(n)} \log \frac{p_{i,j}^{(n)}}{p} + (1 - p_{i,j}^{(n)}) \log \frac{1 - p_{i,j}^{(n)}}{1 - p} \right)$$

which converges to $I_p(h)$ as $n \to \infty$. This proves the lower bound.

3. Conditional distributions

Theorem 2.3 gives estimates of the probabilities of rare events for $G(n, p)$. However, it does not quite answer the following question: given that some particular rare event has occurred, what does the graph look like? Naturally, one might expect that if $G(n, p) \in \widetilde{F}$ for some closed set $\widetilde{F} \subseteq \widetilde{\mathcal{W}}$ satisfying

$$\inf_{\widetilde{h}\in\widetilde{F}^o} I_p(\widetilde{h}) = \inf_{\widetilde{h}\in\widetilde{F}} I_p(\widetilde{h}) > 0, \tag{12}$$

then $G(n, p)$ should resemble one of the minimizers of I_p in $\widetilde{F}$. (Here $\widetilde{F}^o$ denotes the interior of $\widetilde{F}$, as usual.) In other words, given that $G(n, p) \in \widetilde{F}$, one might expect that $\delta_\square(G(n, p), \widetilde{F}^*) \approx 0$, where $\widetilde{F}^*$ is the set of minimizers of I_p in $\widetilde{F}$ and

$$\delta_\square(G(n, p), \widetilde{F}^*) := \inf_{\widetilde{h}\in\widetilde{F}^*} \delta_\square(G(n, p), \widetilde{h}).$$

However, it is not obvious that a minimizer must exist in $\widetilde{F}$. Here is where the compactness of $\widetilde{\mathcal{W}}$ comes to the rescue yet one more time: since the function I_p is lower semicontinuous on $\widetilde{F}$ and $\widetilde{F}$ is closed, therefore a minimizer must necessarily exist. The following theorem formalizes this argument.

Theorem 3.1. *Take any $p \in (0, 1)$. Let $\widetilde{F}$ be a closed subset of $\widetilde{\mathcal{W}}$ satisfying (12). Let $\widetilde{F}^*$ be the subset of $\widetilde{F}$ where I_p is minimized. Then $\widetilde{F}^*$ is non-empty and compact, and for each n, and each $\epsilon > 0$,*

$$\mathbb{P}(\delta_\square(G(n, p), \widetilde{F}^*) \geq \epsilon \mid G(n, p) \in \widetilde{F}) \leq e^{-C(\epsilon,\widetilde{F})n^2}$$

where $C(\epsilon, \widetilde{F})$ is a positive constant depending only on ϵ and $\widetilde{F}$. In particular, if $\widetilde{F}^*$ contains only one element $\widetilde{h}^*$, then the conditional distribution of $G(n, p)$ given $G(n, p) \in \widetilde{F}$ converges to the point mass at $\widetilde{h}^*$ as $n \to \infty$.

Proof. Since $\widetilde{\mathcal{W}}$ is compact and $\widetilde{F}$ is a closed subset, therefore $\widetilde{F}$ is also compact. Since I_p is a lower semicontinuous function on $\widetilde{F}$ (Lemma 2.1) and $\widetilde{F}$ is compact, it must attain its minimum on $\widetilde{F}$. Thus, $\widetilde{F}^*$ is non-empty. By the lower semicontinuity of I_p, $\widetilde{F}^*$ is closed (and hence compact). Fix $\epsilon > 0$ and let

$$\widetilde{F}_\epsilon := \{\widetilde{h} \in \widetilde{F} : \delta_\square(\widetilde{h}, \widetilde{F}^*) \geq \epsilon\}.$$

Then $\widetilde{F}_\epsilon$ is again a closed subset. Observe that

$$\mathbb{P}(\delta_\square(G(n,p), \widetilde{F}^*) \geq \epsilon \mid G(n,p) \in \widetilde{F}) = \frac{\mathbb{P}(G(n,p) \in \widetilde{F}_\epsilon)}{\mathbb{P}(G(n,p) \in \widetilde{F})}.$$

Thus, with

$$I_1 := \inf_{\widetilde{h} \in \widetilde{F}} I_p(\widetilde{h}), \qquad I_2 := \inf_{\widetilde{h} \in \widetilde{F}_\epsilon} I_p(\widetilde{h}),$$

Theorem 2.3 and condition (12) give

$$\limsup_{n\to\infty} \frac{1}{n^2} \log \mathbb{P}(\delta_\square(G(n,p), \widetilde{F}^*) \geq \epsilon \mid G(n,p) \in \widetilde{F}) \leq I_1 - I_2.$$

The proof will be complete if it is shown that $I_1 < I_2$.

Now clearly, $I_1 \leq I_2$. If $I_1 = I_2$, the compactness of $\widetilde{F}_\epsilon$ implies that there exists $\widetilde{h} \in \widetilde{F}_\epsilon$ satisfying $I_p(\widetilde{h}) = I_2$. However, this means that $\widetilde{h} \in \widetilde{F}^*$ and hence $\widetilde{F}_\epsilon \cap \widetilde{F}^* \neq \emptyset$, which is impossible. □

4. Application to triangle counts

4.1. *Brief history of the problem*

Let $T_{n,p}$ be the number of triangles in $G(n, p)$. The primary objective of this section is to compute the large deviation rate function for the upper tail of $T_{n,p}$ when p remains fixed and $n \to \infty$. In other words, given $p \in [0, 1]$ and $\epsilon > 0$, we wish to evaluate the limit

$$\lim_{n\to\infty} \frac{1}{n^2} \log \mathbb{P}(T_{n,p} \geq (1+\epsilon)\mathbb{E}(T_{n,p})) \tag{13}$$

as a function of p and ϵ.

The problem of estimating tail probabilities like $\mathbb{P}(T_{n,p} \geq (1+\epsilon)\mathbb{E}(T_{n,p}))$ has been studied extensively in the random graphs literature, particularly in the case when p is allowed to tend to zero as $n \to \infty$. Computing upper and lower bounds on such tail probabilities that are sharp up to constants in the exponent was a prominent open problem in this area until it was solved recently in [8]. Let us refer to the paper [8] for a survey of the aforementioned literature.

When p is fixed, computing sharp upper and lower bounds is relatively easy. The difficult problem in this case is the exact evaluation of the limit (13). The first progress in this direction was made in [9] where it was shown that, given $p \in (0, 1)$, there exist $p^3/6 < t' \leq t'' < 1/6$ such that for all $t \in (p^3/6, t') \cup (t'', 1/6)$,

$$\lim_{n\to\infty} \frac{1}{n^2} \log \mathbb{P}(T_{n,p} \geq tn^3) = -I_p((6t)^{1/3}), \tag{14}$$

when I_p is the entropy function defined in (6). Explicit formulas for p' and p'' are also given in [9]. Unfortunately, the result does not cover all values of (p, t); and neither is the above formula true for all (p, t), as we shall see below.

There is a related unpublished manuscript by Bolthausen, Comets and Dembo [4] on large deviations for subgraph counts. As of now, to the best of our knowledge, the authors of [4] have only looked at subgraphs that do not complete loops, like 2-stars. Besides [9] and [4], we know of no other papers that attack the exact evaluation of (13) (or equivalently, (14)).

4.2. Exact large deviations for the upper tail

In this subsection, the limit (14) is evaluated for all values of p and t. It comes as the solution of the following variational problem. Let $\mathcal{W}$, $\widetilde{\mathcal{W}}$ and $\delta_\square$ be defined as in Section 1. For each $f \in \mathcal{W}$, let

$$T(f) := \frac{1}{6}\int_0^1\int_0^1\int_0^1 f(x,y)f(y,z)f(z,x)\mathrm{d}x\mathrm{d}y\mathrm{d}z$$

and let $I_p(f)$ be defined as in (7). Note that T can be defined on $\widetilde{\mathcal{W}}$ simply by letting $T(\widetilde{f}) := T(f)$, because T is a continuous map on $\mathcal{W}$ under the $d_\square$ pseudometric (Theorem 3.7 in [6]).

For each $p \in [0,1]$ and $t \in [0, 1/6)$, let

$$\phi(p,t) := \inf\{I_p(f) : f \in \mathcal{W},\ T(f) \geq t\}. \tag{15}$$

For $t \geq 1/6$, let $\phi(p,t) = \infty$. The following result gives the large deviation rate function for the upper tail of $T_{n,p}$. (Note that this is just an illustrative example. Theorem 2.3 can be used to derive large deviations for any subgraph count, or even joint large deviations for the counts of more than one subgraph.)

Theorem 4.1. *Let $G(n,p)$ be the Erdős–Rényi random graph on n vertices with edge probability p. Let $T_{n,p}$ denote the number of triangles in $G(n,p)$. Let ϕ be defined as above. Then for each $p \in (0,1)$ and each $t \geq 0$,*

$$\lim_{n\to\infty}\frac{1}{n^2}\log\mathbb{P}(T_{n,p} \geq tn^3) = -\phi(p,t).$$

Next, take any $p \in (0,1)$ and $t \in (p^3/6, 1/6)$. Let $F^*_{p,t}$ be the set of minimizers for the variational problem (15) and $\widetilde{F}^*_{p,t}$ be its image in $\widetilde{\mathcal{W}}$. Then $\widetilde{F}^*_{p,t}$ is a non-empty compact subset of $\widetilde{\mathcal{W}}$. Moreover, for each $\epsilon > 0$ there exists a positive constant $C(\epsilon,p,t)$ depending only on ϵ, p and t such that for any n,

$$\mathbb{P}(\delta_\square(G(n,p), \widetilde{F}^*_{p,t}) \geq \epsilon \mid T_{n,p} \geq tn^3) \leq \mathrm{e}^{-C(\epsilon,p,t)n^2}.$$

Proof. Let $F := \{f \in \mathcal{W} : T(f) \geq t\}$. By Theorem 3.7 in [6], F is a closed subset of $\mathcal{W}$. Therefore by Theorem 2.3,

$$\begin{aligned}\limsup_{n\to\infty}\frac{1}{n^2}\log\mathbb{P}(T_{n,p} \geq tn^3) &= \limsup_{n\to\infty}\frac{1}{n^2}\log\mathbb{P}(G(n,p)\in F)\\ &\leq -\inf_{h\in F} I_p(h) = -\phi(p,t).\end{aligned}$$

Next, let $U := \{f \in \mathcal{W} : T(f) > t\}$. Again by Theorem 3.7 of [6], U is an open set. Therefore by Theorem 2.3, for each $\epsilon > 0$,

$$\begin{aligned}\liminf_{n\to\infty}\frac{1}{n^2}\log\mathbb{P}(T_{n,p} \geq tn^3) &\geq \liminf_{n\to\infty}\frac{1}{n^2}\log\mathbb{P}(G(n,p)\in U)\\ &\geq -\inf_{h\in U} I_p(h) \geq -\phi(p,t+\epsilon).\end{aligned}$$

In Proposition 4.2 below, it is proved that ϕ is a continuous function of t for every fixed p. This completes the proof of the first assertion of the theorem. The second assertion is merely a corollary of Theorem 3.1. The condition (12) required for Theorem 3.1 can be easily shown to follow from the continuity of ϕ in t, because any $\widetilde{f}$ with $T(\widetilde{f}) > t$ lies in the interior of the set $\{\widetilde{h} : T(\widetilde{h}) \geq t\}$. □

4.3. Properties of the rate function

Given Theorem 2.3, there is a natural desire to understand the rate function ϕ. The following proposition summarizes some basic properties of ϕ. The first property is required in the proof of Theorem 4.1 above.

Proposition 4.2. *For each fixed $p \in (0,1)$, the following hold:*

(i) *The function $\phi(p,t)$ is continuous in t in the interval $[0, 1/6)$.*

(ii) *As a function of* t, $\phi(p,t) = 0$ *in the interval* $[0, p^{3/6}]$ *and strictly increasing in* $(p^3/6, 1/6)$. *Moreover, for* $p^3/6 < t < s < 1/6$,

$$\phi(p,t) < (t/s)^{1/3}\phi(p,s).$$

(iii) *For* $t \in (p^3/6, 1/6)$, $\phi(p,t)$ *can be alternately represented as*

$$\phi(p,t) := \inf\{I_p(f) : f \in \mathcal{W},\ T(f) = t\}.$$

Moreover, if $\{f_n\}_{n\geq 1}$ *is a sequence in* $\mathcal{W}$ *such that* $T(f_n) \geq t$ *for all* n *and* $I_p(f_n) \to \phi(p,t)$, *then* $T(f_n) \to t$. *In particular, the elements of* $\widetilde{F}^*_{p,t}$ *all satisfy* $T(f) = t$.

Proof. Let us start by proving that ϕ is continuous in t. For each $f \in \mathcal{W}$ and $\delta \in [0,1]$, let

$$f^\delta := f + \delta(1-f).$$

By the inequality

$$\begin{aligned}
&(a+\delta(1-a))(b+\delta(1-b))(c+\delta(1-c)) \\
&\quad = abc + \delta((1-a)bc + (1-b)ac + (1-c)ab) \\
&\qquad + \delta^2((1-a)(1-b)c + (1-a)(1-c)b + (1-b)(1-c)a) \\
&\qquad + \delta^3(1-a)(1-b)(1-c) \\
&\quad \geq abc + \delta^3((1-a)bc + (1-b)ac + (1-c)ab) \\
&\qquad + \delta^3((1-a)(1-b)c + (1-a)(1-c)b + (1-b)(1-c)a) \\
&\qquad + \delta^3(1-a)(1-b)(1-c) \\
&\quad = abc + \delta^3(1-abc),
\end{aligned}$$

we see that

$$T(f^\delta) \geq T(f)(1-\delta^3) + \frac{\delta^3}{6}. \tag{16}$$

Take any $t \in [0, 1/6)$ and any f such that $T(f) \geq t$. Suppose $t_n \downarrow t$. Let δ_n be the smallest number in $[0,1]$ such that $T(f^{\delta_n}) \geq t_n$. By (16) it follows that δ_n exists and $\lim_{n\to\infty}\delta_n = 0$. Therefore by the dominated convergence theorem, $\lim_{n\to\infty} I_p(f^{\delta_n}) = I_p(f)$. Thus,

$$\lim_{n\to\infty} \phi(p,t_n) \leq \lim_{n\to\infty} I_p(f^{\delta_n}) = I_p(f).$$

Since this is true for every f such that $T(f) \geq t$ and ϕ is a non-decreasing function in t, this proves the right continuity of ϕ.

Next, take a sequence $t_n \uparrow t$. Let f_n be a sequence of functions such that $T(f_n) \geq t_n$ and

$$\lim_{n\to\infty} I_p(f_n) = \lim_{n\to\infty} \phi(p,t_n).$$

For each n, let δ_n be the smallest number in $(0,1)$ such that $T(f_n^{\delta_n}) \geq t$. By (16), δ_n exists and $\lim_{n\to\infty}\delta_n = 0$. Now, the function I_p on $[0,1]$ (defined in (6)) is uniformly continuous on $[0,1]$. As a consequence,

$$\lim_{\delta\to 0}\sup_{f\in\mathcal{W}} |I_p(f^\delta) - I_p(f)| = 0.$$

In particular,

$$\lim_{n\to\infty} I_p(f_n^{\delta_n}) = \lim_{n\to\infty} \phi(p,t_n).$$

But $\phi(p,t) \leq I_p(f_n^{\delta_n})$ for each n. By the monotonicity of ϕ, this proves left continuity.

Next, note that since $\mathbb{P}(T(G_{n,p}) \geq tn^3) \to 1$ for any $t < p^3/6$, Theorem 4.1 and the continuity of ϕ imply that $\phi(p,t) = 0$ for $t \leq p^3/6$. Let us now show that $\phi(p,t)$ is strictly increasing in t when $t \in (p^3/6, 1/6)$.

Fix $p^3/6 \le t < s < 1/6$. Fix $\epsilon > 0$. Take any $f \in \mathcal{W}$ such that $T(f) \ge s$. For each $\delta \in (0, 1)$, let

$$f_\delta(x, y) := (1 - \delta)f(x, y) + \delta p. \tag{17}$$

By the inequality

$$((1 - \delta)a + \delta p)((1 - \delta)b + \delta p)((1 - \delta)c + \delta p) \ge (1 - \delta)^3 abc$$

we have

$$T(f_\delta) \ge (1 - \delta)^3 T(f) \ge (1 - \delta)^3 s.$$

Thus, if we take δ such that $(1 - \delta)^3 = t/s$, then $T(f_\delta) \ge t$. By the convexity of the function I_p on $[0, 1]$ defined in (6), we see that for any $x \ge p$,

$$I_p((1 - \delta)x + \delta p) \le (1 - \delta)I_p(x) + \delta I_p(p) = (1 - \delta)I_p(x),$$

and therefore,

$$I_p(f_\delta) \le (1 - \delta)I_p(f). \tag{18}$$

Thus, $\phi(p, t) \le I_p(f_\delta) \le (t/s)^{1/3} I_p(f)$. Since this holds for any f with $T(f) \ge s$, it shows that

$$\phi(p, t) \le (t/s)^{1/3}\phi(p, s). \tag{19}$$

To show ϕ is a strictly increasing function of t in the interval $(p^3/6, 1/6)$, it therefore suffices to prove that $\phi(p, t) > 0$ for t in this interval. This follows easily, since the strict convexity of I_p on $[0, 1]$ and Eq. (18) show that equality in (19) can hold only if $f \equiv p$ almost everywhere for some f such that $T(f) \ge s$, which is impossible since $s > p^3/6$.

Next, fix $t \in (p^3/6, 1/6)$ and take any sequence $\{f^{(n)}\}$ in $\mathcal{W}$ such that $T(f^{(n)}) \ge t$ for all n and $I_p(f^{(n)}) \to \phi(p, t)$. Recall the subscript notation introduced in Eq. (17) above. By the continuous mapping theorem, there exist $\delta_n \in [0, 1]$ such that for each n,

$$T(f^{(n)}_{\delta_n}) = t.$$

Therefore by (18),

$$\phi(p, t) \le I_p(f^{(n)}_{\delta_n}) \le (1 - \delta_n)I_p(f^{(n)}),$$

which proves that $\delta_n \to 0$. This proves that

$$\phi(p, t) = \inf\{I_p(f) : f \in \mathcal{W},\ T(f) = t\}.$$

Since $\delta_n \to 0$, this also proves that $T(f^{(n)}) \to t$. This completes the proof of Proposition 4.2. □

4.4. The 'Replica Symmetric' phase

Note that there are two "extreme" functions that satisfy $T(f) = t$. First, there is the constant function

$$c_t(x, y) \equiv (6t)^{1/3}. \tag{20}$$

On the opposite extreme, there is the function χ_t, defined as

$$\chi_t(x, y) := \begin{cases} 1 & \text{if } \max\{x, y\} \le (6t)^{1/3}, \\ 0 & \text{otherwise.} \end{cases} \tag{21}$$

In a limiting sense, c_t represents an Erdős–Rényi random graph with edge probability $(6t)^{1/3}$, while χ_t represents the union of a clique of size $n(6t)^{1/3}$ and a set of isolated vertices of size $n(1 - (6t)^{1/3})$.

It is simple to see that (14) holds if and only if the infimum in (15) is attained at the constant function c_t. The following theorem gives a sufficient condition for this to happen. This extends the main result of [9]. The methods of [9] are closely related to methods from statistical physics; drawing inspiration from this connection, one may call the region where c_t solves the variational problem (15) the 'replica symmetric phase' of the problem.

Theorem 4.3. *For each $0 < p < 1$ and $t \in (p^3/6, 1/6)$, let $h_p(t) := I_p((6t)^{1/3})$. For $t \in [0, p^3/6]$, let $h_p(t) = 0$. Let $\hat{h}_p$ be the convex minorant of h_p (i.e. the maximum convex function lying below h_p). If t is a point in $(p^3/6, 1/6)$ where $h_p(t) = \hat{h}_p(t)$, then the variational problem (15) for this pair (p, t) is uniquely solved by the constant function $c_t \equiv (6t)^{1/3}$. Consequently, $\phi(p, t) = h_p(t)$. Moreover, for such (p, t), for each $\epsilon > 0$*

$$\lim_{n\to\infty} \mathbb{P}(\delta_\square(G(n, p), c_t) \geq \epsilon \mid T_{n,p} \geq tn^3) = 0.$$

Since c_t is the limit of $G(n, (6t)^{1/3})$, this means that for such (p, t) the conditional distribution of $G(n, p)$ given $T_{n,p} \geq tn^3$ is indistinguishable from the law of $G(n, (6t)^{1/3})$ in the large n limit.

Proof. Since h_p is an increasing function, it is easy to see that $h_p(t) = \hat{h}_p(t)$ for $t \leq p^3/6$ (incidentally, this also shows that $\hat{h}_p$ is an increasing function in $[0, 1/6)$, and strictly increasing in $(p^3/6, 1/6)$).

Suppose t is a point in $(p^3/6, 1/6)$ such that $h_p(t) = \hat{h}_p(t)$. We claim that there exists $\beta > 0$ such that

$$t = \arg \max_{x\in[0,1/6]} (\beta x - h_p(x)). \tag{22}$$

To see this, observe that since $\hat{h}_p$ is convex and strictly increasing in the interval $(p^3/6, 1/6)$, there exists $\beta > 0$ and $c \in \mathbb{R}$ such that the line $y = \beta x + c$ lies below the curve $y = \hat{h}_p(x)$ and touches it at $x = t$. But we also know that h_p lies above $\hat{h}_p$ and the two curves touch at t. Thus,

$$t = \arg \max_x (\beta x + c - h_p(x)) = \arg \max_x (\beta x - h_p(x)).$$

This proves the claim.

Now take any $f \in \mathcal{W}$ such that $T(f) \geq t$. Let c_t be the function that is identically equal to $(6t)^{1/3}$. Then $T(c_t) = t$, and by Hölder's inequality and (22),

$$\begin{aligned}\beta t - I_p(c_t) &= \iint \left(\frac{\beta c_t(x, y)^3}{6} - I_p(c_t(x, y))\right) dx\, dy \\ &\geq \iint \left(\frac{\beta f(x, y)^3}{6} - I_p(f(x, y))\right) dxdy \\ &\geq \frac{\beta}{6} \iiint f(x, y)f(y, z)f(z, x)dxdydz - \iint I_p(f(x, y))dxdy \\ &\geq \beta t - I_p(f).\end{aligned}$$

Thus, c_t minimizes $I_p(f)$ among all f such that $T(f) \geq t$. This shows that

$$\phi(p, t) = I_p(c_t) = h_p(t).$$

The uniqueness of the optimizer follows from the Hölder step in the above deduction. Finally, the claim about the conditional distribution follows from Theorem 3.1 and the uniqueness of the minimizer. □

It is easy to show that for any $p > 0$, $h_p(t) = \hat{h}_p(t)$ for all $t \in (p^3/6, t') \cup (t'', 1/6)$ where t' and t'' depend on p. Similarly, given any $t \in (0, 1/6)$, there exists $p' < (6t)^{1/3}$ depending on t such that for all $p \in (p', (6t)^{1/3})$, $h_p(t) = \hat{h}_p(t)$. Thus, there is a nontrivial set of (p, t) where c_t solves the variational problem and consequently $\phi(p, t) = h_p(t)$. As mentioned before, this recovers the main result of [9]. The conclusion about the conditional distribution is a new result.

4.5. Replica symmetry breaking

Given Theorem 4.3, it is quite interesting to note that the variational problem (15) is not solved by constant functions everywhere. From the physical point of view espoused in [9], however, this is not surprising; it is simply the effect of replica symmetry breaking down in the 'low temperature regime'.

The phase transition is very easy to establish using Theorem 4.1, by comparing the performances of c_t and χ_t defined in (20) and (21). A simple computation shows that for any $t \in (0, 1/6)$,

$$\lim_{p\to 0} \frac{I_p(c_t)}{\log(1/p)} = \frac{(6t)^{1/3}}{2} > \frac{(6t)^{2/3}}{2} = \lim_{p\to 0} \frac{I_p(\chi_t)}{\log(1/p)}. \tag{23}$$

Combining the above observation with Theorem 3.1, it follows easily that there are values of (p, t) such that given $T_{n,p} \geq tn^3$, the graph $G(n, p)$ must look different than an Erdős–Rényi graph. Again, compactness is crucial. This is formalized by the following theorem.

Theorem 4.4. *Let $\widetilde{C}$ denote the set of constant functions in $\widetilde{\mathcal{W}}$. For each $t \in (0, 1/6)$, there exists $p' > 0$ such that for all $p < p'$, the variational problem (15) is not solved by the constant function $c_t \equiv (6t)^{1/3}$. Moreover, for such (p, t), $\delta_\square(\widetilde{F}^*_{p,t}, \widetilde{C}) > 0$, where $\widetilde{F}^*_{p,t}$ is the set of minimizers defined in Theorem 4.1. Consequently, there exists $\epsilon > 0$ such that*

$$\lim_{n\to\infty} \mathbb{P}(\delta_\square(G(n, p), \widetilde{C}) > \epsilon \mid T_{n,p} \geq tn^3) = 1.$$

Proof. By (23) we see that for each t, there exists $p' > 0$ such that c_t is not a minimizer for the problem (15) if $p < p'$. Take any such (p, t). By Theorem 4.1, $\widetilde{F}^*_{p,t}$ is non-empty and compact. By part (iii) of Proposition 4.2 and the non-optimality of c_t, it follows that $\widetilde{C}$ and $\widetilde{F}^*_{p,t}$ must be disjoint. But $\widetilde{C}$ and $\widetilde{F}^*_{p,t}$ are both compact subsets of $\widetilde{\mathcal{W}}$. Therefore, $\delta_\square(\widetilde{C}, \widetilde{F}^*_{p,t}) > 0$. The last claim follows by Theorem 4.1. □

4.6. The double phase transition

A combination of Theorems 4.3 and 4.4 shows that for all small enough p, the variational problem (15) has a 'double phase transition'. (Actually, there may be more than two phase transitions, but we show that there is at least two.)

Indeed, for all small enough p, Theorem 4.3 (or the results of [9]) show that there exists $p^3/6 < t' \leq t'' < 1/6$ such that $\phi(p, t) = I_p((6t)^{1/3})$ for all $t \in (p^3/6, t') \cup (t'', 1/6)$. On the other hand by Theorem 4.4, it follows that for all small enough p, the variational problem (15) is not solved by a constant function at the point $(p, 1/2)$. Combining these two observations gives the following theorem.

Theorem 4.5. *There exists $p_0 > 0$ such that if $p \leq p_0$, then there exists $p^3/6 < t' < t'' < 1/6$ such that the variational problem (15) is solved by the constant function $c_t \equiv (6t)^{1/3}$ when $t \in (p^3/6, t') \cup (t'', 1/6)$, but there is a non-empty subset of (t', t'') where all optimizers are non-constant.*

Of course, as shown by Theorems 4.3 and 4.4, the significance of optimizers being constant or non-constant is in whether the conditional behavior of $G(n, p)$ given $T_{n,p} \geq tn^3$ is close to that of an Erdős–Rényi graph or not.

4.7. The small p limit

The last theorem of this paper describes the nature of $\phi(p, t)$ and $\widetilde{F}^*_{p,t}$ when t is fixed and p is very small, tending to zero. The essence of the result, perhaps not surprisingly, is that when t is fixed and $p \to 0$, then conditionally on the event $\{T_{n,p} \geq tn^3\}$ the graph $G(n, p)$ must look like a clique.

Theorem 4.6. *For each $t \in (0, 1/6)$,*

$$\lim_{p\to 0} \frac{\phi(p, t)}{\log(1/p)} = \frac{(6t)^{2/3}}{2}.$$

Moreover, if χ_t is the function defined in (21) and $\widetilde{F}^*_{p,t}$ is defined as in Theorem 4.1, then the set $\widetilde{F}^*_{p,t}$ converges to the point $\widetilde{\chi}_t$ as $p \to 0$, in the sense that

$$\lim_{p\to 0} \sup_{\widetilde{f}\in \widetilde{F}^*_{p,t}} \delta_\Box(\widetilde{f}, \widetilde{\chi}_t) = 0.$$

Consequently, for each $\epsilon > 0$,

$$\lim_{p\to 0}\lim_{n\to\infty} \mathbb{P}(\delta_\Box(G(n,p), \widetilde{\chi}_t) \geq \epsilon \mid T_{n,p} \geq tn^3) = 0.$$

Proof. In this proof, C will denote any constant that does not depend on anything else. The value of C may change from line to line. All integrals will be over the interval $[0, 1]$.

Fix $t \in (0, 1/6)$. For each $p < (6t)^{1/3}$, choose a function $f_p \in F^*_{p,t}$. From the definition of I_p, observe that if $p \leq 1/2$,

$$\left| I_p(f_p) - \frac{1}{2}\iint f_p(x,y)\log\frac{1}{p}\,dx\,dy \right| \leq C. \tag{24}$$

On the other hand, by the definition of $\widetilde{F}^*_{p,t}$,

$$I_p(f_p) \leq I_p(\chi_t) \leq \frac{(6t)^{2/3}}{2}\log\frac{1}{p} + C.$$

Combining the last two inequalities gives

$$\iint f_p(x,y)dxdy \leq (6t)^{2/3} + \frac{C}{\log(1/p)}. \tag{25}$$

Next, let

$$h_p(x,y) := \int f_p(x,z) f_p(z,y)\,dz.$$

Then by two applications of the Cauchy–Schwarz inequality, the inequality (25), and the fact that $f_p(x,y) \in [0, 1]$, we get the following important sequence of inequalities.

$$\begin{cases} (6t)^2 = (6T(f_p))^2 = \left(\iint h_p(x,y) f_p(x,y)\,dx\,dy\right)^2 \\ \leq \iint h_p^2(x,y)\,dx\,dy \iint f_p^2(x,y)\,dx\,dy \\ \leq \int\left(\int f_p^2(x,z)\,dz \int f_p^2(y,z)\,dz\right)dx\,dy \iint f_p^2(x,y)\,dx\,dy \\ = \left(\iint f_p^2(x,y)\,dx\,dy\right)^3 \\ \leq \left(\iint f_p(x,y)\,dx\,dy\right)^3 \leq \left((6t)^{2/3} + \frac{C}{\log(1/p)}\right)^3. \end{cases} \tag{26}$$

A direct consequence of (26), combined with (24), is that

$$\lim_{p\to 0}\frac{\phi(p,t)}{\log(1/p)} = \lim_{p\to 0}\frac{I_p(f_p)}{\log(1/p)} = \lim_{p\to 0}\frac{1}{2}\iint f_p(x,y)\,dx\,dy = \frac{(6t)^{2/3}}{2},$$

which proves the first assertion of the theorem. A second important consequence of (26), to be useful later, is that

$$\iint f_p(x,y)(1 - f_p(x,y))\,dx\,dy \leq \frac{C}{\log(1/p)}. \tag{27}$$

Next, note that for any x, y,

$$\frac{1}{2}\iint (f_p(x,z)f_p(z',y) - f_p(x,z')f_p(z,y))^2\,dz\,dz' = \iint f_p^2(x,z)f_p^2(z',y)\,dz\,dz' - h_p^2(x,y)$$
$$= \int f_p^2(x,z)dz \int f_p^2(y,z)dz - h_p^2(x,y).$$

It follows from this and (26) that

$$\iiiint (f_p(x,z)f_p(z',y) - f_p(x,z')f_p(z,y))^2\,dz\,dz'\,dx\,dy \iint f_p^2(x,y)\,dx\,dy \le \frac{C}{\log(1/p)}.$$

The above inequality and the lower bound on $\iint f_p^2(x,y)dxdy$ from (26) give

$$\iiiint (f_p(x,z)f_p(z',y) - f_p(x,z')f_p(z,y))^2\,dz\,dz'\,dx\,dy \le \frac{C}{t^{2/3}\log(1/p)}. \tag{28}$$

Let $M_p := \iint f_p(x,y)dxdy$. For each x, let $m_p(x) := M_p^{-1/2}\int f_p(x,y)dy$. An application of Jensen's inequality to (28) gives

$$\iint (f_p(x,z)M_p - M_p m_p(x)m_p(z))^2\,dz\,dx \le \frac{C}{t^{2/3}\log(1/p)}.$$

By (26), $M_p \ge (6t)^{2/3}$. Thus, m_p is bounded by $(6t)^{-1/3}$, and

$$\iint (f_p(x,z) - m_p(x)m_p(z))^2\,dz\,dx \le \frac{C}{t^2\log(1/p)}. \tag{29}$$

For each p, let $n_p : [0,1] \to [0,(6t)^{-1/3}]$ be a step function (i.e. a function that is constant on intervals) such that

$$\int (m_p(x) - n_p(x))^2 dx \le p. \tag{30}$$

(Such functions exist because we can approximate m_p by a continuous function to any degree of accuracy by Lusin's Theorem [26], and then approximate the continuous function by a step function.)

Let σ_p be a measure preserving bijection of $[0,1]$ such that $n_p(\sigma_p x)$ is a non-increasing function. Such a bijection is easy to construct because n_p is a step function. Let $\ell_p(x) := n_p(\sigma_p x)$ and $g_p(x,y) := f_p(\sigma_p x, \sigma_p y)$. By the monotonicity and uniform boundedness of ℓ_p there exists a sequence $\{p_i\}_{i\ge 1}$ decreasing to zero such that ℓ_{p_i} converges in L^2 to a limit function ℓ. Therefore by (29) and (30), $g_{p_i} \to g$ in L^2, where

$$g(x,y) := \ell(x)\ell(y).$$

By this and (27), g is a $\{0,1\}$-valued function. It is not difficult to see from this and the non-negativity of ℓ that ℓ must also be $\{0,1\}$-valued. (If $\ell(x) \notin \{0,1\}$ on some set of positive measure, then there may be a set A of positive measure where $\ell(x) \in (0,1)$, or there may be a set A of positive measure where $\ell(x) \in (1,\infty)$. In either case, $g(x,y) \notin \{0,1\}$ on $A \times A$.) Since ℓ is monotone decreasing, it follows that ℓ must be the indicator of an interval of the form $[0,b]$ for some $b \in [0,1]$. Lastly, (26) implies that $\iint g(x,y)dxdy = (6t)^{2/3}$, and therefore $b = (6t)^{1/3}$. Consequently, $g = \chi_t$.

The above argument establishes that for any collection $\{f_p\}_{p>0}$ such that $f_p \in F^*_{p,t}$ for each p, there is a sequence $\{p_i\}_{i\ge1}$ decreasing to zero such that $f_{p_i} \to \chi_t$ in the cut metric. The same argument can be extended to show that for any sequence $\{f_{p_i}\}$ such that $p_i \to 0$ and $f_{p_i} \in F^*_{p_i,t}$ for each i, there is a subsequence converging to χ_t in the cut metric. This proves the second assertion of the theorem. The last claim of the theorem follows from this and Theorem 4.1. □

4.8. Open questions

There are many questions that remain unresolved, even in the simple example of upper tails for triangle counts that has been analyzed in this section. For instance, what is the set of optimal solutions of the variational problem (15) in the broken replica symmetry phase (i.e. where the optimizer is not a constant)? Is the solution unique in the quotient space $\widetilde{\mathcal{W}}$, or can there exist multiple solutions? Is it possible to explicitly compute a nontrivial solution of (15) for at least some value of (p, t)? Is it possible to even numerically evaluate or approximate a solution using a computer? Does Theorem 4.3 characterize the full replica symmetric phase? If not, what is the exact phase transition boundary? What happens in the sparse case where p and t are both allowed to tend to zero? At the time of writing this paper, we do not know how to answer any of these questions.

Acknowledgements

The authors thank Amir Dembo for his crucial role in motivating this research and suggesting the formula for the rate function in the case of subgraph counts, Joel Spencer for hinting that Szemerédi's lemma may be useful, and Persi Diaconis for pointing us to the work of Lovász and coauthors (which helped in cutting down the size of the paper by a half). The authors also thank the referee for a very careful report.

Sourav Chatterjee's research was partially supported by NSF grants DMS-0707054 and DMS-1005312, and a Sloan Research Fellowship

S. R. S. Varadhan's research was partially supported by NSF grants DMS-0904701 and OISE-0730136.

References

[1] D. Aldous, Representations for partially exchangeable arrays of random variables, J. Multivariate Anal. 11 (4) (1981) 581–598.
[2] B. Bollobás, Random Graphs, second edition, Cambridge University Press, Cambridge, 2001.
[3] B. Bollobás, O. Riordan, Metrics for sparse graphs, in: Surveys in Combinatorics, in: LMS Lecture Notes Series, vol. 365, CUP, 2009, pp. 211–287.
[4] E. Bolthausen, F. Comets, A. Dembo, Large deviations for random matrices and random graphs, Private communication (2003).
[5] C. Borgs, J. Chayes, L. Lovász, V.T. Sós, K. Vesztergombi, Counting graph homomorphisms, in: Topics in Discrete Mathematics, in: Algorithms Combin., vol. 26, Springer, Berlin, 2006, pp. 315–371.
[6] C. Borgs, J. Chayes, L. Lovász, V.T. Sós, K. Vesztergombi, Convergent sequences of dense graphs I. subgraph frequencies, metric properties and testing, Adv. Math. 219 (6) (2008) 1801–1851.
[7] C. Borgs, J. Chayes, L. Lovász, V.T. Sós, K. Vesztergombi, Convergent sequences of dense graphs II. Multiway cuts and statistical physics. Preprint. 2007, Available at http://research.microsoft.com/en-us/um/people/borgs/papers/conright.pdf.
[8] S. Chatterjee, The missing log in large deviations for subgraph counts. Preprint. 2010, Available at http://arxiv.org/abs/1003.3498.
[9] S. Chatterjee, P.S. Dey, Applications of Stein's method for concentration inequalities. in Ann. Probab. 2009, Available at http://arxiv.org/abs/0906.1034 (in press).
[10] S. Chatterjee, P. Diaconis, A. Sly, Random graphs with a given degree sequence. Preprint. 2010, Available at.
[11] A. Dembo, O. Zeitouni, Large Deviations Techniques and Applications, second edition, Springer, 1998.
[12] P. Diaconis, S. Janson, Graph limits and exchangeable random graphs, Rend. Mat. Appl. 7 (28) (2008) 33–61. 1.
[13] R. Diestel, Graph Theory, second edition, Springer-Verlag, New York, 2000.
[14] P. Erdős, A. Rényi, On the evolution of random graphs, Publ. Math. Inst. Hungar. Acad. Sci. 5 (1960) 17–61.
[15] M. Freedman, L. Lovász, A. Schrijver, Reflection positivity, rank connectivity, and homomorphism of graphs, J. Amer. Math. Soc. 20 (2007) 37–51.
[16] A. Frieze, R. Kannan, Quick approximation to matrices and applications, Combinatorica 19 (1999) 175–220.
[17] D.N. Hoover, Row-column exchangeability and a generalized model for probability, in: Exchangeability in probability and statistics, North-Holland, Amsterdam-New York, Rome,1981, pp. 281–291.
[18] S. Janson, T. Łuczak, A. Ruciński, Random Graphs, Wiley, New York, 2000.
[19] L. Lovász, The rank of connection matrices and the dimension of graph algebras, Eur. J. Comb. 27 (2006) 962–970.
[20] L. Lovász, Connection matrices, in: G. Grimmett, C. McDiarmid (Eds.), Combinatorics, Complexity, and Chance: A Tribute to Dominic Welsh, Oxford University Press, Oxford, 2007, pp. 179–190.
[21] L. Lovász, V.T. Sós, Generalized quasirandom graphs, J. Comb. Theory B. 98 (1) (2008) 146–163.
[22] L. Lovász, B. Szegedy, Limits of dense graph sequences, J. Combin. Theory Ser. B 96 (6) (2006) 933–957.
[23] L. Lovász, B. Szegedy, Szemerédi's lemma for the analyst, Geom. Funct. Anal. 17 (1) (2007) 252–270.
[24] L. Lovász, B. Szegedy, Testing properties of graphs and functions. in Israel J. Math., 2007 (in press).
[25] L. Lovász, B. Szegedy, Contractors and connectors in graph algebras, J. Graph Theory 60 (1) (2009) 11–30.
[26] W. Rudin, Real and Complex Analysis, third edition, McGraw-Hill Book Co., New York, 1987.
[27] I.N. Sanov, On the probability of large deviations of random variables, Selected Translations in Mathematical Statistics and Probability 1 (1957) 213–224. Russian original: Mat. Sb..
[28] E. Szemerédi, Regular partitions of graphs. Problèmes combinatoires et théorie des graphes, Colloq. Internat. CNRS, Univ. Orsay, Orsay, 1976, Colloq. Internat. CNRS, 260, CNRS, Paris, 1978, pp. 399–401.

Communications on Stochastic Analysis
Vol. 6, No. 1 (2012) 1-13

Serials Publications
www.serialspublications.com

LARGE DEVIATIONS FOR RANDOM MATRICES

SOURAV CHATTERJEE* AND S. R. S. VARADHAN†

Dedicated to Professor K. R. Parthasarathy on the occasion of his 75th birthday

ABSTRACT. We prove a large deviation result for a random symmetric $n \times n$ matrix with independent identically distributed entries to have a few eigenvalues of size n. If the spectrum $\mathcal{S}$ survives when the matrix is rescaled by a factor of n, it can only be the eigenvalues of a Hilbert-Schmidt kernel $k(x, y)$ on $[0, 1] \times [0, 1]$. The rate function for k is $I(k) = \frac{1}{2}\int h(k(x, y)dxdy$ where h is the Cramér rate function for the common distribution of the entries that is assumed to have a tail decaying faster than any Gaussian. The large deviation for $\mathcal{S}$ is then obtained by contraction.

1. Introduction

We are interested in the large deviation behavior of random $n \times n$ symmetric matrices $X(\omega) = \{x_{i,j}(\omega)\}$ where the entries for $j \geq i$ are independent identically distributed real random variables having a common distribution μ. Let $\Lambda_n(\omega) = \{\lambda_j^n(\omega); 1 \leq j \leq n\}$ be the set of n real eigenvalues of this random matrix. If we assume that the mean is 0 and the variance is equal to σ^2, according to a theorem of Wigner [6], when divided by $\sqrt{n}$, the empirical distribution of these eigenvalues converges in probability to the semi-circle law, i.e. for any bounded continuous function $f : \mathbb{R} \to \mathbb{R}$

$$\lim_{n\to\infty} \frac{1}{n} \sum_{j=1}^{n} f\left(\frac{\lambda_j(\omega)}{\sqrt{n}}\right) = \int f(y)\phi_\sigma(y)dy$$

in probability, where

$$\phi_\sigma(y) = \begin{cases} 0 & \text{if } |x| \geq 2\sigma, \\ \frac{1}{2\pi\sigma^2}\sqrt{4\sigma^2 - x^2} & \text{if } |x| \leq 2\sigma. \end{cases}$$

In fact it is known [1] that under suitable additional assumptions, for any $\epsilon > 0$,

$$\mathbb{P}\big[\max_{1\leq j\leq n} |\lambda_j^n| \geq (2\sigma + \epsilon)\sqrt{n}\big] \to 0.$$

Received 2011-6-23; Communicated by K. B. Sinha.
2000 *Mathematics Subject Classification.* 60B20; 15B52.
Key words and phrases. Large deviations, eigenvalues, random matrices.
* Sourav Chatterjee's research was partially supported by NSF grants DMS-0707054 and DMS-1005312, and a Sloan Research Fellowship.
† S. R. S. Varadhan's research was partially supported by NSF grants DMS-0904701 and OISE-0730136.

Large deviation results for the distribution of eigenvalues to be different from the Wigner distribution has been considered by Ben Arous and Guionnet [2]. If we decide to divide by n rather than $\sqrt{n}$, and denote the resulting spectrum by $S_n(\omega) = \frac{1}{n}\Lambda_n(\omega)$, then

$$\mathbb{P}[\sup_{\lambda \in S_n} |\lambda| \geq \epsilon] \to 0.$$

If we drop the assumption that the mean is 0, then there will be one large eigenvalue λ_{max} of size n in Λ_n, with $\frac{\lambda_{max}}{n} \simeq \int x d\mu$ and the remaining eigenvalues will follow the semi-circle law as before when divided by $\sqrt{n}$. It is also possible to consider scaling by n^α for $\frac{1}{2} < \alpha < 1$, but they would likely behave like moderate deviations and exhibit Gaussian behavior with rate $n^{2\alpha}$.

Let $\mathcal{S}$ denote the set of all finite or countable collections $S = \{\lambda_j\}$ of positive as well as negative real numbers, with repetitions allowed, that have the property

$$\sum_{\lambda \in S} |\lambda|^2 = \sum_j |\lambda_j|^2 < \infty.$$

They are all possible spectra of self adjoint Hilbert-Schmidt operators. The topology on $\mathcal{S}$ will be ordinary convergence as real numbers of the corresponding eigenvalues outside any arbitrarily small interval around 0. Equivalently, it is the minimal topology such that for any bounded continuous function f that is 0 in some interval around 0, the sum $\sum_{\lambda \in \mathcal{S}} f(\lambda)$ is continuous as a map of $\mathcal{S} \to \mathbb{R}$. One can easily construct a metric for this topology. We enumerate separately the positive and negative values in S in decreasing order of their absolute values as $\{u_j^+\}, \{u_j^-\}$. If one or both of them is only a finite set or empty we augment it by adding 0's. If two points S, S' in $\mathcal{S}$ are enumerated as $\{u_j^+\}, \{u_j^-\}$ and $\{v_j^+\}, \{v_j^-\}$, we define the distance

$$d(S, S') := \sum_{j=1}^{\infty} \frac{|u_j^+ - v_j^+|}{2^{j+1}} + \sum_{j=1}^{\infty} \frac{|u_j^- - v_j^-|}{2^{j+1}}.$$

The random matrix yields a random spectrum and after normalization by n, we obtain a random element of $\mathcal{S}$. Its distribution yields a sequence $\{P_n\}$ of probability measures on $\mathcal{S}$. We will prove a large deviation result for P_n with rate n^2. We assume that for all $\theta > 0$,

$$\int \exp[\theta\, x^2]\mu(dx) < \infty. \tag{1.1}$$

In particular this implies that the moment generating function $M(\theta) = \int e^{\theta\, x}\mu(dx)$ is finite for all θ and satisfies

$$\limsup_{|\theta| \to \infty} \frac{1}{\theta^2} \log M(\theta) = 0$$

and the conjugate rate function of Cramér

$$h(x) = \sup_\theta [\theta\, x - \log M(\theta)] \tag{1.2}$$

satisfies

$$\liminf_{|x| \to \infty} \frac{h(x)}{x^2} = +\infty.$$

The condition (1.1) is important, because an eigenvalue of size n can be produced by a single entry of size n in the random matrix and we would like this to have probability that is super-exponentially small in the scale n^2.

The random symmetric $n \times n$ matrix $X(\omega) = \{x_{i,j}(\omega)\}$ is first mapped into a symmetric kernel $k(x, y, \omega)$

$$k(x, y, \omega) = \sum_{i,j=1}^{n} x_{i,j}(\omega) \mathbf{1}_{J_i^n}(x) \mathbf{1}_{J_j^n}(y) \tag{1.3}$$

where J_i^n is the interval $[\frac{i-1}{n}, \frac{i}{n}]$. This induces a family of probability measures Q_n on the space of symmetric kernels $k(x, y)$ on $D = [0, 1] \times [0, 1]$. We will restrict ourselves to $\mathcal{K} = \{k : \int\int_D |k(x, y)|^2\, dxdy < \infty\}$. Then $k(x, y)$ defines on $L^2[0, 1]$ a Hilbert-Schmidt operator which has a countable spectrum with 0 as the only limit point. Actually for each fixed n the range of the map $X \to k$ is a finite dimensional subspace of simple functions $\mathcal{K}_n \subset \mathcal{K}$. The nonzero spectrum of $k(x, y, \omega)$ is the same as that of $\{\frac{x_{i,j}}{n}\}$ and we can obtain P_n from Q_n through the natural map $\mathcal{K} \to \mathcal{S}$ that takes any k to its set of eigenvalues. We define on $\mathcal{K}$ the following rate function

$$I(k(\cdot, \cdot)) = \frac{1}{2} \int_0^1 \int_0^1 h(k(x, y))dxdy \tag{1.4}$$

with h given by (1.2).

Any permutation $\sigma \in \Pi(n)$ of the rows and columns of X, mapping $\{x_{i,j}\} \to \{x_{\sigma(i),\sigma(j)}\}$ leaves the set of eigen-values of X invariant and the group G of measure preserving transformations σ of $[0, 1]$ onto itself lifts to an action on $\mathcal{K}$ mapping $k \to \sigma k$ where $\sigma k(x, y) := k(\sigma x, \sigma y)$. The map $k \to \sigma k$ leaves $I(k(\cdot, \cdot))$ as well as the spectrum $S(k)$ of k invariant. For establishing the large deviations of P_n on $\mathcal{S}$ it is therefore enough to prove a large deviation principle for the images $\tilde{Q}_n$ of Q_n on $\tilde{\mathcal{K}} = \mathcal{K}/G$.

We will be working with the space $\mathcal{K}$ of symmetric kernels $k(x, y)$ on D. If $I(k) < \infty$, then the operator defined by $k(x, y)$ on $L^2[0, 1]$ is Hilbert-Schmidt and has a countable spectrum with 0 as the only limit point. We need to show some sort of continuity of the map $\tilde{\mathcal{K}} \to \mathcal{S}$ mapping the G-orbit $\tilde{k}$ of k to its spectrum $S(k)$, in order to transfer the large deviation result from $\tilde{\mathcal{K}}$ to $\mathcal{S}$. This requires a topology on $\tilde{\mathcal{K}}$ that will be inherited from $\mathcal{K}$. The weak topology on $\mathcal{K}$ turns out to be too weak and the strong or L^1 topology too strong. What works is the topology induced by the cut metric

$$d_{\square}(k_1, k_2) = \sup_{\substack{|\phi| \le 1 \\ |\psi| \le 1}} \left| \int (k_1(x, y) - k_2(x, y))\phi(x)\psi(y)dxdy \right|, \tag{1.5}$$

ϕ, ψ being Borel measurable functions on $[0, 1]$. Equivalently,

$$d_{\square}(k_1, k_2) = \sup_{A \times B} \left| \int_{A \times B} (k_1(x, y) - k_2(x, y))dxdy \right|, \tag{1.6}$$

where the supremum is taken over all Borel subsets A, B of $[0,1]$. The induced metric on $\tilde{\mathcal{K}}$ is

$$d_\Box(k_1, k_2) = \inf_\sigma d_\Box(\sigma k_1, k_2) = \inf_\sigma d_\Box(k_1, \sigma k_2) = \inf_{\sigma_1,\sigma_2} d_\Box(\sigma_1 k_1, \sigma_2 k_2)$$

where $\sigma k(x,y) = k(\sigma x, \sigma y)$. We can define on $\mathcal{S}$ the rate function

$$J(S) = \inf_{k:S(k)=S} I(k).$$

Our main result is the following.

Theorem 1.1. *Under assumption (1.1) the sequence of measures P_n on $\mathcal{S}$ satisfies a large deviation property with rate function $J(S)$, i.e for closed $C \subset \mathcal{S}$*

$$\limsup_{n\to\infty} \frac{1}{n^2} \log P_n(C) \leq -\inf_{S\in C} J(S)$$

and for $U \subset \mathcal{S}$ that are open

$$\liminf_{n\to\infty} \frac{1}{n^2} \log P_n(U) \geq -\inf_{S\in U} J(S).$$

This is based on a large deviation principle for $\tilde{Q}_n$ on $\tilde{\mathcal{K}}$ in the cut topology with rate function $I(k)$.

Theorem 1.2. *Under assumption (1.1) the sequence of measures $\tilde{Q}_n$ on $\tilde{\mathcal{K}}$ satisfies a large deviation property with rate function $I(\tilde{k})$, i.e for closed $C \subset \tilde{\mathcal{K}}$*

$$\limsup_{n\to\infty} \frac{1}{n^2} \log \tilde{Q}_n(C) \leq -\inf_{\tilde{k}\in C} I(\tilde{k})$$

and for $U \subset \tilde{\mathcal{K}}$ that are open

$$\liminf_{n\to\infty} \frac{1}{n^2} \log \tilde{Q}_n(U) \geq -\inf_{\tilde{k}\in U} I(\tilde{k}).$$

To make the connection we need the map $k \to S(k)$ to be continuous in the cut topology. It is valid, provided we restrict it to sets $\mathcal{K}^\ell$ of the form $\mathcal{K}^\ell = \{k : |k(x,y)| \leq \ell\}$ for some $\ell < \infty$. This requires truncation at level ℓ and then removing the cut-off. Condition (1.1) provides super-exponential bounds in the Hilbert-Schmidt norm for the error due to cutoff and that is used to complete the proof.

Similar methods were used in [3] to study the large deviation behavior of the number of triangles or other finite subgraphs in random graphs as the number of vertices goes to ∞ but the probability of an edge being connected remains fixed at some $p > 0$. This will correspond to each $x_{i,j}$ taking the values 0 or 1.

2. Some Useful Lemmas.

We will be working with partitions of the unit interval $[0,1]$ into a finite disjoint union of subintervals. We will not worry about the end points. We can adopt any convention that makes it a true partition. For each partition $\mathcal{P}$ of the unit interval into m subintervals $\{J_i\}$ there is a corresponding partition of D into m^2 sub-squares $\{J_i \times J_j\}$.

For each integer m we have the special partition $\mathcal{P}_m$ of the unit interval into m equal subintervals and they will be denoted by $J_i^m = [\frac{i-1}{m}, \frac{i}{m}]$. We denote by $\mathcal{K}_m \subset \mathcal{K}$ the space of symmetric kernels of the form

$$f(x,y) = \sum_{i,j} f_{i,j} \mathbf{1}_{J_i^m}(x) \mathbf{1}_{J_j^m}(y).$$

This provides a faithful representation of symmetric matrices of size $m \times m$ as elements of $\mathcal{K}_m$. If $|f_{i,j}| \leq \ell$, then the corresponding $f \in \mathcal{K}_m^\ell = \mathcal{K}^\ell \cap \mathcal{K}_m$. The following is a simple, but useful lemma.

Lemma 2.1. *If $f(x,y) = \sum f_{i,j} \mathbf{1}_{J_i}(x) \mathbf{1}_{J_j}(y)$ and $g(x,y) = \sum g_{i,j} \mathbf{1}_{J_i}(x) \mathbf{1}_{J_j}(y)$ are both simple functions with respect to the same partition $J_1, \ldots, J_m$ of $[0,1]$, then in calculating the distance*

$$d_\square(f,g) = \sup_{A\times B} \left| \int_{A\times B} (f(x,y) - g(x,y)) dx dy \right|$$

the sets A, B can be restricted to sets of the form $\bigcup_{i\in\mathcal{N}} J_i$ where $\mathcal{N} \subset \{1, 2, \ldots, m\}$.

Proof. First note that

$$\begin{aligned}\int_{A\times B} (f(x,y) - g(x,y)) dx dy &= \sum_{i,j} \int_{A\cap J_i \times B\cap J_j} (f(x,y) - g(x,y)) dx dy \\ &= \sum_{i,j} |A\cap J_i||B\cap J_j|(f_{i,j} - g_{i,j}) \\ &= \sum_{i,j} a_i b_j |J_i||J_j|(f_{i,j} - g_{i,j}),\end{aligned}$$

where $a_i = \frac{|A\cap J_i|}{|J_i|}$ and $b_i = \frac{|B\cap J_i|}{|J_i|}$. It is now clear that the supremum of the absolute value is achieved when each a_i and b_j is either 0 or 1. □

Remark 2.2. If f and g are defined in terms of two different partitions of $[0,1]$, they can both be viewed as defined with respect to the finite partition which is their common refinement.

Since our large deviation result on $\mathcal{S}$ is deduced from a large deviation result on $\mathcal{K}$, we need some continuity property of the map $k \to S(k)$ of $\mathcal{K} \to \mathcal{S}$.

Lemma 2.3. *For any $\ell < \infty$, the map $k \to S(k)$ from $\mathcal{K} \to \mathcal{S}$ is continuous in the cut topology when restricted to $\mathcal{K}^\ell = \{k : |k(x,y)| \leq \ell\}$.*

Proof. If $f(\lambda) = 0$ near 0 then $\lambda^{-3} f(\lambda)$ can be approximated uniformly on $[-\ell, \ell]$ by a polynomial in λ. Therefore $f(\lambda)$ is approximated by a polynomial involving only powers λ^m for $m \geq 3$. Since we have a bound on $\sum_{\lambda\in S} |\lambda|^2$ it is enough to show that

$$k \to \sum_{\lambda \in S(k)} \lambda^m$$

are continuous maps of $\mathcal{K}^\ell \to \mathcal{S}$ for each $m \geq 3$. It is elementary to check that since $|k(x,y)| \leq \ell$, the maps

$$k \to \int_{[0,1]^m} k(x_1,x_2) \cdots k(x_{m-1},x_m) k(x_m,x_1) dx_1 \ldots dx_m = \sum_{\lambda\in S(k)} \lambda^m$$

are continuous maps in the cut topology from $\mathcal{K}^\ell \to \mathcal{S}$ provided $m \geq 3$. We start with

$$\int_{[0,1]^m} k_n(x_1,x_2)\cdots k_n(x_{m-1},x_m)k_n(x_m,x_1)dx_1\ldots dx_m$$

which can be written as

$$\int_{[0,1]^{m-2}}\int dx_3\ldots dx_m \int_{[0,1]^2} k_n(x_1,x_2)\phi_n(x_1,x_3,\ldots,x_n)\psi_n(x_2,x_3,\ldots,x_n)dx_1dx_2$$

where ϕ_n and ψ_n are uniformly bounded. If $d_\square(k_n,k)\to 0$, we can then replace $k_n(x_1,x_2)$ by $k(x_1,x_2)$. This is repeated for each factor. □

We will also need the following lemmas: a multicolor version of Szemerédi's regularity lemma for graphs that can be found in [4] and its consequence.

Lemma 2.4. *Given any $\epsilon > 0$ and integers r and m, there exists M and n_0 such that if the edges of a graph $\mathcal{G}_n$ of size $n \geq n_0$ are colored with any one of r colors, then the vertex set can be partitioned into sets $V_0,\ldots,V_p$ for some p in the range $m \leq p \leq M$ so that $|V_0| \leq \epsilon n$, $|V_1| = |V_2| = \cdots = |V_p| = K$ and all but at most ϵp^2 pairs (V_i, V_j) satisfy the following regularity condition. For any $X \subset V_i, Y \subset V_j$ with $i,j \geq 1$ and $|X|,|Y| \geq \epsilon K$ we have*

$$|d_\nu(X,Y) - d_\nu(V_i,V_j)| < \epsilon,$$

where $d_\nu(X,Y)$ is the proportion of edges between X and Y that are colored with color ν.

Lemma 2.5. *For any $\ell < \infty$ and $\epsilon > 0$, there is a compact set of simple functions $W_{\ell,\epsilon} \subset \mathcal{K}$ and $n_0(\epsilon,\ell)$ such that for $n \geq n_0$ and any $k \in \mathcal{K}_n^\ell$, there exists $f \in W_{\ell,\epsilon}$ and $\sigma \in \Pi(n)$ such that*

$$d_\square(\sigma k, f) < \epsilon.$$

Proof. Let $\epsilon > 0$ be given. Let the integer r be chosen such that $\frac{\ell}{r} < \frac{\epsilon}{2}$. Let $\epsilon' = \frac{\epsilon}{12r\ell}$ and $m = \frac{1}{\epsilon'}$. Let $x_{i,j}$ be the value of k in the rectangle $J_i^n \times J_j^n$. The interval $[-\ell,\ell]$ is divided in to r disjoint equal intervals of length $\frac{2\ell}{r}$ and the edge (i,j) is colored according to the interval into which $x_{i,j}$ falls. The 'color' is defined as the value of the mid point of the interval. The color of the edge (i,j) is then a real number $y_{i,j}$ which can equal any one from the finite set $z_1, z_2, \ldots, z_r$ and $|y_{i,j} - x_{i,j}| \leq \frac{\epsilon}{2}$. We apply the multicolor version of Szemerédi's regularity theorem (Lemma 2.4) with parameters (ϵ', m, r) to obtain a partition $V_0, V_1, \ldots V_p$ of $\{1,2,\ldots,n\}$ with the following properties. $|V_0| = K' \leq \epsilon' n$. $|V_1| = |V_2| = \cdots = |V_p| = K$. If $a_{s,i,j}$ be the proportion of edges between V_i and V_j that have color z_s, then for all but $\epsilon' p^2$ pairs (V_i,V_j), $i,j \geq 1$ $i \neq j$, for any two subsets $X \subset V_i$, $Y \subset V_j$ with $|X| \geq \epsilon' K, |Y| \geq \epsilon' K$ the proportion $b_{s,i,j}$ of edges of color z_s, between X and Y satisfies

$$\sup_s |b_{s,i,j} - a_{s,i,j}| \leq \epsilon'.$$

Let us divide the unit interval into subintervals $J_0, \ldots, J_p$ where $J_0 = [0, \frac{K'}{n}]$ and for $i \geq 1$, $J_i = [\frac{K'+(i-1)K}{n}, \frac{K'+iK}{n}]$. We construct a function $f \in \mathcal{K}^\ell$ as

$$f(x, y) = \sum_{s=1}^{r} z_s f(s, x, y),$$

where

$$f(s, x, y) = \sum_{i,j=0}^{p} a_{s,i,j} \mathbf{1}_{J_i}(x) \mathbf{1}_{J_j}(y).$$

For fixed ℓ and ϵ as long as p remains bounded such functions vary over a compact subset of $L^1(D)$.

For the permutation $\sigma \in \Pi(n)$ of the vertices, we rearrange the order of the vertices, so that those in V_0 corresponds to the first K' indices and for $1 \leq i \leq p$, those in V_i correspond respectively to indices in the range $(K' + (i-1)K + 1, K' + iK)$. We will denote by k' the image of $\{x_{\sigma(i),\sigma(j)}\}$ in $\mathcal{K}_n^\ell$. We define $c(s, i, j) = 1$ if $x_{\sigma(i),\sigma(j)}$ belongs to the interval with mid point z_s. Otherwise it is 0. With

$$k'(s, x, y) = \sum_{i,j=1}^{n} c(s, i, j) J_i^n(x) J_j^n(y)$$

and

$$k'(x, y) = \sum_{i,j=1}^{n} x_{\sigma(i),\sigma(j)} J_i^n(x) J_j^n(y)$$

we have that for each x, y,

$$\left| k'(x, y) - \sum_s z_s k'(s, x, y) \right| \leq \frac{\epsilon}{2}.$$

We need to estimate

$$\left| \int_{A \times B} [k'(x, y) - f(x, y)] dx dy \right| \leq \frac{\epsilon}{2} + r\ell \sup_{1 \leq s \leq r} \left| \int_{A \times B} [k'(s, x, y) - f(s, x, y)] dx dy \right|.$$

For each value of s, we will estimate

$$\int_{A \times B} [k'(s, x, y) - f(s, x, y)] dx dy = \sum_{i,j=0}^{p} \int_{A \cap J_i \times B \cap J_j} [k'(s, x, y) - f(s, x, y)] dx dy.$$

The summation over (i, j) will be split into several groups. $F_0 = \{(i, i) : i \geq 1\}$. $F_1 = \{(i, j) : i = 0\} \cup \{(i, j) : j = 0\}$. $F_2 = \{(i, j)\}$ is the collection of at most $\epsilon' p^2$ exceptional pairs from Lemma 2.4. $F_3 = \{(i, j)\}$ for which either $|A \cap J_i| \leq \epsilon' |J_i|$ or $|B \cap J_j| \leq \epsilon' |J_j|$. F_4 will be the rest. Then:

$$\left| \sum_{(i,j) \in F_0} \int_{A \cap J_i \times B \cap J_j} [k'(s, x, y) - f(s, x, y)] dx dy \right| \leq \frac{pK^2}{n^2} \leq \frac{1}{p} \leq \epsilon',$$

$$\left| \sum_{(i,j) \in F_1} \int_{A \cap J_i \times B \cap J_j} [k'(s, x, y) - f(s, x, y)] dx dy \right| \leq 2\epsilon',$$

$$\left|\sum_{(i,j)\in F_2}\int_{A\cap J_i\times B\cap J_j}[k'(s,x,y)-f(s,x,y)]dxdy\right|\le \epsilon' p^2|J_i||J_j|\le \epsilon' p^2\frac{1}{p^2}=\epsilon',$$

$$\left|\sum_{(i,j)\in F_3}\int_{A\cap J_i\times B\cap J_j}[k'(s,x,y)-f(s,x,y)]dxdy\right|\le \epsilon'\sum_{i,j}|J_i||J_j|\le \epsilon'.$$

Finally in the remaining set F_4, since $|A\cap J_i|\ge \epsilon'|J_i|$ and $|B\cap J_j|\ge \epsilon'|J_j|$, it follows that

$$\left|\int_{A\cap J_i\times B\cap J_j}[k'(s,x,y)-f(s,x,y)]dxdy\right|\le \epsilon'|J_i||J_j|$$

and hence

$$\left|\sum_{(i,j)\in F_4}\int_{A\cap J_i\times B\cap J_j}[k'(s,x,y)-f(s,x,y)]dxdy\right|\le \epsilon'\sum_{i,j}|J_i||J_j|\le \epsilon'.$$

Adding them up gives

$$\left|\int_{A\times B}[k'(x,y)-f(x,y)]dxdy\right|\le \frac{\epsilon}{2}+6r\ell\epsilon'\le \epsilon.$$

□

3. Lower Bound

The lower bound for $\tilde{Q}_n$ on $\tilde{\mathcal{K}}$ can be proved by proving a lower bound for Q_n on $\mathcal{K}$ and it can be done without truncation and under the (weaker) assumption that $\int e^{\theta x^2}d\mu(x)<\infty$ for some θ. This implies a lower bound $h(x)\ge c|x|^2$ when $|x|$ is large.

Theorem 3.1. *Under the weaker assumption that for some $\theta>0$,*

$$\int e^{\theta x^2}d\mu(x)<\infty,$$

it follows that for any $f\in\mathcal{K}$ such that

$$I(f)=\frac{1}{2}\int h(f(x,y))dxdy<\infty$$

and for any $\delta>0$,

$$\liminf_{n\to\infty}\frac{1}{n^2}\log Q_n[d_\square(k,f)<\delta]\ge -I(f).$$

Proof. Since h is a convex function of its argument, for any integer q we can replace f by a simple function g of its averages over $J_r^q\times J_s^q$ so that

$$g(x,y)=\sum_{r,s=1}^{q}g_{r,s}\mathbf{1}_{J_r^q}(x)\mathbf{1}_{J_s^q}(y),$$

where

$$g_{r,s}=q^2\int_{J_r^q\times J_s^q}f(x,y)dxdy.$$

For any q, $I(g) \leq I(f)$ and for large q, $d_\square(f,g) < \frac{\delta}{2}$. It suffices to show that

$$\liminf_{n\to\infty} \frac{1}{n^2} \log Q_n[d_\square(g,k) \leq \frac{\delta}{2}] \geq -I(g).$$

We have

$$k(x,y) = \sum_{i,j=1}^{n} x_{i,j} \mathbf{1}_{J_i^n}(x) \mathbf{1}_{J_j^n}(y)$$

and we can view both k and g as members of $\mathcal{K}_{nq}$. The distance $d_\square(g,k)$ can be computed as

$$d_\square(g,k) = \sup_{A,B} \left| \int_{A\times B} (g(x,y) - k(x,y)) dx dy \right|,$$

where A and B are taken to be unions of sub-collections of intervals of the form $\{[\frac{i-1}{nq}, \frac{i}{nq}]\}$. There are exactly $2^{nq} \times 2^{nq}$ such pairs A, B. We now tilt the measure so that $\{x_{i,j}\}$ remains symmetric and $\{x_{i,j}\}$ for $j \geq i$ are still independent but the distribution of $x_{i,j}$ is tilted from μ to $\mu_{i,j}$ given by

$$\mu_{i,j}(dx) = \frac{1}{M(\theta_{r,s})} \exp[\theta_{r,s}\, x] \mu(dx)$$

for $(r-1)n < iq \leq rn$ and $(s-1)n < jq \leq sn$ where $\theta_{r,s} = h'(g_{r,s})$, or equivalently $g_{r,s} = \frac{M'(\theta_{r,s})}{M(\theta_{r,s})}$ for $1 \leq r, s \leq q$. Let Q_n^g be the law of the new k. The law of large numbers applies to each such pair A, B with uniform (in A and B) exponential error bounds of $e^{-cn^2+o(n^2)}$ for some $c > 0$. Therefore,

$$Q_n^g\left[\left|\int_{A\times B} (g(x,y) - k(x,y)) dx dy\right| \geq \frac{\delta}{2}\right] \leq e^{-c(\delta)n^2+o(n^2)}.$$

Since $2^{2nq} \ll e^{cn^2}$, it follows that

$$Q_n^g[d_\square(g,k) \geq \frac{\delta}{2}] \to 0.$$

The relative entropy of Q_n^g with respect to Q_n is easily computed to be $n^2 I(g)$.

The following entropy lower bound using Jensen's inequality, establishes the large deviation lower bound. Suppose α, β are two probability measures and $\beta \ll \alpha$ with $H = \int \phi \log \phi \, d\alpha < \infty$ where $\phi = \frac{d\beta}{d\alpha}$. Since $y \log y \geq -e^{-1}$ for all $y \geq 0$, $\int |\phi \log \phi| d\alpha \leq H + 2e^{-1}$. Therefore

$$\begin{aligned}
\alpha(A) &\geq \int_A \phi^{-1} d\beta = \int_A \exp[-\log \phi] d\beta \\
&= \beta(A) \frac{1}{\beta(A)} \int_A \exp[-\log \phi] d\beta \\
&\geq \beta(A) \exp\left[-\frac{1}{\beta(A)} \int_A \phi \log \phi d\alpha\right] \\
&\geq \beta(A) \exp\left[-\frac{1}{\beta(A)} \int |\phi \log \phi| d\alpha\right] \\
&\geq \beta(A) \exp\left[-\frac{1}{\beta(A)} [H + 2e^{-1}]\right].
\end{aligned}$$

Taking $\alpha = Q_n$ and $\beta = Q_n^g$ and $A = \{k : d_\square(k,g) \leq \frac{\delta}{2}\}$ we have $H = n^2 I(g)$ and $\beta(A) \simeq 1$. □

4. Upper Bound

We assume that $|x_{i,j}| \leq \ell$. According to a result of Lovász and Szegedy [5], for any finite ℓ, the set $\tilde{\mathcal{K}}^\ell = \{\tilde{f} : \sup_{x,y} |f(x,y)| \leq \ell\}$ is compact in $\tilde{K}$. Therefore in order to prove the large deviation upper bound for $\tilde{Q}_n$ with rate function $I(f)$ it is sufficient to prove the local version of the upper bound.

Theorem 4.1. *Let $\tilde{f} \in \tilde{\mathcal{K}}$. Then*

$$\limsup_{\epsilon \to 0} \limsup_{n \to \infty} \frac{1}{n^2} \log \tilde{Q}_n[\tilde{k} : d_\square(\tilde{k}, \tilde{f}) < \epsilon] \leq -I(\tilde{f}).$$

Proof. The theorem is proved in several steps. Closed balls $B(f,\epsilon)$ in $\mathcal{K}$ of the form $\{k : d_\square(k,f) \leq \epsilon\}$ are weakly closed. It is not hard to prove that

$$\limsup_{\epsilon \to 0} \limsup_{n \to \infty} \frac{1}{n^2} \log Q_n[k : d_\square(k,f) \leq \epsilon] \leq -I(f). \tag{4.1}$$

The argument goes as follows. Assume $I(f) < \infty$. Given $\delta > 0$ pick a nice g such that

$$\frac{1}{2}\left[\int f(x,y)g(x,y)dxdy - \int \log\left[\int e^{z\,g(x,y)}\mu(dz)\right]dxdy\right] \geq I(f) - \delta$$

and apply Cramér type estimate using the moment generating function for the half space

$$H_{f,g,\epsilon} = \left\{k : \langle k, g\rangle \geq \inf_{k' \in B(f,\epsilon)} \langle k', g\rangle\right\}$$

which contains $B(f,\epsilon)$. This gives

$$Q_n[B(f,\epsilon)] \leq Q_n[H_{f,g,\epsilon}] \leq \exp\left[-\frac{n^2}{2} \inf_{k \in B(f,\epsilon)} \langle k, g\rangle\right] \mathbb{E}^{Q_n} \exp\left[\frac{n^2}{2}\langle k, g\rangle\right].$$

It is easy to see (because $x_{i,j} = x_{j,i}$) that

$$\limsup_{n \to \infty} \frac{1}{n^2} \log \mathbb{E}^{Q_n} \exp\left[\frac{n^2}{2}\langle k, g\rangle\right] = \frac{1}{2}\int_D \log\left[\int \exp[zg(x,y)]\mu(dz)\right]dxdy$$

and

$$\lim_{\epsilon \to 0} \inf_{k \in B(f,\epsilon)} \langle k, g\rangle = \langle f, g\rangle.$$

We can let $\delta \to 0$ at the end. If $I(f) = \infty$, pick g such that

$$\frac{1}{2}\left[\int f(x,y)g(x,y)dxdy - \int \log\left[\int e^{z\,g(x,y)}\mu(dz)\right]dxdy\right] \geq L$$

and let $L \to \infty$ in the end.

This proves (4.1). But to prove it for balls in $\tilde{\mathcal{K}}$ we need to estimate the probability of the pre-image in $\mathcal{K}$ of a ball in $\tilde{\mathcal{K}}$. This amounts to showing that

$$\limsup_{\epsilon \to 0} \limsup_{n \to \infty} \frac{1}{n^2} \log Q_n[k : \inf_{\sigma \in G} d_\square(k, \sigma f) < \epsilon] \leq -I(\tilde{f}).$$

We start with the set

$$U(\tilde{f},\epsilon) = \cup_{\sigma\in G} B(\sigma f,\epsilon).$$

We saw that for each $f \in \mathcal{K}$, (4.1) holds. We need to replace the union over $\sigma \in G$ by a union over a finite collection. According to Lemma 2.5 there is a compact set $W_{\ell,\epsilon} \subset \mathcal{K}$ such that its orbit by $\Pi(n)$ nearly covers $\mathcal{K}_n^\ell$ for all sufficiently large n. More precisely for $n \geq n_0(\epsilon,\ell)$,

$$\sup_{k\in\mathcal{K}_n^\ell} \inf_{\sigma\in\Pi(n)} \inf_{g\in W_{\ell,\epsilon}} d_\square(\sigma g, k) < \epsilon.$$

$W_{\ell,\epsilon}$ is compact in the cut topology (and even in L^1) and can be covered by the union of balls $B(g,\epsilon)$ of radius ϵ centered around g from a finite collection $F_{\ell,\epsilon}$. Therefore for sufficiently large n

$$\cup_{\sigma\in\Pi(n)} \cup_{g\in F_{\ell,\epsilon}} B(\sigma g, 2\epsilon) \supset \mathcal{K}_n^\ell.$$

We need to estimate the probability under Q_n of

$$\cup_{\sigma'\in\Pi(n)} \cup_{g\in F_{\ell,\epsilon}} \cup_{\sigma\in G}[B(\sigma' g, 2\epsilon)\cap B(\sigma f,\epsilon)]$$

which is at most $n!$ times

$$\sup_{\sigma'\in\Pi(n)} Q_n[\cup_{g\in F_{\ell,\epsilon}} \cup_{\sigma\in G} [B(\sigma' g, 2\epsilon)\cap B(\sigma f,\epsilon)]].$$

Since $F_{\ell,\epsilon}$ is a finite set independent of n and Q_n is invariant under $\Pi(n) \subset G$, it is enough to show that for each $g \in F_{\ell,\epsilon}$

$$\limsup_{\epsilon\to 0}\limsup_{n\to\infty} \frac{1}{n^2}\log Q_n[B(g,2\epsilon)\cap \cup_{\sigma\in G} B(\sigma f,\epsilon)] \leq -I(f).$$

If the intersection is nonempty, then there is a σ' such that $B(g,2\epsilon) \subset B(\sigma' f, 5\epsilon)$. Thus, from (4.1), the invariance of $I(\sigma f)$ under $\sigma \in G$ and the lower semicontinuity of $I(\cdot)$ on $\tilde{\mathcal{K}}$ we see that there is a $\sigma' \in G$ such that

$$\begin{aligned}
&\limsup_{\epsilon\to 0}\limsup_{n\to\infty} \frac{1}{n^2}\log Q_n[B(g,2\epsilon)\cap \cup_{\sigma\in G} B(\sigma f,\epsilon)]\\
&\qquad \leq \limsup_{\epsilon\to 0}\limsup_{n\to\infty} \frac{1}{n^2}\log Q_n[B(\sigma' f,5\epsilon)]\\
&\qquad = -I(\sigma' f) = -I(f).
\end{aligned}$$

□

5. Truncation

Given $X = \{x_{i,j}\}$ we truncate it at level ℓ. Let $x = f_\ell(x) + g_\ell(x)$ where

$$f_\ell(x) = \begin{cases} x & \text{if } |x| \leq \ell \\ \ell & \text{if } x \geq \ell \\ -\ell & \text{if } x \leq -\ell \end{cases}$$

and $g_\ell(x) = x - f_\ell(x)$. There is a corresponding decomposition of $k = f_\ell(k) + k - f_\ell(k)$ for $k \in \mathcal{K}$. We have the estimate

$$\limsup_{\ell\to\infty}\limsup_{n\to\infty} \frac{1}{n^2}\log \mathbb{E}\exp\Big[\theta\sum_{i,j}[x_{i,j} - f_\ell(x_{i,j})]^2\Big] = 0$$

for any $\theta > 0$. If $\{f_\ell(x_{i,j})\}$ and $\{x_{i,j}\}$ are mapped respectively into k_ℓ and k, we have super exponential estimates on $\Delta(\ell) = \int_D |f_\ell(k(x,y)) - k(x,y)|^2 dxdy$:

$$Q_n[\Delta(\ell) \geq \epsilon] \leq e^{-n^2\theta\epsilon^2} \mathbb{E} \exp\Big[\theta \sum_{i,j} [x_{i,j} - f_\ell(x_{i,j})]^2\Big].$$

Therefore

$$\limsup_{\ell\to\infty} \limsup_{n\to\infty} \frac{1}{n^2} \log Q_n[\Delta(\ell) \geq \epsilon] \leq -\theta\epsilon^2.$$

Since $\theta > 0$ is arbitrary, for any $\epsilon > 0$,

$$\limsup_{\ell\to\infty} \limsup_{n\to\infty} \frac{1}{n^2} \log Q_n[\Delta(\ell) \geq \epsilon] = -\infty.$$

It is easy to complete the proof of Theorem 1.2 using the above identity and the lower-semicontinuity of I.

The difference between the eigenvalues of two kernels k_1 and k_2 can be easily controlled by the Hilbert-Schmidt norm of the difference $k_1 - k_2$. In fact using the variational formula for successive eigenvalues of a compact self adjoint operator,

$$\lambda_{j+1}(A) = \inf_{\substack{\mathcal{M}:\\ \operatorname{codim}(\mathcal{M})=j}} \sup_{\substack{\|z\|=1\\ z\in\mathcal{M}}} \langle Az, z\rangle$$

it is easily seen that if $\lambda_j(A)$ and $\lambda_j(B)$ are the jth largest positive eigenvalues of two compact self adjoint operators A and B

$$|\lambda_j(A) - \lambda_j(B)| \leq \|A - B\| \leq \|A - B\|_{HS}$$

and a similar formula holds for negative eigenvalues as well. Clearly if $k_1, k_2 \in \mathcal{K}$, then $d(S(k_1), S(k_2)) \leq \|k_1 - k_2\| \leq \|k_1 - k_2\|_{HS}$.

If $C \in \mathcal{S}$ is a closed set, and we truncate X at level ℓ and denote by k_ℓ and k the two images in $\mathcal{K}$, then

$$P_n(C) = Q_n[S(k) \in C] \leq Q_n[S(k_\ell) \in C^\epsilon] + Q_n[d(S(k_\ell), S(k)) \geq \epsilon].$$

Applying the upper bound for the truncated version,

$$\limsup_{n\to\infty} \frac{1}{n^2} \log P_n[C] \leq \max\{-\inf_{S\in\overline{C^\epsilon}} J_\ell(S), -c(\ell,\epsilon)\},$$

where

$$c(\ell,\epsilon) = -\limsup_{n\to\infty} \frac{1}{n^2} \log Q_n[d(S(k_\ell), S(k)) \geq \epsilon]$$

and for any $\epsilon > 0$,

$$\lim_{\ell\to\infty} c(\ell,\epsilon) = -\infty.$$

Moreover

$$J_\ell(S) = \inf_{k:S(k)=S} I_\ell(k); \quad I_\ell(k) = \int_D h_\ell(k(x,y))dxdy$$

and

$$h_\ell(z) = \sup[z\theta - \log \int \exp[\theta f_\ell(x)]\mu(dx)].$$

The upper bound is now easily established by letting $\ell \to \infty$ and then letting $\epsilon \to 0$. We observe that the superexponential estimate implies $c(\ell,\epsilon) \to \infty$ as $\ell \to \infty$ for any $\epsilon > 0$. This completes the proof of Theorem 1.1.

References

1. Bai, Z. D.: Methodologies in the spectral analysis of large-dimensional random matrices, a review. *Statist. Sinica* **9** (1999), no. 3, 611–677.
2. Ben Arous, G. and Guionnet, A.: Large deviations for Wigners law and Voiculescu's non-commutative entropy, *Prob. Th. Rel. Fields* **108** (1997) 517–542.
3. Chatterjee, S. and Varadhan, S. R. S.: The large deviation principle for the Erdős-Rényi random graph. *European J. Combin.* **32** (2011), no. 7, 1000–1017
4. Komlós, J. and Simonovits, M.: Szemerédi's regularity lemma and its applications in graph theory. *Bolyai Soc. Math. Stud.,* **2** (1996) 295–352.
5. Lovász, L. and Szegedy, B.: Szemerédi's lemma for the analyst. *Geom. Funct. Anal.* **17** (2007), no. 1, 252–270.
6. Wigner, E. P.: On the distribution of the roots of certain symmetric matrices. *Ann. of Math. (2)* **67** (1958) 325–327.

SOURAV CHATTERJEE: COURANT INSTITUTE, NEW YORK UNIVERSITY, NEW YORK, NY, 10012, USA
E-mail address: sourav@cims.nyu.edu
URL: http://www.cims.nyu.edu/~sourav/

S. R. S. VARADHAN: COURANT INSTITUTE, NEW YORK UNIVERSITY, NEW YORK, NY, 10012, USA
E-mail address: varadhan@cims.nyu.edu
URL: http://www.cims.nyu.edu/~varadhan/